TEXT BOOK OF BACTERIOLOGY

DR. VINOD SINGH
Associate Professor
Department of Microbiology
Barkatullah University,
Bhopal (M.P.) India

Published by

Khushnuma Complex Basement
7, Meerabai Marg (Behind Jawahar Bhawan)
Lucknow 226 001 U.P. (INDIA)
Tel. : 91-522-2209542, 2209543, 2209544, 2209545
Fax : 0522-4045308
E-Mail : ibdco@airtelmail.in

First Edition 2010

ISBN 978-81-8189-498-4 (H.B.)

Composed & Designed at :

Panacea Computers
3rd Floor, Agarwal Sabha Bhawan, Subhash Mohal
Sadar Cantt. Lucknow-226 002
Phone : 0522-2483312, 9335927082
E-mail : prasgupt@rediffmail.com

Printed at:

Salasar Imaging Systems
C-7/5, Lawrence Road Industrial Area
Delhi - 110 035
Tel. : 011-27185653, 9810064311

Preface

Since long I was feeling the need of text book of Bacteriology for the postgraduate as well as undergraduate students of paramedical, medical and basic microbiology of various teaching organizations. A lacuna has always been sensed particularly in the areas of diagnostics and therapeutic bacteriology. Keeping the students demand of biomedical sciences, microbiology, biotechnology, nursing, physiotherapy, dental and other allied courses I took the decision to extend my limited knowledge to students.

I have tried my best to include basic mechanisms of Life cycle, pathogenesis, laboratory diagnosis, immunity and structural aspects of bacterial diversities, detailed ray diagrams and illustration have also been framed to simplify the subject in depth.

I am indebted to many stalwarts in my profession for their valuable guidance, their books and publications in relevant journals which I have consulted freely, if some reference has been omitted due to oversight I would request the readers to please bring this to the attention of the publisher who will ensure that full acknowledgement is made in subsequent editions.

The help provided by my students specially Desh Deepak Singh, Sarika Amdekar, Deepak Dwivedi,PurabiRoy and Sapna Kuswah and other research scholars, colleagues and librarian for preparing this book is acknowledged, without which this difficult task could not have been completed. Suggestion and additional information will always be welcomed from readers for further improvement.

Lastly I would like to extend my special thanks to my family members whose constant support and motivation made this book possible.

Vinod Singh

About the Author

Dr. Vinod Singh is currently Associate Professor with Department of Microbiology of Barkatullah University, Bhopal, M.P,India. His areas of specialization are Microbiology and Immunology. He has a long experience of working in areas of Clinical and Diagnostic Microbiology. He did his Post Graduate in Life Sciences with specialization in Immunology and Clinical Biochemistry from Institute of Life Sciences of CSJM University, Kanpur and Doctorate in Serology and Immunology from Department of Medicine of King George Medical College (KGMC),Lucknow.

He started his Research carrier working in the area of Immunology on the project sponsored by Department of Atomic Energy (DOE) of India on Study of Effect of Chronic and Acute Microwave Radiation on Immunological System of Belgian White Rabbits at Electrical Engineering Department of Indian Institute of Medical Sciences (I.I.T) Kanpur, and after that on the Role of Antinuclear Antibodies (ANA) specifities in the Patients of Systemic Lupus Erythromatosous(SLE) and Rheumatoid Arthritis from Clinical Immunology Unit of Department of All India Institute of Medical Sciences (A.I.I.M.S) New Delhi.

Dr Singh also worked as Senior Project Associate in the International Development Research Council (I.D.R.C) Canada funded project of Development of Low Cost Water filter against Rota Virus at Environmental Engineering Department of Indian Institute of Technology (I.I.T),Kanpur and on the Clinical, Immunological and Virological study in Patients of Dialated Cardio Myopathic Patients caused by Coxsackie virus from Virology Department of All India Institute of Medical Sciences, New Delhi.

He did his postdoctoral work on Walter Reed Army Research Institute,USA funded project on Effect of ã Interferon on murine Malaria and clone drug resistant species of Lesihmania donovani from the division of Microbiology and Parasitology of Central Drug Research Institute (C.D.R.I), Lucknow.

Dr Singh also served as a Senior Assistant Professor at Department of Microbiology of Cancer Hospital and Research Institute, Gwalior and was associated with Laboratory Services of Microbiology and Immunology. There his main focus was on the diagnosis of Pathogenic Bacterial species from Clinical isolates and monitoring of Tumor from various categories of Cancer patients. He was also member of Tumor Board of Cancer Hospital.

He was also involved in teaching the Post Graduate programme of Medical Microbiology and Immunology. He was also Recipient of Best Teacher Award by Gwalior Development Commeettee.

Dr Singh was also an Executive member of Indian Immunology Society (I.I.S), Member of International Union of Immunology Society (I.U.I.S) and presently joint secretary of Society for Immunology and Immunopathology (S.I.I.P). He is also a reviewer of Journal of Rheumatology published by British Society of Rheumatology, U.K, International Dairy Research Journal, Denmark and Journal of Antimicrobial Agent, USA. International Journal of Immunopathology and pharmacology.

He as been the Associate Editor of Journal of Immunology and Immunopathology publisehd by Society of Immunology and Immunpathology(SIIP), and Journal of BioSciences ,ITRC, India . He has published several research papers in field of Microbiology and Immunology in reputed journals.

Contents

Chapter 1

Introduction to Bacteriology

Bacteria are single-celled microorganisms that lack a nuclear membrane, are metabolically active and divide by binary fission. Medically they are a major cause of disease. Superficially, bacteria appear to be relatively simple forms of life; in fact, they are sophisticated and highly adaptable. Many bacteria multiply at rapid rates, and different species can utilize an enormous variety of hydrocarbon substrates, including phenol, rubber, and petroleum. These organisms exist widely in both parasitic and free-living forms. Because they are ubiquitous and have a remarkable capacity to adapt to changing environments by selection of spontaneous mutants, the importance of bacteria in every field of medicine cannot be overstated.

The discipline of bacteriology evolved from the need of physicians to test and apply the germ theory of disease and from economic concerns relating to the spoilage of foods and wine. The initial advances in pathogenic bacteriology were derived from the identification and characterization of bacteria associated with specific diseases. During this period, great emphasis was placed on applying Koch's postulates to test proposed cause-and-effect relationships between bacteria and specific diseases. Today, most bacterial diseases of humans and their etiologic agents have been identified, although importantvariants continue to evolveand sometimes emerge, e.g., Legionnaire's Disease, tuberculosis and toxic shock syndrome.

Major advances in bacteriology over the last century resulted in the development of many effective vaccines (e.g., pneumococcal polysaccharide vaccine, diphtheria toxoid, and tetanus toxoid) as well as of other vaccines (e.g., cholera, typhoid, and plague vaccines) that are less effective or have side effects. Another major advance was the discovery of antibiotics. These antimicrobial substances have not eradicated bacterial diseases, but they are powerful therapeutic tools. Their efficacy is reduced by the emergence of antibiotic resistant bacteria (now an important medical management problem) In reality, improvements in sanitation and water purification have a greater effect on the incidence of bacterial infections in a community than does the availability of antibiotics or bacterial vaccines. Nevertheless, many and serious bacterial diseases remain.

Most diseases now known to have a bacteriologic etiology have been recognized for hundreds of years. Some were described as contagious in the writings of the ancient Chinese, centuries prior to the first descriptions of bacteria by Anton van Leeuwenhoek in 1677. There remain a few diseases (such as chronic ulcerative colitis) that are thought by some investigators to be caused by bacteria but for which no pathogen has been identified. Occasionally, a previously unrecognized diseases is associated with a new group of bacteria. An example is Legionnaire's disease, an acute respiratory infection caused by the previously unrecognized genus, Legionella. Also, a newly recognized pathogen, *Helicobacter,* plays an important role in peptic disease. Another important example, in understanding the etiologies of venereal diseases, was the association of at least 50 percent of the cases of urethritis in male patients with *Ureaplasma urealyticum* or *Chlamydia trachomatis.*

Recombinant bacteria produced by genetic engineering are enormously useful in bacteriologic research and are being employed to manufacture scarce biomolecules (e.g. interferons) needed for research and patient care. The antibiotic resistance genes, while a problem to the physician, paradoxically are indispensable markers in performing genetic engineering. Genetic probes and the polymerase chain reaction (PCR) are useful in the rapid identification of microbial pathogens in patient specimens. Genetic manipulation of pathogenic bacteria continues to be indispensable in defining virulence mechanisms. As more protective protein antigens are identified, cloned, and sequenced, recombinant bacterial vaccines will be constructed that should be much better than the ones presently available. In this regard, a recombinant-based and safer pertussis vaccine is already available in some European countries. Also, direct DNA vaccines hold considerable promise.

In developed countries, 90 percent of documented infections in hospitalized patients are caused by bacteria. These cases probably reflect only a small percentage of the actual number of bacterial infections occurring in the general population, and usually represent the most severe cases. In developing countries, a variety of bacterial infections often exert a devastating effect on the health of the inhabitants. Malnutrition, parasitic infections, and poor sanitation are a few of the factors contributing to the increased susceptibility of these individuals to bacterial pathogens. The World Health Organization has estimated that each year, 3 million people die of tuberculosis, 0.5 million die of pertussis, and 25,000 die of typhoid. Diarrheal diseases, many of which are bacterial, are the second leading cause of death in the world (after cardiovascular diseases), killing 5 million people annually.

Many bacterial diseases can be viewed as a failure of the bacterium to adapt, since a well-adapted parasite ideally thrives in its host without causing significant damage. Relatively nonvirulent (i.e., well-adapted) microorganisms can cause disease under special conditions - for example, if they are present in unusually large numbers, if the host's defenses are impaired, (e.g., AIDS and chemotherapy) or if anaerobic conditions exist. Pathogenic bacteria constitute only a small proportion of bacterial species; many nonpathogenic bacteria are beneficial to humans (i.e. intestinal flora produce vitamin K) and participate in essential processes such as nitrogen fixation, waste breakdown, food production, drug preparation, and environmental bioremediation. This textbook emphasizes bacteria that have direct medical relevance.

In recent years, medical scientists have concentrated on the study of pathogenic mechanisms and host defenses. Understanding host-parasite relationships involving specific pathogens requires familiarity with the fundamental characteristics of the bacterium, the host, and their interactions. Therefore, this section first presents with the basic concepts of the immune response, bacterial structure, taxonomy, metabolism, and genetics. Subsequent chapters emphasize normal relationships among bacteria on external surfaces; mechanisms by which microorganisms damage the host; host defense mechanisms; source and distribution of pathogens (epidemiology); principles of diagnosis; and mechanisms of action of antimicrobial drugs. These chapters provide the basis for the next chapters devoted to specific bacterial pathogens and the diseases they cause. The bacteria in these chapters are grouped on the basis of physical, chemical, and biologic characteristics. These similarities do not necessarily indicate that their diseases are similar; widely divergent diseases may be caused by bacteria in the same group.

Chapter **2**

Structure

General Concepts

Gross Morphology

Bacteria have characteristic shapes (cocci, rods, spirals, etc.) and often occur in characteristic aggregates (pairs, chains, tetrads, clusters, etc.). These traits are usually typical for a genus and are diagnostically useful.

Cell Structure

Prokaryotes have a nucleoid (nuclear body) rather than an enveloped nucleus and lack membrane-bound cytoplasmic organelles. The plasma membrane in prokaryotes performs many of the functions carried out by membranous organelles in eukaryotes. Multiplication is by binary fission.

Surface Structures

Flagella: The flagella of motile bacteria differ in structure from eukaryotic flagella. A basal body anchored in the plasma membrane and cell wall gives rise to a cylindrical protein filament. The flagellum moves by whirling about its long axis. The number and arrangement of flagella on the cell are diagnostically useful.

Pili (Fimbriae): Pili are slender, hairlike, proteinaceous appendages on the surface of many (particularly Gram-negative) bacteria. They are important in adhesion to host surfaces.

Capsules: Some bacteria form a thick outer capsule of high-molecular-weight, viscous polysaccharide gel; others have more amorphous slime layers. Capsules confer resistance to phagocytosis.

Important Chemical Components of Surface Structures

Cell Wall Peptidoglycans: Both Gram-positive and Gram-negative bacteria possess cell wall peptidoglycans, which confer the characteristic cell shape and provide the cell with mechanical protection. Peptidoglycans are unique to prokaryotic organisms and consist of a glycan backbone of muramic acid and glucosamine (both N-acetylated), and peptide chains highly cross-linked with bridges in Gram-positive bacteria (e.g., Staphylococcus aureus) or partially cross-linked in Gram-negative bacteria (e.g.,). The cross-linking transpeptidase enzymes are some of the targets for b-lactam antibiotics.

Teichoic Acids: Teichoic acids are polyol phosphate polymers bearing a strong negative charge. They are covalently linked to the peptidoglycan in some Gram-positive bacteria. They are strongly antigenic, but are generally absent in Gram-negative bacteria.

Lipoteichoic Acids: Lipoteichoic acids as membrane teichoic acids are polymers of amphiphitic glycophosphates with the lipophilic glycolipidand anchored in the cytoplasmic membrane. They are antigenic, cytotoxic and adhesins (e.g., Streptococcus pyogenes).

Lipopolysaccharides: One of the major components of the outer membrane of Gram-negative bacteria is lipopolysaccharide (endotoxin), a complex molecule consisting of a lipid A anchor, a polysaccharide core, and chains of carbohydrates. Sugars in the polysaccharide chains confer serologic specificity.

Wall-Less Forms: Two groups of bacteria devoid of cell wall peptidoglycans are the Mycoplasma species, which possess a surface membrane structure, and the L-forms that arise from either Gram-positive or Gram-negative bacterial cells that have lost their ability to produce the peptidoglycan structures.

Cytoplasmic Structures

Plasma Membrane: The bacterial plasma membrane is composed primarily of protein and phospholipid (about 3:1). It performs many functions, including transport, biosynthesis, and energy transduction.

Organelles: The bacterial cytoplasm is densely packed with 70S ribosomes. Other granules represent metabolic reserves (e.g., poly-b-hydroxybutyrate, polysaccharide, polymetaphosphate, and metachromatic granules).

Endospores: Bacillus and Clostridium species can produce endospores: heat-resistant, dehydrated resting cells that are formed intracellularly and contain a genome and all essential metabolic machinery. The endospore is encased in a complex protective spore coat.

INTRODUCTION

All bacteria, both pathogenic and saprophytic, are unicellular organisms that reproduce by binary fission. Most bacteria are capable of independent metabolic existence and growth, but species of Chlamydia and Rickettsia are obligately intracellular organisms. Bacterial cells are extremely small and are most conveniently measured in microns (10-6 m). They range in size from large cells such as Bacillus anthracis (1.0 to 1.3 μm X 3 to 10 μm) to very small cells such as Pasteurella tularensis (0.2 X 0.2 to 0.7 μm) Mycoplasmas (atypical pneumonia group) are even smaller, measuring 0.1 to 0.2 μm in diameter. Bacteria therefore have a surface-to-volume ratio that is very high: about 100,000.

Bacteria have characteristic shapes. The common microscopic morphologies are cocci (round or ellipsoidal cells, such as Staphylococcus aureus or Streptococcus respectively); rods, such as Bacillus and Clostridium species; long, filamentous branched cells, such as Actinomyces species; and comma-shaped and spiral cells, such as Vibrio cholerae and Treponema pallidum, respectively. The arrangement of cells is also typical of various species or groups of bacteria (Fig. 2-1) . Some rods or cocci characteristically grow in chains; some, such as Staphylococcus aureus, form grapelike clusters of spherical cells; some round cocci form cubic packets. Bacterial cells of other species grow separately. The microscopic appearance is therefore valuable in classification and diagnosis.

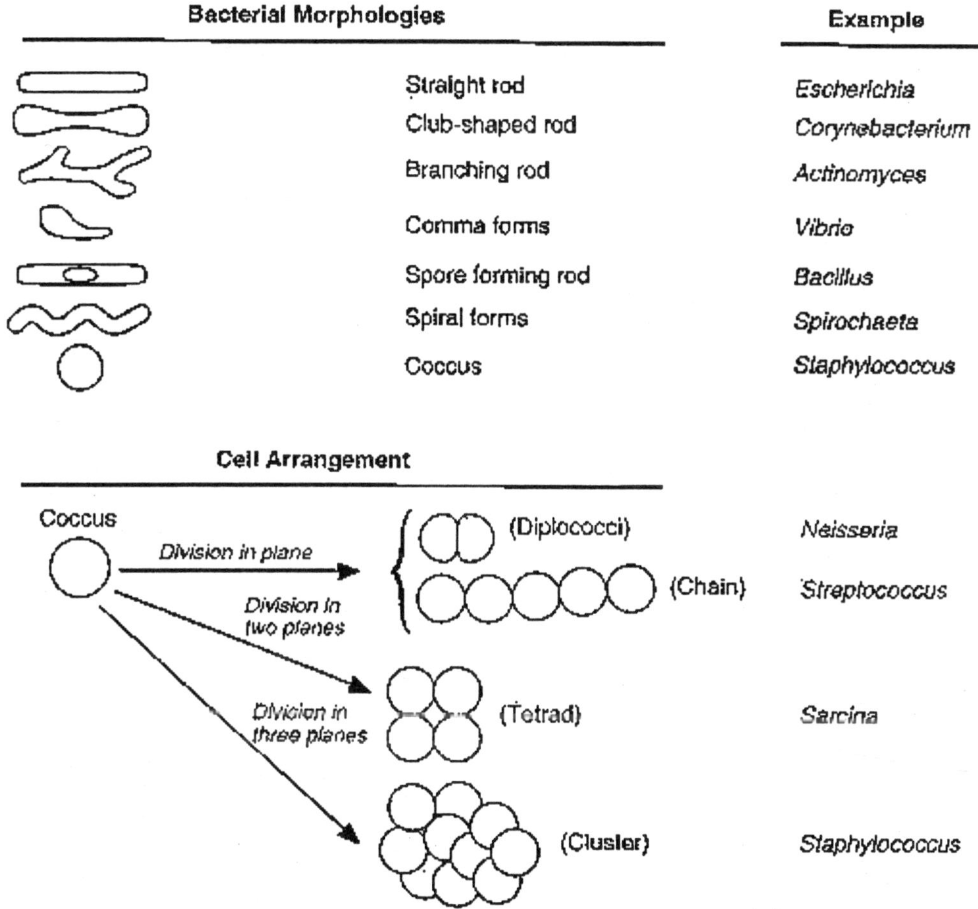

FIGURE 2-1 Typical shapes and arrangements of bacterial cells.

The higher resolving power of the electron microscope not only magnifies the typical shape of a bacterial cell but also clearly resolves its prokaryotic organization (Fig. 2-2).

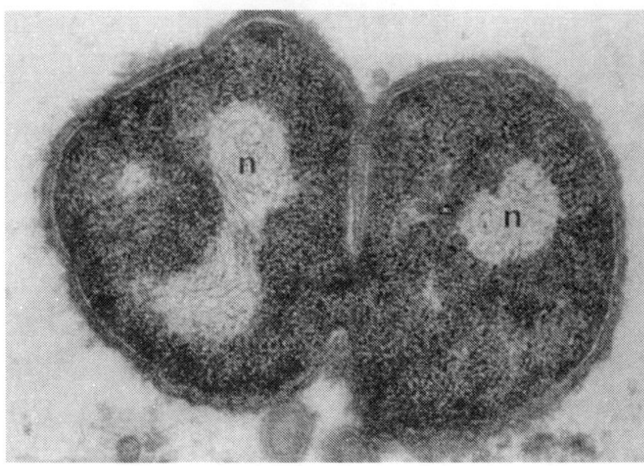

FIGURE 2-2 Electron micrograph of a thin section of Neisseria gonorrhoeae showing the organizational features of prokaryotic cells. Note the electron-transparent nuclear region (n) packed with DNA fibrils, the dense distribution of ribosomal particles in the cytoplasm, and the absence of intracellular membranous organelles.

The Nucleoid

Prokaryotic and eukaryotic cells were initially distinguished on the basis of structure: the prokaryotic nucleoid the equivalent of the eukaryotic nucleusis structurally simpler than the true eukaryotic nucleus, which has a complex mitotic apparatus and surrounding nuclear membrane. As the electron micrograph in Fig. 2-2 shows, the bacterial nucleoid, which contains the DNA fibrils, lacks a limiting membrane. Under the light microscope, the nucleoid of the bacterial cell can be visualized with the aid of Feulgcn staining, which stains DNA. Gentle lysis can be used to isolate the nucleoid of most bacterial cells. The DNA is then seen to be a single, continuous, "giant" circular molecule with a molecular weight of approximately 3 X 109 (see Ch. 5). The unfolded nuclear DNA would be about 1 mm long (compared with an average length of 1 to 2 µm for bacterial cells). The bacterial nucleoid, then, is a structure containing a single chromosome. The number of copies of this chromosome in a cell depends on the stage of the cell cycle (chromosome replication, cell enlargement, chromosome segregation, etc). Although, the mechanism of segregation of the two sister chromosomes following replication is not fully understood, all of the models proposed require that the chromosome be permanently attached to the cell membrane throughout the various stages of the cell cycle.

Bacterial chromatin does not contain basic histone proteins, but low-molecular-weight polyamines and magnesium ions may fulfill a function similar to that of eukaryotic histones. Despite the differences between prokaryotic and eukaryotic DNA, prokaryotic DNA from cells infected with bacteriophage g, when visualized by electron microscopy, has a beaded, condensed appearance not unlike that of eukaryotic chromatin.

Surface Appendages

Two types of surface appendage can be recognized on certain bacterial species: the flagella, which are organs of locomotion, and pili (Latin hairs), which are also known as fimbriae (Latin fringes). Flagella occur on both Gram-positive and Gram-negative bacteria, and their presence can be useful in identification. For example, they are found on many species of bacilli but rarely on cocci. In contrast, pili occur almost exclusively on Gram-negative bacteria and are found on only a few Gram-positive organisms (e.g., Corynebacterium renale).

Some bacteria have both flagella and pili. The electron micrograph in Fig. 2-3 shows the characteristic wavy appearance of flagella and two types of pili on the surface of Escherichia coli.

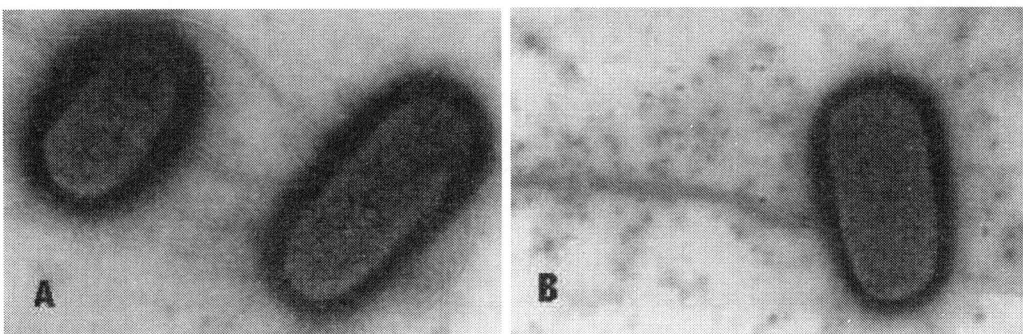

FIGURE 2-3 (A) Electron micrograph of negatively stained E coli showing wavy flagella and numerous short, thinner, and more rigid hairlike structures, the pili. (B) The long sex pilus can be distinguished from the shorter common pili by mixing E coli cells with a male bacteriophage that binds specifically to sex pili.

Flagella

Structurally, bacterial flagella are long (3 to 12 μm), filamentous surface appendages about 12 to 30 nm in diameter. The protein subunits of a flagellum are assembled to form a cylindrical structure with a hollow core. A flagellum consists of three parts: (1) the long filament, which lies external to the cell surface; (2) the hook structure at the end of the filament; and (3) the basal body, to which the hook is anchored and which imparts motion to the flagellum. The basal body traverses the outer wall and membrane structures. It consists of a rod and one or two pairs of discs. The thrust that propels the bacterial cell is provided by counterclockwise rotation of the basal body, which causes the helically twisted filament to whirl. The movement of the basal body is driven by a proton motive force rather than by ATP directly. The ability of bacteria to swim by means of the propeller-like action of the flagella provides them with the mechanical means to perform chemotaxis (movement in response to attractant and repellent substances in the environment). Response to chemical stimuli involves a sophisticated sensory system of receptors that are located in the cell surface and/or periplasm and that transmit information to methyl-accepting chemotaxis proteins that control the flagellar motor. Genetic studies have revealed the existence of mutants with altered biochemical pathways for flagellar motility and chemotaxis.

Chemically, flagella are constructed of a class of proteins called flagellins. The hook and basal-body structures consist of numerous proteins. Mutations affecting any of these gene products may result in loss or impairment of motility. Flagellins are immunogenic and constitute a group of protein antigens called the H antigens, which are characteristic of a given species, strain, or variant of an organism. The species specificity of the flagellins reflects differences in the primary structures of the proteins. Antigenic changes of the flagella known as the phase variation of H1 and H2 occurs in Salmonella typhimurium (see Ch. 21 and Ref. Seifert and So).

The number and distribution of flagella on the bacterial surface are characteristic for a given species and hence are useful in identifying and classifying bacteria. Figure 2-4 illustrates typical arrangements of flagella on or around the bacterial surface. For example, V cholerae has a single flagellum at one pole of the cell (i.e., it is monotrichous), whereas Proteus vulgaris and E coli have many flagella distributed over the entire cell surface (i.e., they are peritrichous). The flagella of a peritrichous bacterium must aggregate as a posterior bundle to propel the cell in a forward direction.

Flagella can be sheared from the cell surface without affecting the viability of the cell. The cell then becomes temporarily nonmotile. In time it synthesizes new flagella and regains motility. The protein synthesis inhibitor chloramphenicol, however, blocks regeneration of flagella.

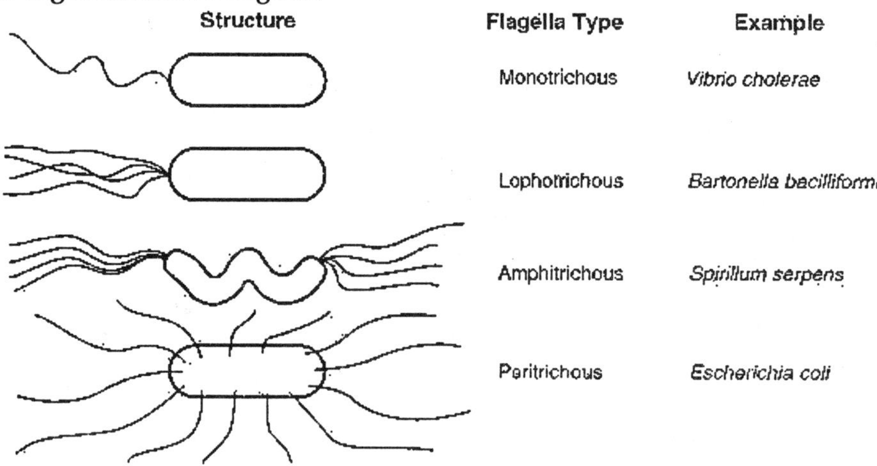

FIGURE 2-4 Typical arrangements of bacterial flagella.

Pili

The terms pili and fimbriae are usually used interchangeably to describe the thin, hairlike appendages on the surface of many Gram-negative bacteria and proteins of pili are referred to as pilins. Pili are more rigid in appearance than flagella (Fig. 2-3). In some organisms, such as Shigella species and E coli, pili are distributed profusely over the cell surface, with as many as 200 per cell. As is easily recognized in strains of E coli, pili can come in two types: short, abundant common pili, and a small number (one to six) of very long pili known as sex pili. Sex pili can be distinguished by their ability to bind male-specific bacteriophages (the sex pilus acts as a specific receptor for these bacteriophages) (Fig. 2-3B). The sex pili attach male to female bacteria during conjugation.

Pili in many enteric bacteria confer adhesive properties on the bacterial cells, enabling them to adhere to various epithelial surfaces, to red blood cells (causing hemagglutination), and to surfaces of yeast and fungal cells. These adhesive properties of piliated cells play an important role in bacterial colonization of epithelial surfaces and are therefore referred to as colonization factors. The common pili found on E coli exhibit a sugar specificity analogous to that of phytohemagglutininsand lectins, in that adhesion and hemagglutinating capacities of the organism are inhibited specifically by mannose. Organisms possessing this type of hemagglutination are called mannose-sensitive organisms. Other piliated organisms, such as gonococci, are adhesive and hemagglutinating, but are insensitive to the inhibitory effects of mannose. Extensive antigenic variations in pilins of gonococci are well known (see Ref. Seifert and So).

Surface Layers

The surface layers of the bacterial cell have been identified by various techniques: light microscopy and staining; electron microscopy of thin-sectioned, freeze-fractured, and negatively stained cells; and isolation and biochemical characterization of individual morphologic components of the cell. The principal surface layers are capsules and loose slime, the cell wall of Gram-positive bacteria and the complex cell envelope of Gram-negative bacteria, plasma (cytoplasmic) membranes, and mesosomal membrane vesicles, which arise from invaginations of the plasma membrane. In bacteria, the cell wall forms a rigid structure of uniform thickness around the cell and is responsible for the characteristic shape of the cell (rod, coccus, or spiral). Inside the cell wall (or rigid peptidoglycan layer) is the plasma (cytoplasmic) membrane; this is usually closely apposed to the wall layer. The topographic relationships of the cell wall and envelope layers to the plasma membrane are indicated in the thin section of a Gram-positive organism (Micrococcus lysodeikticus) in Figure 2-5A and in the freeze-fractured cell of a Gram-negative organism (Bacteroides melaninogenicus) in Figure 2-5B. The latter shows the typical fracture planes seen in most Gram-negative bacteria, which are weak cleavage planes through the outer membrane of the envelope and extensive fracture planes through the bilayer region of the underlying plasma membrane.

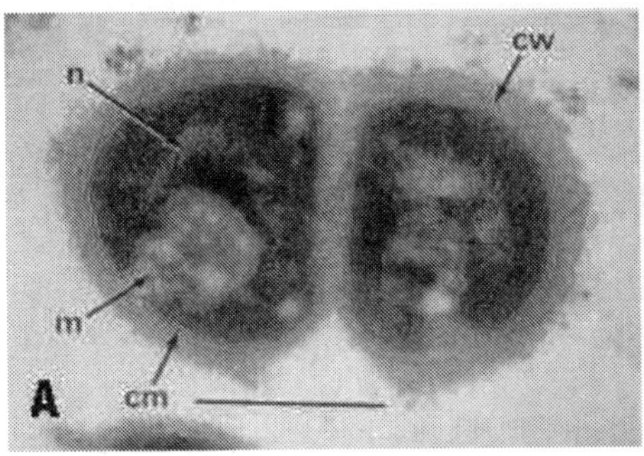

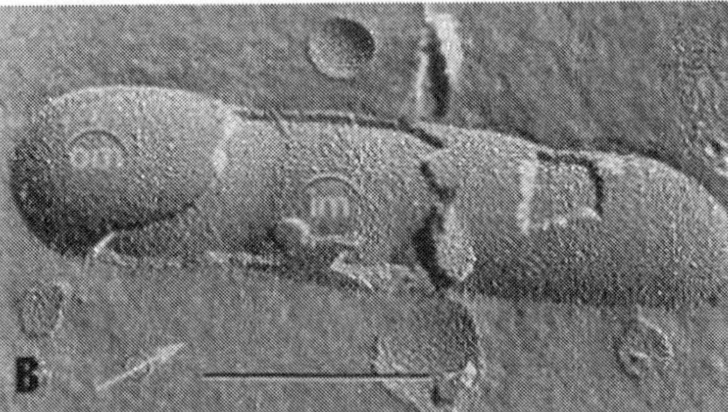

FIGURE 2-5 (A) Electron micrograph of a thin section of the Gram-positive M lysodeikticus showing the thick peptidoglycan cell wall (cw), underlying cytoplasmic (plasma) membrane (cm), mesome (m), and nucleus (n). (B) Freeze-fractured Bacteriodes cell showing typical major convex fracture faces through the inner (im) and outer (om) membranes. Bars = 1 μm; circled arrow in Fig. B indicates direction of shadowing.

Capsules and Loose Slime

Some bacteria form capsules, which constitute the outermost layer of the bacterial cell and surround it with a relatively thick layer of viscous gel. Capsules may be up to 10 μm thick. Some organisms lack a well-defined capsule but have loose, amorphous slime layers external to the cell wall or cell envelope. The a hemolytic Streptococcus mutans, the primary organism found in dental plaque is able to synthesis a large extracellular mucoid glucans from sucrose. Not all bacterial species produce capsules; however, the capsules of encapsulated pathogens are often important determinants of virulence. Encapsulated species are found among both Gram-positive and Gram-negative bacteria. In both groups, most capsules are composed of highmolecular-weight viscous polysaccharides that are retained as a thick gel outside the cell wall or envelope. The capsule of Bacillus anthracis (the causal agent of anthrax) is unusual in that it is composed of a g-glutamyl polypeptide. Table 2-1 presents the various capsular substances formed by a selection of Gram-positive and Gram-negative bacteria. A plasma membrane stage is involved in the biosynthesis and assembly of the capsular substances, which are

TABLE 2-1 Nature of Capsular substances Formed by Various Bacteria

Genus and Species	Capsular substances
Gram-Positive bacteria	
S pneumonia	Polysaccharides: e.g., type 111, glucose, glucoronic acid (celobiuronic acid); other types, various sugars and amino sugars
Streptococcus spp	Polysaccharides: e.g., hyaluronic acid (group A,), others containing amino sugar, uronic acids
B anthracis	γ-Glutamyl polypeptide
Gram-negative bacteria	
H influenzae	Polyrbosephosphaate
Klebsiella spp	Polysaccharides: sugars such as hexoses, fucose, uronic acids
N maningitidis	Polysaccharides: N-acetylmannosamine phosphate polymer (group A); sialic acid polymers (groups B and C)

extruded or secreted through the outer wall or envelope structures. Mutational loss of enzymes involved in the biosynthesis of the capsular polysaccharides can result in the smooth-to-rough variation seen in the pneumococci.

The capsule is not essential for viability. Viability is not affected when capsular polysaccharides are removed enzymatically from the cell surface. The exact functions of capsules are not fully understood, but they do confer resistance to phagocytosis and hence provide the bacterial cell with protection against host defenses to invasion.

Cell Wall and Gram-Negative Cell Envelope

The Gram stain broadly differentiates bacteria into Gram-positive and Gram-negative groups; a few organisms are consistently Gram-variable. Gram-positive and Gram-negative organisms differ drastically in the organization of the structures outside the plasma membrane but below the capsule (Fig. 2-6): in Gram-negative organisms these structures constitute the cell envelope, whereas in Gram-positive organisms they are called a cell wall.

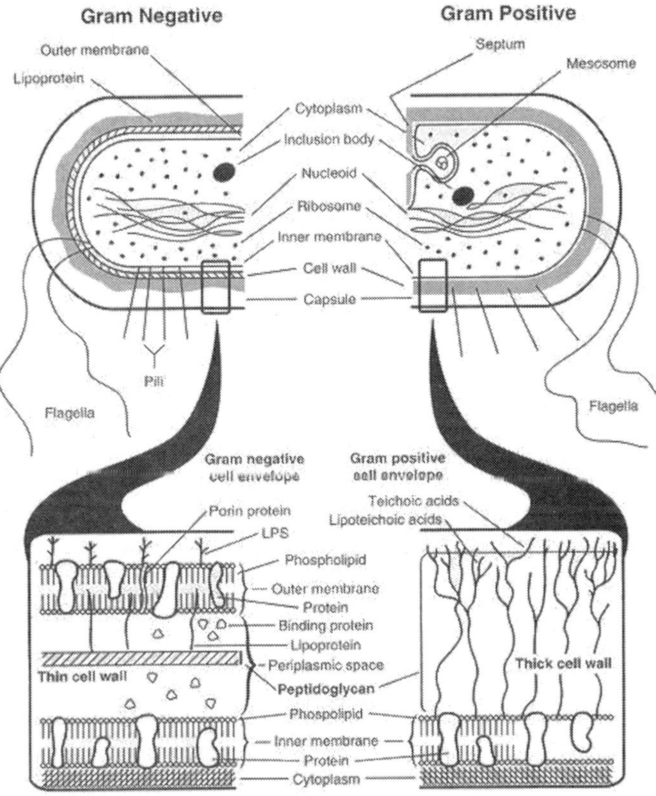

FIGURE 2-6 Comparison of the thick cell wall of Gram-positive bacteria with the comparatively thin cell wall of Gram-negative bacteria. Note the complexity of the Gram-negative cell envelope (outer membrane, its hydrophobic lipoprotein anchor; periplasmic space).

Most Gram-positive bacteria have a relatively thick (about 20 to 80 nm), continuous cell wall (often called the sacculus), which is composed largely of peptidoglycan (also known as mucopeptide or murein). In thick cell walls, other cell wall polymers (such as the teichoic acids, polysaccharides, and peptidoglycolipids) are covalently attached to the peptidoglycan. In contrast, the peptidoglycan layer in Gram-negative bacteria is thin (about 5 to 10 nm thick); in E coli, the peptidoglycan is probably only a monolayer thick. Outside the peptidoglycan layer in the Gram-negative envelope is an outer membrane structure (about 7.5 to 10 nm thick). In most Gram-negative bacteria, this membrane structure is anchored noncovalently to lipoprotein molecules (Braun's lipoprotein), which, in turn, are covalently linked to the peptidoglycan. The lipopolysaccharides of the Gram-negative cell envelope form part of the outer leaflet of the outer membrane structure.

The organization and overall dimensions of the outer membrane of the Gram-negative cell envelope are similar to those of the plasma membrane (about 7.5 nm thick). Moreover, in Gram-negative bacteria such as E coli, the outer and inner membranes adhere to each other at several hundred sites (Bayer patches); these sites can break up the continuity of the peptidoglycan layer. Table 2-2 summarizes the major classes of chemical constituents in the walls and envelopes of Gram-positive and Gram-negative bacteria.

TABLE 2-2 Major classes of Chemical Components in Bacterial Walls and Envelopes

Chemical component	Examples
Gram-positive cell walls	
Puplidoglycan	All species
Poysaccharides Teichoic acid	Streptococcus group A, B, C substances
Ribitol	S aureus, B subtitles, Lactobaillus spp
Gliycerol	S epidermidis, Lactobacillus spp
Telchuronic acids (aminogalacturonic or ainomannuronic acid polymers)	B licheniformis, M lysodeikticus
Peptideglycolipids (Muramylpeptide-polysaccharide-mycolates)	Corynebacterium spp, Mycobacterium spp, Nocardia spp
Glcolipids ("waxes") (polysaccharide-mycolates)	
Gram-negative envelopes	
LPS	All species
Lipoprotein	E coli and marry enteric bacteria, Pseudomonas aeruginosa
Porins (Major outer membrane proteins)	E coli, Salmonella typhimurium
Phospholipids and proteins	All species
Peptidoglycan	Almost all species

The basic differences in surface structures of Gram-positive and Gram-negative bacteria explain the results of Gram staining. Both Gram-positive and Gram-negative bacteria take up the same amounts of crystal violet (CV) and iodine (I). The CV-I complex, however, is trapped inside the Gram-positive cell by the dehydration and reduced porosity of the thick cell wall as a result of the differential washing step with 95 percent ethanol or other solvent mixture. In contrast, the thin peptidoglycan layer and probable discontinuities at the membrane adhesion sites do not impede solvent extraction of the CV-I complex from the Gram-negative cell. The above mechanism of the Gram stain based on the structural differences between the two groups has been confirmed by sophisticated methods of electron microscopy (see Ref. Bereridge and Daries). The sequence of steps in the Gram stain differentiation is illustrated diagrammatically in Figure 2-7. Moreover, mechanical disruption of the cell wall of Gram-positive organisms or its enzymatic removal with lysozyme results in complete extraction of the CV-I complex and conversion to a Gram-negative reaction. Therefore, autolytic wall-degrading enzymes that cause cell wall breakage may account for Gram-negative or variable reactions in cultures of Gram-positive organisms (such as Staphylococcus aureus, Clostridium perfringens, Corynebacterium diphtheriae, and some Bacillus spp).

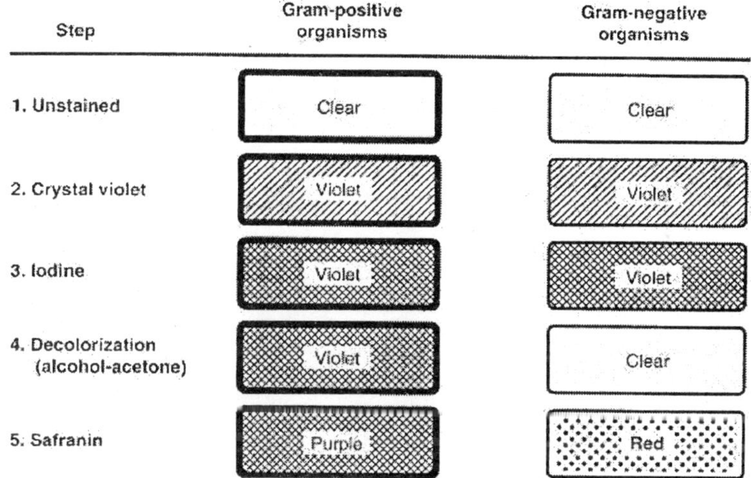

FIGURE 2-7 General sequence of steps in the Gram stain procedure and the resultant staining of Gram-positive and Gram-negative bacteria.

Peptidoglycan

Unique features of almost all prokaryotic cells (except for Halobacterium halobium and mycoplasmas) are cell wall peptidoglycan and the specific enzymes involved in its biosynthesis. These enzymes are target sites for inhibition of peptidoglycan synthesis by specific antibiotics. The primary chemical structures of peptidoglycans of both Gram-positive and Gram-negative bacteria have been established; they consist of a glycan backbone of repeating groups of b1, 4-linked disaccharides of b1,4-N-acetylmuramyl-N-acetylglucosamine. Tetrapeptides of L-alanine-Disoglutamic acid-L-lysine (or diaminopimelic acid)-n-alanine are linked through the carboxyl group by amide linkage of muramic acid residues of the glycan chains; the D-alanine residues are directly cross-linked to the e-amino group of lysine or diaminopimelic acid on a neighboring tetrapeptide, or they are linked by a peptide bridge. In S aureus peptidoglycan, a glycine pentapeptide bridge links the two adjacent peptide structures. The extent of direct or peptide-bridge cross-linking varies from one peptidoglycan to another. The staphylococcal peptidoglycan is highly cross-linked, whereas that of E coli is much less so, and has a more open peptidoglycan mesh. The diamino acid providing the e-amino group for cross-linking is lysine or diaminopimelic acid, the latter being uniformly present in Gram-negative peptidoglycans. The structure of the peptidoglycan is illustrated in Figure 2-8. A peptidoglycan with a chemical structure substantially different from that of all eubacteria has been discovered in certain archaebacteria. Instead of muramic acid, this peptidoglycan contains talosaminuronic acid and lacks the D-amino acids found in the eubacterial peptidoglycans. Interestingly, organisms containing this wall polymer (referred to as pseudomurein) are insensitive to penicillin, an inhibitor of the transpeptidases involved in peptidoglycan biosynthesis in eubacteria.

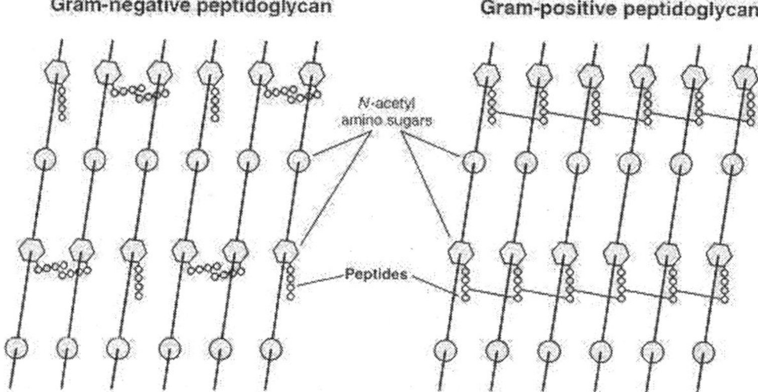

FIGURE 2-8 Diagrammatic representation of peptidoglycan structures with adjacent glycan strands cross-linked directly from the carboxyterminal D-alanine to the e-amino group of an adjacent tetrapeptide or through a peptide cross bridge ,N-acetylmuramic acid; N-acetylglucosamine.

The ß-1,4 glycosidic bond between N-acetylmuramic acid and N-acetylglucosamine is specifically cleaved by the bacteriolytic enzyme lysozyme. Widely distributed in nature, this enzyme is present in human tissues and secretions and can cause complete digestion of the peptidoglycan walls of sensitive organisms. When lysozyme is allowed to digest the cell wall of Gram-positive bacteria suspended in an osmotic stabilizer (such as sucrose), protoplasts are formed. These protoplasts are able to survive and continue to grow on suitable media in the wall-less state. Gram-negative bacteria treated similarly produce spheroplasts, which retain much of the outer membrane structure. The dependence of bacterial shape on the peptidoglycan is shown by the transformation of rod-shaped bacteria to spherical protoplasts (spheroplasts) after enzymatic breakdown of the peptidoglycan. The mechanical protection afforded by the wall peptidoglycan layer is evident in the osmotic fragility of both protoplasts and spheroplasts. There are two groups of bacteria that lack the protective cell wall peptidoglycan structure, the Mycoplasma species, one of which causes atypical pneumonia and some genitourinary tract infections and the L-forms, which originate from Gram-positive or Gram-negative bacteria and are so designated because of their discovery and description at the Lister Institute, London. The mycoplasmas and L-forms are all Gram-negative and insensitive to penicillin and are bounded by a surface membrane structure. L-forms arising "spontaneously" in cultures or isolated from infections are structurally related to protoplasts and spheroplasts; all three forms (protoplasts, spheroplasts, and L-forms) revert infrequently and only under special conditions.

Teichoic Acids

Wall teichoic acids are found only in certain Gram-positive bacteria (such as staphylococci, streptococci, lactobacilli, and Bacillus spp); so far, they have not been found in gram- negative organisms. Teichoic acids are polyol phosphate polymers, with either ribitol or glycerol linked by phosphodiester bonds; their structures are illustrated in Figure 2-9. Substituent groups on the polyol chains can include D-alanine (ester linked), N-acetylglucosamine, N-acetylgalactosamine, and glucose; the substituent is characteristic for the teichoic acid from a particular bacterial species and can act as a specific antigenic determinant. Teichoic acids are covalently linked to the peptidoglycan. These highly negatively charged polymers of the bacterial wall can serve as a cation-sequestering mechanism.

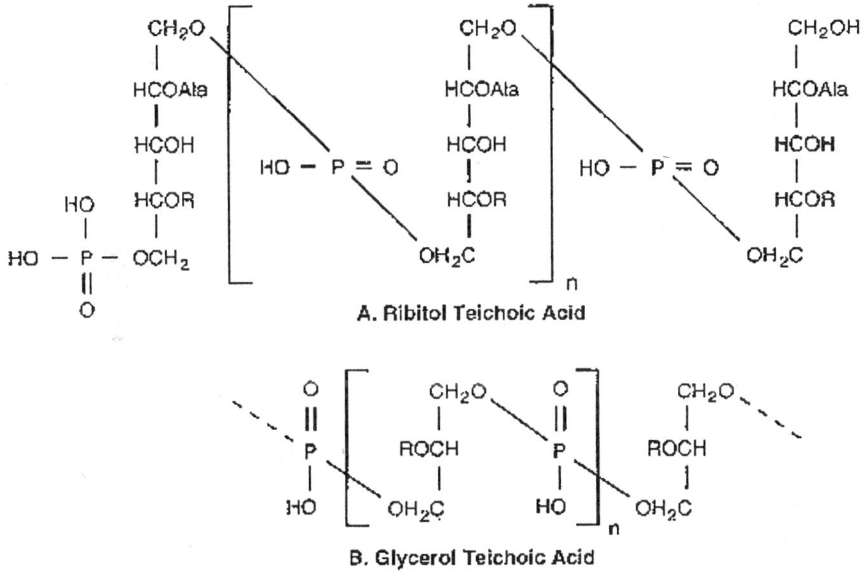

A. Ribitol Teichoic Acid

B. Glycerol Teichoic Acid

FIGURE 2-9 Structures of cell wall teichoic acids. (A) Ribitol teichoic acid with repeating units of 1,5-phosphodiester linkages of D-ribitol and D-alanyl ester on position 2 and glycosyl substituents (R) on position 4. The glycosyl groups may abe N-acetylglucosaminyl (a or b) as in S aureus or a-glucosyl as in B subtilis W23. (B) Glycerol teichoic acid with 1,3-phosphodiester linkages of glycerol repeating units (1,2-linkages in some species). In the glycerol teichoic acid structure shown, the polymer may be unsubstituted (R - H) or substituted (R - D-alanyl or glycosyl).

Accessory Wall Polymers

In addition to the principal cell wall polymers, the walls of certain Gram-positive bacteria possess polysaccharide molecules linked to the peptidoglycan. For example, the C polysaccharide of streptococci confers group specificity. Acidic polysaccharides attached to the peptidoglycan are called teichuronic acids. Mycobacteria have peptidoglycolipids, glycolipids, and waxes associated with the cell wall.

Lipopolysaccharides

A characteristic feature of Gram-negative bacteria is possession of various types of complex macromolecular lipopolysaccharide (LPS). So far, only one Gram-positive organism, Listeria monocytogenes, has been found to contain an authentic LPS. The LPS of this bacterium and those of all Gram-negative species are also called endotoxins, thereby distinguishing these cell-bound, heat-stable toxins from heat-labile, protein exotoxins secreted into culture media. Endotoxins possess an array of powerful biologic activities and play an important role in the pathogenesis of many Gram-negative bacterial infections. In addition to causing endotoxic shock, LPS is pyrogenic, can activate macrophages and complement, is mitogenic for B lymphocytes, induces interferon production, causes tissue necrosis and tumor regression, and has adjuvant properties. The endotoxic properties of LPS reside largely in the lipid A components. Usually, the LPS molecules have three regions: the lipid A structure required for insertion in the outer leaflet of the outer membrane bilayer; a covalently attached core composed of 2-keto-3deoxyoctonic acid (KDO), heptose, ethanolamine, N-acetylglucosamine, glucose, and galactose; and polysaccharide chains linked to the core. The polysaccharide chains constitute the O-antigens of the Gram-negative bacteria, and the individual monosaccharide constituents confer serologic specificity on these components. Figure 2-10 depicts the structure of LPS. Although, it has been known that lipid A is composed of b1,6-linked D-glucosamine disaccharide substituted with phosphomonester groups at positions 4' and 1, uncertainties have existed about the attachment positions of the six fatty acid acyl and KDO groups on the disaccharide. The demonstration of the structure of lipid A of LPS of a heptoseless mutant of Salmonella typhimurium has established that amide-linked hydroxymyristoyl and lauroxymyristoyl groups are attached to the nitrogen of the 2- and 2'-carbons, respectively, and that hydroxymyristoyl and myristoxymyristoyl groups are attached to the oxygen of the 3- and 3'-carbons of the disaccharide, respectively. Therefore, only position 6' is left for attachment of KDO units.

Lipid A	Core	O Antigen
Glucosamine β-hydroxymyristate Fatty acids	Ketodeoxyoctonate Phosphoethanolamine Heptose Glucose, galactose, *N*-acetylglucosamine	Polysaccharide chains: repeating units of species-specific monosaccharides, e.g., galactose, rhamnose, mannose and abequose in *S typhimurium* LPS

FIGURE 2-10 The three major, covalently linked regions that form the typical LPS.

LPS and phospholipids help confer asymmetry to the outer membrane of the Gram-negative bacteria, with the hydrophilic polysaccharide chains outermost. Each LPS is held in the outer membrane by relatively weak cohesive forces (ionic and hydrophobic interactions) and can be dissociated from the cell surface with surface-active agents.

As in peptidoglycan biosynthesis, LPS molecules are assembled at the plasma or inner membrane. These newly formed molecules are initially inserted into the outer-inner membrane adhesion sites.

Outer Membrane of Gram-Negative Bacteria

In thin sections, the outer membranes of Gram-negative bacteria appear broadly similar to the plasma or inner membranes; however, they differ from the inner membranes and walls of Gram-positive bacteria in numerous respects. The lipid A of LPS is inserted with phospholipids to create the outer leaflet of the bilayer structure; the lipid portion of the lipoprotein and phospholipid form the inner leaflet of the outer membrane bilayer of most Gram-negative bacteria (Fig. 2-6).

In addition to these components, the outer membrane possesses several major outer membrane proteins; the most abundant is called porin. The assembled subunits of porin form a channel that limits the passage of hydrophilic molecules across the outer membrane barrier to those having molecular weights that are usually less than 600 to 700. Evidence also suggests that hydrophobic pathways exist across the outer membrane and are partly responsible for the differential penetration and effectiveness of certain b-lactam antibiotics (ampicillin, cephalosporins) that are active against various Gram-negative bacteria. Although, the outer membranes act as a permeability barrier or molecular sieve, they do not appear to possess energy-transducing systems to drive active transport. Several outer membrane proteins, however, are involved in the specific uptake of metabolites (maltose, vitamin B12, nucleosides) and iron from the medium. Thus, outer membranes of the Gram-negative bacteria provide a selective barrier to external molecules and thereby prevent the loss of metabolite-binding proteins and hydrolytic enzymes (nucleases, alkaline phosphatase) found in the periplasmic space. The periplasmic space is the region between the outer surface of the inner (plasma) membrane and the inner surface of the outer membrane (Figure 2-6). Thus, Gram-negative bacteria have a cellular compartment that has no equivalent in Gram-positive organisms. In addition to the hydrolytic enzymes, the periplasmic space holds binding proteins (proteins that specifically bind sugars, amino acids, and inorganic ions) involved in membrane transport and chemotactic receptor activities. Moreover, plasmid-encoded b-lactamases and aminoglycoside-modifying enzymes (phosphorylation or adenylation) in the periplasmic space produce antibiotic resistance by degrading or modifying an antibiotic in transit to its target sites on the membrane (penicillin-binding proteins) or on the ribosomes (aminoglycosides). These periplasmic proteins can be released by subjecting the cells to osmotic shock and after treatment with the chelating agent ethylenediaminetetraacetic acid.

Intracellular Components

Plasma (Cytoplasmic) Membranes

Bacterial plasma membranes, the functional equivalents of eukaryotic plasma membranes, are referred to variously as cytoplasmic, protoplast, or (in Gram-negative organisms) inner membranes. Similar in overall dimensions and appearance in thin sections to biomembranes from eukaryotic cells, they are composed primarily of proteins and lipids (principally phospholipids). Protein-to-lipid ratios of bacterial plasma membranes are approximately 3: 1, close to those for mitochondrial membranes. Unlike eukaryotic cell membranes, the bacterial membrane (except for Mycoplasma species and certain methylotrophic bacteria) has no sterols, and bacteria lack the enzymes required for sterol biosynthesis.

Although, their composition is similar to that of inner membranes of Gram-negative species, cytoplasmic membranes from Gram-positive bacteria possess a class of macromolecules not present in the Gram-negative membranes. Many Gram-positive bacterial membranes contain membrane-bound lipoteichoic acid, and species lacking this component (such as Micrococcus and Sarcina spp) contain an analogous membrane-bound succinylated lipomannan. Lipoteichoic acids are structurally similar to the cell wall glycerol teichoic acids in that they have basal polyglycerol phosphodiester 1-3 linked chains (Fig. 2-9). These chains terminate with the phosphomonoester end of the polymer, which is linked covalently to either a glycolipid or a phosphatidyl glycolipid moiety. Thus, a hydrophobic tail is provided for anchoring in the membrane lipid layers (Fig. 2-6A). As in the cell wall glycerol teichoic acid, the lipoteichoic acids can have glycosidic and D-alanyl ester substituents on the C-2 position of the glycerol.

Both membrane-bound lipoteichoic acid and membrane-bound succinylated lipomannan can be detected as antigens on the cell surface, and the glycerol-phosphate and succinylated mannan chains appear to extend through the cell wall structure (Fig. 2-6). This class of polymer has not yet been found in the cytoplasmic membranes of Gram-negative organisms. In both instances, the lipoteichoic acids and the lipomannans are negatively charged components and can sequester positively charged substances. They have been implicated in adhesion to host cells, but their functions remain to be elucidated.

Multiple functions are performed by the plasma membranes of both Gram-positive and Gram-negative bacteria. Plasma membranes are the site of active transport, respiratory chain components, energy-transducing systems, the H+-ATPase of the proton pump (see Chapter 4), and membrane stages in the biosynthesis of phospholipids, peptidoglycan, LPS, and capsular polysaccharides. In essence, the bacterial cytoplasmic membrane is a multifunction structure that combines the mitochondrial transport and biosynthetic functions that are usually compartmentalized in discrete membranous organelles in eukaryotic cells. The plasma membrane is also the anchoring site for DNA and provides the cell with a mechanism (as yet unknown) for separation of sister chromosomes.

Mesosomes

Thin sections of Gram-positive bacteria reveal the presence of vesicular or tubular-vesicular membrane structures called mesosomes, which are apparently formed by an invagination of the plasma membrane. These structures are much more prominent in Gram-positive than in Gram-negative organisms. At one time, the mesosomal vesicles were thought to be equivalent to bacterial mitochondria; however, many other membrane functions have also been attributed to the mesosomes. At present, there is no satisfactory evidence to suggest that they have a unique biochemical or physiologic function. Indeed, electron-microscopic studies have suggested that the mesosomes, as usually seen in thin sections, may arise from membrane perturbation and fixation artifacts. No general agreement exists about this theory, however, and some evidence indicates that mesosomes may be related to events in the cell division cycle.

Other Intracellular Components

In addition to the nucleoid and cytoplasm (cytosol), the intracellular compartment of the bacterial cell is densely packed with ribosomes of the 70S type (Fig. 2-2). These ribonucleoprotein particles, which have a diameter of 18 nm, are not arranged on a membranous rough endoplasmic reticulum as they are in eukaryotic cells. Other granular inclusions randomly distributed in the cytoplasm of various species include metabolic reserve particles such as poly-b-hydroxybutyrate (PHB), polysaccharide and glycogen-like granules, and polymetaphosphate or metachromatic granules.

Endospores are highly heat-resistant, dehydrated resting cells formed intracellularly in members of the genera Bacillus and Clostridium. Sporulation, the process of forming endospores, is an unusual property of certain bacteria. The series of biochemical and morphologic changes that occur during sporulation represent true differentiation within the cycle of the bacterial cell. The process, which usually begins in the stationary phase of the vegetative cell cycle, is initiated by depletion of nutrients (usually readily utilizable sources of carbon or nitrogen, or both). The cell then undergoes a highly complex, well-defined sequence of morphologic and biochemical events that ultimately lead to the formation of mature endospores. As many as seven distinct stages have been recognized by morphologic and biochemical studies of sporulating Bacillus species: stage 0, vegetative cells with two chromosomes at the end of exponential growth; stage I, formation of axial chromatin filament and excretion of exoenzymes, including proteases; stage II, forespore septum formation and segregation of nuclear material into two compartments; stage III, spore protoplast formation and elevation of tricarboxylic acid and glyoxylate cycle enzyme levels; stage IV, cortex formation and refractile appearance of spore; stage V, spore coat protein formation; stage VI, spore maturation, modification of cortical peptidoglycan, uptake of dipicolinic acid (a unique endospore product) and calcium, and development of resistance to heat and organic solvents; and stage VII, final maturation and liberation of endospores from mother cells (in some species).

When newly formed, endospores appear as round, highly refractile cells within the vegetative cell wall, or sporangium. Some strains produce autolysins that digest the walls and liberate free endospores. The spore protoplast, or core, contains a complete nucleus, ribosomes, and energy generating components that are enclosed within a modified cytoplasmic membrane. The peptidoglycan spore wall surrounds the spore membrane; on germination, this wall becomes the vegetative cell wall. Surrounding the spore wall is a thick cortex that contains an unusual type of peptidoglycan, which is rapidly released on germination. A spore coat of keratin like protein encases the spore contained within a membrane (the exosporium). During maturation, the spore protoplast dehydrates and the spore becomes refractile and resistant to heat, radiation, pressure, desiccation, and chemicals; these properties correlate with the cortical peptidoglycan and the presence of large amounts of calcium dipicolinate.

Recent evidence indicated that the spores of Bacillus spharicus were revived which had been preserved in amber for more than 25 million years. Their claims need to be reevaluated. Figure 2-11 illustrates the principal structural features of a typical endospore (Bacillus megaterium) on initiation of the germination process. The thin section of the spore shows the ruptured, thick spore coat and the cortex surrounding the spore protoplast with the germinal cell wall that becomes the vegetative wall on outgrowth.

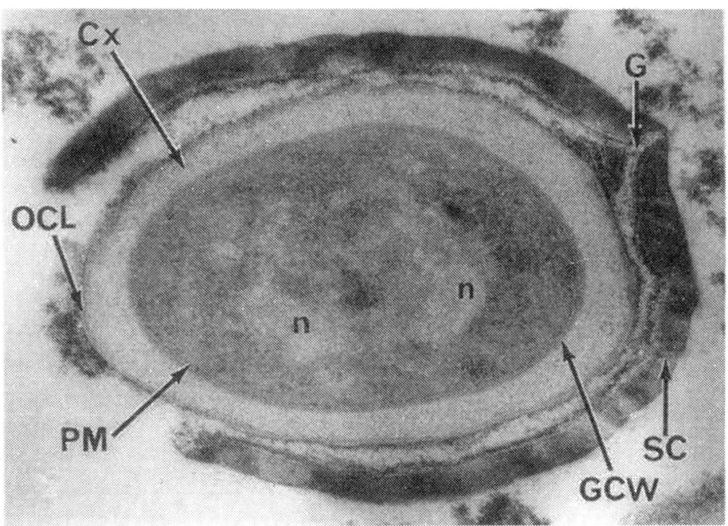

FIGURE 2-11 Electron micrograph of a thin section of a Bacillus megaterium spore showing the thick spore coat (SC), germinal groove (G) in the spore coat, outer cortex layer (OCL) and cortex (Cx) germinal cell wall layer (GCW), underlying spore protoplast membrane (PM), and regions where the nucleoid (n) is visible. (Courtesy of John H Freer, University of Glasgow, Scotland.)

REFERENCES

Beveridge TJ, Davies JA: Cellular responses of Bacillus subtilis and Escherichia coli to the Gram stain. J Bacteriol, 156:846,1983

Costerton JW, Ingram JM, Cheng KJ: Structure and function of the cell envelope of gram-negative bacteria. Bacteriol Rev, 38:87, 1974

Ghuysen J-M, Hakenbeck R: Bacterial cell wall. Elsevier, 1994

Gould GW, Hurst A (eds): The Bacterial Spore. Academic Press, San Diego, 1969

Jawetz E, Melnick JL, Adelberg EA: Medical Microbiology. Appleton & Lange, East Norwalk, CT, 1989

Rogers HJ: Bacterial Cell Structure. American Society for Microbiology, Washington, D.C., 1983

Seifert HS, So M: Genetic mechanisms of bacterial antigenic variation. Microbiological Reviews, Vol. 52:327, 1988

Wright A, Tipper DJ: The outer membrane of gram-negative bacteria. p. 427. In Sokatch JR, Ornston LN (eds): The bacteria. Vol. 7. Academic Press, San Diego, 1979

Chapter 3

Classification

General Concepts

Classification

Bacteria are classified and identified to distinguish among strains and to group them by criteria of interest to microbiologists and other scientists.

Nomenclature

Bacteria are named so that investigators can define and discuss them without the necessity of listing their characteristics.

Species

Species, groups of similar organisms within a genus, are designated by biochemical and other phenotypic criteria and by DNA relatedness, which groups strains on the basis of their overall genetic similarity.

Diagnostic Identification

Bacteria are identified routinely by morphological and biochemical tests, supplemented as needed by specialized tests such as serotyping and antibiotic inhibition patterns. Newer molecular techniques permit species tobe identified by their genetic sequences, sometimes directly from the clinical specimen.

Subtyping

Because of differences in pathogenicity or the necessity to characterize a disease outbreak, strains of medical interest are often classified below the species level by serotyping, enzyme typing, identification of toxins or other virulence factors, or characterization of plasmids, protein patterns, or nucleic acids.

New and Unusual Species

Laboratories have no difficulty identifying most bacteria. Problems develop with atypical strains and rare or newly described species; misidentification can lead to inappropriate patient care. Therefore, laboratory personnel and physicians (at least infectious disease specialists) must remain current regarding changes in taxonomy and the recognition of new species.

Role of the Clinical Laboratory

Clinical laboratory scientists detect, isolate, identify, and determine the antimicrobial susceptibility patterns of clinically relevant microbes at the request of physicians, and interface with public health laboratories.

INTRODUCTION

Bacteria are classified and identified to distinguish one organism from another and to group similar organisms by criteria of interest to microbiologists or other scientists. The most important level of this type of classification is the species level.

A species name should mean the same thing to everyone. Within one species, strains and subgroups can differ by the disease they produce, their environmental habitat, and many other characteristics. Formerly, species were created on the basis of such criteria, which may be extremely important for clinical microbiologists and physicians but which are not a sufficient basis for establishing a species. Verification of existing species and creation of new species should involve biochemical and other phenotypic criteria as well as DNA relatedness. In numerical or phenetic approaches to classification, strains are grouped on the basis of a large number of phenotypic characteristics. DNA relatedness is used to group strains on the basis of overall genetic similarity.

Species are identified in the clinical laboratory by morphological traits and biochemical tests, some of which are supplemented by serologic assessments (e.g., identification of *Salmonella* and *Shigella* species). Because of differences in pathogenicity *(Escherichia coli)* or the necessity to characterize a disease outbreak (*Vibrio cholerae,* methicillin-resistant *Staphylococcus aureus),* strains of medical interest are often classified below the species level by serology or identification of toxins. Pathogenic or epidemic strains also can be classified by the presence of a specific plasmid, by their plasmid profile (the number and sizes of plasmids), or by bacteriophage susceptibility patterns (phage typing). Newer molecular biologic techniques have enabled scientists to identify some species and strains (without the use of biochemical tests) by identifying a specific gene or genetic sequence, sometimes directly from the clinical specimen.

Laboratories have no difficulty in identifying typical strains of common bacteria using commonly available test systems. Problems do arise, however, when atypical strains or rare or newly described species are not in the database. Such difficulties are compounded when the strains are misidentified rather than unidentified, and so laboratory personnel and physicians (at least infectious diseases specialists) should be familiar with taxonomic reference texts and journals that publish papers on new species. Bacterial nomenclature at the genus and species level changes often, based primarily on the use of newer genetic techniques. A species may acquire more than one name. In some cases the recognition of a new species results in a unique correlation with specific clinical problems. For example, recognition of

Porphyromonas gingivalis as a unique species, separate from its previous inclusion within *Bacteroides melaninogenicus* (now known to be composed of several taxonomic groups of black-pigmenting anaerobic gram-negativebacilli), elucidated its role as a key pathogen in adult periodontitis. It is important to understand why these changes and synonyms exist in taxonomy.

The clinical laboratory is concerned with the rapid, sensitive, and accurate identification of microbes involved in producing disease. The number and types of tests done in such a laboratory depend on its size and the population it serves. Highly specialized or rarely performed tests should be done only by reference laboratories. Physicians, clinical laboratory personnel, and reference laboratory personnel must have a good working relationship if patients are to receive first-rate care.

In addition, the physician and the clinical laboratory personnel must know which diseases and isolates are reportable to public health laboratories and how to report them.

Definitions

Taxonomy

Taxonomy is the science of classification, identification, and nomenclature. For classification purposes, organisms are usually organized into subspecies, species, genera, families, and higher orders. For eukaryotes, the definition of the species usually stresses the ability of similar organisms to reproduce sexually with the formation of a zygote and to produce fertile offspring. However, bacteria do not undergo sexual reproduction in the eukaryotic sense. Other criteria are used for their classification.

Classification

Classification is the orderly arrangement of bacteria into groups. There is nothing inherently scientific about classification, and different groups of scientists may classify the same organisms differently. For example, clinical microbiologists are interested in the serotype, antimicrobial resistance pattern, and toxin and invasiveness factors in *Escherichia coli,* whereas geneticists are concerned with specific mutations and plasmids.

Identification

Identification is the practical use of classification criteria to distinguish certain organisms from others, to verify the authenticity or utility of a strain or a particular reaction, or to isolate and identify the organism that causes a disease.

Nomenclature

Nomenclature (naming) is the means by which the characteristics of a species are defined and communicated among microbiologists. A species name should mean the same thing to all microbiologists, yet some definitions vary in different countries or microbiologic specialty groups. For example, the organism known as *Clostridium perfringens* in the United States is called *Clostridium welchii* in England.

Species

A bacterial species is a distinct organism with certain characteristic features, or a group of organisms that resemble one another closely in the most important features of their organization. In the past, unfortunately, there was little agreement about these criteria or about the number of features necessary to distinguish a species. Species were often defined solely by such criteria as host range, pathogenicity, or ability to produce gas during the fermentation of a given sugar. Without a universal consensus, criteria reflected the interests of the investigators who described a particular species. For example, bacteria that caused plant diseases were often defined by the plant from which they were isolated; also, each new *Salmonella* serotype that was discovered was given species status. These practices have been replaced by generally accepted genetic criteria that can be used to define species in all groups of bacteria.

Approaches to Taxonomy

Numerical Approach

In their studies on members of the family Enterobacteriaceae, Edwardsand Ewing established the following principles to characterize, classify, and identify organisms (Lennette et al., 1985):

Classification and identification of an organism should be based on its overall morphologic and biochemical pattern. A single characteristic (pathogenicity, host range, or biochemical reaction), regardless of its importance, is not a sufficient basis for classifying or identifying an organism.

A large and diverse strain sample must be tested to determine accurately the biochemical characteristics used to distinguish a given species.

Atypical strains often are perfectly typical members of a given biogroup within an existing species, but sometimes they are typical members of an unrecognized new species.

In numerical taxonomy (also called computer or phenetic taxonomy) many (50 to 200) biochemical, morphological, and cultural characteristics, as well as susceptibilities to antibiotics and inorganic compounds, are used to determine the degree of similarity between organisms. In numerical studies, investigators often calculate the coefficient of similarity or percentage of similarity between strains (where strain indicates a single isolate from a specimen). A dendrogram or a similarity matrix is constructed that joins individual strains into groups and places one group with other groups on the basis of their percentage of similarity. In the dendrogram in Figure 3-1, group 1 represents three *Citrobacter freundii* strains that are about 95 percent similar and join with a fourth *C freundii* strain at the level of 90 percent similarity. Group 2 is composed of three *Citrobacter diversus* strains that are 95 percent similar, and group 3 contains two *E coli* strains that are 95 percent similar, as well as a third *E coli* strain to which they are 90 percent similar. Similarity between groups 1 and 2 occurs at the 70 percent level, and group 3 is about 50 percent similar to groups 1 and 2.

In some cases, certain characteristics may be weighted more heavily; for example, the presence of spores in *Clostridium* might be weighted more heavily than the organism's ability to use a specific carbon source. A given level of similarity can be equated with relatedness at the genus, species, and, sometimes, subspecies levels. For instance, strains of a given species may cluster at a 90% similarity level, species within a given genus may cluster at the 70 percent level, and different genera in the same family may cluster at the 50 percent or lower level (Fig. 3-1).

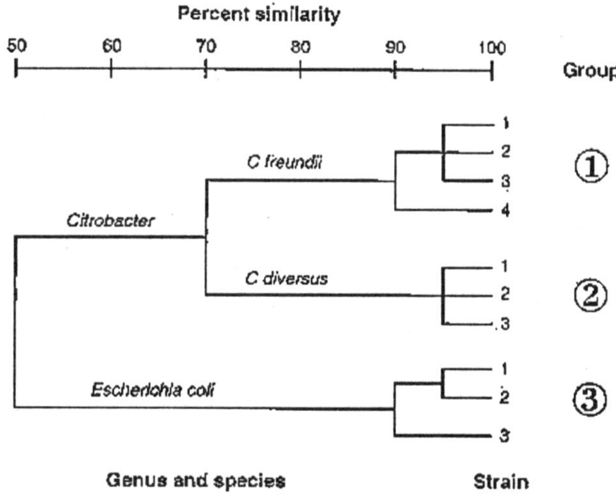

FIGURE 3-1 Example of dendrogram.

When this approach is the only basis for defining a species, it is difficult to know how many and which tests should be chosen; whether and how the tests should be weighted; and what level of similarity should be chosen to reflect relatedness at the genus and species levels.

Most bacteria have enough DNA to specify some 1,500 to 6,000 average-sized genes. Therefore, even a battery of 300 tests would assay only 5 to 20 percent of the genetic potential of a bacterium. Tests that are comparatively simple to conduct (such as those for carbohydrate utilization and for enzymes, presence of which can be assayed colorimetrically) are performed more often than tests for structural, reproductive, and regulatory genes, presence of which is difficult to assay. Thus, major differences may go undetected.

Other types of errors may occur when species are classified solely on the basis of phenotype. For example, different enzymes (specified by different genes) may catalyze the same reaction. Also, even if a metabolic gene is functional, negative reactions can occur because of the inability of the substrate to enter the cell, because of a mutation in a regulatory gene, or by production of an inactive protein. There is not necessarily a one-to-one correlation between a reaction and the number of genes needed to carry out that reaction. For instance, six enzymatic steps may be involved in a given pathway. If an assay for the end product is performed, a positive reaction indicates the presence of all six enzymes, whereas a negative reaction can mean the absence or nonfunction of one to six enzymes. Several other strain characteristics can affect phenotypic characterization; these include growth rate, incubation temperature, salt requirement, and pH. Plasmids that carry metabolic genes can enable strains to carry out reactions atypical for strains of that species.

The same set of "definitive" reactions cannot be used to classify all groups of organisms, and there is no standard number of specific reactions that allows identification of a species. Organisms are identified on the basis of phenotype, but, from the taxonomic standpoint, definition of species solely on this basis is subject to error.

Phylogenetic Approach

The ideal means of identifying and classifying bacteria would be to compare each gene sequence in a given strain with the gene sequences for every known species. This cannot be done, but the total DNA of one organism can be compared with that of any other organism by a method called nucleic acid hybridization or DNA hybridization. This method can be used to measure the number of DNA sequences that any two organisms have in common and to estimate the percentage of divergence within DNA sequences that are related but not identical. DNA relatedness studies have been done for yeasts, viruses, bacteriophages, and many groups of bacteria.

Five factors can be used to determine DNA relatedness: genome size, guanine-plus-cytosine (G+C) content, DNA relatedness under conditions optimal for DNA reassociation, thermal stability of related DNA sequences, and DNA relatedness under conditions supraoptimal for DNA reassociation. Because it is not practical to conduct these genotypic or phylogenetic evaluations in clinical laboratories, the results of simpler tests usually must be correlated with known phylogenetic data. For example, yellow strains of *Enterobacter cloacae* were shown, by DNA relatedness, to form a separate species, *Enterobacter sakazakii,* but were not designated as such until results of practical tests were correlated with the DNA data to allow routine laboratories to identify the new species.

Genome Size

True bacterial DNAs have genome sizes (measured as molecular weight) between 1 X 109 and 8 X 109. Genome size determinations sometimes can distinguish between groups. They were used to distinguish *Legionella pneumophila* (the legionnaire's disease bacterium) from *Bartonella (Rickettsia) quintana,* the agent of trench fever. *L pneumophila* has a genome size of about 3 X 109; that of *B quintana* is about 1 X 109.

Guanine-plus-Cytosine Content

The G+C content in bacterial DNA ranges from about 25 to 75 percent. This percentage is specific, but not exclusive, for a species; two strains with a similar G+C content may or may not belong to the same species. If the G+C contents are very different, however, the strains cannot be members of the same species.

DNA Relatedness under Conditions Optimal for DNA Reassociation

DNA relatedness is determined by allowing single-stranded DNA from one strain to reassociate with single-stranded DNA from a second strain, to form a double-stranded DNA molecule (Figure 3-2). This is a specific, temperature-dependent reaction. The optimal temperature for DNA reassociation is 25 to 30°C below the temperature at which native double-stranded DNA denatures into single strands. Many studies indicate that a bacterial species is composed of strains that are 70 to 100 percent related. In contrast, relatedness between different species is 0 to about 65 percent. It is important to emphasize that the term "related" does not mean "identical" or "homologous." Similar but nonidentical nucleic acid sequences can reassociate.

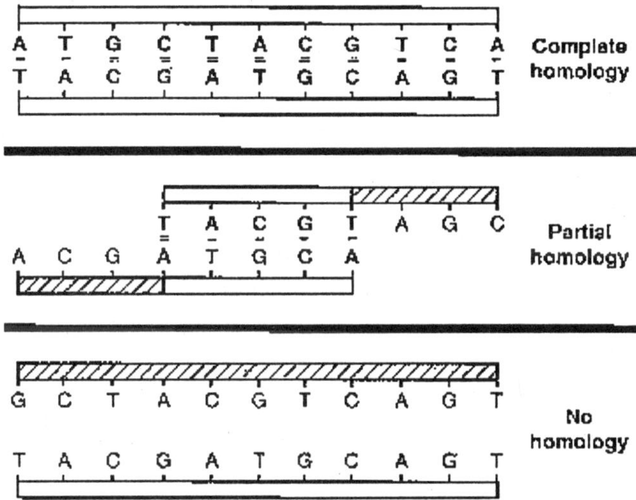

FIGURE 3-2 Diagram of DNA reassociation. DNA is composed of two purine nucleoside bases, adenine (A) and guanine (G), and two pyrimidine nucleoside bases, thymine (T) and cytosine (C). Double-stranded DNA is formed through hydrogen bonds that can occur only between the complementary bases A and T or G and C. (Top) Perfectly reassociated DNA base sequence in which all nucleosides are paired by hydrogen bonds. (Middle) Perfectly paired DNA base sequence in the center with unpaired, single-strand ends on each strand. (Bottom) None of the bases in the sequence (left to right) GCTACGTCAGT on the top strand are complementary to the sequence TACGATGCAGT in the bottom strand.

Thermal Stability of Related DNA Sequences

Each 1 percent of unpaired nucleotide bases in a double-stranded DNA sequence causes a 1 percent decrease in the thermal stability of that DNA duplex. Therefore, a comparison between the thermal stability of a control double-stranded molecule (in which both strands of DNA are from the same organism) and that of a heteroduplex (DNA strands from two different organisms) allows assessment of divergence between related nucleotide sequences.

DNA Relatedness under Supraoptimal Conditions for DNA Reassociation

When the incubation temperature used for DNA reassociation is raised from 25-30° C below the denaturation temperature to only 10-15° C below the denaturation temperature, only very closely related (and therefore highly thermally stable) DNA sequences can reassociate. Strains from the same species are 60 percent or more related at these supraoptimal incubation temperatures.

Defining Species on the Basis of DNA Relatedness

Use of these five factors allows a species definition based on DNA. Thus, *E coli* can be defined as a series of strains with a G+C content of 49 to 52 moles percent, a genome molecular weight of 2.3 X 109 to 3.0 X 109, relatedness of 70 percent or more at an optimal reassociation temperature with 0 to 4 percent divergence in related sequences, and relatedness of 60 percent or more at a supraoptimal reassociation temperature. Experience with more than 300 species has produced an arbitrary phylogenetic definition of a species to which most taxonomists subscribe: "strains with approximately 70% or greater DNA-DNA relatedness and with 5°C or less divergence in related sequences." When these two criteria are met, genome size and G+C content are always similar, and relatedness is almost always 60 percent or more at supraoptimal incubation temperatures. The 70 percent species relatedness rule has been ignored occasionally when the existing nomenclature is deeply ingrained, as is that for *E coli* and the four *Shigella* species. Because these organisms are all 70 percent or more related, DNA studies indicate that they should be grouped into a single species, instead of the present five species in two genera. This change has not been made because of the presumed confusion that would result.

DNA relatedness provides one species definition that can be applied equally to all organisms. Moreover, it cannot be affected by phenotypic variation, mutations, or the presence or absence of metabolic or other plasmids. It measures overall relatedness, and these factors affect only a very small percentage of the total DNA.

Polyphasic Approach

In practice, the approach to bacterial taxonomy should be polyphasic (Fig. 3-3). The first step is phenotypic grouping of strains by morphological, biochemical and any other characteristics of interest. The phenotypic groups are then tested for DNA relatedness to determine whether the observed phenotypic homogeneity (or heterogeneity) is reflected by phylogenetic homogeneity or heterogeneity. The third and most important step is reexamination of the biochemical characteristics of the DNA relatedness groups. This allows determination of the biochemical borders of each group and determination of reactions of diagnostic value for the group. For identification of a given organism, the importance of specific tests is weighted on the basis of correlation with DNA results. Occasionally, the reactions commonly used will not distinguish completely between two distinct DNA relatedness groups. In these cases, other biochemical tests of diagnostic value must be sought.

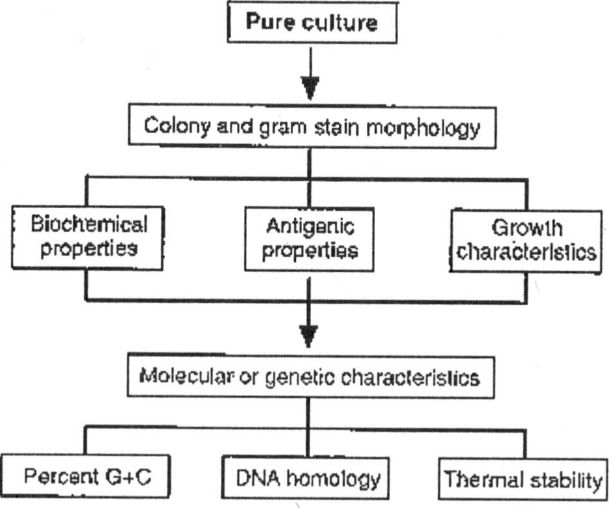

FIGURE 3-3 Bacterial identification.

Phenotypic Characteristics Useful in Classification and Identification

Morphologic Characteristics

Both wet-mounted and properly stained bacterial cell suspensions can yield a great deal of information. These simple tests can indicate the Gram reaction of the organism; whether it is acid-fast; its motility; the arrangement of its flagella; the presence of spores, capsules, and inclusion bodies; and, of course, its shape. This information often can allow identification of an organism to the genus level, or can minimize the possibility that it belongs to one or another group. Colony characteristics and pigmentation are also quite helpful. For example, colonies of several *Porphyromonas* species autofluoresence under long-wavelength ultraviolet light, and *Proteus* species swarm on appropriate media.

Growth Characteristics

A primary distinguishing characteristic is whether an organism grows aerobically, anaerobically, facultatively (i.e., in either the presence or absence of oxygen), or microaerobically (i.e., in the presence of a less than atmospheric partial pressure of oxygen). The proper atmospheric conditions are essential for isolating and identifying bacteria. Other important growth assessments include the incubation temperature, pH, nutrients required, and resistance to antibiotics. For example, one diarrheal disease agent, *Campylobacter jejuni*, grows well at 42° C in the presence of several antibiotics; another, *Y enterocolitica*, grows better than most other bacteria at 4° C. *Legionella, Haemophilus*, and some other pathogens require specific growth factors, whereas *E coli* and most other Enterobacteriaceae can grow on minimal media.

Antigens and Phage Susceptibility

Cell wall (O), flagellar (H), and capsular (K) antigens are used to aid in classifying certain organisms at the species level, to serotype strains of medically important species for epidemiologic purposes, or to identify serotypes of public health importance. Serotyping is also sometimes used to distinguish strains of exceptional virulence or public health importance, for example with *V cholerae* (O1 is the pandemic strain) and *E coli* (enterotoxigenic, enteroinvasive, enterohemorrhagic, and enteropathogenic serotypes).

Phage typing (determining the susceptibility pattern of an isolate to a set of specific bacteriophages) has been used primarily as an aid in epidemiologic surveillance of diseases caused by *Staphylococcus aureus*, mycobacteria, *P aeruginosa, V cholerae*, and *S typhi*. Susceptibility to bacteriocins has also been used as an epidemiologic strain marker. In most cases recently, phage and bacteriocin typing have been supplanted by molecular methods.

Biochemical Characteristics

Most bacteria are identified and classified largely on the basis of their reactions in a series of biochemical tests. Some tests are used routinely for many groups of bacteria (oxidase, nitrate reduction, amino acid degrading enzymes, fermentation or utilization of carbohydrates); others are restricted to a single family, genus, or species (coagulase test for staphylococci, pyrrolidonyl arylamidase test for Gram-positive cocci).

Both the number of tests needed and the actual tests used for identification vary from one group of organisms to another. Therefore, the lengths to which a laboratory should go in detecting and identifying organisms must be decided in each laboratory on the basis of its function, the type of population it serves, and its resources. Clinical laboratories today base the extent of their work on the clinical relevance of an isolate to the particular patient from which it originated, the public health significance of complete identification, and the overall cost-benefit analysis of their procedures. For example, the Centers for Disease Control and Prevention (CDC) reference laboratory uses at least 46 tests to identify members of the Enterobacteriaceae, whereas most clinical laboratories, using commercial identification kits or simple rapid tests, identify isolates with far fewer criteria.

Classification Below and Above the Species Level

Below the Species Level

Particularly for epidemiological purposes, clinical microbiologists must distinguish strains with particular traits from other strains in the same species. For example, serotype O157:H7 *E coli* are identified in stool specimens because of their association with bloody diarrhea and subsequent hemolytic urenic syndrome.

Below the species level, strains are designated as groups or types on the basis of common serologic or biochemical reactions, phage or bacteriocin sensitivity, pathogenicity, or other characteristics. Many of these characteristics are already used and accepted: serotype, phage type, colicin type, biotype, bioserotype (a group of strains from the same species with common biochemical and serologic characteristics that set them apart from other members of the species), and pathotype (e.g., toxigenic *Clostridium difficile,* invasive *E coli,* and toxigenic *Corynebacterium diphtheriae*).

Above the Species Level

In addition to species and subspecies designations, clinical microbiologists must be familiar with genera and families. A genus is a group of related species, and a family is a group of related genera.

An ideal genus would be composed of species with similar phenotypic and phylogenetic characteristics. Some phenotypically homogeneous genera approach this criterion *(Citrobacter, Yersinia,* and *Serratia).* More often, however, the phenotypic similarity is present, but the genetic relatedness is not. *Bacillus, Clostridium,* and *Legionella* are examples of accepted phenotypic genera in which genetic relatedness between species is not 50 to 65 percent, but 0 to 65 percent. When phenotypic and genetic similarity are not both present, phenotypic similarity generally should be given priority in establishing genera. Identification practices are simplified by having the most phenotypically similar species in the same genus. The primary consideration for a genus is that it contain biochemically similar species that are convenient or important to consider as a group separate from other groups of organisms.

The sequencing of ribosomal RNA (rRNA) genes, which have been highly conserved through evolution, allows phylogenetic comparisons to be made between species whose total DNAs are essentially unrelated. It also allows phylogenetic classification at the genus, family, and higher taxonomic levels. The rRNA sequence data are usually not used to designate genera or families unless supported by similarities in phenotypic tests.

Designation of New Species and Nomenclatural Changes

Species are named according to principles and rules of nomenclature set forth in the Bacteriological Code. Scientific names are taken from Latin or Greek. The correct name of a species or higher taxon is determined by three criteria: valid publication, legitimacy of the name with regard to the rules of nomenclature, and priority of publication (that is, it must be the first validly published name for the taxon).

To be published validly, a new species proposal must contain the species name, a description of the species, and the designation of a type strain for the species, and the name must be published in the *International Journal for Systematic Bacteriology (IJSB).* Once proposed, a name does not go through a formal process to be accepted officially; in fact, the opposite is truea validly published name is assumed to be correct unless and until it is challenged officially. A challenge is initiated by publishing a request for an opinion (to the Judicial Commission of the International Association of Microbiological Societies) in the *IJSB.* This occurs only in cases in which the validity of a name is questioned with respect to compliance with the rules of the Bacteriological Code. A question of classification that is based on scientific data (for example, whether a species, on the basis of its biochemical or genetic characteristics, or both, should be placed in a new genus or an existing genus) is not settled by the Judicial Commission, but by the preference and usage of the scientific community. This is why there are pairs of names such as *Providencia rettgeri/Proteus rettgeri, Moraxella catarrhalis/Branhamella catarrhalis,* and *Legionella micdadei/Tatlockia micdadei.* More than one name may thus exist for a single organism. This is not, however, restricted tobacterial nomenclature. Multiple names exist for many antibiotics and other drugs and enzymes.

A number of genera have been divided into additional genera and species have been moved to new or existing genera, such as *Arcobacter* (new genus for former members of *Campylobacter*) and *Burkholderia* species (formerly species of *Pseudomonas*). Two former *Campylobacter* species (*cinaedi* and *fennelliae*) have been moved to the existing genus *Helicobacter* in another example.

The best source of information for new species proposals and nomenclatural changes is the *IJSB.* In addition, the *Journal of Clinical Microbiology* often publishes descriptions of newly described microorganisms isolated from clinical sources. Information, including biochemical reactions and sources of isolation, about new organisms of clinical importance, disease outbreaks caused by newer species, and reviews of clinical significance of certain organisms may be found in the *Annals of Internal Medicine, Journal of Infectious Diseases, Clinical Microbiology Reviews,* and *Clinical Infectious Diseases.* The data provided in these publications supplement and update *Bergey's Manual of Systematic Bacteriology,* the definitive taxonomic reference text.

Assessing Newly Described Bacteria

Since 1974, the number of genera in the family Enterobacteriaceae has increased from 12 to 28 and the number of species from 42 to more than 140, some of which have not yet been named. Similar explosions have occurred in other genera. In 1974, five species were listed in the genus *Vibrio* and four in *Campylobacter;* the genus *Legionella* was unknown. Today, there are at least 25 species in *Vibrio,* 12

Campylobacter species, and more than 40 species in *Legionella*. The total numbers of genera and species continue to increase dramatically.

The clinical significance of the agent of legionnaire's disease was well known long before it was isolated, characterized, and classified as *Legionella pneumophila*. In most cases, little is known about the clinical significance of a new species at the time it is first described. Assessments of clinical significance begin after clinical laboratories adopt the procedures needed to detect and identify the species and accumulate a body of data.

In fact, the detection and even the identification of uncultivatable microbes from different environments are now possible using standard molecular methods. The agents of cat scratch disease (*Bartonella henselae*) and Whipple's disease (*Tropheryma whippelii*) were elucidated in this manner. *Bartonella henselae* has since been cultured from several body sites from numerous patients; *T whippelii* remains uncultivated.

New species will continue to be described. Many will be able to infect humans and cause disease, especially in those individuals who are immunocompromised, burned, postsurgical, geriatric, and suffering from acquired immunodeficiency syndrome (AIDS). With today's severely immunocompromised patients, often the beneficiaries of advanced medical interventions, the concept of "pathogen" holds little meaning. Any organism is capable of causing disease in such patients under the appropriate conditions.

Role of the Clinical Laboratory

Clinical laboratory scientists should be able to isolate, identify, and determine the antimicrobial susceptibility pattern of the vast majority of human disease agents so that physicians can initiate appropriate treatment as soon as possible, and the source and means of transmission of outbreaks can be ascertained to control the disease and prevent its recurrence. The need to identify clinically relevant microorganisms both quickly and cost-effectively presents a considerable challenge.

To be effective, the professional clinical laboratory staff must interact with the infectious diseases staff. Laboratory scientists should attend infectious disease rounds. They must keep abreast of new technology, equipment, and classification and should communicate this information to their medical colleagues. They should interpret, qualify, or explain laboratory reports. If a bacterial name is changed or a new species reported, the laboratory should provide background information, including a reference.

The clinical laboratory must be efficient. A concerted effort must be made to eliminate or minimize inappropriate and contaminated specimens and the performance of procedures with little or no clinical relevance. Standards for the selection, collection, and transport of specimens should be developed for both laboratory and nursing procedure manuals and reviewed periodically by a committee composed of medical, nursing,and laboratory staff. Ongoing dialogues and continuous communication with other health care workers concerning topics such as specimen collection, test selection, results interpretation, and new technology are essential to maintaining high quality microbiological services.

Biochemical and Susceptibility Testing

Most laboratories today use either commercially available miniaturized biochemical test systems or automated instruments for biochemical tests and for susceptibility testing.

The kits usually contain 10 to 20 tests. The test results are converted to numerical biochemical profiles that are identified by using a codebook or a computer. Carbon source utilization systems with up to 95 tests are also available. Most identification takes 4 to 24 hours. Biochemical and enzymatic test systems for which databases have not been developed are used by some reference laboratories.

Automated instruments can be used to identify most Gram-negative fermenters, nonfermenters, and Gram-positive bacteria, but not for anaerobes. Antimicrobial susceptibility testing can be performed for some microorganisms with this equipment, with results expressed as approximate minimum inhibitory drug concentrations. Both tasks take 4 to 24 hours. If semiautomated instruments are used, some manipulation is done manually, and the cultures (in miniature cards or microdilution plates) are incubated outside of the instrument. The test containers are then read rapidly by the instrument, and the results are generated automatically. Instruments are also available for identification of bacteria by cell wall fatty acid profiles generated with gas-liquid chromatography (GLC), analysis of mycolic acids using high performance liquid chromatography (HPLC), and by protein-banding patterns generated by polyacrylamide gel electrophoresis (PAGE). Some other instruments designed to speed laboratory diagnosis of bacteria are those that detect (but do not identify) bacteria in blood cultures, usually faster than manual systems because of continuous monitoring. Also available are many rapid screening systems for detecting one or a series of specific bacteria, including certain streptococci, *N meningitidis,* salmonellae, *Chlamydia trachomatis,* and many others. These screening systems are based on fluorescent antibody, agglutination, or other rapid procedures.

It is important to inform physicians as soon as a presumptive identification of an etiologic agent is obtained so that appropriate therapy can be initiated as quickly as possible. Gram stain and colony morphology; acid-fast stains; and spot indole, oxidase, and other rapid enzymatic tests may allow presumptive identification of an isolate within minutes.

Role of the Reference Laboratory

Despite recent advances, the armamentarium of the clinical laboratory is far from complete. Few laboratories can or should conduct the specialized tests that are often essential to distinguish virulent from avirulent strains. Serotyping is done only for a few species, and phage typing only rarely. Few pathogenicity tests are performed. Not many laboratories can conduct comprehensive biochemical tests on strains that cannot be identified readily by commercially available biochemical systems. Even fewer laboratories are equipped to perform plasmid profiles, gene probes, or DNA hybridization. These and other specialized tests for the serologic or biochemical identification of some exotic bacteria, yeasts, molds, protozoans, and viruses are best done in regional reference laboratories. It is not cost-effective for smaller laboratories to store and control the quality of reagents and media for tests that are seldom run or quite complex. In addition, it is impossible to maintain proficiency when tests are performed rarely. Sensitive methods for the epidemiologic subtyping of isolates from disease outbreaks, such as electrophoretic enzyme typing, rRNA fingerprinting, whole-cell protein electrophoretic patterns, and restriction endonuclease analysis of whole-cell or plasmid DNA, are used only in reference laboratories and a few large medical centers.

Specific genetic probes are now available commercially for identifying virulence factors and many bacteria and viruses. Genetic probes are among the most common methods used for identification of *Mycobacterium tuberculosis* and *M avium* complex in the U.S. today. Probes for *Neisseria gonorrhoeae* and *Chlamydia trachomatis* are now being used directly on clinical specimens with excellent sensitivity and almost universal specificity with same-day results. Mycobacterial probes are also being evaluated for direct specimen testing.

Interfacing with Public Health

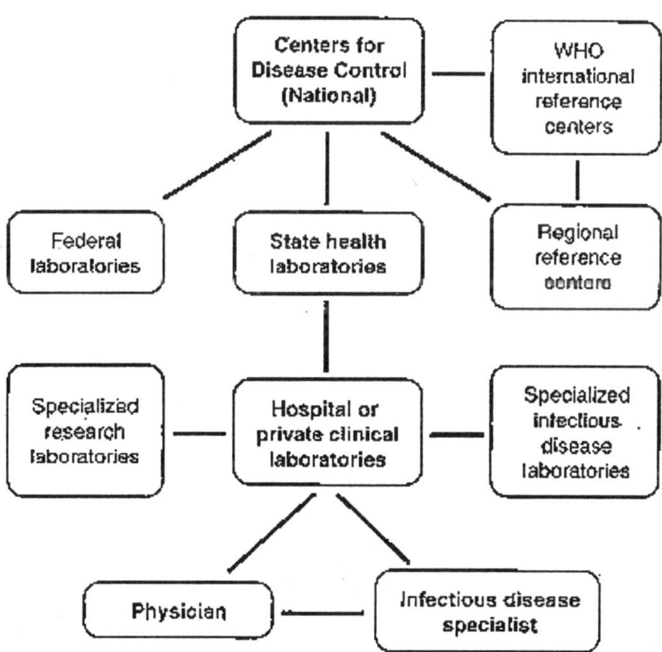

FIGURE 3-4 Pathways for laboratory identification of pathogens and information exchange.

Laboratories

Hospital and local clinical laboratories interact with district, state, and federal public health laboratories in several important ways (Fig. 3-4). The clinical laboratories participate in quality control and proficiency testing programs that are conducted by federally regulated agencies. The government reference laboratories supply cultures and often reagents for use in quality control, and they conduct training programs for clinical laboratory personnel.

All types of laboratories should interact closely to provide diagnostic services and epidemic surveillance. The primary concern of the clinical laboratory is identifying infectious disease agents and studying nosocomial and local outbreaks of disease. When the situation warrants, the local laboratory may ask the state laboratory for help in identifying an unusual organism, discovering the cause or mode of transmission in a disease outbreak, or performing specialized tests not done routinely in clinical laboratories. Cultures should be pure and should be sent on appropriate media following appropriate procedures for transport of biohazardous materials. Pertinent information, including the type of specimen; patient name (or number), date of birth, and sex; clinical diagnosis, associated illness, date of onset, and present condition; specific agent suspected, and any other organisms isolated; relevant epidemiologic and clinical data; treatment of patient; previous laboratory results (biochemical or serologic tests); and necessary information about the submitting party must accompany each request.

These data allow the state laboratory to test the specimen properly and quickly, and they provide information about occurrences within the state. For example, a food-borne outbreak might extend to many parts of the state (or beyond its boundaries). The state laboratory can alert local physicians to the possibility of such outbreaks.

Another necessary interaction between local and state laboratories is the reporting of notifiable diseases by the local laboratory. The state laboratory makes available to local laboratories summaries of the incidence of these diseases. The state laboratories also submit the summaries to the CDC weekly (or, for some diseases, yearly), and national summaries are published weekly in the *Morbidity and Mortality Weekly Report.*

Interaction between the CDC and state and federal laboratories is very similar to that between local and state laboratories. The CDC provides quality control cultures and reagents to state laboratories, and serves as a national reference laboratory for diagnostic services and epidemiologic surveillance. Local laboratories, however, must initially send specimens to the local or state public health laboratory, which, when necessary, forwards them to the CDC. The CDC reports its results back to the state laboratory, which then reports to the local laboratory.

Hazards of Clinical Laboratory Work

Clinical laboratory personnel, including support and clerical employees, are subject to the risk of infection, chemical hazards, and, in some laboratories, radioactive contamination. Such risks can be prevented or minimized by a laboratory safety program.

Radiation Hazards

Personnel who work with radioactive materials should have taken a radioactivity safety course; they should wear radiation monitor badges and be aware of the methods for decontaminating hands, clothing, work surfaces, and equipment. They should wear gloves when working with radioactive compounds. When they work with high-level radiation, they should use a hood and stand behind a radiation shield. Preparative radioactive work should be done in a separate room with access only by personnel who are involved directly in the work.

Chemical Hazards

Chemicals can harm laboratory personnel through inhalation or skin absorption of volatile compounds; bodily contact with carcinogens, acids, bases, and other harmful chemicals; or introduction of poisonous or skin-damagingliquids into the mouth. Good laboratory practices require that volatile compounds be handled only under a hood, that hazardous chemicals never be pipetted by mouth, and that anyone working with skin-damaging chemicals wear gloves, eye guards, and other personal protective equipment as necessary. Workers should be familiar with the materials safety data sheets (MSDS) posted in an accessible place in every laboratory. These forms contain information about chemical hazards and procedures for decontamination should an accident occur.

Biologic Hazards

Microbiologic contamination is the greatest hazard in clinical microbiology laboratories. Laboratory infections are a danger not only to the clinical laboratory personnel but also to anyone else who enters the laboratory, including janitors, clerical and maintenance personnel, and visitors. The risk of infection is governed by the frequency and length of contact with the infectious agent, its virulence, the dose and route of administration, and the susceptibility of the host. The inherent hazard of any infectious agent is affected by factors such as the volume of infectious material used, handling of the material, effectiveness of safety containment equipment, and soundness of laboratory methods. Body fluids from patients, particularly those containing blood, are considered potentially infectious for blood-borne pathogens, and must be handled appropriately.

If possible, agents that are treated differently, such as viruses as opposed to bacteria, or *M tuberculosis* in contrast to *E coli,* should be handled in different laboratories or in different parts of the same laboratory. When the risk category of an agent is known, it should be handled in an area with appropriate containment. All specimens sent for microbiological studies and all organisms sent to the laboratory for identification should be assumed to be potentially infectious. A separate area should be set aside for the receipt of specimens. Personnel should be aware of the potential hazards of improperly packed, broken, or leaking packages and of the proper methods for their handling and decontamination.

To prevent infection, personnel should wear moisture-proof laboratory coats at all times, wash their hands before and after wearing gloves and at the conclusion of each potential exposure to etiologic agents, refrain from mouth pipetting, and not eat, drink, smoke, or apply cosmetics in the laboratory. Immunization may be appropriate for employees who are exposed often to certain infectious agents, including hepatitis B, yellow fever, rabies, polioviruses, meningococci, *Y pestis, S typhi,* and *Francisella tularensis.* Universal precautions, body substance isolation, and other mandated practices involve the use of personal protective equipment and engineering controls to minimize laboratory scientists' exposure to blood-borne pathogens, even when the risk of infection is unknown.

Biosafety Levels

Infectious agents are assigned to a biosafety level from 1 to 4 on the basis of their virulence. The containment levels for organisms should correlate with the biosafety level assigned. Biosafety level 1 is for well-defined organisms not known to cause disease in healthy humans; it includes certain nonvirulent *E coli* strains (such as K-12) and *B subtilis.* Containment level 1 involves standard microbiologic practices, and safety equipment is not needed.

Biosafety level 2, the minimum level for clinical laboratories, is for moderate-risk agents associated with human disease. Containment level 2 includes limited access to the work area, decontamination of all infectious wastes, use of protective gloves, and a biologic safety cabinet for use in procedures that may create aerosols. Examples of biosafety level 2 agents include nematode, protozoan, trematode, and cestode human parasites; all human fungal pathogens except *Coccidioides immitis;* all members of the Enterobacteriaceae except *Y pestis; Bacillus anthracis; Clostridium tetani; Corynebacterium diphtheriae; Haemophilus* species; leptospires; legionellae; mycobacteria other than *M tuberculosis;* pathogenic *Neisseria* species; staphylococci, streptococci, *Treponema pallidum; V cholerae;* and hepatitis and influenza viruses. Clinical specimens potentially containing some biosafety level 3 agents, such as *Brucella* spp., are usually handled using biosafety level 2 containment practices.

Biosafety level 3 is for agents that are associated with risk of serious or fatal aerosol infection. In containment level 3, laboratoryaccess is controlled, special clothing is worn in the laboratory, and containment equipment is used for all work with the agent. *M tuberculosis, Coccidioides immitis, Coxiella burnetii,* and many of the arboviruses are biosafety 3 level agents. Containment level 3 usually is recommended for work with cultures of rickettsiae, brucellae, *Y pestis,* and a wide variety of viruses, including human immunodeficiency viruses.

Biosafety level 4 indicates dangerous and novel agents that cause diseases with high fatality rates. Maximum containment and decontamination procedures are used in containment level 4, which is found in only a few reference and research laboratories. Only a few viruses (including Lassa, Ebola, and Marburg viruses) are classified in biosafety level 4.

REFERENCES

Bergey's Manual of Systematic Bacteriology. Vol. 1-4. Williams & Wilkins, Baltimore, 1984-1989

Center for Infectious Diseases. Reference/Diagnostic Services. Centers for Disease Control and Prevention, Atlanta, Ga.

Centers for Disease Control and Prevention: Morbidity and Mortality Weekly Report.Massachusetts Medical Society, Waltham, Ma.

Fleming DO, Richardson JH, Tulis JI, Vesley D (eds): Laboratory Safety: Principles and Practices. 2nd Ed. ASM Press, Washington, D.C., 1995

Lennette EH, et al (eds): Manual of Clinical Microbiology 4th Ed. American Society for Microbiology, Washington, D.C., 1985

Murray PR, Baron EJ, Pfaller MA, Tenover FC, Yolken RH (eds): Manual of Clinical Microbiology 6th Ed. ASM Press, Washington, D.C., 1995

Richmond JY, McKinney RW (eds): Biosafety in Microbiological and Biomedical Laboratories. 3rd Ed. Centers for Disease Control and Prevention, Atlanta, Ga., and National Institutes of Health, Bethesda, Md., 1993

Skerman VBD, McGowan V, Sneath PHA (eds): Approved lists of bacterial names. Int J Syst Bacteriol 30:225, 1980

Sneath, PHA (ed): International Code of Nomenclature of Bacteria: Bacteriological Code, 1990 Revision. American Society for Microbiology, Washington, D.C., 1990

Chapter 4

Bacterial Metabolism

General Concepts

Heterotrophic Metabolism

Heterotrophic metabolism isthe biologic oxidation of organiccompounds, such as glucose, to yield ATP and simpler organic (or inorganic) compounds, which are needed by the bacterial cell for biosynthetic or assimilatory reactions.

Respiration

Respiration is a type of heterotrophic metabolism that uses oxygen and in which 38 moles of ATP are derived from the oxidation of 1 mole of glucose, yielding 380,000 cal. (An additional 308,000 cal is lost as heat.)

Fermentation

In fermentation, another type of heterotrophic metabolism, an organic compound rather than oxygen is the terminal electron (or hydrogen) acceptor. Less energy is generated from this incomplete form of glucose oxidation, but the process supports anaerobic growth.

Krebs Cycle

The Krebs cycle is the oxidative process in respiration by which pyruvate (via acetyl coenzyme A) is completely decarboxylated to CO_2. The pathway yields 15 moles of ATP (150,000 calories).

Glyoxylate Cycle

The glyoxylate cycle, which occurs in some bacteria, is a modification of the Krebs cycle. Acetyl coenzyme A is generated directly from oxidation of fatty acids or other lipid compounds.

Electron Transport and Oxidative Phosphorylation

In the final stage of respiration, ATP is formed through a series of electron transfer reactions within the cytoplasmic membrane that drive the oxidative phosphorylation of ADP to ATP. Bacteria use various flavins, cytochrome, and non-heme iron components as well as multiple cytochrome oxidases for this process.

Mitchell or Proton Extrusion Hypothesis

The Mitchell hypothesis explains the energy conservation in all cells on the basis of the selective extrusion of H+ ions across a proton-impermeable membrane, which generates a proton motive force. This energy allows for ATP synthesis both in respiration and photosynthesis.

Bacterial Photosynthesis

Bacterial photosynthesis is a light-dependent, anaerobic mode of metabolism. Carbon dioxide is reduced to glucose, which is used for both biosynthesis and energy production. Depending on the hydrogen source used to reduce CO_2, both photolithotrophic and photoorganotrophic reactions exist in bacteria.

Autotrophy

Autotrophy is a unique form of metabolism found only in bacteria. Inorganic compounds are oxidized directly (without using sunlight) to yield energy (e.g., NH_3, NO_2-, S_2, and Fe^{2+}). This metabolic mode also requires energy for CO_2 reduction, like photosynthesis, but no lipid-mediated processes are involved. This metabolic mode has also been called chemotrophy, chemoautotrophy, or chemolithotrophy.

Anaerobic Respiration

Anaerobic respiration is another heterotrophic mode of metabolism in which a specific compound other than 0_2 serves as a terminal electron acceptor. Such acceptor compounds include NO_3-, SO_4^{2-}, fumarate, and even CO_2 for methane-producing bacteria.

The Nitrogen Cycle

The nitrogen cycle consists of a recycling process by which organic and inorganic nitrogen compounds are used metabolically and recycled among bacteria, plants, and animals. Important processes, including ammonification, mineralization, nitrification, denitrification, and nitrogen fixation, are carried out primarily by bacteria.

INTRODUCTION

Metabolism refers to all the biochemical reactions that occur in a cell or organism. The study of bacterial metabolism focuses on the chemical diversity of substrate oxidations and dissimilation reactions (reactions by which substrate molecules are broken down), which normally function in bacteria to generate energy. Also within the scope of bacterial metabolism is the study of the uptake and utilization of the inorganic or organic compounds required for growth and maintenance of a cellular steady state (assimilation reactions). These respective exergonic (energy-yielding) and endergonic (energy-requiring) reactions are catalyzed within the living bacterial cell by integrated enzyme systems, the end result being self-replication of the cell. The capability of microbial cells to live, function, and replicate in an appropriate chemical milieu (such as a bacterial culture medium) and the chemical changes that result during this transformation constitute the scope of bacterial metabolism.

The bacterial cell is a highly specialized energy transformer. Chemical energy generated by substrate oxidations is conserved by formation of high-energy compounds such as adenosine diphosphate (ADP) and adenosine triphosphate

$$(R-\overset{\overset{\displaystyle O}{\|}}{C}\sim S-R)$$, such as acetyl ~ S-coenzyme A

(ATP) or compounds containing the thioester bond (acetyl ~ SCoA) or succinyl ~ SCoA. ADP and ATP represent adenosine monophosphate (AMP) plus one and two high-energy phosphates (AMP ~ P and AMP ~ P~ P, respectively); the energy is stored in these compounds as high-energy phosphate bonds. In the presence of proper enzyme systems, these compounds can be used as energy sources to synthesize the new complex organic compounds needed by the cell. All living cells must maintain steady-state biochemical reactions for the formation and use of such high-energy compounds.

Kluyver and Donker (1924 to 1926) recognized that bacterial cells, regardless of species, were in many respects similar chemically to all other living cells. For example, these investigators recognized that hydrogen transfer is a common and fundamental feature of all metabolic processes. Bacteria, like mammalian and plant cells, use ATP or the high-energy phosphate bond (~ P) as the primary chemical energy source. Bacteria also require the B-complex vitamins as functional coenzymes for many oxidation-reduction reactions needed for growth and energy transformation. An organism such as *Thiobacillus thiooxidans*, grown in a medium containing only sulfur and inorganic salts, synthesizes large amounts of thiamine, riboflavine, nicotinic acid, pantothenic acid, pyridoxine, and biotin. Therefore, Kluyver proposed the unity theory of biochemistry (*Die Einheit in der Biochemie*), which states that all basic enzymatic reactions which support and maintain life processes within cells of organisms, had more similarities than differences. This concept of biochemical unity stimulated many investigators to use bacteria as model systems for studying related eukaryotic, plant and animal biochemical reactions that are essentially "identical" at the molecular level.

From a nutritional, or metabolic, viewpoint, three major physiologic types of bacteria exist: the heterotrophs (or chemoorganotrophs), the autotrophs (or chemolithotrophs), and the photosynthetic bacteria (or phototrophs) (Table 4-1). These are discussed below.

Heterotrophic Metabolism

Heterotrophic bacteria, which include allpathogens, obtain energy from oxidation of organic compounds. Carbohydrates (particularly glucose), lipids, and protein are the most commonly oxidized compounds. Biologic oxidation of these organic compounds by bacteria results in synthesis of ATP as the chemical energy source. This process also permits generation of simpler organic compounds (precursor molecules) needed by the bacteria cell for biosynthetic or assimilatory reactions.

TABLE 4-1 Nutritional Diversity exhibited by Physiologically Different Bacteria

Required Components for bacterial Growth				
Physiologic Type	**Carbon Source**	**Nitrogen Source[a] Source[b]**	**Energy Source**	**Hydrogen Source**
Heterotrophic (chemoorganotrophic)	Organic	Organic or inorganic	Oxidation of organic compounds	-
Autotrophic[c] (chemolitnotrophic)	CO_2	inorganic	Oxidation of inorganic compounds	-
Photosynthetic[d] Photolithotrophic (Bacterial)	CO_2	inorganic	Sunlight	H_2S or H_2
Cyanobacterial	CO_2	inorganic	Sunlight	Photolysis of H_2O
Photoorganotrophic (Bacterial)	CO_2	inorganic	Sunlight	Organic compounds

a Common inorganic nitrogen sources are NO_3; or NH_4 ions; nitrogen fixers can use N_2.
b Many phototrophs and chemotrophs are nitrogen-fixing organisms.
c Results in O_2 evolution (or oxygenic photosynthesis) as commonly occurs in plants.
d Organic acids such as formate, acetate, and succinate can serve as hydrogen donors.

The Krebs cycle intermediate compounds serve as precursor molecules (building blocks) for the energy-requiring biosynthesis of complex organic compounds in bacteria. Degradation reactions that simultaneously produce energy and generate precursor molecules for the biosynthesis of new cellular constituents are called amphibolic.

All heterotrophic bacteria require preformed organic compounds. These carbon- and nitrogen-containing compounds are growth substrates, which are used aerobically or anaerobically to generate reducing equivalents (e.g., reduced nicotinamide adenine dinucleotide; NADH + H^+); these reducing equivalents in turn are chemical energy sources for all biologic oxidative and fermentative systems. Heterotrophs are the most commonly studied bacteria; they grow readily in media containing carbohydrates, proteins, or other complex nutrients such as blood. Also, growth media may be enriched by the addition of other naturally occurring compounds such as milk (to study lactic acid bacteria) or hydrocarbons (to study hydrocarbon-oxidizing organisms).

Respiration

Glucose is the most common substrate used for studying heterotrophic metabolism. Most aerobic organisms oxidize glucose completely by the following reaction equation:

$$C_6H_{12}O_6 + 6O_2 \longrightarrow 6CO_2 + 6H_2O + \text{energy}$$

This equation expresses the cellular oxidation process called respiration. Respiration occurs within the cells of plants and animals, normally generating 38 ATP molecules (as energy) from the oxidation of 1 molecule of glucose. This yields approximately 380,000 calories (cal) per mode of glucose (ATP ~ 10,000 cal/mole). Thermodynamically, the complete oxidation of one mole of glucose should yield approximately 688,000 cal; the energy that is not conserved biologically as chemical energy (or ATP formation) is liberated as heat (308,000 cal). Thus, the cellular respiratory process is at best about 55% efficient.

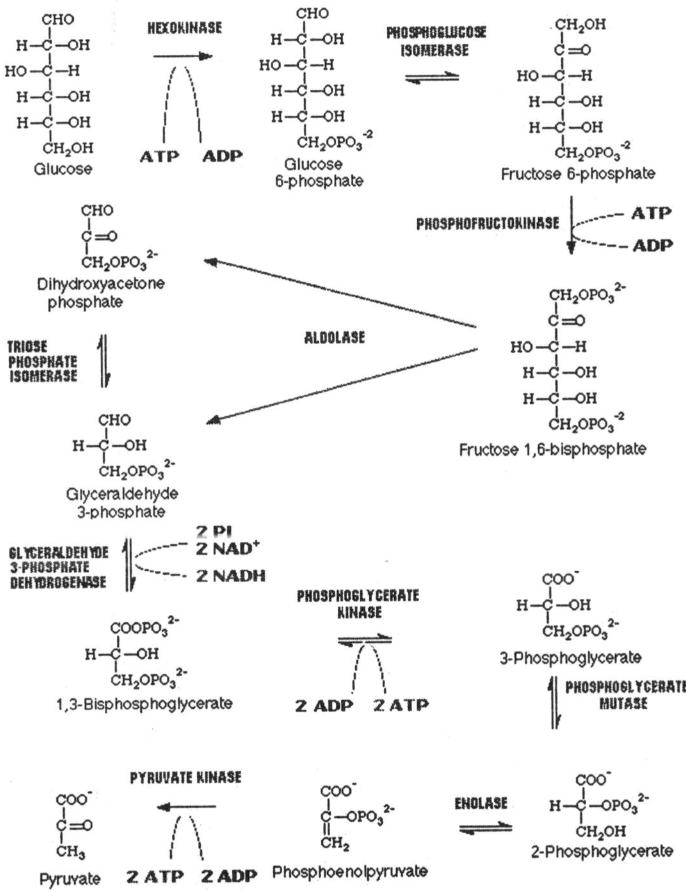

FIGURE 4-1 Glycolytic (EMP) pathway.

Glucose oxidation is the most commonly studied dissimilatory reaction leading to energy production or ATP synthesis. The complete oxidation of glucose may involve three fundamental biochemical pathways. The first is the glycolytic or Embden-Meyerhof-Parnas pathway (Fig. 4-1), the second is the Krebs cycle (also called the citric acid cycle or tricarboxylic acid cycle), and the third is the series of membrane-bound electron transport oxidations coupled to oxidative phosphorylation.

Respiration takes place when any organic compound (usually carbohydrate) is oxidized completely to CO_2 and H_2O. In aerobic respiration, molecular 0_2 serves as the terminal acceptor of electrons. For anaerobic respiration, NO_3-, SO_4^{2-}, CO_2, or fumarate can serve as terminal electron acceptors (rather than 0_2), depending on the bacterium studied. The end result of the respiratory process is the complete oxidation of the organic substrate molecule, and the end products formed are primarily CO_2 and H_2O. Ammonia is formed also if protein (or amino acid) is the substrate oxidized. The biochemical pathways normally involved in oxidation of various naturally occurring organic compounds are summarized in Figure 4-2.

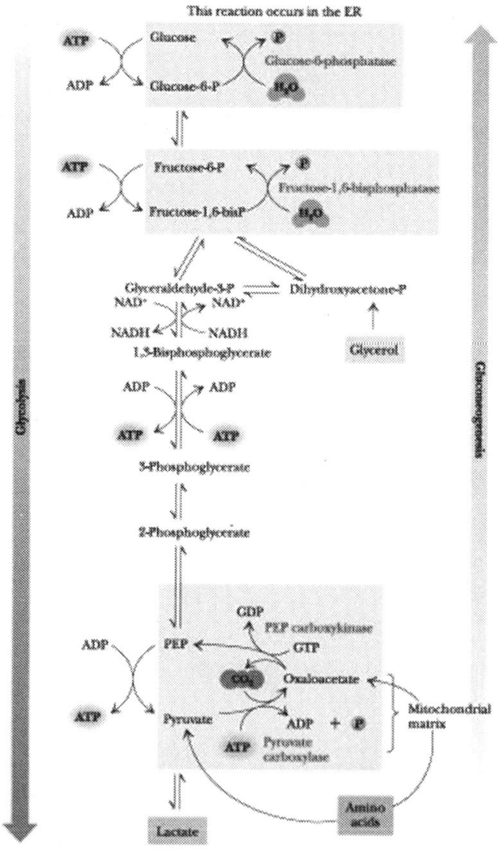

FIGURE 4-2 Heterotrophic metabolism, general pathway.

Metabolically, bacteria are unlike cyanobacteria (blue-green algae) and eukaryotes in that glucose oxidation may occur by more than one pathway. In bacteria, glycolysis represents one of several pathways by which bacteria can catabolically attack glucose. The glycolytic pathway is most commonly associated with anaerobic or fermentative metabolism in bacteria and yeasts. In bacteria, other minor heterofermentative pathways, such as the phosphoketolase pathway, also exist.

In addition, two other glucose-catabolizing pathways are found in bacteria: the oxidative pentose phosphate pathway (hexose monophosphate shunt), (Fig. 4-3) and the Entner-Doudoroff pathway, which is almost exclusively found in obligate aerobic bacteria (Fig. 4-4). The highly oxidative *Azotobacter* and most *Pseudomonas* species, for example, utilize the Entner-Doudoroff pathway for glucose catabolism, because these organisms lack the enzyme phosphofructokinase and hence cannot synthesize fructose 1,6-diphosphate, a key intermediate compound in the glycolytic pathway. (Phospho-fructokinase is also sensitive to molecular 02 and does not function in obligate aerobes). Other bacteria, which lack aldolase (which splits fructose-1,6-diphosphate into two triose phosphate compounds), also cannot have a functional glycolytic pathway. Although, the Entner-Doudoroff pathway is usually associated with obligate aerobic bacteria, it is present in the facultative anaerobe *Zymomonas mobilis* (formerly *Pseudomonas lindneri*). This organism dissimilates glucose to ethanol and represents a major alcoholic fermentation reaction in a bacterium.

Glucose dissimilation also occurs by the hexose monophosphate shunt (Fig. 4-3). This oxidative pathway was discovered in tissues that actively metabolize glucose in the presence of two glycolytic pathway inhibitors (iodoacetate and fluoride). Neither inhibitor had an effect on glucose dissimilation, and NADPH + H+ generation occurred directly from the oxidation of glucose-6-phosphate (to 6-phosphoglucono-d-lactone) by glucose-6phosphate dehydrogenase. The pentose phosphate pathway subsequently permits the direct oxidative decarboxylation of glucose to pentoses. The capability of this oxidative metabolic system to bypass glycolysis explains the term shunt.

The biochemical reactions of the Entner-Doudoroff pathway are a modification of the hexose monophosphate shunt, except that pentose sugars are not directly formed. The two pathways are identical up to the formation of 6-phosphogluconate (see Fig. 4-4) and then diverge. In the Entner-Doudoroff pathway, no oxidative decarboxylation of 6-phosphogluconate occurs and no pentose compound is formed. For this pathway, a new 6 carbon compound intermediate (2-keto-3-deoxy6-phosphogluconate) is generated by the action of 6-phosphogluconate dehydratase (an Fe2+- and glutathione-stimulatedenzyme); this intermediate compound is then directly cleaved into the triose (pyruvate) and a triose-phosphate compound (glyceraldehyde-3-phosphate) by the 2-keto-3-deoxy6-phosphogluconate aldolase. The glyceraldehyde-3-phosphate is further oxidized to another pyruvate molecule by the same enzyme systems that catalyze the terminal glycolytic pathway (see Fig. 4-4).

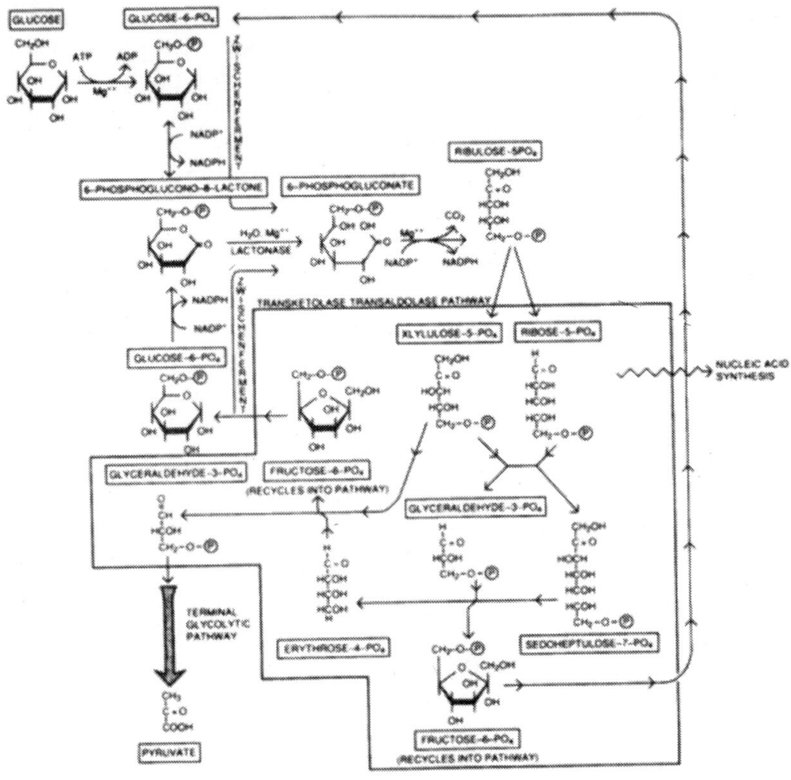

FIGURE 4-3 Hexose monophosphate (HMS) pathway.

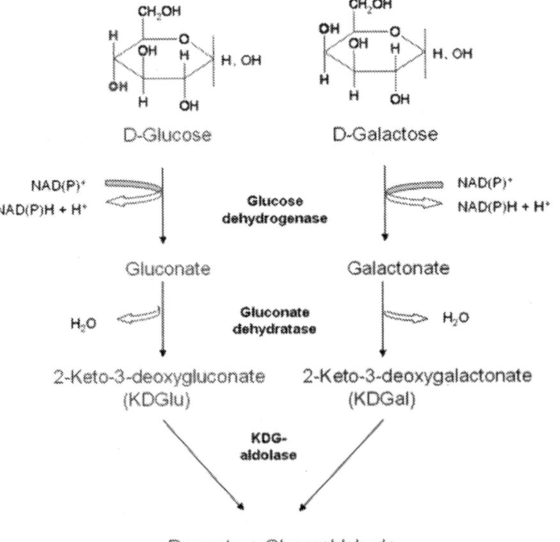

FIGURE 4-4 Entner-Doudoroff (ED) pathway.

The glycolytic pathway may be the major one existing concomitantly with the minor oxidative pentose phosphate- hexose monophosphate shunt pathway; the Entner-Doudoroff pathway also may function as a major pathway with a minor hexose monophosphate shunt. A few bacteria possess only one pathway. All cyanobacteria, *Acetobacter suboxydans*, and *A xylinum* possess only the hexose monophosphate shunt pathway; *Pseudomonas saccharophilia* and *Z mobilis* possess solely the Entner-Doudoroff pathway. Thus, the end products of glucose dissimilatory pathways are as follows:

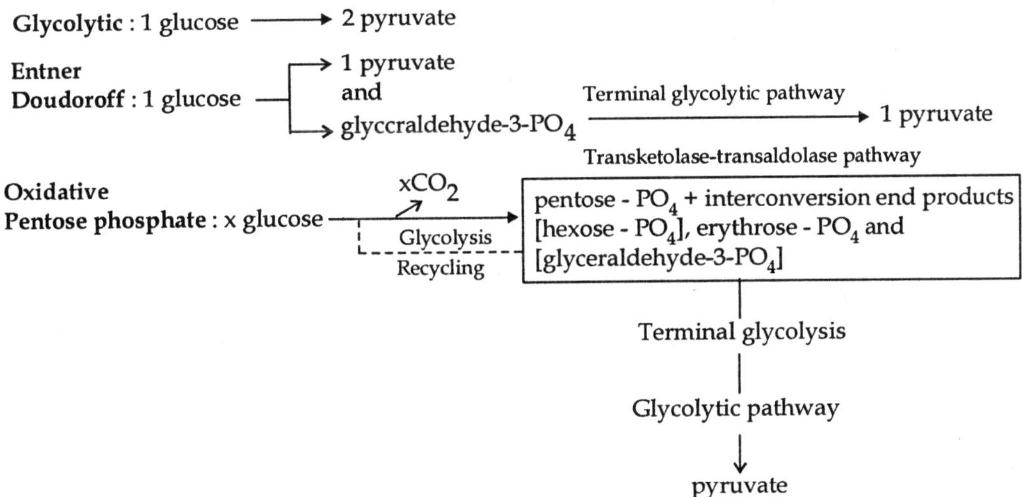

The glucose dissimilation pathways used by specific microorganisms are shown in Table 4-2.

TABLE 4-2 Glucose dissimilation Pathways Utilized by Bacteria, Cyanobacteria, and Yeasts

Bacteria	Glycolytic Pathway	Oxidative pentose Phosphate pathway	Entner- Doudoroff Pathway
Acetobacter suboxydans		Sole	
Acetobacter xylinum		Sole	
Agrobacterium spp			Major
*Azolobacter vinelandil**			Major
Bacilus subtitis	Major	Minor	
Caulobacter spp			Major
Escherlchia coli	Major		Minor
Lactobacillus delbrueckn	Major		
Leuconostoc mesenteroides		Major	
Neissenia gonorrhoeae		Minor	Major
Neissena meningitides		Minor	Major
Neisenia perllave		Major	
Neisseria sicca		Major	

Pseudomonas aeruginose[a]			Major
Pseudomonas saccharophilia			Sole
Rhizobium spp			Major
Sarcina lutea	Major	Minor	
Sprillum spp			Major
Streotococus faecalis			(Major)[b]
Streptomyces griseus	Major	Minor	
Zymomonas anaerobia			Sole
Zymomonas rnobilis			Sole
All gyanobacteria			Sole
All yeasts	Major	Minor	

a Most species utilize the Enter-Doudorafl pathway as major pathway.
b Induced by growth on gluconale.

All major pathways of glucose or hexose catabolism have several metabolic features in common. First, there are the preparatory steps by which key intermediate compounds such as the triose-PO_4, glyceraldehyde-3-phosphate, and/or pyruvate are generated. The latter two compounds are almost universally required for further assimilatory or dissimilatory reactions within the cell. Second, the major source of phosphate for all reactions involving phosphorylation of glucose or other hexoses is ATP, not inorganic phosphate (Pi). Actually, chemical energy contained in ATP must be initially spent in the first step of glucose metabolism (via kinase-type enzymes) to generate glucose-6-phosphate, which initiates the reactions involving hexose catabolism. Third, NADH + H^+ or NADPH + H^+ is generated as reducing equivalents (potential energy) directly by one or more of the enzymatic reactions involved in each of these pathways.

Fermentation

Fermentation, another example of heterotrophic metabolism, requires an organic compound as a terminal electron (or hydrogen) acceptor. In fermentations, simple organic end products are formed from the anaerobic dissimilation of glucose (or some other compound). Energy (ATP) is generated through the dehydrogenation reactions that occur as glucose is broken down enzymatically. The simple organic end products formed from this incomplete biologic oxidation process alsoserve as final electron and hydrogen acceptors. On reduction, these organic end products are secreted into the medium as waste metabolites (usually alcohol or acid). The organic substrate compounds are incompletely oxidized by bacteria, yet yield sufficient energy for microbial growth. Glucose is the most common hexose used to study fermentation reactions.

In the late 1850s, Pasteur demonstrated that fermentation is a vital process associated with the growth of specific microorganisms, and that each type of fermentation can be defined by the principal organic end product formed (lactic acid, ethanol, acetic acid, or butyric acid). His studies on butyric acid fermentation led directly to the discovery of anaerobic microorganisms. Pasteur concluded that oxygen inhibited the microorganisms responsible for butyric acid fermentation because both bacterial mobility and butyric acid formation ceased when air was bubbled into the fermentation mixture. Pasteur also introduced the terms aerobic and anaerobic. His views on fermentation are made clear from his microbiologic studies on the production of beer (from *Etudes sur la Biere,* 1876):

In the experiments which we have described, fermentation by yeast is seen to be the direct consequence of the processes of nutrition, assimilation and life, when these are carried on without the agency of free oxygen. The heat required in the accomplishment of that work must necessarily have been borrowed from the decomposition of the fermentation matter.... Fermentation by yeast appears, therefore, to be essentially connected with the property possessed by this minute cellular plant of performing its respiratory functions, somehow or other, with the oxygen existing combined in sugar.

For most microbial fermentations, glucose dissimilation occurs through the glycolytic pathway (Fig. 4-1). The simple organic compound most commonly generated is pyruvate, or a compound derived enzymatically from pyruvate, such as acetaldehyde, a-acetolactate, acetyl ~ SCoA, or lactyl ~ SCoA (Fig. 4-5). Acetaldehyde can then be reduced by NADH + H+ to ethanol, which is excreted by the cell. The end product of lactic acid fermentation, which occurs in streptococci (e.g., *Streptococcus lactis*) and many lactobacilli (e.g., *Lactobacillus casei, L pentosus*), is a single organic acid, lactic acid. Organisms that produce only lactic acid from glucose fermentation are homofermenters. Homofermentative lactic acid bacteria dissimilate glucose exclusively through the glycolytic pathway. Organisms that ferment glucose to multiple end products, such as acetic acid, ethanol, formic acid, and CO_2, are referred to as heterofermenters. Examples of heterofermentative bacteria include *Lactobacillus, Leuconostoc,* and *Microbacterium* species. Heterofermentative fermentations are more common among bacteria, as in the mixed-acid fermentations carried out by bacteria of the family Enterobacteriaceae (e.g., *Escherichia coli, Salmonella, Shigella,* and *Proteus* species). Many of these glucose fermenters usually produce CO_2 and H_2 with different combinations of acid end products (formate, acetate, lactate, and succinate). Other bacteria such as *Enterobacter aerogenes, Aeromonas, Serratia, Erwinia,* and *Bacillus* species also form CO_2 and H_2 as well as other neutral end products (ethanol, acetylmethylcarbinol [acetoin], and 2,3-butylene glycol). Many obligately anaerobic clostridia (e.g., *Clostridium saccharobutyricum, C thermosaccharolyticum*) and *Butyribacterium* species ferment glucose with the production of butyrate, acetate, CO_2, and H_2, whereas other *Clostridum* species (*C acetobutylicum* and *C butyricum*) also form these fermentation end products plus others (butanol, acetone, isopropanol, formate, and ethanol). Similarly, the anaerobic propionic acid bacteria (*Propionibacterium* species) and the related *Veillonella* species ferment glucose to form CO_2, propionate, acetate, and succinate. In these bacteria, propionate is formed by the partial reversal of the Krebs cycle reactions and involves a CO_2 fixation by pyruvate (the Wood-Werkman reaction) that forms oxaloacetate (a four-carbon intermediate). Oxaloacetate is then reduced to malate, fumarate, and succinate, which is decarboxylated to propionate. Propionate is also formed by another three-carbon pathway in *C propionicum, Bacteroides ruminicola,* and *Peptostreptococcus* species, involving a lactyl ~ SCoA intermediate. The obligately aerobic acetic acid bacteria (*Acetobacter* and the related *Gluconobacter* species) can also ferment glucose, producing acetate and gluconate. Figure 4-5 summarizes the pathways by which the various major fermentation end products form from the dissimilation of glucose through the common intermediate pyruvate.

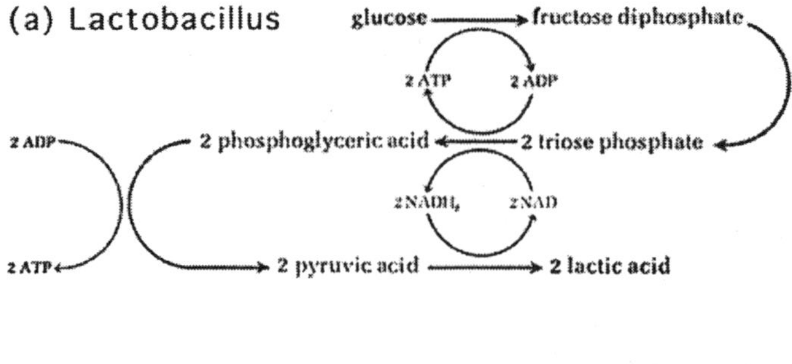

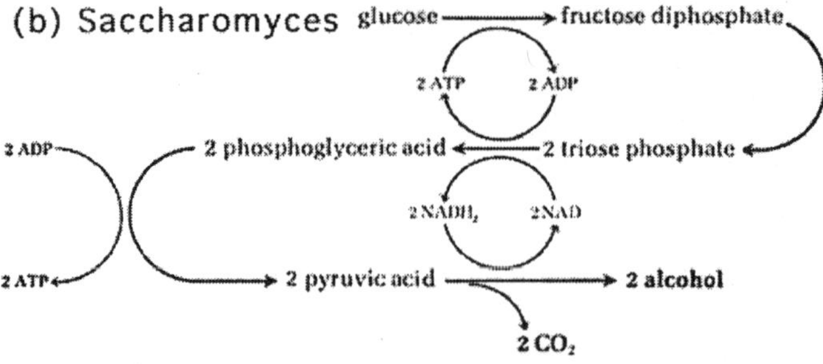

FIGURE 4-5 Fermentative pathways of bacteria and the major end products formed with the organism type carrying out the fermentation.

For thermodynamic reasons, bacteria that rely on fermentative process for growth cannot generate as much energy as respiring cells. In respiration, 38 ATP molecules (or approximately 380,000 cal/mole) can be generated as biologically useful energy from the complete oxidation of 1 molecule of glucose (assuming 1 NAD(P)H = 3 ATP and 1 ATP ADP + Pi = 10,000 cal/mole). Table 4-3 shows comparable bioenergetic parameters for the lactate and ethanolic fermentations by the glycolytic pathway. Although, only 2 ATP molecules are generated by this glycolytic pathway, this is apparently enough energy to permit anaerobic growth of lactic acid bacteria and the ethanolic fermenting yeast, *Saccharomyces cerevisiae.* The ATP-synthesizing reactions in the glycolytic pathway (Fig. 4-1) specifically involve the substrate phosphorylation reactions catalyzed by phosphoglycerokinase and pyruvic kinase. Although, all the ATP molecules available for fermentative growth are believed to be generated by these substrate phosphorylation reactions, some energy equivalents are also generated by proton extrusion reactions (acid liberation), which occur with intact membrane systems and involve the proton extrusion reactions of energy conservation (Fig. 4-9) as it applies to fermentative metabolism.

Krebs Cycle

The Krebs cycle (also called the tricarboxylic acid cycle or citic acid cycle) functions oxidatively in respiration and is the metabolic process by which pyruvate or acetyl ~ SCoA is completely decarboxylated to CO_2. In bacteria, this reaction occurs through acetyl ~ SCoA, which is the first product in the oxidative decarboxylation of pyruvate by pyruvate dehydrogenase. Bioenergetically, the following overall exergonic reaction occurs:

$$CH_3-\underset{\underset{O}{||}}{C}-COOH + 5O \xrightarrow[\text{Electron transport and oxidative phosphorylation}]{\text{Krebs cycle (viz } CH_3-\overset{\overset{O}{||}}{C}-O-SCoA)}$$

$$3CO_2 + 2H_2O + 15ATP\ (\sim 150{,}000\ \text{cal/mole})$$

If 2 pyruvate molecules are obtained from the dissimilation of 1 glucose molecule, then 30 ATP molecules are generated in total. The decarboxylation of pyruvate, isocitrate, and a-ketoglutarate accounts for all CO_2 molecules generated during the respiratory process. Figure 4-6 shows the enzymatic reactions in the Krebs cycle. The chemical energy conserved by the Krebs cycle is contained in the reduced compounds is not available as ATP until the final step of respiration (electron transport and oxidative phosphorylation) occurs.

Table 4-3 Energy obtained from Bacterial Fermentations by Substrate Phosphorylations

		Fermentation	Actual Energy (cal/mole)	Theoretical Energy (cal/mole)	Efficiency (%)
Homolactic		H			
	Glycolysis	$2CH_3-C-COOH +$	~20,000	57,000	35
$C_4H_{12}O_6$ (Glucose)		OH (Lactic acid)			
Alcoholic		H			
	Glycolysis	$2CH_3-C-OH + 2CO +$	~20,000	58,000	34
$C_4H_{12}O_6$ (Glucose)		H (Ethanol)			

The Krebs cycle is therefore another preparatory stage in the respiratory process. If 1 molecule of pyruvate is oxidized completely to 3 molecules of CO_2, generating 15 ATP molecules, the oxidation of 1 molecule of glucose will yield as many as 38 ATP molecules, provided glucose is dissimilatedby glycolysis and the Krebs cycle (further assuming that the electron transport/oxidative phosphorylation reactions are bioenergetically identical to those of eukaryotic mitochondria).

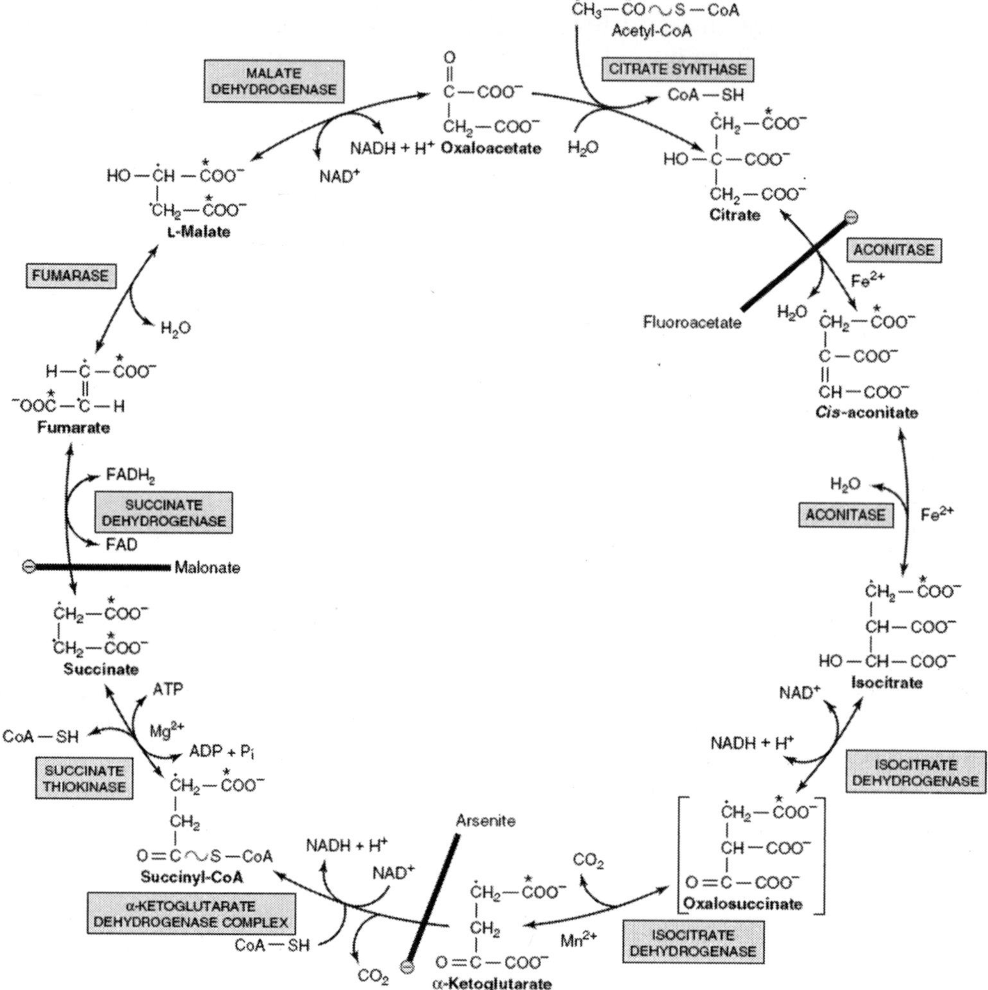

FIGURE 4-6 Krebs cycle (also tricarboxylic acid or citric acid cycle).

Glyoxylate Cycle

In general, the Krebs cycle functions similarly in bacteria and eukaryotic systems, but major differences are found among bacteria. One difference is that in obligate aerobes, L-malate may be oxidized directly by molecular 0_2 via an electron transport chain. In other bacteria, only some Krebs cycle intermediate reactions occur because a-ketoglutarate dehydrogenase is missing.

A modification of the Krebs cycle, commonly called the glyoxylate cycle, or shunt (Fig. 4-7), which exists in some bacteria. This shunt functions similarly to the Krebs cycle but lacks many of the Krebs cycle enzyme reactions. The glyoxylate cycle is primarily an oxidative pathway in which acetyl~SCoA is generated from the oxidation, of acetate, which usually is derived from the oxidation of fatty acids. The oxidation of fatty acids to acetyl~SCoA is carried out by the b-oxidation pathway. Pyruvate oxidation is not directly involved in the glyoxylate shunt, yet this shunt yields sufficient succinate and malate, which are required for energy production (Fig. 4-7). The glyoxylate cycle also generates other precursor compounds needed for biosynthesis (Fig. 4-7). The glyoxylate cycle was discovered as an unusual metabolic pathway during an attempt to learn how lipid (or acetate) oxidation in bacteria and plant seeds could lead to the direct biosynthesis of carbohydrates. The glyoxylate cycle converts oxaloacetate either to pyruvate and CO_2 (catalyzed by pyruvate carboxylase) or to phosphoenolpyruvate and CO_2 (catalyzed by the inosine triphosphate [ITP]-dependent phosphoenolpyruvate

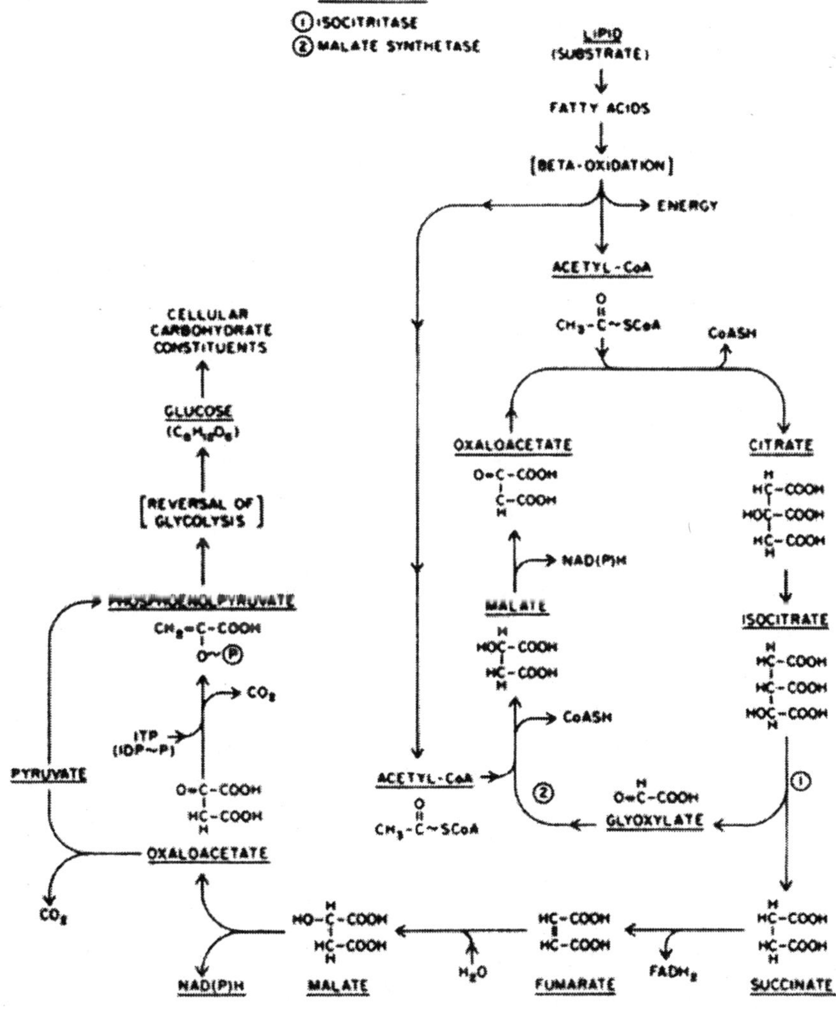

FIGURE 4-7 Glyoxylate shunt.

carboxylase kinase). Either triose compound can then be converted to glucose by reversal of the glycolytic pathway. The glyoxylate cycle is found in many bacteria, including *Azotobacter vinelandii* and particularly in organisms that grow well in media in which acetate and other Krebs cycle dicarboxylic acid intermediates are the sole carbon growth source. One primary function of the glyoxylate cycle is to replenish the tricarboxylic and dicarboxylic acid intermediates that are normally provided by the Krebs cycle. A pathway whose primary purpose is to replenish such intermediate compounds is called anaplerotic.

Electron Transport and Oxidative Phosphorylation

The final stage of respiration occurs through a series of oxidation-reduction electron transfer reactions that yield the energy to drive oxidative phosphorylation; this in turn produces ATP. The enzymes involved in electron transport and oxidative phosphorylation reside on the bacterial inner (cytoplasmic) membrane. This membrane is invaginated to form structures called respiratory vesicles, lamellar vesicles, or mesosomes, which function as the bacterial equivalent of the eukaryotic mitochondrial membrane.

Respiratory electron transport chains vary greatly among bacteria, and in some organisms are absent. The respiratory electron transport chain of eukaryotic mitochondria oxidizes NADH + H+, NADPH + H+, and succinate (as well as the coacylated fatty acids such as acetyl~SCoA). The bacterial electron transport chain also oxidizes these compounds, but it can also directly oxidize, via non-pyridine nucleotide-dependent pathways, a larger variety of reduced substrates such as lactate, malate, formate, a-glycerophosphate, H_2, and glutamate. The respiratory electron carriers in bacterial electron transport systems are more varied than in eukaryotes, and the chain is usually branched at the site(s) reacting with molecular O_2. Some electron carriers, such as nonheme iron centers and ubiquinone (coenzyme Q), are common to both the bacterial and mammalian respiratory electron transport chains. In some bacteria, the naphthoquinones or vitamin K may be found with ubiquinone. In still other bacteria, vitamin K serves in the absence of ubiquinone. In mitochondrial respiration, only one cytochrome oxidase component is found (cytochrome *a* + *a3* oxidase). In bacteria there are multiple cytochrome oxidases, including cytochromes *a, d, o,* and occasionally *a* + *a3* (Fig. 4-8)

In bacteria cytochrome oxidases usually occur as combinations of *a1: d: o* and *a* + *a3*: *o*. Bacteria also possess mixed-function oxidases such as cytochromes P-450 and P-420 and cytochromes *c'* and *c'c'*, which also react with carbon monoxide. These diverse types of oxygen-reactive cytochromes undoubtedly have evolutionary significance. Bacteria were present before O_2 was formed; when O_2 became available as a metabolite, bacteria evolved to use it in different ways; this probably accounts for the diversity in bacterial oxygen-reactive hemoproteins.

Cytochrome oxidases in many pathogenic bacteria are studied by the bacterial oxidase reaction, which subdivides Gram-negative organisms into two major

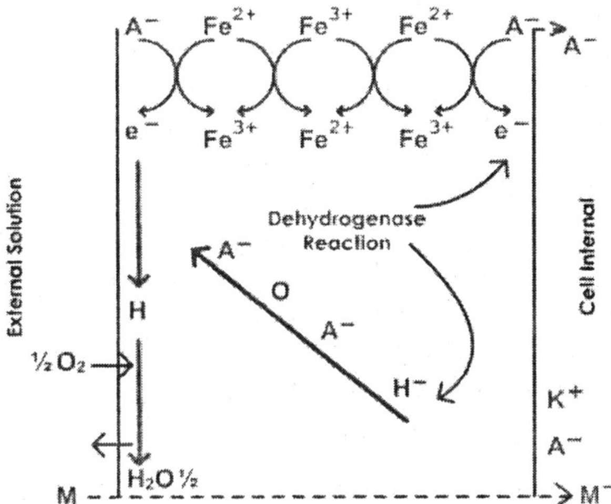

FIGURE 4-8 Respiratory electron transport chains.

groups, oxidase positive and oxidase negative. This oxidase reaction is assayed for by using N,N,N', N'-tetramethyl-*p*-phenylenediamine oxidation (to Wurster's blue) or by using indophenol blue synthesis (with dimethyl-*p*-phenylenediamine and a-naphthol). Oxidase-positive bacteria contain integrated (cytochrome *c* type:oxidase) complexes, the oxidase component most frequently encountered is cytochrome *o*, and occasionally *a* + *a*3. The cytochrome oxidase responsible for the indophenol oxidase reaction complex was isolated from membranes of *Azotobacter vinelandii*, a bacterium with the highest respiratory rate of any known cell. The cytochrome oxidase was found to be an integrated cytochrome *c4:o complex*, which was shown to be present in *Bacillus* species. These *Bacillus* strains are also highly oxidase positive, and most are found in morphologic group II.

Both bacterial and mammalian electron transfer systems can carry out electron transfer (oxidation) reactions with NADH + H^+, NADPH + H^+, and succinate. Energy generated from such membrane oxidations is conserved within the membrane and then transferred in a coupled manner to drive the formation of ATP. The electron transfer sequence is accomplished entirely by membrane-bound enzyme systems. As the electrons are transferred by a specific sequence of electron carriers, ATP is synthesized from ADP + inorganic phosphate (Pi) or orthophosphoric acid (H3PO4) (Fig. 4-8).

In respiration, the electron transfer reaction is the primary mode of generating energy; electrons (2*e*-) from a low-redox-potential compoundsuch as NADH + H+ are sequentially transferred to a specific flavoprotein dehydrogenase or oxidoreductase (flavin mononucleotide [FMN] type for NADH or flavin adenine dinucleotide [FAD] type for succinate); this electron pair is then transferred to a nonheme iron center (FeS) and finally to a specific ubiquinone or a naphthoquinone derivative. This transfer of electrons causes a differential chemical redox potential change so that within the membrane enough chemical energy is conserved to be transferred by a coupling mechanism to a high-energy compound (e.g., ADP + Pi Æ ATP). ATP molecules represent the final stable high-energy intermediate compound formed.

A similar series of redox changes also occurs between ubiquinone and cytochrome c, but with a greater differential in the oxidation-reduction potential level, which allows for another ATP synthesis step. The final electron transfer reaction occurs at the cytochrome oxidase level between reduced cyotchrome c and molecular O_2; this reaction is the terminal ATP synthesis step.

Mitchell or Proton Extrusion Hypothesis

A highly complex but attractive theory to explain energy conservation in biologic systems is the chemiosmotic coupling of oxidative and photosynthetic phosphorylations, commonly called the Mitchell hypothesis. This theory attempts to explain the conservation of free energy in this process on the basis of an osmotic potential caused by a proton concentration differential (or proton gradient) across a proton-impermeable membrane. Energy is generated by a proton extrusion reaction during membrane-bound electron transport, which in essence serve as a proton pump; energy conservation and coupling follow. This represents an obligatory "intact" membrane phenomenon. The energy thus conserved (again within the confines of the membrane and is coupled to ATP synthesis. This would occur in all biologic cells, even in the lactic acid bacteria that lack a cytochrome-dependent electron transport chain but stillpossesses a cytoplasmic membrane. In this hypothesis, the membrane allows for charge separation, thus forming a proton gradient that drives all bioenergization reactions. By such means, electromotive forces can be generated by oxidation-reduction reactions that can be directly coupled to ion translocations, as in the separation of H+ and OH ions in electrochemical systems. Thus, an enzyme or an electron transfer carrier on a membrane that undergoes an oxidation-reduction reaction serves as a specific conductor for OH (or 0_2), and "hydrodehydration" provides electromotive power, as it does in electrochemical cells.

The concept underlying Mitchell'shypothesis is complex, and many modifications have been proposed, but the theory's most attractive feature is that it unifies all bioenergetic conservation principles into a single concept requiring an intact membrane vesicle to function properly. Figure 4-9 shows how the Mitchell hypothesis might be used to explain energy generation, conservation, and transfer by a coupling process. The least satisfying aspect of the chemiosmotic hypothesis is the lack of understanding of how chemical energy is actually conserved within the membrane and how it is transmitted by coupling for ATP synthesis.

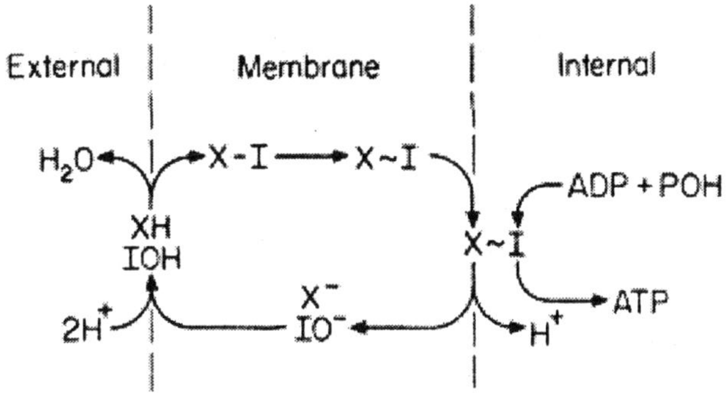

FIGURE 4-9 Mitchell hypotheses, a chemiosmotic model of energy transduction.

Bacterial Photosynthesis

Many prokaryotes (bacteria and cyanobacteria) possess phototrophic modes of metabolism (Table 4-1) . The types of photosynthesis in the two groups of prokaryotes differ mainly in the type of compound that serves as the hydrogen donor in the reduction of CO_2 to glucose (Table 4-1). Phototrophic organisms differ from heterotrophic organisms in that they utilize the glucose synthesized intracellularly for biosynthetic purposes (as in starch synthesis) or for energy production, which usually occurs through cellular respiration.

Unlike phototrophs, heterotrophs require glucose (or some other preformed organic compound) that is directly supplied as a substrate from an exogenous source. Heterotrophs cannot synthesize large concentrations of glucose from CO_2 by specifically using H_2O or (H_2S) as a hydrogen source and sunlight as energy. Plant metabolism is a classic example of photolithotrophic metabolism: plants need CO_2 and sunlight; H_2O must be provided as a hydrogen source and usually NO_3 is the nitrogen source for protein synthesis. Organic nitrogen, supplied as fertilizer, is converted to NO_3 in all soils by bacteria via the process of ammonification and nitrification. Although, plant cells are phototrophic, they also exhibit a heterotrophic mode of metabolism in that they respire. For example, plants use classic respiration to catabolize glucose that is generated photosynthetically. Mitochondria as well as the soluble enzymes of the glycolytic pathway are required for glucose dissimilation, and these enzymes are also found in all plant cells. The soluble Calvin cycle enzymes, which are required for glucose synthesis during photosynthesis, are also found in plant cells. It is not possible to feed a plant by pouring a glucose solution on it, but water supplied to a plant will be "photolysed" by chloroplasts in the presence of light; the hydrogen(s) generated from H_2O is used by Photosystems I and II (PSI and PSII) to reduce NADP+ to NADPH + H+. With the ATP generated by PSI and PSII, these reduced pyridine nucleotides, CO_2 is reduced intracellularly to glucose. This metabolic process is carried out in an integrated manner by Photosystems I and II ("Z" scheme) and by the Calvin cycle pathway. A new photosynthetic, and nitrogen fixing bacterium, *Heliobacterium chlorum*, staining Gram positive was isolated, characterized, and found to contain a new type of chlorophyll, i.e., bacteriochlorophyll'g'. 16S r-RNA sequence analyses showed this organism to be phylogenetically related to members of the family *Bacillaceae,* although all currently known phototrophes are Gram negative (see Table 4.4). A few *Heliobacteriium* strains did show the presence of endospores. Another unusual phototrophe is the Gram negative *Halobacterium halobium* (now named *Halobacterium salinarium*), an archaebacterium growing best at 30°C in 4.0-5.0 M (or 25%, w/v) NaCl. This bacterium is a facultative phototrophe having a respiratory mode; it also possesses a purple membrane within which bacteriorhodopsin serves as the active photosynthetic pigment. This purple membranae possesses a light driven proton translocation pump which mediates photosynthetic ATP synthesis via a proton extrusion reaction (see Mitchell Hypothesis). Table 4-4 summarizes the characteristics of known photosynthetic bacteria.

TABLE 4-4 Characteristics Commonly Exhibited by Phototrophic Bacteria[a]

Photosynthetic Type	Characteristics	Representative Families and Genera
Purple bacteria Sulfur-type (formerly Thiorho daceae) photolinotrophic bacteria	Oligate phototrophs Strict anaerobes H_2S (or H_2) serve as H source Possess S grannules when H_2S used Contain bacteriochlorophyll a or b	Chromatiaceae (Chromatium, Thiosprillum, Thiocapsa)
Non-sulfur-type (formerly Athiorhodaceae), photogrganotrophic bacteria	Facultative phototrophs (have respiratory mechanism and will grow heterotrophically) Oxygen-tolerant anaerobes Most require one or more B vitamins Simple organic compounds serve as H source Contain bactenochlorophyll a or b	Rhodospinllaceae (Rhodopseudomonas, Phodospinllum Rhodomicrobium)
Green bacteria Photolithotrophic bacteria	Obligate phototrophs Strict anaerobes Contains chlorobium chlorophyll which is currently referred to as bacteriochlorophyll type c and d Many require vitamin B_{12} S_2 deposited extracellularly	Chlorobiaceae (Chlorobium Chloropseudomonas)

a All are Gram negative : if motile they exhibit polar flagellation. Most species are anaerobic, although some purple nonsulfur bacteria (family Athiorhodaceae) are facultative phototrophs and can grow as heterotrophs by using the anaerobic respiratory mode of metabolism, they are therefore oxygen tolerant. For further information, see Bergey's Manual of Determinative Bacteriology. 8th ed, part 1.

Autotrophy

Bacteria that grow solely at the expense of inorganic compounds (mineral ions), without using sunlight as an energy source, are called autotrophs, chemotrophs, chemoautotrophs, or chemolithotrophs. Like photosynthetic organisms, all autotrophs use CO_2 as a carbon source for growth; their nitrogen comes from inorganic compounds such as NH_3, NO_3^-, or N_2 (Table 4-1). Interestingly, the energy source for such organisms is the oxidation of specific inorganic compounds. Which inorganic compound is oxidized depends on the bacteria in question (Table 4-5). Many autotrophs will not grow on media that contain organic matter, even agar.

TABLE 4-5 Inorganic Oxidation Reactions Used by Autotrophic Bacteria as Energy Sources

Chemosynthetic Type	Inorganic Compounds Oxidized as Energy (~ E) Source	Representative Families, Genera, and Species[a]	Nitrogen cycle Reaction
NH_3 oxidizers (aerobic)		Nitrobacteriaceae (Nitrosomonas, Nitrosococcus, Nitrosospira)	Nitrification Nitrification Nitrification Nitrification
NO_2 oxidizers (aerobic)		Nitrobacteriaceae (Nitrobacter, Nitrococcus)	Nitrification Nitrification Nitrification
Sulfur oxidizers[b] (aerobic) Iron oxidizers (aerobic)		Thiobacillus thiooxidans, Thiobacillus, ferrooxidans, Ferrobacilius, Leptothrix	
Sulfur-compound oxidizers Dentrification (anaerobic)	S_2 O_2 oxidized; NO_3 reduced	Thiobacillus denitrificans	

a All are Gram-negative species (see Borgey's Manual of beterminative Bacteriology. 8th ed, part 12).

b strict autotrophic modes of metabolism are not present in sulfur and sulfur compound-oxidizing bacteria. For example, heterotrophic sulfur compound oxidizers are known, the aerobic species being able to oxidize.................... (e.g. Beggiatoa and thiothrix species).

Also found among the autotrophic microorganisms are the sulfur-oxidizing or sulfur-compound-oxidizing bacteria, which seldom exhibit a strictly autotrophic mode of metabolism like the obligate nitrifying bacteria (see discussion of nitrogen cycle below). The representative sulfur compounds oxidized by such bacteria are H_2S, S_2, and S_2O_3. Among the sulfur bacteria are two very interesting organisms; *Thiobacillus ferrooxidans,* which gets its energy for autotrophic growth by oxidizing elemental sulfur or ferrous iron, and *T denitrificans,* which gets its energy by oxidizing S_2O_3 anaerobically, using NO_3^- as the sole terminal electron acceptor. T denitrificans reduces NO_3 to molecular N_2, which is liberated as a gas; this biologic process is called denitrification.

All autotrophic bacteria must assimilate CO_2, which is reduced to glucose from which organic cellular matter is synthesized. The energy for this biosynthetic process is derived from the oxidation of inorganic compounds discussed in the previous paragraph. Note that all autotrophic and phototrophic bacteria possess essentially the same organic cellular constituents found in heterotrophic bacteria; from a nutritional viewpoint, however, the autotrophic mode of metabolism is unique, occurring only in bacteria.

Anerobic Respiration

Some bacteria exhibit a unique mode of respiration called anaerobic respiration. These heterotrophic bacteria that will not grow anaerobically unless a specific chemical component, which serves as a terminal electron acceptor, is added to the medium. Among these electron acceptors are NO_3, SO_4^2, the organic compound fumarate, and CO_2. Bacteria requiring one of these compounds for anaerobic growth are said to be anaerobic respirers.

A large group of anaerobic respirers are the nitrate reducers (Table 4-6). The nitrate reducers are predominantly heterotrophic bacteria that possess a complex electron transport system(s) allowing the NO_3 ion to serve anaerobically as a terminal acceptor of electrons (NO_3 2e- NO_2; NO_3 5e- N_2; or NO_3 8e- NH_3). The organic compounds that serve as specific electron donors for these three known nitrate reduction processes are shown in Table 4-6. The nitrate reductase activity is common in bacteria and is routinely used in the simple nitrate reductase test to identify bacteria (see *Bergey's Manual of Deterininative Bacteriology*, 8th ed.).

The methanogens are among the most anaerobic bacteria known, being very sensitive to small concentrations of molecular O_2. They are also archaebacteria, which typically live in unusual and deleterious environments.

All of the above anaerobic respirers obtain chemical energy for growth by using these anaerobic energy-yielding oxidation reactions.

TABLE 4-6 Nitrate Reduces

Physiologic Types of Nitrate Reductases	Electron Donor(s)	Representative
Respiratory	Formate	Escherichia coli
($NO_3^- \rightarrow NO_2^-$)	NADH	Klebsiella acrogenes
Denitrifying	NADH	Pseudomonas aeroginosa
($NO_3^- \rightarrow N_2$)	Pyruvate NADH, succinate	Clostridium perfringens Paracoccus denitrificans
Assimilatory	Lactrate	Staphylococcus aureus
($NO_3^- \rightarrow NH_3$)	H2, formate NADH, succinate NADH NADH, lactale, glycerol phosphate	Vibrio succinogenes Bacillus stearothermophilus Enterobacter aerogenes Escherichia coli

$$4AH_2 + HNO_3 \xrightarrow{\text{Nitrate reduction}} 4A + NH_3 + 3H_2O + \text{energy}$$

(AH_2 = organic substrate, which serves as electron donor)

A second group of anaerobic respirers, the sulfate reducers, utilize SO_4^{2} ion in similar fashion

(SO_4^{2} & H_2S)

$$4AH_2 + H_2SO_4 \xrightarrow{\text{Sulfate reduction}} 4A + H_2S + 4H_2O + \text{energy}$$

The third group, the fumarate respirers are anaerobic bacteria that require exogenous HOOC – CH = CH COOH for growth. Fumarate is reduced to succinate (HOOC – CH_2 – CH_2 COOH), which is secreted as a by-product.

$$AH_2 + HOOC = \overset{H}{\overset{||}{C}} - \overset{H}{\overset{||}{COOH}} \xrightarrow{\text{Fumarate reduction}}$$

$$A + HOOC - CH_2 \quad CH_2 - COOH + \text{energy}$$

Organisms of still another specialized group of anaerobic respirers, the methanogens, produce methane gas CO_2 & CH_4 as a metabolic end product of microbial growth. H_2 gas is the growth. H_2 gas is the growth substrate; CO_2 is the terminal electron acceptor.

$$4H_2 - CO_2 \xrightarrow{CO_2 \text{ reduction}} CH_4 + 2H_2O + \text{energy}$$

The Nitrogen Cycle

Nowhere can the total metabolic potential of bacteria and their diverse chemical-transforming capabilities be more fully appreciated than in the geochemical cycling of the element nitrogen. All the basic chemical elements (S, O, P, C, and H) required to sustain living organisms have geochemical cycles similar to the nitrogen cycle.

The nitrogen cycle is an ideal demonstration of the ecologic interdependence of bacteria, plants, and animals. Nitrogen is recycled when organisms use one form of nitrogen for growth and excrete another nitrogenous compound as a waste product. This waste product is in turn utilized by another type of organism as a growth or energy substrate. Figure 4-10 shows the nitrogen cycle.

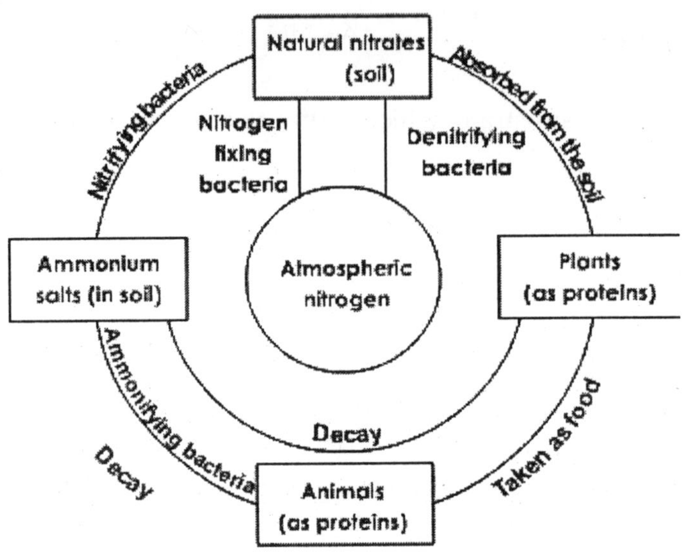

FIGURE 4-10 the nitrogen cycle.

When the specific breakdown of organic nitrogenous compounds occurs, that is, when proteins are degraded to amino acids (proteolysis) and then to inorganic NH_3, by heterotrophic bacteria, the process is called ammonification. This is an essential step in the nitrogen cycle. At death, the organic constituents of the tissues and cells decompose biologically to inorganic constituents by a process called mineralization; these inorganic end products can then serve as nutrients for other life forms. The NH_3 liberated in turn serves as a utilizable nitrogen source for many other bacteria. The breakdown of feces and urine also occurs by ammonification.

The other important biologic processes in the nitrogen cycle include nitrification (the conversion of NH_3 to NO_3 by autotrophes in the soil; denitrification (the anaerobic conversion of NO_3 to N_2 gas) carried out by many heterotrophs); and nitrogen fixation (N_2 to NH_3, and cell protein). The latter is a very specialized prokaryotic process called diazotrophy, carried out by both free-living bacteria (such as *Azotobacter, Derxia, Beijeringeia,* and *Azomona* species) and symbionts (such as *Rhizobium* species) in conjunction with legume plants (such as soybeans, peas, clover, and bluebonnets). All plant life relies heavily on NO_3- as a nitrogen source, and most animal life relies on plant life for nutrients.

REFERENCES

Buchanan RE, Cibbons NE (eds): Bergey's Manual of Determinative Bacteriology. 8th Ed. Williams & Wilkins, Baltimore, 1974

Green DE: A critique of the chemosmotic model of energy coupling. Proc Natl Acad Sci USA, 78:2249, 1981

Haddock BA, Hamilton WA (eds): Microbial energetics. 27th Symposium of the Society of General Microbiology. Cambridge University Press, Cambridge, 1977

Hempfling WP: Microbial Respiration. Benchman Papers in Microbiology no. 13.

Downden, Hutchinson and Ross, Stroudsburg, Pa, 1979

Hill R: The biochemists' green mansions: the photosynthetic electron-transport chain in plants. In Campbell PN, Greville CD (eds): Essays in Biochemistry. Vol.1. Academic Press, New York, 1965

Jurtshuk P, Jr, Liu JK: Cytochrome oxidase and analyses of *Bacillus* strains: existence of oxidase-positive species. Int J Syst Bacterol 33:887, 1983

Jurtshuk P, Jr, Mueller TJ, Acord WC: Bacterial terminal oxidases. Crit Rev Microbiol, 3:359, 1975

Jurtshuk P, Jr, Mueller TJ, Wong TY: Isolation and purification of the cytochrome oxidase of *Azotobacter vinelandii*. Biochim Biophys Acta 637:374, 1981

Jurtshuk P, Jr, Yang TY: Oxygen reactive hemoprotein components in bacterial respiratory systems. In Knowles CJ (ed): Diversity of Bacterial Respiratory Systems. Vol. 1. CRC Press, Boca Raton, FL, 1980

Kamp AF, La Riviere JWM, Verhoeven W (eds): Jan Albert Kluyver: His Life and Work. Interscience, New York, 1959

Kluyver JA, Van Niel CB: The microbe's contribution to biology. Harvard University Press, Cambridge, MA, 1956

Kornberg HL: The role and maintenance of the tricarboxylic acid cycle in *Escherichia coli*. In Goodwin TW (ed): British Biochemistry Past and Present. Biochemistry Society Symposium no. 30. Academic Press, London, 1970

Lemberg R, Barrett J: Bacterial cytochromes and cytochrome oxidases. In Lemberg R, Barrett J: Cytochromes. Academic Press, New York, 1973

Mandelstam J, McQuillen K, Dawes I (eds): Biochemistry of Bacterial Growth. 3rd Ed. Blackwell, Oxford, 1982

O'Leary WM: The chemistry and metabolism of microbial lipids. World Publishing Co, Cleveland, 1967

Schlegel HG, Bowier B (eds): Autotrophic Bacteria. Science Tech, Madison, WI, 1989

Slepecky RA, Leadbetter ER: Ecology and relationships of endospore-forming bacteria: Changing perspectives. In Piggot P, Moran Jr, CP and Youngman P (Eds). Regulation of Bacterial Differentiation. Am Soc Microbiol Press, 1994

Thauer RK, Jungermann K, Decker K: Energy conservation in chemotrophic anaerobic bacteria. Bacteriol Rev 41:100, 1977

Thimann KV: The Life of Bacteria. 2nd Ed. Macmillan, New York, 1966

Chapter 5

Genetics

General Concepts

Genetic Information in Microbes

Genetic information in bacteria and many viruses is encoded in DNA, but some viruses use RNA. Replication of the genome is essential for inheritance of genetically determined traits. Gene expression usually involves transcription of DNA into messenger RNA and translation of mRNA into protein.

Genome Organization

The bacterial chromosome is a circular molecule of DNA that functions as a self-replicating genetic element (replicon). Extrachromosomal genetic elements such as plasmids and bacteriophages are nonessential replicons which often determine resistance to antimicrobial agents, production of virulence factors, or other functions. The chromosome replicates semiconservatively; each DNA strand serves as template for synthesis of its complementary strand.

Mutation and Selection

The complete set of genetic determinants of an organism constitutes its genotype, and the observable characteristics constitute its phenotype. Mutations are heritable changes in genotype that can occur spontaneously or be induced by chemical or physical treatments. Organisms selected as reference strains are called wild type, and their progeny with mutations are called mutants. Selective media distinguish between wild type and mutant strains based on growth; differential media distinguish between them based on other phenotypic properties.

Exchange of Genetic Information

Genetic exchanges among bacteria occur by several mechanisms. In transformation, the recipient bacterium takes up extracellular donor DNA. In transduction, donor DNA packaged in a bacteriophage infects the recipient bacterium. In conjugation, the donor bacterium transfers DNA to the recipient by mating. Recombination is the rearrangement of donor and recipient genomes to form new, hybrid genomes. Transposons are mobile DNA segments that move from place to place within or between genomes.

Recombinant DNA and Gene Cloning

Gene cloning is the incorporation of a foreign gene into a vector to produce a recombinant DNA molecule that replicates and expresses the foreign gene in a recipient cell. Cloned genes are detected by the phenotypes they determine or by specific nucleotide sequences that they contain. Recombinant DNA and gene cloning are essential tools for research in molecular microbiology and medicine. They have many medical applications, including development of new vaccines, biologics, diagnostic tests, and therapeutic methods.

Regulation of Gene Expression

Expression of genes in microbes is often regulated by intracellular or environmental conditions. Regulation can affect any step in gene expression, including transcription initiation or termination, translation, or activity of gene products. An operon is a set of genes that is transcribed as a single unit and expressed coordinately. Specific regulation induces or represses a particular gene or operon. Global regulation affects a set of operons, which constitute a regulon. All operons in the regulon are coordinately controlled by the same regulatory mechanism.

INTRODUCTION

Genetic Information In Microbes

The genetic material of bacteria and plasmids is DNA. Bacterial viruses (bacteriophages or phages) have DNA or RNA as genetic material. The two essential functions of genetic material are replication and expression. Genetic material must replicate accurately so that progeny inherit all of the specific genetic determinants (the genotype) of the parental organism. Expression of specific genetic material under a particular set of growth conditions determines the observable characteristics (phenotype) of the organism. Bacteria have few structural or developmental features that can be observed easily, but they have a vast array of biochemical capabilities and patterns of susceptibility to antimicrobial agents or bacteriophages. These latter characteristics are often selected as the inherited traits to be analyzed in studies of bacterial genetics.

Nucleic Acid Structure

Nucleic acids are large polymers consisting of repeating nucleotide units (Fig. 5-1). Each nucleotide contains one phosphate group, one pentose or deoxypentose sugar, and one purine or pyrimidine base. In DNA the sugar is D-2-deoxyribose; in RNA the sugar is D-ribose. In DNA the purine bases are adenine (A) and guanine (G), and the pyrimidine bases are thymine (T) and cytosine (C). In RNA, uracil (U) replaces thymine. Chemically modified purine and pyrimidine bases are found in some bacteria and bacteriophages. The repeating structure of polynucleotides involves alternating sugar and phosphate residues, with phosphodiester bonds linking the 3'-hydroxyl group of one nucleotide sugar to the 5'-hydroxyl group of the adjacent nucleotide sugar. These asymmetric phosphodiester linkages define the polarity of the polynucleotide chain. A purine or pyrimidine base is linked at the 1'-carbon atom of each sugar residue and projects from the repeating sugar-phosphate backbone. Double-stranded DNA is helical, and the two strands in the helix are antiparallel. The double helix is stabilized by hydrogen bonds between purine and pyrimidine bases on the opposite strands. At each position, A on one strand pairs by two hydrogen bonds with T on the opposite strand, or G pairs by three hydrogen bonds with C. The two strands of double-helical DNA are, therefore, complementary. Because of complementarity, double-stranded DNA contains equimolar amounts of purines (A + G) and pyrimidines (T + C), with A equal to T and G equal to C, but the mole fraction of G + C in DNA varies widely among different bacteria. Information in nucleic acids is encoded by the ordered sequence of nucleotides along the polynucleotide chain, and in double-stranded DNA the sequence of each strand determines what the sequence of the complementary strand must be. The extent of sequence homology between DNAs from different microorganisms is the most stringent criterion for determining how closely they are related.

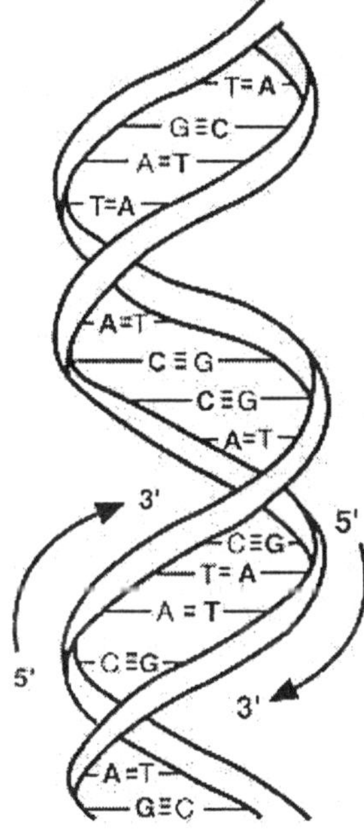

FIGURE 5-1 Double helical structure of DNA. The diagram shows the structure of DNA represented as a helical ladder. The backbone of each polynucleotide strand (represented as a ribbon) consists of alternating phosphate and deoxyribose residues linked by phosphodiester bonds, and the strands have opposite polarities (arrows). The purine or pyrimidine base of each nucleotide on one strand projects toward the complementary base of the corresponding nucleotide from the other strand and is linked to it by hydrogen bonds. The double helix has a diameter of 2 nm. Each full turn of the double helix contains 10 nucleotide pairs and is 3.4 nm in length.

DNA Replication

During replication of the bacterial genome, each strand in double-helical DNA serves as a template for synthesis of a new complementary strand. Each daughter double-stranded DNA molecule thus contains one old polynucleotide strand and one newly synthesized strand. This type of DNA replication is called semiconservative. Replication of chromosomal DNA in bacteria starts at a specific chromosomal site called the origin and proceeds bidirectionally until the process is completed (Fig. 5-2). When bacteria divide by binary fission after completing DNA replication, the replicated chromosomes are partitioned into each of the daughter cells. The origin regions specifically and transiently associate with the cell membrane after DNA replication has been intitiated, leading to a model whereby membrane attachment directs separation of daughter chromosomes (the replicon model). These characteristics of DNA replication during bacterial growth fulfill the requirements of the genetic material to be reproduced accurately and to be inherited by each daughter cell at the time of cell division.

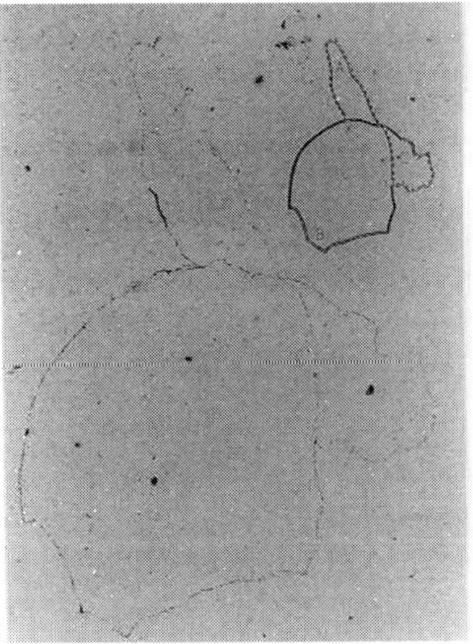

FIGURE 5-2 Autoradiograph of intact replicating chromosome of E coli. Bacteria were radioactively labeled with tritiated thymidine for approximately two generations and were lysed gently. Bacterial DNA was then examined by autoradiography. Insert shows replicating bacterial chromosome in diagrammatic form. The chromosome is circular, and two forks (X and Y) are present in replicating structure. The segments of chromosome represented by double lines had completed two replications in presence of tritiated thymidine, whereas segments represented by a solid line and a dotted line had replicated only once in presence of tritiated thymidine. The density of grains in the autoradiogram was twice as great in the segments of chromosome that had completed two cycles of replication in presence of tritiated thymnidine. Bar, 100 μm. From Cairns, J.P.: Cold Spring Harbor Symposia on Quantitative Biology 28:44, 1963.

Gene Expression

Genetic information encoded in DNA is expressed by synthesis of specific RNAs and proteins, and information flows from DNA to RNA to protein. The DNA-directed synthesis of RNA is called transcription. Because the strands of double-helical DNA are antiparallel and complementary, only one of the two DNA strands can serve as template for synthesis of a specific mRNA molecule. Messenger RNAs (mRNAs) transmit information from DNA, and each mRNA in bacteria functions as the template for synthesis of one or more specific proteins. The process by which the nucleotide sequence of an mRNA molecule determines the primary amino acid sequence of a protein is called translation. Ribosomes, complexes of ribosomal RNAs (rRNAs) and several ribosomal proteins, translate each mRNA into the corresponding polypeptide sequence with the aid of transfer RNAs (tRNAs), amino-acyl tRNA synthesases, initiation factors and elongation factors. All of these components of the apparatus for protein synthesis function in the production of many different proteins. A gene is a DNA sequence that encodes a protein, rRNA, or tRNA molecule (gene product). The genetic code determines how the nucleotides in mRNA specify the amino acids in a polypeptide. Because there are only 4 different nucleotides in mRNA (containing U, A, C and G), single nucleotides do not contain enough information to specify uniquely all 20 of the amino acids. In dinucleotides 16 (4 x 4) arrangements of the four nucleotides are possible, and in trinucleotides 64 (4 x 4 x 4) arrangements are possible. Thus, a minimum of three nucleotides is required to provide at least one unique sequence corresponding to each of the 20 amino acids. The "universal" genetic code employed by most organisms (Table 1) is a triplet code in which 61 of the 64 possible trinucleotides (codons) encode specific amino acids, and any of the three remaining codons (UAG, UAA or UGA) results in termination of translation. The chain-terminating codons are also called nonsense codons because they do not specify any amino acids. The genetic code is described as degenerate, because several codons may be used for a single amino acid, and as nonoverlapping, because adjacent codons do not share any common nucleotides. Exceptions to the "universal" code include the use of UGA as a tryptophan codon in some species of Mycoplasma and in mitochondrial DNA, and a few additional codon differences in mitochondrial DNAs from yeasts, Drosophila, and mammals. Translation of mRNA is usually initiated at an AUG codon for methionine, and adjacent codons are translated sequentially as the mRNA is read in the 5' to 3' direction. The corresponding polypeptide chain is assembled beginning at its amino terminus and proceeding toward its carboxy terminus. The sequence of amino acids in the polypeptide is, therefore, colinear with the sequence of nucleotides in the mRNA and the corresponding gene. Specific enzymatic reactions involved in DNA, RNA, and protein synthesis are beyond the scope of this chapter.

Expression of genetic determinants in bacteria involves the unidirectional flow of information from DNA to RNA to protein. In bacteriophages, either DNA or RNA can serve as genetic material. During infection of bacteriaby RNA bacteriophages,

RNA molecules serve as templates for RNA replication and as mRNAs. Studies with the retrovirus group of animal viruses reveal that DNA molecules can be synthesized from RNA templates by enzymes designated as RNA-dependent DNA polymerases (reverse transcriptases). This reversal of the usual direction for flow of genetic information, from RNA to DNA instead of from DNA to RNA, is an important mechanism for enabling informationfrom retroviruses to be encoded in DNA and to become incorporated into the genomes of animal cells.

TABLE 5-1 The Genetic Code[a]

First Nucleotide of Codon	Second Nucleofide of codon				Third Nucleotide of Codon
	U	C	A	G	
	Pne	Ser	Tyr	Cys	U
U	Pne	Ser	Tyr	Cys	C
	Leu	Ser	Termination	Termination	A
	Leu	Ser	Termination	Trp	G
	Leu	Pro	His	Arg	U
C	Leu	Pro	His	Arg	C
	Leu	Pro	His	Arg	A
	Leu	Pro	His	Arg	G
	Ile	Thr	Asn	Ser	U
A	Ile	Thr	Asn	Ser	C
	Met	Thr	Lys	Arg	G
	Val	Ala	Asp	Gly	U
G	Val	Ala	Asp	Gly	C
	Val	Ala	Glu	Gly	A
	Val	Ala	Glu	Gly	C

aAbbreviations: Ala, alarine; Arg, arginine; asn, asparagine; Asp; aspartic acid; Cys, cysteine; Gln, glutamne; Glu, glutamic acid: Gly, glycine: His, histidine; lle, isoleucne: Pro, proline; Ser, serine; Thr, threonine: Try, Trosine: Trp, tryptophan; Val, valine.

Genome Organization

DNA molecules that replicate as discrete genetic units in bacteria are called replicons. In some Escherichia coli strains, the chromosome is the only replicon present in the cell. Other bacterial strains have additional replicons, such as plasmids and bacteriophages.

Chromosomal DNA

Bacterial genomes vary in size from about 0.4 x 109 to 8.6 x 109 daltons (Da), some of the smallest being obligate parasites (Mycoplasma) and the largest belonging to bacteria capable of complex differentiation such as Myxococcus. The amount of DNA in the genome determines the maximum amount of information that it can encode. Most bacteria have a haploid genome, a single chromosome consisting of a circular, double stranded DNA molecule. However linear chromosomes have been found in Gram-positive Borrelia and Streptomyces spp., and one linear and one circular chromosome is present in the Gram-negative bacterium Agrobacterium tumefaciens. The single chromosome of the common intestinal bacterium E coli is 3 x 109 Da (4,500 kilobase pairs [kbp]) in size, accounting for about 2 to 3 percent of the dry weight of the cell. The E coli genome is only about 0.1% as large as the human genome, but it is sufficient to code for several thousand polypeptides of average size (40 kDa or 360 amino acids).

The chromosome of E coli has a contour length of approximately 1.35 mm, several hundred times longer than the bacterial cell, but the DNA is supercoiled and tightly packaged in the bacterial nucleoid. The time required for replication of the entire chromosome is about 40 minutes, which is approximately twice the shortest division time for this bacterium. DNA replication must be initiated as often as the cells divide, so in rapidly growing bacteria a new round of chromosomal replication begins before an earlier round is completed. At rapid growth rates there may be four chromosomes replicating to form eight at the time of cell division, which is coupled with completion of a round of chromosomal replication. Thus, the chromosome in rapidly growing bacteria is replicating at more than one point. The replication of chromosomal DNA in bacteria is complex and involves many different proteins.

Plasmids

Plasmids are replicons that are maintained as discrete, extrachromosomal genetic elements in bacteria. They are usually much smaller than the bacterial chromosome, varying from less than 5 to more than several hundred kbp, though plasmids as large as 2 Mbp occur in some bacteria. Plasmids usually encode traits that are not essential for bacterial viability, and replicate independently of the chromosome. Most plasmids are supercoiled, circular, double-stranded DNA molecules, but linear plasmids have also been demonstrated in Borrelia and Streptomyces. Closely related or identical plasmids demonstrate incompatibility; they cannot be stably maintained in the same bacterial host. Classification of plasmids is based on incompatibility or on use of specific DNA probes in hybridization tests to identify nucleotide sequences that are characteristic of specific plasmid replicons. Some hybrid plasmids contain more than one replicon.Conjugative plasmids code for functions that promote transfer of the plasmid from the donor bacterium to other recipient bacteria, but nonconjugative plasmids do not. Conjugative plasmids that also promote transfer of the bacterial chromosome from the donor bacterium to other recipient bacteria are called fertility plasmids, and are discussed below. The average number of molecules of a given plasmid per bacterial chromosome is called its copy number. Large plasmids (>40 kilobase pairs) are often conjugative, have small copy numbers (1 to several per chromosome), code for all functions required for their replication, and partition themselves among daughter cells during cell division in a manner similar to the bacterial chromosome. Plasmids smaller than 7.5 kilobase pairs usually are nonconjugative, have high copy numbers (typically 10-20 per chromosome), rely on their bacterial host to provide some functions required for replication, and are distributed randomly between daughter cells at division.

Many plasmids control medically important properties of pathogenic bacteria, including resistance to one or several antibiotics, production of toxins, and synthesis of cell surface structures required for adherence or colonization. Plasmids that determine resistance to antibiotics are often called R plasmids (or R factors). Representative toxins encoded by plasmids include heat-labile and heat-stable enterotoxins of E coli, exfoliative toxin of Staphylococcus aureus, and tetanus toxin of Clostridium tetani. Some plasmids are cryptic and have no recognizable effects on the bacterial cells that harbor them. Comparing plasmid profiles is a useful method for assessing possible relatedness of individual clinical isolates of a particular bacterial species for epidemiological studies. The role of plasmids in the evolution of resistance to antibiotics is discussed below.

Bacteriophages

Bacteriophages (bacterial viruses, phages) are infectious agents that replicate as obligate intracellular parasites in bacteria. Extracellular phage particles are metabolically inert and consist principally of proteins plus nucleic acid (DNA or RNA, but not both). The proteins of the phage particle form a protective shell (capsid) surrounding the tightly packaged nucleic acid genome. Phage genomes vary in size from approximately 2 to 200 kilobases per strand of nucleic acid and consist of double-stranded DNA, single-stranded DNA, or RNA. Phage genomes, like plasmids, encode functions required for replication in bacteria, but unlike plasmids they also encode capsid proteins and nonstructural proteins required for phage assembly. Several morphologically distinct types of phage have been described, including polyhedral, filamentous, and complex. Complex phages have polyhedral heads to which tails and sometimes other appendages (tail plates, tail fibers, etc.) are attached.

A single cycle of phage growth is shown in Fig. 5-3. Infection is initiated by adsorption of phage to specific receptors on the surface of susceptible host bacteria. The capsids remain at the cell surface, and the DNA or RNA genomes enter the target cells (penetration). Because infectivity of genomic DNA or RNA is much less than that of mature virus, there is a time immediately after infection called the eclipse period during which intracellular infectious phage cannot be detected. The infecting phage RNA or DNA is replicated to produce many new copies of the phage genome, and phage-specific proteins are produced. For most phages assembly of progeny occurs in the cytoplasm, and release of the progeny occurs by cell lysis. In contrast, filamentous phages are formed at the cell envelope and released without killing the host cells. The eclipse period ends when intracellular infectious progeny appear. The latent period is the interval from infection until extracellular progeny appear, and the rise period is the interval from the end of the latent period until all phage are extracellular. The average number of phage particles produced by each infected cell, called the burst size, is characteristic for each virus and often ranges between 50 and several hundred. For discussions of structure, multiplication, and classification of animal viruses, see Chapters 41 and 42.

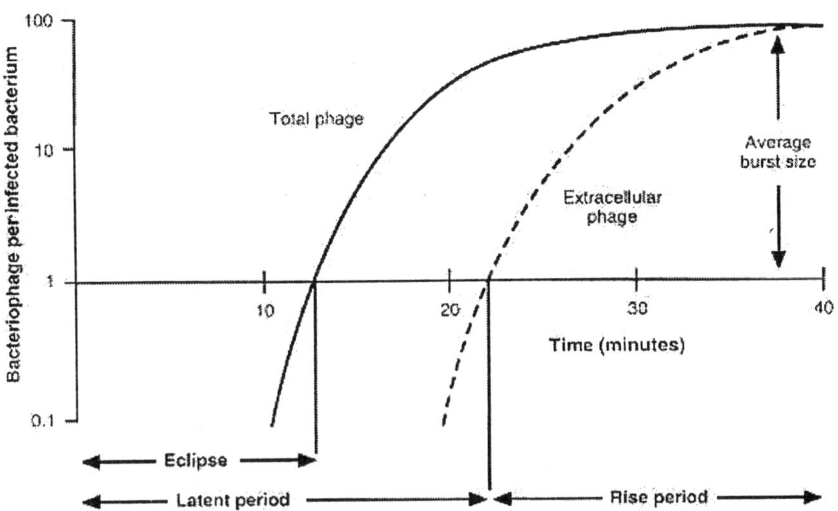

FIGURE 5-3 One-step growth of bacteriophage. A culture of susceptible bacteria is synchronously infected with bacteriophage added at time 0 at low multiplicity of infection. Unabsorbed phage is inactivated shortly thereafter by addition of anti-phage antiserum, and the culture is then diluted to prevent further activity of the antiserum. Samples are taken at intervals for phage assays. Total phage (intracellular plus extracellular) is determined by testing the sample after treating it to disrupt infected bacteria, and extracellular phage is determined by testing supernatant after removal of bacteria by centrifugation or ultrafiltration. Phage titers are as the ratio of phage per infected bacterial cell.

Phages are classified into two major groups: virulent and temperate. Growth of virulent phages in susceptible bacteria destroys the host cells. Infection of susceptible bacteria by temperate phages can have either of two outcomes: lytic growth or lysogeny. Lytic growth of temperate and virulent bacteriophages is similar, leading to production of phage progeny and death of the host bacteria. Lysogeny is a specific type of latent viral infection in which the phage genome replicates as a prophage in the bacterial cell. In most lysogenic bacteria, the genes required for lytic phage development are not expressed, and production of infectious phage does not occur. Furthermore, the lysogenic cells are immune to superinfection by the virus which they harbor as a prophage. The physical state of the prophage is not identical for all temperate viruses. For example, the prophage of bacteriophage l in E coli is integrated into the bacterial chromosome at a specific site and replicates as part of the bacterial chromosome, whereas the prophage of bacteriophage P1 in E coli replicates as an extrachromosomal plasmid.

Lytic phage growth occurs spontaneously in a small fraction of lysogenic cells, and a few extracellular phages are present in cultures of lysogenic bacteria. For some lysogenic bacteria, synchronous induction of lytic phage development occurs in the entire population of lysogenic bacteria when they are treated with agents that damage DNA, such as ultraviolet light or mitomycin C. The loss of prophage from a lysogenic bacterium, converting it to the nonlysogenic state and restoring susceptibility to infection by the phage that was originally present as prophage, is called curing.

Some temperate phages contain genes for bacterial characteristics that are unrelated to lytic phage development or the lysogenic state, and expression of such genes is called phage conversion (or lysogenic conversion). Examples of phage conversion that are important for microbial virulence include production of diphtheria toxin by Corynebacterium diphtheriae, erythrogenic toxin by Streptococcus pyogenes (group A b-hemolytic streptococci), botulinum toxin by Clostridium botulinum, and Shiga-like toxins by E coli. In each of these examples the gene which encodes the bacterial toxin is present in a temperate phage genome. The specificity of O antigens in Salmonella can also be controlled by phage conversion. Phage typing is the testing of strains of a particular bacterial species for susceptibility to specific bacteriophages. The patterns of susceptibility to the set of typing phages provide information about the possible relatedness of individual clinical isolates. Such information is particularly useful for epidemiological investigations.

Mutation and Selection

Mutations are heritable changes in the genome. Spontaneous mutations in individual bacteria are rare. Some mutations cause changes in phenotypic characteristics; the occurrence of such mutations can be inferred from the effects they produce. In microbial genetics specific reference organisms are designated as wild-type strains, and descendants that have mutations in their genomes are called mutants. Thus, mutants are characterized by the inherited differences between them and their ancestral wild-type strains. Variant forms of a specific genetic determinant are called alleles. Genotypic symbols are lower case, italicized abbreviations that specify individual genes, with a (+) superscript indicating the wild type allele. Phenotypic symbols are capitalized and not italicized, to distinguish them from genotypic symbols. For example, the genotypic symbol for the ability to produce b-galactosidase, required to ferment lactose, is lacZ+, and mutants that cannot produce b-galactosidase are lacZ. The lactose-fermenting phenotype is designated Lac+, and inability to ferment lactose is Lac-.

Detection of Mutant Phenotypes

Selective and differential media are helpful for isolating bacterial mutants. Some selective media permit particular mutants to grow, but do not allow the wild-type strains to grow. Rare mutants can be isolated by using such selective media. Differential media permit wild-type and mutant bacteria to grow and form colonies that differ in appearance. Detection of rare mutants on differential media is limited by the total number of colonies that can be observed. Consider a wild-type strain of E coli that is susceptible to the antibiotic streptomycin (phenotype Strs) and can utilize lactose as the sole source of carbon (phenotype Lac+). Spontaneously occurring Strr mutants are rare and are usually found at frequencies of less than one per 109 bacteria in cultures of wild-type E coli. Nevertheless, Strr mutants can be isolated easily by using selective media containing streptomycin, because the wild-type Strs bacteria are killed. Isolation of lactose-negative (phenotype Lac-) mutants of E coli poses a different problem. On minimal media with lactose as the sole source of carbon, Lac+ wild-type strains will grow, but Lac- mutants cannot grow. On differential media such as MacConkey-lactose agar or eosin-methylene blue-lactose agar, Lac+ wild-type and Lac- mutant strains of E coli can be distinguished by their color, but spontaneous Lac- mutants are too rare to be isolated easily. Selective media for Lac- mutants of E coli can be made by incorporating chemical analogs of lactose that are converted into toxic metabolites by Lac+ bacteria but not by Lac- mutants. The Lac- mutants can then grow on such media, but the Lac+ wild-type bacteria are killed.

Mutations that inactivate essential genes in haploid organisms are usually lethal, but such potentially lethal mutations can often be studied if their expression is controlled by manipulation of experimental conditions. For example, a mutation that increases the thermolability of an essential gene product may prevent bacterial growth at 42°C, although the mutant bacterium can still grow at 25°C. Conversely, cold-sensitive mutants express the mutant phenotype at low temperature, but not at high temperature. Temperature-sensitive and cold-sensitive mutations are examples of conditional mutations, as are suppressible mutations described later in this chapter. A conditional lethal phenotype indicates that the mutant gene is essential for viability.

Spontaneous and Induced Mutations

The mutation rate in bacteria is determined by the accuracy of DNA replication, the occurrence of damage to DNA, and the effectiveness of mechanisms for repair of damaged DNA.

For a particular bacterial strain under defined growth conditions, the mutation rate for any specific gene is constant and is expressed as the probability of mutation per cell division. In a population of bacteria grown from a small inoculum, the proportion of mutants usually increases progressively as the size of the bacterial population increases.

Mutations in bacteria can occur spontaneously and independently of the experimental methods used to detect them. This principle was first demonstrated by the fluctuation test (Fig. 5-4). The numbers of phage-resistant mutants of E coli in replicate cultures grown from small inocula were measured and compared with those in multiple samples taken from a single culture. If mutations to phage resistance occurred only after exposure to phage, the variability in numbers of mutants between cultures should be similar under both sets of conditions. In contrast, if phage-resistant mutants occurred spontaneously before exposure of the bacteria to phage, the numbers of mutants should be more variable in the independently grown cultures, because differences in the size of the bacterial population when the first mutant appeared would contribute to the observed variability. The data indicated that the mutations to phage resistance in E coli occurred spontaneously with constant probability per cell division.

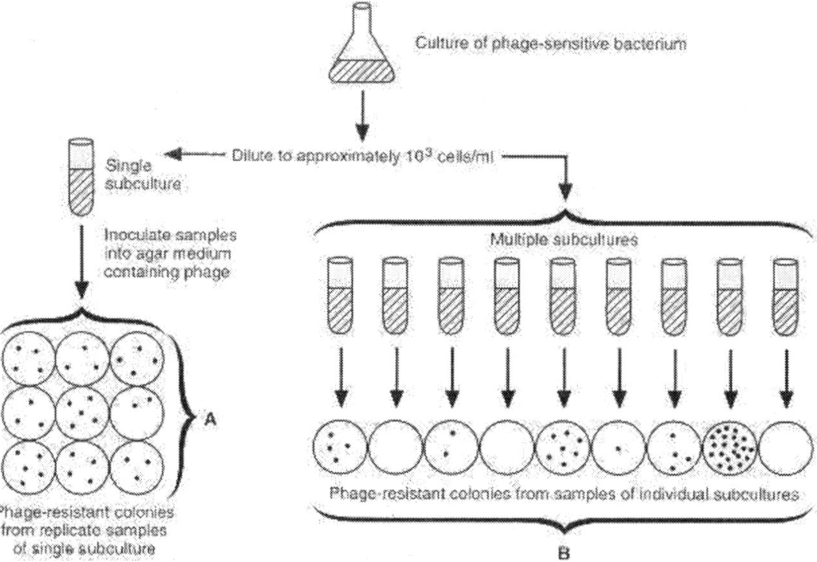

FIGURE 5-4 the fluctuation test. Differences in numbers of colonies of phage-resistant mutants in replicate samples from single subculture were small and reflected only expected fluctuations due to sampling errors. In contrast, numbers of phage-resistant colonies in samples from individual subcultures were more variable and reflected both sampling errors and the independent origins of mutants in individual subcultures. Sizes of clonal populations of mutants in each culture reflected numbers of generations of growth between times that mutations occurred and time of sampling.

Replica plating confirmed that mutations in bacteria can occur spontaneously, without exposure of bacteria to selective agents (Fig. 5-5). For replica plating, a flat, sterile, velveteen surface is used to pick up an inoculum from the surface of an agar master plate and transfer samples to other agar plates. In this manner, samples of the bacterial population from the master plate are transferred to the replica plates without distorting their spatial arrangement. If the replica plates contain selective medium and the master plates do not, the positions of selected mutant colonies on the replica plates can be noted, and bacteria that were not exposed to the selective conditions can be isolated from the same positions on the master plate. Mutants of E coli resistant to bacteriophage T1 or to streptomycin have been isolated in this way, without exposing the wild-type bacteria to the bacteriophage or the antibiotic.

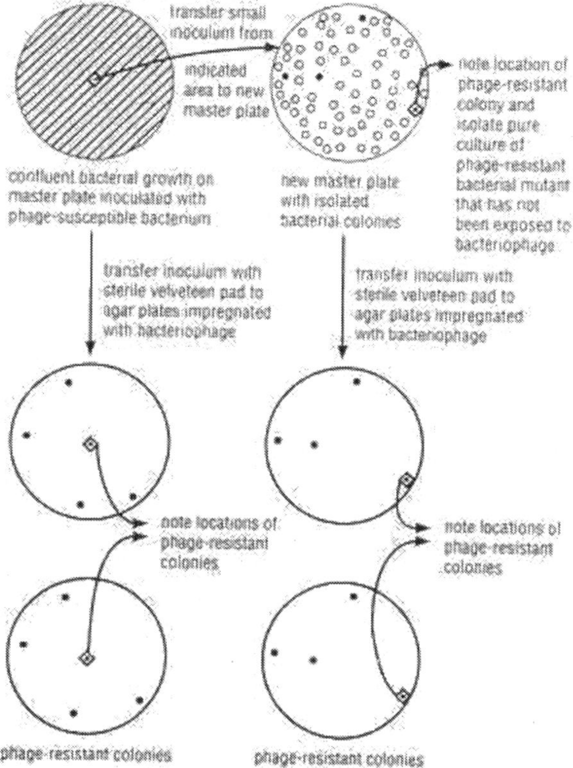

FIGURE 5-5 Detecting preexisting bacterial mutants by replica plating. Master plate was heavily inoculated with sample from pure cultures of phage-susceptible bacterium. After incubation, bacteria from master plate were transferred by replica plating to duplicate agar plates impregnated with bacteriophage. Phage-susceptible bacteria were killed by the bacteriophage. Colonies of phage-resistant bacteria appeared at identical positions on duplicate plates, indicating that phage-resistant bacteria had been transferred to each replica plate from the corresponding locations on master plate. Bacterial inocula selected from appropriate locations on master plate contained a higher proportion of phage-resistant mutants than original bacterial culture. By repeating these procedures several times, it was possible to isolate pure cultures of phage-resistant bacterial mutants that had never been exposed to bacteriophage.

Both environmental and genetic factors affect mutation rates. Exposure of bacteria to mutagenic agents causes mutation rates to increase, sometimes by several orders of magnitude. Many chemical and physical agents, including X-rays and ultraviolet light, have mutagenic activity. Chemicals that are carcinogenic for animals are often mutagenic for bacteria, or can be converted by animal tissues to metabolites that are mutagenic for bacteria. Standardized tests for mutagenicity in bacteria are used as screening procedures to identify environmental agents that may be carcinogenic in humans. Mutator genes in bacteria cause an increase in spontaneous mutation rates for a wide variety of other genes. Expression of these genes, induced by DNA damage (see SOS response later), enables the repair of DNA lesions that would otherwise be lethal, but by an error-prone mechanism that increases the rate of mutation. The overall mutation ratethe probability that a mutation will occur somewhere in the bacterial genome per cell divisionis relatively constant for a variety of organisms with genomes of different sizes and appears to be a significant factor in determining the fitness of a bacterial strain for survival in nature. Most mutations are deleterious, and the risk of adverse mutations for individual bacteria must be balanced against the positive value of mutability as a mechanism for adaptation of bacterial populations to changing environmental conditions.

Molecular Basis of Mutations

Mutations are classified on the basis of structural changes that occur in DNA (Table 2). Some mutations are localized within short segments of DNA (for example, nucleotide substitutions, microdeletions, and microinsertions). Other mutations involve large regions of DNA and include deletions, insertions, or rearrangements of segments of DNA.

TABLE 5-2 Classification of Mutations

Change in DNA	Effect on Polypeptide Structure	Effect on Polypeptide Function	Comments
Nucleotide substitution	1. None	1. None	1. Silent mutation (no phenotypic change)
	2. Amino acid substituting	2. Variable	2. Missense mutation (usually CRM)[a]
	3. Premature termination	3. Usually lost	3. Nonsense mutations (CRM or CRM), estrogenic suppression common
Microdeletion or microsertion	Frameshift mutation	Usually lost	Intragenic suppression common
Large insertions	Altered	Usually lost	See section on transposons
Large deletions	Altered	Usually lost	No reversion

aCRM, Cross-reacting material. Mutant polypeptides are CRM. It they share antigenic determinants with the corresponding wildtype polypeptides.

When a nucleotide substitution occurs in a region of DNA that codes for a polypeptide, one of the three nucleotides within a single codon of a corresponding mRNA molecule will be changed. Silent mutations cause no change in polypeptide structure or function, because one codon in mRNA is changed to another for the same amino acid. Other substitutions cause one amino acid to be replaced by another at the specific position within the polypeptide corresponding to the altered codon. Mutations that result in replacement of one amino acid for another within a polypeptide chain are called missense mutations. The effects of amino acid replacements on the function of a polypeptide gene product vary and depend on the location and the identity of the amino acid replacement. Mutant polypeptides containing amino acid replacements usually share antigenic determinants with the wild-type polypeptide and often have some residual biologic activity. Mutations that result in replacement of an amino acid codon with a termination codon are called nonsense mutations. This results in production of an amino-terminal fragment of the normal polypeptide when the mutant mRNA is translated. Nonsense mutations often result in complete loss of activity of the gene product.

Because of the triplet nature of the genetic code, the consequences of mutations caused by insertions or deletions of small numbers of nucleotides (microinsertions, microdeletions) depend on both the number and sequence of nucleotides involved. Deletion or addition of multiples of three nucleotide pairs does not affect the reading frame, but causes deletion or addition of appropriate numbers of amino acids at one site within the polypeptide. If a new chain-terminating codon is introduced, premature chain termination occurs within the polypeptide. In contrast, addition or deletion of other numbers of nucleotide pairs alters the reading frame for the entire segment of mRNA from the mutation to the distal end of the gene. Therefore, frameshift mutations are likely to cause drastic changes in the structure and activity of polypeptide gene products, and they are often classified as nonsense mutations.

Complementation Tests

To determine if mutations are located in the same gene or different genes, complementation tests are performed with partially diploid bacterial strains (Fig. 5-6). Two copies of the region of the bacterial chromosome harboring a mutation are present in the same bacterium, with each copy containing a different mutation (mutations are in the trans arrangement). A wild-type phenotype indicates that the mutations are in different genes. This phenomenon is called complementation. If a mutant phenotype is observed, a control experiment should be performed with the mutations in the cis arrangement to exclude the possibility that the wild-type alleles cannot be expressed normally in a partially diploid bacterial strain. Complementation tests were originally called "cis-trans" tests, and the term cistron is sometimes used as a synonym for gene. Complementation tests can be performed and interpreted even if the specific biochemical functions of the gene products are unknown.

As an example, consider using a complementation test to characterize two independently derived Lac- mutants of E coli. The biochemical pathway for utilization of lactose requires ß-galactoside permease (genotypic symbol lacY) to transport lactose into the bacterial cell and b-galactosidase (genotypic symbol lacZ) to convert lactose into D-glucose and D-galactose. Mutants that lack b-galactoside permease or b-galactosidase cannot utilize lactose for growth. If the mutations in both Lac- mutants inactivated the same protein (e.g., b-galactoside) then a partial diploid strain containing the lacZ genes from both mutants in the trans arrangement would be unable to utilize lactose. In contrast, if the genotypes of the two mutants were lacZ+ lacY and lacZ lacY+, the partially diploid bacterium would produce

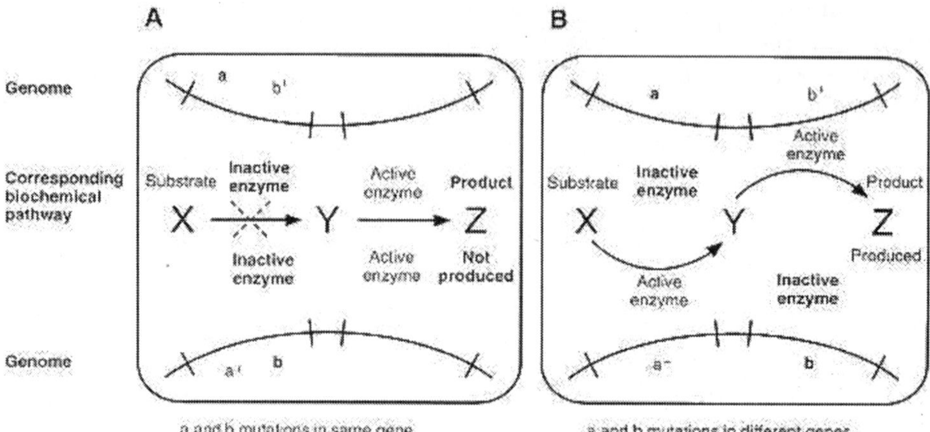

FIGURE 5-6 Complementation is a method to test for functional gene products. Two mutants with similar phenotypes (inability to convert substrate X to product Z) were isolated. Mutations in these strains are designated a and b, respectively, and the wild type alleles are a+ and b+. Partially diploid heterozygous strains were tested to determine if mutations a and b were in the same structural gene (cistron) and inactivated the same gene products. A), If a and b are in the same structural gene (e.g., encoding the enzyme that converts X to Y), neither the a+b nor the ab+ allele codes for an active enzyme, substrate X cannot be utilized, the mutant phenotype is expressed, and no complementation occurs. B), If a and b are in different cistrons (e.g., encoding the enzymes that convert X to Y and Y to Z), the a+ and b+ alleles encode active enzymes, substrate X is converted to product Z, the wild type phenotype is expressed, and complementation occurs.

active b-galactosidase from the lacZ+ determinant and active b-galactoside permease from the lacY+ determinant. Complementation would occur, and the partially diploid strain would utilize lactose.

Reversion and Suppression

Mutations that convert the phenotype from wild-type to mutant are called forward mutations, and mutations that change the phenotype from mutant back to wild-type are called reverse mutations (reversions). Bacterial strains that contain reverse mutations are called revertants. Analysis of mutations that cause phenotypic reversion yields useful information. Reverse mutations that restore the exact nucleotide sequence of the wild-type DNA are true reversions. True revertants are identical to wild-type strains both genotypically and phenotypically. Reverse mutations that do not restore the exact nucleotide sequence of the wild-type DNA are called suppressor mutations (suppressors). Some revertants that harbor suppressor mutations are phenotypically indistinguishable from wild-type strains. Other revertants, called pseudorevertants, can be distinguished phenotypically from wild-type strains, for example, by subtle differences in the characteristics of an enzymatic activity that has been regained (such as specific activity, substrate specificity, kinetic constants, or susceptibility to thermal or chemical inactivation). Recognition of pseudorevertant phenotypes suggests the presence of suppressor mutations.

Suppressor mutations can be intragenic or extragenic. Intragenic suppressors are located in the same gene as the forward mutations that they suppress. The possible locations and nature of intragenic suppressors are determined by the original forward mutation and by the relationships between the primary structure of the gene product and its biologic activity. Extragenic suppressors are located in different genes from mutations whose effects they suppress. The ability of extragenic suppressors to suppress a variety of independent mutations can be tested. Some extragenic suppressors are specific for particular genes, some are specific for particular codons, and some have other specificity patterns. Extragenic suppressors that reverse the phenotypic effects of chain-terminating codons have been well characterized and found to alter the structure of specific tRNAs. A particular suppressor tRNA can permit a specific chain-terminating codon to be translated, resulting in incorporation of a specific amino acid into the nascent polypeptide at the position corresponding to the chain-terminating codon. In a bacterium that has a chain-terminating mutation and an appropriate extragenic suppressor, translation of the mRNA containing the mutant codon can therefore result in formation of a full-length polypeptide. The biologic activity of the full-length polypeptide formed as a consequence of suppression depends both on the amount of protein made and on the functional consequences of the specific amino acid replacement determined by the suppressor tRNA.

Exchange of Genetic Information

The biologic significance of sexuality in microorganisms is to increase the probability that rare, independent mutations will occur together in a single microbe and be subjected to natural selection. Genetic interactions between microbes enable their genomes to evolve much more rapidly than by mutation alone. Representative phenomena of medical importance that involve exchanges of genetic information or genomic rearrangements include the rapid emergence and dissemination of antibiotic resistance plasmids, flagellar phase variation in Salmonella, and antigenic variation of surface antigens in Neisseria and Borrelia.

Sexual processes in bacteria involve transfer of genetic information from a donor to a recipient and result either in substitution of donor alleles for recipient alleles or addition of donor genetic elements to the recipient genome. Transformation, transduction, and conjugation are sexual processes that use different mechanisms to introduce donor DNA into recipient bacteria (Fig. 5-7). Because donor DNA cannot persist in the recipient bacterium unless it is part of a replicon, recombination between donor and recipient genomes is often required to produce stable, hybrid progeny. Recombination is most likely to occur when the donor and recipient bacteria are from the same or closely related species.

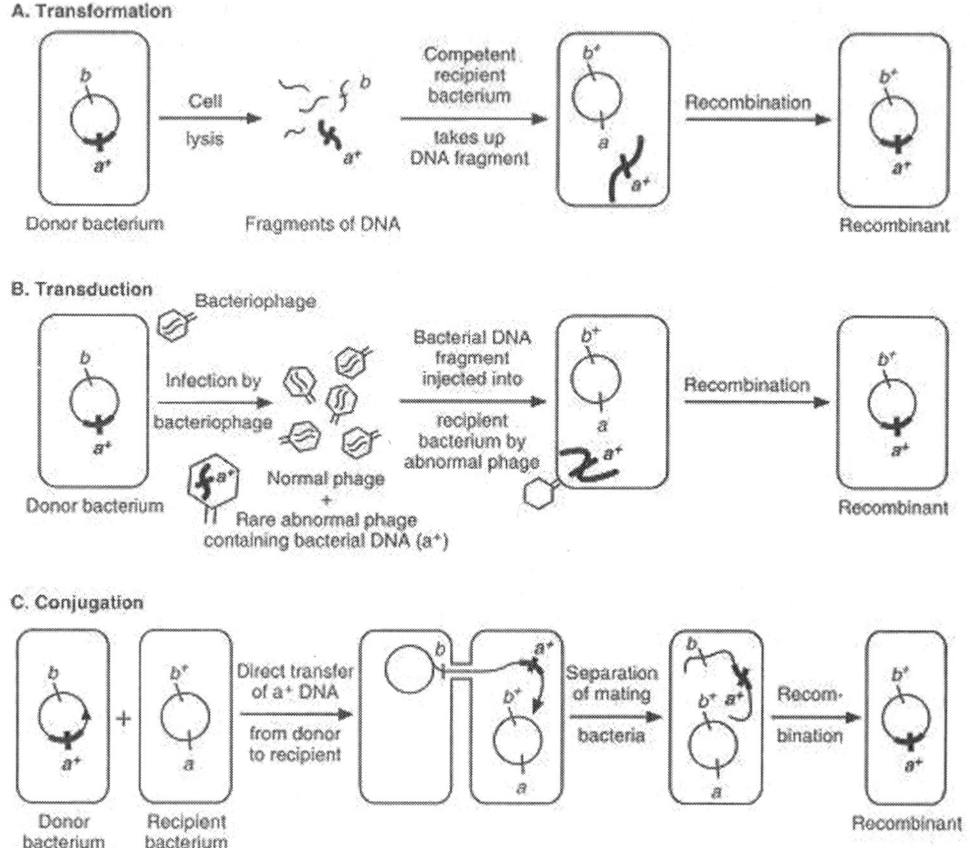

FIGURE 5-7 Exchange of genetic information in bacteria. Transformation, transduction, and conjugation differ in means for introducing DNA from donor cell into recipient cell. A) In transformation, fragments of DNA released from donor bacteria are taken up by competent recipient bacteria. B) In transduction, abnormal bacteriophage particles containing DNA from donor bacteria inject their DNA into recipient bacteria. C) Conjugation occurs by formation of cytoplasmic connections between donor and recipient bacteria, with direct transfer of newly synthesized donor DNA into the recipient cells. In all three cases, recombination between donor and recipient DNA molecules is required for formation of stable recombinant genomes. Bacterial genome is represented diagrammatically as a circular element in bacterial cells. Donor and recipient DNA are indicated by fine lines and heavy lines, respectively. In each recombinant genome, the a+ allele from donor strain has replaced the a allele from recipient strain, and the b+ allele is derived from recipient strain.

For a recombinant to be detected, its phenotype must be different from both parental phenotypes. Growth or cell division may be required before the recombinant phenotype is expressed. Delay in expression of a recombinant phenotype until a haploid recombinant genome has segregated is called segregation lag, and delay until synthesis of products encoded by donor genes has occurred is called phenotypic lag. Testing for linkage (nonrandom reassortment of parental alleles in recombinant progeny) is possible when the parental bacteria have different alleles for several genes. The donor allele of an unselected gene is more likely to be present in recombinants if it is linked to the selected donor gene than if it is not linked to the selected donor gene. Quantitative analysis of linkage permits construction of genetic maps. The genome of E coli is circular (Fig. 5-8), as determined both by genetic linkage and direct biochemical analysis of chromosomal DNA, and the genetic map is colinear with the physical map of the chromosomal DNA. Genetic and physical mapping are also used to analyze extrachromosomal replicons such as bacteriophages and plasmids.

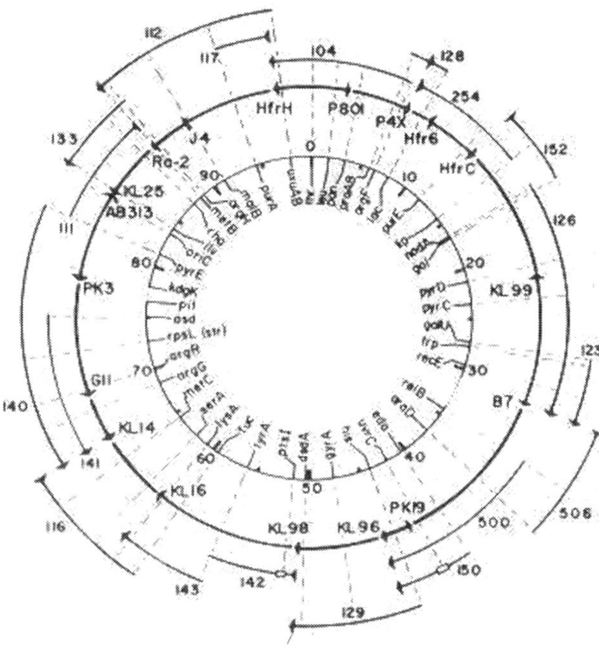

FIGURE 5-8 Circular genetic map of E coli. Positions of representative genes are indicated on inner circle. Distances between genes are calibrated in minutes, based on times required for transfer during conjugation. Position of threonine (thr) locus is arbitrarily designated as 0 minutes, and other assignments are relative to thr. On next circle, symbols and arrowheads identify specific Hfr donor strains of E coli and their characteristics. For each Hfr strain the point of arrowhead is the origin for chromosomal transfer; oriented transfer of chromosome during conjugation proceeds from point of arrowhead, followed immediately by base of the arrowhead, and so on. F′ plasmids are identified by numbers, and the fragment of the E coli chromosome present in each F′ plasmid is represented by an arc corresponding to a specific segment of the circular genetic map. See text for definitions of Hfr and F′ donor strains and for description of the conjugal mating system in E coli. From Bachman, B.M., Low, K.B. Microbiol Rev, 1980;44:31.

Many bacteria have restriction-modification (RM) systems, consisting of modifying enzymes that methylate adenine or cytosine residues at specific sequences in their own DNA and corresponding restriction endonucleases that cleave foreign DNA which does not carry the specific modification at the same target sequences. Some restriction enzymes will only cleave DNA that has been methylated at specific sequences. These restriction systems, which may have evolved to protect bacteria against invasion by phages or plasmids, are an important barrier to genetic exchanges between different bacterial strains or species. Recent evidence suggests that plasmid-borne RM systems may be a way for the plasmid to ensure its carriage in a host strain, since cells that lose the plasmid (and the corresponding protective methylase gene) are killed by the action of the more stable restriction enzyme, which attacks the newly replicated but unmodified chromosomal DNA.

Transformation

In transformation, pieces of DNA released from donor bacteria are taken up directly from the extracellular environment by recipient bacteria. Recombination occurs between single molecules of transforming DNA and the chromosomes of recipient bacteria. To be active in transformation, DNA molecules must be at least 500 nucleotides in length, and transforming activity is destroyed rapidly by treating DNA with deoxyribonuclease. Molecules of transforming DNA correspond to very small fragments of the bacterial chromosome. Cotransformation of genes is unlikely, therefore, unless they are so closely linked that they can be encoded on a single DNA fragment. Transformation was discovered in Streptococcus pneumoniae and occurs in other bacterial genera including Haemophilus, Neisseria, Bacillus, and Staphylococcus. The ability of bacteria to take up extracellular DNA and to become transformed, called competence, varies with the physiologic state of the bacteria. Many bacteria that are not usually competent can be made to take up DNA by laboratory manipulations, such as calcium shock or exposure to a high-voltage electrical pulse (electroporation). In some bacteria (including Haemophilus and Neisseria) DNA uptake depends on the presence of specific oligonucleotide sequences in the transforming DNA, but in others (including Streptococcus pneumoniae) DNA uptake is not sequence-specific. Competent bacteria may also take up intact bacteriophage DNA (transfection) or plasmid DNA, which can then replicate as extrachromosomal genetic elements in the recipient bacteria. In contrast, a piece of chromosomal DNA from a donor bacterium usually cannot replicate in the recipient bacterium unless it becomes part of a replicon by recombination. Historically, characterization of "transforming principle" from S pneumoniae provided the first direct evidence DNA is genetic material.

Transduction

In transduction, bacteriophages function as vectors to introduce DNA from donor bacteria into recipient bacteria by infection. For some phages, called generalized transducing phages, a small fraction of the virions produced during lytic growth are aberrant and contain a random fragment of the bacterial genome instead of phage DNA. Each individual transducing phage carries a different set of closely linked genes, representing a small segment of the bacterial genome. Transduction mediated by populations of such phages is called generalized transduction, because each part of the bacterial genome has approximately the same probability of being transferred from donor to recipient bacteria. When a generalized transducing phage infects a recipient cell, expression of the transferred donor genes occurs. Abortive transduction refers to the transient expression of one or more donor genes without formation of recombinant progeny, whereas complete transduction is characterized by production of stable recombinants that inherit donor genes and retain the ability to express them. In abortive transduction the donor DNA fragment does not replicate, and among the progeny of the original transductant only one bacterium contains the donor DNA fragment. In all other progeny, the donor gene products become progressively diluted after each generation of bacterial growth until the donor phenotype can no longer be expressed. On selective medium upon which only bacteria with the donor phenotype can grow, abortive transductants produce minute colonies that can be distinguished easily from colonies of stable transductants. The frequency of abortive transduction is typically one to two orders of magnitude greater than the frequency of generalized transduction, indicating that most cells infected by generalized transducing phages do not produce recombinant progeny.

Specialized transduction differs from generalized transduction in several ways. It is mediated only by specific temperate phages, and only a few specific donor genes can be transferred to recipient bacteria. Specialized transducing phages are formed only when lysogenic donor bacteria enter the lytic cycle and release phage progeny. The specialized transducing phages are rare recombinants which lack part of the normal phage genome and contain part of the bacterial chromosome located adjacent to the prophage attachment site. Many specialized transducing phages are defective and cannot complete the lytic cycle of phage growth in infected cells unless helper phages are present to provide missing phage functions. Specialized transduction results from lysogenization of the recipient bacterium by the specialized transducing phage and expression of the donor genes. Phage conversion and specialized transduction have many similarities, but the origin of the converting genes in temperate converting phages is unknown.

Conjugation

In conjugation, direct contact between the donor and recipient bacteria leads to establishment of a cytoplasmic bridge between them and transfer of part or all of the donor genome to the recipient. Donor ability is determined by specific conjugative plasmids called fertility plasmids or sex plasmids.

The F plasmid (also called F factor) of E coli is the prototype for fertility plasmids in Gram-negative bacteria. Strains of E coli with an extrachromosomal F plasmid are called F+ and function as donors, whereas strains that lack the F plasmid are F- and behave as recipients. The conjugative functions of the F plasmid are specified by a cluster of at least 25 transfer (tra) genes which determine expression of F pili, synthesis and transfer of DNA during mating, interference with the ability of F+ bacteria to serve as recipients, and other functions. Each F+ bacterium has 1 to 3 F pili that bind to a specific outer membrane protein (the ompA gene product) on recipient bacteria to initiate mating. An intercellular cytoplasmic bridge is formed, and one strand of the F plasmid DNA is transferred from donor to recipient, beginning at a unique origin and progressing in the 5' to 3' direction. The transferred strand is converted to circular double-stranded F plasmid DNA in the recipient bacterium, and a new strand is synthesized in the donor to replace the transferred strand. Both of the exconjugant bacteria are F+, and the F plasmid can therefore spread by infection among genetically compatible populations of bacteria. In addition to the role of the F pili in conjugation, they also function as receptors for donor-specific (male-specific) phages.

The F plasmid in E coli can exist as an extrachromosomal genetic element or be integrated into the bacterial chromosome (Fig. 5-9). Because the F plasmid and the bacterial chromosome are both circular DNA molecules, reciprocal recombination between them produces a larger DNA circle consisting of F plasmid DNA inserted linearly into the chromosome. E coli contains multiple copies of several different genetic elements called insertion sequences (see section on transposons for more detail), at various locations in its chromosome and in the F plasmid. Homologous recombination between insertion sequences in the chromosome and the F plasmid leads to preferential integration of the F plasmid at chromosomal sites where insertion sequences are located. The chromosomal sites where insertion sequences are found vary, however, among strains of E coli.

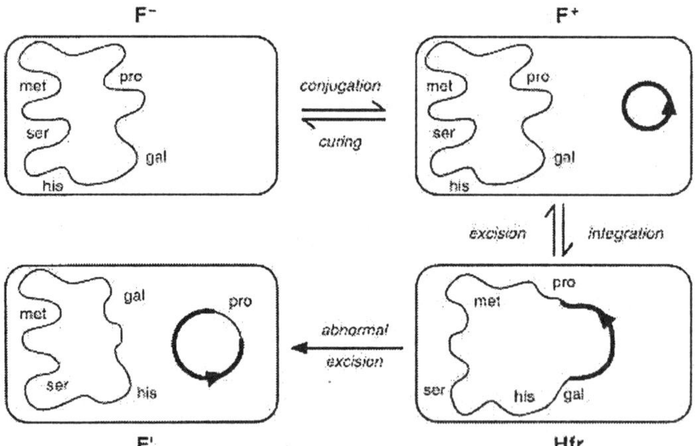

FIGURE 5-9 Role of F plasmid in determining donor and recipient states of E coli. The F plasmid is representative of specific conjugative plasmids that control donor ability in E coli. F- strains lack the F plasmid and are genetic recipients. F+ strains harbor the F plasmid as a cytoplasmic element, express F pili, and are genetic donors. The F plasmid can become integrated into bacterial chromosome at various locations to produce Hfr (high-frequency recombination) donor strains. Abnormal excision of F plasmid can result in formation of F′ plasmids that contain segments of bacterial chromosome and the corresponding bacterial genes. The arrowhead in F plasmid defines origin for transfer of DNA during conjugation. F plasmid and chromosomal DNA are indicated by heavy and fine lines, respectively. For additional data concerning the genomes of Hfr and F′ strains, see Fig. 5-8.

An E coli strain with an integrated F plasmid retains its ability to function as a donor in conjugal matings. Because donor strains with integrated F factors can transfer chromosomal genes to recipients with high efficiency, they are called Hfr (High frequency recombination) strains. Transfer of single-stranded DNA from an Hfr donor to a recipient begins from the origin within the F plasmid and proceeds as described above, except that the transferred DNA is the hybrid replicon consisting of F plasmid integrated into the bacterial chromosome. Transfer of this entire replicon, including the bacterial chromosome, requires approximately 100 minutes. The identity of the first chromosomal gene to be transferred and the polarity of chromosomal transfer are determined by the site of integration of the F plasmid and its orientation with respect to the bacterial chromosome. Because the mating bacteria usually separate spontaneously before the entire chromosome is transferred, conjugation typically transfers only a fragment of the donor chromosome into the recipient. The probability that a donor gene will enter the recipient bacterium during conjugation decreases, therefore, as its distance from the F origin (and therefore the time of its transfer) increases. Mating cells can also be broken apart experimentally by subjecting them to strong shearing forces in a mechanical blender; this is called interrupted mating. Formation of recombinant progeny requires recombination between the transferred donor DNA and the genome of the recipient bacterium. Analysis of progeny from matings that are interrupted after different intervals demonstrates which chromosomal genes are transferred first by particular donor strains, the sequential times of entry for genes that are transferred subsequently, and the progressively lower probability that genes transferred later will appear in recombinant progeny. The circularity of the genetic map of E coli was originally deduced from the overlapping, circularly permuted groups of linked genes that were transferred early by individual donor strains in which the F factor was integrated at different chromosomal locations.

In matings between F+ and F- bacteria, only the F plasmid is transferred with high efficiency to recipients. Chromosomal genes are transferred with very low efficiency, and it is the spontaneous Hfr mutants in F+ populations that mediate transfer of donor chromosomal genes. In matings between Hfr and F- strains, the segment of the F plasmid containing the tra region is transferred last, after the entire bacterial chromosome has been transferred. Most recombinants from matings between Hfr and F- cells fail to inherit the entire set of F plasmid genes and are phenotypically F-. In matings between F+ and F- strains, the F plasmid spreads rapidly throughout the bacterial population, and most recombinants are F+.

Integrated F plasmids in Hfr strains can sometimes be excised from the bacterial chromosome. If excision precisely reverses the integration process, F+ cells are produced. On rare occasions, however, excision occurs by recombinations involving insertion sequences or other genes on the bacterial chromosome that are located at some distance from the original integration site. In such cases segments of the bacterial chromosome can become incorporated into hybrid F plasmids that are called F′ plasmids (see Fig. 5-9). By similar processes, segments of the bacterial chromosome can sometimes become incorporated into R plasmids to produce hybrid R′ plasmids. Conjugative R′ plasmids can function as fertility plasmids because they can integrate into the bacterial chromosome by homologous recombination and mediate transfer of chromosomal genes during matings with recipient bacteria. F′ plasmids, R′ plasmids, specialized transducing phages, and recombinant plasmids or phages constructed by gene cloning (described below) are hybrid replicons that can include segments of the bacterial chromosome. Therefore, any of these genetic elements can be used to construct the partially diploid bacterial strains that are required for complementation tests and other purposes.

Conjugation also occurs in Gram-positive bacteria. Gram-positive donor bacteria produce adhesins that cause them to aggregatewith recipient cells, but sex pili are not involved. In some Streptococcus species, recipient bacteria produce extracellular sex pheromones that cause the donor phenotype to be expressed by bacteria that harbor an appropriate conjugative plasmid, and the conjugative plasmid prevents the donor cells from producing the corresponding pheromone.

Recombination

Recombination involves breakage and joining of parental DNA molecules to form hybrid, recombinant molecules. Several distinct kinds of recombination have been identified that depend on different features of the participating genomes and require the activities of different gene products. Specific enzymes that act on DNA (for example, exonucleases, endonucleases, polymerases, ligases) participate in recombination. Detailed discussion of the biochemical events in recombination is beyond the scope of this chapter.

Generalized recombination involves donor and recipient DNA molecules that have homologous nucleotide sequences. Reciprocal exchanges can occur between any homologous donor and recipient sites. In E coli, the product of the recA gene is essential for generalized recombination, but other gene products also participate.

Site-specific recombination involves reciprocal exchanges only between specific sites in donor and recipient DNA molecules. The recA gene product is not required for site-specific recombination. Integration of the temperate bacteriophage l into the chromosome of E coli is a well-studied example of site-specific recombination (Fig. 5-10). The specific attachment (att) sites on the E coli chromosome and l phage DNA have a common core sequence of 15 nucleotides, within which reciprocal recombination occurs, flanked by adjacent sequences that are not homologous in the phage and bacterial genomes. In phage l the product of the int gene (integrase) is required for the site-specific integration event in lysogenization; the products of the int and xis (excisionase) genes are both needed for the complementary site-specific excision event that occurs during induction of lytic phage development in lysogenic cells.

Illegitimate recombination is the term used to describe nonhomologous, aberrant recombination events such as those involved in formation of specialized transducing phages. The mechanisms of illegitimate recombination are unknown.

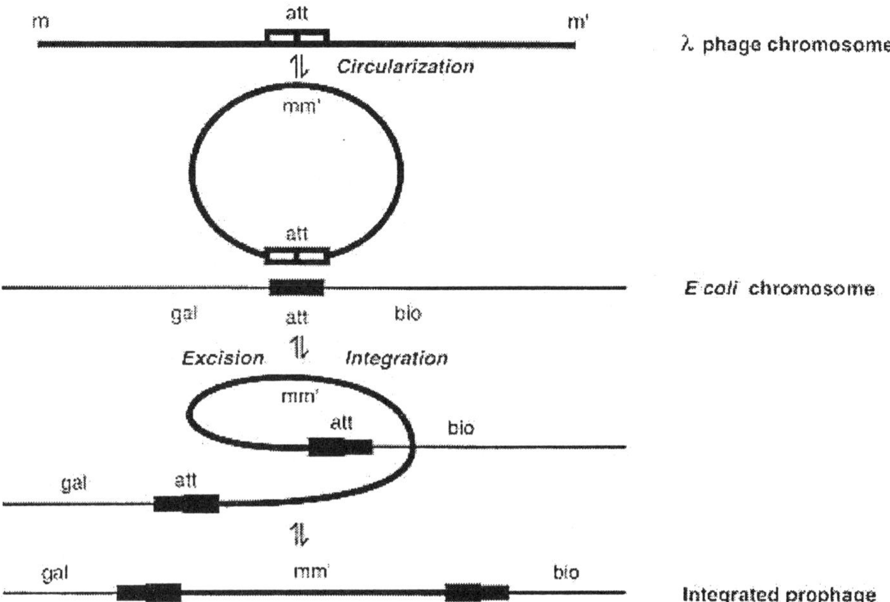

FIGURE 5-10 Integration and excision of bacteriophage l are examples of site-specific recombination. l DNA is shown by thin lines and chromosomal DNA by thick lines. Attachment (att) sites are closed boxes for the bacterial chromosome and open boxes for the l chromosome. The gal and bio operons, which determine utilization of galactose and biosynthesis of biotin, are located adjacent to the bacterial attachment site. In an infected E coli the l DNA becomes circular by joining ends m and m′, and site-specific recombination between phage and bacterial att sites results in insertion of the l genome into the bacterial chromosome. The arrangement of the prophage DNA (m and m′ located internally) is, therefore, a circular permutation of l virion DNA (m and m′ located terminally).

Transposons

Transposons are segments of DNA that can move from one site in a DNA molecule to other target sites in the same or a different DNA molecule. The process is called transposition and occurs by a mechanism that is independent of generalized recombination. Transposons are important genetic elements because they cause mutations, mediate genomic rearrangements, function as portable regions of genetic homology, and acquire new genes and contribute to their dissemination within bacterial populations. Insertion of a transposon often interrupts the linear sequence of a gene and inactivates it. Transposons have a major role in causing deletions, duplications, and inversions of DNA segments as well as fusions between replicons. Transposons are not self-replicating genetic elements, however, and they must integrate into other replicons to be maintained stably in bacterial genomes.

Most transposons share a number of common features. Each transposon encodes the functions necessary for its transposition, including a transposase enzyme that interacts with specific sequences at the ends of the transposon. During transposition a short sequence of target DNA is duplicated, and the transposon is inserted between the directly repeated target sequences. The length of this short duplication varies, but is characteristic for each transposon. The duplication is presumed to involve asymmetric cleavage of DNA at the target site, followed by synthesis of new complementary strands corresponding to the region between the cleavage sites. Some transposons insert into almost any target sequence, whereas others have relatively stringent target specificity. Two types of transposition are recognized. Excision of the transposon from a donor site followed by its insertion into a target site is called nonreplicative transposition. If the transposon at a donor site is replicated and a copy is inserted into the target site, however, the process is called replicative transposition. The process of replicative transposition can involve formation of a cointegrate, a single circular DNA molecule consisting of two replicons joined with copies of the transposon in an alternating sequence. Resolution of the cointegrate into its component replicons is often accomplished by a transposon-encoded resolvase that catalyzes site-specific recombination between the transposons. Generalized recombination between homologous transposons can also lead to the formation or resolution of cointegrates. Transposition differs from site-specific recombination by duplicating a segment of the target sequence and by using a variety of different target sequences for a single donor sequence.

Most transposons in bacteria can be separated into three major classes (Fig. 5-11). Insertion sequences and related composite transposons comprise the first class. Insertion sequences are simplest in structure and encode only the functions needed for transposition. The known insertion sequences vary in length from approximately 780 to 1500 nucleotide pairs, have short (15-25 base pair) inverted repeats at their ends, and are not closely related to each other. The DNA between the inverted terminal repeats contains one (or rarely two) transposase genes and does not encode a resolvase. Complex transposons vary in length from about 2,000 to more than 40,000 nucleotide pairs and contain insertion sequences (or closely related sequences) at each end, usually as inverted repeats. The entire complex element can transpose as a unit. The DNA between the terminal insertion sequences of complex transposons encodes multiple functions that are not essential for transposition. In medically important bacteria, genes that determine production of adherence antigens, toxins, or other virulence factors, or specify resistance to one or more antibiotics, are often located in complex transposons. Well-known examples of complex transposons are Tn5 and Tn10, which determine resistance to kanamycin and tetracycline, respectively. The complex transposons probably evolve by transposition of homologous insertion sequences to nearby sites within a DNA molecule.

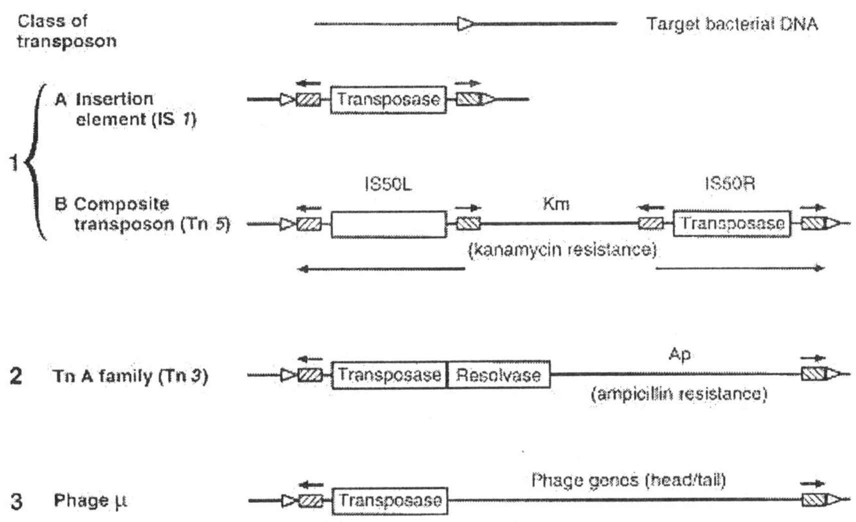

FIGURE 5-11 Features of representative transposons (heavy lines) integrated into the bacterial chromosome (fine lines). Transposons are important genetic elements because they cause mutations, mediate genomic rearrangements, function as portable regions of genetic homology, and acquire new genes and contribute to their dissemination within bacterial populations. 1A) IS1 insertion sequence (786 base pairs) has transposase gene flanked by inverted terminal repeats (hatched bars with arrows above them). The IS1 element is flanked by copies of target site (open arrows) with same orientation. 1B) Composite transposon Tn5 (5816 base pairs) consists of kanamycin resistance determinant flanked by inverted copies of IS50 insertion element. 2), Transposon TnA (4957 base pairs) contains ampicillin resistance determinant, transposase and resolvase genes between terminal inverted repeat sequences (hatched bars with arrows above them), flanked by direct repeats of target site (open arrows). 3) Phage Mu (37 kilobase pairs) encodes transposase that catalyzes recombination between the ends of Mu DNA and target DNA. Direct repeats of the target site (open arrows) flank the integrated Mu genome. Mu virion DNA is longer than Mu prophage and contains chromosomal sequences at both ends, reflecting the process by which prophage Mu is excised and packaged.

The second class of transposons consists of the highly homologous TnA family. These transposons have longer (35 to 40 base pair) terminal inverted repeats than the complex transposons described above, but they lack terminal insertion sequences. All members of the family encode both transposase and resolvase functions. Well known examples from the TnA transposon family include the ampicillin resistance transposon Tn3 and Tn1000 (the gamma-delta transposon) found in the F plasmid. The TnA family has an important place in the history of medical microbiology. The development of high-level resistance to ampicillin in Haemophilus influenzae and Neisseria gonorrhoeae during the 1970s, which severely limited the usefulness of ampicillin for treatment of gonorrhea and Haemophilus infections in areas where such strains became prevalent, was caused by dissemination of ampicillin resistance determinants from TnA transposons in plasmids of the Enterobacteriaceae to plasmids in Haemophilus and Neisseria.

The third class of transposons consists of bacteriophage Mu and related temperate phages. The entire phage genome functions as a transposon, and replication of the phage DNA during vegetative growth occurs by replicative transposition. Prophage integration can occur at many different sites in the bacterial chromosome and often causes mutations. For that reason Mu and related phages are sometimes called mutator phages.

A fourth class of transposons, discovered in Gram-positive bacteria and represented by Tn917, consists of conjugative transposons that are completely different from the transposons described above. The conjugative transposon does not generate a duplication of the target sequence into which it inserts, and in Gram-positive bacteria the host strain carrying the transposon can act as a conjugal donor. Recipient bacteria need not be closely related to the donor bacterium. The transposon is excised from the chromosome of the donor and transmitted by conjugation to the recipient, where it integrates randomly into the chromosome. Tn917 encodes tetracycline resistance, but other larger conjugative transposons may encode additional antibiotic resistances. Conjugative transposons appear to be a major cause of the spread of antibiotic resistance in Gram-positive bacteria.

Some roles of transposons in bacterial evolution are illustrated by considering enteric Gram-negative bacteria and the structure of their plasmids. Bacteria collected during the pre-antibiotic era contained many plasmids, but they usually lacked resistance determinants. Many of the R plasmids from current clinical isolates belong to the same incompatibility groups as plasmids found previously, but they also determine resistance to multiple antibiotics. The close relationships between their replicons provide strong evidence that many current R plasmids evolved from the older plasmids by acquisition of resistance determinants. Some of the multiple antibiotic resistant plasmids have individual transposons with several resistance determinants, others have multiple resistance transposons located at separate sites, and still others contain complex hybrid resistance transposons formed by integration of one transposon into another. The stepwise acquisition of resistance determinants can lead, in some cases, to the formation of composite transposons that encode multiple resistance determinants. Therapeutic use of antibiotics and their incorporation into animal feeds provide selective advantages for bacteria with R plasmids, whereas conjugation, transformation and transfection provide means for dissemination of R plasmids within and between bacterial species. After a plasmid carrying a transposon is introduced into a new bacterial host, the transposon and its determinants can jump into the chromosome or indigenous plasmids of the new host. Therefore, stability of the mobilizing plasmid in a new bacterial host is not essential for persistence of genetic determinants located on a transposon.

Recombination DNA and Gene Cloning

Many methods are available to make hybrid DNA molecules in vitro (recombinant DNA) and to characterize them. Such methods include isolating specific genes in hybrid replicons, determining their nucleotide sequences, and creating mutations at designated locations (site-directed mutagenesis). A clone is a population of organisms or molecules derived by asexual reproduction from a single ancestor. Gene cloning is the process of incorporating foreign genes into hybrid DNA replicons. Cloned genes can be expressed in appropriate host cells, and the phenotypes that they determine can be analyzed. Some key concepts underlying representative methods are summarized here.

The first step in gene cloning is to make fragments of the donor DNA by mechanical or enzymatic methods. Certain restriction endonucleases, designated as class II, are particularly useful for preparing defined fragments of DNA molecules. They cleave both strands of double-stranded DNA molecules at specific, palindromic sequences (restriction sites) that usually vary from four to eight nucleotides in length, and the resulting DNA fragments are called restriction fragments. Some restriction endonucleases cleave at coincident sites to create blunt-ended DNA fragments, and others cut at staggered positions to create DNA fragments with short, self-complementary, single-stranded 5' or 3' ends (see Table 3). The random probability that n adjacent nucleotides in a DNA strand will correspond to a specific restriction site is approximately 1/4n. Sites for enzymes that recognize unique 4, 6, or 8 nucleotide targets are likely to occur about once in every 256, 4096, or 65,536 nucleotides, respectively. By choosing appropriate restriction enzymes, specific DNA molecules, including bacterial chromosomes, plasmids, and phage genomes, can be digested into sets of restriction fragments that have appropriate sizes for specific applications.

Table 5-3 Specificities of Representative Class II Restriction Endonucleases

Enzyme	Isolated from	Recognition site Length (bp)	Sequence (5-3)	End structure of Restriction Fragment
Sau3A	*Staphylococcus aureus*	4	↓ GATC	4-base 5' extension
Nlall	***Neisseria lactamica***	4	CATG ↓	4-base 3' extension
Dpnl	*Diplococcus pneumoniae*	4	GA ↓ TC	Blunt
Sspl	*Sphaerotilus natans*	6	AAT ↓ ATT	Blant
Pstl	*Providenica stuartii*	6	CTGCA ↓ G	4-base 3' extension
EcoRl	*Escherichia coli*	6	G ↓ AATTC	4-base 5' extension
Clal	*Caryopharon latum*	6	AT ↓ CGAT	2-base 5' extension
Norl	*Nocardia otitidis-caviarum*	8	GC ↓ GFCCFC	4-base 5' extension

A restriction map identifies the positions of target sites for specific restriction endonucleases in a DNA molecule. Restriction maps are available for many cloned DNA fragments, plasmids and phage genomes, as well as for the entire chromosome of E coli and several other bacteria.

The second step in gene cloning is to create hybrid replicons consisting of donor DNA fragments and a cloning vector (Fig. 5-12). Cloning vectors are small plasmid or phage replicons that have one or more restriction sites into which foreign DNA can be inserted. Hybrid replicons are produced by using DNA ligase to join the restricted vector DNA with donor DNA fragments that have compatible ends, or, alternatively, synthetic oligonucleotides are used as linkers to create compatibility between donor and vector DNA molecules with different ends. Ligating a vector to a heterogeneous set of DNA fragments from a donor genome is called shotgun cloning, and the collection of recombinant DNA molecules that contains the various fragments is called a genomic library. If a specific DNA fragment is available, it can be incorporated into a recombinant replicon by direct cloning into an appropriate vector chosen from the wide variety of vectors available. Plasmid and phage vectors are used mainly to clone small inserts usually less than 10 kbp. Examples of more special purpose vectors include cosmids, which are plasmid vectors that can be packaged into phage capsids (lambda cosmids accept inserts up to 30-40 kbp), and phagemids, which are plasmid-phage hybrid replicons that can exist either as plasmids or as single-stranded DNA phages under different experimental conditions. Phage P1 cosmids can accept inserts up to 100 kbp, and still larger DNA molecules can be cloned in yeast artificial chromosomes (YACs) which can stably maintain inserts up to and exceeding 1 Mbp in size. Other specialized vectors detect promoters, transcription termination signals, or other regulatory elements within foreign DNA inserts or, conversely, provide promoters from which transcription of cloned genes can be initiated.

The final steps in gene cloning are to introduce hybrid replicons into appropriate recipient cells and test them for expression of donor genes of interest. Prokaryotic cells (including bacteria) or eukaryotic cells (including yeast, animal or plant cells) can be used as recipients, but they differ with respect to their permissiveness for specific replicons, the transcriptional signals that they recognize, and the post-translational modifications of protein structure that they can accomplish. Recombinant DNA molecules produced in vitro can be introduced directly into recipient cells by transformation or transfection. In addition, clones in cosmid or phage vectors can be packaged into phage coats and introduced into susceptible recipient cells by transduction. By using specialized vectors (shuttle vectors) that can replicate in multiple cell types, genes from any organism can be cloned and manipulated in a convenient bacterial system and subsequently reintroduced into cells of the original organism for analysis in their natural environment.

Many methods are available to identify bacteria that contain recombinant DNA molecules. Most cloning vectors have genes for traits that can be positively selected,

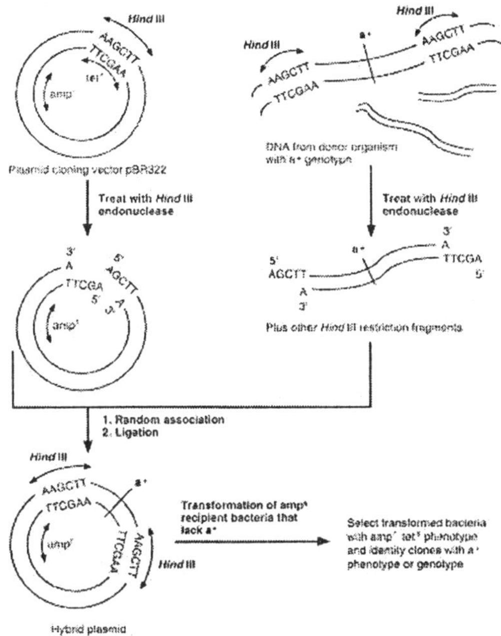

FIGURE 5-12 Diagrammatic representation of gene cloning experiment. Plasmid cloning vector pBR322 is 4.36 kilobase pairs in size, has genes for resistance to ampicillin (ampr) and to tetracycline (tetr), and has only one HindIII restriction site that is located within the tetr locus. HindIII is used to treat samples of DNA from plasmid pBR322 and from a donor organism with a gene, designated a+, to be cloned. The donor can be a prokaryotic or a eukaryotic organism. If HindIII restriction sites are located adjacent to a+ in donor DNA, but do not occur within a+, a restriction fragment carrying intact a+ marker can be generated from donor DNA. Hybrid plasmids can be formed by random association and ligation of the HindIII-treated donor and vector DNA fragments. Although pBR322 is tetr, hybrid plasmids will be tests because the donor DNA fragments are inserted at the HindIII restriction site within the tetr locus. After transformation of amps-recipient bacteria that also lack a+, transconjugants with hybrid plasmids can be selected by their ampr tests phenotypes. Strains in which the a+ gene is present can then be identified by expresion of a+ or by testing for the polynucleotide sequence corresponding to a+. The pBR322 plasmid contains other uniuqe restriction sites that can also be used for cloning (e.g., PstI in ampr and BamHI in tetr). Many other cloning vectors and restriction endonucleases have also been used for gene cloning experiments.

such as resistance to antibiotics. Furthermore, it is often possible to introduce foreign DNA into the cloning vector at a site that inactivates a nonessential, but easily recognizable, vector function. If both of these conditions are fulfilled, bacteria that contain recombinant molecules can be selected and distinguished easily from bacteria that contain only the vector. Bacteria in a genomic library that contain a particular cloned gene can be identified by using biochemical or immunologic methods to test for the desired gene product. Alternatively, the cloned gene of interest can be detected directly by using nucleic acid hybridization methods, provided that a specific DNA or RNA probe is available. Because insertion of foreign DNA into a cloning vector at an appropriate site does not inactivate its ability to replicate in appropriate recipient cells, hybrid replicons of interest can be amplified by replication, and the recombinant DNA molecules or their gene products can be purified and studied. The ability to purify specific DNA molecules made it feasible to develop enzymatic and chemical methods for determining their nucleotide sequences, and current methods for introducing mutations at defined sites in cloned genes are based on knowing their restriction maps or nucleotide sequences.

Recombinant DNA methods make it feasible to clone specific DNA fragments from any source into vectors that can be studied in well-characterized bacteria, in eukaryotic cells, or in vitro. Applications of DNA cloning are expanding rapidly in all fields of biology and medicine. In medical genetics such applications range from the prenatal diagnosis of inherited human diseases to the characterization of oncogenes and their roles in carcinogenesis. Pharmaceutical applications include large-scale production from cloned human genes of biologic products with therapeutic value, such as polypeptide hormones, interleukins, and enzymes. Applications in public health and laboratory medicine include development of vaccines to prevent specific infections and probes to diagnose specific infections by nucleic acid hybridization or polymerase chain reaction (PCR). The latter process uses oligonucleotide primers and DNA polymerase to amplify specific target DNA sequences during multiple cycles of synthesis in vitro, making it possible to detect rare target DNA sequences in clinical specimens with great sensitivity.

Regulation of Gene Expression

The phenotypic properties of bacteria are determined by their genotypes and growth conditions. For bacteria in pure culture, changes in growth conditions often result in predictable physiological adaptations in all members of the population. Typically, essential gene products are made in amounts that permit fastest growth in the given environment, and products required under special circumstances are made only when they are needed.

Physiological adaptations are often associated with changes in metabolic activities. The flow of metabolites through particular biochemical pathways can be controlled both by regulating the synthesis of specific enzymes and by altering the activities of existing enzymes. Mechanisms that regulate expression of genes by affecting synthesis of specific gene products are discussed here.

Specific regulation involves a gene or group of genes involved in a particular metabolic process. Induction and repression enable bacteria to regulate production of specific gene products in response to appropriate signals. Generally catabolic enzymes are induced when the substrate for the pathway is present in the growth medium, and biosynthetic enzymes are repressed by the product of the pathway. Enzymes that participate in a single biochemical pathway often occupy adjacent positions on the bacterial chromosome and are coordinately induced or repressed. They form an operon, a group of contiguous genes that is transcribed as a single unit and translated to produce the corresponding gene products. Organization into an operon is an important strategy for coordinately regulating the expression of genes in bacteria. Operons that can be induced or repressed are controlled by binding of specific regulatory proteins to particular nucleotide sequences that function as regulatory sites within the operon. Comparison of the amino acid sequences of many of these different regulatory proteins showed that they could be grouped together into families of regulators (e.g. the lysR family of proteins) that may have evolved from common ancestoral genes. Members of the lysR family include regulators of such diverse phenomena as lysine, cysteine and methionine metabolism in E coli and iron repression in V cholerae.

Global regulation simultaneously alters expression of a group of genes and operons, collectively called a regulon, that are controlled by the same regulatory signal. Global regulation determines responses of bacteria to basic nutrients such as carbon, nitrogen or phosphate, reactions to stresses such as DNA damage or heat shock, and synthesis by pathogens of specific virulence factors during growth in their host animals.

The amount of a specific protein in a bacterial cell can vary from none to many thousands of molecules. This wide range is often determined by the combined action of several regulatory mechanisms that affect expression of the corresponding structural gene. Regulation is achieved by determining how often a gene is transcribed into functional mRNA, how efficiently the mRNA is translated into protein, how rapidly the mRNA is degraded, how rapidly the protein product turns over, and whether the activity of the protein product can be altered by allosteric effects or covalent modifications.

mRNAs as Transcriptional Units

Gene expression begins with DNA-dependent RNA polymerase (RNA polymerase) catalyzing the transcription of specific mRNA from one strand of a DNA template. Binding of RNA polymerase to DNA occurs at specific sites called promoters, and transcription begins adjacent to the promoter. Strong promoters can interact efficiently with RNA polymerase and initiate transcription at a high rate; weak promoters initiate transcription at slow rates. In either case, mRNA is synthesized from its 5' end toward its 3' end at an approximately constant rate until the RNA polymerase recognizes another specific site called a terminator. RNA polymerase then dissociates from the template, and transcription of the mRNA is completed.

Individual mRNA molecules may code for one or more polypeptides. Transcription of an operon produces a polycistronic mRNA that codes for several polypeptides. Translation of polycistronic mRNAs leads to coordinate synthesis of the encoded polypeptides, but each polypeptide is synthesized as a separate molecule. A specific ribosome binding site is located just upstream from the start of each coding sequence on the mRNA molecule.

Messenger RNAs in bacteria are degraded rapidly with an average half life of several minutes, in contrast to tRNAs and rRNAs which are much more stable. Although mRNAs represent about half of the newly synthesized RNA, they represent only a small fraction of the total RNA. The short half-life of mRNAs has important consequences for gene expression. If the synthesis of a specific mRNA is prevented, production of the corresponding polypeptides declines rapidly.

Control of gene expression occurs by regulating one or more of the steps in the pathway from the DNA template to the active gene product. Simultaneous regulation at several levels permits greater control over gene expression than would be possible with a single regulatory mechanism. The most common way to regulate gene expression in bacteria is to control the production of specific mRNAs. Since the rate of elongation of an RNA molecule is approximately constant, the major factors that control mRNA synthesis are the rate of initiation and the probability that a full length transcript will be produced.

Regulation of Transcription Initiation

Some mRNAs in bacteria are synthesized at constant rates, resulting in constitutive production of the encoded polypeptides. The amounts of specific mRNAs and polypeptides produced from different constitutive genes vary greatly, however, and often reflect differences in strength of the promoters for those genes.

Transcription of many operons is regulated in response to changing environmental conditions. The promoters determine the maximum rate of transcription initiation for such operons, but regulatory proteins participate in controlling transcription. Nucleotide sequences in operons to which specific regulatory proteins bind are called regulatory sites or operators. Operators and promoters are located close together within operons and may have overlapping DNA sequences. The binding of regulatory proteins to operators can either increase (positive regulation) or decrease (negative regulation) the frequency of transcription initiation. Proteins that function as negative regulators are usually called repressors. Because regulatory proteins can diffuse through the cytoplasm, the structural genes for regulatory proteins do not have to be linked to the target operons.

The ability to sense the presence or absence of specific compounds and change the rates of synthesis of appropriate gene products are central to the control of gene expression. Regulatory proteins offer one solution to this problem of stimulus-response coupling. Many regulatory proteins are bifunctional and bind not only to appropriate operators but also to specific effectors, which are small molecules such as particular sugars, amino acids, and other metabolites. Furthermore, regulatory proteins are allosteric, meaning that they can exist in different conformations which exhibit different binding affinities for their cognate operators and effectors. A sufficient concentration of effector favors formation of the regulatory protein-effector complex, which has either high or low affinity for the operator in any specific case. In negatively regulated systems the effector functions as a corepressor if the regulatory protein-effector complex is the active repressor, and the effector functions as an inducer and causes derepression if the free regulatory protein is the active repressor. Conversely, in positively regulated systems, the effector stimulates expression of the operon if the regulatory protein-effector complex is the positive regulator, and the effector inhibits expression of the operon if the free regulatory protein is the positive regulator.

The lactose (lac) operon of E coli is an example of an inducible, negatively regulated operon (Fig. 5-13). The lacI gene codes for a repressor that binds to the lac operator and prevents transcription from the lac promoter. The structural gene for this repressor is separate from the lac operon, and the repressor is synthesized constitutively at a low rate. When inducer binds to the lac repressor, the complex cannot bind to the operator and cannot prevent binding of the RNA polymerase to the promoter. If other conditions are favorable, the lac operon is expressed, resulting in synthesis of b-galactosidase, b-galactoside permease and b-galactoside transacetylase. The lac operon can be induced by lactose or by structurally related compounds such as isopropyl-b-D-thiogalactoside (IPTG). IPTG is called a gratuitous inducer because it induces the lac operon, but is not a substrate for b-galactosidase. Negative regulation also occurs in many biosynthetic operons in E coli. In such operons a product of the biosynthetic pathway functions as the effector for the negative regulatory system.

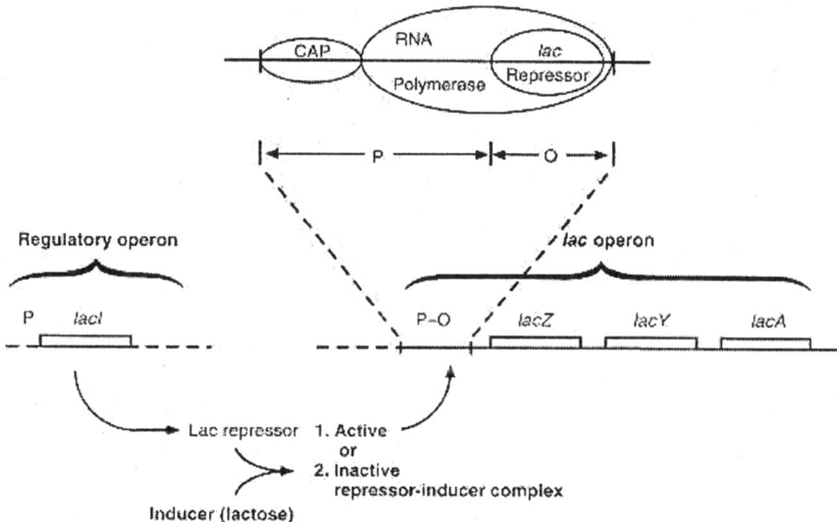

FIGURE 5-13 Regulation of lac operon in *E coli*. Structural genes lacZ, lacY, and lacA code for b-galactosidase, b-galactoside permease, and b-galactoside transacetylase, respectively. The physiologic role of lacA is unknown. The lac repressor is product of lacI gene in separate regulatory operon. Transcription of mRNA encoding lacZ, lacY, and lacA is negatively regulated. and binding of lac repressor to operator lacO prevents initiation of transcription at promoter lacP. Inducer binds to lac repressor and inactivates it. Catabolite activator protein (CAP) forms a complex with cyclic AMP, and binding of the complex to a site immediately adjacent to the lac promoter stimulates transcription of the lac operon by RNA polymerase. An expanded diagram of the lac operator-promoter region shows the binding sites for CAP, RNA polymerase, and lac repressor.

The arabinose (ara) operon in E coli is both positively and negatively regulated. In the presence of arabinose the regulatory protein stimulates transcription of the ara operon. In the absence of arabinose, however, the regulatory protein represses the ara operon.

Operons are often controlled by more than one mechanism. When E coli is grown in a medium containing glucose and an alternative carbon source such as lactose or arabinose, induction of the lac or ara operon and utilization of the lactose or arabinose are delayed until the glucose has been consumed. This phenomenon is called diauxic growth. The failure to induce the lac or ara operon in the presence of glucose is an example of catabolite repression. The lac and ara operons are positively regulated by cyclic-3',5'-adenosine monophosphate (cAMP) and the catabolite gene activator (CAP) protein (the product of the crp gene). The cAMP-CAP complex interacts with CAP binding sites in the regulatory regions of some operons, including the lac and ara operons, and stimulates transcription from the corresponding promoters. The level of intracellular cAMP in E coli is high during growth in the absence of glucose, and low during growth in the presence of glucose. Catabolite repression is due, therefore, to lack of activation of cAMP-dependent operons when the bacteria are grown in the presence of glucose or certain other rapidly metabolizable carbon sources.

Regulation of Transcription Termination

Attenuation is a mechanism for regulating operons by terminating transcription of mRNA prematurely. Attenuation is common in biosynthetic operons, including the trp, histidine (his), threonine (thr), isoleucine-valine (ilv), and phenylalanine (phe) operons. The trp operon in E coli is controlled both by repression and attenuation. In the presence of excess tryptophan, initiation of transcription from the trp promoter is repressed. In addition, however, those transcripts that are initiated from the trp promoter are usually terminated before any of the structural genes of the trp operon are transcribed. The concentration of intracellular tryptophan required to maintain repression exceeds that needed for attenuation. Such dual control enables the cell to fine tune the expression of the trp operon in response to decreasing concentrations of tryptophan.

The secondary structure of mRNA has an important role in the mechanism of attenuation. All mRNAs have a leader sequence between the transcriptional start site and the beginning of the coding sequence for the first structural gene. For amino acid biosynthetic operons that are subject to attenuation, the mRNA leader sequence has two distinctive features. It encodes a short peptide containing the amino-acid produced by the regulated pathway, and it can form alternative, mutually incompatible, double-stranded RNA structures that participate in regulatory events. For example, the peptide encoded by the trp mRNA leader sequence contains two adjacent tryptophan residues, and the peptide encoded by the his mRNA leader sequence has a series of seven consecutive histidine residues. Fig. 5-14 shows the trp operon and illustrates alternative secondary structures in the leader sequence of trp mRNA. There are three possible secondary structures for this region, called the pause site (segments 1+2), the anti-terminator (segments 2+3), and the attenuator (segments 3+4). Segment 1 of the pause site overlaps with the coding region for the trpL peptide. Which secondary structures are formed depends on efficiency of translation of the trpL peptide. When segments 1 and 2 are transcribed, they immediately anneal and cause the RNA polymerase to pause temporarily. Subsequent initiation of translation of the trpL peptide disrupts the pause site and allows RNA polymerase to continue transcription. If tryptophan is present, transcription of segments 3 and 4 and formation of the attenuator structure occurs while the ribosome is blocking segment 2, causing the RNA polymerase to terminate transcription. If tryptophan is deficient, however, tryptophanyl-tRNA is also deficient, and the ribosome stalls at the tryptophan codons in segment 1. This allows segment 2 to anneal with newly synthesized segment 3 to form the antiterminator, thereby making segment 3 unavailable to anneal with segment 4. Formation of the attenuator is therefore prevented, and the RNA polymerase transcribes the entire trp operon. In this manner depletion of tryptophan (actually the supply of tryptophanyl-tRNA) is coupled to regulation of transcription of the biosynthetic operon for tryptophan.

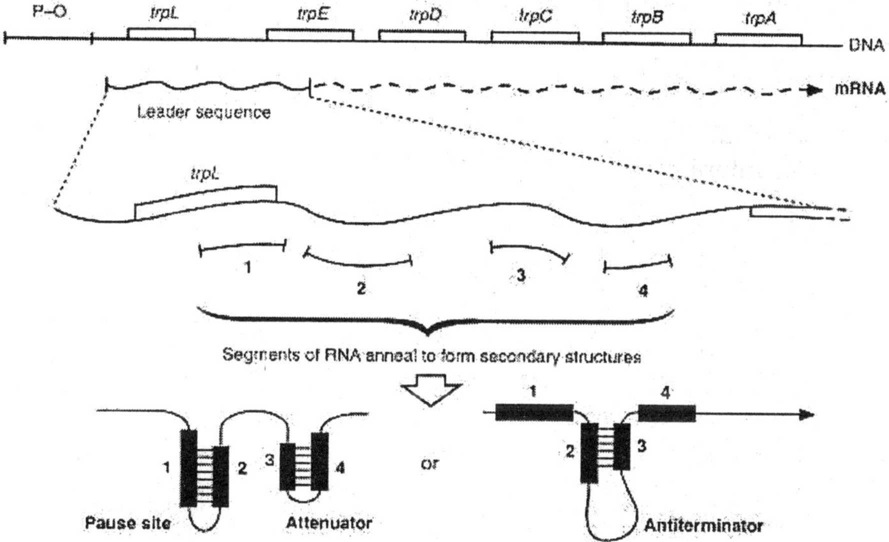

FIGURE 5-14 Regulation of trp operon in E coli. The organization of the trp operon is shown at the top of the figure. The five structural genes trpE, trpD, trpC, trpB, and trpA encode enzymes that catalyze terminal sequence of reactions in tryptophan formation. Transcription initiation is controlled at the promoter-operator (p-o) locus, and signals within the 162 nucleotide trp mRNA leader sequence control termination of transcription fcby attenuation. The leader sequence of trp mRNA is expanded to show locations of the trpL coding sequence, the complementary segments 1, 2, 3, and 4, and their possible alternative secondary structures which function as pause site, anti-terminator, or attenuator (see text).

In E coli transcription and translation are functionally coupled. Nonsense mutations that cause premature termination of translation often cause decreased transcription of more distal genes in the same operon. This phenomenon is called polarity. Ribosomes usually initiate translation of a growing mRNA molecule prior to completion of transcription, and such translation masks sites that would otherwise cause the RNA polymerase to terminate transcription. Premature termination of translation by a nonsense codon dissociates the ribosomes from the mRNA and enables RNA polymerase to interact with the unmasked transcription termination sites.

In some biological systems, including phage lambda, antitermination is used as a positive regulatory mechanism to control gene expression. Immediately after infection of E coli by lambda, RNA polymerase binds to two promoters in lambda DNA and initiates divergent primary transcripts which terminate at specific sites on the lambda genome. A protein encoded by one of the primary transcripts interacts with RNA polymerase and enables it to continue transcription through the primary termination sites, thereby expressing a second set of lambda genes. One of the products encoded by a secondary transcript blocks termination of another mRNA and activates expression of a third set of genes. Antitermination has a key role, therefore, in controlling the cascade of gene expression during lytic growth of phage lambda. Antitermination is also involved in the regulation of E coli rRNA operons.

Regulation of Translation

The ribosome binding site on mRNA is complementary to a sequence at the 3' end of 16S rRNA. Interaction between these sequences facilitates formation of the initiation complex for protein synthesis. Both the extent of homology with 16S rRNA and the spacing of the ribosome binding site from the initiation codon affect the efficiency of translation initiation. Codon usage in mRNA also influences translation efficiency. Messenger RNAs for proteins that are required in large amounts tend to use codons that are translated by the most abundant species of tRNA, and the converse is also true.

Translational control is important for regulation of synthesis of ribosomal proteins. Production of ribosomes involves a high metabolic cost for bacteria, and at high growth rates ribosomes can constitute nearly one-half of the cell weight. Most ribosomal proteins and rRNAs are found assembled into ribosomes, and the pool of free ribosomal subunits is very small. The genes for ribosomal proteins are organized into several operons. Certain of the free ribosomal proteins directly inhibit the translation of the polycistronic mRNAs that encode them, thereby ensuring that synthesis of ribosomal proteins is balanced with the requirement for their utilization.

Regulons and signal transducing proteins

A regulon is a group of genes or operons controlled bya common regulator. There are several advantages to placing different operons under control by the same regulator. It enables the sensing of a single stimulus to be coupled to expression of a large number of genes that may be needed for an appropriate response, and it eliminates the requirement for the coordinately regulated genes to be linked on the bacterial chromosome. The stimulus to which the regulon responds can be an intracellular component or an environmental signal. Individual operons may also be subject to regulation by several different mechanisms and expressed under conditions that differ from those affecting the whole regulon.

More than 40 different regulons have been identified in E coli. Specific examples of regulons that respond to intracellular components include the cAMP-CAP regulon described previously and the regulons controlled by the stringent response and the SOS response. When ribosomes encounter uncharged tRNA molecules during protein synthesis, the stringent response is activated and results in prompt cessation of rRNA synthesis. A novel nucleotide called guanosine-3'-diphosphate-5'-diphosphate (ppGpp) accumulates during amino-acid starvation. The ppGpp produced by idling ribosomes appears to be a mediator of the stringent response, but the precise mechanism causing inhibition of rRNA synthesis is unknown. The SOS response is associated with damage to DNA and involves induction of more than 20 genes involved in several DNA repair pathways. The product of the recA gene detects inhibition of DNA synthesis and initiates events leading to proteolytic cleavage and inactivation of the repressor for the SOS pathway, encoded by the lexA gene.

Some regulons are induced by specific environmental stimuli, such as nutrient limitation or osmotic stress. Often operons from more than one regulon may be induced, and the term stimulon has been used to describe the set of genes so induced. Typically, bacteria sense such environmental conditions by two component systems. The first component is a membrane-spanning protein with extracellular and intracellular domains. Its extracellular domain detects the environmental stimulus, and its cytoplasmic domain transmits the signal. The second component is a bifunctional cytoplasmic protein. It has a receiver domain that interacts with the transmitter module of the first component, as well as an effector domain that controls expression of the corresponding regulon. The transmitter and receiver modules of the two component regulatory systems from a wide variety of regulons are genetically related and share amino-acid homology. The signal-detecting and effector domains of the proteins from different regulons vary, however, and determine the signal that is detected and the operons that are activated or repressed in response to that signal.

Global regulation has an important role in the physiology of pathogenic bacteria. For example, Vibrio cholerae and Bordetella pertussis express many of their virulence determinants under the control of signal transducing systems that are related to the two component systems described above. The expression of proteins needed for the invasive phenotype is controlled by temperature in Shigella. Yersinia enterocolitica senses both the environmental temperature and the concentration of calcium ions and couples these signals to the expression of genes and cellular location of the gene products that are appropriate for an intracellular or extracellular environment. In host tissues the concentration of free iron is extremely low, and most pathogenic bacteria have high affinity iron transport systems that are induced under low-iron conditions. The synthesis of diphtheria toxin by C diphtheriae, Shiga toxin by Shigella dysenteriae, exotoxin A by Pseudomonas aeruginosa, and other specific proteins in many pathogenic bacteria is induced under conditions of iron-limited growth. These examples illustrate how environmental factors can regulate the expression of virulence genes in pathogenic bacteria.

REFERENCES

Dorman CJ: The genetics of bacterial virulence. Blackwell Scientific Press, Oxford, England, 1994

Drlica K, Riley M (eds): The bacterial chromosome. American Society for Microbiology, Washington, DC, 1990

Harwood AJ (ed): Protocols for gene analysis. Methods in Molecular Biology vol. 31. Human Press, NJ, 1993

Holloway BW: Genetics for all bacteria. Annu Rev Microbiol 47:659, 1993

Lewin B: Genes V. Oxford University Press, Oxford, England, 1994

Miller JH: A short course in bacterial genetics: a laboratory manual and handbook for Escherichia coli and related bacteria. Cold Spring Harbor Laboratory Press, NY, 1992

Miller VL, Kaper JB, Portnoy DA et al. (eds): Molecular genetics of bacterial pathogenesis. American Society for Microbiology, Washington DC, 1994

Saylers AA, Whitt DD: Bacterial pathogenesis: a molecular approach. American Society for Microbiology, Washington DC, 1994

Singer M, Berg P. Genes and genomes: a changing perspective. University Science Books, Mill Valley, CA, 1991

Chapter 6

Specific Acquired Immunity

General Concepts

Basis of Acquired Resistance

Specific acquired immunity against infectious diseases may be mediated by antibodies and/or T lymphocytes. Immunity mediated by these two factors may be manifested by a direct effect upon a pathogen, such as (1) antibody-initiated, complement-dependent bacteriolysis or (2) opsonophagocytosis and killing, as occurs for some bacteria, (3) neutralization of viruses or toxins, or (4) by T lymphocytes which will kill a cell parasitized by a microorganism.

Primary vs Opportunistic Pathogens

Among the almost infinite varieties of microorganisms, relatively few are capable of causing a disease in an otherwise normal or healthy individual. These disease-causing microorganisms are conveniently classified as primary pathogens. A disease may also be caused by organisms ordinarily in contact with the host, such as bacteria or fungi in the colon or in the upper respiratory tract (opportunistic pathogens), following an injury (whether mechanical, such as an open fracture, following a disease with immunosuppressive activity, such as measles or malaria, or induced by cytotoxic chemotherapy).

Protective Antigens

In general, specific acquired immunity to human pathogens is directed to only one or a few protective antigens. The immunologic properties of the protective antigen are important determinants of the human protective defense.

Protein Antigens

Proteins, in most instances, are defined as T cell dependent antigens. T cells can be activated by protein antigens and host cells parasitized by microorganisms that are intracellular parasites. The activated antigen-specific T cells release cytokines that cause the plasma cell to divide and to increase its secretion of specific antibody. Protein protective antigens are highly specific and are unique to each pathogen. Acquisition of antibodies to protein protective antigens follows either infection with the pathogen or vaccination.

Polysaccharides

The surface polysaccharides of pathogens may serve as protective antigens. These polysaccharides may be capsular or constitute the outermost domain of bacteria. In contrast to a protein, a polysaccharide is multivalent for each epitope. This multivalency explains why a polysaccharide can crosslink the receptors of plasma cells, resulting in their aggregation and activation.

The protective epitopes of polysaccharides, in contrast to proteins, are widely shared in nature, and natural immunity or antibody synthesis in the absence of the homologous organism occurs in most individuals during development. Polysaccharides are T cell independent.

Immune Mechanisms

Antibody and secondary biologic activities

For some pathogens such as meningococci (Gram-negative), the antigen-antibody configuration will activate the serum complement protein cascade resulting in lysis or phagocytosis of many bacteria. Gram-positive organisms are killed by antibody-initiated complement-dependent opsonophagocytosis and intracellular killing. The exact protective mechanism of viral-specific antibodies is not fully known and may be unique for each pathogen. Antibodies that neutralize bacterial toxins (antitoxins), such as tetanus, diphtheria and pertussis toxins, are highly protective and therapeutic.

Preventive Immunity

Disease-acquired

Convalescence from most infectious diseases confers immunity. This immunity, in most instances, may be transferred for a limited period to non-immune individuals by injection of serum IgG, passively-acquired maternal serum IgG or by milk. The best explanation for the preventive action of antibodies is that they kill or inactivate the inoculum of the pathogen which also results in decreased transmission of the pathogen.

Natural immunity

Acquisition of serum antibodies to the surface polysaccharides of human pathogens is age-related and often occurs without the individual encountering the homologous organism. The stimulus for these natural and protective antibodies is probably cross-reacting polysaccharides of the enteric and respiratory tract floras.

Vaccination-induced active and passive immunity

Vaccines are heterogenous according to the nature of the immunizing substance. Inert and injected vaccines, such as tetanus toxoid, elicit mostly serum antibodies. Living vaccines elicit secretory antibodies and sensitized T cells.

INTRODUCTION

Basis of Acquired Resistance

Acquired resistance is mediated by antigen-specific immune mechanisms. This specificity may be acquired following a disease, by asymptomatic carriage of the pathogen, by harboring an organism with a similar structure (crossreacting,) or by vaccination.

Specific acquired immunity against infectious diseases may be mediated by antibodies and/or T lymphocytes. Immunity mediated by these two factors may be manifested by a direct effect upon a pathogen, such as (1) antibody-initiated complement-dependent bacteriolysis, (2) opsonophagocytosis and killing, as occurs for some bacteria, (3) neutralization of viruses so that these organisms cannot enter cells, or (4) by T lymphocytes which will kill a cell parasitized by a microorganism.

Primary vs Opportunistic Pathogens

Among the almost infinite varieties of microorganisms, relatively few cause a disease in an otherwise normal or healthy individual. These highly virulent microorganisms are conveniently classified as primary pathogens. Opportunistic infectious disease may be caused by organisms that are ordinarily in contact with the host, such as bacteria or fungi in the colon or in the upper respiratory tract; following an injury, whether mechanical (such as a open fracture); or by a disease with immunosuppressive activity (such as measles or malaria, or one induced by cytotoxic chemotherapy). Organisms, which cause an infectious disease in a host with depressed resistance, are classified as opportunistic pathogens. Primary pathogens may also cause more explosive disease in a host with depressed resistance. Our knowledge of the protective antigens and specific acquired host immune factors is more complete for primary pathogens.

Most primary pathogens are inhabitants of, and pathogens for, humans only. Opportunistic pathogens, in contrast, may cause disease in many species of mammals. Some exceptions are tetanus, anthrax and rabies which may inhabit and cause disease in many animal species including humans.

Protective Antigens

Microorganisms adapt to cause disease by many mechanisms (see Ch. 49). Many bacteria, for example, produce macromolecules that cause (1) inflammation, (2) adherence to human tissues, or (3) are toxins that chemically alter host metabolism. But, in general, specific acquired immunity to human pathogens is directed to one (protective) antigen.

Protein antigens

Proteins, in most instances, are defined as T cell dependent antigens by two properties. First, proteins are hydrolyzed to peptides by intracellular proteases. Second, proteins may have multiple but unique specificities (epitopes). These two properties permit recognition by receptors (membrane immunoglobulin) of antibody-producing cells (plasma cells) and internalization and proteolysis by enzymes. Activation requires that the peptide fragment of the protein antigen interact with the plasma cell histocompatibility antigen to form a complex that attracts and activates T cells. The activated antigen-specific T cells release cytokines that cause plasma cells to divide and to increase their secretion of specific antibody.

The same mechanism serves to activate T cells by contact with host cells parasitized by intracellular microorganisms. Proteins secreted by intracellular organisms interact with histocompatibility antigens of the host cell (T cell epitopes) and provide a specific site for T cells that kill the parasitized host cell (cytotoxic T lymphocytes).

Some protective epitopes, especially those of viruses, are expressed only on the intact organism (conformation epitope). An example is the neutralizing epitope of polioviruses (D antigen) that requires the intact capsid to elicit neutralizing antibodies.

The protective protein antigens are highly specific and are unique to each pathogen. Acquisition of antibodies to protein protective antigens either follows infection with the pathogen or vaccination.

Polysaccharide antigens

Surface polysaccharides of pathogens may serve as protective antigens (Table 8-1). These polysaccharides may be capsular polysaccharides, present on either Gram-negative or Gram-positive organisms or the outermost domain of the lipopolysaccharide of Gram-negative organisms. Polysaccharides have simple structures composed of identical repeating units so that each molecule will have relatively fewer epitopes than a protein. But in contrast to a protein, a polysaccharide is multivalent for each epitope. This multivalency explains why a polysaccharide can crosslink the receptors of plasma cells, resulting in activation and multiplication of plasma cells, both of which increase secretion of antibodies.

TABLE 8-1 The surface Polysaccharides of Primary Bacterial Pathogens Causing Systemic Infections

Name of pathogen	Polysaccharide designation
	Capsular polysccharides:
Streptococcus pneumoriae	Type
Haemophilus influenzae	Type
Nelsseria meringitiis	Group
Group B streptococcus	Group
Esherichia coli	K antigen
Salmonella type	V_1

	Lipopolysaccharides :
Salmonella	Group
Shigella	Type
Enteroinvasive	Type
Vibrio cholerae 01	Serotype

The protective epitopes of polysaccharides, in contrast to proteins, are widely shared in nature, and natural immunity: antibody synthesis in the absence of the homologous organism, occurs in almost every individual during development. Similarly, disease and often asymptomatic carriage will also stimulate serum polysaccharide antibodies. Lastly, because they do not interact with T cells, polysaccharides are designated as T cell independent.

Each bacterial species, such as pneumococci, may have many capsular polysaccharides but only a fraction of these will be associated with a disease. For example, there are now 89 reported types but most systemic pneumococcal infections are caused by about 23 types. In infants and children, most pneumococcal infections are caused by only 8 types. Similarly, of the six types of Haemophilus influenzae, almost all systemic infections, especially meningitis, are caused by type b.

It is important to understand that immunity may be directed towards the intact pathogens such as bacteria, viruses, protozoa, or fungi, or to individual extracellular antigens such as toxins (antitoxin).

Immune Mechanisms

Antibody and secondary biologic activities

Serum antibodies are the signal and specific component of a complex inactivation system. Serum IgM and IgG antibodies exert their protective effect directly upon bacteria whose surface polysaccharides or proteins are protective antigens. This signal is generated by the configuration of the non-antibodycombining site region of the heavy polypeptide chain after binding of antibody with an epitope. For some pathogens such as meningococci (Gram-negative), the antigen-antibody configuration will activate the serum complement protein cascade with deposition of a C8,9 peptide probe that drills itself through the outer membrane. The resultant lesion causes release of intracellular components and lysis of the meningococci. In addition to bacteriolysis, other Gram-negatives, such as Haemophilus influenzae type b, also may be inactivated by serum antibody, affixed to the polysaccharide of this pathogen, that attracts and activates serum complement proteins to form C3 and C5 complexes. The latter complexes attract and activate phagocytic cells that engulf and digest the pathogen (opsonophagocytosis). This antibody-initiated, complement-dependent phagocytosis and killing are required for Gram-positive bacteria whose cell wall is not susceptible to lysis by complement.

Serum antibodies may confer immunity by binding directly to viral pathogens. The effect of this simple interaction is neutralization or inactivation of the virus. The exact protective mechanism of viral-specific antibodies is not known and may be unique for each pathogen (see Ch. 50). In many cases, antibody binding renders the viral pathogen incapable of infecting a host cell by preventing penetration of cells. As an example, antibodies to the fusion (F) protein of measles prevent the integration of virus with the cell membrane of the host. Experimental proof for this direct antiviral effect is that monovalent fragments of IgG, unable to activate complement, exert similar neutralization activity as the intact antibody. Some larger viral pathogens, coated with specific antibody, may be phagocytized and digested (See Chapter 50).

Antibodies that neutralize bacterial toxins (antitoxins), such as tetanus, diphtheria and pertussis toxins, are protective and therapeutic. The protective effects of antitoxins are varied. Antitoxin does not exert antibacterial action upon Clostridium tetani. Rather, antitoxin inactivates the functions of tetanus toxin that facilitates its migration up the neural sheath to the synapse, and antibodies to the enzymatic region inhibit its alteration of the synapse. The neutralizingactivities of diphtheria and pertussis antitoxins, in contrast, exert secondary antibacterial actions upon their respective pathogens. Both toxins serve to condition the respiratory epithelium to permit colonization by Corynebacterium diphtheriae tox+ and Bordetella pertussis: the former by its cytotoxicity and the latter by its inactivation of the function of phagocytic cells. Antitoxins block these actions and facilitate the function of phagocytic cells.

Cell-mediated immunity

Antigen-specific activation of T cells has been described (vide supra). The targets for activated T cells are parasitized host cells (also see Ch. 50). The secondary or inactivation mechanisms invoked by activated T cell phagocytic host cell complexes are not clearly understood. One important mechanism is the release of nitrous oxide that results in killing of the host cell and of the pathogen. Cell-mediated immunity is largely, if not exclusively, a curative mechanism.

Preventive Immunity

Immune resistance to an infectious disease requires a critical level of either antigen-specific antibodies and/or T cells when the host encounters the pathogen. Prevention of an infection requires immune mechanisms to kill or inactivate the inoculum of the pathogen. This immunity may be expressed as a protective level of antibodies so that resistance to specific infections may be reliably predicted by a serologic assay such as the level of neutralizing antibodies to measles, mumps or Groups A, B, Y and W135 meningococci. These assays can predict resistance to a disease for individuals or can be used to assess the immune status of communities. Quantitation of antigen-specific T cells to predict immunity on a clinical basis is, as yet, an investigative tool.

Disease-acquired Immunity

Convalescence from most infectious diseasesconfers immunity. This immunity, in most instances, may be transferred for a limited period to non-immune individuals by injection of IgG as FDA-licensed immunoglobulin(Table 8-2) or to the newborn by passively acquired maternal serum IgG. There is also evidence that secretory IgA acquired by breast feeding confers immunity to newborns. These findings indicate that critical levels of antibodies are sufficient to prevent infectious diseases. Their preventive action is best explained by antibodies killing or inactivating the inoculum of the pathogen on epithelial surfaces. Herd immunity follows vaccination with Haemophilus type b conjugates, diphtheria toxoid and measles virus vaccines. The resulting immunity causes a decreased transmission of the pathogen. Since there is no animal vector for these pathogens, the incidence of the disease in the entire community is far below that percentage of the population that has been vaccinated (herd immunity). Antibody-mediated inactivation of the inoculum may also occur in the blood stream in the case of pathogens inoculated directly into the tissues or blood stream, such as hepatitis B or malaria. There is yet no evidence in humans that antigen-specific T lymphocytes can prevent infectious diseases.

TABLE 8-2 U.S. Licensed Immunoglobulin For Passive Immunization

Disease	Biologic	Indication*
Botulism	Specific equine Ig	Treatmont
CMV	Hyperimmune human IV Ig	Prophylaxis
Diphtheria	Specific equine Ig	Treatmont
Hepatitis A, measles	Pooled human Ig	Prophylaxis
Ig deficiency* , ITP, Kawasaki disease	Pooled human IgG	Treatmont
Hepatitis B Ig	Immune human Ig	Prophylaxis
Rabies (HRIG)	Immune human Ig	Prophylaxis
Tetanus Ig (TIG)	Immune human Ig	Treatmont
Vaccina	Immune human Ig	Treatmont
Varicella-zoster	Immune human Ig	Prophylaxis

* The positive effect passively administered immunoglobulin has been established for the immunodeficiency disease, X-linked hypogammaglobulinemia, in which bacterial diseases, such as otitis media, pneumonia and meningitis are prevented. Prior to the advent of viral vaccines, passively administered immunoglobulin was routinely used for prevention of measles, rubella, mumps, poliomyelitis, and varicella. Human immune syncytial virus infections.

Natural immunity

Acquisition of serum antibodies to surface polysaccharides of pathogens is age-related and often occurs without the individual encountering the homologous organism. An example is group A meningococci, the cause of epidemic meningitis.

Despite the absence of this pathogen in the United States for about 50 years, either as a cause of meningitis or in asymptomatic carriers, most adults have antibodies to this capsular polysaccharide. Antigenic stimuli for Group A meningococcal antibodies are likely due to exposure to several Gram-positive and Gram-negative bacterial species in human stools. Another example is Shigella dysenteriae type 1, the cause of epidemic dysentery. Adults in Sweden and the United States have antibodies to the LPS of S dysenteriae type 1, despite the virtual absence of this pathogen in these countries during the past 50 years. The stimulus for these natural and protective antibodies is probably cross-reacting polysaccharides of the enteric and respiratory tract floras.

Natural serum antibodies to surface polysaccharides confer specific protection to adults and are transmitted to newborns. The highest attack rate and mortality occur during childhood when these maternally-acquired anti-polysaccharide antibodies have waned and adult levels have not been reached. Acquisition of natural antibodies is not uniform and many adults remain non-immune. Vaccination with polysaccharide-based vaccines increases the percentage of adults with protective levels to almost 100%. These principles are elegantly illustrated by the development of groups A and C meningococcal polysaccharide vaccines during the 1960's when outbreaks of meningitis caused by this pathogen occurred in armed forces recruits during the Vietnam conflict. Introduction of these polysaccharide vaccines rapidly eliminated these outbreaks.

Vaccination-induced active and passive immunity

Vaccines are heterogeneous according to (1) the nature of the immunizing substance, whether they are inert or living, and (2) by the method of their administration. Vaccines include living attenuated strains of viruses (poliovirus) or bacteria (BCG), inactivated viruses (yellow fever) and bacteria (anthrax), purified polysaccharides (pneumococcal 23 valent) or polysaccharide-protein conjugates (Haemophilus type b conjugate). Inert and injected vaccines, such as tetanus toxoid, elicit mostly serum antibodies. Living vaccines, such as attenuated strains of viruses (poliovirus), elicit secretory antibodies and sensitized T cells.

To date, the FDA regulates vaccines and seroepidemiologic studies to assess the immune status of populations by measurement of biologically active antibodies only. Thus, the status of diphtheria immunity is evaluated by the percentage of the population with protective levels of neutralizing antibodies to diphtheria toxin (antitoxin). Similarly, the immune status to measles is evaluated by measurement of the percentage of the population with protective levels of neutralizing antibodies. For some vaccines, such as BCG, there is as yet no measure of immunity to assess their effectiveness.

Some investigational vaccines utilize recombinant DNA technology to mobilize genes governing the synthesis of protective antigens. These specific genes may be inserted into a virulent vectors. Administration of vectors is designed to stimulate the comprehensive immunity that follows disease with the pathogen itself. Naked DNA may be incorporated into plasmids that infect somatic cells and continually induce synthesis of protective antigens that stimulate antibodies and activated T cells.

Curative Immunity

Infection with most pathogens does not result in death of the host and the offending organism is ultimately cleared after the symptoms of the disease have waned. The basis of the curative process of patients is not well understood, but it is likely mediated by expansion of both specific immune and effector mechanisms. Quantitative increases of specific antibodies and activated T cells is accompanied by increases in serum complement levels and phagocytic cells. Also, infection increases the levels of cytokines and serum proteins known as acute phase reactants, such as C-reactive protein, alpha 1 trypsin inhibitor, and transferrin, that serve as scavengers or inhibitors of bacterial debris. Cure of infectious diseases is most likely the prolonged interaction of maximal levels of host specific and non-specific (effector) (see Ch. 49) mechanisms with the pathogen.

REFERENCES

Centers for Disease Control: MMWR, Volume 43, 1994

Chanock RM, Crowe JE Jr, Murphy BR, et al: Human monoclonal antibody Fab fragments cloned from combinatorial libraries: Potential usefulness in prevention and/or treatment of major human viral diseases. Infect Agents Dis 2:118, 1993

Goldschneider I, Gotschlich EC, Artenstein MS, et al: Human immunity to the meningococcus. I. The role of humoral antibodies. J Exp Med 129:1307, 1969

Nicholson A, Lepow IH: Host defense against Neisseria meningitidis require a complement-dependent bactericidal activity. Science 205:298, 1979.

Pappenheimer AM, Jr: Diphtheria. p. 1. In Germanier R (ed): Bacterial Vaccines. Academic Press, Inc. New York, 1984

Robbins JB, Schneerson R, Szu SC: Perspective: Hypothesis: Serum IgG antibody is sufficient to confer protection against infectious diseases by inactivating the inoculum. J Inf Dis 171:1387, 1995

Sutton A, Schneerson R, Kendall-Morris S, et al: Differential complement resistance mediates virulence of Haemophilus influenzae type b. Infect Immun 35:95, 1982

Young JD-E: Killing of target cells by lymphocytes: A mechanistic view. Physiol Rev 69:250, 1989

Chapter 7

Normal Flora

General Concepts

Significance of the Normal Flora

The normal flora influences the anatomy, physiology, susceptibility to pathogens, and morbidity of the host.

Skin Flora

The varied environment of the skin results in locally dense or sparse populations, with Gram-positive organisms (e.g., staphylococci, micrococci, diphtheroids) usually predominating.

Oral and Upper Respiratory Tract Flora

A varied microbial flora is found in the oral cavity, and streptococcal anaerobes inhabit the gingival crevice. The pharynx can be a point of entry and initial colonization for Neisseria, Bordetella, Corynebacterium, and Streptococcus spp.

Gastrointestinal Tract Flora

Organisms in the stomach are usually transient, and their populations are kept low (10^3 to 10^6/g of contents) by acidity. Helicobacter pylori is a potential stomach pathogen that apparently plays a role in the formation of certain ulcer types. In normal hosts the duodenal flora is sparse (0 to 10^3/g of contents). The ileum contains a moderately mixed flora (10^6 to 10^8/g of contents). The flora of the large bowel is dense (10^9 to 10^{11}/g of contents) and is composed predominantly of anaerobes. These organisms participate in bile acid conversion and in vitamin K and ammonia production in the large bowel. They can also cause intestinal abscesses and peritonitis.

Urogenital Flora

The vaginal flora changes with the age of the individual, the vaginal pH, and hormone levels. Transient organisms (e.g., Candida spp) frequently cause vaginitis. The distal urethra contains a sparse mixed flora; these organisms are present in urine specimens (10^4/ml) unless a clean-catch, midstream specimen is obtained.

Conjunctival Flora

The conjunctiva harbors few or no organisms. Haemophilus and Staphylococcus are among the genera most often detected.

Host Infection

Many elements of the normal flora may act as opportunistic pathogens, especially in hosts rendered susceptible by rheumatic heart disease, immunosuppression, radiation therapy, chemotherapy, perforated mucous membranes, etc. The flora of the gingival crevice causes dental caries in about 80 percent of the population.

INTRODUCTION

A diverse microbial flora is associated with the skin and mucous membranes of every human being from shortly after birth until death. The human body, which contains about 10^{13} cells, routinely harbors about 10^{14} bacteria (Fig. 6-1). This bacterial population constitutes the normal microbial flora. The normal microbial flora is relatively stable, with specific genera populating various body regions during particular periods in an individual's life. Microorganisms of the normal flora may aid the host (by competing for microenvironments more effectively than such pathogens as Salmonella spp or by producing nutrients the host can use), may harm the host (by causing dental caries, abscesses, or other infectious diseases), or may exist as commensals (inhabiting the host for long periods without causing detectable harm or benefit). Even though most elements of the normal microbial flora inhabiting the human skin, nails, eyes, oropharynx, genitalia, and gastrointestinal tract are harmless in healthy individuals, these organisms frequently cause disease in compromised hosts. Viruses and parasites are not considered members of the normal microbial flora by most investigators because they are not commensals and do not aid the host.

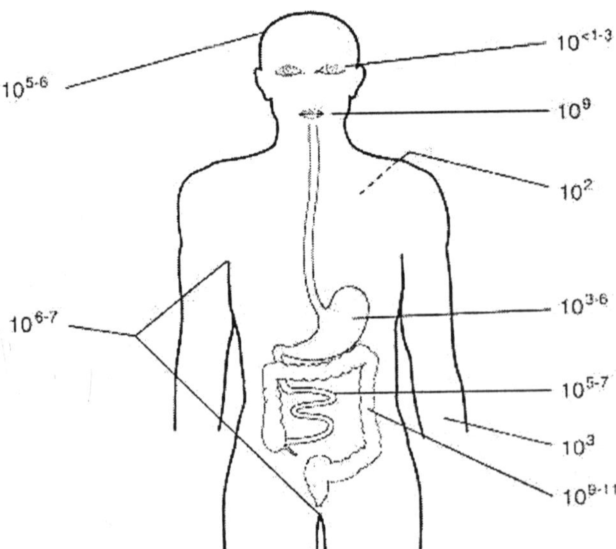

FIGURE 6-1 Numbers of bacteria that colonize different parts of the body. Numbers represent the number of organisms per gram of homogenized tissue or fluid or per square centimeter of skin surface.

Significance of the Normal Flora

The fact that the normal flora substantially influences the well-being of the host was not well understood until germ-free animals became available. Germ-free animals were obtained by cesarean section and maintained in special isolators; this allowed the investigator to raise them in an environment free from detectable viruses, bacteria, and other organisms. Two interesting observations were made about animals raised under germ-free conditions. First, the germ-free animals lived almost twice as long as their conventionally maintained counterparts, and second, the major causes of death were different in the two groups. Infection often caused death in conventional animals, but intestinal atonia frequently killed germ-free animals. Other investigations showed that germ-free animals have anatomic, physiologic, and immunologic features not shared with conventional animals. For example, in germ-free animals, the alimentary lamina propria is underdeveloped, little or no immunoglobulin is present in sera or secretions, intestinal motility is reduced, and the intestinal epithelial cell renewal rate is approximately one-half that of normal animals (4 rather than 2 days).

Although, the foregoing indicates that bacterial flora may be undesirable, studies with antibiotic treated animals suggest that the flora protects individuals from pathogens. Investigators have used streptomycin to reduce the normal flora and have then infected animals with streptomycin-resistant Salmonella. Normally, about 10^6 organisms are needed to establish a gastrointestinal infection, but in streptomycin-treated animals whose flora is altered, fewer than 10 organisms were needed to cause infectious disease. Further studies suggested that fermentation products (acetic and butyric acids) produced by the normal flora inhibited Salmonella growth in the gastrointestinal tract. Figure 6-2 shows some of the factors that are important in the competition between the normal flora and bacterial pathogens.

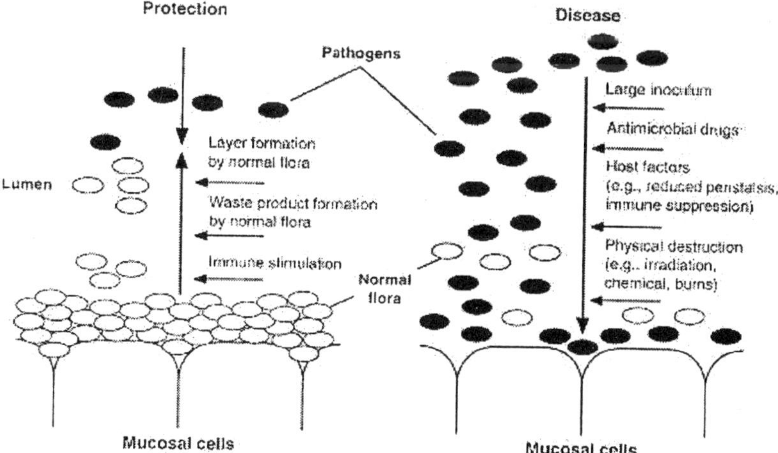

FIGURE 6-2 Mechanisms by which the normal flora competes with invading pathogens. Compare this schematic with Figure 6-3.

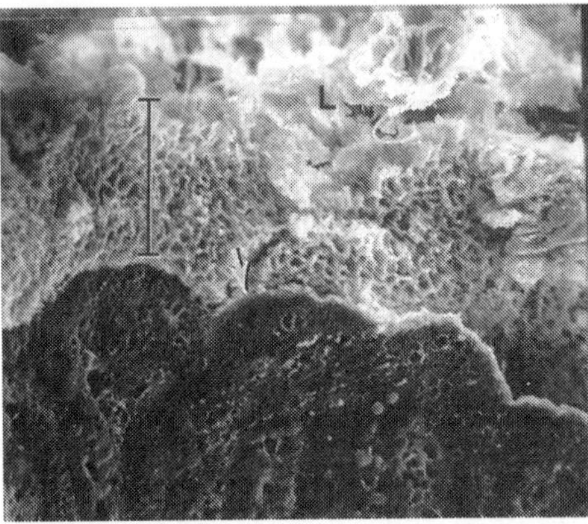

FIGURE 6-3 (A) Scanning electron micrograph of a cross-section of rat colonic mucosa. The bar indicates the thick layer of bacteria between the mucosal surface and the lumen (L) (X 262,)

FIGURE 6-3 (B) Higher magnification of the area indicated by the arrow in Fig. A, showing a mass of bacteria (B) immediately adjacent to colonized intestinal tissue (T), (X2,624.) (Figure from Davis CP: Preservation of bacteria and their microenvironmental association in the rat by freezing. Appl Environ Microbiol 31:310,1976, with permission.)

The normal flora in humans usually develops in an orderly sequence, or succession, after birth, leading to the stable populations of bacteria that make up the normal adult flora. The main factor determining the composition of the normal flora in a body region is the nature of the local environment, which is determined by pH, temperature, redox potential, and oxygen, water, and nutrient levels. Other factors such as peristalsis, saliva, lysozyme secretion, and secretion of immunoglobulins also play roles in flora control. The local environment is like a concerto in which one principal instrument usually dominates. For example, an infant begins to contact organisms as it moves through the birth canal. A Gram-positive population (bifidobacteria arid lactobacilli) predominates in the gastrointestinal tract early in life if the infant is breast-fed. This bacterial population is reduced and displaced somewhat by a Gram-negative flora (Enterobacteriaceae) when the baby is bottle-fed. The type of liquid diet provided to the infant is the principal instrument of this flora control; immunoglobulins and, perhaps, other elements in breast milk may also be important.

What, then, is the significance of the normal flora? Animal and some human studies suggest that the flora influences human anatomy, physiology, lifespan, and, ultimately, cause of death. Although, the causal relationship of flora to death and disease in humans is accepted, of her roles of the human microflora need further study.

Normal Flora of Skin

Skin provides good examples of various microenvironments. Skin regions have been compared to geographic regions of Earth: the desert of the forearm, the cool woods of the scalp, and the tropical forest of the armpit. The composition of the dermal microflora varies from site to site according to the character of the microenvironment. A different bacterial flora characterizes each of three regions of skin: (1) axilla, perineum, and toe webs; (2) hand, face and trunk; and (3) upper arms and legs. Skin sites with partial occlusion (axilla, perineum, and toe webs) harbor more microorganisms than do less occluded areas (legs, arms, and trunk). These quantitative differences may relate to increased amount of moisture, higher body temperature, and greater concentrations of skin surface lipids. The axilla, perineum, and toe webs are more frequently colonized by Gram-negative bacilli than are drier areas of the skin.

The number of bacteria on an individual's skin remains relatively constant; bacterial survival and the extent of colonization probably depend partly on the exposure of skin to a particular environment and partly on the innate and species-specific bactericidal activity in skin. Also, a high degree of specificity is involved in the adherence of bacteria to epithelial surfaces. Not all bacteria attach to skin; staphylococci, which are the major element of the nasal flora, possess a distinct advantage over viridans streptococci in colonizing the nasal mucosa. Conversely, viridans streptococci are not seen in large numbers on the skin or in the nose but dominate the oral flora.

The microbiology literature is inconsistent about the density of bacteria on the skin; one reason for this is the variety of methods used to collect skin bacteria. The scrub method yields the highest and most accurate counts for a given skin area. Most microorganisms live in the superficial layers of the stratum corneum and in the upper parts of the hair follicles. Some bacteria, however, reside in the deeper areas of the hair follicles and are beyond the reach of ordinary disinfection procedures. These bacteria are a reservoir for recolonization after the surface bacteria are removed.

Staphylococcus epidermidis

S epidermidis is a major inhabitant of the skin, and in some areas it makes up more than 90 percent of the resident aerobic flora.

Staphylococcus aureus

The nose and perineum are the most common sites for S aureus colonization, which is present in 10 percent to more than 40 percent of normal adults. S aureus is prevalent (67 percent) on vulvar skin. Its occurrence in the nasal passages varies with age, being greater in the newborn, less in adults. S aureus is extremely common (80 to 100 percent) on the skin of patients with certain dermatologic diseases such as atopic dermatitis, but the reason for this finding is unclear.

Micrococci

Micrococci are not as common as staphylococci and diphtheroids; however, they are frequently present on normal skin. Micrococcus luteus, the predominant species, usually accounts for 20 to 80 percent of the micrococci isolated from the skin.

Diphtheroids (Coryneforms)

The term diphtheroid denotes a wide range of bacteria belonging to the genus Corynebacterium. Classification of diphtheroids remains unsatisfactory; for convenience, cutaneous diphtheroids have been categorized into the following four groups: lipophilic or nonlipophilic diphtheroids; anaerobic diphtheroids; diphtheroids producing porphyrins (coral red fluorescence when viewed under ultraviolet light); and those that possess some keratinolytic enzymes and are associated with trichomycosis axillaris (infection of axillary hair). Lipophilic diphtheroids are extremely common in the axilla, whereas nonlipophilic strains are found more commonly on glabrous skin.

Anaerobic diphtheroids are most common in areas rich in sebaceous glands. Although, the name Corynebacterium acnes was originally used to describe skin anaerobic diphtheroids, these are now classified as Propionibacterium acnes and as P granulosum. P acnes is seen eight times more frequently than P granulosum in acne lesions and is probably involved in acne pathogenesis. Children younger than 10 years are rarely colonized with P acnes. The appearance of this organism on the skin is probably related to the onset of secretion of sebum (a semi-fluid substance composed of fatty acids and epithelial debris secreted from sebaceous glands) at puberty. P avidum, the third species of cutaneous anaerobic diphtheroids, is rare in acne lesions and is more often isolated from the axilla.

Streptococci

Streptococci, especially ß-hemolytic streptococci, are rarely seen on normal skin. The paucity of ß-hemolytic streptococci on the skin is attributed at least in part to the presence of lipids on the skin, as these lipids are lethal to streptococci. Other groups of streptococci, such as a-hemolytic streptococci, exist primarily in the mouth, from where they may, in rare instances, spread to the skin.

Gram-Negative Bacilli

Gram-negative bacteria make up a small proportion of the skin flora. In view of their extraordinary numbers in the gut and in the natural environment, their scarcity on skin is striking. They are seen in moist intertriginous areas, such as the toe webs and axilla, and not on dry skin. Desiccation is the major factor preventing the multiplication of Gram-negative bacteria on intact skin. Enterobacter, Klebsiella, Escherichia coli, and Proteus spp are the predominant Gram-negative organisms found on the skin. Acinetobacter spp also occurs on the skin of normal individuals and, like other Gram-negative bacteria, is more common in the moist intertriginous areas.

Nail Flora

The microbiology of a normal nail is generally similar to that of the skin. Dust particles and other extraneous materials may get trapped under the nail, depending on what the nail contacts. In addition to resident skin flora, these dust particles may carry fungi and bacilli. Aspergillus, Penicillium, Cladosporium, and Mucor are the major types of fungi found under the nails.

Oral and Upper Respiratory Tract Flora

The oral flora is involved in dental caries and periodontal disease, which affect about 80 percent of the population in the Western world. The oral flora, its interactions with the host, and its response to environmental factors are thoroughly discussed in another Chapter. Anaerobes in the oral flora are responsible for many of the brain, face, and lung infections that are frequently manifested by abscess formation.

The pharynx and trachea contain primarily those bacterial genera found in the normal oral cavity (for example, alpha-and ß-hemolytic streptococci); however, anaerobes, staphylococci, neisseriae, diphtheroids, and others are also present. Potentially pathogenic organisms such as Haemophilus, mycoplasmas, and pneumococci may also be found in the pharynx. Anaerobic organisms also are reported frequently. The upper respiratory tract is so often the site of initial colonization by pathogens (Neisseria meningitides, C diphtheriae, Bordetella pertussis, and many others) and could be considered the first region of attack for such organisms. In contrast, the lower respiratory tract (small bronchi and alveoli) is usually sterile, because particles the size of bacteria do not readily reach it. If bacteria do reach these regions, they encounter host defense mechanisms, such as alveolar macrophages, that are not present in the pharynx.

Gastrointestinal Tract Flora

The stomach is a relatively hostile environment for bacteria. It contains bacteria swallowed with the food and those dislodged from the mouth. Acidity lowers the bacterial count, which is highest (approximately 10^3 to 10^6 organisms/g of contents) after meals and lowest (frequently undetectable) after digestion. Some Helicobacter species can colonize the stomach and are associated with type B gastritis and peptic ulcer disease. Aspirates of duodenal or jejunal fluid contain approximately 10^3 organisms/ml in most individuals. Most of the bacteria cultured (streptococci, lactobacilli, Bacteroides) are thought to be transients. Levels of 10^5 to about 10^7 bacteria/ml in such aspirates usually indicate an abnormality in the digestive system (for example, achlorhydria or malabsorption syndrome). Rapid peristalsis and the presence of bile may explain in part the paucity of organisms in the upper gastrointestinal tract. Further along the jejunum and into the ileum, bacterial populations begin to increase, and at the ileocecal junction they reach levels of 10^6 to 10^8 organisms/ml, with streptococci, lactobacilli, Bacteroides, and bifidobacteria predominating.

Concentrations of 10^9 to 10^{11} bacteria/g of contents are frequently found in human colon and feces. This flora includes a bewildering array of bacteria (more than 400 species have been identified); nonetheless, 95 to 99 percent belong to anaerobic genera such as Bacteroides, Bifidobacterium, Eubacterium, Peptostreptococcus, and Clostridium. In this highly anaerobic region of the intestine, these genera proliferate, occupy most available niches, and produce metabolic waste products such as acetic, butyric, and lactic acids. The strict anaerobic conditions, physical exclusion (as is shown in many animal studies), and bacterial waste products are factors that inhibit the growth of other bacteria in the large bowel.

Although, the normal flora can inhibit pathogens, many of its members can produce disease in humans. Anaerobes in the intestinal tract are the primary agents of intra-abdominal abscesses and peritonitis. Bowel perforations produced by appendicitis, cancer, infarction, surgery, or gunshot wounds almost always seed the peritoneal cavity and adjacent organs with the normal flora. Anaerobes can also cause problems within the gastrointestinal lumen. Treatment with antibiotics may allow certain anaerobic species to become predominant and cause disease. For example, *Clostridium difficile,* which can remain viable in a patient undergoing antimicrobial therapy, may produce pseudomembranous colitis. Other intestinal pathologic conditions or surgery can cause bacterial overgrowth in the upper small intestine. Anaerobic bacteria can then deconjugate bile acids in this region and bind available vitamin B12 so that the vitamin and fats are malabsorbed. In these situations, the patient usually has been compromised in some way; therefore, the infection caused by the normal intestinal flora is secondary to another problem.

More information is available on the animal than the human microflora. Research on animals has revealed that unusual filamentous microorganisms attach to ileal epithelial cells and modify host membranes with few or no harmful effects. Microorganisms have been observed in thick layers on gastrointestinal surfaces (Fig. 6-3) and in the crypts of Lieberkuhn. Other studies indicate that the immune response can be modulated by the intestinal flora. Studies of the role of the intestinal flora in biosynthesis of vitamin K and other host-utilizable products, conversion of bile acids (perhaps to cocarcinogens), and ammonia production (which can play a role in hepatic coma) show the dual role of the microbial flora in influencing the health of the host. More basic studies of the human bowel flora are necessary to define their effect on humans.

Urogenital Flora

The type of bacterial flora found in the vagina depends on the age, pH, and hormonal levels of the host. Lactobacillus spp predominate in female infants (vaginal pH, approximately 5) during the first month of life. Glycogen secretion seems to cease from about I month of age to puberty. During this time, diphtheroids, *S epidermidis*, streptococci, and *E coli* predominate at a higher pH (approximately pH 7). At puberty, glycogen secretion resumes, the pH drops, and women acquire an adult flora in which L acidophilus, corynebacteria, peptostreptococci, staphylococci, streptococci, and Bacteroides predominate. After menopause, pH again rises, less glycogen is secreted, and the flora returns to that found in prepubescent females. Yeasts (Torulopsis and Candida) are occasionally found in the vagina (10 to 30 percent of women); these sometimes increase and cause vaginitis.

In the anterior urethra of humans, *S epidermidis*, enterococci, and diphtheroids are found frequently; *E coli*, Proteus, and Neisseria (nonpathogenic species) are reported occasionally (10 to 30 percent). Because of the normal flora residing in the urethra, care must be taken in clinically interpreting urine cultures; urine samples may contain these organisms at a level of 10^4/ml if a midstream (clean-catch) specimen is not obtained.

Conjunctival Flora

The conjunctival flora is sparse. Approximately 17 to 49 percent of culture samples are negative. Lysozyme, secreted in tears, may play a role in controlling the bacteria by interfering with their cell wall formation. When positive samples show bacteria, corynebacteria, neisseriae, and moraxellae are cultured. Staphylococci and streptococci are also present, and recent reports indicate that Haemophilus parainfluenzae is present in 25 percent of conjunctival samples.

Host Infection by Elements of the Normal Flora

This chapter has briefly described the normal human flora; however, the pathogenic mechanisms of various genera or the clinical syndromes in which they are involved was not discussed. Although, such material is presented in other chapters, note that a breach in mucosal surfaces often results in infection of the host by members of the normal flora. Caries, periodontal disease, abscesses, foul-smelling discharges, and endocarditis are hallmarks of infections with members of the normal human flora (Fig. 6-4). In addition, impairment of the host (for example, those with heart failure or leukemia) or host defenses (due to immunosuppression, chemotherapy, or irradiation) may result in failure of the normal flora to suppress transient pathogens or may cause members of the normal flora to invade the host themselves. In either situation, the host may die.

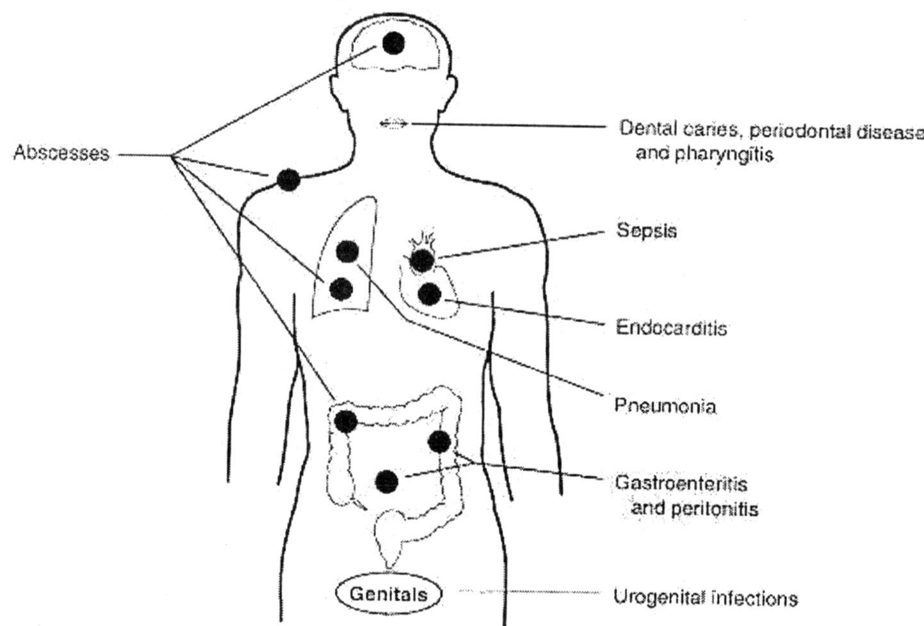

FIGURE 6-4 Clinical conditions that may be caused by members of the normal flora.

REFERENCES

Bitton G, Marshall KC: Adsorption of Microorganisms to Surfaces. John Wiley & Sons, New York, 1980

Draser BS, Hill MJ: Human Intestinal Flora. Academic Press, London, 1974.

Freter R, Brickner J, Botney M, et al: Survival and implantation of Escherichia coli in the intestinal tract. Infect Immun 39:686, 1983

Hentges DJ, Stein AJ, Casey SW, Que JU: Protective role of intestinal flora against Pseudomonas aeruginosa in mice: influence of antibiotics on colonization resistance. Infect Immun 47:118, 1985

Herthelius M, Gorbach SL, Mollby R, et al: Elimination of vaginal colonization with Escherichia coli by administration of indigenous flora. Infect Immun 57:2447, 1989

Maibach H, Aly R: Skin Microbiology: Relevance to Clinical Infection. Springer-Verlag, New York, 1981

Marples MJ: Life in the skin. Sci Am 220:108, 1969

Savage DC: Microbial ecology of the gastrointestinal tract. Annu Rev Microbiol 31:107, 1977

Tannock GW: Normal Microflora. Chapman and Hall,London, UK, 1995

Chapter 8

Bacterial Pathogenesis

General Concepts

Host Susceptibility

Resistance to bacterial infections is enhanced by phagocytic cells and an intact immune system. Initial resistance is due to nonspecific mechanisms. Specific immunity develops over time. Susceptibility to some infections is higher in the very young and the very old and in immunosuppressed patients.

Bacterial Infectivity

Bacterial infectivity results from a disturbance in the balance between bacterial virulence and host resistance. The "objective" ofbacteria is to multiply rather than to cause disease; it is in the best interest of the bacteria not to kill the host.

Host Resistance

Numerous physical and chemical attributes of the host protect against bacterial infection. These defenses include the antibacterial factors in secretions covering mucosal surfaces and rapid rate of replacement of skin and mucosal epithelial cells. Once the surface of the body is penetrated, bacteria encounter an environment virtually devoid of free iron needed for growth, which requires many of them to scavenge for this essential element. Bacteria invading tissues encounter phagocytic cells that recognize them as foreign, and through a complex signaling mechanism involving interleukins, eicosanoids, and complement, mediate an inflammatory response in which many lymphoid cells participate.

Genetic and Molecular Basis for Virulence

Bacterial virulence factors may beencoded on chromosomal, plasmid, transposon, or temperate bacteriophage DNA; virulence factor genes on transposons or temperate bacteriophage DNA may integrate into the bacterial chromosome.

Host-Mediated Pathogenesis

In certain infections (e.g., tuberculosis), tissue damage results from the toxic mediators released by lymphoid cells rather than from bacterial toxins.

Intracellular Growth

Some bacteria (e.g., Rickettsia species) can grow only within eukaryotic cells, whereas others (e.g., *Salmonella* species) invade cells but do not require them for growth. Most pathogenic bacteria multiply in tissue fluids and not in host cells.

Virulence Factors

Virulence factors help bacteria to (1) invade the host, (2) cause disease, and (3) evade host defenses. The following are types of virulence factors:

Adherence Factors: Many pathogenic bacteria colonize mucosal sites by using pili (fimbriae) to adhere to cells.

Invasion Factors: Surface components that allow the bacterium to invade host cells can be encoded on plasmids, but more often are on the chromosome.

Capsules: Many bacteria are surrounded by capsules that protect them from opsonization and phagocytosis.

Endotoxins: The lipopolysaccharide endotoxins on Gram-negative bacteria cause fever, changes in blood pressure, inflammation, lethal shock, and many other toxic events.

Exotoxins: Exotoxins include several types of protein toxins and enzymes produced and/or secreted from pathogenic bacteria. Major categories include cytotoxins, neurotoxins, and enterotoxins.

Siderophores: Siderophores are iron-binding factors that allow some bacteria to compete with the host for iron, which is bound to hemoglobin, transferrin, and lactoferrin.

INTRODUCTION

Infection is the invasion of the host by microorganisms, which then multiply in close association with the host's tissues. Infection is distinguished from disease, a morbid process that does not necessarily involve infection (diabetes, for example, is a disease with no known causative agent). Bacteria can cause a multitude of different infections, ranging in severity from inapparent to fulminating. Table 7-1 lists these types of infections.

TABLE 7-1 Types of Bacterial Infections

Type of infection	Description	Examples
Inapparent (subclinical)	No detectable clinical symptoms of infection	Asymptomic gonorrhea in women and men
Dormaint (fatent)	Carner state	Typhoid carrier
Accidental	Zoonosis or environmental or inadvertent exposures	Anthrax, cryptococcal infection, and laboratory exposure, respectively

Opportunistic	Infection caused by normal host defenses are compromised	Serrati or candlde infection of the genitourinary
Primary	Clinically apparent (e.g. invasion and multiplication of microbes in body tissues, causing local tissue injury)	Shigella dysentery
Secondary	Microbial invasion subsequent to primary infection	Bacterial pneumonia following viral lung infection
Mixed	Two or more microbes infecting the same tissue	Anaerobic abscess (E coli and Bacterides fragils)
Acute	Rapid onset (hours or days); brief duration (days or weeks)	Diphthone
Chronic	Prolonged duration (months or years)	Mycobacterial diseases (tuberculosis and leprosy)
Localized	Confined to a small area or to an organ	Staphylococcal boil
Generalized	Disseminated to many body regions (goncoccernia)	Gram-negative bacteremia
Pyognic	Pus-forming	Staphyloccal and streptococcal infection
Retrograde	Microbes ascending in a duet or tube against the flow of secretions or excretions	E coli urinary tract infection
fulminant	Infections that occur suddenly and intensely	Airborne Yesinia pests (pneumonic)

The capacity of a bacterium to cause disease reflects its relative pathogenicity. On this basis, bacteria can be organized into three major groups. When isolated from a patient, frank or primary pathogens are considered to be probable agents of disease (e.g., when the cause of diarrheal disease is identified by the laboratory isolation of *Salmonella* spp from feces). Opportunistic pathogens are those isolated from patients whose host defense mechanisms have been compromised. They may be the agents of disease (e.g., in patients who have been predisposed to urinary tract infections with *Escherichia coli* by catheterization). Finally, some bacteria, such as *Lactobacillus acidophilus,* are considered to be nonpathogens, because they rarely or never cause human disease. Their categorization as nonpathogens may change, however, because of the adaptability of bacteria and the detrimental effect of modern radiation therapy, chemotherapy, and immunotherapy on resistance mechanisms. In fact, some bacteria previously considered to be nonpathogens are now known to cause disease. *Serratia marcescens,* for example, is a common soil bacterium that causes pneumonia, urinary tract infections, and bacteremia in compromised hosts.

Virulence is the measure of the pathogenicity of an organism. The degree of virulence is related directly to the ability of the organism to cause disease despite host resistance mechanisms; it is affected by numerous variables such as the number of infecting bacteria, route of entry into the body, specific and nonspecific host defense mechanisms, and virulence factors of the bacterium. Virulence can be measured experimentally by determining the number of bacteria required to cause animal death, illness, or lesions in a defined period after the bacteria are administered by a designated route. Consequently, calculations of a lethal dose affecting 50 percent of a population of animals (LD_{50}) or an effective dose causing a disease symptom in 50 percent of a population of animals (ED_{50}) are useful in comparing the relative virulence of different bacteria.

Pathogenesis refers both to the mechanism of infection and to the mechanism by which disease develops. The purpose of this chapter is to provide an overview of the many bacterial virulence factors and, where possible, to indicate how they interact with host defense mechanisms and to describe their role in the pathogenesis of disease. It should be understood that the pathogenic mechanisms of many bacterial diseases are poorly understood, while those of others have been probed at the molecular level. The relative importance of an infectious disease to the health of humans and animals does not always coincide with the depth of our understanding of its pathogenesis. This information is best acquired by reading each of the ensuing chapters on specific bacterial diseases, infectious disease texts, and public health bulletins.

Host Susceptibility

Susceptibility to bacterial infections depends on the physiologic and immunologic condition of the host and on the virulence of the bacteria. Before increased amounts of specific antibodies or T cells are formed in response to invading bacterial pathogens, the "nonspecific" mechanisms of host resistance (such as polymorphonuclear neutrophils and macrophage clearance) must defend the host against the microbes. Development of effective specific immunity (such as an antibody response to the bacterium) may require several weeks (Fig. 7-1). The normal bacterial flora of the skin and mucosal surfaces also serves to protect the host against colonization by bacterial pathogens. In most healthy individuals, bacteria from the normal flora that occasionally penetrate the body (e.g., during tooth extraction or routine brushing of teeth) are cleared by the host's cellular and humoral mechanisms. In contrast, individuals with defective immune responses are prone to frequent, recurrent infections with even the least virulent bacteria. The best-known example of such susceptibility is acquired immune deficiency syndrome (AIDS), in which the $CD4^+$ helper lymphocytes are progressively decimated by human immunodeficiency virus (HIV). However, resistance mechanisms can be altered by many other processes. For example, aging often weakens both nonspecific and specific defense systems so that they can no longer effectively combat the challenge of bacteria from the environment. Infants are also especially susceptible to certain pathogens (such as group B streptococci because their immune systems are not yet fully developed and cannot mount a protective immune response to important bacterial antigens. In addition, some individuals have genetic defects of the complement system or cellular defenses (e.g., inability of polymorphonuclear neutrophils to kill bacteria). Finally, a patient may develop granulocytopenia as a result of a predisposing disease, such as cancer, or immunosuppressive chemotherapy for organ transplants or cancer.

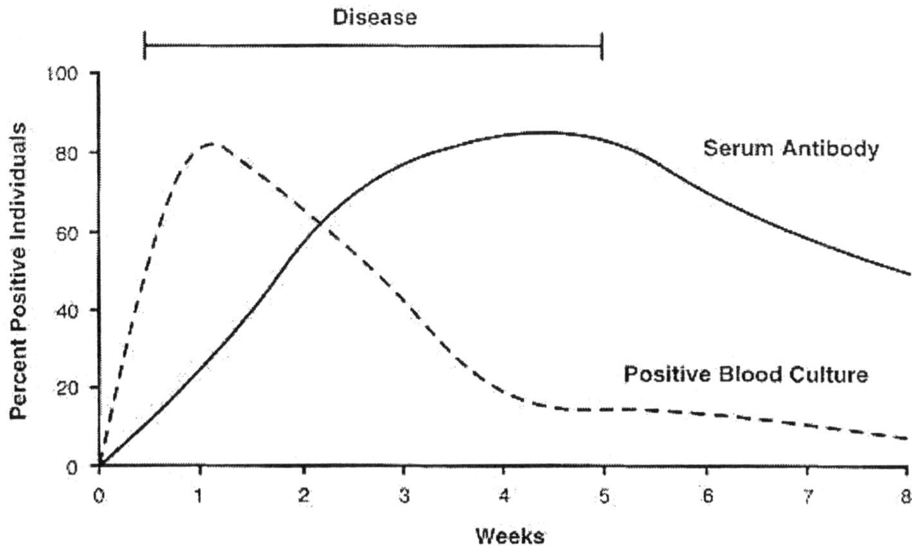

FIGURE 7-1 Serum antibody response to Salmonella typhi during typhoid fever and its relationship to septicemia.

Host resistance can be compromised by trauma and by some underlying diseases. An individual becomes susceptible to infection with a variety of bacteria if the skin or mucosa is breached, particularly in the case of severe wounds such as burns or contaminated surgical wounds. Cystic fibrosis patients, who have poor ciliary function and consequently cannot clear mucus efficiently from the respiratory tract, are abnormally susceptible to infection with mucoid strains of *Pseudomonas aeruginosa*, resulting in serious respiratory distress. Ascending urinary tract infections with *Escherichia coli* are common in women and are particularly troublesome in patients with urinary tract obstructions. A variety of routine medical procedures, such as tracheal intubation and catheterization of blood vessels and the urethra, increase the risk of bacterial infection. The plastic devices used in these procedures are readily colonized by bacteria from the skin, which migrate along the outside of the tube to infect deeper tissues or enter the bloodstream. Because of this problem, it is standard practice to change catheters frequently (e.g., every 72 hours for peripheral intravenous catheters).

Many drugs have been developed to treat bacterial infections. Antimicrobial agents are most effective, however, when the infection is also being fought by healthy phagocytic and immune defenses. Some reasons for this situation are the poor diffusion of antibiotics into certain sites (such as the prostate gland), the ability of many bacteria to multiply or survive inside cells (where many antimicrobial agents have little or no effect), the bacteriostatic rather than bactericidal action of some drugs, and the capacity of some organisms to develop resistance to multiple antibiotics.

Many bacterial pathogens are transmitted to the host by a vector, usually an arthropod. For example, Rocky Mountain spotted fever and Lyme disease are both vectored by ticks, and bubonic plague is spread by fleas. Susceptibility to these diseases depends partly on the host's contact with the vector.

Pathogenic Mechanisms

Bacterial Infectivity

Factors that are produced by a microorganism and evoke disease are called virulence factors. Examples are toxins, surface coats that inhibit phagocytosis, and surface receptors that bind to host cells. Mostfrank (as opposed to opportunistic) bacterial pathogens have evolved specific virulence factors that allow them to multiply in their host or vector without being killed or expelled by the host's defenses. Many virulence factors are produced only by specific virulent strains of a microorganism. For example, only certain strains of *E coli* secrete diarrhea-causing enterotoxins.

Virulence factors should never be considered independently of the host's defenses; the clinical course of a disease often depends on the interaction of virulence factors with the host's response. An infection begins when the balance between bacterial pathogenicity and host resistance is upset. In essence, we live in an environment that favors the microbe, simply because the growth rate of bacteria far exceeds that of most eukaryotic cells. Furthermore, bacteria are much more versatile than eukaryotic cells in substrate utilization and biosynthesis. The high mutation rate of bacteria combined with their short generation time results in rapid selection of the best-adapted strains and species. In general, bacteria are much more resistant to toxic components in the environment than eukaryotes, particularly when the major barriers of eukaryotes (skin and mucous membranes) are breached.

From a practical standpoint, bacteria can be said to have a single objective to multiply. Only a few of the vast number of bacterial species in the environment consistently cause disease in a given host. From a teleologic standpoint, it is not in the best interest of the pathogen to kill the host, because in most cases the death of the host means the death of the pathogen. The most highly evolved or adapted pathogens are the ones that acquire the necessary nutritional substances for growth and dissemination with the smallest expenditure of energy and least damage to the host. For example, *Rickettsia akari*, the etiologic agent of rickettsialpox, causes a mild, self-limited infection consisting of headache, fever, and a papulovesicular rash. Other members of the rickettsial group, such as *R rickettsii*, the agent of Rocky Mountain spotted fever, elicit more severe, life-threatening infections. Some bacteria that are poorly adapted to the host synthesize virulence factors (e.g., tetanus and diphtheria toxin) so potent that they threaten the life of the host.

Host Resistance

Although easily damaged, the skin represents one of the most important barriers of the body to the microbial world, which contains a diverse array of bacteria in enormous numbers. Fortunately, most bacteria in the environment are relatively benign to individuals with normal immune systems. However, patients who are immunosuppressed, such as individuals receiving cancer chemotherapy or have AIDS, opportunistic microbial pathogens can establish life-threatening infections. Normally, microbes in the environment are prevented from entering the body by the skin and mucous membranes. The outermost surface of the skin consists of squamous cell epithelium, largely comprised of dead cells that are sloughed off as new cells are formed below them. In addition to the skin barrier, mucous membranes of the respiratory, gastrointestinal, and urogenital systems represent other portals through which bacteria can gain access to the body. Like the squamous epithelial cells of the skin, the mucosal epithelial cells divide rapidly, and as the cells mature, they are pushed laterally toward theintestinal lumen and shed. The entire process is reported to require only 36-48 hours for complete replacement of the epithelium, which diminishes the number of bacteria associated with the epithelium. The skin surface is a dry, acidic environment, and the temperature is less than 37° C. The pores and crevices of the skin also are colonized by the "normal bacterial flora", which ensure competition for pathogens to which the skin is exposed. Similarly, the mucous layer that covers the epithelia contains hostile substances to microbial colonization. Protective levels of lysozyme, lactoferrin, and lactoperoxidase in the mucus either kill bacteria or restrict their growth. In addition, the mucus contains secretory immunoglobulins (predominantly sIgA) synthesized by plasma cells resident in the submucosal tissue. During the normal course of life, individuals develop local antibodies specific for a variety of intestinal bacteria that colonize mucosal surfaces.

Another mechanism of restricting growth of bacteria that penetrate the skin and mucous membranes is competition for iron. Typically, the amount of free iron in tissues and blood available to bacteria is very low, since plasma transferrin binds virtually all iron in the blood. Similarly, hemoglobin in the erythrocytes binds iron. Without free iron, bacterial growth is restricted unless the bacteria synthesize siderophores or receptors for iron containing molecules that compete for transferrin-bound iron. Such siderophores strip iron from transferrin and present it to the bacteria, which enables them to grow. The phagocytic cells of the body patrol the blood and tissues for foreign substances, including bacteria. This task is assumed predominantly by polymorphonuclear neutrophils; however, monocytes, macrophages, and eosinophils also participate. After phagocytosis, these bacterial cells usually are killed unless their numbers are excessive or they possess virulence factors, that enable them to survive the lysosomal enzymes and acidic pH. In some instances, the bacteria kill the phagocyte or multiply within the macrophage, escaping the hostile extracellular environment. When inflammation occurs, phagocytic cells, along with lymphocytes, play an important role in innate immunity to bacterial infections. During the interaction of bacterial cells with macrophages, T cells, and B cells, specific antibody responses and/or cell-mediated immunity develop to protect against reinfection.

Genetic and Molecular Basis for Virulence

Virulence factors in bacteria may be encoded on chromosomal DNA, bacteriophage DNA, plasmids, or transposons in either plasmids or the bacterial chromosome (Fig. 7-2; Table 7-2). For example, the capacity of the Shigella species to invade cells is a property encoded in part on a 140-mega-dalton plasmid. Similarly, the heat-labile enterotoxin (LTI) of *E coli* is plasmid encoded, whereas the heat-labile toxin (LTII) is encoded on the chromosome. Other virulence factors are acquired by bacteria following infection by a particular bacteriophage, which integrates its genome into the bacterial chromosome by the process of lysogeny (Fig. 7-2). Temperate bacteriophages often serve as the basis of toxin production in pathogenic bacteria. Examples include diphtheria toxin production by Corynebacterium diphtheriae, erythrogenic toxin formation by Streptococcus pyogenes, Shiga-like toxin synthesis by *E coli*, and production of botulinum toxin (types C and D) by *Clostridium botulinum*. Other virulence factors are encoded on the bacterial chromosome (e.g., cholera toxin, *Salmonella enterotoxin*, and *Yersinia* invasion factors).

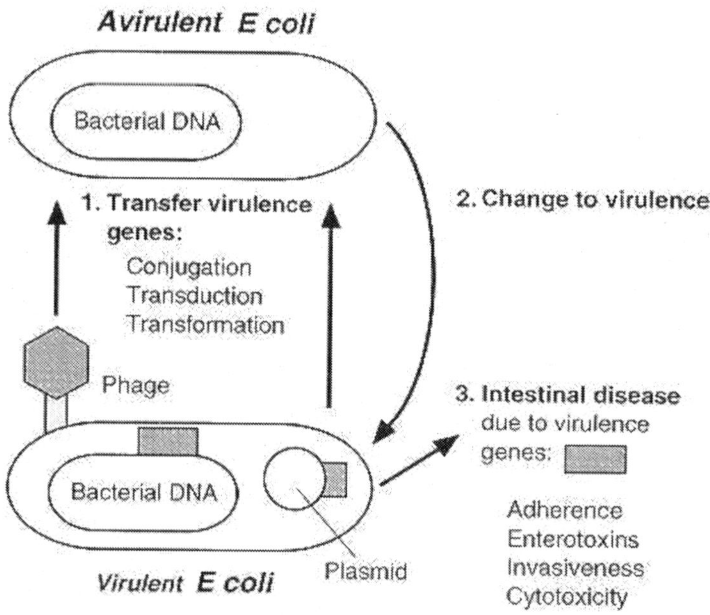

FIGURE 7-2 Mechanisms of acquiring bacterial virulence genes.

TABLE 7-2 Genetic Basis for Virulence of Selected Bacterial Pathogens

Gene(a) Encoded on	Bacterial Pathogen	Virulence Factor
Chromosome	Vibrid cholerae	Enterotoxin
	Salmonella lyphinurium	Enterotoxin, invasion factors
	Shigella spp	Enterotoxin, invasion factors
	Aeromonas hydrophille	Enterotoxin, aerolysin
	Pseuodmonas aenuginose	Exotoxin, A
	Staphylococcus aureus	Enterotoxin, B
	Yersinia enterocofica	Invasion factors
	Yersinia pesudotuberculosis	Invasion factors
	Escherchia coli	Enterotoxin (LTH)
Plasmid	Shiglla spp	Invasion factors, colonization factor and enterotoxin
	Escherichia coli	Invasion factors, colonization factor and enterotoxin (LTI)
	Stapfrylococcus aureus	Exfoliative toxin
	Baciflus anthracis	Anthrax toxin
Bacteriophage	Corynetracferlum diphthorae	Diphtheria toxin
	Sterptococcus pyogenes	Erythrogenic toxin
	Escherichia coli	Shiga-like entrotoxin
	Clostridium botulioum	Botuflrum toxin (C, D)
Transposons*	Escherichia coli	Enterotoxin(STA and STB), iron acquisition hemolysin

* Transposable genetic elements located on plasmids that often insert into the chromosome

The transfer of genes for antibiotic resistance among bacteria is a significant medical problem, although none of these properties actually confers increased virulence to the bacterium. Rather, they provide the opportunity for resistant bacteria to proliferate and produce other virulence factors in patients who are being treated with an inappropriate antibiotic. Resistance factors are discussed fully in Chapter 5.

An intriguing question regarding most bacterial protein toxins is the purpose they serve for the bacteriophage or the bacterium carrying them. Several bacterial toxins are enzymes. For example, cholera toxin, diphtheria toxin, Pseudomonas exotoxin A, and pertussis toxin all are NAD+ glycohydrolases that also act as ADP-ribosyltransferases. The toxic effect of these bacterial enzymes on the host is integral to the pathogenesis of the bacterial infections, but the function of the enzymes in the normal bacterial physiology is not known. Of all the protein toxins synthesized by pathogenic bacteria, there are few instances in which the function of the protein to the bacterium is known. It would be unlikely for the bacterium or infecting bacteriophage to expend the energy necessary to synthesize these relatively high-molecular-weight and complex molecules if they offered it no advantage. Frequently the toxicity of these substances is "unintentional" as far as the bacteria are concerned, considering that the primary goal of the microorganisms is to acquire nutrients and multiply rather than to harm the host.

Host-Mediated Pathogenesis

The pathogenesis of many bacterial infections cannot be separated from the host immune response, for much of the tissue damage is caused by the host response rather than by bacterial factors. Classic examples of host response-mediated pathogenesis are seen in diseases such as Gram-negative bacterial sepsis, tuberculosis, and tuberculoid leprosy. The tissue damage in these infections is caused by toxic factors released from the lymphocytes, macrophages, and polymorphonuclear neutrophils infiltrating the site of infection (Fig. 7-3). Often the host response is so intense that host tissues are destroyed, allowing resistant bacteria to proliferate. In lepromatous leprosy, in contrast, the absence of a cellular response to Mycobacterium leprae allows the bacteria to multiply to such large numbers in the skin that they become tightly packed and replace healthy tissue. The molecular basis for this specific immune energy is poorly understood.

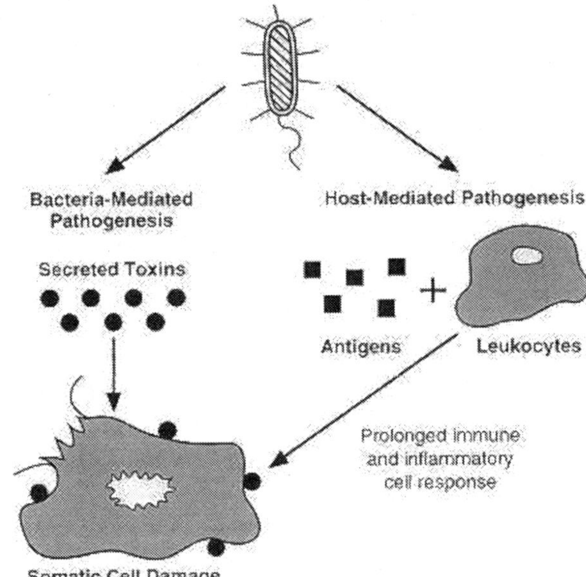

FIGURE 7-3 Generalized mechanisms of bacterial pathogenesis: bacteria-induced toxicity or host-mediated damage.

Intracellular Growth

In general, bacteria that can enter and survive within eukaryotic cells are shielded from humoral antibodies and can be eliminated only by a cellular immune response. However, these bacteria must possess specialized mechanisms to protect them from the harsh effects of the lysosomal enzymes encountered within the cell (see Ch. 1). Pathogenic bacteria can be grouped into three categories on the basis of their invasive properties for eukaryotic cells (Fig. 7-4; Table 7-3). Although, some bacteria (e.g., *Rickettsia, Coxiella,* and *Chlamydia*) grow only inside host cells, others (e.g.,

Salmonella, Shigella, and *Yersinia*) are facultative intracellular pathogens, invading cells when it gives them a selective advantage in the host.

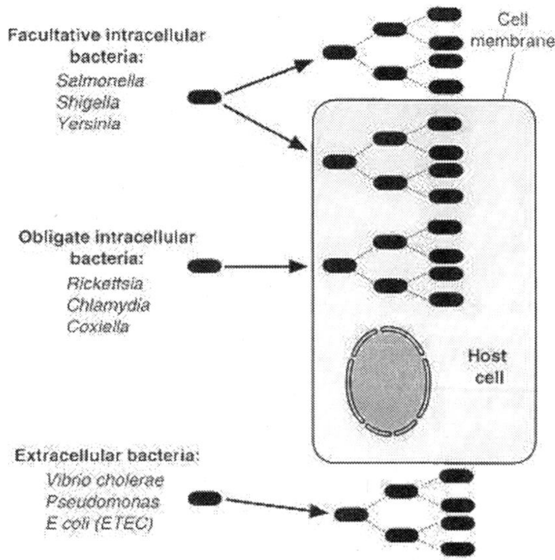

FIGURE 7-4 Examples of pathogenic bacteria, indicating their preferred growth phase within the host. (ETEC:enterotoxigenic *E coli*)

TABLE 7-3 Intracellular of Extracellular Growth Preference Relatinve to Eukaryotic Cells

Category	Bacterial Pathogen
Obligate intracellular	Richttsia spp
	Coxiella burhetll
	Chlarnyolia spp
Faculrative Intracellular	Salmonella spp
	Shigella spp
	Legoinella pneumophils
	Invasive Escharichia coli
	Neisseria spp
	Mycobacterium spp
	Listeria monocytogentes
	Bordetelia perfusis
Predominantly extracellular	Mycoplasme spp
	Pseudomonas aeruginosa
	Enterotoxigenic Escherichia coli
	Vibrio cholerae
	staphylococcus aursus
	Streptococcus pyogenes
	haemophillus influenzae
	Bacillus arthracis

Some bacteria survive the intracellular milieu by producing phospholipases to dissolve the phagocytic vesicle surrounding them. This appears to be the case for R rickettsii, which destroys the phagosomal membrane with which the lysosomes fuse. Legionella pneumophila, which prefers the intracellular environment of macrophages for growth, appears to induce its own uptake and blocks lysosomal fusion by undefined mechanisms. Other bacteria have evolved to the point that they prefer the low-pH environment within the lysosomal granules, as may be the case for *Coxiella burnetii,* a highly resistant member of the rickettsial group. Salmonella and Mycobacterium species also appear to be very resistant to intracellular killing by phagocytic cells, but their mechanisms of resistance are not yet fully understood. Certainly, the capacity of bacteria to survive and multiply within host cells has great impact on the pathogenesis of the respective infections.

Most bacterial pathogens do not invade cells, proliferating instead in the extracellular environment enriched by body fluids. Some of these bacteria (e.g., V cholerae and Bordetella pertussis) do not even penetrate body tissues, but, rather, adhere to epithelial surfaces and cause disease by secreting potent protein toxins. Although, bacteria such as *E coli* and *P aeruginosa* are termed noninvasive, they frequently spread rapidly to various tissues once they gain access to the body. All bacteria could at some point be considered intracellular once they become ingested by polymorphonuclear neutrophilsand macrophages, but these organisms are not renowned for their capacity to survive the intracellular environment or to induce their own uptake by most host cells.

Specific Virulence Factors

The virulence factors of bacteria can be divided into a number of functional types. These are discussed in the following sections:

Adherence and Colonization Factors

To cause infection, many bacteria must first adhere to a mucosal surface. For example, the alimentary tract mucosa is continually cleansed by the release of mucus from goblet cells and by the peristaltic flow of the gut contents over the epithelium. Similarly, ciliated cells in the respiratory tract sweep mucus and bacteria upward. In addition, the turnover of epithelial cells at these surfaces is fairly rapid. The intestinal epithelial cell monolayer is continually replenished, and the cells are pushed from the crypts to the villar tips in about 48 hours. To establish an infection at such a site, a bacterium must adhere to the epithelium and multiply before the mucus and extruded epithelial cells are swept away. To accomplish this, bacteria have evolved attachment mechanisms, such as pili (fimbriae), that recognize and attach the bacteria to cells (see Ch. 2). Colonization factors (as they are often called) are produced by numerous bacterial pathogens and constitute an important part of the pathogenic mechanism of these bacteria. Some examples of piliated, adherent bacterial pathogens are *V cholerae, E coli, Salmonella* spp, *N gonorrheae, N meningitidis,* and *Streptococcus pyogenes.*

Invasion Factors

Mechanisms that enable a bacterium to invade eukaryotic cells facilitate entry at mucosal surfaces. Some of these invasive bacteria (such as Rickettsia and Chlamydia species) are obligate intracellular pathogens, but most are facultative intracellular pathogens (Fig. 7-4). The specific bacterial surface factors that mediate invasion are not known in most instances, and often, multiple gene products are involved. Some *Shigella* invasion factors are encoded on a 140 megadalton plasmid, which, when conjugated into *E coli,* gives these noninvasive bacteria the capacity to invade cells. Other invasion genes have also recently been identified in *Salmonella* and *Yersinia* pseudotuberculosis. The mechanisms of invasion of *Rickettsia,* and *Chlamydia* species are not well known.

Capsules and Other Surface Components

Bacteria have evolved numerous structural and metabolic virulence factors that enhance their survival rate in the host. Capsule formation has long been recognized as a protective mechanism for bacteria (see Ch. 2). Encapsulated strains of many bacteria (e.g., pneumococci) are more virulent and more resistant to phagocytosis and intracellular killing than are nonencapsulated strains. Organisms that cause bacteremia (e.g., *Pseudomonas*) are less sensitive than many other bacteria to killing by fresh human serum containing complement components, and consequently are called serum resistant. Serum resistance may be related to the amount and composition of capsular antigens as well as to the structure of the lipopolysaccharide. The relationship between surface structure and virulence is important also in *Borrelia* infections. As the bacteria encounter an increasing specific immune response from the host, the bacterial surface antigens are altered by mutation, and the progeny, which are no longer recognized by the immune response, express renewed virulence. *Salmonella typhi* and some of the paratyphoid organisms carry a surface antigen, the *Vi antigen,* thought to enhance virulence. This antigen is composed of a polymer of galactosamine and uronic acid in 1,4-linkage. Its role in virulence has not been defined, but antibody to it is protective.

Some bacteria and parasites have the ability to survive and multiply inside phagocytic cells. A classic example is Mycobacterium tuberculosis,whose survival seems to depend on the structure and composition of its cell surface. The parasite *Toxoplasma gondii* has the remarkable ability to block the fusion of lysosomes with the phagocytic vacuole. The hydrolytic enzymes contained in the lysosomes are unable, therefore, to contribute to the destruction of the parasite. The mechanism(s) by which bacteria such as *Legionella pneumophila, Brucella abortus,* and *Listeria* monocytogenes remain unharmed inside phagocytes are not understood.

Endotoxins

Endotoxin is comprised of toxic lipopolysaccharide components of the outer membrane of Gram-negative bacteria (see Ch. 2). Endotoxin exerts profound biologic effects on the host and may be lethal. Because it is omnipresent in the environment, endotoxin must be removed from all medical supplies destined for injection or use during surgical procedures. The term endotoxin was coined in 1893 by Pfeiffer to distinguish the class of toxic substances released after lysis of bacteria from the toxic substances (exotoxins) secreted by bacteria. Few, if any, other microbial products have been as extensively studied as bacterial endotoxins. Perhaps it is appropriate that a molecule with such important biologic effects on the host, and one produced by so many bacterial pathogens, should be the subject of intense investigation.

Structure of Endotoxin

Figure 7-5 illustrates the basic structure of endotoxin. Endotoxin is a molecular complex of lipid and polysaccharide; hence, the alternate name lipopolysaccharide. The complex is secured to the outer membrane by ionic and hydrophobic forces, and its strong negative charge is neutralized by Ca^{2+} and Mg^{2+} ions.

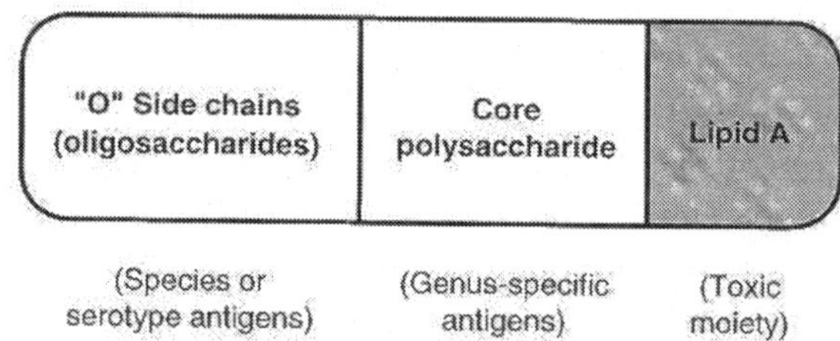

FIGURE 7-5 Basic structure of endotoxin (lipopolysaccharide) from Gram-negative bacteria.

The structure of endotoxin molecules from *Salmonella* spp and *E coli* is known in detail. Enough data on endotoxin from other Gram-negative organisms have been gathered to reveal a common pattern with genus and species diversity. Although, all endotoxin molecules are similar in chemical structure and biologic activity, some diversity has evolved. Purified endotoxin appears as large aggregates. The molecular complex can be divided into three regions (Fig. 7-5): (1) the O-specific chains, which consist of a variety of repeating oligosaccharide residues, (2) the core polysaccharide that forms the backbone of the macromolecule, and (3) lipid A, composed usually of a glucosamine disaccharide with attached long-chain fatty acids and phosphate. The polysaccharide portions are responsible for antigenic diversity, whereas the lipid A moiety confers toxicity. Dissociation of the complex has revealed that the polysaccharide is important in solubilizing the toxic lipid A component, and in the laboratory it can be replaced by carrier proteins (e.g., bovine serum albumin).

Members of the family Enterobacteriaceae exhibit O-specific chains of various lengths, whereas *N gonorrhoeae, N meningitidis,* and *B pertussis* contain only core polysaccharide and lipid A. Some investigators working on the latter forms of endotoxin prefer to call them lipooligosaccharides to emphasize the chemical difference from the endotoxin of the enteric bacilli. Nevertheless, the biologic activities of all endotoxin preparations are essentially the same, with some being more potent than others.

Biologic Activity of Endotoxin

The biologic effects of endotoxin have been extensively studied. Purified lipid A (conjugated to bovine serum albumin) and endotoxin elicit the same biologic responses. Table 7-4 lists some of the biologic effects of endotoxin. The more pertinent toxic effects include pyrogenicity, leukopenia followed by leukocytosis, complement activation, depression in blood pressure, mitogenicity, induction of prostaglandin synthesis, and hypothermia. These events can culminate in sepsis and lethal shock. However, it should be noted from Table 7-4 that not all effects of endotoxin are necessarily detrimental; several induce responses potentially beneficial to the host, assuming the stimulation is not excessive. These include:

1. mitogenic effects on B lymphocytes that increase resistance to viral and bacterial infections
2. induction of gamma interferon production by T lymphocytes, which may enhance the antiviral state, promote rejection of tumor cells, and activate macrophages and natural killer cells
3. activation of the complement cascade with the formation of C3a and C5a
4. induction of the formation of interleukin-1 by macrophages and interleukin-2 and other mediators by T lymphocytes.

TABLE 7-4 Multiple Biologic activities Exhibited by the Lipid a Component of Endotoxin

Pyrogenicity
leukopenla, leukocytosis
Complement activtioDepression of blood pressure
Hageman factor activation
platelet activtion
Induction of plasminogen activator
Bone marrow necrosis
Hypothermia in mice
Lethal toxicity in mice
shwartzman reaction
Induction of prostaglandin synthesis
limulus lysate gelation
*Induction of nonspecific resistance to infection
*Induction of endotoxin tolerance
*Aduvant activity
*Miloganic activity for lymohocytes
*Macrophage activation
*Induction of interferon synthesis
*Induction of tumor necrosis factor synthesis

*Potentially beneficial stimulatory effects of endotoxin in low doses.

Current research focuses on exploiting some of the potential beneficial effects of "nontoxic" endotoxin derivatives and holds promise for development of future treatment regimens for stimulating the immune response. For example, the toxicity of endotoxin is largely attributed to lipid A, attached to a polysaccharide carrier. The toxicity of lipid A is markedly reduced after hydrolysis of a phosphate group or deacylation of one or more fatty acids from the lipid A molecule. Clinical trials are in progress to test a monophosphoryl lipid A for its potential of inducing low dose tolerance to endotoxin. Tolerance to endotoxin can be achieved by pretreatment of an animal with low doses of endotoxin or a detoxified lipid A derivative before challenge with high doses of endotoxin. Experimental studies have demonstrated that induction of tolerance to endotoxin reduces the dangerous effects of endotoxin. It is hoped that these relatively nontoxic lipid A derivatives may be useful in reducing the severity of bacterial sepsis in which bacterial endotoxin produces a life-threatening clinical course.

Endotoxin, which largely accumulates in the liver following injection of a sublethal dose by the intravenous route, can be devastating because of its ability to affect a variety of cell and host proteins. Kupffer cells, granulocytes, macrophages, platelets, and lymphocytes all have a cell receptor on their surface called CD14, which binds endotoxin. Endotoxin binding to the CD14 receptor on macrophages is enhanced by interaction with a host protein madein the liver (i.e., LPS-binding protein). The extent of involvement of each cell type probably depends on the level of endotoxin exposure. The effects of endotoxin on such a wide variety of host cells result in a complex array of host responses that can culminate in the serious condition gram-negative sepsis, which often leads to shock and death. The effects of endotoxin on host cells are known to stimulate prostaglandin synthesis and to activate the kallikrein system, the kinin system, the complement cascade via the alternative pathway, the clotting system, and the fibrinolytic pathways. When these normal host systems are activated and operate out of control, it is not surprising that endotoxin can be lethal. Although, it is difficult to comprehend the mechanisms of all the cell responses and the myriad sequelae of the cell mediators released rather indiscriminately in the host following exposure to endotoxin, it does seem clear that the host cellular response to endotoxin, rather than a direct toxic effect of endotoxin, plays the major role in causing tissue damage (Fig. 7-3).

Detection of Endotoxin in Medical Solutions

Endotoxin is omnipresent in the environment. It is found in most deionized-water lines in hospitals and laboratories, for example, and affects virtually every biologic assay system ever examined. It tends to be a scapegoat for all biologic problems encountered in the laboratory, and, many times, this reputation is deserved. Because of its pyrogenic and destructive properties, extreme care must be taken to avoid exposing patients to medical solutions containing endotoxin. Even though all supplies should be sterile, solutions for intravenous administration can become contaminated with endotoxin-containing bacteria after sterilization as a result of improper handling. Furthermore, water used in the preparation of such solutions must be filtered through ion exchange resins to remove endotoxin, because it is not removed by either autoclave sterilization or filtration through bacterial membrane filters. If endotoxin-containing solutions were used in such medical procedures as renal dialysis, heart bypass machines, blood transfusions, or surgical lavage, the patient would suffer immediate fever accompanied by a rapid and possibly lethal alterations in blood pressure.

Solutions for human or veterinary use are prepared under carefully controlled conditions to ensure sterility and to remove endotoxin. Representative samples of every manufacturing batch are checked for endotoxin by one of two procedures: the Limulus lysate test or the rabbit pyrogenicity test. The rabbit pyrogenicity test is based on the exquisite sensitivity of rabbits to the pyrogenic effects of endotoxin. A sample of the solution to be tested usually is injected intravenously into the ear veins of adult rabbits while the rectal temperature of the animal is monitored. Careful monitoring of the temperature responses provides a sensitive and reliable indicator of the presence of endotoxin and, importantly, one measure of the safety of the solution for use in patients.

The Limulus lysate test is more common and less expensive. This test, which is based on the ability of endotoxin to induce gelation of lysates of amebocyte cells from the horseshoe crab Limulus polyphemus,is simple, fast, and sensitive (about 1 ng/ml). It is so sensitive, however, that trace quantities of endotoxin in regular deionized water often obscure the results. It can be used for rapid detection of certain Gram-negative infections (e.g., of cerebrospinal fluid); however, blood contains inhibitors that prevent gelation. Test kits are commercially available. The amebocyte is the sole phagocytic immune cell of the horseshoe crab, and the gelation reaction is believed to be involved in sequestering invading Gram-negative bacteria.

Exotoxins

Exotoxins, unlike the lipopolysaccharide endotoxin, are protein toxins released from viable bacteria. They form a class of poisons that is among the most potent, per unit weight, of all toxic substances. Most of the higher molecular-sized exotoxin proteins are heat labile; however, numerous low molecular-sized exotoxins are heat-stable peptides. Unlike endotoxin, which is a structural component of all Gram-negative cells, exotoxins are produced by some members of both Gram-positive and Gram-negative genera. The functions of these exotoxins for the bacteria are usually unknown, and the genes for most can be deleted with no noticeable effect on bacterial growth. In contrast to the extensive systemic and immune-system effects of endotoxin on the host, the site of action of most exotoxins is more localized and is confined to particular cell types or cell receptors. Tetanus toxin, for example, affects only internuncial neurons. In general, exotoxins are excellent antigens that elicit specific antibodies called antitoxins. Not all antibodies to exotoxins are protective, but some react with important binding sites or enzymatic sites on the exotoxin, resulting in complete inhibition of the toxic activity (i.e., neutralization).

Exotoxins can be grouped into several categories (e.g., neurotoxins, cytotoxins, and enterotoxins) based on their biologic effect on host cells. Neurotoxins are best exemplified by the toxins produced by *Clostridium* spp, for example, the botulinum toxin formed by *C botulinum*. This potent neurotoxin acts on motor neurons by preventing the release of acetylcholine at the myoneural junctions, thereby preventing muscle excitation and producing flaccid paralysis. The cytotoxins constitute a larger, more heterogeneous grouping with a wide array of host cell specificities and toxic manifestations. One cytotoxin is diphtheria toxin, which is produced by *Corynebacterium diphtheriae*. This cytotoxin inhibits protein synthesis in many cell types by catalyzing the ADP-ribosylation of elongation factor II, which blocks elongation of the growing peptide chain.

Enterotoxins stimulate hypersecretion of water and electrolytes from the intestinal epithelium and thus produce watery diarrhea. Some enterotoxins are cytotoxic (e.g., shiga-like enterotoxin from *E coli*), while others perturb eukaryotic cell functions and are cytotonic (e.g., cholera toxin). Enterotoxins also can disturb normal smooth muscle contraction, causing abdominal cramping and decrease transit time for water absorption in the intestine. Enterotoxigenic *E coli* and *V cholerae* produce diarrhea after attaching to the intestinal mucosa, where they elaborate enterotoxins. Neither pathogen invades the body in substantial numbers, except in the case of *E coli* species that have acquired an invasion plasmid. Importantly, cholera toxin and *E coli* heat-labile enterotoxins I and II cause ADP-ribosylation of cell proteins in a manner similar to diphtheria toxin, except that the primary target is the regulatory protein (G_{s-}) of adenylate cyclase, resulting in increased levels of cyclic 3',5'-adenosine monophosphate (cAMP) (see Ch. 25). In contrast, the organisms responsible for shigellosis (*Shigella dysenteriae, S boydii, S flexneri,* and *S sonnei*) penetrate the mucosal surface of the colon and terminal ileum to proliferate and cause ulcerations that bleed into the intestinal lumen. Despite causing extensive ulceration of the mucosa, the pathogens rarely enter the bloodstream. The Shiga enterotoxin produced by *Shigella* species and the Shiga-like enterotoxin elaborated by many isolates of *E coli* inhibit protein synthesis in eukaryotic cells. It is not clear how this cytotoxic enterotoxin causes hypersecretion of water and electrolytes from the intestinal epithelium. These enterotoxins differ from those secreted by *V cholerae* and *E coli* in that the Shiga toxins are cytotoxic and lethal, whereas the cholera toxin-like enterotoxins are not. The latter enterotoxins cause no structural damage to cells, and are described as cytotonic. The ensuing inflammatory response to the invading bacteria and/or their toxins appears to activate neurologic control mechanisms (e.g., prostaglandins, serotonin) that normally regulate water and electrolyte transport.

Siderophores

Both animals and bacteria require iron for metabolism and growth, and the control of this limited resource is often used as a tactic in the conflict between pathogen and host. Animals have evolved mechanisms of "withholding" iron from tissue fluids in an attempt to limit the growth of invading bacteria. Although, blood is a rich source of iron, this iron is not readily available to bacteria since it is not free in solution. Most of the iron in blood is bound either to hemoglobin in erythrocytes or to transferrin in plasma. Similarly, the iron in milk and other secretions (e.g., tears, saliva, bronchial mucus, bile, and gastrointestinal fluid) is bound to lactoferrin. Some bacteria express receptors for eukoyotic iron-binding proteins (e.g., transferrin-binding outer membrane proteins on the surface of *Neisserira* spp). Via these specialized receptors iron acquisition is facilitated, providing the esssential element for bacterial growth.

Other bacteria have evolved elaborate mechanisms to extract the iron from host proteins (Fig. 7-6). Siderophores are substances produced by many bacteria (and some plants) to capture iron from the host. The absence of iron triggers transcription of the genes coding for the enzymes that synthesize siderophores, as well as for a set of surface protein receptors that recognize siderophores carrying bound iron. The binding constants of the siderophores for iron are so high that even iron bound to transferrin and lactoferrin is confiscated and taken up by the bacterial cells. An example of a bacterial siderophore is enterochelin, which is produced by *Escherichia* and *Salmonella* species. Classic experiments have demonstrated that Salmonella mutants that have lost the capacity to synthesize enterochelin lose virulence in an assay of lethality in mice. Injection of purified enterochelin along with the Salmonella mutants restores virulence to the bacteria. Therefore, siderophore production by many pathogenic bacteria is considered an important virulence mechanism.

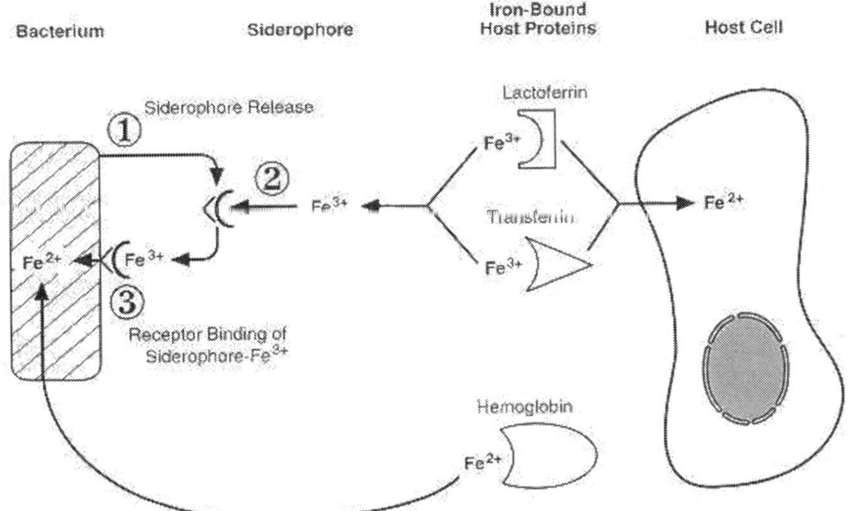

FIGURE 7-6 Competition between host cells and bacterial pathogens for iron, illustrating the importance of siderophores. Since free iron is scarce in tissue fluids and blood, bacterial siderophores compete effectively for Fe^{3+} bound to lactoferrin and transferrin.

Epilogue

Many factors determine the outcome of the bacterium-host relationship. The host must live in an environment filled with a diverse population of microorganisms. Because of the magnitude of the infectious-disease problem, we strive to understand the natural immune mechanisms of the host so that future improvements in resistance to bacterial infections may be possible. Similarly, massive research efforts are being expended to identify and characterize the virulence factors of pathogenic bacteria and hence allow us to interrupt the pathogenic mechanisms of virulent bacteria. The availability of an array of antibiotics and vaccines has provided the medical profession with powerful tools to control or cure many infections. Unfortunately, these drugs and vaccines have eliminated no bacterial disease from the human or animal populations, and bacterial infections and drug resistance remain a serious medical problem.

REFERENCES

Astiz ME, Rackow EC, Still JG, et al: Pretreatment of normal humans with monophosphoryl lipid A induces tolerance to endotoxin: A prospective, double-blind, randomized, controlled trial. Crit Care Med 23:9, 1995

Berry LJ: Bacterial toxins. Crit Rev Toxicol 5: 239, 1977

Eisenstein TK, Actor P, Friedman H: Host Defenses to Intracellular Pathogens. Plenum Publishing Co, New York, 1983

Finlay BB, Falkow S: Common themes in microbial pathogenicity. Microbiol Rev 53:210, 1989

Foster TJ: Plasmid-determined resistance to antimicrobial drugs and toxic metal ions in bacteria. Microbiol Rev 47:361, 1983

Hardegree MC, Tu AT (eds): Handbook of Natural Toxins. Vol.4: Bacterial Toxins. Marcel Dekker, New York, 1988

Iglewski BH, Clark VL (eds): Molecular Basis of Bacterial Pathogenesis. Vol. XI of The Bacteria: A Treatise on Structure and Function. Academic Press, Orlando, FL, 1990

Luderitz O, Galanos C: Endotoxins of gram-negative bacteria. p.307. In Dorner F, Drews J (eds): Pharmacology of Bacterial Toxins. International Encyclopedia of Pharmacology and Therapeutics, Section 119. Pergamon, Elmsford, NY, 1986

Mims CA: The Pathogenesis of Infectious Disease. Academic Press, London, 1976

Payne SM: Iron and virulence in the family Enterobacteriaceae. Crit Rev Microbiol 16:81, 1988

Sack RB: Human diarrheal disease caused by enterotoxigenic *Escherichia coli*. Annu Rev Microbiol 29:333, 1975

Salyers, AA, Whitt DD: Bacterial Pathogenesis - A Molecular Approach ASM Press, 1994

Smith H: Microbial surfaces in relation to pathogenicity. Bacteriol Rev 41:475, 1977

Smith H, Turner JJ (eds): The Molecular Basis of Pathogenicity. Verlag Chemie, Deerfield Beach, FL, 1980

Weinberg ED: Iron withholding: a defense against infection and neoplasia. Physiol Rev 64:65, 1984

Chapter 9

Epidemiology

General Concepts

Definitions

Epidemiology is the study of the determinants, occurrence, and distribution of health and disease in a defined population. Infection is the replication of organisms in host tissue, which may cause disease. A carrier is an individual with no overt disease who harbors infectious organisms. Dissemination is the spread of the organism in the environment.

Chain of Infection

There are three major links in disease occurrence: the etiologic agent, the method of transmission (by contact, by a common vehicle, or via air or a vector), and the host.

Epidemiologic Methods

Epidemiologic studies may be (1) descriptive, organizing data by time, place, and person; (2) analytic, incorporating a case-control or cohort study; or (3) experimental. Epidemiology utilizes an organized approach to problem solving by: (1) confirming the existence of an epidemic and verifying the diagnosis; (2) developing a case definition and collating data on cases; (3) analyzing data by time, place, and person; (4) developing a hypothesis; (5) conducting further studies if necessary; (6) developing and implementing control and prevention measures; (7) preparing and distributing a public report; and (8) evaluating control and preventive measures.

INTRODUCTION

This chapter reviews the general concepts of epidemiology, which is the study of the determinants, occurrence, distribution, and control of health and disease in a defined population. Epidemiology is a descriptive science and includes the determination of rates, that is, the quantification of disease occurrence within a specific population. The most commonly studied rate is the attack rate: the number of cases of the disease divided by the population among whom the cases have occurred. Epidemiology can accurately describe a disease and many factors concerning its occurrence before its cause is identified. For example, Snow described many aspects of the epidemiology of cholera in the late 1840s, fully 30 years before

Koch described the bacillus and Semmelweis described puerperal fever in detail in 1861 and recommended appropriate control and prevention measures a number of years before the streptococcal agent was fully described. One goal of epidemiologic studies is to define the parameters of a disease, including risk factors, in order to develop the most effective measures for control. This chapter includes a discussion of the chain of infection, the three main epidemiologic methods, and how to investigate an epidemic (Table 9-1).

Table 9-1 Epidemiologic Methodes and Investigation

Descriptive	Analytic	Experimental
Time	Case control	Manipulate cause and note effect
Secular	Cohort	
Periodic		
Seasonal		
Epidemic		
Place		
Person		

Proper interpretation of disease-specific epidemiologic data requires information concerning past as well as present occurrence of the disease. An increase in the number of reported cases of a disease that is normal and expected, representing a seasonal pattern of change in host susceptibility, does not constitute an epidemic. Therefore, the regular collection, collation, analysis, and reporting of data concerning the occurrence of a disease is important to properly interpret short-term changes in occurrence.

A sensitive and specific surveillance program is important for the proper interpretation of disease occurrence data. Almost every country has a national disease surveillance program that regularly collects data on selected diseases. The quality of these programs varies, but, generally, useful data are collected that are important in developing control and prevention measures. There is an international agreement that the occurrence of three diseases cholera, plague, and yellow fever will be reported to the World Health Organization in Geneva, Switzerland. In the United States, the Centers for Disease Control and Prevention (CDC), U.S. Public Health Service, and the state health officers of all 50 states have agreed to report the occurrence of 51 diseases weekly and of another 10 diseases annually from the states to the CDC. Many states have regulations or laws that mandate reporting of these diseases and often of other diseases of specific interest to the state health department.

The methods of case reporting vary within each state. Passive reporting is one of the main methods. In such a case, physicians or personnel in clinics or hospitals report occurrences of relevant diseases by telephone, postcard, or a reporting form, usually at weekly intervals. In some instances, the report may be initiated by the public health or clinical laboratory where the etiologic agent is identified. Some diseases, such as human rabies, must be reported by telephone as soon as diagnosed. In an active surveillance program, the health authority regularly initiates the request for reporting. The local health department may call all or some health care providers at regular intervals to inquire about the occurrence of a disease or diseases. The active system may be used during an epidemic or if accurate data concerning all cases of a disease are desired.

The health care provider usually makes the initial passive report to a local authority, such as a city or county health department. This unit collates its data and sends a report to the next highest health department level, usually the state health department.

The number of cases of each reportable disease are presented weekly, via computer linkage, by the state health department to the CDC. Data are analyzed at each level to develop needed information to assist public health authorities in disease control and prevention. For some diseases, such as hepatitis, the CDC requests preparation of a separate case reporting form containing more specific details.

In addition, the CDC prepares and distributes routine reports summarizing and interpreting the analyses and providing information on epidemics and other appropriate public health matters. Most states and some county health departments also prepare and distribute their own surveillance reports. The CDC publishes *Morbidity and Mortality Weekly Report,* which is available for a small fee from the Massachusetts Medical Society. The CDC also prepares more detailed surveillance reports for specific diseases, as well as an annual summary report, all of which can also be obtained through the Massachusetts Medical Society.

Infection is the replication of organisms in the tissue of a host; when defined in terms of infection, disease is overt clinical manifestation. In an inapparent (subclinical) infection, an immune response can occur without overt clinical disease. A carrier (colonized individual) is a person in whom organisms are present and may be multiplying, but who shows no clinical response to their presence. The carrier state may be permanent, with the organism always present; intermittent, with the organism present for various periods; or temporary, with carriage for only a brief period. Dissemination is the movement of an infectious agent from a source directly into the environment; when infection results from dissemination, the source, if an individual, is referred to as a dangerous disseminator.

Infectiousness is the transmission of organisms from a source, or reservoir (see below), to a susceptible individual. A human may be infective during the preclinical, clinical, postclinical, or recovery phase of an illness. The incubation period is the interval in the preclinical period between the time at which the causative agent first infects the host and the onset of clinical symptoms; during this time the agent is replicating. Transmission is most likely during the incubation period for some diseases such as measles; in other diseases such as shigellosis, transmission occurs during the clinical period. The individual may be infective during the convalescent phase, as in diphtheria, or may become an asymptomatic carrier and remain infective for a prolonged period, as do approximately 5% of persons with typhoid fever.

The spectrum of occurrence of disease in a defined population includes sporadic (occasional occurrence); endemic (regular, continuing occurrence); epidemic (significantly increased occurrence); and pandemic (epidemic occurrence in multiple countries).

Chain of Infection

The chain of infection includes the three factors that lead to infection: the etiologic agent, the method of transmission, and the host (Fig. 9-1). These links should be characterized before control and prevention measures are proposed. Environmental factors that may influence disease occurrence must be evaluated.

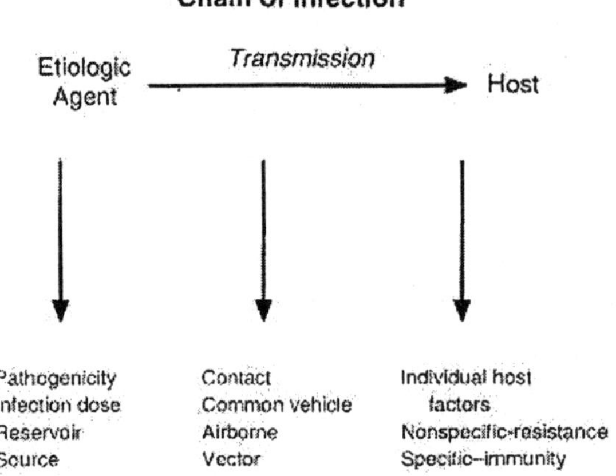

FIGURE 9-1 Summary of important aspects involved in the chain of any infection.

Etiologic Agent

The etiologic agent may be any microorganism that can cause infection. The pathogenicity of an agent is its ability to cause disease; pathogenicity is further characterized by describing the organism's virulence and invasiveness. Virulence refers to the severity of infection, which can be expressed by describing the morbidity (incidence of disease) and mortality (death rate) of the infection. An example of a highly virulent organism is *Yersinia pestis*, the agent of plague, which almost always causes severe disease in the susceptible host.

The invasiveness of an organism refers to its ability to invade tissue. *Vibrio cholerae* organisms are noninvasive, causing symptoms by releasing into the intestinal canal an exotoxin that acts on the tissues. In contrast, *Shigella* organisms in the intestinal canal are invasive and migrate into the tissue.

No microorganism is assuredly avirulent. An organism may have very low virulence, but if the host is highly susceptible, as when therapeutically immunosuppressed, infection with that organism may cause disease. For example, the poliomyelitis virus used in oral polio vaccine is highly attenuated and thus has low virulence, but in some highly susceptible individuals it may cause paralytic disease.

Other factors should be considered in describing the agent. The infecting dose (the number of organisms necessary to cause disease) varies according to the organism, method of transmission, site of entrance of the organism into the host, host defenses, and host species. Another agent factor is specificity; some agents (for example, *Salmonella typhimurium)* can infect a broad range of hosts; others have a narrow range of hosts. *S typhi,* for example, infects only humans. Other agent factors include antigenic composition, which can vary within a species (as in influenza virus or *Streptococcus* species); antibiotic sensitivity; resistance transfer plasmids (see Ch. 5); and enzyme production.

The reservoir of an organism is the site where it resides, metabolizes, and multiplies. The source of the organism is the site from which it is transmitted to a susceptible host, either directly or indirectly through an intermediary object. The reservoir and source can be different; for example, the reservoir for *S typhi* could be the gall bladder of an infected individual, but the source for transmission might be food contaminated by the carrier. The reservoir and source can also be the same, as in an individual who is a permanent nasal carrier of *S aureus* and who disseminates organisms from this site. The distinction can be important when considering where to apply control measures.

Method of Transmission

The method of transmission is the means by which the agent goes from the source to the host. The four major methods of transmission are by contact, by common vehicle, by air or via a vector.

In contact transmission the agent is spread directly, indirectly, or by airborne droplets. Direct contact transmission takes place when organisms are transmitted directly from the source to the susceptible host without involving an intermediate object; this is also referred to as person-to-person transmission. An example is the transmission of hepatitis A virus from one individual to another by hand contact. Indirect transmission occurs when the organisms are transmitted from a source, either animate or inanimate, to a host by means of an inanimate object. An example is transmission of *Pseudomonas* organisms from one individual to another by means of a shaving brush. Droplet spread refers to organisms that travel through the air very short distances, that is, less than 3 feet from a source to a host. Therefore, the organisms are not airborne in the true sense. An example of a disease that may be spread by droplets is measles.

Common-vehicle transmission refers to agents transmitted by a common inanimate vehicle, with multiple cases resulting from such exposure. This category includes diseases in which food or water as well as drugs and parenteral fluids are the vehicles of infection. Examples include food-borne salmonellosis, waterborne shigellosis, and bacteremia resulting from use of intravenous fluids contaminated with a gram-negative organism.

The third method of transmission, airborne transmission, refers to infection spread by droplet nuclei or dust. To be truly airborne, the particles should travel more than 3 feet through the air from the source to the host. Droplet nuclei are the residue from the evaporation of fluid from droplets, are light enough to be transmitted more than 3 feet from the source, and may remain airborne for prolonged periods. Tuberculosis is primarily an airborne disease; the source may be a coughing patient who creates aerosols of droplet nuclei that contain tubercle bacilli. Infectious agents may be contained in dust particles, which may become resuspended and transmitted to hosts. An example occurred in an outbreak of salmonellosis in a newborn nursery in which *Salmonella*-contaminated dust in a vacuum cleaner bag was resuspended when the equipment was used repeatedly, resulting in infections among the newborns.

The fourth method of transmission is vector borne transmission, in which arthropods are the vectors. Vector transmission may be external or internal. External, or mechanical, transmission occurs when organisms are carried mechanically on the vector (for example, *Salmonella* organisms that contaminate the legs of flies). Internal transmission occurs when the organisms are carried within the vector. If the pathogen is not changed by its carriage within the vector, the carriage is called harborage (as when a flea ingests plague bacilli from an infected individual or animal and contaminates a susceptible host when it feeds again; the organism is not changed while in the flea). The other form of internal transmission is called biologic. In this form, the organism is changed biologically during its passage through the vector (for example, malaria parasites in the mosquito vector).

An infectious agent may be transmitted by more than one route. For example, *Salmonella* may be transmitted by a common vehicle (food) or by contact spread (human carrier). *Francisella tularensis* may be transmitted by any of the four routes.

Host

The third link in the chain of infection is the host. The organism may enter the host through the skin, mucous membranes, lungs, gastrointestinal tract, or genitourinary tract, and it may enter fetuses through the placenta. The resulting disease often reflects the point of entrance, but not always: meningococci that enter the host through the mucous membranes may nonetheless cause meningitis. Development of disease in a host reflects agent characteristics (see above) and is influenced by host defense mechanisms, which may be nonspecific or specific.

Nonspecific defense mechanisms include the skin, mucous membranes, secretions, excretions, enzymes, the inflammatory response, genetic factors, hormones, nutrition, behavioral patterns, and the presence of other diseases. Specific defense mechanisms or immunity may be natural, resulting from exposure to the infectious agent, or artificial, resulting from active or passive immunization (see Ch. 8).

The environment can affect any link in the chain of infection. Temperature can assist or inhibit multiplicationof organisms at their reservoir; air velocity can assist the airborne movement of droplet nuclei; low humidity can damage mucous membranes; and ultraviolet radiation can kill the microorganisms. In any investigation of disease, it is important to evaluate the effect of environmental factors. At times, environmental control measures are instituted more on emotional grounds than on the basis of epidemiologic fact. It should be apparent that the occurrence of disease results from the interaction of many factors (Table 9-2). Some of these factors are outlined here.

TABLE 9-2 General Factors That Influence the Occurrence of Infectious Disease

Pathogenic Agent	**Host**
Growth characteristics	Incubation poriod
Stability	Nonspecific defense mechanisms
Ability to form spores	Age
Possession of antibiotic resistance plasmids	Sex
Expression of antigens	Skin
Enzyme production	Secretions
Pathogencity-ability to induce disease	Cough
virulence-influencing disease severity, morbidly and mortally	Ciliary function Peristalsis
Invasoveness	Inflammation
Dose	Nutrition
Reservoir	Genetic factors
Source	Hormones
Mode of dissemination	Personal education Personal hygiene
Host specificity	Behavior patterns Chronic disease
Disease Transmission	
Contact	Specific defense mechanisms (immunity)
Direct	Natural
Indirect	Active-apparent, inapparent
Droplets	Passive-Iransplacental antibody
Common vehicle	Artificial
Food	Active-vaccine, toxoid
Water	Passive-immune serum globulin
Medication	
Solution	Environment
Airborne	Temperature
Drople1 nuclei	Rainfall
Dus1	Humidity
Skin squames	Radiation
Vectorborne-arthropods	Air currents
External Internal-harborage, true biologic transmission	

Epidemiologic Methods

The three major epidemiologic techniques are descriptive, analytic, and experimental. Although, all three can be used in investigating the occurrence of disease, the method used most is descriptive epidemiology. Once the basic epidemiology of a disease has been described, specific analytic methods can be used to study the disease further, and a specific experimental approach can be developed to test a hypothesis.

Descriptive Epidemiology

In descriptive epidemiology, data that describe the occurrence of the disease are collected by various methods from all relevant sources. The data are then collated by time, place, and person. Four time trends are considered in describing the epidemiologic data. The secular trend describes the occurrence of disease over a prolonged period, usually years; it is influenced by the degree of immunity in the population and possibly nonspecific measures such as improved socioeconomic and nutritional levels among the population. For example, the secular trend of tetanus in the United States since 1920 shows a gradual and steady decline.

The second time trend is the periodic trend. A temporary modification in the overall secular trend, the periodic trend may indicate a change in the antigenic characteristics of the disease agent. For example, the change in antigenic structure of the prevalent influenza A virus every 2 to 3 years results in periodic increases in the occurrence of clinical influenza caused by lack of natural immunity among the population. Additionally, a lowering of the overall immunity of a population or a segment thereof (known as herd immunity) can result in an increase in the occurrence of the disease. This can be seen with some immunizable diseases when periodic decreases occur in the level of immunization in a defined population. This may then result in an increase in the number of cases, with a subsequent rise in the overall level of herd immunity. The number of new cases then decreases until the herd's immunity is low enough to allow transmission to occur again and new cases then appear.

The third time trend is the seasonal trend. This trend reflects seasonal changes in disease occurrence following changes in environmental conditions that enhance the ability of the agent to replicate or be transmitted. For example, food-borne disease outbreaks occur more frequently in the summer, when temperatures favor multiplication of bacteria. This trend becomes evident when the occurrence of salmonellosis is examined on a monthly basis (Fig. 9-2).

The fourth time trend is the epidemic occurrence of disease. An epidemic is a sudden increase in occurrence due to prevalent factors that support transmission.

A description of epidemiologic data by place must consider three different sites: where the individual was when disease occurred; where the individual was when

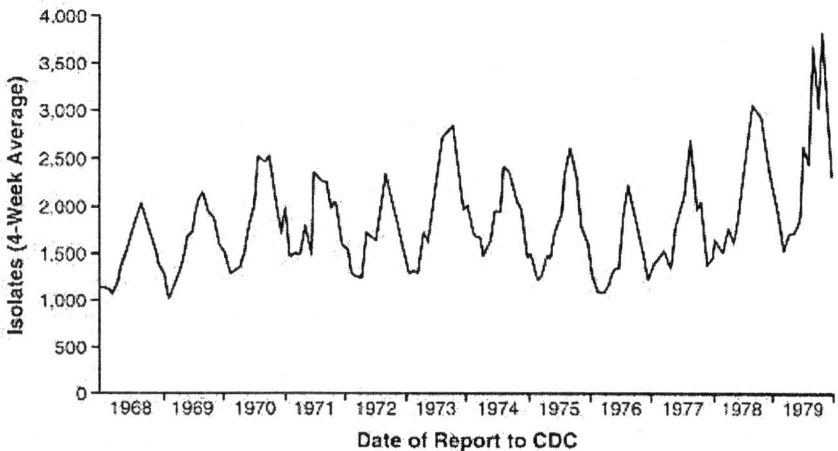

FIGURE 9-2 An example of a disease showing a seasonal trend. Reported human *Salmonella* isolations, by 4-week average, in the United States from 1968 to 1980.

he or she became infected from the source; and where the source became infected with the etiologic agent. Therefore, in anoutbreak of food poisoning, the host may become clinically ill at home from food eaten in a restaurant. The vehicle may have been undercooked chicken, which became infected on a poultry farm. These differences are important to consider in attempting to prevent additional cases.

The third focus of descriptive epidemiology is the infected person. All pertinent characteristics should be noted: age, sex, occupation, personal habits, socioeconomic status, immunization history, presence of underlying disease, and other data.

Once the descriptive epidemiologic data have been analyzed, the features of the epidemic should be clear enough that additional areas for investigation are apparent.

Analytic Epidemiology

The second epidemiologic method is analytic epidemiology, which analyzes disease determinants for possible causal relations. The two main analytic methods are the case-control (or case-comparison) method and the cohort method. The case-control method starts with the effect (disease) and retrospectively investigates the cause that led to the effect. The case group consists of individuals with the disease; a comparison group has members similar to those of the case group except for absence of the disease. These two groups are then compared to determine differences that would explain the occurrence of the disease. An example of a case-control study is selecting individuals with meningococcal meningitis and a comparison group matched for age, sex, socioeconomic status, and residence, but without the disease, to see what factors may have influenced the occurrence in the group that developed disease.

The second analytic approach is the cohort method, which prospectively studies two populations: one that has had contact with the suspected causal factor under study and a similar group that has had no contact with the factor. When both groups are observed, the effect of the factor should become apparent. An example of a cohort approach is to observe two similar groups of people, one composed of individuals who received blood transfusions and the other of persons who did not. The occurrence of hepatitis prospectively in both groups permits one to make an association between blood transfusions and hepatitis; that is, if the transfused blood was contaminated with hepatitis B virus, the recipient cohort should have a higher incidence of hepatitis than the nontransfused cohort.

The case-control approach is relatively easy to conduct, can be completed in a shorter period than the cohort approach, and is inexpensive and reproducible; however, bias may be introduced in selecting the two groups, it may be difficult to exclude subclinical cases from the comparison group, and a patient's recall of past events may be faulty. The advantages of a cohort study are the accuracy of collected data and the ability to make a direct estimate of the disease risk resulting from factor contact; however, cohort studies take longer and are more expensive to conduct.

Another analytic method is the cross-sectional study, in which a population is surveyed over a limited period to determine the relationship between a disease and variables present at the same time that may influence its occurrence.

Experimental Epidemiology

The third epidemiologic method is the experimental approach. A hypothesis is developed and an experimental model is constructed in which one or more selected factors are manipulated. The effect of the manipulation will either confirm or disprove the hypothesis. An example is the evaluation of the effect of a new drug on a disease. A group of people with the disease is identified, and some members are randomly selected to receive the drug. If the only difference between the two is use of the drug, the clinical differences between the groups should reflect the effectiveness of the drug.

Epidemic Investigation

An epidemic investigation describes the factors relevant to an outbreak of disease; once the circumstances related to the occurrence of disease are defined, appropriate control and prevention measures can be identified. In an epidemic investigation, data are collected, collated according to time, place, and person, and analyzed and inferences are drawn.

In the investigation, the first action should be to confirm the existence of the epidemic by noting from past surveillance data the number of cases suspected and comparing this with the number of cases initially reported. Additionally, the investigator should discuss the occurrence of the disease with physicians or others who have seen or reported cases after examining patients and reviewing laboratory and hospital records. These diagnoses should then be verified. A case definition should be developed to differentiate patients who represent actual cases, those who represent suspected or presumptive cases, and those who should be omitted from further study. Additional cases may be sought or additional patient data obtained, and a rough case count made.

This initial phase consists basically of collecting data, which then must be organized according to time, place, and person. The population at risk should be identified and a hypothesis developed concerning the occurrence of the disease. If appropriate, specimens should be collected and transported to the laboratory. More specific studies may be indicated. Additional data from these studies should be analyzed and the hypothesis confirmed or altered. After analysis, control and prevention measures should be developed and, as far as possible, implemented. A report containing this information should be prepared and distributed to those involved in investigating the outbreak and in implementing control and/or prevention measures. Continued surveillance activities may be appropriate to evaluate the effectiveness of the control and prevention measures.

In the United States, the CDC assists state health departments by providing epidemiologic and laboratory support services on request. Its assistance supports disease investigations and diagnostic laboratory activities and includes various training programs conducted in the states and at the CDC. A close working relationship exists between the CDC and state health departments. Additionally, physicians frequently consult with CDC personnel on a variety of health-related problems and attend public health training programs.

The use of epidemiology to characterize a disease before its etiology has been identified is exemplified by the initial studies of acquired immune deficiency syndrome (AIDS). The first cases came to the attention of the CDC late in 1981 when an increase was observed in requests for pentamidine for treatment of *Pneumocystis carinii* pneumonia. This initiated specific surveillance activities and epidemiologic studies that provided important information about this newly diagnosed disease.

Initial symptoms include fever, loss of appetite, weight loss, extreme fatigue, and enlargement of lymph nodes. A severe immune deficiency then develops, which appears to be associated with opportunistic infections. These infections include *P carinii* pneumonia, diagnosed in 52 percent of cases; Kaposi sarcoma in 26 percent of cases; and both *P carinii* pneumonia and Kaposi sarcoma in 7 percent of cases. The remaining 15 percent of AIDS patients have other parasitic, fungal, bacterial, or viral infections associated with immunodeficiencies. Among the first 2,640 cases reported to the CDC, there were 1,092 deaths, a case-fatality rate of 41 percent. Approximately 95 percent of the cases were male; 70 percent were 20 to 49 years of age at the time of diagnosis. Approximately 40 percent of the cases were reported from New York City, 12 percent from San Francisco, 8 percent from Los Angeles, and the remainder from 32 other states. Cases were reported from at least 16 other countries. Among the 90 percent of patients who were categorized according to possible risk factors, those at highest risk were homosexuals or bisexuals (70 percent), intravenous drug abusers (17 percent), Haitian entrants into the United States (9.5 percent), and persons with hemophilia (1 percent).

Analysis of these initial data, collected before the etiologic agent of AIDS was identified, supported the hypothesis that transmission occurred primarily by sexual contact, receipt of contaminated blood or blood products, or contact with contaminated intravenous needles. Spread through casual contact did not seem likely. The epidemiologic data indicated that AIDS was an infectious disease. It has now been determined that AIDS results from infection with a retrovirus of the human T cell leukemia/lymphoma virus family, which has been designated human immunodeficiency virus type I (HIV-l). The initial hypotheses have been proven as shown by analysis of data subsequently collected.

REFERENCES

Beaglehole R, Bonita R, Kjellstrom T: Basic Epidemiology. World Health Organization, Geneva, Switzerland, 1993

Benenson A: Control of Communicable Disease Manual 16th Ed., American Public Health Association, Washington, DC, 1995

Bennett JV, Brachman PS: Hospital Infections. 3rd Ed., Little, Brown, Boston, 1992

Evans AS, Brachman PS: Bacterial Infections of Humans. Epidemiology and Control.2nd Ed. Plenum New York, 1991

Fox JP, Hall CE, Elveback LR: Epidemiology, Man and Disease. Macmillan, New York, 1970

Hennekens CH, Buring JE: Epidemiology in Medicine. Little, Brown, Boston, 1987

Langmuir AD: The surveillance of communicable diseases of national importance. N Engl J Med 268: 182, 1963

Lilienfeld DE, Stolley PD: Foundations of Epidemiology. 3rd Ed. Oxford University Press, New York, 1994

MacMahon B, Pugh TF: Epidemiology Principles and Methods. Little, Brown, Boston, 1970

Mandell GL, Douglas RG, Jr, Bennett JE: Principles and Practice of Infectious Diseases, 3rd Ed. Churchill Livingstone, New York, 1990

Smith DM, Haupt BJ: Hospital discharge data used as feedback in planning research and education for primary care. Public Health Rep 98:457, 1983

World Health Organization: The surveillance of communicable diseases. WHO Chron 22:439, 1968

Chapter **10**

Principles of Diagnosis

General Concepts

Manifestations of Infection

The clinical presentation of an infectious disease reflects the interaction between the host and the microorganism. This interaction is affected by the host immune status and microbial virulence factors. Signs and symptoms vary according to the site and severity of infection. Diagnosis requires a composite of information, including history, physical examination, radiographic findings, and laboratory data.

Microbial Causes of Infection

Infections may be caused by bacteria, viruses, fungi, and parasites. The pathogen may be exogenous (acquired from environmental or animal sources or from other persons) or endogenous (from the normal flora).

Specimen Selection, Collection, and Processing

Specimens are selected on the basis of signs and symptoms, should be representative of the disease process, and should be collected before administration of antimicrobial agents. The specimen amount and the rapidity of transport to the laboratory influence the test results.

Microbiologic Examination

Direct Examination and Techniques: Direct examination of specimens reveals gross pathology. Microscopy may identify microorganisms. Immunofluorescence, immuno-peroxidase staining, and other immunoassays may detect specific microbial antigens. Genetic probes identify genus- or species-specific DNA or RNA sequences.

Culture: Isolation of infectious agents frequently requires specialized media. Nonselective (noninhibitory) media permit the growth of many microorganisms. Selective media contain inhibitory substances that permit the isolation of specific types of microorganisms.

Microbial Identification: Colony and cellular morphology may permit preliminary identification. Growth characteristics under various conditions, utilization of carbohydrates and other substrates, enzymatic activity, immunoassays, and genetic probes are also used.

Serodiagnosis: A high or rising titer of specific IgG antibodies or the presence of specific IgM antibodies may suggest or confirm a diagnosis.

Antimicrobial Susceptibility: Microorganisms, particularly bacteria, are tested in vitro to determine whether they are susceptible to antimicrobial agents.

INTRODUCTION

Some infectious diseases are distinctive enough to be identified clinically. Most pathogens, however, can cause a wide spectrum of clinical syndromes in humans. Conversely, a single clinical syndrome may result from infection with any one of many pathogens. Influenza virus infection, for example, causes a wide variety of respiratory syndromes that cannot be distinguished clinically from those caused by streptococci, mycoplasmas, or more than 100 other viruses.

Most often, therefore, it is necessary to use microbiologic laboratory methods to identify a specific etiologic agent. Diagnostic medical microbiology is the discipline that identifies etiologic agents of disease. The job of the clinical microbiology laboratory is to test specimens from patients for microorganisms that are, or may be, a cause of the illness and to provide information (when appropriate) about the in vitro activity of antimicrobial drugs against the microorganisms identified (Fig. 10-1).

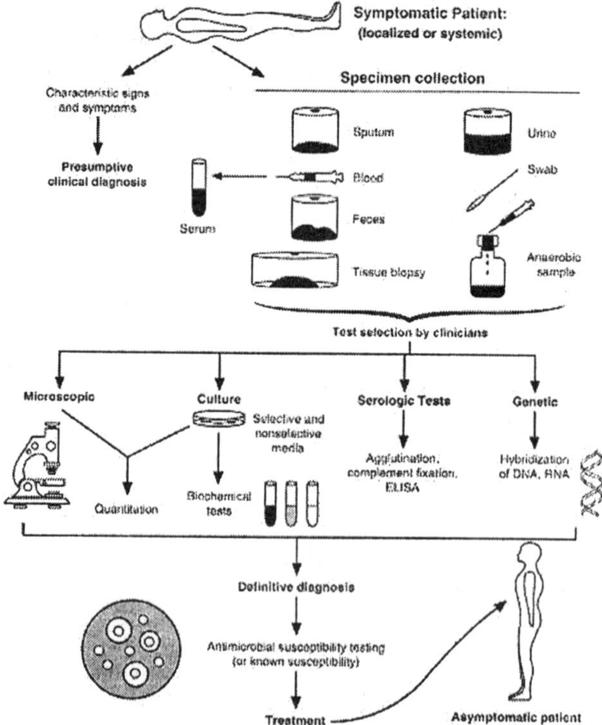

FIGURE 10-1 Laboratory procedures used in confirming a clinical diagnosis of infectious disease with a bacterial etiology.

The staff of a clinical microbiology laboratory should be qualified to advise the physician as well as process specimens. The physician should supply salient information about the patient, such as age and sex, tentative diagnosis or details of the clinical syndrome, date of onset, significant exposures, prior antibiotic therapy, immunologic status, and underlying conditions. The clinical microbiologist participates in decisions regarding the microbiologic diagnostic studies to be performed, the type and timing of specimens to be collected, and the conditions for their transportation and storage. Above all, the clinical microbiology laboratory, whenever appropriate, should provide an interpretation of laboratory results.

Manifestations of Infection

The manifestations of an infection depend on many factors, including the site of acquisition or entry of the microorganism; organ or system tropisms of the microorganism; microbial virulence; the age, sex, and immunologic status of the patient; underlying diseases or conditions; and the presence of implanted prosthetic devices or materials. The signs and symptoms of infection may be localized, or they may be systemic, with fever, chills, and hypotension. In some instances the manifestations of an infection are sufficiently characteristic to suggest the diagnosis; however, they are often nonspecific.

Microbial Causes of Infection

Infections may be caused by bacteria (including mycobacteria, chlamydiae, mycoplasmas, and rickettsiae), viruses, fungi, or parasites. Infection may be endogenous or exogenous. In endogenous infections, the microorganism (usually a bacterium) is a component of the patient's indigenous flora. Endogenous infections can occur when the microorganism is aspirated from the upper to the lower respiratory tract or when it penetrates the skin or mucosal barrier as a result of trauma or surgery. In contrast, in exogenous infections, the microorganism is acquired from the environment (e.g., from soil or water) or from another person or an animal. Although, it is important to establish the cause of an infection, the differential diagnosis is based on a careful history, physical examination, and appropriate radiographic and laboratory studies, including the selection of appropriate specimens for microbiologic examination. Results of the history, physical examination, and radiographic and laboratory studies allow the physician to request tests for the microorganisms most likely to be the cause of the infection.

Specimen Selection, Collection and Processing

Specimens selected for microbiologic examination should reflect the disease process and be collected in sufficient quantity to allow complete microbiologic examination. The number of microorganisms per milliliter of a body fluid or per gram of tissue is highly variable, ranging from less than 1 to 10^8 or 10^{10} colony-forming units (CFU). Swabs, although popular for specimen collection, frequently yield too small a specimen for accurate microbiologic examination and should be used only to collect material from the skin and mucous membranes.

Because skin and mucous membranes have a large and diverse indigenous flora, every effort must be made to minimize specimen contamination during collection. Contamination may be avoided by various means. The skin can be disinfected before aspirating or incising a lesion. Alternatively, the contaminated area may be bypassed altogether. Examples of such approaches are transtracheal puncture with aspiration of lower respiratory secretions or suprapubic bladder puncture with aspiration of urine. It is often impossible to collect an uncontaminated specimen, and decontamination procedures, cultures on selective media, or quantitative cultures must be used (see above).

Specimens collected by invasive techniques, particularly those obtained intraoperatively, require special attention. Enough tissue must be obtained for both histopathologic and microbiologic examination. Histopathologic examination is used to distinguish neoplastic from inflammatory lesions and acute from chronic inflammations. The type of inflammation present can guide the type of microbiologic examination performed. If, for example, a caseous granuloma is observed histopathologically, microbiologic examination should include cultures for mycobacteria and fungi. The surgeon should obtain several samples for examination from a single large lesion or from each of several smaller lesions. If an abscess is found, the surgeon should collect several milliliters of pus, as well as a portion of the wall of the abscess, for microbiologic examination. Swabs should be kept out of the operating room.

If possible, specimens should be collected before the administration of antibiotics. Above all, close communication between the clinician and the microbiologist is essential to ensure that appropriate specimens are selected and collected and that they are appropriately examined.

Microbiologic Examination

Direct Examination

Direct examination of specimens frequently provides the most rapid indication of microbial infection. A variety of microscopic, immunologic, and hybridization techniques have been developed for rapid diagnosis (Table 10-1).

TABLE 10-1 Rapid Tests Commonly Used to Detect Microorganisms in Specimens

Specimen	Test	Application
Blood	Gelmsa	Plasmodia, microfilarie
	EIA	Hepatitis A and B virus, human immunodeficiency virus
Cerebrospinal	Gram stain	Bacteria
	LA; COA	Haemophilus influenzae, Neisseria

	India ink wet mount or LA	Streoticiccus pneumoniae
Wound exudates, pus	Gram stain	Bacteria
Respiratory secretions	Gram stain	Bacteria
	Acid-fast stain	Mycobacteria, nocardiae
	IFA or genetic probe	Legionella species, Streptococcus pyogenes
	KOH wet mount	Fungi
	Gormori methenamine silver stain	Fungi, Pneumocystis carnil
	FA, EIA	Respiratory syncytrial virus
Urine	Gram stain	Bacteria
Urethral or cervical scrapings or exudates	Gram stain, EIA, IFA, EIA, or genetic probe	Neisseria gonorrhoeae Chlamydia trachomatis, papilomaviruses
Genital ulcer	FA, EIA, or genetic probe	Herpes simplex virus
Feces	Methylene blue stain	Leukocytes
	Eosin wet mount, trichrome stain	Parasites
	EM, LA, EIA	Rotavirses
	EIA	Adenoviruses, Clostridium difficife

Abbreviations: COA, coagglutination; EIA, enzyme immunoassay: IFA, immunofluorescent antibody: LA, latex agglutination.

Sensitivity and Specificity

The sensitivity of a technique usually depends on the number of microorganisms in the specimen. Its specificity depends on how morphologically unique a specific microorganism appears microscopically or how specific the antibody or genetic probe is for that genus or species. For example, the sensitivity of Gram stains is such that the observation of two bacteria per oil immersion field (X 1,000) of a Gram-stained smear of uncentrifuged urine is equivalent to the presence of $\geq 10^5$ CFU/ml of urine. The sensitivity of the Gram-stained smear for detecting Gram-negative coccobacilli in cerebrospinal fluid from children with *Haemophilus influenzae* meningitis is approximately 75 percent because in some patients the number of colony-forming units per milliliter of cerebrospinal fluid is less than 10^4. At least 10^4 CFU of tubercle bacilli per milliliter of sputum must be present to be detected by an acid-fast smear of decontaminated and concentrated sputum.

An increase in the sensitivity of a test is often accompanied by a decrease in specificity. For example, examination of a Gram-stained smear of sputum from a patient with pneumococcal pneumonia is highly sensitive but also highly nonspecific if the criterion for defining a positive test is the presence of any Gram-positive cocci. If, however, a positive test is defined as the presence of a preponderance of Gram-positive, lancet-shaped diplococci, the test becomes highly specific but has a sensitivity of only about 50 percent. Similar problems related to the number of microorganisms present affect the sensitivity of immunoassays and genetic probes for bacteria, chlamydiae, fungi and viruses. In some instances, the sensitivity of direct examination tests can be improved by collecting a better specimen. For example, the sensitivity of fluorescent antibody stain for Chlamydia trachomatis is higher when endocervical cells are obtained with a cytobrush than with a swab. The sensitivity may also be affected by the stage of the disease at which the specimen is collected. For example, the detection of herpes simplex virus by immunofluorescence, immunoassay, or culture is highest when specimens from lesions in the vesicular stage of infection are examined. Finally, sensitivity may be improved through the use of an enrichment or enhancement step in which microbial or genetic replication occurs to the point at which a detection method can be applied.

Techniques

For microscopic examination it is sufficient to have a compound binocular microscope equipped with low-power (1OX), high-power (40X), and oil immersion (1OOX) achromatic objectives, 10X wide-field oculars, a mechanical stage, a substage condenser, and a good light source. For examination ofwet-mount preparations, a darkfield condenser or condenser and objectives for phase contrast increases image contrast. An exciter barrier filter, darkfield condenser, and ultraviolet light source are required for fluorescence microscopy.

For immunologic detection of microbial antigens, latex particle agglutination, coagglutination, and enzyme-linked immunosorbent assay (ELISA) are the most frequently used techniques in the clinical laboratory. Antibody to a specific antigen is bound to latex particles or to a heat-killed and treated protein A-rich strain of Staphylococcus aureus to produce agglutination (Fig. 10-2). There are several approaches to ELISA; the one most frequently used for the detection of microbial antigens uses an antigen-specific antibody that is fixed to a solid phase, which may be a latex or metal bead or the inside surface of a well in a plastic tray. Antigen present in the specimen binds to the antibody as in Fig. 10-2. The test is then completed by adding a second antigen-specific antibody bound to an enzyme that can react with a substrate to produce a colored product. The initial antigen antibody complex forms in a manner similar to that shown in Figure 10-2. When the enzyme-conjugated antibody is added, it binds to previously unbound antigenic sites, and the antigen is, in effect, sandwiched between the solid phase and the enzyme-conjugated antibody. The reaction is completed by adding the enzyme substrate.

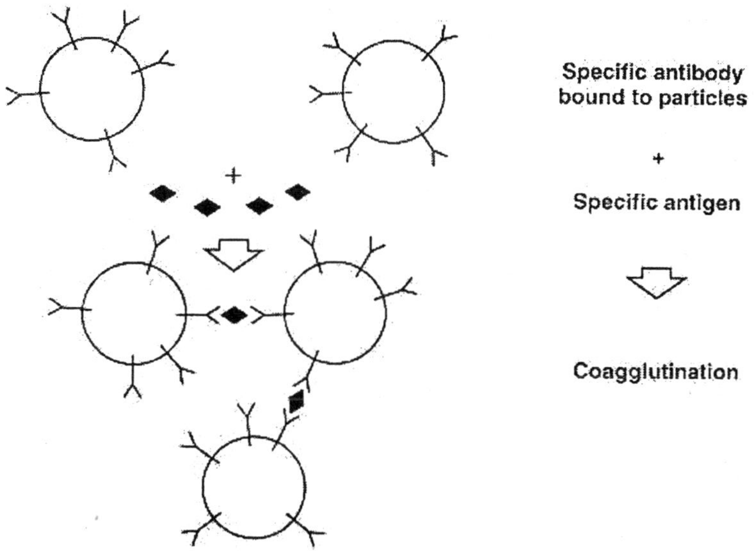

Specific antibody bound to particles

+

Specific antigen

Coagglutination

FIGURE 10-2 Agglutination test in which inert particles (latex beads or heat-killed *S aureus* Cowan 1 strain with protein A) are coated with antibody to any of a variety of antigens and then used to detect the antigen in specimens or in isolated bacteria.

Genetic probes are based on the detection of unique nucleotide sequences with the DNA or RNA of a microorganism. Once such a unique nucleotide sequence, which may represent a portion of a virulence gene or of chromosomal DNA, is found, it is isolated and inserted into a cloning vector (plasmid), which is then transformed into Escherichia coli to produce multiple copies of the probe. The sequence is then reisolated from plasmids and labeled with an isotope or substrate for diagnostic use. Hybridization of the sequence with a complementary sequence of DNA or RNA follows cleavage of the double-stranded DNA of the microorganism in the specimen.

The use of molecular technology in the diagnoses of infectious diseases has been further enhanced by the introduction of gene amplication techniques, such as the polymerase chain reaction (PCR) in which DNA polymerase is able to copy a strand of DNA by elongating complementary strands of DNA that have been initiated from a pair of closely spaced oligonucleotide primers. This approach has had major applications in the detection of infections due to microorganisms that are difficult to culture (e.g. the human immunodeficiency virus) or that have not as yet been successfully cultured (e.g. the Whipple's disease bacillus).

Culture

In many instances, the cause of an infection is confirmed by isolating and culturing microorganism either in artificial media or in a living host. Bacteria (including mycobacteria and mycoplasmas) and fungi are cultured in either liquid (broth) or on solid (agar) artificial media. Liquid media provide greater sensitivity for the isolation of small numbers of microorganisms; however, identification of mixed cultures growing in liquid media requires subculture onto solid media so that isolated colonies can be processed separately for identification. Growth in liquid media also cannot ordinarily be quantitated. Solid media, although somewhat less sensitive than liquid media, provide isolated colonies that can be quantified if necessary and identified. Some genera and species can be recognized on the basis of their colony morphologies.

In some instances one can take advantage of differential carbohydrate fermentation capabilities of microorganisms by incorporating one or more carbohydrates in the medium along with a suitable pH indicator. Such media are called differential media (e.g., eosin methylene blue or MacConkey agar) and are commonly used to isolate enteric bacilli. Different genera of the Enterobacteriaceae can then be presumptively identified by the color as well as the morphology of colonies.

Culture media can also be made selective by incorporating compounds such as antimicrobial agents that inhibit the indigenous flora while permitting growth of specific microorganisms resistant to these inhibitors. One such example is Thayer-Martin medium, which is used to isolate *Neisseria gonorrhoeae*. This medium contains vancomycin to inhibit Gram-positive bacteria, colistin to inhibit most Gram-negative bacilli, trimethoprim-sulfamethoxazole to inhibit *Proteus* species and other species that are not inhibited by colistin and anisomycin to inhibit fungi. The pathogenic *Neisseria* species, *N gonorrhoeae* and *N meningitidis*, are ordinarily resistant to the concentrations of these antimicrobial agents in the medium.

The number of bacteria in specimens may be used to define the presence of infection. For example, there may be small numbers ($< 10^3$ CFU/ml) of bacteria in clean-catch, midstream urine specimens from normal, healthy women; with a few exceptions, these represent bacteria that are indigenous to the urethra and periurethral region. Infection of the bladder (cystitis) or kidney (pyelone-phritis) is usually accompanied by bacteriuria of about $> 10^4$ CFU/ml. For this reason, quantitative cultures (Fig. 10-3) of urine must always be performed. For most other specimens a semiquantitative streak method (Fig. 10-3) over the agar surface is sufficient. For quantitative cultures, a specific volume of specimen is spread over the agar surface and the number of colonies per milliliter is estimated. For semiquantitative cultures, an unquantitated amount of specimen is applied to the agar and diluted by being streaked out from the inoculation site with a sterile bacteriologic loop (Fig. 10-3). The amount of growth on the agar is then reported semiquantitatively as many, moderate, or few (or 3+, 2+, or 1+), depending on how far out from the inoculum site colonies appear. An organism that grows in all streaked areas would be reported as 3+.

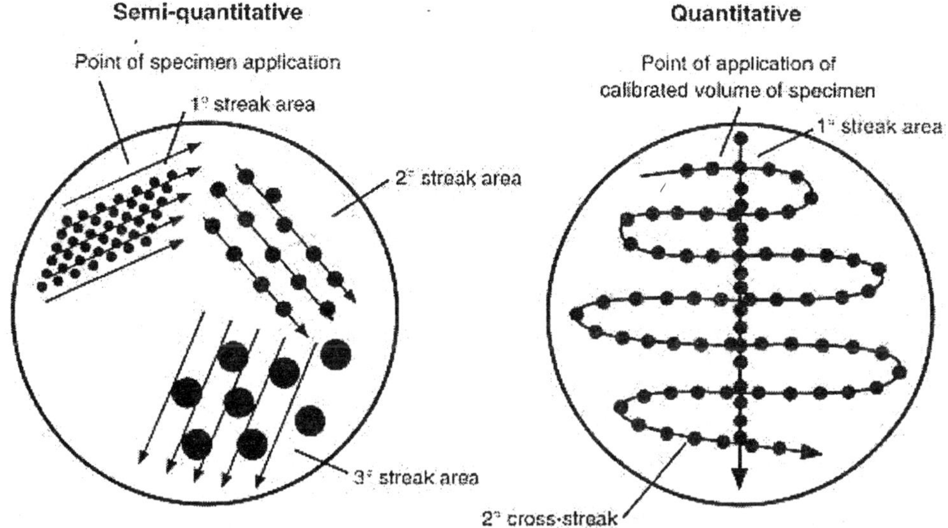

FIGURE 10-3 Quantitative versus semiquantitative culture, revealing the number of bacteria in specimens.

Chlamydiae and viruses are cultured in cell culture systems, but virus isolation occasionally requires inoculation into animals, such as suckling mice, rabbits, guinea pigs, hamsters, or primates. Rickettsiae may be isolated with some difficulty and at some hazard to laboratory workers in animals or embryonated eggs. For this reason, rickettsial infection is usually diagnosed serologically. Some viruses, such as the hepatitis viruses, cannot be isolated in cell culture systems, so that diagnosis of hepatitis virus infection is based on the detection of hepatitis virus antigens or antibodies.

Cultures are generally incubated at 35 to 37°C in an atmosphere consisting of air, air supplemented with carbon dioxide (3 to 10 percent), reduced oxygen (microaerophilic conditions), or no oxygen (anaerobic conditions), depending upon requirements of the microorganism. Since clinical specimens from bacterial infections often contain aerobic, facultative anaerobic, and anaerobic bacteria, such specimens are usually inoculated into a variety of general purpose, differential, and selective media, which are then incubated under aerobic and anaerobic conditions (Fig. 10-4).

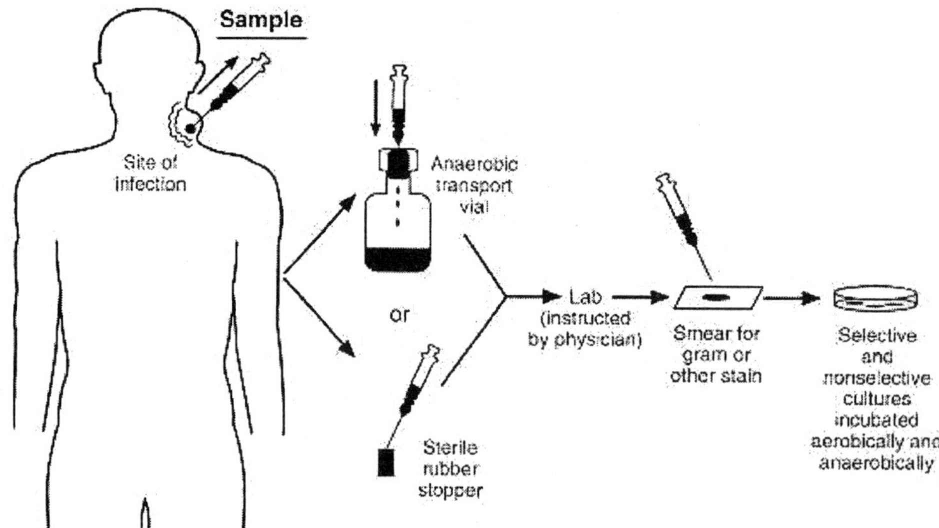

FIGURE 10-4 General procedure for collecting and processing specimens for aerobic and/or anaerobic bacterial culture. The duration of incubation of cultures also varies with the growth characteristics of the microorganism. Most aerobic and anaerobic bacteria will grow overnight, whereas some mycobacteria require as many as 6 to 8 weeks.

Microbial Identification

Microbial growth in cultures is demonstrated by the appearance of turbidity, gas formation, or discrete colonies in broth; colonies on agar; cytopathic effects or inclusions in cell cultures; or detection of genus- or species-specific antigens or nucleotide sequences in the specimen, culture medium, or cell culture system.

Identification of bacteria (including mycobacteria) is based on growth characteristics (such as the time required for growth to appear or the atmosphere in which growth occurs), colony and microscopic morphology, and biochemical, physiologic, and, in some instances, antigenic or nucleotide sequence characteristics. The selection and number of tests for bacterial identification depend upon the category of bacteria present (aerobic versus anaerobic, Gram-positive versus Gram-negative, cocci versus bacilli) and the expertise of the microbiologistexamining the culture. Gram-positive cocci that grow in air with or without added CO_2 may be identified by a relatively small number of tests (see Ch. 12). The identification of most Gram-negative bacilli is far more complex and often requires panels of 20 tests for determining biochemical and physiologic characteristics. The identification of filamentous fungi is based almost entirely on growth characteristics and colony and microscopic morphology. Identification of viruses is usually based on characteristic cytopathic effects in different cell cultures or on the detection of virus- or species-specific antigens or nucleotide sequences.

Interpretation of Culture Results

Some microorganisms, such as *Shigella dysenteriae*, *Mycobacterium tuberculosis*, *Coccidioides immitis*, and influenza virus, are always considered clinically significant. Others that ordinarily are harmless components of the indigenous flora of the skin and mucous membranes or that are common in the environment may or may not be clinically significant, depending on the specimen source from which they are isolated. For example, coagulase-negative staphylococci are normal inhabitants of the skin, gastrointestinal tract, vagina, urethra, and the upper respiratory tract (i.e., of the nares, oral cavity, and pharynx). Therefore, their isolation from superficial ulcers, wounds, and sputum cannot usually be interpreted as clinically significant. They do, however, commonly cause infections associated with intravascular devices and implanted prosthetic materials. However, because intravascular devices penetrate the skin and since cultures of an implanted prosthetic device can be made only after incision, the role of coagulase-negative staphylococci in causing infection can usually be surmised only when the microorganism is isolated in large numbers from the surface of an intravascular device, from each of several sites surrounding an implanted prosthetic device, or, in the case of prosthetic valve endocarditis, from several separately collected blood samples. Another example, *Aspergillus fumigatus*, is widely distributed in nature, the hospital environment, and upper respiratory tract of healthy people but may cause fatal pulmonary infections in leukemia patients or in those who have undergone bone marrow transplantation. The isolation of *A fumigatus* from respiratory secretions is a nonspecific finding, and a definitive diagnosis of invasive aspergillosis requires histologic evidence of tissue invasion.

Physicians must also consider that the composition of microbial species on the skin and mucous membranes may be altered by disease, administration of antibiotics, endotracheal or gastric incubation, and the hospital environment. For example, potentially pathogenic bacteria can often be cultured from the pharynx of seriously ill, debilitated patients in the intensive care unit, but may not cause infection.

Serodiagnosis

Infection may be diagnosed by an antibody response to the infecting microorganism. This approach is especially useful when the suspected microbial agent either cannot be isolated in culture by any known method or can be isolated in culture only with great difficulty. The diagnosis of hepatitis virus and Epstein-Barr virus infections can be made only serologically, since neither can be isolated in any known cell culture system. Although, human immunodeficiency virus type 1 (HIV-1) can be isolated in cell cultures, the technique is demanding and requires special containment facilities. HIV-1 infection is usually diagnosed by detection of antibodies to the virus.

The disadvantage of serology as a diagnostic tool is that there is usually a lag between the onset of infection and the development of antibodies to the infecting microorganism. Although, IgM antibodies may appear relatively rapidly, it is usually necessary to obtain acute- and convalescent-phase serum samples to look for a rising titer of IgG antibodies to the suspected pathogen. In some instances the presence of a high antibody titer when the patient is initially seen is diagnostic; often, however, the high titer may reflect a past infection, and the current infection may have an entirely different cause. Another limitation on the use of serology as a diagnostic tool is that immunosuppressed patients may be unable to mount an antibody response.

Antimicrobial Susceptibility

The responsibility of the microbiology laboratory includes not only microbial detection and isolation but also the determination of microbial susceptibility to antimicrobial agents. Many bacteria, in particular, have unpredictable susceptibilities to antimicrobial agents, and their susceptibilities can be measured in vitro to help guide the selection of the most appropriate antimicrobial agent.

Antimicrobial susceptibility tests are performed by either disk diffusion or a dilution method. In the former, a standardized suspension of a particular microorganism is inoculated onto an agar surface to which paper disks containing various antimicrobial agents are applied. Following overnight incubation, any zone diameters of inhibition about the disks are measured and the results are reported as indicating susceptibility or resistance of the microorganism to each antimicrobial agent tested. An alternative method is to dilute on a log2 scale each antimicrobial agent in broth to provide a range of concentrations and to inoculate each tube or, if a microplate is used, each well containing the antimicrobial agent in broth with a standardized suspension of the microorganism to be tested. The lowest concentration of antimicrobial agent that inhibits the growth of the microorganism is the minimal inhibitory concentration (MIC). The MIC and the zone diameter of inhibition are inversely correlated (Fig. 10-5). In other words, the more susceptible the microorganism is to the antimicrobial agent, the lower the MIC and the larger the zone of inhibition. Conversely, the more resistant the microorganism, the higher the MIC and the smaller the zone of inhibition.

The term susceptible means that the microorganism is inhibited by a concentration of antimicrobial agent that can be attained in blood with the normally recommended dose of the antimicrobial agent and implies that an infection caused by this microorganism may be appropriately treated with the antimicrobial agent. The term resistant indicates that the microorganism is resistant to concentrations of the antimicrobial agent that can be attained with normal doses and implies that an infection caused by this microorganism could not be successfully treated with this antimicrobial agent.

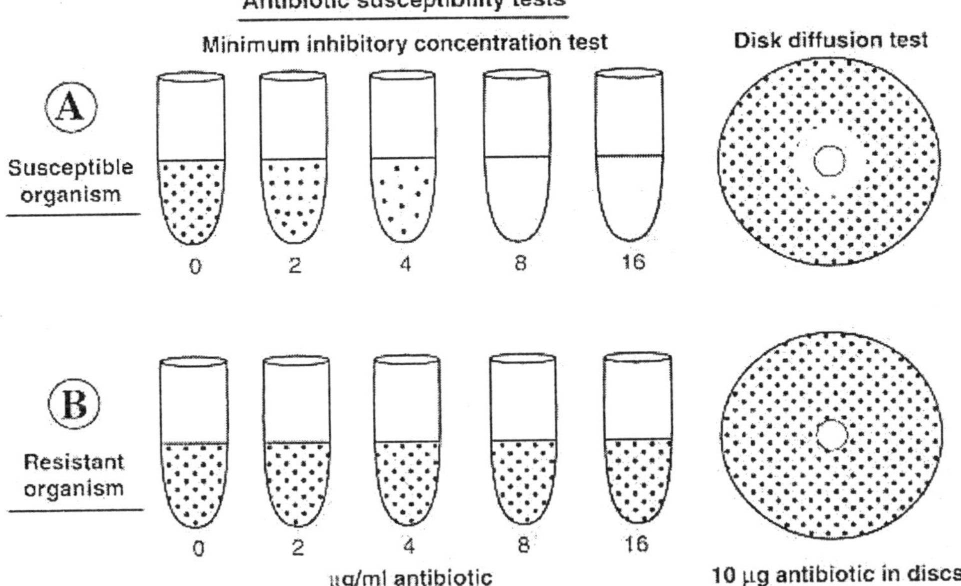

FIGURE 10-5 Two methods for performing antibiotic susceptibility tests. (A) Disk diffusion method. (B) Minimum inhibitory concentration (MIC) method. In the example shown, two different microorganisms are tested by both methods against the same antibiotic. The MIC of the antibiotic for the susceptible microorganism is 8 µg/ml. The corresponding disk diffusion test shows a zone of inhibition surrounding the disk. In the second sample, a resistant microorganism is not inhibited by the highest antibiotic concentration tested (MIC > 16 µg/ml) and there is no zone of inhibition surrounding the disk. The diameter of the zone of inhibition is inversely related to the MIC.

REFERENCES

Baron EJ, Pererson LR, Finegold SM (eds): Bailey and Scott's Diagnostic Microbiology. 9th ed. CV Mosby, St. Louis, 1994

Koneman EW, Allen SD, Schreckenberg PC, Winn WC (eds): Atlas and Textbook of Diagnostic Microbiology. 4th ed. JB Lippincott, Philadelphia, 1992

Kunin CM: Detection, Prevention and Management of Urinary Tract Infections. 4th ed. Lea & Febiger, Philadelphia, 1987

Murray PR, Baron EJ, Pfaller MA, Tenover PC, Yolken RH (eds): Manual of Clinical Microbiology. 6th ed. American Society for Microbiology, Washington, DC, 1995

Pennington JE (ed): Respiratory Infections: Diagnosis and Management. 3rd ed. Raven Press, New York, 1994

Woods GL, Washington JA: The Clinician and the Microbiology Laboratory. Mandell GL, Bennett JE, Dolin R (eds): Principles and Practice of Infectious Diseases. 4th ed. Churchill Livingstone, New York, 1995

Chapter **11**

Antimicrobial Chemotherapy

General Concepts

Basis of Antimicrobial Action

Various antimicrobial agents act by interfering with (1) cell wall synthesis, (2) plasma membrane integrity, (3) nucleic acid synthesis, (4) ribosomal function, and (5) folate synthesis.

Action of Specific Agents

Cell wall synthesis is inhibited by ß-lactams, such as penicillins and cephalosporins, which inhibit peptidoglycan polymerization, and by vancomycin, which combines with cell wall substrates. Polymyxins disrupt the plasma membrane, causing leakage. The plasma membrane sterols of fungi are attacked by polyenes (amphotericin) and imidazoles. Quinolones bind to a bacterial complex of DNA and DNA gyrase, blocking DNA replication. Nitroimidazoles damage DNA. Rifampicin blocks RNA synthesis by binding to DNA directed RNA polymerase. Aminoglycosides, tetracycline, chloramphenicol, erythromycin, and clindamycin all interfere with ribosome function. Sulfonamides and trimethoprim block the synthesis of the folate needed for DNA replication.

Bacterial Resistance

Bacteria can evolve resistance to antibiotics. Resistance factors can be encoded on plasmids or on the chromosome. Resistance may involve decreased entry of the drug, changes in the receptor (target) of the drug, or metabolic inactivation of the drug.

Effects of Combination Therapy

Combinations of antibiotics may act synergistically-producing an effect stronger than the sum of the effects of the two drugs alone or antagonistically, if one agent inhibits the effect of the other.

Adverse Effects of Antimicrobial Agents

Many antibiotics are toxic to the host. Alterations of the normal intestinal flora caused by antibiotics may result in diarrhea or in superinfection with opportunistic pathogens.

INTRODUCTION

The earliest evidence of successful chemotherapy is from ancient Peru, where the Indians used bark from the cinchona tree to treat malaria. Other substances were used in ancient China, and we now know that many of the poultices used by primitive peoples contained antibacterial and antifungal substances. Modern chemotherapy has been dated to the work of Paul Ehrlich in Germany, who sought systematically to discover effective agents to treat trypanosomiasis and syphilis. He discovered p-rosaniline, which has antitrypanosomal effects, and arsphenamine, which is effective against syphilis. Ehrlich postulated that it would be possible to find chemicals that were selectively toxic for parasites but not toxic to humans. This idea has been called the "magic bullet" concept. It had little success until the 1930s, when Gerhard Domagk discovered the protective effects of prontosil, the forerunner of sulfonamide. Ironically, penicillin G was discovered fortuitously in 1929 by Fleming, who did not initially appreciate the magnitude of his discovery. In 1939 Florey and colleagues at Oxford University again isolated penicillin. In 1944 Waksman isolated streptomycin and subsequently found agents such as chloramphenicol, tetracyclines, and erythromycin in soil samples. By the 1960s, improvements in fermentation techniques and advances in medicinal chemistry permitted the synthesis of many new chemotherapeutic agents by molecular modification of existing compounds. Progress in the development of novel antibacterial agents has been great, but the development of effective, nontoxic antifungal and antiviral agents has been slow. Amphotericin B, isolated in the 1950s, remains an effective antifungal agent, although newer agents such as fluconazole are now widely used. Nucleoside analogs such as acyclovir have proved effective in the chemotherapy of selected viral infections.

Biochemical Basis of Antimicrobial Action

Bacterial cells grow and divide, replicating repeatedly to reach the large numbers present during an infection or on the surfaces of the body. To grow and divide, organisms must synthesize or take up many types of biomolecules. Antimicrobial agents interfere with specific processes that are essential for growth and/or division (Fig. 11-1). They can be separated into groups such as inhibitors of bacterial and fungal cell walls, inhibitors of cytoplasmic membranes, inhibitors of nucleic acid synthesis, and inhibitors of ribosome function (Table 11-1). Antimicrobial agents may be either bactericidal, killing the target bacterium or fungus, or bacteriostatic, inhibiting its growth. Bactericidal agents are more effective, but bacteriostatic agents can be extremely beneficial since they permit the normal defenses of the host to destroy the microorganisms.

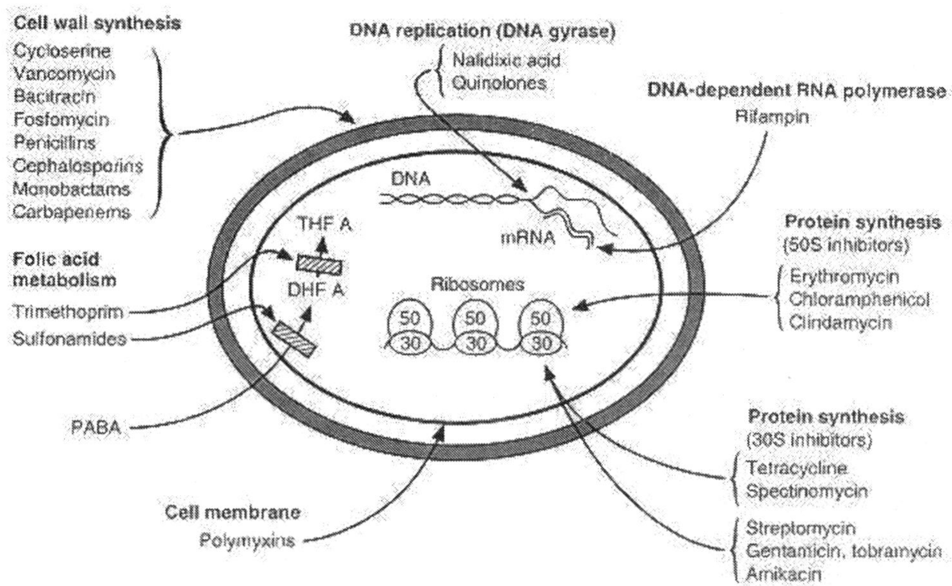

FIGURE 11-1 Sites of action of different antimicrobial agents. PABA, paraminobenzoic acid; DHFA, dihydrofolic acid; THFA, tetrahydrofolic acid.

Table 11-1 Mechanisms of Action of Antimicrobial Agents

Inhibitors of Bacterial Cell Wall Synthesis	Inhibitors of Ribosome Function
Drugs that inhibit biosynthetic enzymes	Inhibitors of 30S units
Fosfomycin	Streptomycin
Cysloserine	Kanamycin, gentamicin, amikacin
Drugs that combine with carrier molecules	Spectinomycin
Bactracin	Tetracyclines
Drugs that combine with cell wall substrates	Inhibitors of 50S units
Vancomycin	Chloramphenicol
Drugs that inhibit polymerization and attachment of new pepidoglycan to cell wall	Clindamycin
Penicillins	Erythromycin
Cephalosporins	Fusidic acid
Carbapenems	Inhibitors of Folate Metabolism
Monobactams	Inhibitor of pteroic acid synthesis
Inhibitors of Cytoplasmis Membranes	Sulfonamides
Drugs that disorganize the cytoplasmic membrane	**Inhibitors of dihydrofolate reductase**
Tyrocidins	Trimephorim
Polymyxins	

Drugs that produce pores in membranes
  Gramicidins
Drugs that alter structure of fungi
  Polyenes (amphotericin)
  Imidazoles (ketoconazole, fluconazole)

Inhibitors of Nucleic Acid Synthesis

Inhibitors of nucleotide metabolism
  Adenosine arabinoside (viruses)
  Acyclovir (viruses)
  Flucytosine (fungi)
Agents that impair DNA template function
  Intercalating agents
  Chloroquine (parasites)
Inhibitors of DNA replication
  Quinolones
  Nitroimdazoles
Inhibitors of RNA polymerase
  Rifampin

Inhibition of Bacterial Cell Wall Synthesis

As noted in earlier chapters, bacteria are classified as Gram-positive and Gram-negative organisms on the basis of staining characteristics. Gram-positive bacterial cell walls contain peptidoglycanand teichoic or teichuronic acid, and the bacterium may or may not be surrounded by a protein or polysaccharide envelope. Gram-negative bacterial cell walls contain peptidoglycan, lipopolysaccharide, lipoprotein, phospholipid, and protein (Fig. 11-2). The critical attack site of anti-cell-wall agents is the peptidoglycan layer. This layer is essential for the survival of bacteria in hypotonic environments; loss or damage of this layer destroys the rigidity of the bacterial cell wall, resulting in death.

Peptidoglycan synthesis occurs in three stages. The first stage takes place in the cytoplasm, where the low-molecular-weight precursors UDP-GlcAc and UDP-MurNAc-L-Ala-D-Glu-meso-Dap-D-Ala-D-Ala are synthesized. A number of antimicrobial agents interfere with these early steps in cell wall biosynthesis. UTP and N-acetylglucosamine alpha-1-P are converted to UDP-N-acetylglucosamine, which is subsequently converted by the enzyme phosphoenolpyruvate: UDP-GlcNAc-3-enol-pyruvyltransferase. Fosfomycins block this transfer by a direct nucleophilic attack on the enzyme. Because mammalian enzymes such as enolase, pyruvate kinase, carboxykinases, and the shikimate enlaces are not inhibited by these compounds, fosfomycins have no effect on the host. Three amino acids are added to the muramyl peptide to yield a tripeptide to which two more amino acids will be linked. The dipeptide D-alanyl-D-alanine is synthesized from two molecules of D-alanine by the enzyme D-alanyl-D-alanine synthetase. D-Alanine

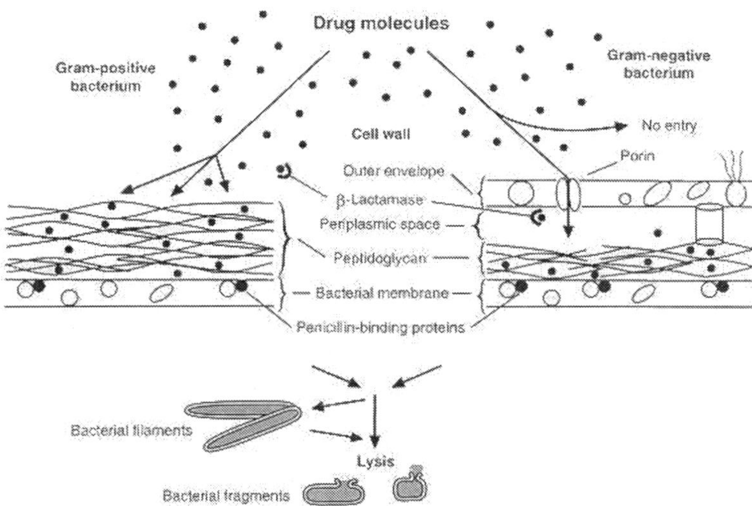

FIGURE 11-2 Outer wall of Gram-positive and Gram-negative species and detail of porin channels of Gram-negative bacteria. Antimicrobial agents diffuse easily through the loose outer wall of Gram-positive bacteria, but must go through the narrow channels of the Gram-negative species.

is produced from L-alanine by an alanine racemase. Cycloserine inhibits both alanine racemase and D-alanyl-D-alanine synthetase owing to the structural similarity of cycloserine and D-alanine and to the fact that cycloserine actually binds to the enzymes better than the D-alanine.

The second stage of cell wall synthesis is catalyzed by membrane-bound enzymes. The nonnucleotide portion of the precursor molecules previously made are transferred sequentially to a carrier in the cytoplasmic membrane. This carrier is a phosphorylated undecaprenyl alcohol. The lipid carrier functions as a point of attachment to the membrane for the precursors and allows for transport of the subunits across the hydrophobic interior of the cytoplasmic membrane to the outside surface. Bacitracin is a peptide antibiotic that specifically interacts with the pyrophosphate derivate of the undecaprenyl alcohol, preventing further transfer of the muramylpentapeptide from the precursor nucleotide to the nascent peptidoglycan.

The third stage of cell wall synthesis involves polymerization of the subunits and the attachment of nascent peptidoglycan to the cell wall. Polymerization occurs by transfer of the new peptidoglycan chain from its carrier in the membrane to the nonreducing N-acetylglucosamine of the new saccharide-peptide that is attached to the membrane. The new peptidoglycan is attached to preexisting cell wall peptidoglycan by a transpeptidase reaction that involves peptide chains in both polymers, one of which must possess a D-alanyl-D-alanine terminus. It is believed that the transpeptidase enzyme cleaves the peptide bond between two D-alanyl residues in the pentapeptide and become acylated via the carbonyl group of the penultimate D-alanine residue. This final reaction is inhibited by ß-lactam antibiotics. These antibiotics contain a critical four-membered ring, which undergoes an acylation reaction with the transpeptidases that cross-link the polymers mentioned above. The ß-lactam antibiotics are the penicillins (penams), cephalosporins (including oxacephems and cephamycins), penems, thienamycins (carbapenems), and aztreonam (monobactams) (Fig. 11-3). The enzymes involved in this final process of cell wall formation are called penicillin-binding proteins since they were discovered by labeling with radioactive penicillin G. The enzymes are different in Gram-positive and Gram-negative bacteria and in anaerobic species. Differences in the penicillin-binding proteins explain, to some extent, differences in antibacterial activity of the ß-lactam antibiotics. The penicillin-binding protein, to which a particular ß-lactam antibiotic binds, affects the morphologic response of the bacterium to the agent. For example, some antibiotics bind to a penicillin-binding protein that is involved in forming the septum between dividing cells; as a result, the bacteria continue to grow into long filaments, which eventually die. Binding to another penicillin-binding protein results in rapid lysis of a bacterium because the wall bulges and the bacterium bursts. ß-Lactams such as mecillinam (an amidino penicillin) do not bind to the penicillin-binding proteins of Gram-positive bacteria and therefore do not affect these bacteria. Aztreonam binds only to Gram-negative penicillin-binding proteins and does not inhibit Gram-positive or anaerobic species.

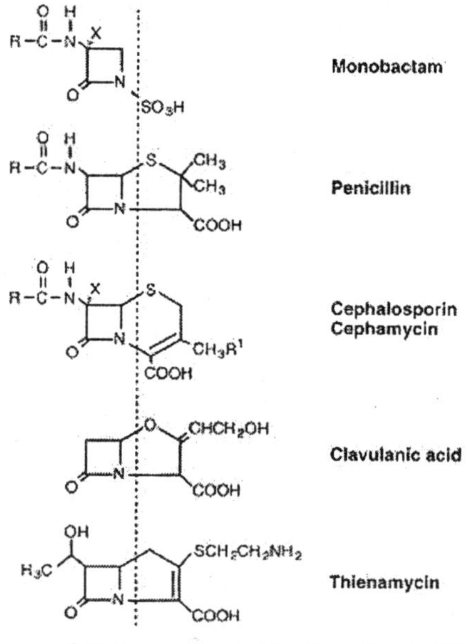

Monobactam

Penicillin

Cephalosporin
Cephamycin

Clavulanic acid

Thienamycin

FIGURE 11-3 Basic structures of ß-lactam antibiotics. Penicillins and cephalosporins/ cephamycins are widely used to inhibit both Gram-positive and Gram-negative bacilli. Monobactams inhibit only aerobic Gram-negative bacilli, clavulanic acid acts as a ß-lactamase inhibitor, and thienamycin inhibits a wide range of aerobic and anaerobic species. R and R' represent various carbon groups. X can be either hydrogen or a mehoxy group.

Vancomycin interrupts cell wall synthesis by forming a complex with the C-terminal D-alanine residues of peptidoglycan precursors. Complex formation at the outer surface of the cytoplasmic membrane prevents the transfer of the precursors from a lipid carrier to the growing peptidoglycan wall by transglycosidases. Biochemical reactions in the cell wall catalyzed by transpeptidases and D,D-carboxypeptidases are also inhibited by vancomycin and other glycopeptide antimicrobials. Because of its large size and complex structure (Fig. 11-4), vancomycin does not penetrate the outer membrane of gram-negative organisms. With resistance to beta-lactams increasing in frequency among staphylococci and enterococci, glycopeptides such as vancomycin remain important therapeutic agents against such bacteria.

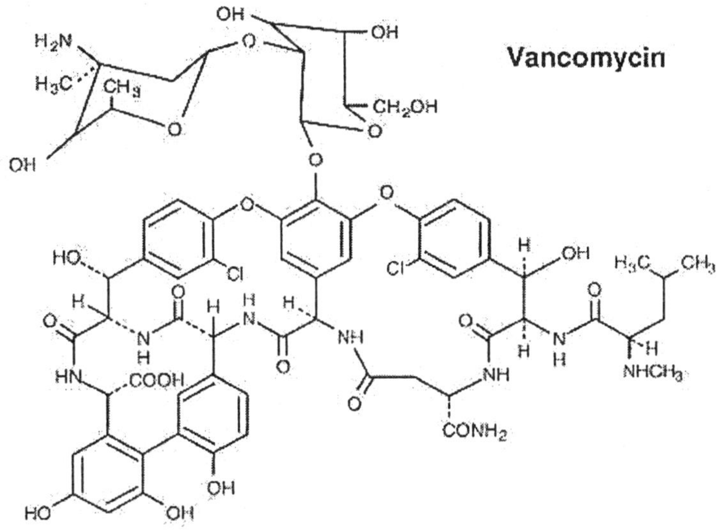

FIGURE 11-4 Structure of vancomycin.

Antibiotics that Affect the Function of Cytoplasmic Membranes

Bacterial Cytoplasmic Membranes

Biologic membranes are composed basically of lipid, protein, and lipoprotein. The cytoplasmic membrane acts as a diffusion barrier for water, ions, nutrients, and transport systems. Most workers now believe that membranes are a lipid matrix with globular proteins randomly distributed to penetrate through the lipid bilayer. A number of antimicrobial agents can cause disorganization of the membrane. These agents can be divided into cationic, anionic, and neutral agents. The best-known compounds are polymyxin B and colistemethate (polymyxin E). These high-molecular-weight octapeptides inhibit Gram-negative bacteria that have negatively charged lipids at the surface. Since the activity of the polymyxins is antagonized by Mg^{2+} and Ca^{2+}, they probably competitively displace Mg^{2+} or Ca^{2+} from the negatively charged phosphate groups on membrane lipids. Basically, polymyxins disorganize membrane permeability so that nucleic acids and cations leak out and the cell dies. The polymyxins are of virtually no use as systemic agents since they bind to various ligands in body tissues and are potent toxins for the kidney and nervous system. Gramicidins are also membrane-active antibiotics that appear to act by producing aqueous pores in the membranes. They also are used only topically.

Fungal Membranes

Fungal membranes contain sterols, whereas bacterial membranes do not. The polyene antibiotics, which apparently act by binding to membrane sterols, contain a rigid hydrophobic center and a flexible hydrophilic section. Structurally, polyenes are tightly packed rods held in rigid extension by the polyene portion. They interact with fungal cells to produce a membrane-polyene complex that alters the membrane permeability, resulting in internal acidification of the fungus with exchange of K+ and sugars; loss of phosphate esters, organic acids, nucleotides; and eventual leakage of cell protein. In effect, the polyene makes a pore in the fungal membrane and the contents of the fungus leak out. Prokaryotic cells neither bind to nor are inhibited by polyenes. Although, numerous polyene antibiotics have been isolated, only amphotericin B is used systemically (Fig. 11-5). Nystatin is used as a topical agent and primaricin as an ophthalmic preparation.

A number of other agents interfere with the synthesis of fungal lipid membranes. These agents belong to a class of compounds referred to as imidazoles: miconazole, ketoconazole, clotrimazole, and fluconazole. These compounds inhibit the incorporation of subunits into ergosterol and may also directly damage the membrane.

Amphotericin B

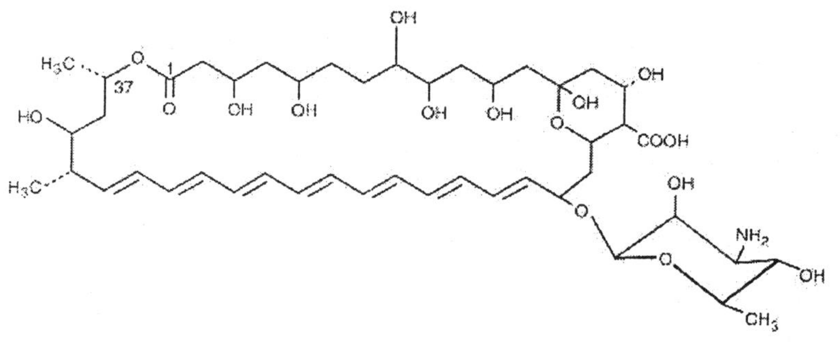

H_3C 37 1 O HO OH OH OH OH OH OH OH COOH H_3C O OH NH_2 O OH CH_3

FIGURE 11-5 Structure of amphotericin B.

Antibiotics that Inhibit Nucleic Acid Synthesis

Antimicrobial agents can interfere with nucleic acid synthesis at several different levels. They can inhibit nucleotide synthesis or interconversion; they can prevent DNA from functioning as a proper template; and they can interfere with the polymerases involved in the replication and transcription of DNA.

Interference with Nucleotide Synthesis

A large number of agents interfere with purine and pyrimidine synthesis or with the interconversion or utilization of nucleotides. Other agents act as nucleotide analogs that are incorporated into polynucleotides.

Flucytosine (5-fluorocytosine) is an antifungal agent that inhibits yeast species. It is converted in the fungal cell to 5-fluorouracfl, which inhibits thymidylate synthetase resulting in a deficit of thymine nucleotides and impaired DNA synthesis. Adenosine arabinoside inhibits viruses. It is phosphorylated in virus-infected cells and acts as a competitive analog of DATP, inhibiting the incorporation of DATP into DNA. Acyclovir is a nucleoside analog that, after being converted to a triphosphate, inhibits the thymidine kinase and DNA polymerase of herpes viruses. Zidovudine (AZT) inhibits human immunodeficiency virus (HM replication by interfering with viral RNA-dependent DNA polymerase (reverse transcriptase).

Agents That Impair the Template Function of DNA

A number of substances bind to DNA by intercalation. None of them is useful as an antibacterial agent; however, chloroquine and miracil D (lucanthone) inhibit plasmodia and schistosomes, respectively. These agents are thought to intercalate into the DNA and thereby to inhibit further nucleic acid synthesis. Acridine dyes such as proflavine act by this intercalation mechanism, but because they are toxic and carcinogenic in mammals they cannot be used as antibacterial agents.

Inhibition of DNA-Directed DNA Polymerase

Rifampcins are a class of antibiotics that inhibit DNA-directed RNA polymerase (Fig. 11-6). Polypeptide chains in RNA polymerase attach to a factor that confers specificity for the recognition of promoter sites that initiate transcription of the DNA. Rifampin binds noncovalently but strongly to a subunit of RNA polymerase and interferes specifically with the initiation process. However, it has no effect once polymerization has begun.

Rifampin

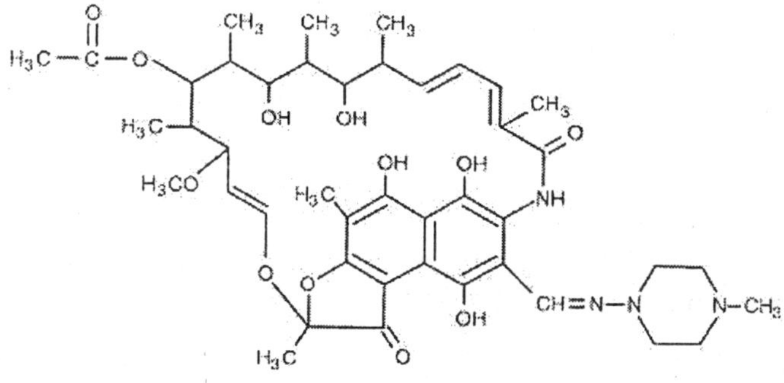

FIGURE 11-6 Structure of rifampin, which inhibits the DNA-directed RNA polymerase.

Inhibition of DNA Replication

DNA gyrase and topoisomerase I act in concert to maintain an optimum supercoiling state of DNA in the cell. In this capacity, DNA gyrase is essential for relieving torsional strain during replication of circular chromosomes in bacteria. The enzyme is a tetrameric protein composed of two A and two B subunits. A transient, covalent bond between the A subunit and DNA occurs during the double strand passage reaction catalyzed by gyrase. Quinolones such as nalidixic acid (Fig. 11-7), bind to the cleavage complex composed of DNA and gyrase during this strand passage. This interaction of quinolone acts to stabilize the cleavage intermediate which has a detrimental effect on the normal DNA replication process. The effects of this inhibition result in the death of the bacterial cell. The newer fluoroquinolones such as ciprofloxacin, norfloxacin, and ofloxacin also interact with DNA gyrase and possess a broad spectrum of antimicrobial activity.

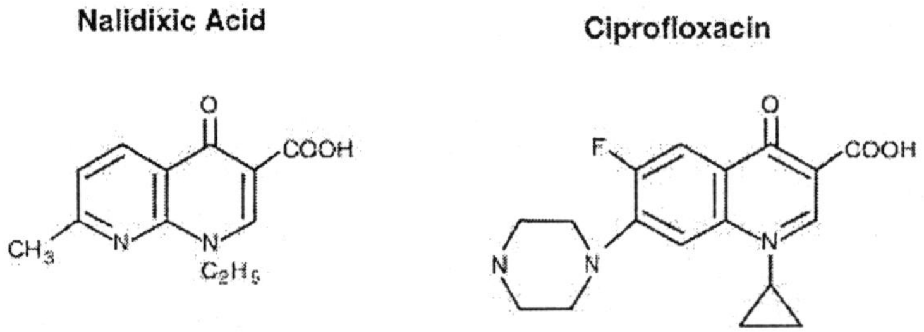

Nalidixic Acid

Ciprofloxacin

FIGURE 11-7 Structure of quinolone antibiotica. Nalidixic inhibits only aerobic Gram-negative species. In ciprofloxacin, the flourine provides Gram-positive activity, the piperazine group increases activity against members of the *Enterovacteriaceae*, and the piperazine and cylopropyl groups give activity against *Pseudomonas* species.

Nitroimidazoles such as metronidazole inhibit anaerobic bacteria and protozoa. The nitro group of the nitrosohydroxyl amino moiety is reduced by an electron transport protein in anaerobic bacteria (Fig. 11-8). The reduced drug causes strand breaks in the DNA. Mammalian cells are unharmed because they lack enzymes to reduce the nitro group of these agents.

Antimicrobial Inhibitors of Ribosome Function.

The basic structure and function of ribosomes are presented in Fig. 11-1. A number of antibacterial agents act by inhibiting ribosome function. Bacterial ribosomes contain two subunits, the 50S and 30S subunits, and it is possible to localize the action of antibiotics to one or both subunits. It is also possible to isolate the specific ribosomal proteins to which an agent binds and to isolate bacterial mutants that lack a specific ribosomal protein and therefore show resistance to a particular agent.

Aminoglycosides act by binding to specific ribosomal subunits. Aminoglycosides are complex sugars connected in glycosidic linkage (Fig. 11-9). They differ both in the molecular nucleus, which can be streptidine or 2-deoxystreptidine, and in the

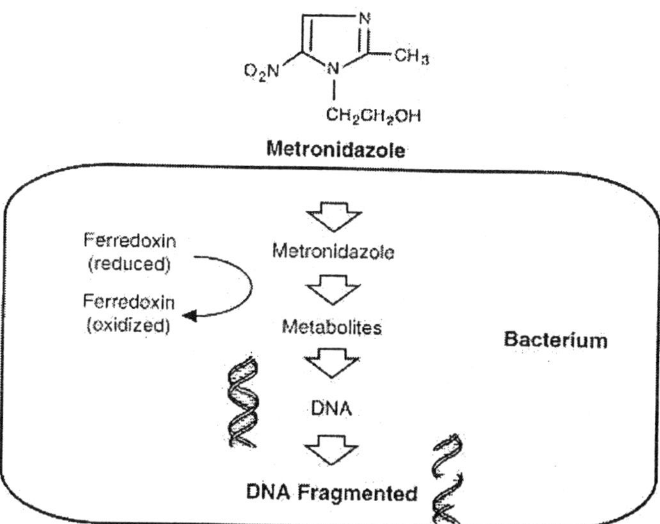

FIGURE 11-8 Structure of metronidazole and its mechanism of action. Metronidazole enters an aerobic bacterium where, via the electron transport protein ferrodoxin, it is reduced. The drug then binds to DNA, and DNA breakage occurs.

aminohexoses linked to the nucleus. Essential to the activity of these agents are free NH, and OH groups by which aminoglycosides bind to specific ribosomal proteins. Streptomycin, the first aminoglycoside studied, was a useful tool in elucidating protein synthesis. However, it is rarely used clinically today except to treat tuberculosis, and its mode of action differs to some extent from that of the other clinically useful aminoglycosides, which are 2-deoxystreptidine derivatives such as gentamicin, tobramycin, and amikacin. Streptomycin binds to a specific S12 protein in the 30S ribosomal subunit (Fig. 11-10) and causes the ribosome to misread the genetic code. Other aminoglycosides bind not only to the S12 protein of the 30S ribosome, but also to some extent to the L6 protein of the 50S ribosome. This latter binding is quite important in terms of the resistance of bacteria to aminoglycosides. Indeed, the aminoglycoside-type drugs can combine with other binding sites on 30S ribosomes, and theykill bacteria by inducing the formation of aberrant, nonfunctional complexes as well as by causing misreading (Fig. 11-11).

Spectinomycin is an aminocylitol antibiotic that is closely related to the aminoglycosides. It binds to a different protein in the ribosome and is bacteriostatic but not bactericidal. It is used to treat penicillin-resistant gonorrhea.

Other agents that bind to 30S ribosomes are the tetracyclines (Fig. 11-12). These agents appear to inhibit the binding of aminoacyl-tRNA into the A site of the bacterial ribosome. Tetracycline binding is transient, so these agents are bacteriostatic. Nonetheless, they inhibit a wide variety of bacteria, chlamydias, and mycoplasmas and are extremely useful antibiotics.

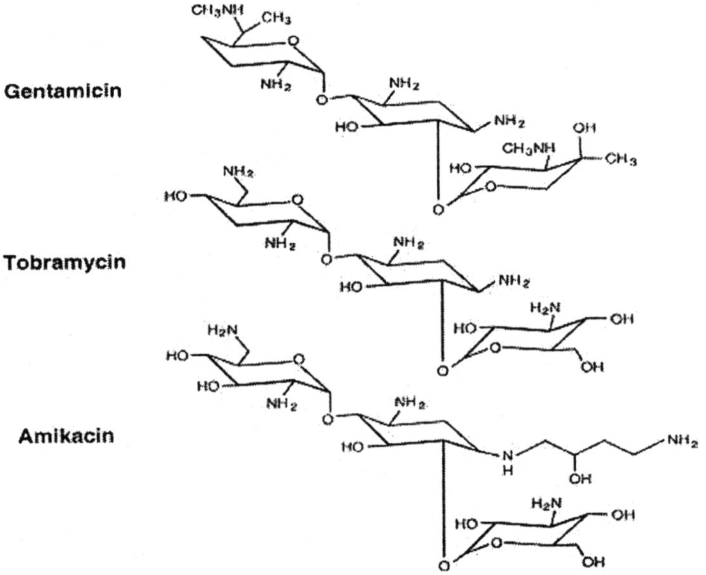

Gentamicin

Tobramycin

Amikacin

FIGURE 11-9 Structures of three aminoglycoside antibiotics used clinically. Critical aspects of the molecules are the amino and hydroxy groups that bind to proteins in the ribosomes.

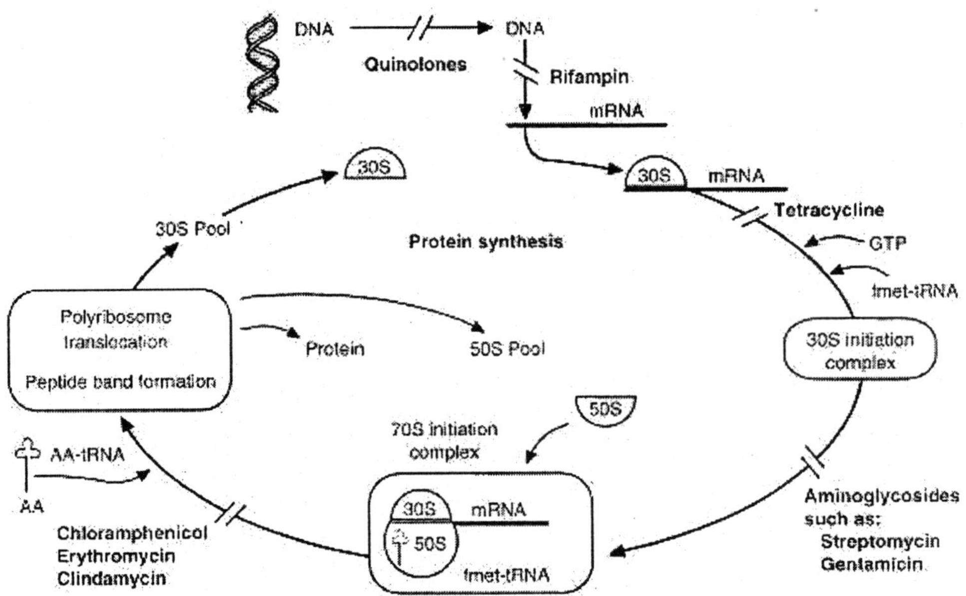

FIGURE 11-10 Diagrammatic representation of inhibition sites of protein biosynthesis by various antibiotics that bind to the 30S and 50S ribosomes.

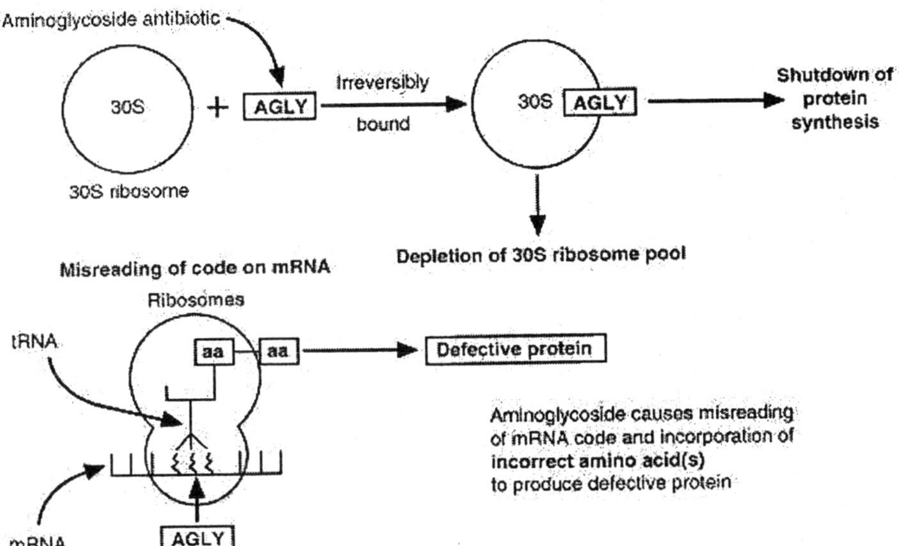

FIGURE 11-11 Inhibition of protein biosynthesis by aminoglycosides.

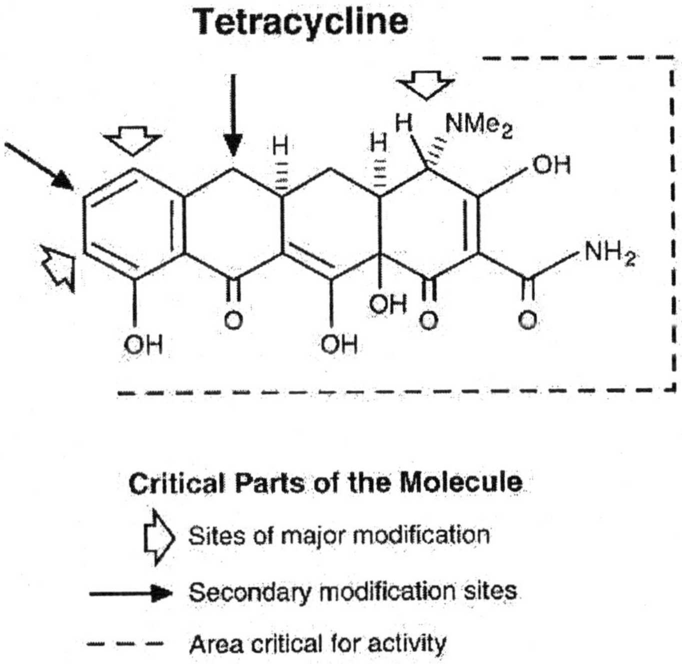

Tetracycline

H NMe₂ H H OH NH₂ OH O OH OH O O

Critical Parts of the Molecule

- Sites of major modification
- Secondary modification sites
- Area critical for activity

FIGURE 11-12 Structure of tetracycline showing the area critical for activity and major and minor points of modification.

There are three important classes of drugs that inhibit the 50S ribosomal subunit. Chloramphenicol (Fig. 11-13) is a bacteriostatic agent that inhibits both Gram-positive and Gram-negative bacteria. It inhibits peptide bond formation by binding to a peptidyltransferase enzyme on the 50S ribosome. Macrolides are large lactone ring compounds that bind to 50S ribosomes and appear to impair a peptidyltransferase reaction or translocation, or both. The most important macrolide is erythromycin, which inhibits Gram-positive species and a few Gram-negative species such as *Haemophilus, Mycoplasma, Chlamydia, and Legionella.*

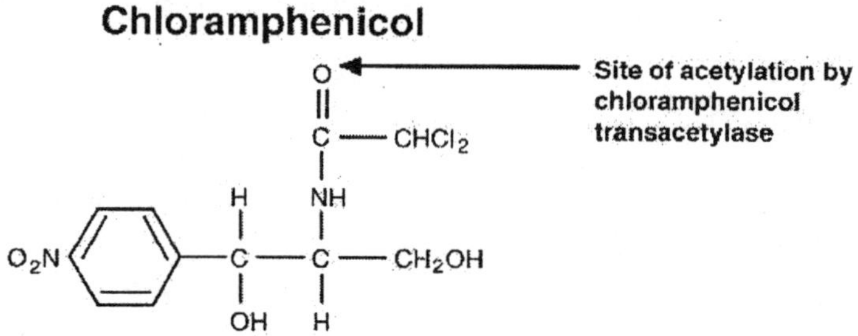

FIGURE 11-13 Structure of chloramphenicol.

New molecules such as azithromycin and clarithromycin have greater activity than erythromycin against many of these pathogens. Lincinoids, of which the most important is clindamycin, have a similar site of activity (Fig. 11-14). Both macrolides and lincinoids are generally bacteriostatic, inhibiting only the formation of new peptide chains.

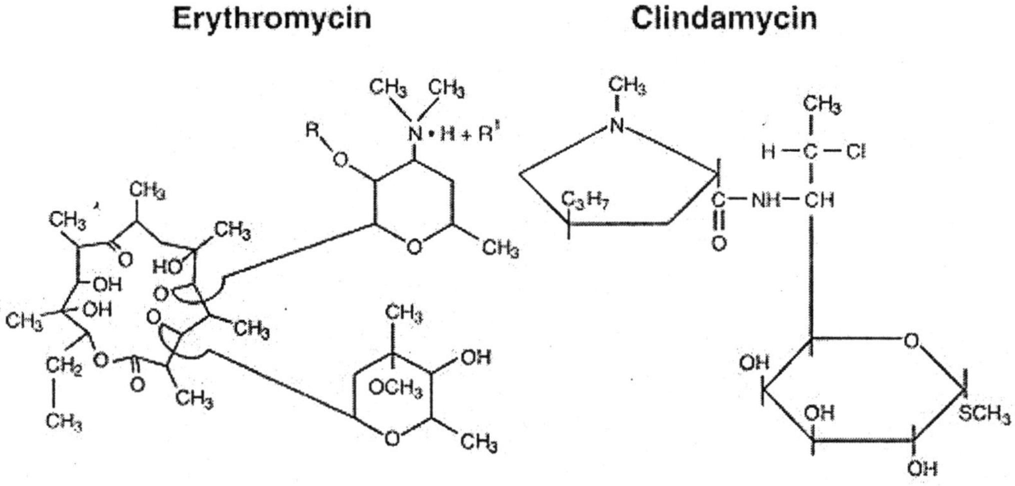

FIGURE 11-14 Structure of erythromycin (prototype or macrolide) and clindamycin. Although extremely different in structure, both compounds inhibit protein synthesis by binding to 50S ribosome.

Drugs that Inhibit Other Biochemical Targets

Both trimethoprim and the sulfonamides interfere with folate metabolism in the bacterial cell by competitively blocking the biosynthesis of tetrahydrofolate, which acts as a carrier of one-carbon fragments and is necessary for the ultimate synthesis of DNA, RNA and bacterial cell wall proteins (Fig. 11-15). Unlike mammals, bacteria and protozoan parasites usually lack a transport system to take up preformed folic acid from their environment. Most of these organisms must synthesize folates, although some are capable of using exogenous thymidine, circumventing the need for folate metabolism.

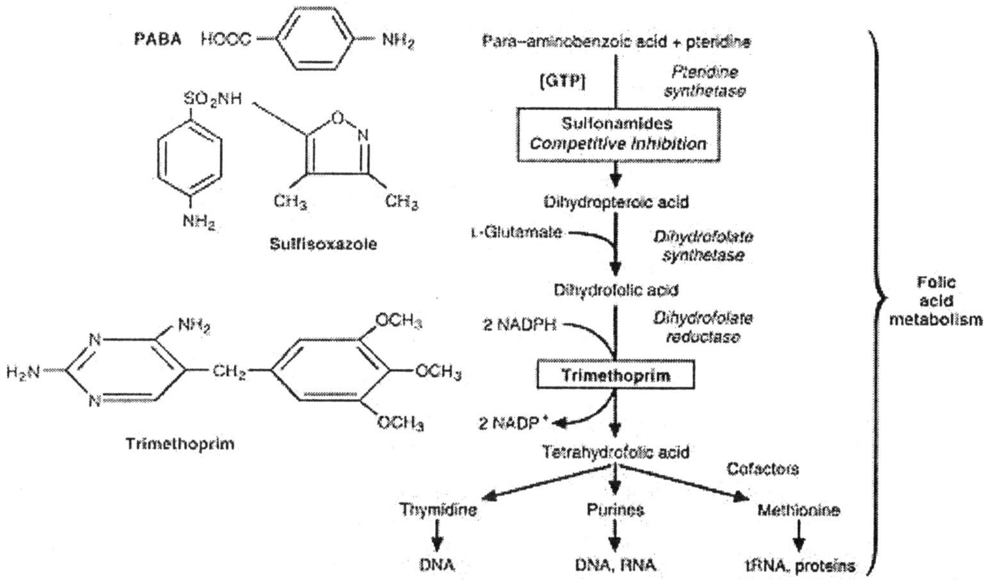

FIGURE 11-15 Structure of sulfonamide and trimethoprim with sites of inhibition of folic metabolism.

Sulfonamides competitively block the conversion of pteridine and p-aminobenzoic acid (PABA) to dihydrofolic acid by the enzyme pteridine synthetase. Sulfonamides have a greater affinity than p-aminobenzoic acid for pteridine synthetase. Trimethoprim has a tremendous affinity for bacterial dihydrofolate reductase (10,000 to 100,000 times higher than for the mammalian enzyme); when bound to this enzyme, it inhibits the synthesis of tetrahydrofolate.

Antibacterial Agents that Affect Mycobacteria

Isoniazid is a nicotinamide derivative that inhibits mycobacteria. Its precise mode of action is not known, but it affects the synthesis of lipids, nucleic acids, and the mycolic acid of the cell walls of these species. Ethambutol is also an antimycobacterial agent whose mechanism of action is unknown. It is mycostatic, whereas isoniazid is mycocidal. The other antituberculosis drugs, rifampin and streptomycin, affect mycobacteria in the same manner that they inhibit bacteria. Pyrazinamide is a synthetic analog of nicotinamide. It is bactericidal, but its exact mechanism is unknown.

Bacterial Resistance

Bacteria have proved adept at developing resistance to new antimicrobial agents. There are a number of ways in which bacteria can become resistant (Table 11-2). Most of the early studies of bacterial resistance focused on single-step mutational events of chromosomal origin. Resistance to the early sulfonamides, for example, was the result of a single amino acid change in the enzyme pteridine synthetase that caused sulfonamides to bind less well than p-aminobenzoic acid. Similarly, a single step mutation that altered a ribosomal protein conferred resistance to streptomycin. In the late 1950s, Japanese workers found that enteric bacteria such as Shigella dysenteriae had become resistant not only to sulfonamides but also to the tetracyclines and chloramphenicol. This resistance was due not to a chromosomal change, but rather to the presence of extrachromosomal DNA that was transmissible. This type of resistance is called plasmid-mediated resistance.

Table 11-2 Mechanisms of Resistance

Alteration of Target

- Modification to insensitivity to inhibitor
- Reduction in physiologic inportance of target
- Synthesis of new target enzyme that duplicates function of inhibited target

Prevention of access to target

- Efflux of more drug than enters cell
- Failure of modified drug to enter cell

Inactivation of agent

- Destruction of the agent
- Modification of the agent so it fails to bind to target

Failure to convert an inactive precursor agent to its active form

Resistance-conferring plasmids are present in virtually all bacteria (Table 11-3). For example, resistance to ampicillin appeared in *Haemophilus influenzae* in 1974 and in *Neisseria gonorrhoeae* in 1976. In the last several years, organisms such as enterococci have been shown to contain plasmids that confer resistance to drugs such as ampicillin and aminoglycosides.

Table 11-2 R-Plasmid -Mediated Resistance

Antibiotic	Mechanism	Organisms
Penicillin, ampicillin, carbenicillin, etc.	β-Lactamase hydrolysis	Staphylococci, enterococci (rate), Enterobacteriaceae, pseudomonads, bacterioides
Oxacillin, methicillin etc.	β -Lactamase hydrolysis	Enterobacteriaceae, pseudomonads
Cephalosporins	β -Lactamase	Staphylococci, Enterobacteriaceae, pseudomonads, bacteroides
Chloramphenicol	Acetylation	Staphylococci, enterococci, streptococci, Enterobacteriaceae, pseudomonads
Tetracyclines	Permeability block	Staphylococci, enterococci, streptococci, Enterobacteriaceae, pseudomonads, bacteroides
Aminoglycosides Streptomycin Neomycin Kanamycin Tobramycin Amikacin	Acetylation Phosphorylation Adenylation (alters binding to ribosomes and uptake of drug)	Staphylococci, enterococci, Enterobacteriaceeae, pseudomonads
Macrolides-Lincinoids Erythromycin	Alterated 23S RNA	Staphylococci, enterococci, streptococci, bacteroides
Clindamycin	Alteraed 23S RNA	
Trimethoprim	Alterated dihydrofolate reductase	Staphylococci, Enterobacteriaceae
Sulfonamides	Altered tetraphydropteroic synthetase	Staphylococci, enterococci, streptococci, Enterobacteriaceae, pseudomonads
Fostomycin	Altered glucose	Staphylococci, Enterobacteriaceae
Vancomycin	New protein	Enterococci

Bacteria also contain transposons, which can insert into plasmids and also into the chromosome (see Ch. 5). Transposon-mediated resistance to most of the major antibiotics has been found in the past few years.

Antimicrobial agents exert a strong selective pressure on the development of both chromosomal and plasmid-mediated resistance, as discussed below. Administration of an antibiotic destroys the susceptible bacteria in a population, but may permit resistant ones to proliferate. From an epidemiologic viewpoint, plasmid-mediated resistance is the most important type, since it is transmissible, is usually highly stable, confers resistance to many different classes of antibiotics simultaneously, and often is associated with other characteristics that enable a microorganism to colonize and invade a susceptible host.

Mechanisms of Resistance

The basic mechanisms by which a microorganism can resist an antimicrobial agent are (1) to alter the receptor for the drug (the molecule on which it exerts its effect);

(2) to decrease the amount of drug that reaches the receptor by altering entry or increasing removal of the drug; (3) to destroy or inactivate the drug; and (4) to develop resistant metabolic pathways. Bacteria can possess one or all of these mechanisms simultaneously.

Resistance Due to Altered Receptors

ß-Lactam Resistance. The ability to analyze changes in receptors for ß-lactams by competition experiments in which [^{14}C]penicillin is inhibited from binding to penicillin-binding proteins has explained a number of cases of bacterial resistance to penicillins and cephalosporins. In 1977, *Streptococcus pneumoniae* strains resistant to penicillin G were encountered in South Africa. Plasmids were not the cause of the resistance. Penicillin-resistant *S pneumoniae* cells have altered penicillin-binding proteins, which bind penicillin less well. Resistance of *S pneumoniae* to penicillin has been increasing, and there are now relatively resistant isolates (minimal inhibitory concentration [MIC], 0.1 to 1 mg/ml) in many parts of the world.

Altered penicillin-binding proteins also explain the resistance of some *Staphylococcus aureus* strains to ß-lactamase-stable penicillins (the so-called methicillin-resistant strains). The ß-lactams induce synthesis of a new penicillin-bindingprotein, PBP2a, which does not bind any ß-lactam. The ß-lactam resistance of coagulase-negative staphylococci is also the result of altered penicillin-binding proteins. Staphylococcal organisms resistant to methicillin are resistant to all penicillins, cephalosporins, and carbapenems.

The resistance of group D streptococci to ß-lactam antibiotics appears to be the result of lower affinity of the penicillin-binding proteins for the penicillins. Enterococci are resistant to all cephalosporins because of failure to bind to the penicillin-binding proteins. One Gram-negative species for which resistance to ß-lactam antibiotics can be correlated withdiminished affinity of the target enzymes is *N gonorrhoeae.*

Vancomycin Resistance. Certain transposable genetic elements encode special cell wall-synthesizing enzymes which change the structure of the normal D-Ala-D-Ala side chain in the peptidoglycan assembly pathway. The altered side chain (D-Ala-D-Lac) does not bind vancomycin and allows normal peptidoglycan polymerization to occur in the presence of the drug. Depending upon the nature of the vancomycin resistance gene, high-level resistance can occur to glycopeptides. Thus far, this type of resistance has been found in enterococci but not in multi-resistant isolates of *Staphylococcus aureus.*

Macrolide-lincomycin Resistance. Macrolide-lincomycin resistance in clinical isolates of staphylococci and streptococci has been recognized for several decades. The resistance is due to methylation of two adenine nucleotides in the 23S component of 50S RNA. This resistance is plasmid mediated, and the resistance is encoded on transposons. Resistance results from induction of an enzyme that is normally repressed. The methylated RNA binds macrolidelincomycin-type drugs less well than unmethylated RNA does. Induction of resistance varies by species, and in most Gram-positive species erythromycin is a more effective inducer of resistance than is clindamycin. The plasmids that mediate macrolide-lincomycin resistance in streptococci and staphylococci have extensive structural similarity, indicating that these plasmids readily pass between these species.

Rifampin Resistance. The resistance of bacteria to rifampin is caused by an alternation of one amino acid in DNA-directed RNA polymerase, which results in reduced binding of rifampin. The degree of resistance is related to the degree to which the enzyme is changed, but does not correlate strictly with enzyme inhibition. This form of resistance occurs at a low level in any population of bacteria so that resistance develops by natural selection during a course of therapy. Naturally resistant organisms are more common among members of the Enterobacteriaceae, explaining why agents of urinary tract infections rapidly became resistant to rifampin. The resistance of *Neisseria meningitides* to rifampin appeared in closed military settings in which rifampin has been used for prophylaxis.

Sulfonamide-trimethoprim Resistance. Sulfonamide can be rendered ineffective by altered or new dihydropteroicsynthetase that has poor affinity for sulfonamides and preferentially binds p-aminobenzoic acid. Sulfonamide resistance of this type can result from a point mutation or from acquisition of a plasmid that causes synthesis of the new enzyme. A most serious resistance problem is an increase in resistance to trimethoprim. This plasmid- and transposon-mediated resistance is due to production of an altered dihydrofolate reductase that has markedly reduced affinity for trimethoprim.

Quinolone Resistance. Resistance to quinolones can be caused by mutations in DNA gyrase subunits A or B, reduced outer membrane permeability in gram-negative cells, or to active efflux transporters found in many bacteria. The highest level of resistance to the newer fluoroquinolones is most frequently associated with chromosomal mutations, causing amino acid substitutions in a highly conserved region in the A subunit of DNA gyrase. Multiple-mechanisms of resistance can occur in a single isolate of bacteria, leading to a higher level of resistance to many fluoroauinolones.

Resistance Due to Decreased Entry of a Drug

Tetracycline Resistance. The uptake of tetracycline by members of the Enterobacteriaceae is biphasic. In an initial energy-independent rapid phase, tetracycline binds to cell surface layers and passes by diffusion through the outer layers of the cell. In the second, energy-dependent phase, tetracycline crosses the cytoplasmic membrane, probably by means of a proton-motive force. The precise transport system has not been identified.

Tetracycline resistance is common in both Gram-positive and Gram-negative bacteria. In most cases it is plasmid encoded and inducible; however, chromosomal, constitutive resistance is found in some organisms such as *Proteus* species. Many plasmid-encoded specified tetracycline resistance determinants have been found in enteric bacteria. The most common of these determinants, TetB, is also present in *H influenzae*. Tetracycline resistance in *Staphylococcus aureus* is due primarily to small multicopy plasmids; chromosomal resistance is rare. Tetracycline resistance is found on nonconjugative plasmids in *Streptococcus faecalis* and on the chromosome of *S pneumoniae, S agalactiae* (group B streptococci), and oral streptococci. *Clostridium* species such as *C difficile* harbor chromosomal genes for tetracycline resistance.

Basically, tetracycline resistance is due to a decrease in the levels of drug accumulation. Decreased uptake and increased efflux both probably participate. Resistant bacteria bind less tetracycline, and the tetracycline they do accumulate is lost by an energy-dependent process when they are in a drug-free milieu.

Plasmid-mediated resistance to tetracyclines can be partially overcome in Gram-positive species by modifying the tetracycline nucleus. Hence, achievable concentrations of minocycline and doxycycline, in particular, will inhibit some tetracycline-resistant streptococci such as *S pneumoniae,* and some *S aureus* strains. Molecular modification has not been successful in overcoming the tetracycline resistance of members of the Enterobacteriaceae or *Pseudomonas* or most *Bacteroides* species.

Tetracycline resistance is a major concern because it is located on plasmids near insertion sites, and these plasmids readily acquire other genetic information to enlarge the spectrum of resistance. The widespread use of tetracycline in animal feeds may be a factor in the extensive, worldwide resistance of members of the Enterobacteriaceae, particularly enteric species such as *Salmonella,* to tetracyclines and subsequently to many other drugs. Not only can tetracycline resistance move among members of the Enterobacteriaceae on plasmids, but plasmids mediating tetracycline resistance have moved between *S aureus, S epidermidis, S pyogenes, S pneumoniae,* and *S faecalis.*

Fosfomycin Resistance: Fosfomycin and fosmidomycin, which inhibit cell wall synthesis, enter bacteria by means of a glycerol-phosphate or glucose-6-phosphate transport system. Gram-positive bacteria in which the glucose-6-phosphate transport system is poorly developed do not take up these drugs in concentrations adequate to inhibit the cell wall synthesis. This resistance usually is chromosomal. The resistance of Gram-negative bacteria to these agents is related primarily to the presence in the population of some bacteria that can function without the transport system. Plasmids and transposons that transfer resistance to fosfomycin have been found in bacteria such as *Serratia marcescens.*

Aminoglycoside Resistance: In the most important form of aminoglycoside resistance, the compound is modified outside the cell and resistance is due partly to poor uptake of the altered compound. Also, all aminoglycosides have free amino and hydroxy groups that are essential for binding to ribosomal proteins. A number of enzymes can acetylate the amino groups and phosphorylate or adenylate the hydroxyl groups (Fig. 11-9). Other forms of resistance, such as altered binding site on 30S ribosomes, are much less common.

In members of the Enterobacteriaceae and in *Pseudomonas* species, the aminoglycosides pass through the cell wall via channels designed to admit cationic molecules to the periplasmic space. These channels, called porin channels, are lined by the porin protein. Aminoglycosides are then translocated across the cell membrane by an energy-dependent proton-motive force and, in the cytoplasm, bind to ribosomes located just below the membrane. Aminoglycosides bind only to ribosomes actively engaged in protein synthesis. Binding to the ribosomes induces a protein involved in the uptake of the aminoglycosides.

Bacteria may contain in the periplasmic space enzymes that acetylate, phosphorylate, or adenylate aminoglycosides to various degrees. It is not clear whether the enzymes are free in the periplasmic space or bound to the cytoplasmic membrane. The modified aminoglycosides do not bind well to ribosomes, and hence uptake is poor or absent (Fig. 11-16).

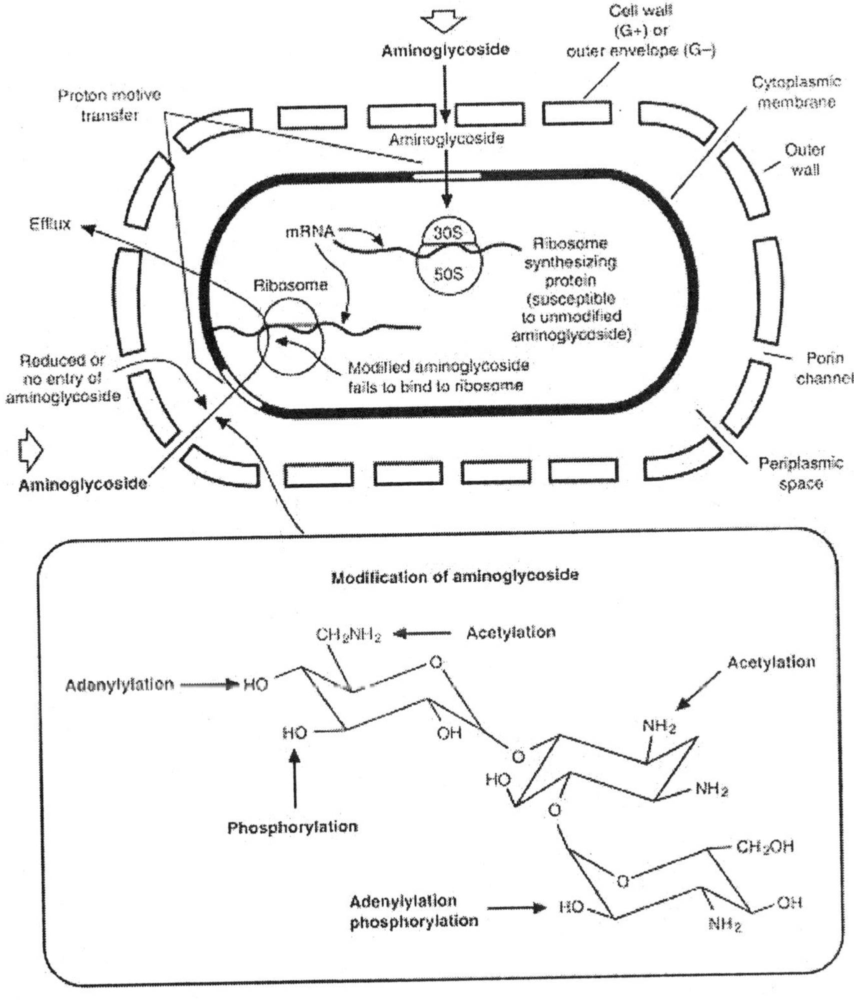

FIGURE 11-16 Diagrammatic representation of transfer and transfer reduction of aminoglycoside across the bacterial cell wall. If it is modified by acetylation, adenylation, or phosphorylation (see box), the drug will not bind to ribosomes and will leave the bacterial cell.

Aminoglycoside-modifying enzymes have been found in Gram-positive species such as *S aureus, S faecalis, S pyogenes,* and enzymes are particularly prevalent in members of the Enterobacteriaceae and *P aeruginosa. S pneumoniae.* These Many of the genes for aminoglycoside-modifying enzymes are carried on transposons

Anaerobic organisms such as *Bacteroides* species are resistant to aminoglycosides because they lack an oxygen-dependent transport system to move the drugs across the cytoplasmic membrane. Although, most resistance of *S aureus* to aminoglycosides is due to aminoglycoside-modifying enzymes, small-colony variants of staphylococci also show resistance, which may be due to a defect in adenylate cyclase or in cyclic adenosine 5'-monophosphate (cAMP)binding proteins such that cells with a reduced growth rate do not transport aminoglycosides into the cytoplasm. Some members of the Enterobacteriaceae and *P aeruginosa* appear to be resistant because of altered porin channels, since these bacteria do not take up any drug and do not have aminoglycoside-inactivating enzymes.

Resistance Due to Destruction or Inactivation of a Drug

Chloramphenicol Resistance: Many Gram-positive and Gram-negative bacteria, including some recently discovered *H influenzae* strains, are resistant to chloramphenicol because they possess the enzyme chloramphenicol transacetylate, which acetylates hydroxyl groups on the chloramphenicol structure. This enzyme, unlike the aminoglycoside-inactivating enzymes and ß-lactamases, is an intracellular enzyme of higher molecular weight and subunit structure. Acetylated chloramphenicol binds less well to the 50S ribosome.

ß-Lactam Resistance: The best-known mechanism of bacterial resistance is the resistance to ß-lactams, which is mediated by penicillinase enzymes. Resistance of *E coli* to penicillin was recognized in 1940, before sufficient penicillin was made to be clinically useful. In the 1940s, resistance of staphylococci was shown to be due to a penicillinase. As these enzymes also attack other ß-lactam compounds such as cephalosporins, carbapenems, and monobactams, they would be more appropriately designated ß-lactamases. The most important activity of these enzymes is alteration of the ß-lactam nucleus (Fig. 11-17). ß-lactamases are widely distributed in nature and are usually classified on the basis of the principal compounds they destroy (e.g., as penicillinases or cephalosporinases) (Fig. 11-18). ß-Lactamases may be chromosomally or plasmid mediated, and they may be constitutive or inducible.

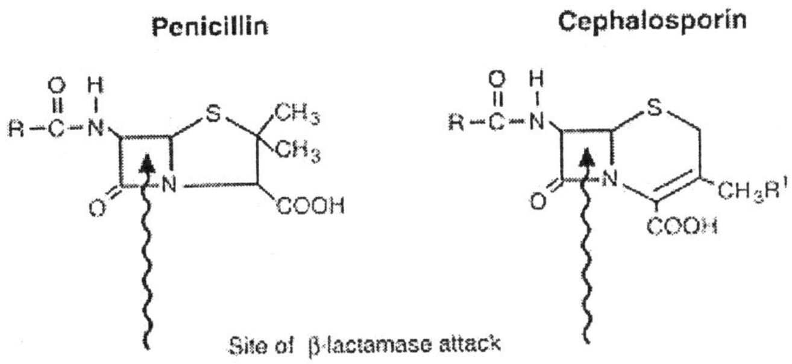

Penicillin

Cephalosporin

Site of β-lactamase attack

FIGURE 11-17 Site of ß-lactamase attack in penicillins and cephalosporins.

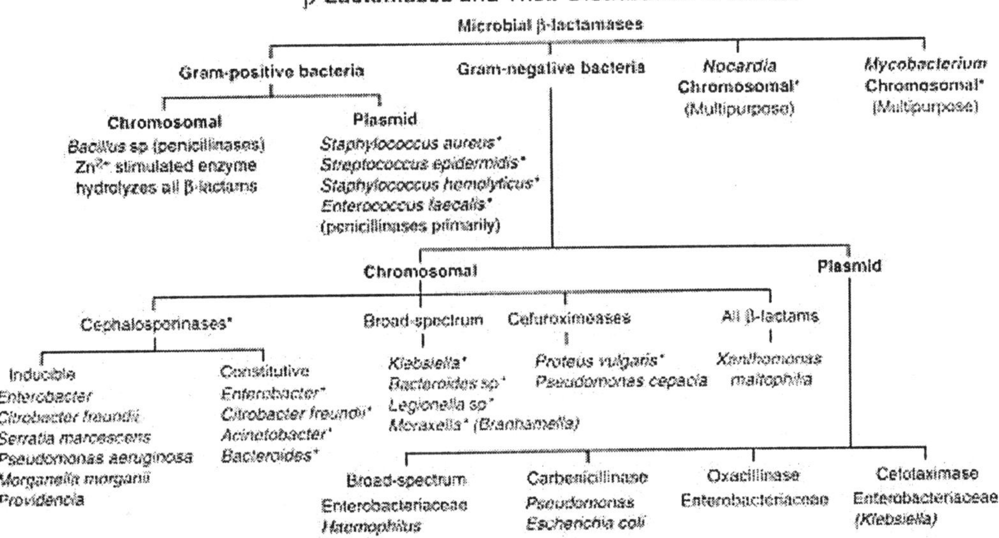

*Inhibited by clavulanate, sulbactam

FIGURE 11-18 β-Lactamase found in bacteria and their classification and synthesis, whether chromosomally or plasmid mediated.

In Gram-positive species, ß-lactamases are primarily exoenzymes; that is, they are excreted into the milieu around the bacteria. Virtually all hospital isolates of staphylococci, both *S aureus* and *S epidermidis*, have β-lactamases, and 50 to 80 percent of community-acquired staphylococcal isolates produce β-lactamases. In Gram-negative species, both aerobic and anaerobic, β-lactamases are contained in the periplasmic space, thus effectively protecting the penicillin-binding proteins.

Resistance of staphylococci to β-lactams was soon overcome with the antistaphylococcal penicillins and the cephalosporins. Some strains of *S aureus* produce more β-lactamase constitutively and can destroy some of the cephalosporins.

In 1974, *H influenzae* was shown to possess a plasmid mediated ß-lactamase. At present 10 to 35 percent of *H influenzae* strains in the United States produce ß-lactamases. The TnA transposon has become more widespread, and the resistance of *Haemophilus* species to penicillin G and ampicillin seems to be increasing yearly. The *Haemophilus* ß-lactamase is the same structurally as the enzyme found in *E coli, Salmonella, Shigella,* and *N gonorrhoeae.* The enzyme has generally been called the TEM enzyme after the initials of the Greek girl from whom an *E coli* strain containing a plasmid ß-lactamase was first isolated. These enzymes are also called Richmond-Sykes class IIIa enzymes from a classification proposed by Richmond and Sykes in 1973. By far the most common plasmid ß-lactamase found in nature is TEM-1, which accounts for 75 to 80 percent of plasmid-mediated ß-lactamase resistance worldwide. Recently new ß-lactamases have been found that hydrolyze compounds such as inomethoxy cephalosporin, which were not destroyed by other plasmid-encoded ß-lactamases. The new ß-lactamases have an altered amino acid composition, which permits binding to the cephalosporin and subsequent hydrolysis. How common these new enzymes will become is unknown.

Chromosomally mediated ß-lactamases are present in many *Enterobacter, Citrobacter, Proteus-Providencia,* and *Pseudomonas* species. All *Klebsiella* species possess a ß-lactamase, which acts primarily as a penicillinase and is chromosomally mediated. Constitutively produced ß-lactamases are also present in many anaerobic species.

Table 11-3 lists the major ß-lactamases of clinical importance. ß-Lactamases vary in their ability to destroy penicillins and cephalosporins. ß-Lactamase activity is only one component of the ß-lactam resistance of Gram-negative bacteria, since resistance to ß-lactams is a combination of decreased entry, ß-lactamase stability and affinity of the compounds for penicillin-binding proteins.

Table 11-3 R-Plasmid -Mediated Resistance

Antibiotic	Mechanism	Organisms
Penicillin, ampicillin, carbenicillin, etc.	β-Lactamase hydrolysis	Staphylococci, enterococci (rate), Enterobacteriaceae, pseudomonads, bacterioides
Oxacillin, methicillin etc.	β -Lactamase hydrolysis	Enterobacteriaceae, pseudomonads
Cephalosporins	β -Lactamase	Staphylococci, Enterobacteriaceae, pseudomonads, bacteroides
Chloramphenicol	Acetylation	Staphylococci, enterococci, streptococci, Enterobacteriaceae, pseudomonads
Tetracyclines	Permeability block	Staphylococci, enterococci, streptococci, Enterobacteriaceae, pseudomonads, bacteroides

Aminoglycosides Streptomycin Neomycin Kanamycin Tobramycin Amikacin	Acetylation Phosphorylation Adenylation (alters binding to ribosomes and uptake of drug)	Staphylococci, enterococci, Enterobacteriaceeae, pseudomonads
Macrolides-Lincinoids Erythromycin	Alterated 23S RNA	Staphylococci, enterococci, streptococci, bacteroides
Clindamycin	Alteraed 23S RNA	
Trimethoprim	Alterated dihydrofolate reductase	Staphylococci, Enterobacteriaceae
Sulfonamides	Altered tetraphydropteroic synthetase	Staphylococci, enterococci, streptococci, Enterobacteriaceae, pseudomonads
Fostomycin	Altered glucose	Staphylococci, Enterobacteriaceae
Vancomycin	New protein	Enterococci

Synthesis of Resistant Metabolic Pathway

No synthesis of a new type of cell wall resistant to β-lactams has occurred, but some bacteria, particularly some streptococci, lack the hydrolytic enzymes necessary for forming a new cell wall, and so β-lactams do not lyse these bacteria. An altered hydrolytic system thus converts a bactericidal antibioticinto a bacteriostatic agent. Whether such resistance occurs in Gram-negative species is not clear.

Some thymidine-requiring streptococci are not inhibited by trimethoprim and sulfonamides and so are not killed by these agents. These organisms are a rare cause of urinary tract infections. Other bacteria produce adequate deoxyribosylthymine 5'-monophosphate (DTMP) by alternativemethods and, as a result. survive exposure to these folate inhibitors.

Certain *Candida* or *Cryptococcus* yeasts are resistant to flucytosine because they cannot convert it to its active component, fluorouracil. Other fungi can resist the polyenes and imidazoles because they synthesize membrane components by different metabolic mechanisms.

Combinations of Antimicrobial Agents

Antibiotics are frequently used in combination for the following reasons: (1) to treat a life-threatening infection; (2) to prevent emergence of bacterial resistance; (3) to treat mixed infections of aerobic and anaerobic bacteria; (4) to enhance antibacterial activity (synergy); and (5) to use lower doses of a toxic drug. Combined treatment is reasonable when the precise agents of a serious infection are unknown. Use of two or more drugs to prevent the emergence of resistance is effective for tuberculosis and for therapy of some chronic infections. The use of combinations to achieve synergy is more complicated. Synergy occurs when a combination of two drugs causes inhibition or killing when used at a fourfold-lower concentration than that of either component drug used separately (Fig. 11-19). However, indifference or antagonism may occur instead. Indifference means that the combined action is the same as with either component; antagonism refers to a reduction in the activity of one or both components in the presence of the other (Fig. 11-19).

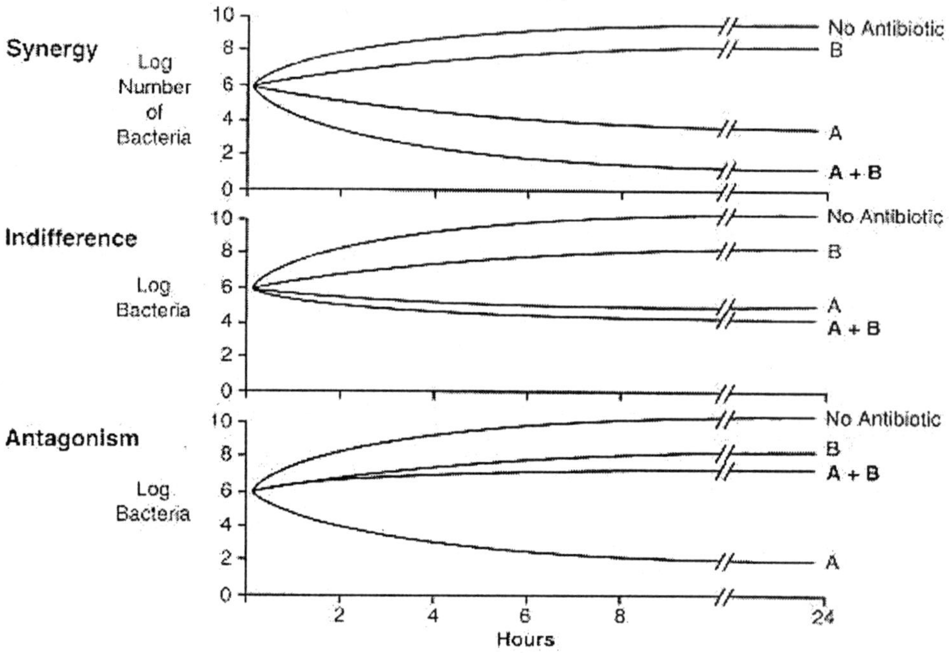

FIGURE 11-19 Example of how two antibiotics (A and B) may interact with synergy, indifference, or antagonism.

Important examples of bacterial synergy include (1) combinations of anti-cell wall agents with aminoglycosides, (2) use of ß-lactamase inhibitors with ß-lactamase-susceptible antibiotics, and (3) combinations of drugs that act on sequential steps in bacterial metabolic or synthetic pathways. Examples of synergy that have proved clinically important are the combination of penicillin and streptomycin to treat *Enterococcus faecalis* endocarditis and the combination of carbenicillin and gentamicin to treat P aeruginosa infections. Recently the ß-lactamase inhibitor clavulanic acid or sulbactam has been combined with amino penicillins to inhibit *S aureus, Klebsiella pneumoniae, H influenzae,* and anaerobic organisms such as *Bacteroides* species, all of which are resistant to amoxicillin or ampicillin when they contain ß-lactamases. The combination of sulfamethoxazole and trimethoprim attacks two parts of the folic acid cycle and synergistically inhibits many bacteria. Finally, combinations of two penicillins that affect different stages of cell wall synthesis in Gram-negative bacteria are synergistic. This is true for the combination of mecillinam (amdinocillin), which binds to PBP2 of bacteria, and penicillins or cephalosporins, which bind to PBP lb or 3.

Antagonism can occur when a bacteriostatic agent is combined with a bactericidal agent. The classic example has been the combination of chlortetracycline and penicillin in treatment of pneumococcal meningitis. This effect has not been explained from a molecular standpoint, and tetracyclines and penicillins or cephalosporins are used to treat mixed infections such as pelvic inflammatory disease due to *N gonorrhoeae* and *Chlamydia.* Some ß-lactam antibiotics can induce ß-lactamases that inactivate other ß-lactams, and antagonism can be shown in the test tube, but the relevance for clinical infections is not established.

Toxicology of Antimicrobial Agents

Antimicrobial agents can be directly toxic, can interact with other drugs to increase their toxicity, or can alter microbial flora to cause infection by organisms that are normally saprophytic. Allergic reactions can becaused by an agent, but penicillins can produce either immediate, IgE-mediated, or delayed hypersensitivity reactions. Cutaneous reactions have been reported with every class of antimicrobial agent. Hematologic reactions can range from the life-threatening blood dyscrasia that occurs in 1 in 60,000 individuals who receive chloramphenicol to hemolytic anemia due to sulfonamides in individuals who lack the enzyme glucose-6-phosphate dehydrogenase. Depression of blood platelet activity has occurred with many agents. By altering the gastrointestinal flora, almost all antibiotics can cause overgrowth of *Clostridium difficile,* which produces a toxin that causes diarrhea and even pseudomembranous colitis. Alteration of intestinal flora by antibiotics can also result in overgrowth of *Candida* in the mouth, vagina, or gastrointestinal tract. Since a number of antibiotics are metabolized in the liver, damage to the liver can occur. This has been of particular concern with isoniazid, which is used to treat tuberculosis. Damage to the kidneys can follow the use of aminoglycosides. Neurologic toxicity is fortunately fairly uncommon, but the aminoglycosides can damage the auditory or vestibular apparatus if the dosage is not closely monitored.

Bacteria continue to evolve new mechanisms of resistance to old and to new antimicrobial agents. Some bacteria such as *P aeruginosa* are particularly adept at utilizing a number of different mechanisms simultaneously to become resistant to agents in virtually every class and those with such diverse sites of action as cell wall, protein biosynthesis, or DNA and RNA synthesis. Progress in medicine will keep patients alive who have nosocomial infections with resistant pathogens.

Mechanism to Reduce Bacterial Resistance

Proper selection of new antibiotics will be a major force in slowing the development of antimicrobial resistance. Proper hygiene practices will reduce plasmid transfer and the establishment of multiple drug-resistant bacteria in the hospital and will delay the appearance of such species in the community. Table 11-4 lists a number of mechanisms to prevent bacterial resistance. The health care provider must be continually alert to the appearance of antibiotic resistance within the hospital and community.

Table 11-4 Mechanisms to Reduce Antibiotic Resistance

1. Control, reduce, or cycel antibiotic usage.
2. Improve hyginene in hospitals and among hospital presonnel and reduce movement of patients to eliminate the dissemination of resistant organisms within hospitals.
3. Discover or develop new antibiotics.
4. Modify existing antibiotics chemically to produce compounds next to known mechanisms of resistance.
5. Develop inhibitors of antibiotic-modifying enzymes.
6. Define agents that would "cure" resistance plasmids.

REFERENCES

Arthur M: Genetics and mechanisms of glycopeptide resistance in enterococci. Antimicrob Agents Chemother 37:1563, 1993

Gale EF, Cundliffe E, Reynolds PE et al: The Molecular Basis of Antibiotic Action. 2nd Ed. John Wiley & Sons, New York, 1981

Kucers A, Bennett N: The Use of Antibiotics. 4th Ed. JB Lippincott, Philadelphia, 1985. Lorian V (ed): Antibiotics in Laboratory Medicine. 3rd Ed. Williams & Wilkins, Baltimore, 1991

Murray B: New Aspects of antimicrobial resistance and the resulting therapeutic dilemmas. J Infect Dis 163:1185, 1991

Neu HC (ed): Update on antibiotics. 1. Med Clin N Am 71:1051, 1987

Neu HC (ed): Update on antibiotics. 11. Med Clin N Am 72:555, 1988

Neu H: The crisis in antibiotic resistance. Science 257:1064, 1992

Norrby SR, Bergan T, Holm SE et a] (eds): Evaluation of new beta-lactam antibiotics. Rev Infect Dis 8 (Suppl. 3):S235, 1986

Schaberg D: Resistant gram-positive organisms. Ann Emergency Med 24(3):462, 1994

Waxman DJ, Strominger JL: Beta-lactam antibiotics: biochemical modes of action. p. 210. In Morin RB, Gorman M (eds): Chemistry and Biology of Beta-Lactam Antibiotics. Academic Press, San Diego, 1982

Wolfson JS, Hooper DC (eds): Quinolone Antimicrobial Agents. 2nd Ed. American Society for Microbiology, Washington, 1993

Chapter 12

Staphylococcus

General Concepts

Clinical Manifestations

Staphylococci can cause many forms of infection. (1) *S aureus* causes superficial skin lesions (boils, styes) and localized abscesses in other sites. (2) *S aureus* causes deep-seated infections, such as osteomyelitis and endocarditis and more serious skin infections (furunculosis). (3) *S aureus* is a major cause of hospital acquired (nosocomial) infection of surgical wounds and, with *S epidermidis*, causes infections associated with indwelling medical devices. (4) *S aureus* causes food poisoning by releasing enterotoxins into food. (5) *S aureus* causes toxic shock syndrome by release of superantigens into the blood stream. (6) *S saprophiticus* causes urinary tract infections, especially in girls. (7) Other species of staphylococci (*S lugdunensis, S haemolyticus, S warneri, S schleiferi, S intermedius*) are infrequent pathogens.

Structure

Staphylococci are Gram-positive cocci 1 μm in diameter. They form clumps.

Classification

S aureus *and* S intermedius *are coagulase positive. All other staphylococci are coagulase negative. They are salt tolerant and often hemolytic. Identification requires biotype analysis.*

Natural Habitat

S aureus colonizes the nasal passage and axillae. S epidermidis is a common human skin commensal. Other species of staphylococci are infrequent human commensals. Some are commensals of other animals.

Pathogenesis

S aureus expresses many potential virulence factors. (1) Surface proteins that promote colonization of host tissues. (2) Factors that probably inhibit phagocytosis (capsule, immunoglobulin binding protein A). (3) Toxins that damage host tissues and cause disease symptoms. Coagulase-negative staphylococci are normally less virulent and express fewer virulence factors. S epidermidis readily colonizes implanted devices.

Host Defenses

Phagocytosis is the major mechanism for combatting staphylococcal infection. Antibodies are produced which neutralize toxins and promote opsonization. The capsule and protein A may interfere with phagocytosis. Biofilm growth on implants is impervious to phagocytosis.

Treatment

Infections acquired outside hospitals can usually be treated with penicillinase-resistant ß-lactams. Hospital acquired infection is often caused by antibiotic resistant strains and can only be treated with vancomycin.

Antibiotic Resistance

Multiple antibiotic resistance isincreasingly common in *S aureus* and *S epidermidis*. Methicillin resistance is indicative of multiple resistance. Methicillin-resistant *S aureus* (MRSA) causes outbreaks in hospitals and can be epidemic.

Epidemiology

Epidemiological tracing of *S aureus* is traditionally performed by phage typing, but has limitations. Molecular typing methods are being tested experimentally.

Diagnosis

Diagnosis is based on performing tests with colonies. Tests for clumping factor, coagulase, hemolysins and thermostable deoxyribonuclease are routinely used to identify *S aureus*. Commercial latex agglutination tests are available. Identification of *S epidermidis* is confirmed by commercial biotyping kits.

Control

Patients and staff carrying epidemic strains, particularly MRSA, should be isolated. Patients may be given disinfectant baths or treated with a topical antibiotic to eradicate carriage of MRSA. Infection control programs are used in most hospitals.

INTRODUCTION

Bacteria in the genus Staphylococcus are pathogens of man and other mammals. Traditionally they were divided into two groups on the basis of their ability to clot blood plasma (the coagulase reaction). The coagulase-positive staphylococci constitute the most pathogenic species *S aureus*. The coagulase-negative staphylococci (CNS) are now known to comprise over 30 other species. The CNS are common commensals of skin, although some species can cause infections. It is now obvious that the division of staphylococci into coagulase positive and negative is artificial and indeed, misleading in some cases. Coagulase is a marker for *S aureus* but there is no direct evidence that it is a virulence factor. Also, some natural isolates of *S aureus* are defective in coagulase. Nevertheless, the term is still in widespread use among clinical microbiologists.

S aureus expresses a variety of extracellular proteins and polysaccharides, some of which are correlated with virulence. Virulence results from the combined effect of many factors expressed during infection. Antibodies will neutralize staphylococcal toxins and enzymes, but vaccines are not available. Both antibiotic treatment and surgical drainage are often necessary to cure abscesses, large boils and wound infections. Staphylococci are common causes of infections associated with indwelling medical devices. These are difficult to treat with antibiotics alone and often require removal of the device. Some strains that infect hospitalized patients are resistant to most of the antibiotics used to treat infections, vancomycin being the only remaining drug to which resistance has not developed.

Taxonomy

DNA-ribosomal RNA (rRNA) hybridization and comparative oligonucleotide analysis of 16S rRNA has demonstrated that staphylococci form a coherent group at the genus level. This group occurs within the broad Bacillus-Lactobacillus-Streptococcus cluster defining Gram-positive bacteria with a low G + C content of DNA.

At least 30 species of staphylococci have been recognized by biochemical analysis and in particular by DNA-DNA hybridization. Eleven of these can be isolated from humans as commensals. *S aureus* (nares) and *S epidermidis* (nares, skin) are common commensals and also have the greatest pathogenic potential. *S saprophyticus* (skin, occasionally) is also a common cause of urinary tract infection. *S haemolyticus, S simulans, S cohnii, S warneri* and *S lugdunensis* can also cause infections in man.

Identification of Staphylococci in the Clinical laboratory

Structure

Staphylococci are Gram-positive cocci about 0.5 - 1.0 μm in diameter. They grow in clusters, pairs and occasionally in short chains. The clusters arise because staphylococci divide in two planes. The configuration of the cocci helps to distinguish micrococci and staphylococci from streptococci, which usually grow in chains. Observations must be made on cultures grown in broth, because streptococci grown on solid medium may appear as clumps. Several fields should be examined before deciding whether clumps or chains are present.

Catalase Test

The catalase test is important in distinguishing streptococci (catalase-negative) staphylococci which are catalase positive. The test is performed by flooding an agar slant or broth culture with several drops of 3% hydrogen peroxide. Catalase-positive cultures bubble at once. The test should not be done on blood agar because blood itself will produce bubbles.

Isolation and Identification

The presence of staphylococci in a lesion might first be suspected after examination of a direct Gram stain. However, small numbers of bacteria in blood preclude microscopic examination and require culturing first.

The organism is isolated by streaking material from the clinical specimen (or from a blood culture) onto solid media such as blood agar, tryptic soy agar or heart infusion agar. Specimens likely to be contaminated with other microorganisms can be plated on mannitol salt agar containing 7.5% sodium chloride, which allows the halo-tolerant staphylococci to grow. Ideally a Gram stain of the colony should be performed and tests made for catalase and coagulase production, allowing the coagulase-positive *S aureus* to be identified quickly. Another very useful test for *S aureus* is the production of thermostable deoxyribonuclease. *S aureus* can be confirmed by testing colonies for agglutination with latex particles coated with immunoglobulin G and fibrinogen which bind protein A and the clumping factor, respectively, on the bacterial cell surface. These are available from commercial suppliers (e.g., Staphaurex). The most recent latex test (Pastaurex) incorporates monoclonal antibodies to serotype 5 and 8 capsular polysaccharide in order to reduce the number of false negatives. (Some recent clinical isolates of*S aureus* lack production of coagulase and/or clumping factor, which can make identification difficult.)

The association of *S epidermidis* (and to a lesser extent of other coagulase-negative staphylococci) with nosocomial infections associated with indwelling devices means that isolation of these bacteria from blood is likely to be important and not due to chance contamination, particularly if successive blood cultures are positive. Nowadays, identification of S *epidermidis* and other species of *Staphylococcus* is performed using commercial biotype identification kits, such as API Staph Ident, API Staph-Trac, Vitek GPI Card and Microscan Pos Combo. These comprise preformed strips containing test substrates.

Epidemiology of *Staphylococcus Aureus* Infections

Because *S aureus* is a major cause of nosocomial and community-acquired infections, it is necessary to determine the relatedness of isolates collected during the investigation of an outbreak. Typing systems must be reproducible, discriminatory, and easy to interpret and to use. The traditional method for typing *S aureus* is phage-typing. This method is based on a phenotypic marker with poor reproducibility. Also, it does not type many isolates (20% in a recent survey at the Center for Disease Control and Prevention), andit requires maintenance of a large number of phage stocks and propagating strains and consequently can be performed only by specialist reference laboratories.

Many molecular typing methods have been applied to the epidemiological analysis of *S aureus*, in particular, of methicillin-resistant strains (MRSA). Plasmid analysis has been used extensively with success, but suffers the disadvantage that plasmids can easily be lost and acquired and are thus inherently unreliable. Methods designed to recognize restriction fragment length polymorphisms (RFLP) using a variety of gene probes, including rRNA genes (ribotyping), have had limited success in the epidemiology of MRSA. In this technique, the choice of restriction enzyme used to cleave the genomic DNA, as well as the probes, is crucial. Random primer PCR offers potential for discriminating between strains but a suitable primer has yet to be identified for *S aureus*. The method currently regarded as the most reliable is pulsed field gel electrophoresis, where genomic DNA is cut with a restriction enzyme that generates large fragments of 50-700 kb.

Clinical Manifestations of *S Aureus*

S aureus is notorious for causing boils, furuncles, styes, impetigo and other superficial skin infections in humans (Figure 12-1). It may also cause more serious infections, particularly in persons debilitated by chronic illness, traumatic injury, burns or immunosuppression. These infections include pneumonia, deep abscesses, osteomyelitis, endocarditis, phlebitis, mastitis and meningitis, and are often associated with hospitalized patients rather than healthy individuals in the community. *S aureus* and *S epidermidis* are common causes of infections associated with indwelling devices such as joint prostheses, cardiovascular devices and artificial heart valves (Fig. 12-2).

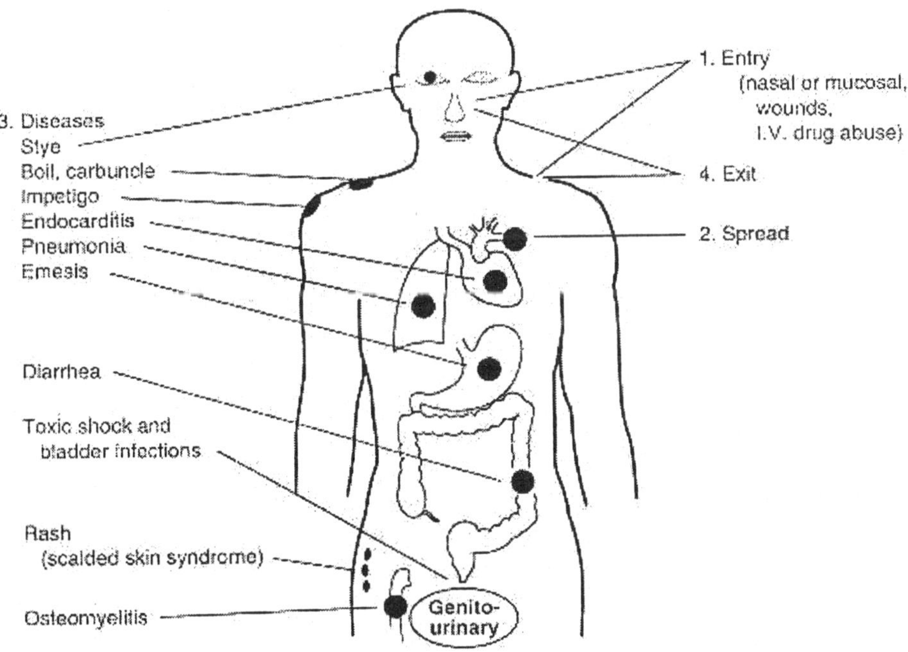

FIGURE 12-1 Pathogenesis of staphylococcal infections

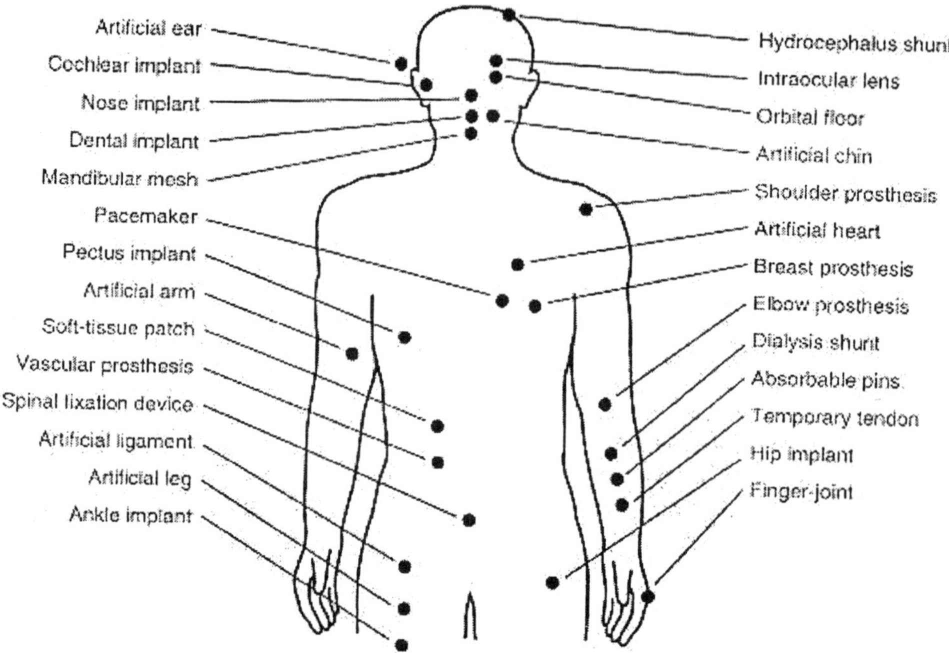

FIGURE 12-2 Infections associated with indwelling devices

Pathogenesis of *S aureus* Infections

S aureus expresses many cell surface-associated and extracellular proteins that are potential virulence factors. For the majority of diseases caused by this organism, pathogenesis is multifactorial. Thus it is difficult to determine precisely the role of any given factor. This also reflects the inadequacies of many animal models for staphylococcal diseases.

However, there are correlations between strains isolated from particular diseases and expression of particular factors, which suggests their importance in pathogenesis. With some toxins, symptoms ofa human disease can be reproduced in animals with pure proteins. The application of molecular biology has led to recent advances in the understanding of pathogenesis of staphylococcal diseases. Genes encoding potential virulence factors have been cloned and sequenced and proteins purified. This has facilitated studies at the molecular level on their modes of action, both in in vitro and in model systems. In addition, genes encoding putative virulence factors have been inactivated, and the virulence of the mutants compared to the wild-type strain in animal models. Any diminution in virulence implicates the missing factor. If virulence is restored when the gene is returned to the mutant then "Molecular Koch's Postulates" have been fulfilled. Several virulence factors of *S aureus* have been confirmed by this approach.

Adherence

In order to initiate infection the pathogen must gain access to the host and attach to host cells or tissues.

S aureus Adheres to Host Proteins

S aureus cells express on their surface proteins that promote attachment to host proteins such as laminin and fibronectin that form part of the extracellular matrix (Figure 12-3). Fibronectin is present on epithelial and endothelial surfaces as well as being a component of blood clots. In addition, most strains express a fibrinogen/fibrin binding protein (the clumping factor) which promotes attachment to blood clots and traumatized tissue. Most strains of *S aureus* express fibronectin and fibrinogen-binding proteins.

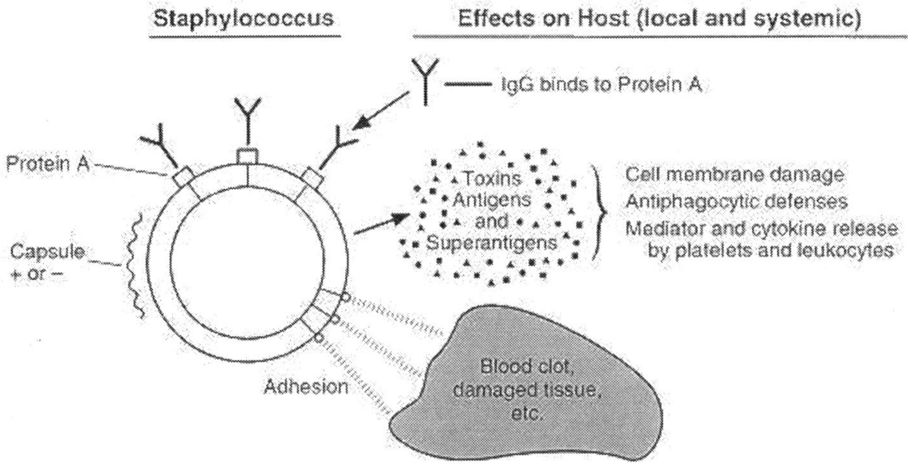

FIGURE 12-3 Summary of virulence factors of *Staphylococcus aureus*

The receptor which promotes attachment to collagen is particularly associated with strains that cause osteomyelitis and septic arthritis. Interaction with collagen may also be important in promoting bacterial attachment to damaged tissue where the underlying layers have been exposed.

Evidence that these staphylococcal matrix-binding proteins are virulence factors has come from studying defective mutants in vitro adherence assays and in experimental infections. Mutants defective in binding to fibronectin and to fibrinogen have reduced virulence in a rat model for endocarditis, suggesting that bacterial attachment to the sterile vegetations caused by damaging the endothelial surface of the heart valve is promoted by fibronectin and fibrinogen. Similarly, mutants lacking the collagen-binding protein have reduced virulence in a mouse model for septic arthritis. Furthermore, the soluble ligand-binding domain of the fibrinogen, fibronectin and collagen-binding proteins expressed by recombinant methods strongly blocks interactions of bacterial cells with the corresponding host protein.

Role of Adherence in Infections Associated with Medical Devices

Infections associated with indwelling medical devices ranging from simple intravenous catheters to prosthetic joints and replacement heart valves can be caused by *S aureus* and *S epidermidis* (Figure 12-2). Very shortly after biomaterial is implanted in the human body it becomes coated with a complex mixture of host proteins and platelets. In one model system involving short-term contact between biomaterial and blood, fibrinogen was shown to be the dominant component and was primarily responsible for adherence of *S aureus* in subsequent in vitro assays. In contrast, with material that has been in the body for longer periods (e.g., human intravenous catheters) the fibrinogen is degraded and no longer promotes bacterial attachment. Instead, fibronectin, which remains intact, becomes the predominant ligand promoting attachment.

Adherence to Endothelial Cells

S aureus can adhere to the surface of cultured human endothelial cells and become internalized by a phagocytosis-like process. It is not clear if attachment involves a novel receptor or a known surface protein of *S aureus*. Some researchers think that *S aureus* can initiate endocarditis by attaching to the undamaged endothelium. Others feel that trauma of even a very minor nature is required to promote attachment of bacteria.

Avoidance of Host Defenses

S aureus expresses a number of factors that have the potential to interfere with host defense mechanisms. However, strong evidence for a role in virulence of these factors is lacking.

Capsular Polysaccharide

The majority of clinical isolates of *S aureus* express a surface polysaccharide of either serotype 5 or 8. This has been called a microcapsule because it can be visualized only by electron microscopy after antibody labeling, unlike the copious capsules of other bacteria which are visualized by light microscopy. *S aureus* isolated from infections expresses high levels of polysaccharide but rapidly loses it upon laboratory subculture. The function of the capsule is not clear. It may impede phagocytosis, but in in vitro tests this was only demonstrated in the absence of complement. Conversely, comparing wild-type and a capsule defective mutant strain in an endocarditis model suggested that polysaccharide expression actually impeded colonization of damaged heart valves, perhaps by masking adhesins.

Protein A

Protein A is a surface protein of *S aureus* which binds immunoglobulin G molecules by the Fc region (Fig. 12-3). In serum, bacteria will bind IgG molecules the wrong way round by this non-immune mechanism. In principle this will disrupt opsonization and phagocytosis. Indeed mutants of *S aureus* lacking protein A are more efficiently phagocytozed in vitro, and studies with mutants in infection models suggest that protein A enhances virulence.

·Leukocidin

S aureus can express a toxin that specifically acts on polymorphonuclear leukocytes. Phagocytosis is an important defense against staphylococcal infection so leukocidin should be a virulence factor. This toxin is discussed in more detail in the next section.

Damage to the Host

S aureus can express several different types of protein toxins which are probably responsible for symptoms during infections. Some damage the membranes of erythrocytes, causing hemolysis; but it is unlikely that hemolysis is relevant in vivo. The leukocidin causes membrane damage to leukocytes and is not hemolytic. Systemic release of a-toxin causes septic shock, while enterotoxins and TSST-1 cause toxic shock.

Membrane Damaging Toxins

(a) α-toxin

The best characterized and most potent membrane-damaging toxin of *S aureus* is α-toxin. It is expressed as a monomer that binds to the membrane of susceptible cells. Subunits then oligomerize to form hexameric rings with a central pore through which cellular contents leak.

Susceptible cells have a specific receptor for a-toxin which allows low concentrations of toxin to bind, causing small pores through which monovalent cations can pass. At higher concentrations, the toxin reacts non-specifically with membrane lipids, causing larger pores through which divalent cations and small molecules can pass. However, it is doubtful if this is relevant under normal physiological conditions.

In humans, platelets and monocytes are particularly sensitive to α-toxin. They carry high affinity sites which allow toxin to bind at concentrations that are physiologically relevant. A complex series of secondary reactions ensue, causing release of eicosanoids and cytokines which trigger production of inflammatory mediators. These events cause the symptoms of septic shock that occur during severe infections caused by *S aureus*.

The notion that α-toxin is a major virulence factor of *S aureus* is supported by studies with the purified toxin in animals and in organ culture. Also, mutants lacking α-toxin are less virulent in a variety of animal infection models.

(b) β-toxin

ß-toxin is a sphingomyelinase which damages membranes rich in this lipid. The classical test for ß-toxin is lysis of sheep erythrocytes. The majority of human isolates of *S aureus* do not express ß-toxin. A lysogenic bacteriophage is inserted into the gene that encodes the toxin. This phenomenon is called negative phage conversion. Some of the phages that inactivate the ß-toxin gene carry the determinant for an enterotoxin and staphylokinase (see below).

In contrast the majority of isolates from bovine mastitis express ß-toxin, suggesting that the toxin is important in the pathogenesis of mastitis. This is supported by the fact that ß-toxin-deficient mutants have reduced virulence in a mouse model for mastitis.

(c) δ-toxin

The δ-toxin is a very small peptide toxin produced by most strains of *S aureus*. It is also produced by *S epidermidis* and *S lugdunensis*. The role of δ-toxin in disease is unknown.

(d) γ-toxin and leukocidin

The γ-toxin and the leukocidins are two-component protein toxins that damage membranes of susceptible cells. The proteins are expressed separately but act together to damage membranes. There is no evidence that they form multimers prior to insertion into membranes. The γ-toxin locus expresses three proteins. The B and C components form a leukotoxin with poor hemolytic activity, whereas the A and B components are hemolytic and weakly leukotoxic.

The classical Panton and Valentine (PV) leukocidin is distinct from the leukotoxin expressed by the γ-toxin locus. It has potent leukotoxicity and, in contrast to γ-toxin, is non-hemolytic. Only a small fraction of *S aureus* isolates (2% in one survey) express the PV leukocidin, whereas 90% of those isolated from severe dermonecrotic lesions express this toxin. This suggests that PV leukocidin is an important factor in necrotizing skin infections.

PV-leukocidin causes dermonecrosis when injected subcutaneously in rabbits. Furthermore, at a concentration below that causing membrane damage, the toxin releases inflammatory mediators from human neutrophils, leading to degranulation. This could account for the histology of dermonecrotic infections (vasodilation, infiltration and central necrosis).

Superantigens: enterotoxins and toxic shock syndrome toxin

S aureus can express two different types of toxin with superantigen activity, enterotoxins, of which there are six serotypes (A, B, C, D, E and G) and toxic shock syndrome toxin (TSST-1). Enterotoxins cause diarrhea and vomiting when ingested and are responsible for staphylococcal food poisoning. When expressed systemically, enterotoxins can cause toxic shock syndrome (TSS) - indeed enterotoxins B and C cause 50% of non-menstrual TSS. TSST-1 is very weakly related to enterotoxins and does not have emetic activity. TSST-1 is responsible for 75% of TSS, including all menstrual cases. TSS can occur as a sequel to any staphylococcal infection if an enterotoxin or TSST-1 is released systemically and the host lacks appropriate neutralizing antibodies. Tampon-associated TSS is not a true infection, being caused by growth of *S aureus* in a tampon and absorption of the toxin into the blood stream. TSS came to prominence with the introduction of super-absorbent tampons; and although the number of such cases has decreased dramatically, they still occur despite withdrawal of certain types of tampons from the market.

Superantigens stimulate T cells non-specifically without normal antigenic recognition (Figure 12-4). Up to one in five T cells may be activated, whereas only 1 in 10,000 are stimulated during antigen presentation. Cytokines are released in large amounts, causing the symptoms of TSS. Superantigens bind directly to class II major histocompatibility complexes of antigen-presenting cells outside the conventional antigen-binding grove. This complex recognizes only the Vb element of the T cell receptor. Thus any T cell with the appropriate Vb element can be stimulated, whereas normally antigen specificity is also required in binding.

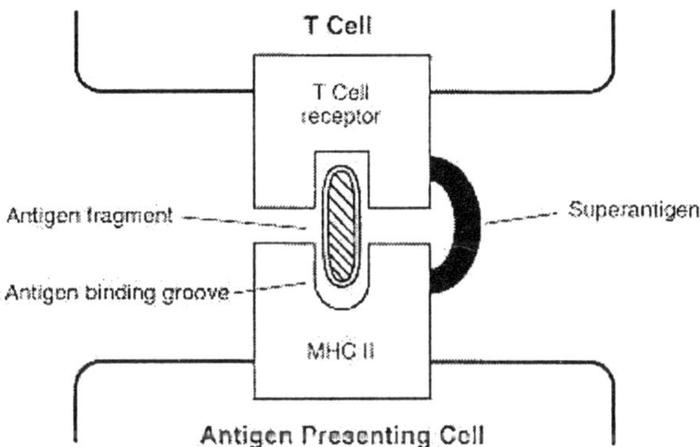

FIGURE 12-4 Superantigens and the non-specific stimulation of T cells

Epidermolytic (exfoliative) toxin (ET)

This toxin causes the scalded skin syndrome in neonates, with widespread blistering and loss of the epidermis. There are two antigenically distinct forms of the toxin, ETA and ETB. There is evidence that these toxins have protease activity. Both toxins have a sequence similarity with the *S aureus* serine protease, and the three most important amino acids in the active site of the protease are conserved. Furthermore, changing the active site of serine to a glycine completely eliminated toxin activity. However, ETs do not have discernible proteolytic activity but they do have esterase activity. It is not clear how the latter causes epidermal splitting. It is possible that the toxins target a very specific protein which is involved in maintaining the integrity of the epidermis.

Other Extracellular Proteins

Coagulase

Coagulase is not an enzyme. It is an extracellular protein which binds to prothrombin in the host to form a complex called staphylothrombin. The protease activity characteristic of thrombin is activated in the complex, resulting in the conversion of fibrinogen to fibrin. This is the basis of the tube coagulase test, in which a clot is formed in plasma after incubation with the *S aureus* broth-culture supernatant. Coagulase is a traditional marker for identifying *S aureus* in the clinical microbiology laboratory. However, there is no evidence that it is a virulence factor, although it is reasonable to speculate that the bacteria could protect themselves from host defenses by causing localized clotting. Notably, coagulase deficient mutants have been tested in several infection models but no differences from the parent strain were observed.

There is some confusion in the literature concerning coagulase and clumping factor, the fibrinogen-binding determinant on the *S aureus* cell surface. This is partly due to loose terminology, with the clumping factor sometimes being referred to as bound coagulase. Also, although coagulase is regarded as an extracellular protein, a small fraction is tightly bound on the bacterial cell surface where it can react with prothrombin. Finally, it has recently been shown that the coagulase can bind fibrinogen as well as thrombin, at least when it is extracellular. Genetic studies have shown unequivocally that coagulase and clumping factor are distinct entities. Specific mutants lacking coagulase retain clumping factor activity, while clumping factor mutants express coagulase normally.

Staphylokinase

Many strains of *S aureus* express a plasminogen activator called staphylokinase. The genetic determinant is associated with lysogenic bacteriophages. A complex formed between staphylokinase and plasminogen activates plasmin-like proteolytic activity which causes dissolution of fibrin clots. The mechanism is identical to streptokinase, which is used in medicine to treat patients suffering from coronary thrombosis. As with coagulase there is no evidence that staphylokinase is a virulence factor, although it seems reasonable to imagine that localized fibrinolysis might aid in bacterial spreading.

Enzymes

S aureus can express proteases, a lipase, a deoxyribonuclease (DNase) and a fatty acid modifying enzyme (FAME). The first three probably providenutrients for the bacteria, and it is unlikely that they have anything but a minor role in pathogenesis. However, the FAME enzyme may be important in abscesses, where it could modify anti-bacterial lipids and prolong bacterial survival. The thermostable DNase is an important diagnostic test for identification of *S aureus*.

Coagulase Negative Staphylococci

Staphylococci other than *S aureus* can cause infections in man. *S epidermidis* is the most important coagulase-negative staphylococcus (CNS) species and is the major cause of infections associated with prosthetic devices and catheters. CNS also cause peritonitis in patients receiving continuous ambulatory peritoneal dialysis and endocarditis in those with prosthetic valves. These infections are not usually nosocomially acquired. Other species such as *S haemolyticus, S warneri, S hominis, S capitis, S intermedius, S schleiferi* and *S simulans* are infrequent pathogens. *S lugdunesis* is a newly recognized species. It is probably more pathogenic than are other CNS species, with cases of endocarditis and other infections being reported. It is likely that the incidence of infections caused by these organisms is underestimated because of difficulties in identification.

Diagnosis of CNS infections is difficult. Infections are often indolent and chronic with few obvious symptoms. This is due to the smaller array of virulence factors and toxins compared to those in the case of *S aureus*. *S epidermidis* is a skin commensal and is one of the most common contaminants of samples sent to the diagnostic laboratory, while *S lugdunensis* is often confused with *S aureus*. Precise identification of CNS species requires the use of expensive test kits, such as the API-Staph.

In contrast to *S aureus*, little is known about mechanisms of pathogenesis of *S epidermidis* infections. Adherence is obviously a crucial step in the initiation of foreign body infections. Much research has been done on the interaction between *S epidermidis* and plastic material used in implants, and a polysaccharide adhesion (PS/A) has been identified. Mutants lacking PS/A are less virulent in an animal model for foreign body infection, and immunization with purified PS/A is protective. Bacteria-plastic interactions are probably important in colonization of catheters through the point of entry. However, host proteinsare quickly deposited on implants. *S epidermidis* does not bind to fibrinogen but most isolates bind fibronectin, albeit less avidly than *S aureus*. However, it is not known if a protein analogous to the fibronectin binding protein of *S aureus* is involved.

A characteristic of clinical isolates of *S epidermidis* is the production of "slime." This is a controversial topic. Some feel that slime is an in vitro manifestation of the ability to form a biofilm in vivo, for example on the surface of a prosthetic device, and is thus a virulence marker. In vitro, slime is formed during growth in broth as a biofilm on the surface of the growth vessel. The composition of this slime is probably influenced by the growth medium. One study with defined medium showed that the slime was predominantly secreted teichoic acid, a polymer normally found in the cell wall of staphylococci. Some polysaccharides in slime from bacteria grown on solid medium are derived from the agar.

Resistance of Staphylococci to Antimicrobial Drugs

Hospital strains of *S aureus* are often resistant to many different antibiotics. Indeed strains resistant to all clinically useful drugs, apart from the glycopeptides vancomycin and teicoplanin, have been described. The term MRSA refers to methicillin resistance and most methicillin-resistant strains are also multiply resistant. Plasmid-associated vancomycin resistance has been detected in some enterococci and the resistance determinant has been transferred from enterococci to *S aureus* in the laboratory and may occur naturally. *S epidermidis* nosocomial isolates are also often resistant to several antibiotics including methicillin. In addition, *S aureus* expresses resistance to antiseptics and disinfectants, such as quaternary ammonium compounds, which may aid its survival in the hospital environment.

Since the beginning of the antibiotic era *S aureus* has responded to the introduction of new drugs by rapidly acquiring resistance by a variety of genetic mechanisms including (1) acquisition of extrachromosomal plasmids or additional genetic information in the chromosome via transposons or other types of DNA insertion and (2) by mutations in chromosomal genes (Table 12-1).

TABLE 12-1 Antimicrobial Resistance

Antimicrobial	Resistance Mechanism	Genetic Basis
Penicillin	β-laclamase, Enzymatic Inactivation of penicillin	Plasmid
Methicillin	Expression of new penicillin-resistant penicillin-binding protein. Bypass	Novel chromosomal locus acquired form unknown source
Tetracycline	1. Efflux from cell	Plasmid
	2. Modification of ribosome	Novel chromosomal locus acquired form unknown source
Chloramphenicol	Enzymatic inactivation	Plasmid
Erythromycin	Enzymatic modification of ribosomal RNA, Prevents drug binding to ribosome	Plasmid Transposon in chromosome
Streptomycin	1. Mutation in ribosomal protein. Prevents drug binding	Mutation in chromosomal gene encoding drug target
	2. Enzymatic inactivation	Plasmid
Kanamycin	Enzymatic inactivation	Plasmid
Gentamicin		Transposon in chromosome
Trimathoprim	Alternative dihydofolate reductase. Bypass	Plasmid
Mupirocin	Alternative isoleucyl tRNA synthase. Bypass	Plasmid
Fluoroquinalones	1. Altered DNA	Mutation in chromosomal gene encoding drug target
	2. Efflux	Mutation increases expression of natural afflux mechanism
Antiseptics	Efflux	Plasmid

Many plasmid-encoded determinants have recently become inserted into the chromosome at a site associated with the methicillin resistance determinant. There may be an advantage to the organism having resistance determinants in the chromosome because they will be more stable. There are essentially four mechanisms of resistance to antibiotics in bacteria: (1) enzymatic inactivation of the drug, (2) alterations to the drug target to prevent binding, (3) accelerated drug efflux to prevent toxic concentrations accumulating in the cell, and (4) a by-pass mechanism whereby an alternative drug-resistant version of the target is expressed (Table 12-1).

Future Prospects

Antimicrobial Drugs

Ever since the first use of penicillin, *S aureus* has shown a remarkable ability to adapt. Resistance has developed to new drugs within a short time of their introduction. Some strains are now resistant to most conventional antibiotics. It is worrisome that there do not seem to be any new antibiotics on the horizon. Any recent developments have been modifications to existing drugs.

The original strategy used by the pharmaceutical industry to find antimicrobial drugs was to screen natural products and synthetic chemicals for antimicrobial activity. The mechanism of action was then investigated.

New approaches are being adopted to find the next generation of antimicrobials. Potential targets such as enzymes involved in an essential function (e.g., in cell division) are identified based on knowledge of bacterial physiology and metabolism. Screening methods are then developed to identify inhibitors of a specific target molecule. In addition, with detailed molecular knowledge of the target molecule, specific inhibitors can be designed.

Vaccines and New Approaches to Combatting Nosocomial Infections

No vaccine is currently available to combat staphylococcal infections. There may now be a case for considering methods to prevent disease, particularly in hospitalized patients.

Hyperimmune serum from human volunteer donors or humanized monoclonal antibodies directed towards surface components (e.g., capsular polysaccharide or surface protein adhesions) could both prevent bacterial adherence and also promote phagocytosis of bacterial cells. Indeed a prototype vaccine based on capsular polysaccharide from *S aureus* has been administered to volunteers to raise hyperimmune serum, which could be given to patients in hospital before surgery. A vaccine based on fibronectin binding protein induces protective immunity against mastitis in cattle and might also be used as a vaccine in humans.

When the molecular basis of the interactions between the bacterial surface proteins and the host matrix protein ligands are known it might be possible to design compounds that block the interactions and thus prevent bacterial colonization. These could be administered systemically or topically.

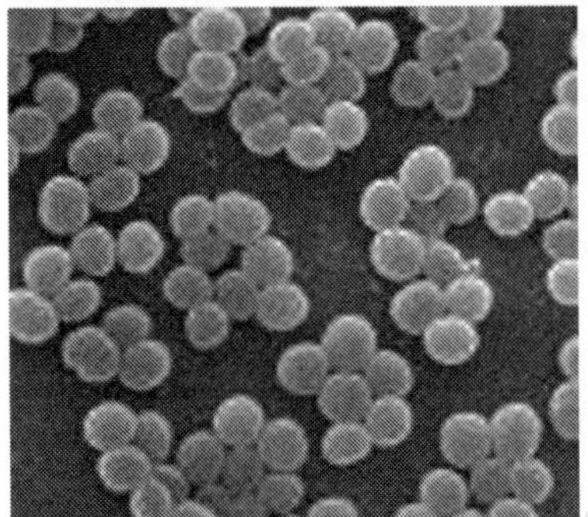
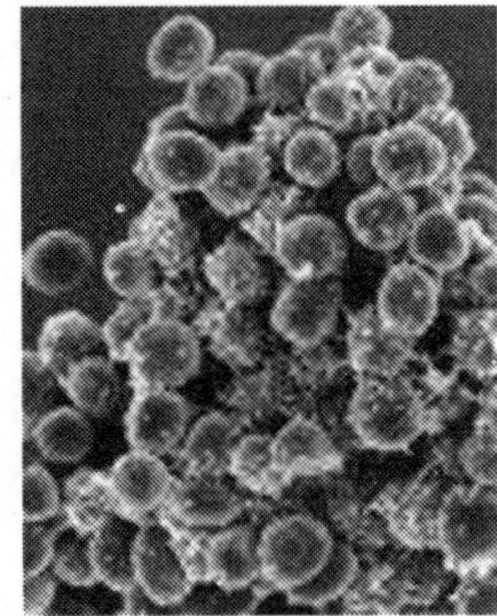
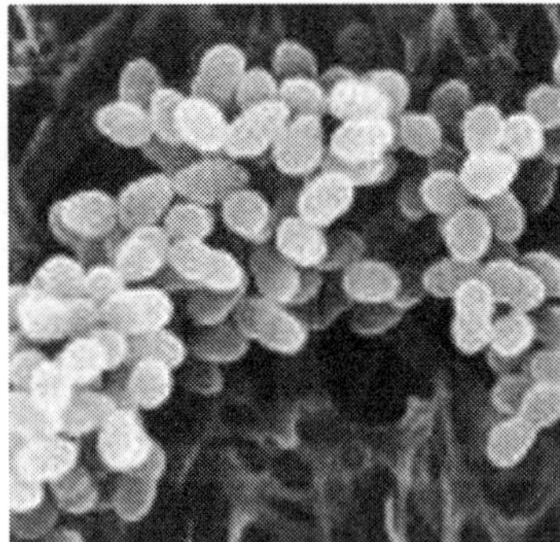

Colony Morphology of *Staphylococci*

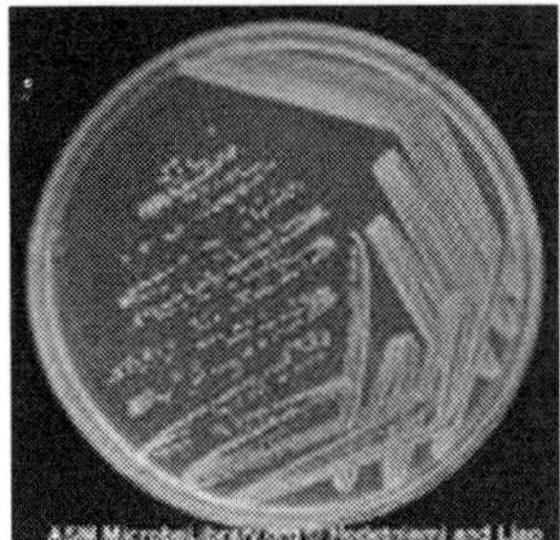

Culture appearance of *Staphylococci*

REFERENCES

Bhakdi S, Tranum-Jensen J: Alpha-toxin of *Staphylococcus aureus*. Microbiol Rev 55:733, 1991

Easmon CSF, Adlam C: Staphylococci and staphylococcal infections. Vols 1 and 2. Academic Press, London, 1983

Foster TJ: Potential for vaccination against infections caused by *Staphylococcus aureus*. Vaccine 9:221, 1991

Foster TJ, McDevitt D: Molecular basis of adherence of staphylococci to biomaterials. p. 31, In Bisno AL, Waldvogel FA (eds): Infections Associated with Indwelling Medical Devices, 2nd Edition. American Society for Microbiology, Washington, D.C., 1994.

Lyon BR, Skurray R: Antimicrobial resistance in *Staphylococcus aureus*: genetic basis. Microbiol Reviews 51:88, 1987

Prevost G, Couppie P, Prevost P et al: Epidemiological data on *Staphylococcus aureus* strains producing synergohymenotropic toxins. J Med Microbiol 42:237, 1995

Rupp ME, Archer GL: Coagulase-negative staphylococci: pathogens associated with medical progress. Clin Infect Dis 19:231, 1994

Schlievert PM: Role of superantigens in human disease. J Infect Dis 167:997, 1993

Skinner GRB, Ahmad, A: Staphylococcal vaccines - present status and future prospects. p. 537. In Mollby R, Flock JI, Nord CE, Christensson B (eds): Staphylococci and Staphylococcal Infections. Zbl. Bakt. Suppl. 26, Fischer Verlag, Stuttgart, 1994

Tenover F, Arbeit R, Archer, G et al: Comparison of traditional and molecular methods of typing isolates of *Staphylococcus aureus*. J Clin Microbiol 32:407, 1994

Vaudaux PE, Lew DP, Waldvogel FA: Host factors predisposing to and influencing therapy of foreign body infections. p. 1. In Bisno AL, Waldvogel FA (eds): Infections Associated with Indwelling Medical Devices. 2nd Ed. American Society for Microbiology, Washington, D.C., 1994

Chapter **13**

Streptococcus

General Concepts

Streptococcus pyogenes, *other* Streptococci, *and* Enterococcus

Clinical Manifestations

Acute *Streptococcus pyogenes* infections may take the form of pharyngitis, scarlet fever (rash), impetigo, cellulitis, or erysipelas. Invasive infections can result in necrotizing fasciitis, myositis and streptococcal toxic shock syndrome. Patients may also develop immune-mediated sequelae such as acute rheumatic fever and acute glomerulonephritis. *S agalactiae* may cause meningitis, neonatal sepsis, and pneumonia in neonates; adults may experience vaginitis, puerperal fever, urinary tract infection, skin infection, and endocarditis. Viridans streptococci can cause endocarditis, and Enterococcus is associated with urinary tract and biliary tract infections. Anaerobic streptococci participate in mixed infections of the abdomen, pelvis, brain, and lungs.

Structure

Streptococci are Gram-positive, nonmotile, nonsporeforming, catalase-negative cocci that occur in pairs or chains. Older cultures may lose their Gram-positive character. Most streptococci are facultative anaerobes, and some are obligate (strict) anaerobes. Most require enriched media (blood agar). Group A streptococci have a hyaluronic acid capsule.

Classification and Antigenic Types

Streptococci are classified on the basis of colony morphology, hemolysis, biochemical reactions, and (most definitively) serologic specificity. They are divided into three groups by the type of hemolysis on blood agar: b-hemolytic (clear, complete lysis of red cells), a hemolytic (incomplete, green hemolysis), and g hemolytic (no hemolysis). Serologic grouping is based on antigenic differences in cell wall carbohydrates (groups A to V), in cell wall pili-associated protein, and in the polysaccharide capsule in group B streptococci.

Pathogenesis

Streptococci are members of the normal flora. Virulence factors of group A streptococci include (1) M protein and lipoteichoic acid for attachment; (2) a hyaluronic acid capsule that inhibits phagocytosis; (3) other extracellular products, such as pyrogenic (erythrogenic) toxin, which causes the rash of scarlet fever; and (4) streptokinase, streptodornase (DNase B), and streptolysins. Some strains are nephritogenic. Immune-mediated sequelae do not reflect dissemination of bacteria. Nongroup A strains have no defined virulence factors.

Host Defenses

Antibody to M protein gives type-specific immunity to group A streptococci. Antibody to erythrogenic toxin prevents the rash of scarlet fever. Immune mechanisms are important in the pathogenesis of acute rheumatic fever. Maternal IgG protects the neonate against group B streptococci.

Epidemiology

Group A ß-hemolytic streptococci are spread by respiratory secretions and fomites. The incidence of both respiratory and skin infections peaks in childhood. Infection can be transmitted by asymptomatic carriers. Acute rheumatic fever was previously common among the poor; susceptibility may be partly genetic. Group B streptococci are common in the normal vaginal flora and occasionally cause invasive neonatal infection.

Diagnosis

Diagnosis is based on cultures from clinical specimens. Serologic methods can detect group A or B antigen; definitive antigen identification is by the precipitin test. Bacitracin sensitivity presumptively differentiates group A from other b-hemolytic streptococci (B, C, G); group B streptococci typically show hippurate hydrolysis; group D is differentiated from other viridans streptococci by bile solubility and optochin sensitivity. Acute glomerulonephritis and acute rheumatic fever are identified by anti-streptococcal antibody titers. In addition, acute rheumatic fever is diagnosed by clinical criteria.

Control

Prompt penicillin treatment of streptococcal pharyngitis reduces the antigenic stimulus and therefore prevents glomerulonephritis and acute rheumatic fever. Vancomycin resistance among the enterococci is an emerging microbial threat. Vaccines are under development.

Streptococcus pneumoniae

Clinical Manifestations

S pneumoniae causes pneumonia, meningitis, and sometimes occult bacteremia.

Structure

Pneumococci are lancet-shaped, catalase-negative, capsule-forming, a-hemolytic cocci or diplococci. Autolysis is enhanced by adding bile salts.

Classification and Antigenic Types

There are more than 85 antigenic types of *S pneumoniae*, which are determined by capsule antigens. There is no Lancefield group antigen.

Pathogenesis

S pneumoniae is a normal member of the respiratory tract flora; invasion results in pneumonia. The best defined virulence factor is the polysaccharide capsule, which protects the bacterium against phagocytosis.

Host Defenses

Protection against infection depends on a normal mucociliary barrier and intact phagocytic and T-independent immune responses. Type-specific anti-capsule antibody is protective.

Epidemiology

Pneumococcal pneumonia is most common in elderly, debilitated, or immunosuppressed individuals. The disease often sets in after a preceding viral infection damages the respiratory ciliated epithelium; incidence therefore peaks in the winter.

Diagnosis

Diagnosis is based on a sputum Gram stain and culture; blood or cerebrospinal fluid may also be cultured. Capsular antigen can be detected serologically. *Pneumococci* are distinguished from viridans streptococci by the quellung (capsular swelling) reaction, bile solubility, and optochin inhibition.

Control

Treatment is usually with penicillin. However, strains resistant to penicillin and multiple antibiotics are rapidly emerging. A vaccine is available.

INTRODUCTION

The genus *Streptococcus*, a heterogeneous group of Gram-positive bacteria, has broad significance in medicine and industry. Various streptococci are important ecologically as part of the normal microbial flora of animals and humans; some can also cause diseases that range from subacute to acute or even chronic. Among the significant human diseases attributable to streptococci are scarlet fever, rheumatic heart disease, glomerulonephritis, and pneumococcal pneumonia. Streptococci are essential in industrial and dairy processes and as indicators of pollution.

The nomenclature for streptococci, especially the nomenclature in medical use, has been based largely on serogroup identification of cell wall components rather than on species names. For several decades, interest has focused on two major species that cause severe infections: *S pyogenes* (group A streptococci) and *S pneumoniae* (pneumococci). In 1984, two members were assigned a new genus - the group D enterococcal species (which account for 98% of human enterococcal infections) became *Enterococcus faecalis* (the majority of human clinical isolates) and *E faecium* (associated with a remarkable capacity for antibiotic resistance).

In recent years, increasing attention has been given to other streptococcal species, partly because innovations in serogrouping methods have led to advances in understanding the pathogenetic and epidemiologic significance of these species. A variety of cell-associated and extracellular products are produced by streptococci, but their cause-effect relationship with pathogenesis has not been defined. Some of the other medically important streptococci are *S agalactiae* (group B), an etiologic agent of neonatal disease; *E faecalis* (group D), a major cause of endocarditis, and the viridans streptococci. Particularly for the viridans streptococci, taxonomy and nomenclature are not yet fully reliable or consistent. Important members of the viridans streptococci, normal commensals, include *S mutans* and *S sanguis* (involved in dental caries), *S mitis* (associated with bacteremia, meningitis, periodontal disease and pneumonia), and "*S milleri*" (associated with suppurative infections in children and adults). There remains persistent taxonomic confusion regarding "*S milleri*." These and other streptococci of medical importance are listed in Table 13-1 by serogroup designation, normal ecologic niche, and associated disease.

Clinical Manifestations

In humans, diseases associated with the streptococci occur chiefly in the respiratory tract, bloodstream, or as skin infections. Human disease is most commonly associated with Group A streptococci. Acute group A streptococcal disease is most often a respiratory infection (pharyngitis or tonsillitis) or a skin infection (pyoderma). Also medically significant are the late immunologic sequelae, not directly attributable to dissemination of bacteria, of group A infections (rheumatic fever following respiratory infectionand glomerulonephritis following respiratory or skin infection) which remain a major worldwide health concern. Much effort is being directed toward clarifying the risk and mechanisms of these sequelae and identifying rheumatogenic and nephritogenic strains. *S pneumoniae* remains a primary cause of serious focal and systemic infections, the first most common cause of community acquired pneumonia in the United States and of fatal bacterial pneumonia in developing countries. Hemorrhagic shock in association with *S pneumoniae* sepsis in previously healthy children has been reported recently in the United States. Of major biologic importance is a renewed interest in safe and effective streptococcal vaccines.

Structure

Both *S pyogenes* and *S pneumoniae* are Gram-positive cocci, nonmotile, and nonsporulating; they usually require complex culture media. *S pyogenes* characteristically is a round-to-ovoid coccus 0.6-1.0 µm in diameter (Fig. 13-1). They divide in one plane and thus occur in pairs, or (especially in liquid media or clinical material) in chains of varying lengths. S *pneumoniae* appears as a 0.5-1.25 µm diplococcus, typically described as lancet-shaped but sometimes difficult to distinguish morphologically from other streptococci. Streptococcal cultures older than the logarithmic phase, which is the most active growth period of a culture, may lose their Gram-positive staining characteristics.

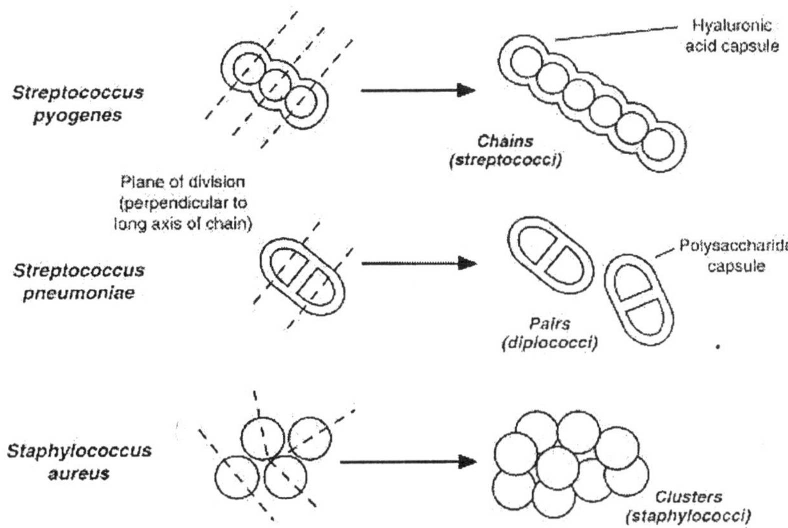

FIGURE 13-1 Morphology of the streptococci in comparison with staphylococci. Streptococci divide in a single plane and tend not to separate, causing chain formation. Capsules are antiphagocytic.

Unlike *Staphylococcus* (Chapter 12), all streptococci lack the enzyme catalase. Most are facultative anaerobes but some are obligate anaerobes. Streptococci often have a mucoid or smooth colonial morphology, and *S pneumoniae* colonies exhibit a central depression caused by rapid partial autolysis. As *S pneumoniae* colonies age, viability is lost during fermentative growth in the absence of catalase and peroxidase because of the accumulation of peroxide. Some group B and D streptococci produce pigment. Recently, nutritionally deficient streptococci (also known as wall-deficient, L form, thiol-requiring, satelliting, or pyridoxal-dependent) have been recovered from a variety of clinical sources, including blood, abscesses, and oral and urethral ulcers. These variants demonstrate bizarre pleomorphism microscopically and do not grow on routine subculture.

Classification, Antigenic Types and Extracellular Growth Products

The type of hemolytic reaction displayed on blood agar has long been used to classify the streptococci. b-Hemolysis is associated with complete lysis of red cells surrounding the colony, whereas a-hemolysis is a partial or "greening" hemolysis associated with reduction of red cell hemoglobin. Nonhemolytic colonies have been termed g-hemolytic. Hemolysis is affected by the species and age of red cells as well as by other properties of the base medium. Use of the hemolytic reaction in classification is not completely satisfactory. Some group A streptococci appear nonhemolytic; group B can manifest a-, b-, or even g-hemolysis; most *S pneumoniae* are a-hemolytic but can cause ß-hemolysis during anaerobic incubation. The viridans group, although linked by the property of a-hemolysis, is actually an extremely diverse group of organisms that does not usually react with Lancefield grouping sera. The taxonomy and biochemical and genetic relationships of these organisms continue to be clarified (Table 13-1).

TABLE 13-1 Medically Important Streptococci

Type species	Lancefield serogroup Normal habitat	Normal habitat	significant human disease
S pyogenes	A	Humans,	Acute pharyngitis and others
S agalactiae	B	Cattle, humans	Neonatal meningitis and sepsis and infections in adults
S equisimllis	C	Wide human and animal distribution	Endocarditis, bacteremia, pneumonia, meningitis, mild upper respiratory infection
E laecalis			
S bovis (nonenterococcus)	D	Human and animal intestinal tracts, dairy products bacteremia	Biliary or urinary tract infection, endocardistis
S anginosus	F, G[a]	Humans, animals	Subcutaneous or organ abscesses, endocarditis, mild upper respiratory infection
S sanguis[b]	H	Humans	Endocarditis, caries
S salivarius	K	Humans	Endocarditis, caries
None	O	Humans	Endocarditis
S suis	R	Swine	Meningitis
[c]Viridans S mitis,			
S mutans[b]	None identified	Humans	Caries, endocarditis
Anaerobic or micro aerophilic	None identified	Wide human and animal distribution	Brain and pulmonary abscesses, gynecologia infections
S pneumoniae	None identified	Humans	Lobar pneumonia and others

a Strains of the S millerl group (S cnstallatus, s intermedius , sanginosus, minule strains) may pssess antigens of groups A, C, F, or G or no identifiable lancefield group antigens: a heterogeneous group, genetically related but with a wide variety of phenotypic and biochemical characteristics.

b Disparate grouping undergoing further definition.

c Other viridans streptococci (S sanguls, s salivrius S miller, S bovis) have identified group antigens(s); nutritionally variant streptococci may be included in this diverse category.

Antigenic Types

The cell wall structure of group A streptococci is among the most studied of any bacteria (Fig. 13-2). The cell wall is composed of repeating units of N-acetylglucosamine and N-acetylmuramic acid, the standard peptidoglycan. For decades, the definitive identification of streptococci has rested on the serologic reactivity of cell wall polysaccharide antigens originally delineated by Rebecca Lancefield. Eighteen group-specific antigens were established. The group A polysaccharide is a polymer of N-acetylglucosamine and rhamnose. Some group antigens are shared by more than one species; no Lancefield group antigen has been identified for *S pneumoniae* or for some other a- or g-streptococci. With advances in serologic methods, other streptococci have been shown to possess several established group antigens.

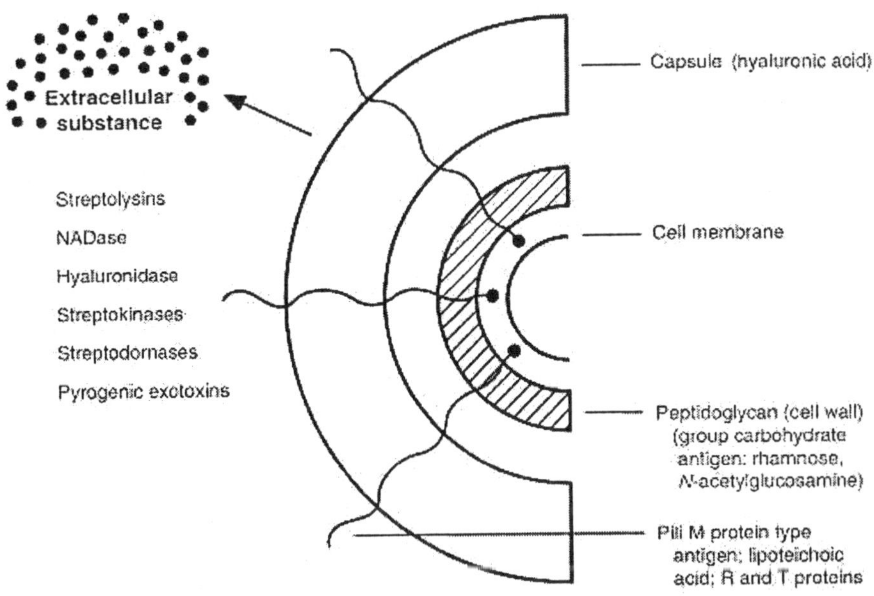

FIGURE 13-2 Cell surface structure of *S pyogenes* and extracellular substances.

The cell wall also consists of several structural proteins (Figure 13-2). In group A streptococci, the R and T proteins may serve as epidemiologic markers, but the M proteins are clearly virulence factors associated with resistance to phagocytosis. More than 50 types of *S pyogenes* M proteins have been identified on the basis of antigenic specificity. Both the M proteins and lipoteichoic acid are supported externally to the cell wall on fimbriae, and the lipoteichoic acid, in particular, appears to mediate bacterial attachment to host epithelial cells. M protein, peptidoglycan, N-acetylglucosamine, and group-specific carbohydrate portions of the cell wall have antigenic epitopes similar in size and charge to those of mammalian muscle and connective tissue. Recently emerging strains of increased virulence are distinctly mucoid, rich in M protein and highly encapsulated.

The capsule of *S pyogenes* is composed of hyaluronic acid, which is chemically similar to that of host connective tissue and is therefore nonantigenic. In contrast, the antigenically reactive and chemically distinct capsular polysaccharide of *S pneumoniae* allows the single species to be separated into more than 80 serotypes. The antiphagocytic *S pneumoniae* capsule is the most clearly understood virulence factor of these organisms; type 3 *S pneumoniae,* which produces copious quantities of capsular material, are the most virulent. Unencapsulated *S pneumoniae* are avirulent. The polysaccharide capsule in *S agalactiae* allows differentiation into types Ia, Ib, Ic, II and III.

Finally, the cytoplasmic membrane of *S pyogenes* has antigens similar to those of human cardiac, skeletal, and smooth muscle, heart valve fibroblasts, and neuronal tissues, resulting in a molecular mimicry.

Extracellular Growth Products

The importance of the interaction of streptococcal products with mammalian blood and tissue components is becoming widely recognized. The soluble extracellular growth products or toxins of the streptococci, especially of *S pyogenes* (see Fig. 13-2), have been studied intensely. Streptolysin S is an oxygen-stable cytolysin; Streptolysin O is a reversibly oxygen-labile cytolysin. Both are leukotoxic, as is NADase. Hyaluronidase (spreading factor) can digest host connective tissue hyaluronic acid as well as the organism's own capsule. Streptokinases participate in fibrin lysis. Streptodornases A-D possess deoxyribonuclease activity; B and D possess ribonuclease activity as well. Protease activity similar to that in Staph aureus has been shown in strains causing soft tissue necrosis or toxic shock syndrome. This large repertoire of products may be important in the pathogenesis of *S pyogenes* by enhancing virulence; however, antibodies to these products appear not to protect the host even though they have diagnostic importance.

Three pyrogenic exotoxins of *S pyogenes* (SPEs) are recognized: types A, B, C. These toxins act as superantigens by a mechanism similar to those described for staphylococci, not requiring processing by antigen presenting cells. Rather, they stimulate T cells by binding class II MHC molecules directly and nonspecifically. With superantigens about 20% of T cells may be stimulated (vs 1/10,000 T cells stimulated by conventional antigens) resulting in massive detrimental cytokine release. When *S pyogenes* is lysogenized by certain bacteriophages, the SPEs A or C are produced; nonlysogenized strains are atoxic. SPE B is encoded by the bacterial chromosome. Re-emergence in the late 1980's of these exotoxin-producing strains has been associated with a toxic shock-like syndrome similar in pathogenesis and manifestation to staphylococcal toxic shock syndrome (Ch.12) and other forms of invasive disease associated with severe tissue destruction. SPE's have also been identified from non group A streptococci (groups B, C, F. G) in association with the toxic shock-like syndrome.

Virulence factors in the other streptococcal species, including the enterococci, are less well identified. In group B streptococci, carbohydrate surface antigens associated with antiphagocytosis have been identified, as has neuraminidase, which may play a role in pathogenesis. Among the viridans streptococci, production of the exopolysaccharide (glycocalyx) is associated with the ability to adhere to the cardiac valves and to form vegetations on the valve leaflets.

Pathogenesis

Streptococcus pyogenes *and* Streptococcus pneumoniae

Streptococci vary widely in pathogenic potential. Despite the remarkable array of cell-associated and extracellular products previously described (Fig.13- 2), no clear scheme of pathogenesis has been worked out. *S pneumoniae* and, to a lesser extent, *S pyogenes* are part of the normal human nasopharyngeal flora. Their numbers are usually limited by competition from the nasopharyngeal microbial ecosystem and by nonspecific host defense mechanisms, but failure of these mechanisms can result in disease. More often disease results from the acquisition of a new strain following alteration of the normal flora. *S pyogenes* causes inflammatory purulent lesions at the portal of entry, often the upper respiratory tract or the skin. Some strains of streptococci show a predilection for the respiratory tract; others, for the skin. Generally, streptococcal isolates from the pharynx and respiratory tract do not cause skin infections.

Invasion of other portions of the upper or lower respiratory tracts results in infections of the middle ear (otitis media), sinuses (sinusitis), or lungs (pneumonia). In addition, meningitis can occur by direct extension of infection from the middle ear or sinuses to the meninges or by way of bloodstream invasion from the pulmonary focus. Bacteremia can also result in infection of bones (osteomyelitis) or joints (arthritis).

S pyogenes (a group A streptococcus) is the leading cause of uncomplicated bacterial pharyngitis and tonsillitis (Fig. 13-3). Indeed, only group A streptococci are sought routinely in cases of pharyngitis, although groups B, C, and G are sometimes identified. *S pyogenes* infections can also result in sinusitis, otitis, mastoiditis, pneumonia with empyema, joint or bone infections, necrotizing fasciitis or myositis, and, more infrequently, in meningitis or endocarditis. *S pyogenes* infections of the skin can be superficial (impetigo) or deep (cellulitis). Although, scarlet fever was formerly a severe complication of streptococcal infection, because of antibiotic therapy it is now little more than streptococcal pharyngitis accompanied by rash. Similarly, erysipelas, a form of cellulitis accompanied by fever and systemic toxicity, is less common today. There has, however, been an apparent recent increase in variety, severity and sequelae of *S pyogenes* infections. Because cases of streptococcal disease are not reported to national disease clearinghouses in the US, absolute numbers are not available. However, the recent resurgence of severe invasive infections has prompted descriptions of "flesh eating bacteria" in the news media.

There has been no major change in susceptibility of *S pyogenes* to commonly used antibiotics but rather in the strain variations described above (antigenic types and extracellular growth products). However, a complete explanation for the decline and resurgence is not yet available.

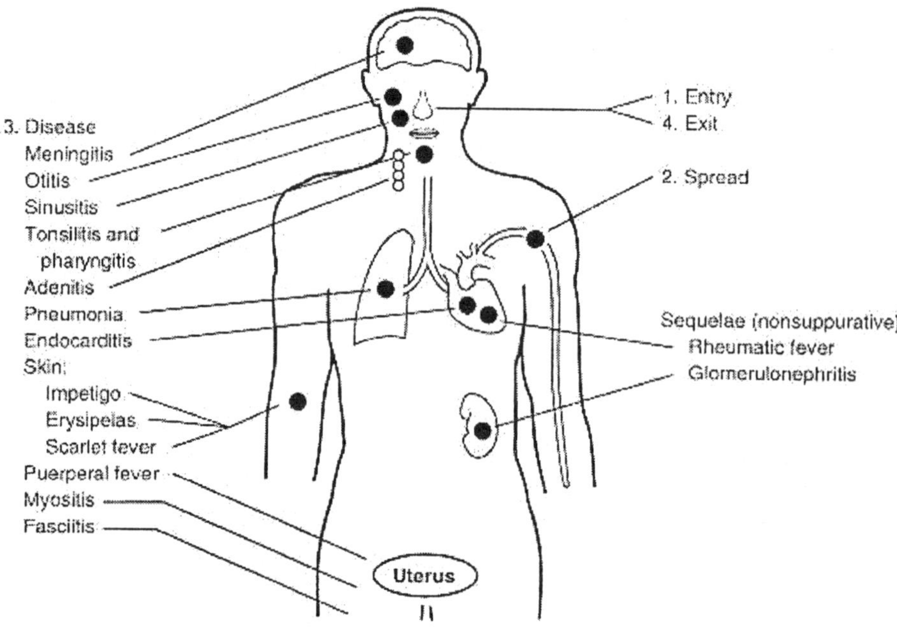

FIGURE 13-3 Pathogenesis of *S pyogenes* infections.

The capsule of *S pneumoniae* renders it resistant to phagocytosis. The ability to evade this important host defense mechanism allows *S pneumoniae* to survive, multiply, and spread to various organs (Fig.13-4). The cell wall of S *pneumoniae* contains teichoic acid. The inflammatory response induced by Gram-positive cell walls differs from that induced by the endotoxin of Gram-negative organisms, but does include recruitment of polymorphonuclear neutrophils, changes in permeability and perfusion, cytokine release, and stimulation of platelet-activating factor. The role of other S pneumoniae moieties in virulence is less clear: protein A, pneumolysin, and peptide permeases. *S pneumoniae* is the leading cause of bacterial pneumonia beyond the neonatal period. Pleural effusion is the most common and empyema (pus in the pleural space) one of the most serious complications of *S pneumoniae*. This organism is also the most common cause of sinusitis, acute bacterial otitis media, and conjunctivitis beyond early childhood. Dissemination from a respiratory focus results in serious disease: outpatient bacteremia in children, meningitis, occasionally acute septic arthritis and bone infections in patients with sickle cell disease and, more rarely, peritonitis (especially in patients with nephrotic syndrome) or endocarditis.

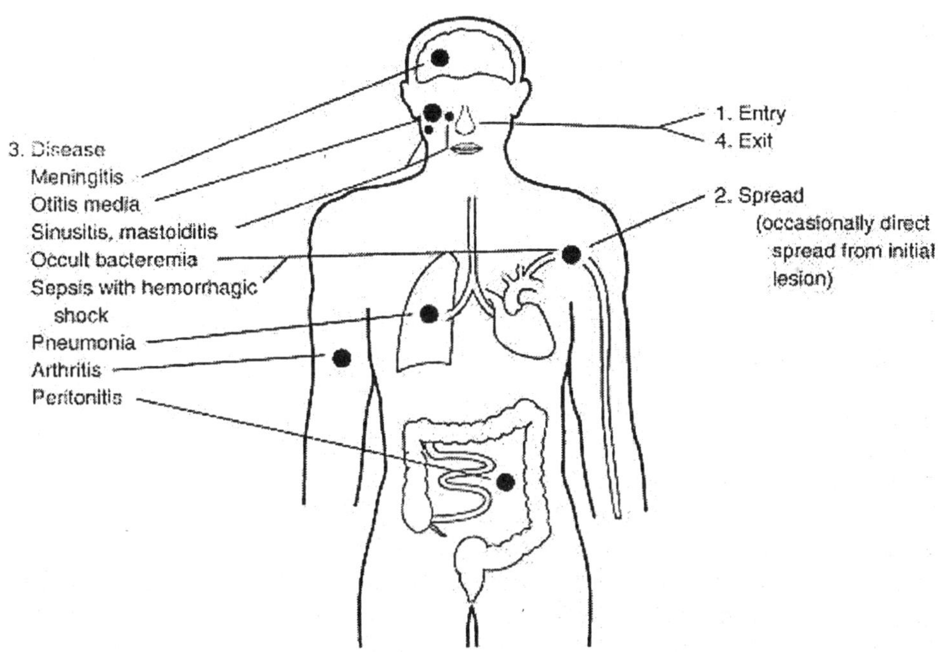

FIGURE 13-4 Pathogenesis of *S pneumoniae* infections.

Postinfectious Sequelae

Infection with *S pyogenes* (but not *S pneumoniae*) can give rise to serious nonsuppurative sequelae: acute rheumatic fever and acute glomerulonephritis. These sequelae begin 1-3 weeks after the acute illness, a latent period consistent with an immune-mediated rather than pathogen-disseminated etiology. Whether all *S pyogenes* strains are rheumatogenic is still controversial; however, clearly not all are nephritogenic. These differences in pathogenic potential are not yet understood.

Acute rheumatic fever is a sequela only of pharyngeal infections, but acute glomerulonephritis can follow infections of the pharynx or the skin. Although, there is no adequate explanation for the precise pathogenesis of acute rheumatic fever or for its failure to occur after streptococcal pyoderma, an abnormal or enhanced immune response seems essential. Also, persistence of the organism, due perhaps in part to the greater avidity with which the organism adheres to host pharyngeal cells, is associated with an increased likelihood of rheumatic fever. Acute glomerulonephritis results from deposition of antigen-antibody-complement complexes on the basement membrane of kidney glomeruli. The antigen may be streptococcal in origin or it may be a host tissue species with antigenic determinants similar to those of streptococcal antigen (cross-reactive epitopes for endocardium, sarcolemma, vascular smooth muscle). In the United States, the incidence of acute rheumatic fever had decreased dramatically. Although, several areas reported a resurgence in cases in the late 1980's, subsequently, a slow, steady decline continued. Acute rheumatic fever can result in permanent damage to the heart valves. Less than 1% of sporadic streptococcal pharyngitis infections result in acute rheumatic fever; however, recurrences are common, and life-long antibiotic prophylaxis is recommended following a single case. The incidence of acute glomerulonephritis in the United States is more variable, perhaps due to cycling of nephritogenic strains, but appears to be decreasing; recurrences are uncommon, and prophylaxis following an initial attack is unnecessary.

Other Streptococcal Species

Lancefield Group Streptococci

Streptococcal groups B, C, and G initially were recognized as animal pathogens (see Table 13-1) and as part of the normal human flora. Recently, the pathogenic potential for humans of some of these non-group-A streptococci has been clarified. Group B streptococci, a major cause of bovine mastitis, are a leading cause of neonatal septicemia and meningitis, accounting for a significant changing clinical spectrum of diseases in both pregnant women and their infants. Mortality rates in full-term infants range from 2-8% but in pre-term infants are approximately 30%. Early-onset neonatal disease (associated with sepsis, meningitis and pneumonia at 6d life) is thought to be transmitted vertically from the mother; late-onset (from 7d to 3 mos age) meningitis is acquired horizontally, in some instances as a nosocomial infection. Group B organisms also have been associated with pneumonia in elderly patients. They are part of the normal oral and vaginal flora and have also been isolated in adult urinary tract infection, chorioamnionitis and endometritis, skin and soft tissue infection, osteomyelitis, meningitis, bacteremia without focus, and endocarditis. Infection in patients with HIV can occur at any age.

Streptococci of groups C and G are associated with mild, as well as severe human disease. None of these groups has been implicated in acute rheumatic fever or acute glomerulonephritis. Group D streptococci are important etiologic agents of urinary tract infections and infections associated with biliary tract procedures, as well as cases of disseminated infection, bacteremia, and endocarditis. Streptococcus bovis bacteremia has been recognized more often in cases of bowel disease.

Group F streptococci are associated with abscess formation and purulent disease. Group R streptococci, well-documented causes of meningitis and septicemia in pigs, also pose a serious health hazard to workers in the pork industry.

Viridans Streptococci

The biochemically and antigenically diverse group of organisms classified as viridans streptococci, as well as other non-groupable streptococci of the oral and gastrointestinal cavities and urogenital tract, include important etiologic agents of bacterial endocarditis. Dental manipulation and dental disease with the associated transient bacteremia are the most common predisposing factors in bacterial endocarditis, especially if heart valves have been damaged by previous rheumatic fever or by congenital cyanotic heart disease. *S mutans* and *S sanguis* are odontopathogens responsible for the formation of dental plaque, the dense adhesive microbial mass that colonizes teeth and is linked to caries and other human oral disease (see Ch. 99). *S mutans* is the more cariogenic of the two species, and its virulence is directly related to its ability to synthesize glucan from fermentable carbohydrates as well as to modify glucan in promoting increased adhesiveness.

Anaerobes

Like their aerobic counterparts, anaerobic streptococci are part ofthe normal flora, particularly of the mouth and intestinal tract; they are also part of the normal flora of the upper respiratory and genital tracts and the skin. These anaerobic organisms are linked to a wide variety of serious mixed infections of the female genital tract as well as to brain, pulmonary, and abdominal abscesses.

Host Defenses

The streptococci are part of the endogenous microbial flora of the nasopharynx. Disease may result from circumvention of the normal specific or nonspecific host defense mechanisms. More often, both *S pyogenes* and *S pneumoniae* are exogenous secondary invaders following viral disease or disturbances in the normal bacterial flora.

In the normal host, nonspecific defense mechanisms prevent organisms from penetrating beyond the superficial epithelium of the upper respiratory tract. These mechanisms include mucociliary movement and the cough, sneeze and epiglottal reflexes. The host phagocytic system is a second line of defense against pathogens.

Organisms can be opsonized by activation of the classical or alternate complement pathway or by specific immunoglobulin binding.

The capsules of both *S pyogenes* and *S pneumoniae* allow the organisms to evade opsonization. The hyaluronic acid outer surface of *S pyogenes* is only weakly antigenic; however, protective immunity results from the development of type-specific antibody to the M protein of the fimbriae, which protrude from the cell wall through the capsular structure. This antibody, which follows respiratory and skin infections, is persistent. Presumably, IgA in the respiratory secretions and serum IgG are the important protective antibody classes. *S pyogenes* is rapidly killed following phagocytosis enhanced by specific antibody. Prompt, effective antibiotic treatment of streptococcal infections may preclude development of this persistent antibody. Evidence has shown that antibody to the erythrogenic toxin involved in scarlet fever is also long lasting. This is the basis of the Dick test, as in vivo skin test, rarely used today, which measures host antitoxin. The capsular polysaccharides of *S pneumoniae* are highly antigenic and type-specific. Type-specific anticapsular antibodies to these T-independent antigens result in effective opsonization and host recovery. In untreated *S pneumoniae* infections, recovery clearly is due to opsonizing antibody. Even when adequate and appropriate antibiotic therapy is given, opsonizing antibody probably contributes significantly to recovery from pneumococcal disease. The normal host is somewhat resistant to *S pneumoniae* disease, but compromised hosts of several types are highly susceptible to serious infections: alcoholics, the semicomatose, very young, and very old individuals, patients who have undergone splenectomy, and patients with underlying diseases (specifically, chronic cardiac, pulmonary, or renal disease; sickle cell anemia; leukopenia; multiple myeloma; cirrhosis; and diabetes).

Cross-reactive antigens, especially of *S pyogenes* and various mammalian tissues, help explain the autoimmune responses that develop following some infections. The level of humoral response to infection with *S pyogenes* is greater in patients with rheumatic fever than in patients with uncomplicated pharyngitis. In addition, cell-mediated immunity may play a significant role in acute rheumatic fever.

Neonatal susceptibility to group B streptococci may result from immature neonatal phagocytic function, humoral immunity, or cell-mediated immunity, or from lack of passively acquired maternal antibody.

Evidence from the Rhesus monkey animal model in dental research shows that IgG may be a more important antibody class than IgA or IgM in protection against caries. Part of the reason may be that IgG is the antibody isotype most efficient at enhancing phagocytosis of *S mutans*. Cell-mediated immunity appears to participate in the protective host response against caries.

Epidemiology

The streptococci are widely distributed in nature and frequently form part of the normal human flora (see Table 29-1). Approximately 5-15% of humans carry *S pyogenes* or *S agalactiae* in the nasopharynx. *S pneumoniae* infects humans exclusively, and no reservoir is found in nature. The carrier rate of *S pneumoniae* in the normal human nasopharynx is 20-40%.

All ages, races, and sexes are susceptible to streptococcal disease. Because *S pneumoniae* is a particularly labile organism, sensitive to heat, cold, and drying, horizontal transmission requires close person-to-person contact. Infection is more likely at the extremes of life (<2 yr, > 65 yr), when host resistance is reduced, as described in the preceding section, or after the introduction of a more virulent strain. In the United States, pneumococcal disease is most prevalent during winter, coinciding with increased rates of acquisition but not necessarily of carriage. Alaskan natives have higher rates of invasive pneumococcal disease than do other American populations. The reason for this is unclear.

The incidence of respiratory disease attributed to *S pyogenes* peaks at about 6 years of age, and then again at 13 years of age, and is most common during late winter and early spring in temperate climates. Skin infections are more common among preschool-age children, and are most prevalent in late summer and early fall in temperate climates (when hot, humid weather prevails), and at all times in tropical climates. *S pyogenes* is spread by respiratory droplets or by contact with fomites used by the index individual, either patient or carrier. Skin infections often follow minor skin irritation, such as insect bites. There are occasional reports of streptococcal disease traced to rectal carriers, and of food-borne and vector-born outbreaks. In children, invasive disease with *S pyogenes* may follow varicella, or be associated with burns or malignancy; in adults with surgical or nonsurgical wounds or underlying medical problems, i.e., diabetes, cirrhosis, underlying peripheral vascular disease, or malignancy.

The world prevalence of the serious late sequelae of *S pyogenes* infections (acute rheumatic fever and acute glomerulonephritis) has shifted from temperate to tropical climates. In particular, acute rheumatic fever had ceased to be a major health concern in the US., despite no concomitant decline in group A streptococcal pharyngitis. These diseases previously affected persons with a low standard of living and limited access to medical care. Since 1985, there have been scattered outbreaks of acute rheumatic fever in some regions of the United States. Temporal and geographic clustering provides further evidence for "rheumatogenic" strains. Whether ethnic or racially determined factors affect this shift is not known.

Other streptococcal groups show striking epidemiologic features. An increasing prevalence of non-group-A as compared to group A streptococci in throats has been reported. Studies of the vaginal flora among women of child-bearing age show a *S agalactiae* carrier rate of 15-40%. Vertical transmission of the organisms to neonates of vaginally infected mothers ranges from 40-73%, but the incidence of neonates with disease (in contrast to colonized, healthy neonates) is low, 1-2%. *S suis* has been linked to meningitis among meat handlers. Isolation of *S milleri* or *S bovis* from the bloodstream should raise suspicion of immunosuppression or underlying disease visceral abscess formation or other bowel disease (including colon carcinoma).

In the United States, enterococci are the second most common nosocomial pathogenes associated with both endogenous colonization and patient-to-patient spread. A wide variety of infections results, especially urinary tract and surgical wound infections, with a marked propensity for antibiotic resistance. The widespread usage of newer cephalosporins, which have poor activity against enterococci, allows "break through" of enterococci as clinically significant isolates, the development of resistance in areas of heavy antibiotic use, and a selective advantage to these organisms.

Diagnosis

Clinical

It is not usually possible to diagnose streptococcal pharyngitis or tonsillitis on clinical grounds alone. Accurate differentiation from viral pharyngitis is difficult even for the experienced clinician, and therefore the use of bacteriologic methods is essential. However, distinguishing acute streptococcal pharyngitis from the carrier state may be difficult. When documented streptococcal pharyngitis is accompanied by an erythematous punctiform rash (Fig.13-4), the diagnosis of scarlet fever can be made. With streptococcal toxic shock syndrome, unlike staphylococcal toxic shock syndrome where the organism is elusive, there is often a focal infection or bacteremia. Criteria for diagnosis of streptococcal toxic shock syndrome include hypotension and shock, isolation of *S pyogenes*, as well as 2 or more of the following: ARDS, renal impairment, liver abnormality, coagulopathy, rash with desquamating soft tissue necrosis. The invasive, potentially fatal *S pyogenes* infections require early recognition, definitive diagnosis, and early aggressive treatment.

Rheumatic fever is a late sequela of pharyngitis and is marked by fever, polyarthritis, and carditis. A combination of clinical and laboratory criteria (Table 13-2) is used in the diagnosis of acute rheumatic fever. Since the original Jones criteria were published in 1944, these have been modified (1955), revised (1965, 1984) and updated (1992). The other late sequela, acute glomerulonephritis, is preceded by pharyngitis or pyoderma; is characterized by fever, blood in the urine (hematuria), and edema; and is sometimes accompanied by hypertension and elevated blood urea nitrogen (azotemia). Pneumococcal pneumonia is a life-threatening disease, often characterized by edema and rapid lobar consolidation.

TABLE 13-2 Jones Diagnostic criteria for Acute Rheumatic Fever[a]

Major	Minor
Carditis	Clinical
Polyarthritis	Fever
Erythema marginatum	Arthralgia
Subcutaneous nodules	
chorea	Laboratory
	Increased erythrocyte sedimentation rate
	Increased C - reactive protein level
	ECG
	Prolonged PR interval

a in conjunction with culture or serologic evidence of recent streptococcal infection. (From American Heart Association, jones criteria, updated: JAMA 268:2069, 1992, with permission)

Specimens For Direct Examination And Culture

S pyogenes is usually isolated from throat cultures. In cases of cellulitis or erysipelas thought to be caused by *S pyogenes*, aspirates obtained from the advancing edge of the lesion may be diagnostic. *S pneumoniae* is usually isolated from sputum or blood. Precise streptococcal identification is based on the Gram stain and on biochemical properties, as well as on serologic characteristics when group antigens are present.

Table 13-3 shows biochemical tests that provide sensitive group-specific characteristics permitting presumptive identification of Gram-positive, catalase-negative cocci.

TABLE 13-3 Characteristics for the Presumptive Identification of Streptococci of Human clinical Importance

	Results of Group						
Procedure	**A**	**B**	**D(enterococcus)**	**D(nonenterococcus)**	**Non-A,B,D**	**S pneumonlae**	**viridans**
Hemolysis[a]	β	β,α,γ	α,β,γ	α,γ,β	β.α.γ	α	α
Bacitracin sensitivity	+	-(+)[b]	-	-	-(+)	±	-
CAMP test	-	+	-	-	-	-	-
Growth at 45°C	-	-	+	+	-	-	+
Optochin sensitivity	-(+)	-(+)	-	-	-(+)	+	-
Hydrolysis of sodium hippurate	-	+	-(+)	-	-	-	-(+)
Tolerance to 6.5 percents NaCl	-	-	+	-	-	-	-
Hydrolysis of esculin presence of 40 percent bile	-	-	+	+	-	-	-(+)
Hydrolysis of pyrrolidonyl napgthylamide	+	-	+	-	-	-	-

a In general order of frequency.

b Signs in parentheses indicate occasional result,

Identification

Hemolysis should not be used as a stringent identification criterion. Bacitracin susceptibility is a widely used screening method for presumptive identification of *S pyogenes*; however, some *S pyogenes* are resistant to bacitracin (up to 10%) and some group C and G streptococci (about 3-5%) are susceptible to bacitracin. Some of the group B streptococci also may be bacitracin sensitive, but are presumptively identified by their properties of hippurate hydrolysis and CAMP positivity. *S pneumoniae* can be separated from other a-hemolytic streptococci on the basis of sensitivity to surfactants, such as bile or optochin (ethylhydrocupreine hydrochloride). These agents activate autolytic enzymes in the organisms that hydrolyze peptidoglycan.

In many instances, presumptive identification is not carried further. Serologic grouping has not been performed as often as it might be because of the lack of available methods and the practical constraints of time and cost; however, only serologic methods, as listed in Table 13-4, provide definitive identification of the streptococci. The Lancefield capillary precipitation test is the classical serologic method. *S pneumoniae,* which lacks a demonstrable group antigen by the Lancefield test, is conventionally identified by the quellung or capsular swelling test that employs type-specific anticapsular antibody. Inspection of Gram-stained sputum remains a reliable predictor for initial antibiotic therapy in community-acquired pneumonia.

TABLE 13-4 Methods of Serogrouping Streptococci

Nature of streptococcal Antigen	Techniques
Whole	Fluorescent antibody direct bacterial agglufination Coagglutination with slaphlococcal protein A Carrier agglutination (antibody-coated latex particles) Quellung reaction (for S pneumoniae)
soluble extract	Procipitation (classical capillary of counterimmunoelectrohoresis) Coagglutination with staphylococcal protein A Carrier agglutination (antibody-coated latex particles) Enzyme-linked immunosorbent assay (ELISA) antigen capture assays Optical immunoassay

New methods for serogrouping that show sensitivity and specificity now are being explored. Organisms from throat swabs, incubated for only a few hours in broth, can be examined for the presence of *S pyogenes* using the direct fluorescent antibody or enzyme-linked immunosorbent technique. Additional rapid antigen detection systems for the group carbohydrate have become increasingly popular. However, the sensitivity (70-90%) of these currently available rapid tests for group A streptococcal carbohydrate does not allow exclusion of streptococcal pharyngitis without conventional throat culture (sensitivity of a single throat culture is 90-99%). A third generation assay, the optical immunoassay, is currently being evaluated. *S pneumoniae* can be identified rapidly by counterimmuno-electrophoresis, a modification of the gel precipitin method. The coagglutination test, described in Ch.12, is a more sensitive modification ofthe conventional direct bacterial agglutination test. The Fc portion of group-specific antibody binds to the protein A of dead staphylococci, leaving the Fab portion free to react with specific streptococcal antigen. The attachment of antibody to other carrier particles in suspension (for example, latex) also is used. The fact that whole streptococcal cells can be used in recently developed methods circumvents the difficulties involved in extracting components that retain appropriate antigenic reactivity. These newer serogrouping methods should make it more practical to identify not only b-hemolytic isolates from the blood or normally sterile sites, but also a-and nonhemolytic strains. It has become increasingly important to identify more of these strains to avoid simply misclassifying them as contaminants. Such information will expand our understanding of the importance of non-group-A streptococci.

Serologic Titers

Antibodies to some of the extracellular growthproducts of the streptococci are not protective but can be used in diagnosis. The antistreptolysin O (ASO) titer which peak 2-4 wks after acute infection and anti-NADase titers (which peaks 6-8 weeks after acute infection) are more commonly elevated after pharyngeal infections than after skin infections. In contrast, antihyaluronidase is elevated after skin infections, and anti-DNase B rises after both pharyngeal and skin infections. Titers observed during late sequelae (acute rheumatic fever and acute glomerulonephritis) reflect the site of primary infection. Although, it is notas well known as the ASO test, the anti-DNase B test appears superior because high-titer antibody is detected following skin and pharyngeal infections and during the late sequelae. Those titers should be interpreted in terms of the age of the patient and geographic locale.

Although, not used in diagnosis, bacteriocin production and phage typing of streptococci are employed in research and epidemiologic studies.

Control

Antibiotic Treatment

Penicillin remains the drug of choice for *S pyogenes*. It is safe, inexpensive, and of narrow spectrum, and there is no direct or indirect evidence of loss of efficacy. Prior to the 1990's, *S pneumoniae* was also uniformly sensitive to penicillin but a recent abrupt shift in the usefulness of penicillin has occurred. The group D enterococci are resistant to penicillins, including penicillinase-resistant penicillins such as methicillin, nafcillin, dicloxacillin, and oxacillin, and are becoming increasingly resistant to many other antibiotics. Group B streptococci are often resistant to tetracycline but remain sensitive to the clinically achievable blood levels of penicillin, even though they have penicillin minimal inhibitory concentrations (MIC) considerably higher than those of *S pyogenes*. Although, the duration of penicillin therapy varies with the degree of invasiveness, streptococcal pharyngitis is generally adequately treated with 10 days of antibiotic therapy, and pneumococcal pneumonia with 7-14 days. If penicillin allergy occurs, an alternative drug for treating pharyngitis is erythromycin, although sporadic erythromycin and tetracycline resistance has been reported, leaving clindamycin or the newer macrolides as possible treatments. The most important goal of therapy in acute streptococcal pharyngitis is still to prevent rheumatic fever. However, therapy also hastens clinical recovery, avoids suppurative complications and renders the patient non-infectious for others. In addition to antibiotics, the patient with *S pyogenes* myositis or necrotizing fasciitis requires surgical debridement. Lifelong prophylaxis against recurrences of rheumatic fever is achieved with long-acting penicillin or erythromycin. Sulfonamides will not eradicate the streptococcus and thus are not acceptable therapy for streptococcal pharyngitis, but sulfadiazine is effective for preventing recurrent attacks of rheumatic fever. Additional prophylactic coverage before some dental and surgical procedures is necessary in the presence of rheumatic heart disease or prosthetic heart valves. Although, streptococcal pharyngitis is usually a benign, self-limited disease, therapy is important to prevent rheumatic fever. There is no convincing evidence that antibiotic therapy prevents glomerulonephritis. Disconcertingly, some patients in recent outbreaks of acute rheumatic fever do not give a history of preceding pharyngitis.

Methods of treating the asymptomatic pharyngeal carrier of *S pyogenes* remain controversial. Recent evidence suggests that up to 20% of children and young adults are carriers, the carrier state involves no risk to the carrier or to others, and it is frequently difficult to eradicate despite the exquisite sensitivity of the organism to penicillin in vitro. A similar failure of antibiotic therapy to eradicate nasopharyngeal carriage or to prevent reinfection with *S pneumoniae* also occurs.

Although, antibiotic resistance in *S pneumoniae* is common in many parts of the world, in the United States such strains previously had a geographically limited focus. Recent widespread emergence of *S pneumoniae* resistant to penicillin and other antibiotics has become a microbial threat in the United States as well. Even cefotaxime and ceftriaxone resistance has been documented. Isolates must be carefully screened for susceptibility by oxacillin disc testing, with definitive MIC determination by the E test (A B Biodisk NA, Piscataway, NJ), a convenient and reliable method for detection of resistance to penicillin and extended spectrum cephalosporins.

It is inappropriate to universally treat of pregnant women who are carriers of group B streptococci, or their colonized neonates, for several reasons: the high carrier rate; cost; the associated high risk of penicillin hypersensitivity; the potential increase in infections with penicillin-resistant organisms; the difficulty in altering colonization of women (even when their sexual partners were also treated); and the low risk of neonatal disease. The controversy continues despite recent recommendations for universal screening of pregnant women and selective intrapartum chemoprophylaxis for screen-positive mothers with preterm labor, premature or prolonged rupture of membranes, fever in labor, multiple births or previous infants with group B streptococcal disease.

Clearly, penicillin has reduced the severe morbidity and mortality associated with *S pneumoniae*. The emergence of resistance has now forced re-evaluation of empiric therapy. Clinicians must report clusters of *S pneumoniae* infection and be aware of local patterns of resistance. Penicillin susceptible organisms show MICs 0.06 mg/ml, intermediate strains 0.1-1.0 mg/ml and high level resistant strains 2 mg/ml.

For nonmeningeal infection by intermediate strains, parenteral penicillin at high dose can probably be used since the mechanism of resistance involves alteration in penicillin binding proteins (PBP) and saturation. For meningeal infection with intermediate strains or any infection by high level resistant strains only ceftriaxone and cefotaxime retain sufficient activity. Resistance even to these extended spectrum cephalosporins was first reported for the US in 1991. At this writing only vancomycin remains uniformly effective but as discussed below, its use incurs potential for selection of vancomycin resistant enterococci (VRE) or risk of transferring vancomycin resistance from enterococci to *S pneumoniae*.

Currently, no single agent is reliably bactericidal against enterococci. Serious infections with group D enterococci often require a classic synergistic regime combining penicillin or ampicillin with an aminoglycoside, designed to weaken the cell wall with the ß-lactam and facilitate entry of the bacteriocidal aminoglycoside. Other ß-lactam drugs with good activity against enterococci include piperacillin and imipenem. An alternate drug of choice is vancomycin, but vancomycin-resistant strains of enterococci have been isolated. Nosocomial acquisition of these resistant organisms is of grave concern.

This antibiotic resistance among the streptococci/enterococci is an increasing problem. Studies show that in vitro exchange of resistant DNA can occur in conjugation via plasmids and transposons, or in transduction with bacteriophages. The mechanisms involved in the in vivo genetic exchange are not clearly defined. Evidence is accumulating that other streptococci may be the important donors of resistance markers. Transposon transfer is thought to be the most likely mechanism in *S pneumoniae*, although point mutations also occur. In the setting of heavy ß-lactam use, selective pressure is important in emergence of resistant strains. The first penicillin-resistant *S pneumoniae* were reported in 1967 in Australia and in 1974 in North America. In New Guinea, where the first penicillin-resistant strains were reported in 1971, one-third of *S pneumoniae* isolates from patients with severe pneumococcal disease were resistant by 1978. In Hungary in 1992, 69% of *S pneumoniae* isolates were penicillin resistant. This resistance is not ß-lactamase mediated but due to alteration in PBP which results in decreased binding of penicillin by the organism, rendering the drug less effective and requiring higher concentrations for saturation. Some strains resistant to erythromycin or tetracycline also have been reported, as well as some multiply resistant strains. In South Africa, outbreaks of infection with strains of *S pneumoniae* resistant to ß-lactam antibiotics (penicillins and cephalosporins) as well as to tetracycline, chloramphenicol, erythromycin, streptomycin, clindamycin, sulfonamides, and rifampin were reported in 1977. Although, antibiotic resistance among *S pneumoniae* was infrequent in the United States, a major shift occurred from 1988 to 1990, resulting in the present situation of 15-25% of *S pneumoniae* intermediately or completely resistant to penicillin. Communities with "low prevalence" have 5-10% resistance. Single or multiply resistant strains are transmitted person to person, especially in settings of frequent salivary exchange, antibiotic use and hand-to-hand transmission (as in day care centers) or of crowding (corrections facilities, homeless shelters, nursing homes, military training groups). Control of the problem of emerging, antibiotic-resistant *S pneumoniae* is multifactorial: 1) surveillance for clusters of invasive disease, resistance and prevalent serotypes; 2) education of physicians and the public about antibiotic use (decrease unnecessary antibiotic use for obviously viral infections and decrease antibiotic prophylaxis for otitis by use of intermittent or expectant dosing or of non ß-lactam based prophylaxissulfa. Use topical treatment for impetigo, and short course therapies and narrow spectrum antibiotics); 3) adherence to infection control strategies in day care centers; 4) aggressive promotion of the current 23-valent *S pneumoniae* vaccine and support of efforts to design a new vaccine effective in those <2 years of age, analogous to the eminently successful Haemophilus influenzae type B vaccine (see Ch. 30) where bacterial polysaccharide is conjugated to protein to elicit a T cell-dependent response.

Among the enterococci, resistance to a wide variety of common antibiotics has emerged, with some strains resistant to all currently available antibiotics. There is no clinically proven treatment effective against enterococci multiply resistant to lactams, aminoglycosides, and vancomycin. The emergence of such organisms poses a stunning management dilemma. Resistance among the enterococci can be either intrinsic or acquired (by de novo genetic mutation or acquisition of DNA from resistant organisms). Enterococcal resistance to lactams is also mediated by altered PBP as in pneumococci, allowing cell wall synthesis even in the presence of antibiotic, or much less commonly by ß-lactamase. Resistance to aminoglycosides is mediated by decreased uptake or aminoglycoside modifying enzymes, and to vancomycin by decreased cell wall affinity for glycopeptide antibiotics. Further research into the mechanisms of resistance and new class(es) of antibiotics is essential.

A final concern about emerging resistance among the enterococci is the potential for genetic transfer of resistance genes to more virulent pathogens: Staph aureus, *S pneumoniae* and even Gram-negative organisms. So significant is this threat of emerging enterococcal resistance that the Centers for Disease Control and Prevention has issued a document addressing national guidelines. These include recommendations for 1) education of physicians and the public about the impact of vancomycin resistant enterococci (VRE), 2) vigilant surveillance for and detection of VRE, 3) strict enforcement of infection control strategies in hospitals, and 4) prudent vancomycin use or monotherapeutic use of extended spectrum cephalosporins. In a recent study of vancomycin use in US hospitals, use was about equally divided for treatment of a specific isolate, for prophylaxis, and for empiric coverage. The recommendations discourage vancomycin use for routine surgical prophylaxis, empiric prophylaxis in the patient with febrile neutropenia, the low birth weight infant or patients with vascular or peritoneal catheters, treatment of a single blood culture positive for coagulase-negative staphylococci, primary treatment of antibiotic-associated colitis, attempted eradication of colonization by methicillin-resistant Staph aureus (MRSA), or selective decontamination of the gastrointestinal tract.

Vaccination

As chemotherapeutic management becomes more difficult because of the threat of resistance, prevention becomes more important. With the introduction of antibiotics, previously successful pneumococcal vaccines fell into disuse. However, although prompt treatment with antibiotics has reduced the serious consequences of *S pneumoniae* infections (pre-antibiotic mortality rate of 30%), the disease incidence remains unchanged, and attention has been redirected to vaccines for *S pneumoniae* as well as for other streptococci. Pneumococcal vaccines (containing the pneumococcal polysaccharides of the most prevalent serotypes) have been licensed in several countries, including the United States. Initial use shows them to be useful and safe, but they remain under-utilized. The spectre of multidrug resistant *S pneumoniae* may provide a new incentive for their use. In 1983, the United States Food and Drug Administration licensed a vaccine containing 23 serotypes, representing coverage against nearly 89% of the pneumococcal isolates submitted to the CDC in the 1987-1988 National Surveillance Study. The population target of pneumococcal vaccines includes those at high risk for serious pneumococcal disease: the elderly (65 and older) and children (2 years of age and older) with sickle cell anemia, with an immunocompromised state (lymphoma, asplenia, myeloma, acquired immunodeficiency syndrome), with nephrotic syndrome, or with chronic cardiopulmonary disease. Vaccines for the other streptococci remain experimental.

Vaccine production for the streptococci presents several formidable problems. For both *S pyogenes* and *S pneumoniae,* a large number of serotypes must be included in effective vaccines since successful selection of a common epitope remains elusive. Continuing surveillance to determine prevalent serotypes is necessary to insure that the vaccine formulations remain appropriate. For *S pyogenes,* it is critical to determine rheumatogenic and nephritogenic strains to limit the required multivalency of the vaccines. Alternatively a newly described conserved portion of M protein is a distant goal. Toxicity has been associated with M protein preparations, but lack of immunogenicity in highly purified preparations of antigens is still a problem. With streptococcal vaccines, the potential risk of antigenic cross-reactivity with cardiac tissue and an associated increased risk of acute rheumatic fever must be appreciated.

In group B neonatal disease chemoprophylaxis does not appear as practical as vaccine control. Passive immunity in group B streptococcal neonatal infection appears protective. Polyvalent hyperimmune gamma globulin and human monoclonal IgM antibody which reacts with multiple serotypes are undergoing efficacy studies. Active immunization of pregnant women with undegraded sialic acid-containing polysaccharide group B antigens is another important aspect of control.

The streptococci are ubiquitous, and their significance in medicine is remarkable. Exciting advances are being made in diagnosis and in understanding the mechanisms of pathogenesis, as well as in control of these well-known organisms. Problems with antibiotic resistance must preclude complacency in dealing with these common pathogens.

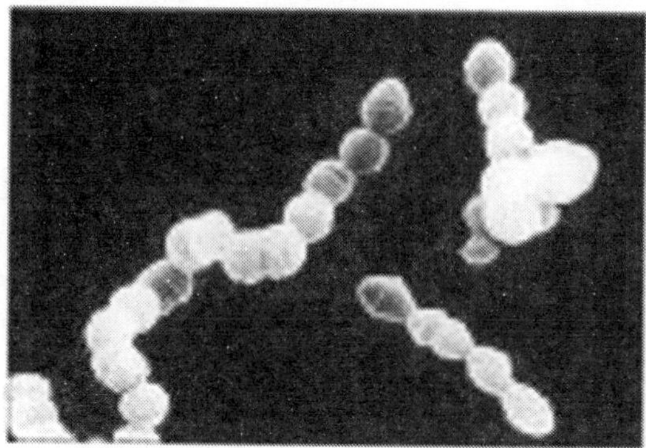

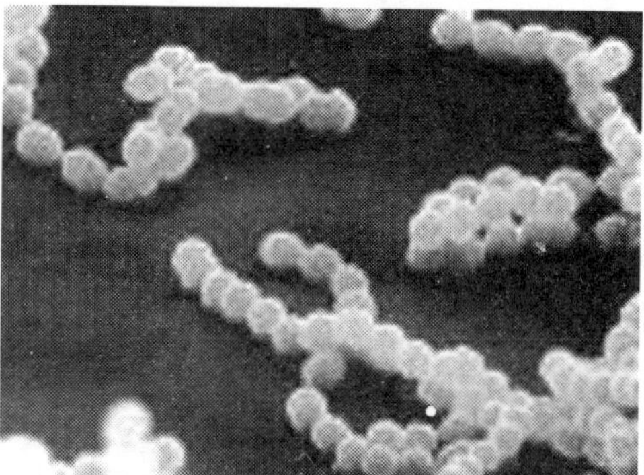

Colony morphology of *Streptococc*

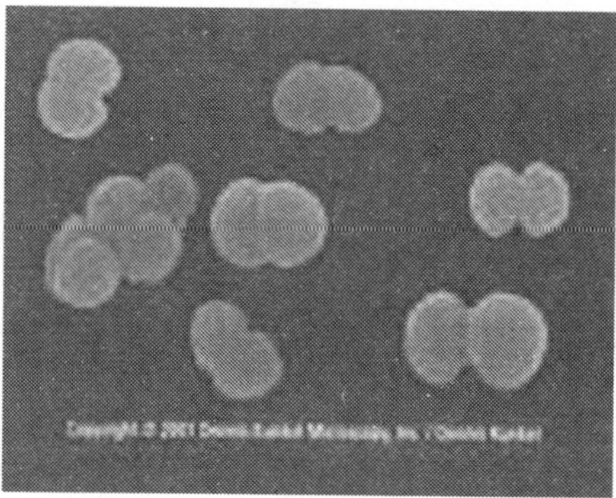

Colony morphology of *Pneumococci*

REFERENCES

Awada A, van der Auwera P, Meunier F, et al. Streptococcal and enterococcal bacteremia in patients with cancer. Clin. Infect. Dis 1992; 15:33-48.

Bisno AL Medical progress: group A streptococcal infections and acute rheumatic fever. N. Engl. J. Med. 1991; 325:783-793.

Butler JC, Breiman RF, Campbell JF, et al. Pneumococcal polysaccharide vaccine efficacy: an evaluation of current recommendations. JAMA 1993; 270:1826-1831.

CDC. Addressing emerging infectious disease threats: a prevention strategy for the United States. Atlanta: US Dept Health & Human Serv. Pub Hlth Serv. CDC 1994. CDC.

Preventing the spread of vancomycin resistance report from the Hospital Infection Control Practices Advisory Committee. Fed Reg. 1994; 59:25757-25763.

Coykendall AL. Classification and identification of the viridans streptococci. Clin Microbiol Rev 1989; 2:315-328.

Dillon HC. Post-streptococcal glomerulonephritis following pyoderma. Rev Infec Dis 1979; 1:935-943.

Denny FW, Wannamaker LW, Brink WR, et al. Prevention of rheumatic fever: treatment of the preceding streptococcal infection. JAMA 1950; 143:151-153.

Farley MM, Harvey RC, Stull T, et al. A population-based assessment of invasive disease due to group B streptococcus in nonpregnant adults. N. Engl. J. Med. 1993; 328:1807-1811.

Hoge CW, Schwartz B, Talkington DF, et al. The changing epidemiology of invasive group A streptococcal infections and the emergence of streptococcal toxic shock-like syndrome. JAMA 1993; 269:384-389.

Kaplan EL. Global assessment of rheumatic fever and rheumatic heart disease at the close of the century. Circulation 1993; 88:1964-1972.

McCracken G. Emergence of resistant Streptococcus pneumoniae: a problem in pediatrics. Pediatr. Infect. Dis. J. 1995; 14:424-428.

Moellering RC. Emergence of Enterococcus as a significant pathogen. Clin. Infect. Dis. 1992; 14:1173-1178.

Ruoff KL. Streptococcus anginosus ("Streptococcus milleri"). The unrecognized pathogen. Clin. Microbiol. Rev. 1988; 1:102-108.

Special Writing Group of the Committee on Rheumatic Fever, Endocarditis, and Kawasaki Disease of the Council on Cardiovascular Disease in the Young, of the American Heart Association. Guidelines for the diagnosis of rheumatic fever. Jones criteria, 1992 update. JAMA 1992; 268:2069-2073.

Stevens DL, Tanner MH, Winship J, et al. Severe group A streptococcal infections associated with a toxic shock-like syndrome and scarlet fever toxin A. N Engl J Med 1989; 321:1-8.

Tanz RR, Poncher JR, Corydon KE, et al. Clindamycin treatment of chronic pharyngeal carriage of group A streptococci. J. Pediatr. 1991; 119:123-128

Wannamaker LW. Changes and changing concepts in the biology of group A streptococci and the epidemiology of streptococcal infections. Rev Infect Dis 1979; 1:967-973.

Wessels MR, Kasper DL. The changing spectrum of group B streptococcal disease. N Engl J Med 1993; 328:1843-1844.

Zabriskie JB. Rheumatic fever: the interplay between host, genetics and microbe. Circulation 1985; 71:1077-1086.

Chapter **14**

Neisseria, Moraxella, Kingella and Eikenella

General Concepts

Neisseria gonorrhoeae

Clinical Manifestations

Symptomatic or asymptomatic localized infections include urethritis, cervicitis, proctitis, pharyngitis, and conjunctivitis. Disseminated infections occur either by extension to adjacent organs (pelvic inflammatory disease, epididymitis) or by bacteremic spread (skin lesions, tenosynovitis, septic arthritis, endocarditis, and meningitis).

Structure

Cells are Gram-negative cocci, usually seen in pairs with the adjacent sides flattened.

Classification and Antigenic Types

N gonorrhoeae strains have been typed on the basis of their growth requirements (auxotyping) or by antigenic differences in the porin protein (serotyping). More recently, restriction fragment length polymorphisms in genes encoding ribosomal RNA (ribotyping) and the separation of large DNA fragments by pulsed- field gel electrophoresis have been used to type isolates.

Pathogenesis

Gonorrhea is usually acquired by sexual contact. Gonococci adhere to columnar epithelial cells, penetrate them, and multiply on the basement membrane. Adherence is facilitated through pili and opa proteins. Gonococcal lipopolysaccharide stimulates the production of tumor necrosis factor, which causes cell damage. Gonococci may disseminate via the bloodstream. Strains that cause disseminated infections are usually resistant to serum and complement.

Host Defenses

Infection stimulates inflammation and local immunity; however, it is not known whether the secretory immune response is protective. Serum antibodies also appear. Individuals with genetic defects in late-acting complement components are at increased risk for disseminated infections. Protection, if it exists, may be strain specific.

Epidemiology

Gonorrhea is a sexually transmitted disease of worldwide importance. The highest attack rate in both men and women occurs between 15 and 29 years of age. Host-related factors such as the number of sexual partners, contraceptive practices, sexual preference, and population mobility contribute to the incidence of gonorrhea.

Diagnosis

Gonorrhea cannot be diagnosed solely on clinical grounds. For men, a Gram-stained smear of urethral exudate showing intracellular Gram-negative diplococci is diagnostic. For women, and for men when a direct smear is not definitive, culturing on selective medium is often required. N gonorrhoeae must be differentiated from other Neisseria species. Where appropriate, isolates should be examined for antibiotic resistance. A non-amplified DNA probe test is commercially available. This test does not require viable organisms and is useful where maintenance of viability during specimen transport is a problem; however, it is not as sensitive as culture. Serologic tests are not recommended for uncomplicated infections.

Control

Recommended treatment for uncomplicated infections is a third-generation cephalosporin or a fluoroquinolone plus an antibiotic (e.g., doxycycline) effective against possible coinfection with Chlamydia trachomatis. Sex partner(s) should be referred and treated. No effective vaccine yet exists. Condoms are effective in preventing gonorrhea.

Neisseria meningitidis

Clinical Manifestations

Infection with N meningitidis has two presentations, meningococcemia, characterized by skin lesions, and acute bacterial meningitis. Fulminant disease (with or without meningitis) is characterized by multisystem involvement and high mortality.

Structure

Cell morphology is identical to that of N gonorrhoeae. The antiphagocytic polysaccharide capsule is a prominent feature.

Classification and Antigenic Types

N meningitidis is grouped, on the basis of capsular polysaccharides, into 12 serogroups, some of which are subdivided according to the presence of outer membrane protein and lipopolysaccharide antigens.

Pathogenesis

Infection is by aspiration of infective particles, which attach to epithelial cells of the nasopharyngeal and oropharyngeal mucosa, cross the mucosal barrier, and enter the bloodstream. If not cleared blood-borne bacteria may enter the central nervous system and cause meningitis.

Host Defenses

Meningococci establish systemic infections only in individuals who lack serum bacterial antibodies directed against the capsular or noncapsular antigens of the invading strain, or in patients deficient in the late-acting complement components.

Epidemiology

Asymptomatic carriage of meningococci in the nasopharynx provides a reservoir for infection but also enhances host immunity. Attack rates peak in infants 3 months to 1 year old. Meningococcal meningitis occurs both sporadically (mainly groups B and C meningococci) and in epidemics (mainly group A meningococci), with the highest incidence during late winter and early spring.

Diagnosis

Symptoms are suggestive; diagnosis is confirmed by identifyingN meningitidis in specimens of blood, cerebrospinal fluid, and nasopharyngeal secretions collected before antibiotic administration.

Control

Penicillin is the drug of choice. Household contacts require chemoprophylaxis with rifampin. Groups A, C, AC, and ACYW135 capsular polysaccharide vaccines are available. In children under 1 year old, antibody levels decline rapidly after immunization. Routine vaccination is not recommended.

Other Genera and Species

Moraxella is an oxidase-positive bacterium, sometimes mistaken for Neisseria, that may be isolated from eye infections and respiratory tract infections. M catarrhalis causes lower respiratory infection in adults with chronic lung disease and is a common cause of otitis media, sinusitis, and conjunctivitis in children. Kingella and Eikenella species are short bacilli or coccoid bacteria that act as opportunistic pathogens. They are sometimes secondary invaders of damaged tissues.

INTRODUCTION

The family Neisseriaceae comprises the genera Neisseria, Moraxella, Kingella, and Acinetobacter. The only significant human pathogens are N gonorrhoeae, the agent of gonorrhea, and N. meningitidis, an agent of acute bacterial meningitidis. N gonorrhoeae infections have a high prevalence and low mortality, whereas N meningitidis infections have a low prevalence and high mortality.

Gonococcal infections are acquired by sexual contact and usually affect mucous membranes of the urethra in men and the endocervix in women, although the infection may disseminate to a variety of tissues. The pathogenic mechanism involves the attachment of the gonococci to nonciliated epithelial cells via pili (fimbriae) and the production of cytotoxic factors (endotoxin). Similarly, the lipopolysaccharide of meningococci is highly toxic, but an additional virulence factor is the antiphagocytic capsule. Both pathogens produce proteases that cleave and inactivate human immunoglobulin A1 (IgA1), a major mucosal immunoglobulin of humans. Many normal individuals harbor meningococci, whereas gonococci are present only if sexual contact with an infected person has occurred. Epidemics of meningococcal meningitis occur sporadically. Gonococcal infections occur frequently and affect large numbers of sexually active people. Other species in this genus are primarily parasites on mucosal surfaces of humans and other animals. Human disease caused by these organisms usually is associated with opportunistic infections in compromised patients.

Nesseria gonorrhoeae

Clinical Manifestations

Gonorrheal infection is generally limited to superficial mucosal surfaces lined with columnar epithelium. The areas most frequently involved are the cervix, urethra, rectum, pharynx, and conjunctiva (Fig. 14-1). Squamous epithelium, which lines the adult vagina, is not susceptible to infection by the gonococcus. However, the prepubertal vaginal epithelium, which has not been keratinized under the influence of estrogen, may be infected. Hence, gonorrhea in young girls may present as vulvovaginitis. Mucosal infections are usually characterized by a marked local neutrophilic response (purulent discharge).

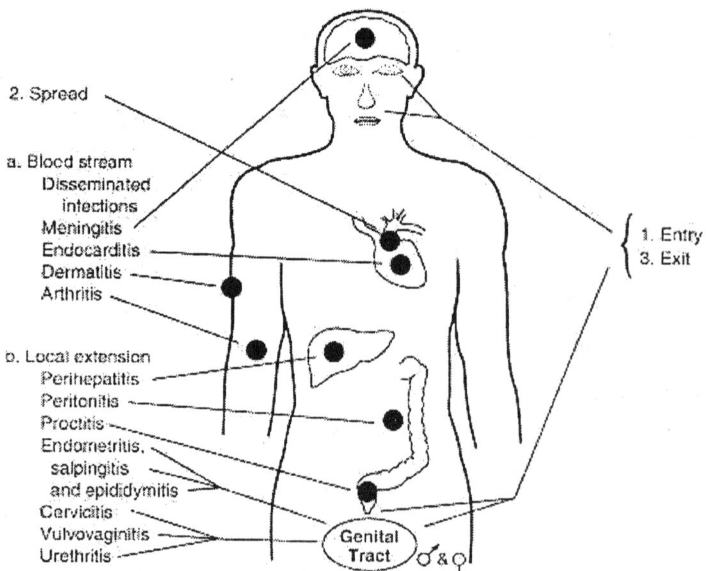

FIGURE 14-1 Clinical manifestations of N gonorrhoeae infection.

The most common symptom of uncomplicated gonorrhea is a discharge that may range from a scanty, clear, or cloudy fluid to one that is copious and purulent. Dysuria is often present. Men with asymptomatic urethritis are an important reservoir for transmission. In addition, such men and those who ignore their symptoms are at increased risk for developing complications.

Endocervical infection is the most common form of uncomplicated gonorrhea in women. Such infections are usually characterized by vaginal discharge and sometimes by dysuria (because of coexistent urethritis). The cervical as may be erythematous and friable, with a purulent exudate. About 50 percent of women with cervical infections are asymptomatic. Local complications include abscesses in Bartholin's and Skene's glands.

Rectal infections with N gonorrhoeae occur in about one-third of women with cervical infection. They most often result from autoinoculation with cervical discharge and are rarely symptomatic. Rectal infections in homosexual men usually result from anal intercourse and are more often symptomatic. The symptoms and signs of gonococcal proctitis range from mild burning on defecation to itching to severe tenesmus and from mucopurulent discharge to frank blood in the stools.

Pharyngeal infections are diagnosed most often in women and homosexual men with a history of fellatio. Such infections may be a focal source of gonococcemia.

Ocular infections can have serious consequences (corneal scarring or perforation); prompt diagnosis and treatment are therefore important. Ocular infections (ophthalmia neonatorum) occur most commonly in newborns who are exposed to infected secretions in the birth canal. Keratoconjunctivitis is occasionally seen in adults as a result of autoinoculation.

Disseminated gonococcal infections result from gonococcal bacteremia. Asymptomatic infections of the pharynx, urethra, or cervix often serve as focal sources for bacteremia. The most common form of disseminated gonococcal infection is the dermatitis-arthritis syndrome. It is characterized by fever, chills, skin lesions, and arthralgias (usually involving the hands, feet, and elbows), which are due to periarticular inflammation of the tendon sheaths. Occasionally, a patient develops a septic joint with effusion. Skin lesions may be mascular, pustular, centrally necrotic, or hemorrhagic. Rarely, disseminated gonococcal infection causes endocarditis or meningitis. Gonococci may ascend from the endocervical canal through the endometrium to the fallopian tubes and ultimately to the pelvic peritoneum, resulting in endometritis, salpingitis, and finally, peritonitis. Women usually present with pelvic and abdominal pain, fever, chills, and cervical motion tenderness. This complex of signs and symptoms is referred to as pelvic inflammatory disease (PID). This disease may also be caused by other sexually transmitted organisms (e.g., Chlamydia trachomatis) as well as by non-sexually transmitted bacteria that are part of the normal vaginal flora. Complications of pelvic inflammatory disease include tubo-ovarian abscesses, pelvic peritonitis, or

Fitz-Hugh and Curtis syndrome, which is an inflammation of Glisson's capsule of the liver. As many as 15 percent of women with uncomplicated cervical infections may develop pelvic inflammatory disease. The disease may have serious consequences, including an increased probability of infertility and ectopic pregnancy.

Structure

Neisseria species are Gram-negative cocci, 0.6 to 1.0 µm in diameter. The organisms are usually seen in pairs with the adjacent sides flattened. Pili, hairlike filamentous appendages extend several micrometers from the cell surface and have a role in adherence. The outer membrane is composed of proteins, phospholipids, and lipopolysaccharide (LPS). Features that distinguish gonococcal LPS from enteric LPS are the highly branched basal oligosaccharide structure and the absence of repeating O-antigen subunits. For these reasons gonococcal LPS, as well as that of other mucosal pathogens, is referred to as lipooligosaccharide (LOS). Gonococci characteristically release outer membrane fragments (blebs) during growth. These blebs contain LOS and may have a role in pathogenesis.

Classification and Antigenic Types

The gonococcus is an obligate human pathogen. It is one of two Neisseria species that cause significant human infections. The genus also includes several nonpathogenic species (Table 14-1), which may be part of the normal flora and therefore can be confused with *N. gonorrhoeae*. Gonococcal strains can be characterized according to their nutritional requirements (auxotyping). A panel of monoclonal antibodies specific for epitopes on protein I have also been used to type strains. Strains exhibiting specific reaction patterns are termed serovars. A combined auxotype-serovar classification provides greater resolution among gonococcal isolates and is useful in epidemiologic investigations.

Pathogenesis

Our knowledge of the molecular basis of gonococcal pathogenesis is incomplete (Fig. 14-2). Attachment of gonococci to mucosal cells is mediated in part by pili, although, nonspecific factors such as surface charge and hydrophobicity may be important. Pili undergo both phase and antigenic variation. Opa proteins (protein II), which are located in the outer membrane, are also involved in attachment to host cells. Gonococci attach only to microvilli of nonciliated columnar epithelial cells; attachment to ciliated cells is not observed.

Much of our knowledge of gonococcal invasion comes from studies with tissue culture cells and human fallopian tube organ culture. After gonococci attach to the nonciliated epithelial cells of the fallopian tube, they are surrounded by the microvilli, which draw them to the surface of the mucosal cell. The gonococci appear to enter the epithelial cells by a process called parasite-directed endocytosis. This process seems to be initiated by microbial factors because it does not occur unless

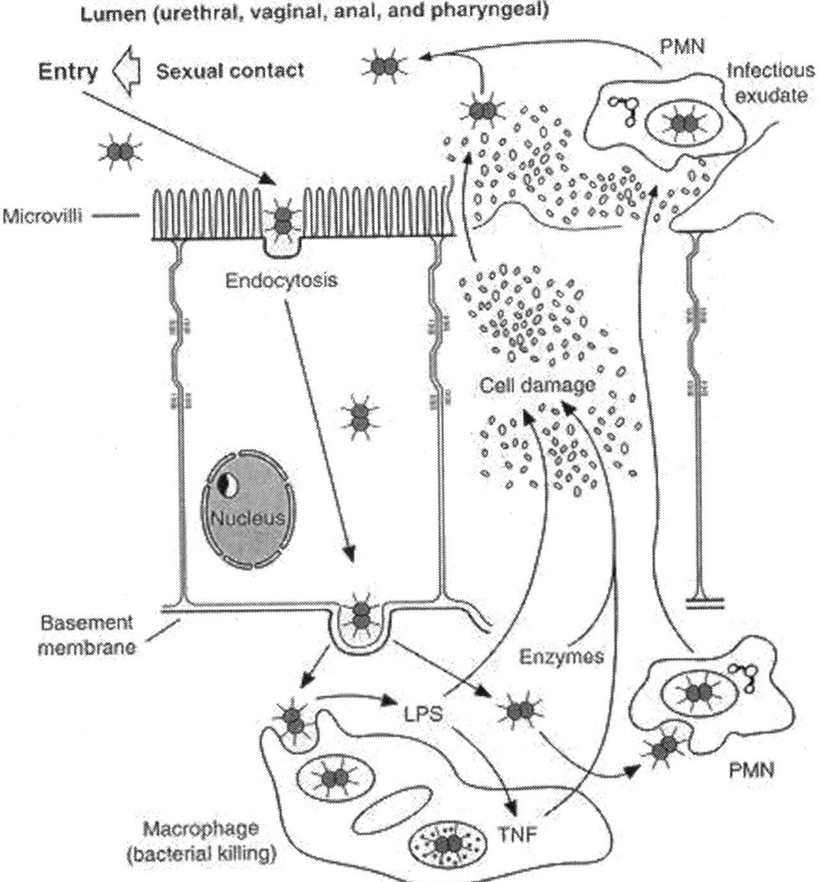

FIGURE 14-2 Pathogenesis of uncomplicated gonorrhea. Gonococci can invade columnar epithelial cells, although they do not invade ciliated columnar epithelium of the genitourinary tract.

the gonococci are viable and because it involves host cells that are not normally phagocytic. An unidentified factor in serum enhances engulfment of gonococci. The process is inhibited by drugs that block the actions of the microtubule (demecolcine) and microfilament (cytochalasin B) systems. During endocytosis the membrane of the mucosal cell retracts, pinching off a membrane-bound vacuole that contains gonococci; this vacuole is rapidly transported to the base of the cell, where gonococci are released by exocytosis into the subepithelial tissue. Gonococci are not destroyed within the phagocytic vacuole; it is not clear whether they replicate in the vacuoles.

The major porin protein of the gonococcal outer membrane, Por (protein I), has been proposed as a candidate invasin (a substance that helps mediate invasion into a host cell). The insertion of Por into neutrophils treated with the chemotactic peptide, fMLP and leukotriene B4, inhibits degranulation but not the generation of the superoxide anion. The significance of these observations with respect to the pathogenesis of gonorrhea remains to be determined. Each gonococcal strain expresses only one type of Por; however, the Por of different strains may exhibit antigenic differences.

Gonococci can produce one or several outer membrane proteins called Opa proteins (proteins II). These proteins are subject to phase variation and are usually found on cells from colonies possessing an opaque phenotype (O+). At any one time, a gonococcus may express zero, one, or several different Opa proteins, though each strain has 10 or more genes for different Opas. Trypsin-like proteases present in cervical mucus may help select for protease-resistant transparent (O-) colony phenotypes. O+ colony phenotypes (protease sensitive) predominate in cultures taken during the middle portion of the menstrual cycle. Cervical proteases increase during the second half of the cycle, resulting in an increase in the O- phenotype. The O- colony types can be isolated from tubal as well as endocervical cultures; O+ colony phenotypes have been isolated more often from endocervical cultures than from tubal cultures.

Rmp (Protein III) is an outer membrane protein found in all strains of N gonorrhoeae. It does not undergo phase variation and is found in a complex with Por and LOS. It shares partial homology with the Omp A protein of Escherichia coli. Antibodies to Rmp, induced either by a neisserial infection or by colonization with E coli, block bactericidal antibodies directed against Por and LOS. Rmp antibodies may facilitate infection with N gonorrhoeae.

LOS has a profound effect on the virulence and pathogenesis of N gonorrhoeae. Gonococci can express several antigenic types of LOS and can alter the type of LOS they express by an as yet unknown mechanism. Gonococcal LOS produces mucosal damage in fallopian tube organ cultures and brings about the release of enzymes, such as proteases and phospholipases, that may be important in pathogenesis. More recent evidence suggests that gonococcal LOS stimulates the production of tumor necrosis factor (TNF) in fallopian tube organ cultures; inhibition of tumor necrosis factor with specific antiserum prevents tissue damage. Thus, gonococcal LOS appears to have an indirect role in mediating tissue damage. Gonococcal LOS is also involved in the resistance of N gonorrhoeae to the bactericidal activity of normal human serum. Oligosaccharides containing epitopes defined by specific monoclonal antibodies are associated with a serum-resistant phenotype.

Gonococci can utilize host-derived cytidine monophospho-N-acetylneuraminic acid (CMP-NANA) in vivo to sialylate the oligosaccharide component of its LOS, converting a serum-sensitive organism to a serum-resistant one. When such organisms are grown in vitro without CMP-NANA, their resistance to killing by normal human serum is rapidly lost. Organisms with non-sialylated LOS are more invasive than those with sialylated LOS. There is antigenic similarity between neisserial LOS and antigens present on human erthyrocytes. This similarity to self may preclude an effective immune response to these LPS antigens.

Gonococci are highly autolytic and release peptidoglycan fragments during growth. These fragments, released by bacterial and/or host peptidoglycan hydrolases, are toxic for fallopian tube mucosa and may contribute to the intense inflammatory reactions characteristic of gonococcal disease.

N gonorrhoeae is highly efficient at utilizing transferrin-bound iron for in vitro growth; many strains can also utilize lactoferrin-bound iron. Gonococci (and meningococci) bind only human transferrin and lactoferrin. This specificity is thought to be the reason these organisms are exclusively human pathogens. Nevertheless, the role of transferrin- and lactoferrin-bound iron in in vivo growth is unknown. Gonococci express several new proteins when grown under iron-restricted conditions similar to the conditions occurring in the host. Some of these proteins function as receptors for transferrin, lactoferrin, heme, and hemoglobin; others function in the transport of iron into the cell. Gonococci cannot grow anaerobically unless low concentrations of the alternative electron acceptor nitrite are present. Under these conditions they produce novel proteins. These proteins are apparently produced during an infection because antibodies against them are present in the serum specimens of patients with uncomplicated gonorrhea, disseminated gonococcal infection, or pelvic inflammatory disease. These data suggest that some gonococci in the host are growing under anaerobic conditions. Further studies will determine the relevance of these proteins to pathogenesis.

Strains of N gonorrhoeae (and N meningitidis) produce two distinct extracellular IgA1 proteases, which cleave the heavy chain of human immunoglobulinA1 (IgA1) at different points within the hinge region. Type 1 protease cleaves a prolyl-seryl peptide bond and type 2 protease cleaves a prolyl-threonylbond in the hinge region of the heavy chain. This region is missing in human IgA2, and so this isotype is not susceptible to cleavage. Each gonococcal or meningococcal isolate elaborates only one of these two enzymes. Split products of IgA1 have been found in the genital secretions of women with gonorrhea, suggesting that the gonococcal IgA1 protease is present and active during genital infection. Fab fragments of IgA1 may bind to the gonococcal cell surface and block the Fc-mediated functions of intact immunoglobulins.

Host Defenses

Not everyone exposed to N gonorrhoeae acquires the disease. This may be due to variations in the size or virulence of the inoculum, to nonspecific resistance, or to specific immunity. A 50 percent infective dose (ID50) of about 1,000 organisms has been established, based on the experimental urethral inoculation of male volunteers. There is no reliable ID50 for women, although it is assumed to be similar.

Nonspecific factors have been implicated in natural resistance to gonococcal infection. In women, changes in the genital pH and hormones may increase resistance to infection at certain times of the menstrual cycle. Urinary solutes exhibit bactericidal and bacteriostatic activity of N gonorrhoeae. Factors in urine that seem to be important are pH, osmolarity, and the concentration of urea. The variability in the susceptibility of gonococcal strains to the bactericidal and bacteriostatic properties of urine is thought to be one of the reasons some men do not develop a gonococcal infection when exposed.

Most uninfected individuals have serum antibodies that react with gonococcal antigens. These antibodies probably result from colonization or infection with various Gram-negative bacteria that possess cross-reactive antigens. These "natural" antibodies differ, both qualitatively and quantitatively, from person to person, but may be important in an individual's natural resistance or susceptibility to infection.

Infection with N gonorrhoeae stimulates both mucosal and systemic antibodies to a variety of gonococcal antigens. Mucosal antibodies are primarily IgA and IgG. In genital secretions, antibodies have been identified that react with Por, Opa and LOS, and some of the iron-regulated proteins. Vaccine trials have suggested that antipilus antibodies inhibit the pilus-mediated attachment of the homologous gonococcal strain. Complement is present in endocervical secretions, but in a much lower concentration than in blood. However, there is little evidence to support a role for a complement-mediated bactericidal defense mechanism on the genital mucosa. In general, the IgA response is brief and declines rapidly after treatment; IgG levels decline more slowly.

More information is available about the function of systemic humoral immune mechanisms in gonococcal infection. Gonococcal antigens such as pili, Por, Opa, Rmp, and LOS elicit a serum antibody response during an infection. Antipilus antibody levels tend to be higher in women than in men and are related to the number of previous infections. The predominant IgG subclass that reacts with a variety of gonococcal antigens is IgG3, followed by IgG1 and IgG4. IgG2 is minimal, suggesting that polysaccharides are not important in the immune response to gonococcal infection. Anti-Por antibodies may be bactericidal for the gonococcus. IgG that reacts with Rmp blocks the bactericidal activity of antibodies directed against Por and LOS. Genital infection with N gonorrhoeae stimulates a serum antibody response against the LOS of the infecting strain. Disseminated gonococcal infection results in higher levels of anti-LOS antibody than do genital infections.

Strains that cause uncomplicated genital infections usually are killed by normal human serum and are termed serum sensitive. This bactericidal activity is mediated by IgM and IgG that recognize sites on the LOS. Strains that cause disseminated infections are not killed by most normal human serum and are referred to as serum resistant. Resistance is mediated, in part, by IgA that blocks the IgG-mediated bactericidal activity of the serum. Serum specimens from convalescent patients with disseminating infections contain bactericidal IgG to the LOS of the infecting strain.

Individuals with inherited complement deficiencies have a markedly increased risk of acquiring systemic neisserial infections and are subject to recurring episodes of systemic gonococcal and meningococcal infections, indicating that the complement system is important in host defense. Gonococci activate complement by both the classic and alternative pathways. Complement activation by gonococci leads to the formation of the C5b-9 complex (membrane attack complex) on the outer membrane. In normal human serum, similar numbers of C5b-9 complexes are deposited on serum-sensitive and serum-resistant organisms, but the membrane attack complex is not functional on serum-resistant organisms. Other complement-mediated functions, such as opsonophagocytosis and chemotaxis, are more efficient with serum-sensitive than with serum-resistant gonococci. This may be a significant factor in the pathogenesis of disseminated gonococcal infection and probably contributes to the relative lack of genital symptoms observed with this disease.

Normal human serum contains opsonic anti-Por IgG. Antibodies to various surface-exposed antigens are also present in cervical and urethral secretions of patients with gonorrhea and probably contribute to the opsonophagocytosis of the organism. Opa is important in gonococcus-neutrophil interactions. Gonococci expressing certain Opas interact with neutrophils in the absence of antibodies. Once phagocytosed, gonococci are killed by both oxygen-dependent and oxygen-independent mechanisms. The survival of gonococci within neutrophils has been the subject of considerable controversy, with no clear-cut answer yet available. The opsonization and phagocytosis of gonococci are comparatively more important in mucosal infections than in protection from systemic gonococcal (and meningococcal) infections.

Epidemiology

The only natural host for N gonorrhoeae is the human. Gonorrhea has all but disappeared in Scandinavia and several other European countries. In the United States, gonorrhea remains the most frequently reported infectious disease. Between 1977 and 1993, the number of reported cases decreased 56 percent, from 1 million to 439,673 cases per year. The Centers for Disease Control (CDC) estimates that there are two unreported cases for every reported case of gonorrhea. Gonorrhea is transmitted almost exclusively by sexual contact. The highest rates occur in women between the ages of 15 and 19 years and in men 20 and 24 years of age. Persons who have multiple sex partners are at highest risk. Rates of gonorrhea are higher in males and in minority and inner-city populations.

Gonorrhea is usually contracted from a sex partner who is either asymptomatic or has only minimal symptoms. It is estimated that the efficiency of transmission after one exposure is about 35 percent from an infected woman to an uninfected man and 50 to 60 percent from an infected man to an uninfected woman. More than 90 percent of men with urethral gonorrhea will develop symptoms within 5 days; fewer than 50 percent of women with anogenital gonorrhea will do so. Women with asymptomatic infections are at higher risk of developing pelvic inflammatory disease and disseminated gonococcal infection.

Diagnosis

Gonococcal infection produces several common clinical syndromes that have multiple causes or that mimic other conditions. Laboratory testsare often required to differentiate among the etiologic agents causing urethritis or cervicitis. The etiologic diagnosis of salpingitis and pelvic peritonitis is quite difficult because mixed infections are common and laparoscopy is required to obtain appropriate cultures. Gonococcal perihepatitis may mimic acute cholecystitis. All of the above syndromes are also caused by C trachomatis, a sexually transmitted bacterium that causes more infections in the United States than N gonorrhoeae. The gonococcal arthritis-dermatitis syndrome, must be, differentiated from meningococcemia and Reiter syndrome, in particular, and from other causes of septic arthritis.

Customarily, the laboratory diagnosis of gonorrhea is made presumptively and then confirmed; the latter process involves identifying characteristics that distinguish N gonorrhoeae from other Neisseria spp. that may be present in the specimen. Nonpathogenic Neisseriaare normal inhabitants of the oropharynx and nasopharynx and occasionally are isolated from other sites infected by N gonorrhoeae. A presumptive diagnosis of gonorrhea may be made from Gram-stained smears of urethral, cervical, and rectal specimens if Gram-negative diplococci are observed within leukocytes; it is equivocal if only extracellular Gram-negative diplococci are seen and negativeif no Gram-negative diplococci are seen. Gram stain diagnosis has a sensitivity and specificity of >95 percent in men with symptomatic urethritis. The specificity of Gram stain diagnosis in women is also high if the cervix is wiped clean to remove cervical secretions before collecting the specimen; however, the sensitivity is only about 50 percent. The sensitivity and specificity of the Gram stain for rectal specimens are lower than with cervical specimens.

Specimens for the laboratory diagnosis of gonorrhea should be collected before treating the patient. Ideally, specimens should be inoculated onto appropriate media and incubated immediately after collection at 35 to 36.5°C in a CO_2-enriched atmosphere, which can be obtained by using a candle extinction jar or a CO_2 incubator. Urethral specimens are normally obtained from heterosexual men; urethral, rectal, and pharyngeal specimens are normally obtained from homosexual men; and cervical and rectal specimens are normally obtained from women. Specimens are collected with cotton, polyester, or calcium alginate swabs. When appropriate, specimens may also be obtained from the urethra and from Bartholin's and Skene's glands of infected women. Blood cultures should be performed for patients with suspected disseminated infection. Synovial fluid cultures should be performed for patients with septic arthritis.

Urethral, cervical, and pharyngeal specimens are inoculated onto selective medium such as modified Thayer-Martin, Martin-Lewis, or NYC medium. These are complex media that contain antimicrobial and antifungal agents to inhibit the growth of unwanted organisms. Rectal specimens should be inoculated onto modified Thayer-Martin medium which contains trimethoprim lactate to inhibit the growth and swarming of Proteus species. Specimens collected from normally sterile sites such as blood, synovial fluid, and conjunctivae may be inoculated onto a nonselective medium such as chocolate agar.

The combination of oxidase-positive colonies and Gram-negative diplococci provides a presumptive identification of N gonorrhoeae. Fluorescent-antibody staining, coagglutination, specific biochemical tests (Table 14-1), and DNA probes may be used for confirmation. DNA probes have also been used to detect gonococci in urethral and cervical specimens. A commercial test based on this approach is available. Serologic tests for uncomplicated gonorrhea have not proved satisfactory.

TABLE 14-1 Differential characteristics of Neisseria and related species of human origin[a]

Species	Grwoth on:			Acid From:					Reduction of NO_3	Polysaccharide from SUC	DNase
	MTM, ML, and NYC media	Chocolate or blood agar (22°C)	Nutrient agar (35°C)	GLU	MAL	LAC	SUC	FRU			
	+	0	0	+	0	0	0	0	0	0	0
	+	0	V	+	+	0	0	0	0	0	0
	+	V	+	+	+	+	0	0	0	0	0
	V	0	+	0	0	0	0	0	0	0	0
	V	0	+	+	+	0	0	0	0	+	0
	V	+	+	+	+	0	V	V	0	V	0
	0	+	+	+	+	0	+	+	0	+	0
	0	+	+	+	+	0	+	+	+	+	0
	0	+	+	0	0	0	0	0	0	+	0
	V	+	+	0	0	0	0	0	+	0	+
	+	NT	+	+	0	0	0	0	+	0	0

Control

There is no effective vaccine to prevent gonorrhea. Candidate vaccines consisting of pilus protein or Por are of little benefit. The developmentof an effective vaccine has been hampered by the lack of a suitable animal model and the fact that an effective immune response has never been demonstrated. Condoms are effective in preventing the transmission of gonorrhea.

Contact tracing to identify source contacts (i.e., those who infected the index patient) has been useful in identifying asymptomatic individuals or those with ignored symptoms. Contact tracing has also been used to identify contacts who were exposed to the index patient and who may have become infected.

The evolution of antimicrobial resistance in N gonorrhoeae may ultimately affect the control of gonorrhea. Strains with multiple chromosomal resistance to penicillin, tetracycline, erythromycin, and cefoxitin have been identified in the United States and most other parts of the world. Sporadic high-level resistanceto spectinomycin and fluoroquinolones have been reported.

Penicillinase-producing strains of N gonorrhoeae were first described in 1976. Five related ß-lactamase plasmids of different sizes have been identified in these strains. The strains cause more than one-half of all gonococcal infections in parts of Africa and Asia. Their prevalence has increased dramatically in the United States since 1984 and has affected nearly every major metropolitan area.

Plasmid-mediated high-level resistance of N gonorrhoeae to tetracycline was first described in 1986 and has now been reported in most parts of the world. This resistance is due to the presence of the streptococcal tetM determinant on a gonococcal conjugative plasmid.

The current CDC Treatment Guidelines recommend treatment of all gonococcal infections with antibiotic regimens effective against resistant strains. The recommended antimicrobial agents are ceftriaxone, cefixime, ciprofloxacin, or oflaxacin. Since a significant proportion of patients with gonorrhea are also infected with C trachomatis, doxycycline or erythromycin has been added to treat this concomitant infection.

Neisseria meningitidis

Clinical Presentation

N meningitidis infection results from the bloodborne dissemination (meningococcemia) of the meningococcus, usually following an asymptomatic or mildly symptomatic nasopharyngeal carrier state or a mild rhinopharyngitis (Fig. 14-3). The mildest form is a transient bacteremic illness characterized by a fever and malaise; symptoms resolve spontaneously in 1 to 2 days. Acute meningococcemia is more serious and is often complicated by meningitis. The manifestations of meningococcal meningitis are similar to acute bacterial meningitis caused by organisms such as Streptococcus pneumoniae, Haemophilus influenzae, and E coli. The manifestations result from both infection and increased intracranial pressure. Chills, fever, malaise, and headache are the usual manifestations of infection; headache, vomiting, and rarely, papilledema may result from increased intracranial pressure. Signs of meningeal inflammation are also present. The onset of meningococcal meningitis may be abrupt or insidious.

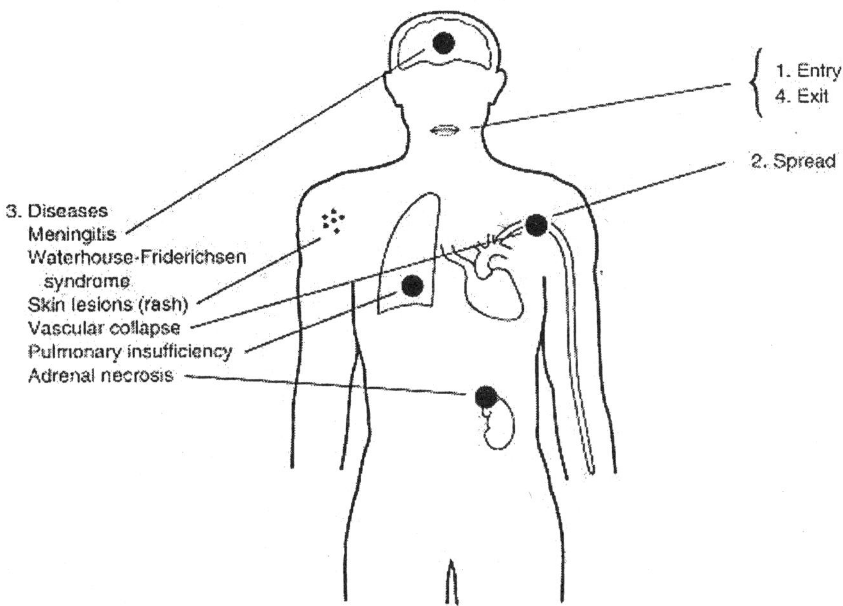

FIGURE 14-3 Clinical manifestations of N meningitidis infection.

Infants with meningococcal meningitis rarely display signs of meningeal irritation. Irritability and refusal to take food are typical; vomiting occurs early in the disease and may lead to dehydration. Fever is typically absent in children younger than 2 months of age. Hypothermia is more common in neonates. As the disease progresses, apneic episodes, seizures, disturbances in motor tone, and coma may develop.

In older children and adults, specific symptoms and signs are usually present, with fever and altered mental status the most consistent findings. Headacheis an early, prominent complaint and is usually very severe. Nausea, vomiting, and photophobia are also common symptoms.

Neurologic signs are common; approximately one-third of patients have convulsions or coma when first seen by a physician. Signs of meningeal irritation such as cervical rigidity (Brudzinski sign), thoracolumbar rigidity, hamstring spasm (Kernig sign), and exaggerated reflexes are common.

Petechiae (minute hemorrhagic spots in the skin) or purpura (hemorrhages into the skin) occurs from the first to the third day of illness in 30 to 60 percent of patients with meningococcal disease, with or without meningitis. The lesions may be more prominent in areas of the skin subjected to pressure, such as the axillary folds, the belt line, or the back.

Fulminant meningococcemia (Waterhouse-Friderichsen syndrome) occurs in 5 to 15 percent of patients with meningococcal disease and has a high mortality rate. It begins abruptly with sudden high fever, chills, myalgias, weakness, nausea, vomiting, and headache. Apprehension, restlessness, and frequently, delirium occur within the next few hours. Widespread purpuric and ecchymotic skin lesions appear suddenly. Typically, no signs of meningitis are present. Pulmonary insufficiency develops within a few hours, and many patients die within 24 hours of being hospitalized despite appropriate antibiotic therapy and intensive care.

Structure

The only distinguishing structural feature between N meningitidis and N gonorrhoeae is the presence of a polysaccharide capsule in the former. The capsule is antiphagocytic and is an important virulence factor.

Classification and Antigenic Types

Meningococcal capsular polysaccharides provide the basis for grouping these organisms. Twelve serogroups have been identified (A, B, C, H, I, K, L, X, Y, Z, 29E, and W135). The most important serogroups associated with disease in humans are A, B, C, Y, and W135. The chemical composition of these capsular polysaccharides, where known, is listed in Table 14-2. The prominent outer membrane proteins of N meningitidis have been designated class 1 through class 5. The class 2 and 3 proteins function as porins and are analogous to gonococcal Por. The class 4 and 5 proteins are analogous to gonococcal Rmp and Opa, respectively. Serogroup B and C meningococci have been further subdivided on the basis of serotype determinants located on the class 2 and 3 proteins. A few serotypes are associated with most cases of meningococcal disease, whereas other serotypes within the same serogroup rarely caused disease. All known group A strains have the same protein serotype antigens in the outer membrane. Another serotyping system is based on the antigenic diversity of meningococcal LOS. The LOS types are independent of the protein serotypes, although certain combinations frequently occur together.

TBLE 14-2 Chemical composition of N. meningitdis Capsular Polysaccharides

Serogroup	Structural Repeating Unit[a]
Serogroup A[b] (homopolymer)	
Serogroup B (homopolymer)	
Serogroup C[c] (homopolymer)	
Serogroup H (monosaccharide-glycerol repeating unit)	
Serogroup I (disaccharide repeating unit)	
Serogroup K (disaccharide repeating unit)	
Serogroup L (disaccharide repeating unit)	
Serogroup W135 (disaccharide repeating unit)	
Serogroup X (homopolymer)	
Serogroup Y (BO) (disaccharide repeating unit)	
Serogroup Z (monosaccharide glycerol repeating unit)	
Serogroup 29E (disaccharide repeating)	

Pathogenesis

The human nasopharynx is the only known reservoir of N meningitidis. Meningococci are spread via respiratory droplets, and transmission requires aspiration of infective particles. Meningococci attach to the nonciliated columnar epithelial cells of the nasopharynx. Attachment is mediated by pili and possibly by other outer membrane components. Invasion of the mucosal cells occurs by a mechanism similar to that observed with gonococci. However, once internalized, meningococci remain in an apical location within the epithelial cell; the route by which they gain access to the subepithelial space remains unclear. Trimers of class 2 and 3 proteins have the ability to translocate from intact cells and insert into eukaryotic cell membranes to form voltage-dependent channels. This process may be important in invasion.

Purified meningococcal LOS is highly toxic and is as lethal for mice as the LOS from E coli or Salmonella typhimurium; however, meningococcal LOS is 5 to 10 times more effective than enteric LPS in eliciting a dermal Shwartzman reaction in rabbits. Meningococcal LPS suppresses leukotriene B4 synthesis in human polymorphonuclear leukocytes. The loss of leukotriene B4 deprives the leukocytes of a strong chemokinetic and chemotactic factor.

The events after bloodstream invasion are unclear. Relatively little information is known about how the meningococcus enters the central nervous system.

Host Defenses

The integrity of the pharyngeal and respiratory epithelium may be important in protection from invasive disease. Chronic irritation of the mucosa due to dust or low humidity, or damage to the mucosa resulting from a concurrent viral or mycoplasmal upper respiratory infection, may be predisposing factors for invasive disease.

The presence of serum bactericidal IgG and IgM is probably the most important host factor in preventing invasive disease. These antibodies are directed against both capsular and noncapsular surface antigens. The antibodies are produced in response to colonization with carrier strains of N meningitidis, N lactamica, or other nonpathogenic Neisseria species. Protective antibodies are also stimulated by cross-reacting antigens on other bacterial species. The role of bactericidal antibodies in prevention of invasive disease explains why high attack rates are seen in infants from 6 to 9 months old, the age at which maternally acquired antibodies are being lost.

The immunity conferred by specific antibody may not be absolute. Illness has been documented in individuals with levels ofantibodies considered to be protective. It has been postulated that the activity of the bactericidal antibodies might be blocked by IgA, induced by other meningococcal strains, or by cross-reacting antigens on enteric or other respiratory bacteria. Since IgA does not bind complement, it may block binding sites for the bactericidal IgG and IgM. Persons with complement deficiencies (C5, C6, C7, or C8) may develop meningococcemia despite protective antibody. This underscores the importance of the complement system in protection from meningococcal disease.

Epidemiology

The meningococcus usually inhabits the human nasopharynx without causing detectable disease. This carrier state may last for a few days to months and is important because it not only provides a reservoir for meningococcal infection but also enhances host immunity. Between 5 and 30 percent of normal individuals are carriers at any given time, yet few develop meningococcal disease. Even during epidemics of meningococcal meningitis in military recruits, when the carrier rate may reach 95 percent, the incidence of systemic disease is less than 1 percent. Meningococcal carriage rates are highest in older children and young adults, but the attack rates are higher in children, peaking at 5 years of age (group B) and 4 to 14 years of age (group C). The low incidence of disseminated disease following colonization suggests that host rather than bacterial factors play an important determining role.

Meningococcal meningitis occurs sporadically and in epidemics, with the highest incidence during late winter and early spring. Most epidemics are caused by group A strains, but small outbreaks have occurredwith group B and C strains. Sporadic cases generally are caused by group B, C, and Y strains. Whenever group A strains become prevalent in the population, the incidence of meningitis increases markedly.

Diagnosis

The most characteristic manifestation of meningococcemia is the skin rash, which is essential for its recognition. Petechiae are the most common type of skin lesion. Ill-defined pink macules and maculopapular lesions also occur. Lesions are sparsely distributed over the body. They tend to occur in crops and on any part of the body; however, the face is usually spared and involvement of the palms and soles is less common. The skin rash may progress from a few ill-defined lesions to a widespread eruption within a few hours.

Acute bacterial meningitis has characteristic signs and symptoms. Except in epidemic situations, it is difficult to identify the causative agent without laboratory tests.

In cases of suspected meningococcal disease, specimens of blood, cerebrospinal fluid, and nasopharyngeal secretions should be collected before administration of any antimicrobial agents and examined for the presence of N meningitidis. Success in isolation is reduced by prior therapy; however, the microscopic diagnosis is not significantly affected. The cerebrospinal fluid should be concentrated by centrifugation and a portion of the sediment cultured on chocolate or blood agar. The plates should be incubated in a candle jar or CO_2 incubator. The presence of oxidase-positive colonies and Gram-negative diplococci provides a presumptive identification of N meningitidis. Production of acid from glucose and maltose but not sucrose, lactose, or fructose may be used for confirmation (Table 14-1). The serologic group may be determined by a slide agglutination test, using first polyvalent and then monovalent antisera.

Nasopharyngeal specimens must be obtained from the posterior nasopharyngeal wall behind the soft palate and then should be inoculated onto a selective medium such as Thayer-Martin medium and processed as above.

Blood specimens are inoculated in 10- to 15-ml aliquots onto each of three blood bottles to give a final concentration of 10% (vol/vol). Evacuated bottles should be vented. Some strains of N meningitidis are inhibited by the sodium polyanetholsulfonate contained in blood medium. Toxicity may be overcome by the addition of gelatin. Sodium amylosulfate is not toxic for the meningococcus. Blood cultures are subcultured blindly onto chocolate or blood agar for confirmation.

Gram-stained smears of cerebrospinal fluid may be diagnostic; however, finding neisseriae in these smears is often more difficult than finding the strains that cause pneumococcal meningitis. Quellung tests may be of value.

Control

Group A, C, Y, and W135 capsular polysaccharide vaccines are available and can be used to control outbreaks due to the meningococcal serogroups covered by the vaccine. The A, C, AC, and ACYW135 polysaccharide formulations are currently licensed in the United States. The polysaccharide vaccines are ineffective in young children, and the duration of protection is limited in children vaccinated at 1 to 4 years of age. Routine vaccination of the civilian population in industrialized countries is not currently recommended because the risk of infection is low and most endemic disease occurs in young children. The group B capsular polysaccharide is a homopolymer of sialic acid and is not immunogenic in humans. A group B meningococcal vaccine consisting of outer membrane protein antigens has recently been developed but is not licensed in the United States.

Meningococcal disease arises from association with infected individuals, as evidenced by the 500- to 800-fold greater attack rate among household contacts than among the general population. Because such household members are at high risk, they require chemoprophylaxis. Sulfonamides were the chemoprophylactic agent of choice until the emergence of sulfonamide-resistant meningococci. At present, approximately 25 percent of clinical isolates of N meningitidis in the United States are resistant to sulfonamides; rifampin is therefore the chemoprophylactic agent of choice.

Penicillin is the drug of choice to treat meningococcemia and meningococcal meningitis. Although, penicillin does not penetrate the normal blood-brain barrier, it readily penetrates the blood-brain barrier when the meninges are acutely inflamed. Either chloramphenicol or a third-generation cephalosporin such as cefotaxime or ceftriaxone is used in persons allergic to penicillins.

Moraxella

Moraxella species are parasites of the mucous membranes of humans and other warm-blooded animals. Many species are nonpathogenic. M lacunata can be isolated from the eyes and may cause conjunctivitis in humans living under conditions of poor hygiene. M nonliquefaciens is found in the upper respiratory tract, especially the nose, and may be a secondary invader in respiratory infections. M urethralis can be isolated from urine and the female genital tract. Some strains formerly designated as Mima polymorpha subsp oxidans belong in this species. These organisms can be mistaken for N gonorrhoeae unless appropriate biochemical characteristics are determined.

M catarrhalis organisms are cocci that morphologically resemble Neisseria cells. Other relevant characteristics are presented in Table 14-1. M catarrhalis was formerly placed in the genus Neisseria; however, studies of DNA base content, fatty acid composition, and genetic transformation showed that this organism did not belong in that genus. M catarrhalis is a member of the normal flora in 40-50% of normal school children; however, it should be considered more than a harmless commensal of the mucous membranes of humans. It is an infrequent, yet significant, cause of severe systemic infections such as pneumonia, meningitis, and endocarditis. It is an important cause of lower respiratory tract infections in adults with chronic lung disease and a common cause of otitis media, sinusitis, and conjunctivitis in otherwise healthy children and adults. M catarrhalis may cause clinical syndromes indistinguishable from those caused by gonococci, and so it is important to distinguish these organisms from one another. Many strains produce ß-lactamase.

Kingella

Kingella kingae and K denitrificans are oxidase-positive non-motile organisms that are hemolytic when grown on blood agar. They are gram-negative rods, but may resemble coccobacilli or diplococci. They are part of the normal oral flora and occasionally cause infections of bone, joints, and tendons. The organism may enter the circulation with minor oral trauma such as tooth brushing. It is susceptible to penicillin, ampicillin, and erythromycin.

Eikenella

Eikenella corrodens is a small oxidase-positive, fastidious gram-negative rod, which requires carbon dioxide for growth. Many isolates form pits in agar during growth on solid medium. E. corrodens is part of the gingival and bowel flora in 40-70% of humans and may be found in mixed flora infections associated with contamination from these sites. It occurs frequently in infections from human bites. E corrodens is resistant to clindamycin, but susceptible to ampicillin and third generation cephalosporins.

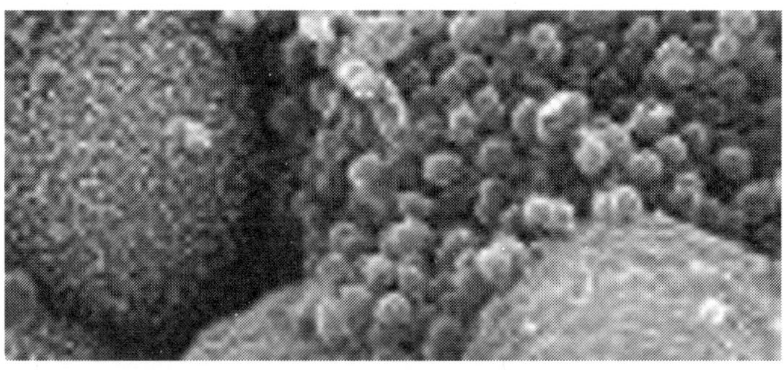

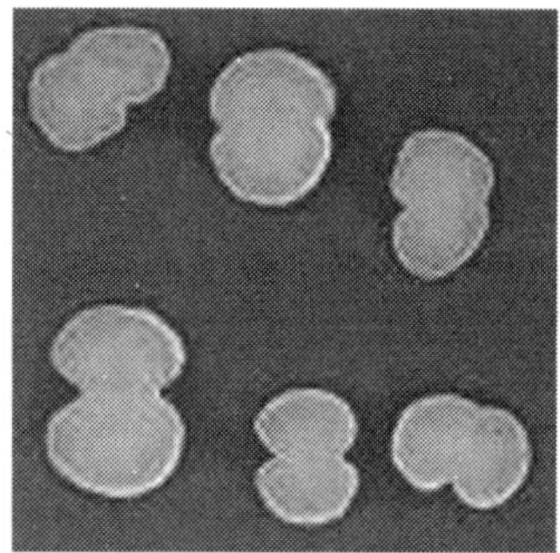

Structural and colony morphology of *Neisseria*

REFERENCES

Brooks GF, Donegan EA: Gonococcal Infection. Edward Arnold, London, 1985.

Centers for Disease Control and Prevention: Sexually transmitted diseases treatment guidelines. MMWR, Vol. 42 (No. RR-14), 1993.

Conde-Glez CJ, Morse S, Rice P, et al (eds): Pathobiology and immunobiology of Neisseriaceae. Instituto Nacional de Salud Publica, Cuernavaca, 1994.

DeVoe IW: The meningococcus and mechanisms of pathogenicity. Microbiol Rev 46:162, 1982.

Holmes KK, Mardh PA, Sparling PF,(eds): Sexually Transmitted Diseases. 2nd Ed McGraw Hill, New York, 1990.

Morse SA, Broome CV, Cannon J, et al (eds): Perspectives on pathogenic neisseriae. Clin Microbiol Rev 2:S1, 1989.

Salyers AA, Whitt DD: Bacterial pathogenesis: a molecular approach. American Society for Microbiology, Washington, D.C., 1994.

Chapter **15**

Pseudomonas

General Concepts

Clinical Manifestations

Pseudomonas aeruginosa and P maltophilia account for 80 percent of opportunistic infections by pseudomonads. Pseudomonas aeruginosa infection is a serious problem in patients hospitalized with cancer, cystic fibrosis, and burns; the case fatality is 50 percent. Other infections caused by Pseudomonas species include endocarditis, pneumonia, and infections of the urinary tract, central nervous system, wounds, eyes, ears, skin, and musculoskeletal system.

Structure, Classification, and Antigenic Types

Pseudomonas species are Gram-negative, aerobic bacilli measuring 0.5 to 0.8, µm by 1.5 to 3.0 µm. Motility is by a single polar flagellum. Species are distinguished by biochemical and DNA hybridization tests. Antisera to lipopolysaccharide and outer membrane proteins show cross-reactivity among serovars.

Pathogenesis

Neutropenia in cancer patients and others receiving immunosuppressive drugs contributes to infection. Pseudomonas aeruginosa has several virulence factors, but their roles in pathogenesis are unclear. An alginate is antiphagocytic, and most strains isolated produce toxin A, a diphtheria-toxin-like exotoxin. All strains have endotoxin, which is a major virulence factor in bacteremia and septic shock.

Host Defenses

Phagocytosis by polymorphonuclear leukocytes is important in resistance to Pseudomonas infections. Antibodies to somatic antigens and exotoxins also contribute to recovery.

Humoral immunity is normally the primary immune mechanism against Pseudomonas infection but does not seem to resolve infection in cystic fibrosis patients despite high levels of circulating antibodies.

Epidemiology

Pseudomonas species normally inhabit soil, water, and vegetation and can be isolated from the skin, throat, and stool of healthy persons. They often colonize hospital food, sinks, taps, mops, and respiratory equipment. Spread is from patient to patient via contact with fomites or by ingestion of contaminated food and water.

Diagnosis

Pseudomonas can be cultured on most general-purpose media and identified with biochemical media.

Control

The spread of Pseudomonas is best controlled by cleaning and disinfecting medical equipment. In burn patients, topical therapy of the burn with antimicrobial agents such as silver sulfadiazine, coupled with surgical debridement, has markedly reduced sepsis. Antibiotic susceptibility testing of clinical isolates is mandatory because of multiple antibiotic resistance; however, the combination of gentamicin and carbenicillin can be very effective in patients with acute P aeruginosa infections.

INTRODUCTION

The genus Pseudomonas contains more than 140 species, most of which are saprophytic. More than 25 species are associated with humans. Most pseudomonads known to cause disease in humans are associated with opportunistic infections. These include P aeruginosa, P fluorescens, P putida, P cepacia, P stutzeri, P maltophilia, and P putrefaciens. Only two species, P mallei and P pseudomallei, produce specific human diseases: glanders and melioidosis. Pseudomonas aeruginosa and P maltophilia account for approximately 80 percent of pseudomonads recovered from clinical specimens. Because of the frequency with which it is involved in human disease, P aeruginosa has received the most attention. It is a ubiquitous free-living bacterium and is found in most moist environments. Although, it seldom causes disease in healthy individuals, it is a major threat to hospitalized patients, particularly those with serious underlying diseases such as cancer and burns. The high mortality associated with these infections is due to a combination of weakened host defenses, bacterial resistance to antibiotics, and the production of extracellular bacterial enzymes and toxins.

Clinical Manifestations

Pseudomonas aeruginosa causes various diseases (Fig. 27-1). Localized infection following surgery or burns commonly results in a generalized and frequently fatal bacteremia. Urinary tract infections following introduction of P aeruginosa on catheters or in irrigating solutions are not uncommon. Furthermore, most cystic fibrosis patients are chronically colonized with P aeruginosa. Interestingly, cystic fibrosis patients rarely have P aeruginosa bacteremia, probably because of high levels of circulating P aeruginosa antibodies. However, most cystic fibrosis patients ultimately die of localized P aeruginosa infections. Necrotizing P aeruginosa pneumonia may occur in other patients following the use of contaminated respirators. Pseudomonas aeruginosa can cause severe corneal infections following eye surgery or injury. It is found in pure culture, especially in children with middle ear infections. It occasionally causes meningitis following lumbar puncture and endocarditis following cardiac surgery. It has been associated with some diarrheal disease episodes. Since the first reported case of P aeruginosa infection in 1890, the organism has been increasingly associated with bacteremia and currently accounts for 15 percent of cases of Gram-negative bacteremia. The overall mortality associated with P aeruginosa bacteremia is about 50 percent. Some infections (e.g., eye and ear infections) remain localized; others, such as wound and burn infections and infections in leukemia and lymphoma patients, result in sepsis. The difference is most probably due to altered host defenses.

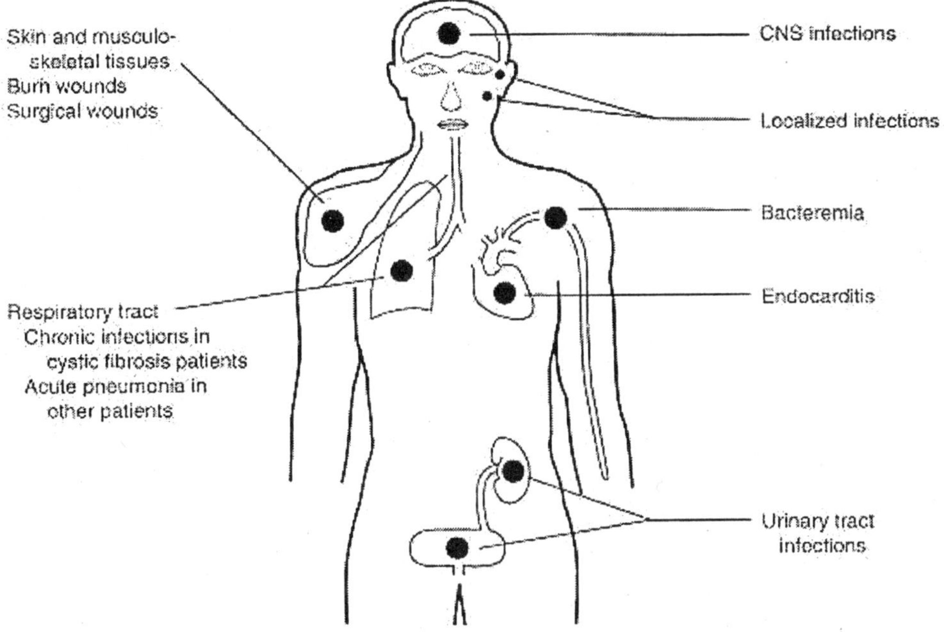

FIGURE 27-1 Diverse sites of infection by P aeruginosa. This opportunistic pathogen may infect virtually any tissue. Infection is facilitated by the presence of underlying disease (e.g., cancer, cystic fibrosis) or by a breakdown in nonspecific host defenses (as in burns).

Pseudomonas maltophilia is the second most frequently isolated pseudomonad species in clinical laboratories. In nature, P maltophilia is found in water and in both raw and pasteurized milk. It has been associated with a variety of opportunistic infections in humans, including pneumonia, endocarditis, urinary tract infections, wound infections, septicemia, and meningitis. Pseudomonas cepacia, although primarily a plant pathogen (onion bulb rot), also is an opportunist. Most human infections caused by P cepacia are nosocomial and include endocarditis, necrotizing vasculitis, pneumonia, wound infections, and urinary tract infections. Pseudomonas cepacia causes chronic lung infections in cystic fibrosis patients. These infections differ from those caused by P aeruginosa in that P cepacia has become systemic in a number of cystic fibrosis patients, whereas P aeruginosa infections remain confined to the lungs. Pseudomonas cepacia is highly resistant to aminoglycosides and other antibiotics, making it very difficult to control.

Unlike most pseudomonads, P mallei and P pseudomallei can cause disease in otherwise healthy individuals. Pseudomonas mallei is the agent of glanders, a disease primarily of equines. Humans generally become infected by inhalation or by direct contract through abraded skin. These infections are frequently fatal within 2 weeks of onset, although chronic infections also have been reported. Today, P mallei infections of equines are controlled and are rarely encountered in the western world. Similarly, melioidosis, an endemic glanders like disease of animals and a human pulmonary infection caused by P pseudomallei, is rare in the western hemisphere. Melioidosis is still found in Southeast Asia, and travelers returning from that area are sometimes infected.

Structure, Classification, and Antigenic Types

Pseudomonas aeruginosa is a Gram-negative rod measuring 0.5 to 0.8 µm by 1.5 to 3.0 µm. Almost all strains are motile by means of a single polar flagellum, and some strains have two or three flagella (Fig. 27-2). The flagella yield heat-labile antigens (H antigen). The significance of antibody directed against these antigens, aside from its value in serologic classification, is unknown. Clinical isolates usually have pili, which may be antiphagocytic and probably aids in bacterial attachment, thereby promoting colonization.

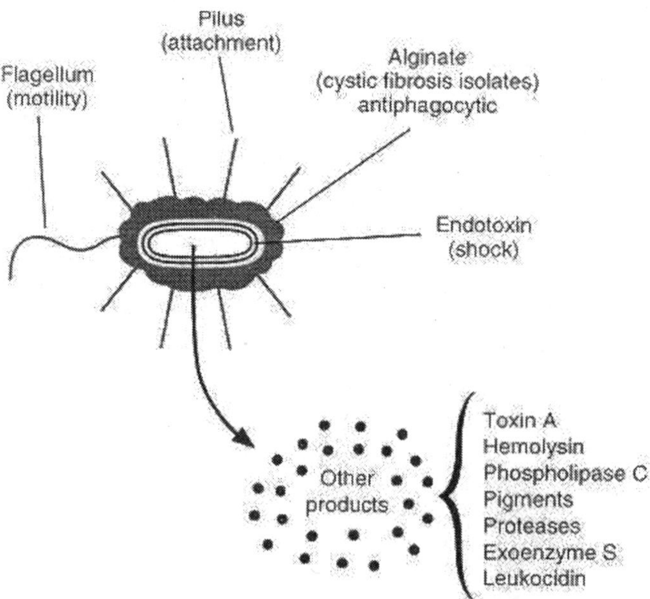

FIGURE 27-2 Structure and pathogenic mechanisms of P aeruginosa. The proposed role of other products is listed in Table 27-1.

The cell envelope of P aeruginosa, which is similar to that of other Gram-negative bacteria, consists of three layers: the inner or cytoplasmic membrane, the peptidoglycan layer, and the outer membrane. The outer membrane is composed of phospholipid, protein, and lipopolysaccharide (LPS). The LPS of P aeruginosa is less toxic than that of other Gram-negative rods. The LPS of most strains of P aeruginosa contains heptose, 2-keto-3-deoxyoctonic acid, and hydroxy fatty acids, in addition to side-chain and core polysaccharides. Recent evidence suggests that the LPS of a large percentage of strains isolated from patients with cystic fibrosis may have little or no polysaccharide side chain (O antigen), and that this finding correlates with the polyagglutinability of these strains with typing sera.

Studies of isolated outer membranes suggest strong conservation of many of the outer membrane proteins of P aeruginosa. Although, numerous serologic types exist (based on evaluations of O-specific antigens), many of the outer membrane proteins from these strains are antigenically crossreactive.

Pseudomonas aeruginosa is a nonfermentative aerobe that derives its energy from oxidation rather than fermentation of carbohydrates. Although, able to use more than 75 different organic compounds, it can grow on media supplying only acetate for carbon and ammonium sulfate for nitrogen. Furthermore, although an aerobe, it can grow anaerobically, using nitrate as an electron acceptor. This organism grows well at 25° C to 37° C, but can grow slowly or at least survive at higher and lower temperatures. Indeed, the ability to grow at 42° C distinguishes it from many other Pseudomonas species. In addition to its nutritional versatility, P aeruginosa resists high concentrations of salt, dyes, weak antiseptics, and many commonly used antibiotics. These properties help explain its ubiquitous nature and contribute to its preeminence as a cause of nosocomial infections.

Pathogenesis

Pseudomonas aeruginosa produces many factors that may contribute to its virulence. Table 27-1 lists some of them. Almost all strains of P aeruginosa are hemolytic on blood agar plates, and several different hemolysins have been described. A heat-stable hemolytic glycolipid consisting of two molecules each of L-rhamnose and 1-b-hydroxydecenoic acid has been purified. Although, this hemolytic glycolipid is not very toxic to animals (5 mg injected intraperitoneally is required to kill a mouse), it is toxic to alveolar macrophages. Furthermore, P aeruginosa strains isolated from respiratory tract infections produce more hemolysin than do environmental strains, suggesting that this glycolipid hemolysin may play a role in P aeruginosa pulmonary infections. Correlation of hemolysin production with infections of other sites has not been reported.

TABLE 27-1 Products of P aeruginosa strains

Product	Incidence of production (%)	LD_{52} In Mlce	Proposed Role
Endotoxin	100	300 μg/TV	Terminal shock
heat-stable hemolysin	95	5 mg/IP	toxic to alveolar macrophages
Leukocidin	4	0.4 μg/IP	Depression of host defenses
Phospholipase C	70	?	Hydrolysis of lecithin
Pigments (pyccyanin and fluorescein)	90	?	Antibacterial agent
Proteases (elastase and alkaline protease)	90	200μg/IP	Local tissue necrosis and spreading factor
Toxin A	90	0.2 μg/IP	lethality and inhibition of host defenses
Exoenzyme S	90	?	Local and systemic toxicity

Several heat-labile protein hemolysins also have been described. One of these hemolysins may be identical to phospholipase C, which is produced by approximately 70 percent of all clinical strains of P aeruginosa. Phospholipase C, which hydrolyzes lecithin, is of unknown toxicity, and its role in P aeruginosa infections also remains unknown. Some strains of P aeruginosa produce a thermolabile protein (leukocidin), which lyses leukocytes from many species including humans but is nonhemolytic. This leukocidin (also called cytotoxin) damages lymphocytes and various tissue culture cells and is very toxic to mice (minimum lethal dose is 1 μg). Despite its toxicity, the role of leukocidin remains unknown.

Some strains of P aeruginosa produce large amounts of extracellular polysaccharide. These mucoid strains usually are isolated only from patients with cystic fibrosis. The role of these polysaccharides inthe pathogenesis of P aeruginosa chronic lung infections is unknown, but they may impede phagocytosis and impair diffusion of antibiotics and thus facilitate colonization and persistence. Interestingly, mucoid strains are frequently deficient in production of elastase, toxin A, and flagella, and their LPS lacks long polysaccharide side chains.

Most strains of P aeruginosa also produce one or more pigments, the most common being pyocyanin (a phenazine pigment) and fluorescein. These pigments are nontoxic in animals. Pyocyanin, however, retards the growth of some other bacteria and thus may facilitate colonization by P aeruginosa. One or more of these pigments appear to function in iron acquisition by P aeruginosa. Additional work is needed to clarify the role of these pigments in P aeruginosa infections.

Approximately 90 percent of P aeruginosa strains produce extracellular protease. Three separate proteases have been purified that differ in pH optimum, isoelectric point, and substrate specificity. Although, all are capable of digesting casein, one of them, protease II, also digests elastin. When injected into the skin of animals, purified P aeruginosa proteases induce formation of hemorrhagic lesions, which become necrotic within 24 hours. These proteases also cause rapid tissue destruction when injected into the cornea of animal eyes or into rabbit lungs; they also probably contribute to the tissue destruction that accompanies P aeruginosa eye or lung infections and may aid bacteria in tissue invasion. Their effects, however, appear to be localized, and they are not highly toxic to animals (LD_{50} = approximately 200 µg/mouse) (Table 27-1).

Toxin A

Toxin A, the most toxic known extracellular protein of P aeruginosa, is produced by 90 percent of all strains. The median lethal dose of pure toxin A is about 0.2 µg/ mouse. Its toxicity has been attributed to its ability to inhibit protein synthesis in susceptible cells. It achieves this by catalyzing the transfer of the ADP-ribosyl moiety of nicotinamide adenine dinucleotide (NAD) onto elongation factor 2 (EF-2) according to the following reaction:

$$\text{NAD} + \text{EF-2} \underset{\rightarrow}{\overset{\text{toxin A}}{\leftarrow}} \text{ADP-ribosyl- EF-2} + \text{nicotiamide} + H^{+}$$

The resultant ADP-ribosyl-EF-2 complex is inactive in protein synthesis. This intracellular mechanism of action of toxin A is identical to that of diphtheria toxin fragment A (see Ch. 32). Also like diphtheria toxin, Pseudomonas toxin A is released by P aeruginosa as a proenzyme. Toxin A is toxic to animals and cultured cells, but the proenzyme has little or no enzymatic activity. Table 27-2 shows the relationship between the various forms of toxin A and their enzymatic activity and mouse toxicity. Evidence suggesting that toxin A may be a major virulence factor of P aeruginosa includes observations that toxin A-deficient mutants are less virulent in several animal models than their toxin A-producing parental strains, as well as the observation that most patients surviving P aeruginosa sepsis have elevated levels of antitoxin A antibody or are infected with strains that produce little or no detectable toxin A in vitro. These studies need to be expanded before firm conclusions can be reached.

TABLE 27-2 Comparison of the structure and Function of Toxin A and Its Fragments

Toxin Moss (Da)	**Molecular**	**Mouse LD_{60} Activity per µg of Protein (%)**	**Maximum ADP-Ribosy ltransferase**
Native toxin	70,000	0.2 µg	1-10
Reduced and denatured toxin*	70,000	>5.0 µg	35
Fragment A	27,000	>5.0 µg	100
Fragment B	43,000	Not tested	<1

*Toxin a was preincubated in 4 M urea plus 1 percent dithiothreitol for 15 minules

Exoenzyme S

A second ADP-ribosyltransferase, exoenzyme S, has been described. Exoenzyme S catalyzes the transfer of ADP-ribose onto a number of GTP-binding proteins, including the product of the proto-oncogene c-H-ras (p21C-H-ras); however, it does not modify elongation factor 2. Exoenzyme S is produced by about 90 percent of clinical isolates of P aeruginosa. Transposon-induced S-deficient mutants are less virulent in several animal models than is their S-producing parental strain; thus, exoenzyme S may be involved in the pathogenesis of some P aeruginosa infections.

Host Defenses

Although, 85 percent of P aeruginosa isolates are resistant to serum alone, addition of polymorphonuclear leukocytes results in bacterial killing. Killing is most efficient in the presence of type-specific opsonizing antibodies, directed primarily at the antigenic determinants of LPS. This suggests that phagocytosis is an important defense and that opsonizing antibody is the principal functioning antibody in protecting from P aeruginosa infections; however, once a P aeruginosa infection is established, other antibodies, such as antitoxin, may be important in preventing death. Although, evidence suggests interaction between P aeruginosa and the cellular immune system, patients with diseases characterized by impaired cellular immune responses (e.g., Hodgkin's disease) do not have an increased incidence of severe P aeruginosa infections. However, patients with diminished antibody responses caused by underlying disease or its associated therapy, have more serious P aeruginosa infections. This underscores the importance of the humoral response in controlling P aeruginosa infections. Cystic fibrosis is the exception. Most cystic fibrosis patients have high levels of circulating antibodies to many bacterial antigens, but are unable to clear P aeruginosa efficiently from their lungs.

Epidemiology

Pseudomonas aeruginosa commonly inhabits soil, water, and vegetation. It is found in the skin of some healthy persons and has been isolated from the throat (5 percent) and stool (3 percent) of nonhospitalized patients. The gastrointestinal carriage rates increase in hospitalized patients to 20 percent within 72 hours of admission. Within the hospital, P aeruginosa finds numerous reservoirs: disinfectants, respiratory equipment, food, sinks, taps, and mops. Furthermore, it is constantly reintroduced into the hospital environment on fruits, plants, vegetables, and patients transferred from other facilities. Spread occurs from patient to patient on the hands of hospital personnel, by direct patient contact with contaminated reservoirs, and by the ingestion of contaminated foods and water.

Several different typing systems are available for epidemiologic studies: serologic, phage, pyocin, and DNA fingerprinting. In the pyocin system, pyocins (bacteriocins or aeruginocins) produced by the test strain are assayed for bactericidal activity against a series of indicator strains. A number of different serologic typing systems are used. Some employ combinations of heat-stable and heat-labile antigens, whereas others use only heat-stable antigens. No system is universally accepted. Recently, DNA fingerprinting has identified probes that are useful in typing P aeruginosa strains.

Diagnosis

Diagnosis of P aeruginosa depends on its isolation and laboratory identification. It grows well on most laboratory media and commonly is isolated on blood agar plates or eosin-methylthionine blue agar. It is identified on the basis of its Gram morphology, inability to ferment lactose, a positive oxidase reaction, its fruity odor, and its ability to grow at 4 2° C. Fluorescence under ultraviolet radiation helps in early identification of P aeruginosa colonies and also is useful in suggesting its presence in wounds. Other pseudomonads are identified by specific laboratory tests.

Control

The spread of P aeruginosa can best be controlled by observing proper isolation procedures, aseptic technique, and careful cleaning and monitoring of respirators, catheters, and other instruments. Topical therapy of burn wounds with antibacterial agents such as mafenide or silver sulfadiazine, coupled with surgical debridement, has dramatically reduced the incidence of P aeruginosa sepsis in burn patients.

Pseudomonas aeruginosa is frequently resistant to many commonly used antibiotics. Although, many strains are susceptible to gentamicin, tobramycin, colistin, and amikacin, resistant forms have developed, making susceptibility testing essential. The combination of gentamicin and carbenicillin is frequently used to treat severe Pseudomonas infections, especially in patients with leukopenia. Several types of vaccines are being tested, but none is currently available for general use.

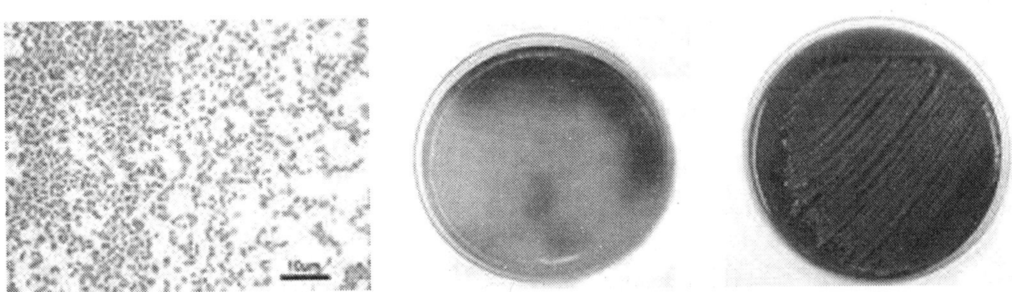

Colony and Cultural morphology of *Pseudomonas*

REFERENCES

Brown MRW (ed): Resistance of a Pseudomonas aeruginosa. John Wiley & Sons, New York, 1975

Clarke PH, Richman MN (eds): Genetics and Biochemistry of Pseudomonas. John Wiley & Sons, New York, 1975

Coburn J, Wyatt RT, Iglewski BH, Gill DM: Several GTP-binding proteins, including o24 C-H-ras, are preferred substrates of Pseudomonas aeruginosa exoenzyme S.J Biol Chem 264:9004, 1989

Cross AS, Sadoff JC, Iglewski BH, Sokol PA: Evidence for the role of toxin A in the pathogenesis of infections with Pseudomonas aeruginosa in humans. J Infect Dis 142:538, 1980

Dunn M, Wunderink RG: Ventilator-associated pnemonia caused by Pseudomonas infection. (Review) Clinics of Chest Medicine. (16):95, 1995

Hancock REW, Mutharia LM, Chan L, et al: Pseudomonas aeruginosa isolates from patients with cystic fibrosis: a class of serum-sensitive, nontypable strains deficient in lipopolysaccharide O side chain. Infect Immun 42: 170, 1983

Liu PV: Extracellular toxins of Pseudomonas aeruginosa. J Infect Dis, suppl.130:S95, 1974

Mutharia LM, Nicas TI, Hancock REW: Outer membrane proteins of Pseudomonas aeruginosa serotyping strains. J Infect Dis 46:770, 1982

Poole K, Bacterial multidrug resistanceemphasis on efflux mechanisms and Pseudomonas aeruginosa. (Review) J Antimicrobial Chemotherapy 34(4):453, 1994.

Pritchard AE, Vasal ML: Possible insertion sequences in a mosaic genomeorganization upstream of the exotoxin A gene in Pseudomonas aeruginosa. J Bacteriol 172:2020, 1990

Woods DE, Iglewski BH: Toxins of Pseudomonas aeruginosa: new perspectives. Rev Infect Dis, suppl. 5:S715, 1983

Chapter **16**

Bacillus

General Concepts

Clinical Manifestations

Anthrax is caused by Bacillus anthracis. Humans acquire the disease directly from contact with infected herbivores or indirectlyvia their products. Theclinical forms include (1) cutaneous anthrax (eschar with edema), from handling infected material (this accounts for more than 95 percent of cases); (2) intestinal anthrax, from eating infected meat; and (3) pulmonary anthrax, from inhaling spore-laden dust. Several other Bacillus spp, in particular B cereus and to a lesser extent B subtilis and B licheniformis, are periodically associated with bacteremia/septicemia, endocarditis, meningitis, and infections of wounds, the ears, eyes, respiratory tract, urinary tract, and gastrointestinal tract. Bacillus cereus causes two distinct food poisoning syndromes: a rapid-onset emetic syndrome characterized by nausea and vomiting, and a slower-onset diarrheal syndrome.

Structure and Classification

Bacillus species are rod-shaped, endospore-forming aerobic or facultatively anaerobic, Gram-positive bacteria; in some species cultures may turn Gram-negative with age. The many species of the genus exhibit a wide range of physiologic abilities that allow them to live in every natural environment. Only one endospore is formed per cell. The spores are resistant to heat, cold, radiation, desiccation, and disinfectants. Bacillus anthracis needs oxygen to sporulate; this constraint has important consequences for epidemiology and control. In vivo, B anthracis produces a polypeptide (polyglutamic acid) capsule that protects it from phagocytosis. The genera Bacillus and Clostridium constitute the family Bacillaceae. Species are identified by using morphologic and biochemical criteria.

Pathogenesis

The virulence factors of B anthracis are its capsule and three-component toxin, both encoded on plasmids. Bacillus cereus produces numerous enzymes and aggressins. The principal virulence factors are a necrotizing enterotoxin and a potent hemolysin (cereolysin). Emetic food poisoningprobably results from the release of emetic factors from specific foods by bacterial enzymes.

Host Defenses

The reasons for marked differences in susceptibility to anthrax among different animal species are not known. The protective actions of the live-spore animal vaccine or the human chemical vaccines are based on induction of humoral and cell-mediated immunity to the protective antigen component of anthrax toxin.

Epidemiology

Individuals at risk for anthrax include those in contact with infected animals or animal products. Episodes of B cereus food poisoning occur sporadically worldwide and result from ingestion of contaminated food in which the bacteria have multiplied to high levels under conditions of improper storage after cooking.

Diagnosis

Cutaneous anthrax is diagnosed on the basis of the characteristic papule (early) or eschar (later) with extensive surrounding edema, backed by a history of exposure to animals or their products. Diagnosis is confirmed by observation of characteristic encapsulated bacilli in polychrome methylene blue-stained smears of blood, exudate, lymph, cerebrospinal fluid, etc., and/or by culture. Other Bacillus infections are diagnosed by culture of the bacteria.

Control

Anthrax: Control in animals is essential for control in humans. In endemic areas, animals that die suddenly should be handled cautiously and livestock should be vaccinated annually. A human vaccine is available for individuals in high-risk occupations. Anthrax is readily treated with antibiotics (e.g., penicillin, tetracycline, chloramphenicol, gentamicin, or erythromycin).

Other Bacillus Infections: Control is by good hygiene. Treatment is with non-ß-lactam antibiotics for Gram-positive bacteria. Food poisoning is controlled by adequate cooking, avoidance of recontamination of cooked food, and proper storage (efficient refrigeration).

Pharmaceutical, Agricultural, and Industrial Importance

Many of the physiologic properties and specialized metabolites of Bacillus species are used in the pharmaceutical, agricultural, and food industries. On the other hand, the resistance of the spores to sterilization and disinfection makes them problem contaminants in foods, medical supplies, surgical procedures, etc.

INTRODUCTION

Bacillus species are aerobic, sporulating, rod-shaped bacteria that are ubiquitous in nature. Bacillus anthracis, the agent of anthrax, is the only obligate Bacillus pathogen in vertebrates. Bacillus larvae, B lentimorbus, B popilliae, B sphaericus, and B thuringiensis are pathogens of specific groups of insects. A number of other species, in particular B cereus, are occasional pathogens of humans and livestock, but the large majority of Bacillus species are harmless saprophytes.

Anthrax has afflicted humans throughout recorded history. The fifth and sixth plagues of Egypt described in Exodus are widely believed to have been anthrax. The disease was featured in the writings of Virgil in 25 BC and was familiar in medieval times as the Black Bane. It was from studies on anthrax that Koch established his famous postulates in 1876, and vaccines against anthrax the best known being that of Pasteur (1881)were among the first bacterial vaccines developed.

Bacillus species are used in many medical, pharmaceutical, agricultural, and industrial processes that take advantage of their wide range of physiologic characteristics and their ability to produce a host of enzymes, antibiotics, and other metabolites. Bacitracin and polymyxin are two well-known antibiotics obtained from Bacillus species. Several species are used as standards in medical and pharmaceutical assays.

The spores of the obligate thermophile *B stearothermophilus* are used to test heat sterilization procedures, and B subtilis subsp globigii, which is resistant to heat, chemicals, and radiation, is widely used to validate alternative sterilization and fumigation procedures. Certain Bacillus species are important in the natural or artificial degradation of waste products. Some Bacillus insect pathogens are used as the active ingredients of insecticides.

Because the spores of many Bacillus species are resistant to heat, radiation, disinfectants, and desiccation, they are difficult to eliminate from medical and pharmaceutical materials and are a frequent cause of contamination. Bacillus species are well known in the food industries as troublesome spoilage organisms.

Clinical Manifestions

Although, anthrax remains the best-known Bacillus disease, in recent years other Bacillus species have been increasingly implicated in a wide range of infections including abscesses, bacteremia/septicemia, wound and burn infections, ear infections, endocarditis, meningitis, ophthalmitis, osteomyelitis, peritonitis, and respiratory and urinary tract infections. Most of these occur as secondary or mixed infections or inimmunodeficient or otherwise immunocompromised hosts (such as alcoholics and diabetics), but a significant proportion are primary infections in otherwise healthy individuals. Some of these infections are severe or lethal. Of the species listed in Table 15-1, most frequently implicated in these types of infection is *B cereus,* followed by *B. licheniformis and B. subtilis. Bacillus alvei, B. brevis, B. circulans, B. coagulans, B. macerans, B. pumilus, B. sphaericus, and B. thuringiensis cause* occasional infections. As secondary invaders, Bacillus species may exacerbate preexisting infections by producing either tissue-damaging toxins or metabolites such as penicillinase that interfere with treatment.

Bacillus cereus is well known as an agent of food poisoning, and a number of other Bacillus species, particularly *B subtilis and B licheniformis*, are also incriminated periodically in this capacity.

Anthrax

Anthrax is primarily a disease of herbivores. Humans acquire it as a result of contact with infected animals or animal products. In humans the disease takes one of three forms, depending on the route of infection. Cutaneous anthrax, which accounts for more than 95 percent of cases worldwide, results from infection through skin lesions; intestinal anthrax results from ingestion of spores, usually in infected meat; and pulmonary anthrax results from inhalation of spores.

Cutaneous anthrax usually occurs through contamination of a cut or abrasion, although in some countries biting flies may also transmit the disease. After a 2- to 3-day incubation period, a small pimple or papule appears at the inoculation site. A surrounding ring of vesicles develops. Over the next few days, the central papule ulcerates, dries, and blackens to form the characteristic eschar (Fig. 15-1). The lesion is painless and is surrounded by marked edema that may extend for some distance. Pus and pain appear only if a pyogenic organism infects the lesion. Similarly, marked lymphangitis and fever usually point to a secondary infection. In most cases, the disease remains limited to the initial lesion and resolves spontaneously. The main dangers are that a lesion on the face or neck may swell to occlude the airway or may give rise to secondary meningitis. If host defenses fail to contain the infection, however, fulminating septicemia develops. Approximately 20 percent of untreated cases of cutaneous anthrax progress to fatal septicemia. However, *B anthracis* is susceptible to penicillin and other common antibiotics, so effective treatment is almost always available.

Intestinal anthrax is analogous to cutaneous anthrax but occurs on the intestinal mucosa. As in cutaneous anthrax, the organisms probably invade the mucosa through a preexisting lesion. Organisms spread from the mucosal lesion to the lymphatic system. In pulmonary anthrax, inhaled spores are transported by alveolar macrophages to the mediastinal lymph nodes, where they germinate and multiply to initiate systemic disease. Gastrointestinal and pulmonary anthrax are both more dangerous than the cutaneous form because they are usually identified too late for treatment to be effective.

Herbivorous animals, the primary hosts of *B anthracis*, contract the infection by ingesting spores on forage plants; the spores are derived from soil or dust or are deposited on leaves by flies after feeding on an anthrax-infected carcass. If the spores enter a lesion in the gastrointestinal mucosa, they germinate and are taken into the bloodstream and lymphatics, finally producing systemic anthrax, which is usually fatal.

Symptoms prior to fulminant systemic anthrax may be absent or mild, consisting,

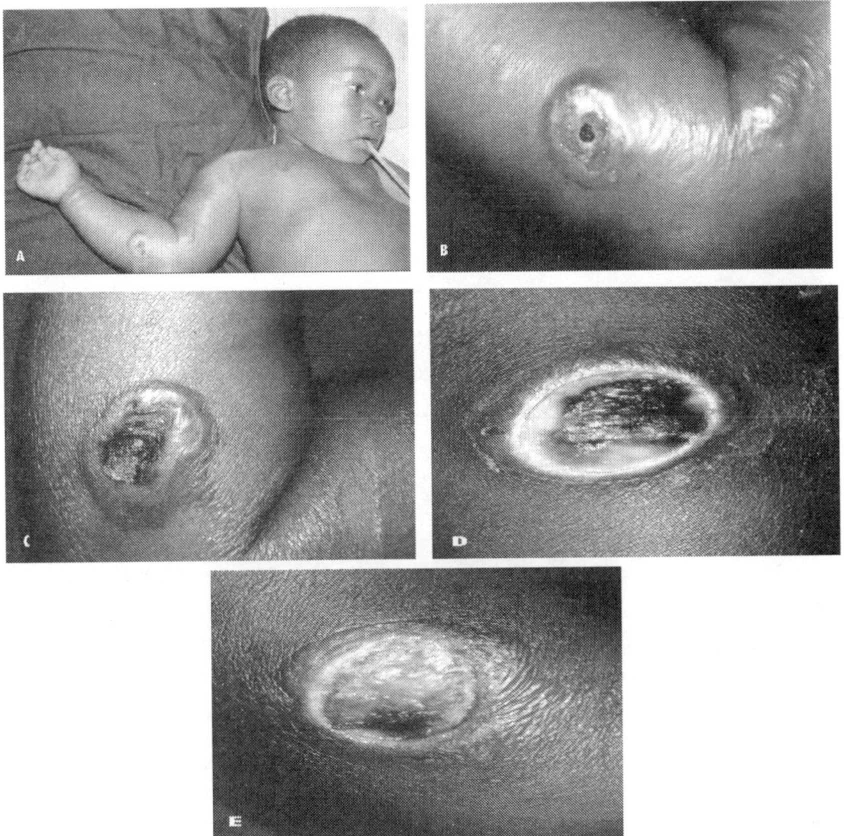

FIGURE 15-1 Evolution of an anthrax eschar in a 4-year-old boy. (A&B) the lesion when first seen (day 0). Note the arm swollen from the characteristic edema. (C) Day 6. (D) Day 10. (E) Day 15. Although penicillin treatment was begun immediately and the lesion was sterile by about 24 hours, it continued to evolve and resolve as seen. (Photographs kindly supplied by W.E. Kobuch, M.D., St. Luke's Hospital, Lupane, Bulawayo, Zimbabwe.)

for example, of malaise, low fever, and mild gastrointestinal symptoms in the case of gastrointestinal disease. During this phase, the organism is multiplying and producing toxin in the regional lymph nodes and spleen. Released toxin causes breakdown of these organs probably of the spleen in particular. This causes the sudden onset of hyperacute illness with dyspnea, cyanosis, high fever, and disorientation, which progress in a few hours to shock, coma, and death. Although, symptoms vary somewhat with the host species, this final acute phase is marked by a high-grade bacteremia. In humans, blood cultures are not always positive.

Bacillus Food Poisoning

Bacillus cereus can cause two distinct types of food poisoning. The diarrheal type is characterized by diarrhea and abdominal pain occurring 8 to 16 hours after consumption of the contaminated food. It is associated with a variety of foods, including meat and vegetable dishes, sauces, pastas, desserts, and dairy products. In emetic disease, on the other hand, nausea and vomiting begin 1 to 5 hours after the contaminated food is eaten. Boiled rice that is held for prolonged periods at ambient temperature and then quick-fried before serving is the usual offender, although dairy products or other foods are occasionally responsible. The symptoms of food poisoning caused by other Bacillus species (*B subtilis, B licheniformis,* and others) are less well defined. Diarrhea and/or nausea occurs 1 to 14 hours after consumption of the contaminated food. A wide variety of food types have proved responsible in recorded instances.

A Bacillus food poisoning episode usually occurs because spores survive cooking or pasteurization and then germinate and multiply when the food is inadequately refrigerated. The symptoms of B cereus food poisoning are caused by a toxin or toxins produced in the food during this multiplication. Toxins have not yet been identified for other Bacillus species that cause food poisoning.

Structure and Classification

The family Bacillaceae, consisting of rod-shaped bacteria that form endospores, has two principal subdivisions: the anaerobic spore-forming bacteria of the genus Clostridium, and the aerobic or facultatively anaerobic spore-forming bacteria of the genus Bacillus frequently known as ASB (aerobic spore-bearers). Bacterial cells of Bacillus cultures are Gram positive when young, but in some species become Gram negative as they age.

Most Bacillus species are saprophytes. Table 15-1 lists the identifying characteristics of some of the species most likely to be encountered by the physician. Not only are Bacillus endospores resistant to hostile physical and chemical conditions, but also various species have unusual physiologic properties that enable them to survive or thrive in harsh environments, ranging from desert sands and hot springs to Arctic soils and from fresh waters to marine sediments. The genus includes thermophilic, psychrophilic, acidophilic, alkaliphilic, halotolerant, and halophilic representatives, which are capable of growing at temperatures, pH values, and salt concentrations at which few other organisms could survive.

Figure 15-2 shows the structure of a generalized Bacillus endospore (details of the structure differ from species to species). One spore is produced per vegetative cell. The central protoplast, or germ cell, carries the constituents of the future vegetative cell, accompanied by dipicolinic acid, which is essential to the heat resistance of the spore. Surrounding the protoplast is a cortex consisting largely of peptidoglycan (murein), which is also important in the heat and radiation resistance of the spore. The inner layer, the cortical membrane or protoplast wall, becomes the cell wall of the new vegetative cell when the spore germinates. The spore coats, which constitute up to 50 percent of the volume of the spore, protect it from chemicals, enzymes, etc.

Table 15-1 Basic Characteristics for identification of Selected Bacillus Species

Species	Mobility	Catalase Production	Parasporasal Bodies	Lipid Globules in Protoplasm	Leclhovitellin Reaction	Citrate Utilization	Anaerobic Growth	V.P. Relation	pH in V.P Medium <6.0	Growth at 50ºC	Growth at 60 ºC	Growth in 7% NaCl	Acid from AS Glucose	Acid + Gas from AS Glucose	Nitrate Reduction	Casein Hydrolysis	Starch Hydrolysis	Propionate Utilization
Morphologic group 1																		
B megaterium	V	+	-	+	-	+	-	-	V	-	+	+	+	-	V	+	+	n
B cereus	+	+	-	+	+	+	+	+	+	-	+	+	+	-	+	+	+	n
B cereus subsp *mycoides*	-	+	-	+	+	+	+	+	+	-	+	+	+	-	+	+	+	n
B anthracis	-	+	-	+	+	V	+	+	+	-	+	+	+	-	+	+	+	n
B thuringiensis	+	+	-	+	+	+	+	+	+	-	+	+	+	-	+	+	+	n
B licheniformis	+	+	-	-	-	+	+	+	+	+	+	+	+	-	+	+	+	+
B subtilis	+	+	-	-	-	+	-	+	V	V	+	+	+	-	+	+	+	-
B pumilus	+	+	-	-	-	+	-	+	+	V	+	+	+	-	-	+	-	-
B firmus	V	+	-	-	-	-	-	-	-	-	+	+	+	-	-	+	+	-
B coagulans	+	+	-	-	-	V	+	+	+	+	-	-	+	-	V	V	+	-
Morphologic group 2																		
B polymyxa	+	+	-	-	-	-	+	+	V	-	-	-	+	+	+	+	+	n
B macerans	+	+	-	-	-	V	+	-	+	+	-	-	+	+	+	-	+	n
B circulans	V	+	-	-	-	V	V	-	+	V	-	V	+	-	V	V	+	n
B stearothermophilus	+	V	-	-	-	-	-	-	+	+	+	-	+	-	V	V	+	n
B alvei	+	+	-	-	-	-	+	+	+	-	-	-	+	-	-	+	+	n
B laterosporus	+	+	-	-	(+)	-	+	-	+	-	-	-	+	-	+	+	-	n
B brevis	+	+	-	-	-	V	-	-	-	V	V	-	+	-	V	+	-	n
Morphologic group 3																		
B sphaeocus	+	+	-	-	-	V	-	-	-	-	-	V	-	-	-	V	-	n

[a] V-P, Voges-Proskauer; AS, ammonium salt; +, more than 85 precent of strains examined by Gordon et al, (See References) were positive; +, more than 85 percent of strains negative; V, variable; n, test not applicable; (+), under colony which must be scraped off to see positive reaction

[b] Species grouped according to the classification scheme of Grodon et al, Morphologic group 1: sporangium not swollen by spore; spore ellipsoidal or cylindrical, central or terminal; Gram positive. Morphologic group 2: sporangium swollen by ellipsoidal spore; spore centrol or terminal; Gram variable. 3: sporangium swollen by spore; spore spherical subterminal, or terminal; Gram variable.

[c] Sporangium and spore have characteristic cance shape.

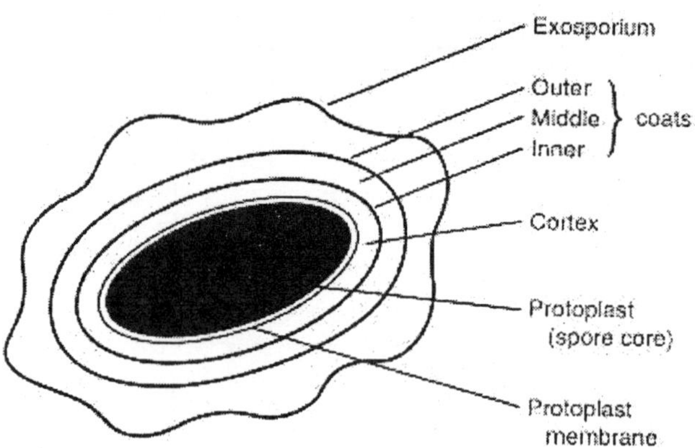

FIGURE 15-2 Cross section of a Bacillus spore.

The events involved in sporulation of vegetative cells and in germination of spores are complex and are influenced by factors such as temperature, pH, and the availability of certain divalent cations and carbon- and nitrogen-containing compounds. Spores formed under different conditions have different stabilities and degrees of resistance to heat, radiation, chemicals, desiccation, and other hostile conditions.

Pathogenisis

The pathogenicity of B anthracis depends on two virulence factors: a poly-y-D-glutamic acid polypeptide capsule, which protects it from phagocytosis by the defensive phagocytes of the host, and a toxin produced in the log phase of growth. This toxin consists of three proteins: protective antigen (PA) (82. 7 kDa), lethal factor (LF) (90.2 kDa), and edema factor (EF) (88.9 kDa). Host proteases in the blood and on the eukaryotic cell surface activate protective antigen by cutting off a 20-kDa segment, exposing a binding site for LF and EF. The activated 63 kDa PA polypeptide binds to specific receptors on the host cell surface, thereby creating a secondary binding site for which LF and EF compete. The complex (PA+LF or PA+EF) is internalized by endocytosis and, following acidification of the endosome, the LF or EF cross the membrane into the cytosol via PA-mediated ion-conductive channels. This is analogous to the A-B structure-function model of cholera toxin with PA behaving as the B (binding) moiety (Fig.15-3). EF, responsible for the characteristic edema of anthrax, is a calmodulin-dependent adenylate cyclase. (Calmodulin is the major intracellular calcium receptor in eukaryotic cells.) The only other known bacterial adenylate cyclase is produced by *Bordetella pertussis* (see Ch 31), but the two toxins share only minor homologies. LF appears to be a zinc-dependent metalloprotease though its substrate and mode of action have yet to be elucidated.

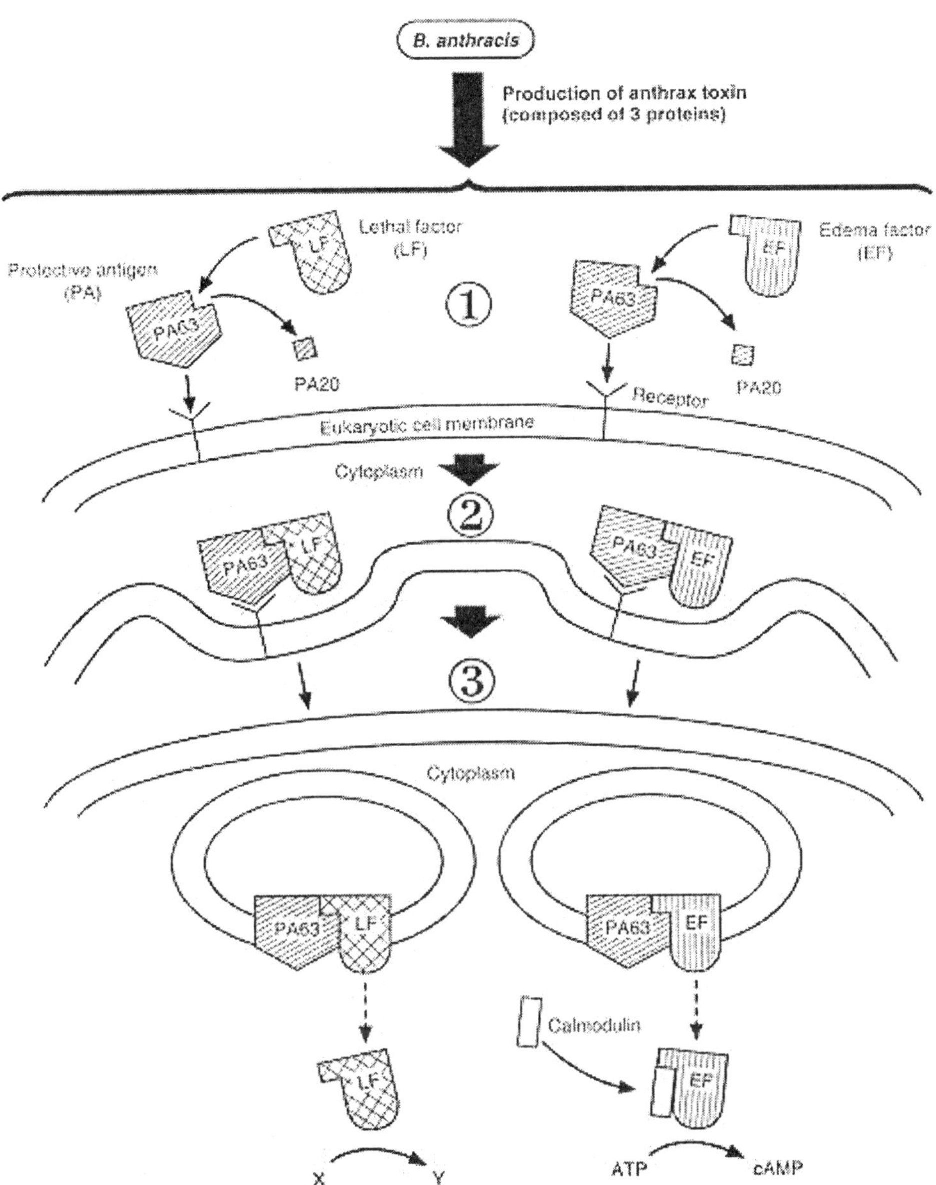

FIGURE 15-3 Mechanism of action of the anthrax toxin. The toxin is composed of three proteins. Protective antigen (PA) binds to an appropriate site on the host cell membrane. A cell surface protease cleaves off a 20-kDa piece from the protective antigen and thereby exposes a secondary binding site for which lethal factor (LF) and edema factor (EF) compete. The complex (PA+LF or PA+EF) is internalized by receptor-mediated endocytosis, and acidification of the endosome results in the transfer of the LF or EF across the endosome membrane into the cytosol where they carry out their catalytic actions. (Model by S.H. Leppla, Ph.D., Laboratory of Microbial Ecology, National Institutes of Health, Bethesda, MD.)

The toxin and capsule of *B anthracis* are encoded on two large plasmids called pXO 1 (110 MDa) and pX02 (60 MDa), respectively. Strains lacking either of these plasmids have greatly reduced virulence (Fig.15-4). The attenuated live vaccine strain developed by Sterne in 1937, which is still the basis of most anthrax vaccines for livestock, lacks pX02 and is therefore Cap- Tox+. The protection afforded by such vaccines apparently is related primarily to antibodies specific for the protective antigen component of the toxin. In contrast, the attenuated vaccine strains developed by Pasteur 110 years ago were inadvertently cured of pXO1 (by subculturing at 42° to 43°C); these Pasteur strains are therefore Cap+ Tox-. Strains of this type do not induce protective immunity; the partial effectiveness of Pasteur's vaccines is now believed to have been due to the residual uncured (Cap+ Tox+) cells they contained, and this would also explain the partial virulence of these strains.

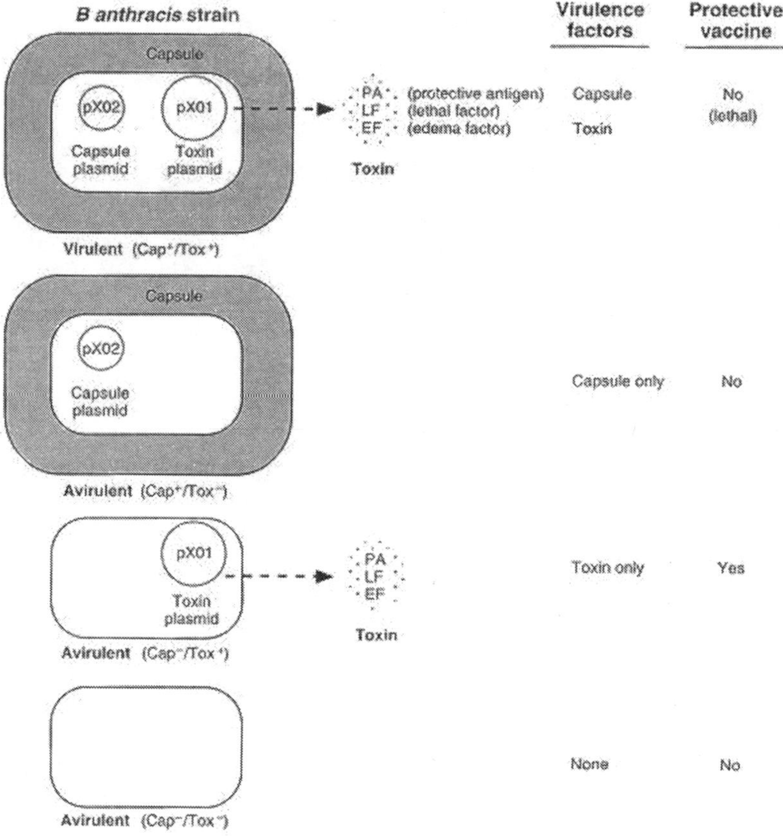

FIGURE 15-4 Genetics of virulence factor production by *B. anthracis*. Plasmids pX01 and pX02 encode, respectively, the anthrax toxin and capsule. Curing the bacteria of pX01 produces an encapsulated, nontoxigenic strain that is nonprotective. Curing of pX02 produces a toxigenic nonencapsulating strain that can be used as a protective vaccine. Production of protective antigen is essential for a strain to be protective.

The only other Bacillus species for which virulence factors have been identified is B cereus. A 38 to 46-kDa protein complex has been shown in animal models to cause necrosis of the skin or intestinal mucosa (Fig. 15-5), to induce fluid accumulation in the intestine, and to be a lethal toxin. This protein is believed to be responsible for the necrotic and toxemic nature of severe B cereus infections and for the diarrheal form of food poisoning. Bacillus cereus also produces two hemolysins; one of these, cereolysin (58 kDa), is a potent necrotic and lethal toxin. Although this toxin is neutralized by serum cholesterol, it probably contributes to the pathogenesis of B cereus infections. Little is known about the other hemolysin at present. Phospholipases produced by B cereus may act as exacerbating factors by degrading host cell membranes following exposure of their phospholipid substrates in wounds or other infections. The agent responsible for the emetic type of B cereus food poisoning has not been clearly identified. The emesis may be induced by breakdown products resulting from the action of one or more B cereus enzymes on the food.

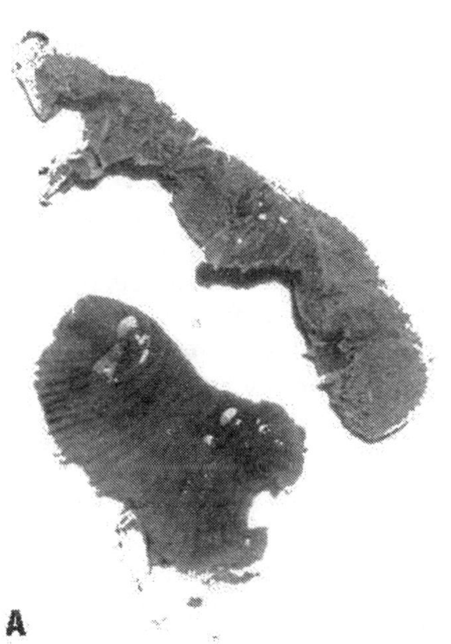

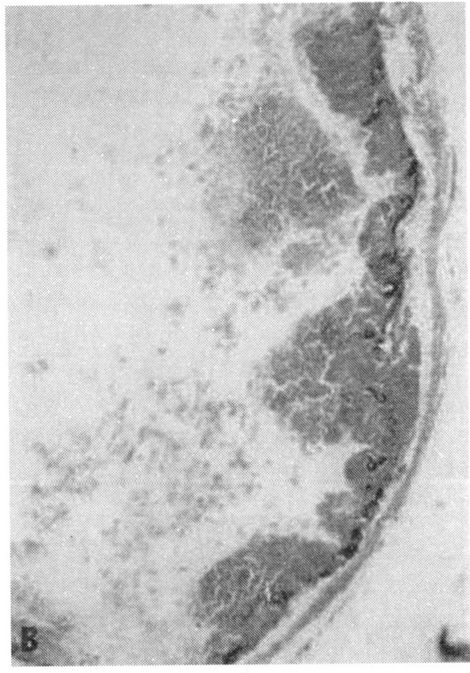

FIGURE 15-5 Necrosis of rabbit ileal mucosa 4 hours after introduction of a toxigenic cell-free culture filtrate of B cereus. (A) Gross appearance of the luminal surface of the ileum compared with a section of control ileum. (B) Histologic appearance of a cross-section of the toxin-exposed ileum. (From Turnbull PCB: Studies on the production of enterotoxins by Bacillus cereus. J Clin Pathol 29:941, 1976, with permission.)

Host Defenses

Anthrax has been documented in a wide variety of warm-blooded animals. Some species, such as rats, chickens, and dogs, are quite resistant to the disease, whereas others (notably herbivores such as cattle, sheep, and horses) are very susceptible. Humans have intermediate susceptibility. The specific mechanisms of resistance in the more resistant species are not known.

Protective immunity against anthrax requires antibodies against components of anthrax toxin, primarily protective antigen. Both the noncellular human vaccines and live-spore animal vaccines confer protection by eliciting antibodies to protective antigen. The poly-g-D-glutamic acid capsule of*B anthracis* is poorly immunogenic, and antibodies to the polysaccharide and other components of the cell wall are not protective.

Nothing is known about immune responses to food poisoning or other types of infections with Bacillus species other than *B anthracis*. These types of infection are rare, and effective vaccines against them have not been developed.

Epidemiology

The ultimate reservoir of *B anthracis* is contaminated soil, in which spores remain viable for long periods. Herbivores, the primary hosts, become infected when foraging in a contaminated region. Because the organism does not depend on an animal reservoir, it cannot readily be eradicated from a region, and anthrax remains endemic in many countries. Humans become infected almost exclusively through contact with infected animals or animal products. Human anthrax is traditionally classified as either nonindustrial or industrial anthrax, depending on whether the disease is acquired directly from animals or indirectly during handling of contaminated animal products. Nonindustrial anthrax usually affects people who work with animals or animal carcasses, such as farmers, veterinarians, knackers, and butchers, and is almost always cutaneous. Industrial anthrax, acquired from handling contaminated hair, hides, wool, bone meal, or other animal products, has a higher chance of being pulmonary as a result of the inhalation of spore-laden dust.

The development of an effective animal vaccine in the 1930s, together with improved factory hygiene, introduction of procedures for sterilizing imported animal products, replacement of animal products with man-made alternatives, and the availability since the mid-1960s of a human vaccine, has resulted in a greatly reduced incidence of the disease in North America. Human anthrax is now very rare in the United States. However, major epidemics still break out in endemic countries, normally following an outbreak in livestock. Nonendemic countries must remain alert for episodes of anthrax arising from imported animal products.

Diagnosis

The clinical diagnosis of anthrax is confirmed by directly visualizing or culturing the anthrax bacilli. Fresh smears of vesicular fluid, fluid from under the eschar, blood, lymph node or spleen aspirates, or (in meningitic cases) cerebrospinal fluid are stained with polychrome methylene blue (M'Fadyean's stain) and examined for the characteristic square-ended, blue-black bacilli surrounded by a pink capsule (Fig. 15-6). (It should be remembered that B anthracis organisms are not invariably detected in stained blood smears of humans dying of anthrax.) Alternatively, the bacilli may be cultured from these specimens and checked for sensitivity to the anthrax gamma phage, for penicillin sensitivity, and for capsule formation. Colonies grown overnight at 37°C on blood agar are gray or white, nonhemolytic, with a dry, ground-glass appearance; they are at least 3 mm in diameter and sometimes have tails (Fig. 15-7). Capsules can be seen in polychrome methylene blue-stained smears of cultures grown on nutrient agar containing 0.7 percent sodium bicarbonate and incubated overnight under CO_2 (e.g., in a candle jar); encapsulated colonies are mucoid. Alternatively, 2 ml of blood (such as commercial defibrinated horse blood) inoculated with a pinhead quantity of material from a suspected colony and incubated at 37°C yields readily demonstrable encapsulated bacilli in 6 hours. Culturing may be unsuccessful if the patient has been treated with antibiotics.

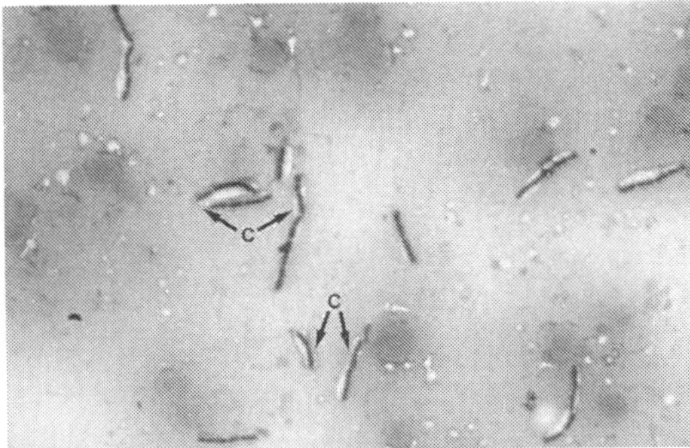

FIGURE 15-6 Blood smears from a guinea pig that died of anthrax, stained with M'Fadyean stain (polychrome methylene blue). The capsule (C) is pink around the dark-blue bacilli. Although not obvious from this photograph, anthrax bacilli frequently have square ends.

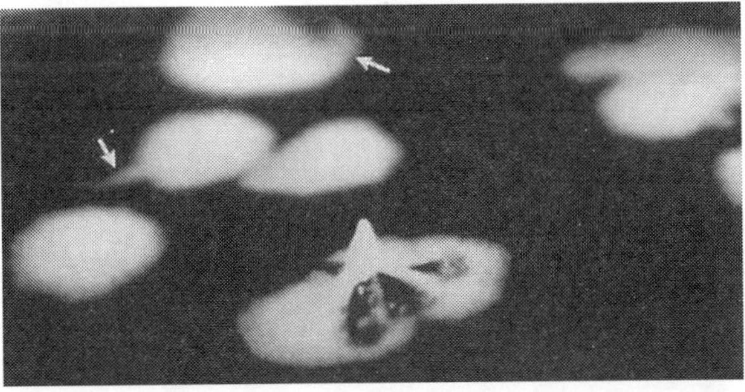

FIGURE 15-7 Colonies of *B anthracis* on a blood agar plate. Note the characteristic tackiness of colonies that allow them to be teased upright with a loop (foreground) and the characteristic tailing seen in the background (arrows). (Photograph kindly supplied by R.W. Charlton. From Turnbull PCB, Kramer JM, Melling J: Bacillus. p. 187. In Parker MT, Duerden BI (eds): Systematic Bacteriology. Topley and Wilson's Principles of Bacteriology, Virology and Immunity. Vol. 2. Edward Arnold, Sevenoaks, England, 1990, with permission.)

Isolation of B *anthracis* from old specimens or from animal or environmental material being examined for public health purposes is more difficult, particularly if, as is often the case, B cereus or other Bacillus species are present in substantial numbers. The specimen should be examined both unheated and heated to 60°C to 65°C for 15 min with subculture to both blood or nutrient agar and specialized selective agars. Very rarely it may be necessary to use mouse or guinea pig inoculation to isolate B *anthracis*. Up to about 0.2 ml of the specimen (or an aqueous extract of the specimen) is injected subcutaneously into a mouse, or intramuscularly or subcutaneously in a guinea pig (more sensitive than a mouse); the encapsulated bacilli can be seen in a smear of blood aspirated from the heart of the animal at death, and the bacteria are readily observed in and isolated from this blood. If soil samples are being used, the animals should be injected 24 hours earlier with tetanus and gas gangrene antitoxin.

When a specimen from an individual not suspected clinically of having anthrax yields substantial numbers of Gram-positive bacilli, the specimen should be cultured and tested as shown in Figure 15-8 to determine the Bacillus species present. The most common Bacillus species may be identifiedby the characteristics in Table 15-1. Incrimination of a Bacillus species as the cause of an infection is usually based on its presence in large numbers at the infection site, especially in the absence of other known pathogens. Since Bacillus species are common environmental organisms, their presence in small numbers is not generally considered significant. For this reason, the use of selective or enrichment systems for isolating clinically relevant, nonanthrax Bacillus species is confined to just a few situations, such as the retrospective examination of feces several days after a food poisoning incident (by which time the offending Bacillus organism may be present in only small numbers).

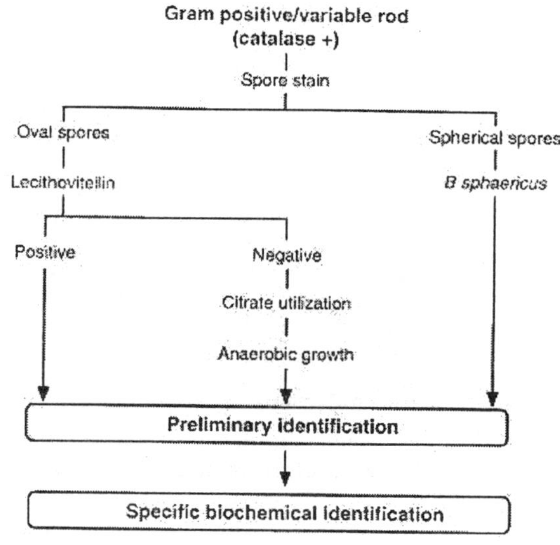

FIGURE 15-8 Flow chart for identification of principal Bacillus species.

Control

To comprehend the strategies used to control anthrax, it is important to understand the cycle of infection in susceptible animals. As a susceptible animal with anthrax approaches death, its blood contains as many as 10^9 bacilli/ml (depending on the species). Necrosis of the walls of small blood vessels during the acute phase of the illness leads to hemorrhages and to characteristic bloody exudations from the mouth, nose, and nausea highly diagnostic sign. These exudates carry vast numbers of the bacilli, which sporulate on exposure to air and produce a heavily contaminated environmental site that is potentially capable of infecting other animals for many years.

Because sporulation of B anthracis requires oxygen and therefore does not occur inside a closed carcass, regulations in most countries forbid postmortem examination of animals when anthrax is suspected. The vegetative cells in the carcass are killed in a few days by the process of putrefaction. Nevertheless, in the case of livestock, legislation invariably requires that the carcass be burned or buried in quicklime (calcium oxide). However, it is becoming increasingly apparent in the new era of sensitivity about environmental contamination that implementation in the past (sometimes many decades) of the order to bury in quicklime has left us a legacy of burial sites which are contaminated with viable anthrax spores and it is to be hoped that this instruction will be removed from veterinary public health orders.

Livestock in endemic areas are effectively protected by yearly inoculations with a vaccine made from spores of a live attenuated strain (see above). Noncellular vaccines for human use are available for individuals in high-risk occupations. They appear to have contributed to the decline in incidence of industrial anthrax since they became available in the 1960s, but animal studies suggest that there are limitations to their ability to protect against anthrax. The human vaccine available in the United States is an aluminum hydroxide-adsorbed cell-free filtrate of a B anthracis culture grown to maximize the yield of protective antigen and minimize the quantities of lethal factor, edema factor, and other unwanted metabolites.

Bacillus anthracis is susceptible to penicillin and to almost all other broad-spectrum antibiotics. Because it is easily recognized, cutaneous anthrax is almost always treated early and cured. Gastrointestinal and pulmonary anthrax infections are difficult to identify before the fulminant phase and therefore carry a high mortality. In uncomplicated anthrax cases, adequate treatment consists of 500 mg of penicillin V taken orally every 6 hours for 5 days, or 600 mg (1 million units) of procaine penicillin administered intramuscularly every 12 to 24 hours for 5 days. In severe cases, 1,200 mg (2 million units) of penicillin G should be administered intravenously every 6 hours, reverting to the intramuscular regime of 600 mg every 12 to 24 hours once recovery starts. If pulmonary anthrax is suspected, continuous-drip administration is advisable. Tetracyclines (tests in animals indicate doxycycline is good), chloramphenicol, gentamicin, or erythromycin may be used if the patient has penicillin hypersensitivity. The fluoroquinolone, ciprofloxacin, has also been shown to be effective in monkeys and guinea pigs and would be expected to be effective in treatment of cases of human anthrax.

Avoidance of other types of Bacillus infections is largely a matter of observing proper hygiene. Bacillus cereus and its close relatives B thuringiensis and B mycoides produce potent ß-lactamases and thus are not responsive to penicillin, ampicillin, or the cephalosporins. They are mostly resistant to trimethoprim as well. These species are generally sensitive to standard empirical treatment with an aminoglycoside combined with vancomycin and to chloramphenicol, erythromycin, tetracycline, clindamycin, and sulfonamides.

Bacillus food poisoning, like all types of food poisoning, can largely be prevented by proper food handling. Food should be cooked adequately; cooked food should not be recontaminated from uncooked food (separate utensils and cutting surfaces should be used for cooked and uncooked food); and, of particular importance, cooked food should be stored under proper refrigeration.

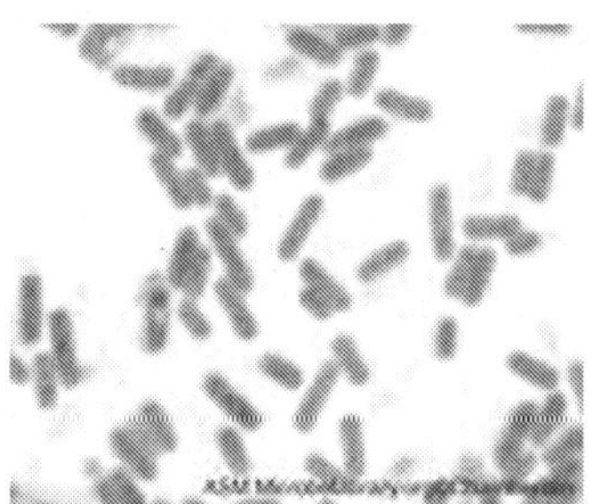

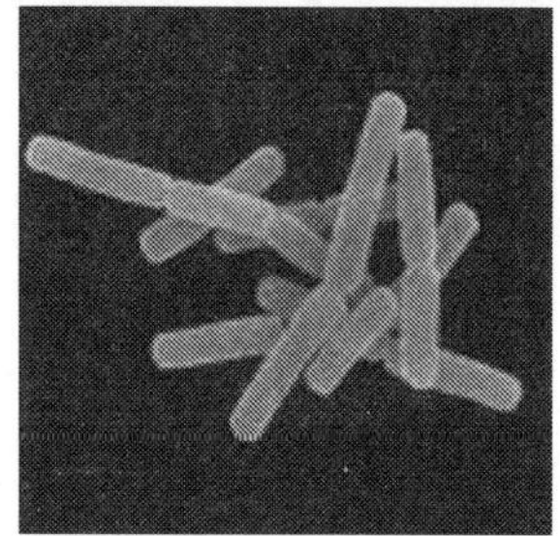

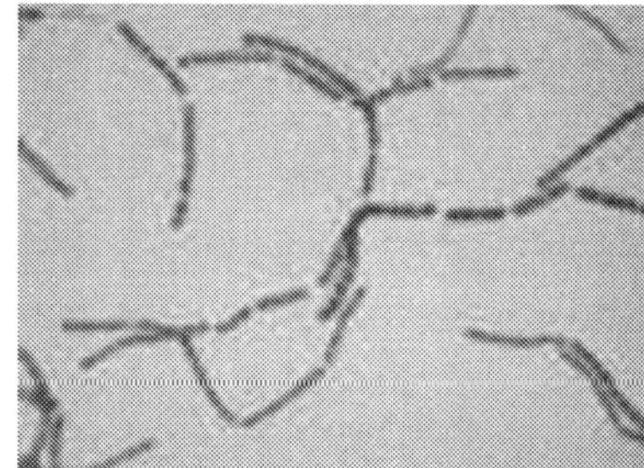

Microscopic and structural appearance of *Bacillus*

REFERENCES

Claus D, Berkeley RCW: Genus Bacillus Cohn 1872, 174AL. p. 1105. In Sneath PHA, Mair NS, Sharpe ME, Holt JG (eds): Bergey's Manual of Systematic Bacteriology. Vol. 2. Williams & Wilkins, Baltimore, 1986

Gordon RE, Haynes WC, Pang CH-N: The genus Bacillus. U.S. Department of Agriculture Agricultural Handbook no. 427. U.S. Department of Agriculture, Washington DC, 1973.

Klimpel KR, Arora N, Leppla SH: Anthrax toxin lethal factor contains a zinc metalloprotease consensus sequence which is required for lethal toxin activity. Molecular Microbiol 13:1093, 1994

Kochi SK, Schiavo G, Mock M, Montecucco C: Zinc content of the Bacillus anthracis lethal factor. FEMS Microbiol Lett 124:343, 1994

Kramer JM, Gilbert RJ: Bacillus cereus and other Bacillus species. p.21. In Doyle MP (ed): Foodborne Bacterial Pathogens. Marcel Dekker, New York, 1989

Leppla SH: The Anthrax toxin complex. p. 277. In Alouf JE, Freer JH (eds): Sourcebook of Bacterial Proteins. Academic Press, New York

Norris JR, Berkeley RCW, Logan NA et al: The genera Bacillus and Sporolactobacillus. p. 1711. In Starr MP, Stolp H, Truper HG (eds): The Prokaryotes A Handbook on Habitats, Isolation and Identification of Bacteria. Vol. 2. Springer-Verlag, New York, 1981

Parry JM, Turnbull PCB, Gibson JR: A Colour Atlas of Bacillus Species. Wolfe Medical Atlas no. 19. Wolfe Publishing, London 1983

Turnbull PCB: Studies on the production of enterotoxins by Bacillus cereus. J Clin Pathol 29:941, 1976

Turnbull PCB: Anthrax. p. 364. In Smith GR, Easmon CR (eds): Bacterial Diseases. Topley and Wilson's Principles of Bacteriology, Virology and Immunity. Vol 3. Edward Arnold. Sevenoaks, England, 1990

Turnbull PCB, Kramer JM: Bacillus. p. 349. In Murray PR, et al. (eds): Manual of Clinical Microbiology, 6th ed. American Society for Microbiology, Washington, DC

Turnbull PCB, Kramer JM, Melling J: Bacillus. p. 187. In Parker MT, Duerden BI (eds): Systematic Bacteriology. Topley and Wilson's Principles of Bacteriology, Virology and Immunity. Vol. 2. Edward Arnold, Sevenoaks, England, 1990.

Chapter **17**

Clostridia: Sporeforming Anaerobic Bacilli

General Concepts

Clostridia are strictly anaerobic to aerotolerant sporeforming bacilli found in soil as well as in normal intestinal flora of man and animals. There are both gram-positive and gram-negative species, although the majority of isolates are gram-positive. Exotoxin(s) play an important role in disease pathogenesis.

Gas Gangrene and Related Clostridial Wound Infections

Clinical Manifestations

Patients may present with a wound infection. Severity varies from invasion of live tissue with systemic toxemia to relatively benign superficial contamination of already necrotic tissue.

Structure

The clostridia that cause gas gangrene are anaerobic, spore-forming bacilli, but some species may not readily sporulate, e.g., *C perfringens.*

Classification and Antigenic Types

Clostridial wound infections are typically polymicrobic. The primary pathogens are various clostridial species, including *C perfringens, C novyi, C septicum,* and others.

Pathogenesis

Wounds are contaminated by clostridia from the environment or the host's normal flora. The anaerobic tissue environment facilitates replication of clostridia and secretion of toxins.

Host Defenses

Host defenses are essentially absent. There is little, if any, innate immunity.

Epidemiology

Clostridial wound infections are found worldwide. Clostridia are ubiquitous in the soil and in the normal microbial flora of humans and animals.

Diagnosis

These infections are diagnosed by recognition of a characteristic lesion coupled with tissue Gram stains and bacterial culture.

Control

Wound infections are controlled by administration of antimicrobial agents (e.g., penicillin, chloramphenicol) coupled with tissue debridement (for more severe forms of clostridial wound infections).

Tetanus and *Clostridium Tetani*

Clinical Manifestations

Tetanus is characterized by twitching of muscles around a wound, pain in neck and jaw muscles (trismus), and around the wound. Patients have no fever, but sweat profusely and exhibit muscle rigidity and spasms.

Structure

These organisms are bacilli with terminal spores.

Classification and Antigenic Types

C tetani is the only species. There are no serotypes.

Pathogenesis

The infection is initiated as a result of contamination of a wound with *C tetani*. The anaerobic tissue environment facilitates *C tetani* replication and secretion of exotoxins. A spasmogenic toxin, tetanospasmin, fixes to inhibitory neurons and blocks the release of neurotransmitters, glycine and gamma-aminobutyric acid.

Host Defenses

Host defenses are essentially absent. There is little, if any, inate immunity and the disease does not produce immunity in the patient. Active immunity follows vaccination with tetanus toxoid.

Epidemiology

C tetani is found worldwide. Ubiquitous in soil, it is occasionally found in intestinal flora of humans and animals.

Diagnosis

Diagnosis is primarily by the clinical symptoms (above). The wound may not be obvious. Furthermore, *C tetani* is recovered from only one-third of all implicated wounds.

Control

The administration of tetanus toxoid is a preventive measure. *C tetani* infection is treated with antimicrobial agents (metronidazole or penicillin) and by local wound debridement. Other measures include tetanus immunoglobulin and supportive therapy.

Botulism and *Clostridium Botulinum*

Clinical Manifestations

These infections may have early gastrointestinal symptoms. The cranialnerves are initially affected, followed by descending, symmetric paralysis of motor nerves, with critical involvement of the respiratory tree. Muscle paralysis may occur.

Structure

These organisms are bacilli with oval, subterminal spores.

Classification and Antigenic Types

C botulinum consists of several biochemically distinct groups of organisms that produce botulinum toxin. Seven types of neurotoxins are designated A, B, C, D, E, F, and G, some of which have been shown to be encoded on bacteriophage DNA.

Pathogenesis

There are three forms: (1) adult botulism, caused by ingestion of preformed toxin in food; (2) infant botulism, in which the organism replicates and secretes toxin in the intestinal tract; and (3) wound botulism, in which the organism replicates in the wound and secretes toxin. Toxin binds to neuromuscular junctions of parasympathetic nerves and interferes with acetylcholine release, causing flaccid muscle paralysis.

Host Defenses

No host defenses are known.

Epidemiology

C botulinum is distributed worldwide, and is ubiquitous in soil. Improper heating of canned foods is a major factor in botulism food poisoning.

Diagnosis

Diagnosis is from the clinical symptoms (above), especially gastrointestinal and neurological symptoms, coupled with laboratory confirmation. A finding of normal spinal fluid helps to eliminate the possible diagnosis of numerous other central nervous system disorders.

Control

The best means of control is to eliminate the toxin source via proper food handling. Once the food poisoning is diagnosed, treatment measures should include an attempt to neutralize unbound toxin. Supportive care is of primary importance.

Antibotic-Associated Diarrhea, Pseudomembranous Colitis, and *Clostridium Difficile*

Clinical Manifestations

Patients can present with a spectrum of disease that varies from uncomplicated antibiotic-associated diarrhea to antibiotic-associated pseudomembranous colitis that may be fatal.

Structure

This species consists of bacilli with large, oval, subterminal spores.

Classification and Antigenic Types

C difficile is the only species. There are no defined serotypes. Toxigenic and nontoxigenic strains exist. The former produce varying amounts of toxin A (enterotoxin) and toxin B (cytotoxin).

Pathogenesis

Broad spectrum antibiotic therapy eliminates much competing normal flora, permitting intestinal overgrowth of toxigenic *C difficile.*

Host Defenses

There are no defined host defenses.

Epidemiology

C difficile is a component of the normal intestinal flora of a small percentage of healthy adults and of a relatively largepercentage of healthy neonates. It also may be found in the environment, especially in hospitals.

Diagnosis

The presence of antibiotic therapy, diarrhea, and pseudomembranes by colonoscopy help establish the severity of disease, coupled with the demonstration of organisms and/or toxin in feces.

Control

Metronidazole and vancomycin should be used therapeutically. However, relapses can occur. Supportive therapy may be needed.

Other Pathogenic Clostridia

Clostridium perfringens causes food poisoning and necrotizing enteritis. *C sordellii* causes bacteremia, endometritis and nonbacteremic infections. *C septicum* is correlated with the presence of cancer. *C tertium* is associated with bacteremia.

INTRODUCTION

Of the anaerobes that infect humans, the clostridia are the most widely studied. They are involved in a variety of human diseases, the most important of which are gas gangrene, tetanus, botulism, pseudomembranous colitis and food poisoning. In most cases, clostridia are opportunistic pathogens; that is, one or more species establishes a nidus of infection in a particular site in a compromised host. All pathogenic clostridial species produce protein exotoxins (such as botulinum and tetanus toxins) that play an important role in pathogenesis.

Most generalizations about Clostridium have exceptions. The clostridia are classically anaerobic rods, but some species can become aerotolerant on subculture; a few species (*C carnis, C histolyticum, and C tertium*) can grow under aerobic conditions. Most species are Gram-positive, but a few are Gram-negative. Also, many Gram-positive species easily lose the Gram reaction, resulting in Gram-negative cultures.

The clostridia form characteristic spores, the position of which is useful in species identification; however, some species do not sporulate unless exposed to exacting cultural conditions. Many clostridia are transient or permanent members of the normal flora of the human skin and the gastrointestinal tracts of humans and animals. Unlike typical members of the human bacterial flora, most clostridia can also be found worldwide in the soil.

Because clostridia are ubiquitous saprophytes, many isolated from clinical specimens are accidental contaminants and not involved in a disease process. Because these organisms are normally found on the skin, even a pure culture of clostridia isolated from blood may have no clinical significance. In determining the importance of a clinical isolate of clostridia, the clinician should consider the frequency of isolation of the species, the presence of other microbes of pathogenic potential, and the clinical symptoms of the patient. Many clostridial infections can be controlled by antibiotic therapy (e.g., penicillin, chloramphenicol, vancomycin, metronidazole) accompanied, in some cases, by tissue debridement. Antitoxin therapy and toxoid immunization are clearly useful in some clostridial infections, such as tetanus.

Gas Gangrene and Related Clostridial Wound Infections

Clinical Manifestations

Clostridial wound infections may be divided into three categories: gas gangrene or clostridial myonecrosis, anaerobic cellulitis, and superficial contamination. Gas gangrene can have a rapidly fatal outcome and requires prompt, often severe, treatment. The more common clostridial wound infections are much less acute and require much less radical treatment; however, they may share some characteristics with gas gangrene and must be included in the differential diagnosis.

Gas gangrene is an acute disease with a poor prognosis and often fatal outcome (Fig. 18-1). Initial trauma to host tissue damages muscle and impairs blood supply. This lack of oxygenation causes the oxidation-reduction potential to decrease and allows the growth of anaerobic clostridia. Initial symptoms are generalized fever and pain in the infected tissue. As the clostridia multiply, various exotoxins (including hemolysins, collagenases, proteases, and lipases) are liberated into the surrounding tissue, causing more local tissue necrosis and systemic toxemia. Infected muscle is discolored (purple mottling) and edematous and produces a foul-smelling exudate; gas bubbles form from the products of anaerobic fermentation. As capillary permeability increases, the accumulation of fluid increases, and venous return eventually is curtailed. As more tissue becomes involved, the clostridia multiply within the increasing area of dead tissue, releasing more toxins into the local tissue and the systemic circulation. Because ischemia plays a significant role in the pathogenesis of gas gangrene, the muscle groups most frequently involved are those in the extremities served by one or two major blood vessels.

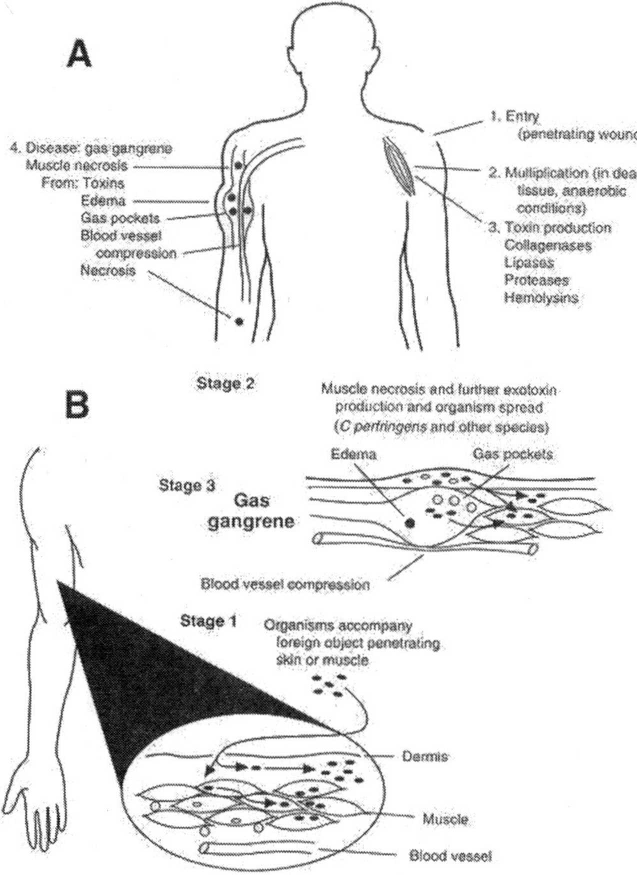

Figure 18-1 Pathogenesis of gas gangrene caused by *C perfringens*. A = macroscopic, B = microscopic.

Clostridial septicemia, although rare, may occur in the late stages of the disease. Severe shock with massive hemolysis and renal failure is usually the ultimate cause of death. The incubation period, from the time of wounding until the establishing of gas gangrene, varies with the infecting clostridial species from 1 to 6 days, but it may be as long as 6 weeks. Average incubation times for the three most prevalent infecting organisms are as follows: *C perfringens*, 10-48 hours; *C septicum*, 2-3 days; and *C novyi*, 5-6 days. Because the organisms need time to establish a nidus of infection, the time lag between wounding and the appropriate medical treatment is a significant factor in the initiation of gas gangrene.

Like gas gangrene, clostridial cellulitis is an infection of muscle tissue, but here the infecting organisms invade only tissue that is already dead; the infection does not spread to healthy, undamaged tissue. Clostridial cellulitis has a more gradual onset than gas gangrene and does not include the systemic toxemia associated with gas gangrene. Pain is minimal, and although only dead tissue is infected, the disease can spread along the planes between muscle groups, causing the surrounding tissue to appear more affected than it actually is. Anaerobic cellulitis may cause formation of many gas bubbles, producing infected tissue that looks similar to the gaseous tissue of gas gangrene. Some tissue necrosis does occur, but it is caused by decreased blood supply and not invasion by the infecting organism. With adequate treatment, anaerobic cellulitis has a good prognosis.

Superficial contamination, the least serious of the clostridial wound infections, involves infection of only necrotic tissue. Usually, the patient experiences little pain, and the process of wound healing proceeds normally; however, occasionally an exudate may form and the infection may interfere with wound healing. Superficial wound contamination caused by clostridia usually involves *C perfringens*, with staphylococci or streptococci, or both, as frequent co-isolates.

Structure

The clostridia that cause gas gangrene are anaerobic, spore-forming bacilli, but some species may not readily sporulate, e.g., *C perfringens*.

Classification and Antigenic Types

Clostridial wound infections usually are polymicrobic because the source of wound contamination (feces, soil) is polymicrobic. In gas gangrene and anaerobic cellulitis, the primary pathogen can be any one of various clostridial species including *C perfringens* (80%), *C novyi* (40%), *C septicum* (20%), and, occasionally, *C bifermentans*, *C histolyticum*, or *C fallax*. Other bacterial isolates may be any of a wide number and variety of organisms (for example, *Proteus, Bacillus, Escherichia, Bacteroides, Staphylococcus*). The distinctive or unique properties of the causative agents of gas gangrene are difficult to list; morphologic characteristics and biochemical reactions vary among these species, and a reliable laboratory manual should be consulted for their proper identification. Isolation of 107 or more clostridia per milliliter of wound exudate is strong evidence for a clostridial wound infection.

The most frequently isolated pathogen, *C perfringens*, has five types, designated A, B, C, D, and E. Each of these types produces a semi-unique spectrum of protein toxins. Alpha-toxin (a lecithinase, also called phospholipase-C) and theta-toxin (oxygen-labile cytolysin) are both considered important in the disease pathology. Alpha-toxin is lethal and necrotizing; it lyses cell membrane lecithins, disrupting cell membranes and causing cell death. Theta-toxin also contributes to rapid tissue destruction by several mechanisms. At the site of infection, theta-toxin acts as a cytolysin, promoting direct vascular injury; lower toxin concentrations activate polymorphonuclear leukocytes and endothelial cells, promoting distal vascular injury by stimulating leukocyte adherence to the endothelium. The result is leukostasis, thrombosis, decreased perfusion, and tissue hypoxia. Theta-toxin also mediates the production of shock through induction of inflammatory mediators such as platelet activating factor, tumor necrosis factor, interleukin 1 and interleukin 6.

Pathogenesis

All clostridial wound infections occur in an anaerobic tissue environment caused by an impaired blood supply secondary to trauma, surgery, foreign bodies, or malignancy. Contamination of the wound by clostridia from the external environment or from the host's normal flora produces the infection. The detailed pathogenesis of the disease is intimately associated with the clinical presentation as described above (Fig. 18-1).

Host Defenses

Host defenses against gas gangrene and other clostridial wound infections are mostly ineffective. Even repeated episodes of clostridial wound infection do not seem to produce effective immunity.

Epidemiology

Clostridial spores are ubiquitous in the soil, on human skin, and in the gastrointestinal tracts of humans and animals. Thus, the causative agents of clostridial wound infections are not environmentally restricted. Even operating theaters can be habitats for infecting clostridial organisms and spores. The incidence of clostridial wound infections has declined with the advance of prompt, adequate medical treatment. Historically, war casualties have had the greatest incidence of gas gangrene; however, the prompt evacuation and medical attention given United States casualties in the Vietnam war greatly decreased the incidence of gas gangrene in these soldiers, emphasizing the importance of prompt medical treatment.

Diagnosis

Diagnosis of clostridial wound infections is based on clinical symptoms coupled with Gram stains and bacterial culture of clinical specimens. Gas gangrene, once initiated, may spread and cause death within hours. By the time the typical lesions of gas gangrene are evident, the disease usually is firmly established and the physician must treat the patient on a clinical basis without waiting for laboratory confirmation. Characteristic lesions and the presence of large numbers of Gram-positive bacilli (with or without spores) in a wound exudate provide strong presumptive evidence. In contrast to tissue infections caused by Staphylococcus aureus, there is typically an absence of polymorphonuclear leukocytes at the site of infection, likely due to the presence of clostridial toxins.Spores are rare in cultures of *C perfringens*, the most common etiologic agent of these diseases. A commonly used laboratory test for presumptive identification of *C perfringens* is the Nagler reaction which detects the presence of alpha-toxin (phospholipase-C), one of the most prominent toxins produced by *C perfringens*. However, several other species of clostridia also have a positive Nagler reaction, and thus this test is not entirely specific for *C perfringens*.

Discussion of the differential diagnosis of clostridial wound infections appropriately includes streptococcal myositis, as this disease can be characterized by an edematous, necrotizing, often gaseous lesion. Like anaerobic cellulitis and superficial contamination with clostridia, streptococcal myositis is a relatively localized disease, but its later stages may include some systemic toxicity that mimics the toxemia of gas gangrene.

Control

Correction of the anaerobic conditions combined with antibiotic treatment form the basis for therapy. Penicillin is the drug of choice for all clostridial wound infections; chloramphenicol is a second-choice antibiotic. Successful treatment of the less severe forms of clostridial wound infections includes local debridement and antibiotic therapy; after these measures are taken, patient recovery usually proceeds along a steady, positive course. Treatment of gas gangrene includes radical surgical debridement coupled with high doses of antibiotics. Blood transfusions and supportive therapy for shock and renal failure also may be indicated.

The usefulness of gas gangrene antitoxin is currently a disputed matter. Some physicians maintain that the efficacy of this polyvalent antitoxin has been proved in the past, but better medical care now may have eliminated the need for its use. Others believe that because of insufficient data, antitoxin should be administered systemically as early as possible after diagnosis, and that the antitoxin should be injected locally into tissue that cannot be excised.

Obviously, prevention of wound contamination is the single most important factor in controlling clostridial wound infections. In the past, immunization has been considered a possible preventive measure for gas gangrene; however, several factors have discouraged the use of active immunization, including difficulty in preparing a suitable antigenic toxoid, availability of prompt wound treatment, and accessibility of effective therapeutic agents.

Tetanus and *Clostridium Tetani*

Clinical Manifestations

Tetanus is a severe disease caused by the toxin of *C tetani* (Fig. 18-2). This organism grows in a wound and secretes a toxin that invades systemically and causes muscle spasms. The initial symptom is cramping and twitching of muscles around a wound. The patient usually has no fever but sweats profusely and begins to experience pain, especially in the area of the wound and around the neck and jaw muscles (trismus). Portions of the body may become extremely rigid, and opisthotonos (a spasm in which the head and heels are bent backward and the body bowed forward) is common. Complications include fractures, bowel impaction, intramuscular hematoma, muscle ruptures, and pulmonary, renal, and cardiac problems.

Structure and Classification

C tetani is an anaerobic gram-positive rod that forms terminal spores, giving it a characteristic tennis racquet appearance. Some strains do not sporulate readily, and spores may not appear until the third or fourth day of culture. Most strains are motile with peritrichous flagella; colonies often swarm on agar plates, but some strains are nonflagellated and nonmotile. The presence of *C tetani* should be suspected on isolation of a swarming rod that produces indole and has terminal spherical spores, but does not produce acid from glucose. Toxigenic *C tetani* contains a plasmid that produces a toxin called tetanospasmin, but nontoxigenic strains also exist. Tetanospasmin is responsible for the infamous toxemia called tetanus. The two animal species most susceptible to this toxemia are horses and humans.

Pathogenesis

As with all clostridial wound infections, the initial event in tetanus is trauma to host tissue, followed by accidental contamination of the wound with *C tetani* (Fig. 18-2). Tissue damage is needed to lower the oxidation-reduction potential and provide an environment suitable for anaerobic growth. Once growth is initiated, the organism itself is not invasive and remains confined to the necrotic tissue, where the vegetative cells of *C tetani* elaborate the lethal toxin. The incubation period from the time of wounding to the appearance of symptoms varies from a few days to several weeks, depending on the infectious dose and the site of the wound (the more peripheral the wound, the longer the incubation time).

Tetanus can be initiated in two different ways, resulting in either generalized or local tetanus. In generalized tetanus (also called descending tetanus), all of the toxin cannot be absorbed by local nerve endings; therefore, it passes into the blood and lymph with subsequent absorption by motor nerves. The most susceptible centers are the head and neck; the first symptom is usually trismus (lockjaw), with muscle spasms descending from the neck to the trunk and limbs. As the disease progresses, the spasms increase in severity, becoming very painful and exhausting. During spasms, the upper airway can become obstructed, resulting in respiratory

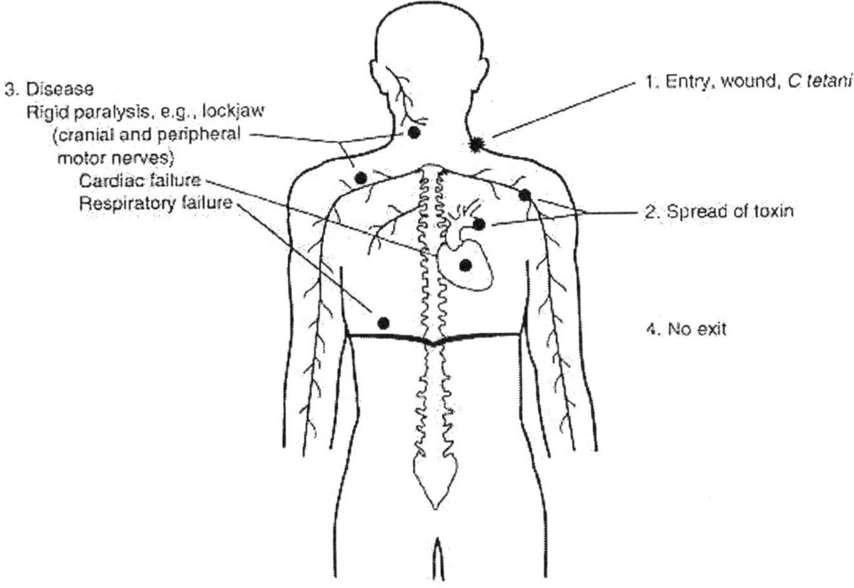

Figure 18-2 Pathogenesis of tetanus caused by *C tetani.*

failure. Spasms often are initiated by environmental stimuli that may be as insignificant as the flash of a light or the sound of a footstep. In the localized form of tetanus (also called ascending tetanus), toxins travel along the neural route (peripheral nerves), causing a disease confined to the extremities and seen most often in inadequately immunized persons. Localized tetanus may last for months but usually resolves spontaneously. Another unusual form of tetanus is called cephalic tetanus which results from head wounds and affects the face, most commonly the muscles innervated by lower cranial nerves. Curiously, cephalic tetanus can occur in fully immunized persons; the outcome is typically poor, but mild cases (often associated with otitis media) have more favorable outcomes. Neonatal tetanus is seen in newborns when the mother lacks immunity and the umbilical stump becomes contaminated with *C tetani* spores.

C tetani actually produces two toxins: tetanolysin, a hemolysin that is inactivated by cholesterol and has no role in pathogenesis, and tetanospasmin, a spasmogenic toxin responsible for the classical symptoms of the disease.

The actions of tetanospasmin are complex and involve three components of the nervous system: central motor control, autonomic function, and the neuromuscular junction. Toxin enters the nervous system primarily through the neuromuscular junction of alpha motor neurons. The toxin is then transported to the other neurons, most importantly presynaptic inhibitory cells, where it is no longer accessible to be neutralized by antitoxin. (This retrograde transport to the central nervous system is similar to that utilized by some viruses, such as herpes virus and rabies.) The toxin also spreads hematogenously, but it still must enter the central nervous system via retrograde transport from peripheral neuronal processes. Once the toxin gains access to inhibitory neurons, it blocks the release of the neurotransmitters glycine and gamma-aminobutyric acid. The absence of this inhibition permits the simultaneous spasms of both agonist and antagonist muscles, producing muscle rigidity and convulsions. Tetanospasmin also acts on the autonomic nervous system and is associated with elevated plasma catacholamine levels; respiratory failure is a frequent complication of the disease. Peripherally, there is a failure of transmission at the neuromuscular junction, involving defective release of acetylcholine in a manner similar to that seen with *C botulinum.* Tetanospasmin may be as potent as the toxin of *C botulinum;* as little as 130 μg constitutes a lethal dose for humans. In untreated tetanus, the fatality rate is 90% for the newborn and 40% for adults. However, with aggressive hospital care, these fatality rates can be substantially reduced. The ultimate cause of death is usually pulmonary or cardiac failure.

Host Defenses

Although, there are scattered reports that tetanus antibodies can be acquired by natural, presumably enteric, infection with *C tetani,* innate immunity to tetanus toxin does not typically exist. In addition, one or more episodes of tetanus do not produce immunity to future attacks. The reason for the lack of immune response may be two fold: the toxin is potent, and the amount released may be too small to trigger immune mechanisms but still be enough to cause symptoms and, because the toxin binds firmly to neural tissue, it may not interact effectively with the immune system.

Epidemiology

C tetani can be isolated from the soil in almost every environment throughout the world. The organism can be found in the gastrointestinal flora of humans, horses, and other animals. Isolation of *C tetani* from the intestinal flora of horses, coupled with the high frequency of equine tetanus, led to the erroneous assumption that the horse was the animal reservoir of *C tetani.*

Generalized outbreaks of tetanus do not occur, but certain populations can be considered at risk. Historically, wounded soldiers have had a high incidence of tetanus, but this phenomenon declined with widespread use of immunizations. Umbilical tetanus (tetanus neonatorum) usually is a generalized, fulminating, fatal disease that occurs with the neonates of unimmunized mothers who have given birth under unsanitary conditions. In the United States, intravenous drug abusers have become another population with an increasing incidence of clinical tetanus. One million cases of tetanus occur annually in the world. Tetanus is rare in most developed countries. The United States has about one case per million per year, most often seen in the elderly with declining immune status due to failure to receive timely tetanus booster vaccinations. In some less developed countries, tetanus is still one of the ten leading causes of death, and neonatal tetanus accounts for approximately one-half of the cases worldwide. In less developed countries, approximate mortality rates remain 85% for neonatal tetanus and 50% for nonneonatal tetanus. This is an unfortunate situation because with adequate immunization, tetanus is a completely preventable disease.

Diagnosis

Diagnosis of tetanus is obvious in advanced cases; however, successful treatment depends on early diagnosis before a lethal amount of toxin becomes fixed to neural tissue. The patient should be treated on a clinical basis without waiting for laboratory data. *C tetani* can be recovered from the wound in only about one-third of the cases, and a wound is not even evident in 10-20% of cases. It is important for the clinician to be aware that toxigenic strains of *C tetani* can grow actively in the wound of an immunized person, but the presence of antitoxin antibodies prevents initiation of tetanus. Also, because tetanus is commonly found in the soil, the mere presence of tetanus in a wound does not imply that the organism is actively replicating and secreting toxin.

Numerous syndromes, includingrabies and meningitis, have symptoms similar to those of tetanus and must be considered in the differential diagnosis. Ingestion of strychnine (found in rat poison) can cause symptoms that closely resemble those of generalized tetanus. Trismus can occur in encephalitis, phenothiazine reactions, and diseases involving the jaw.

Control

Injections of tetanus toxoid are prophylactic. Currently, booster doses are recommended only every 10 years by the CDC. More frequent boosters are unnecessary and may cause local reactions resembling the Arthus phenomenon or a delayed hypersensitivity reaction. It has been noted that, because of their immunodeficiency state, AIDS patients may not respond to prophylactic injections of tetanus toxoid. An antibody titer above 0.01 international units (IU) per ml is usually considered protective. Human tetanus immunoglobulin (HTIG) in a dose of 250 IU intramuscularly should be considered for those with questionable immune status.

Treatment of diagnosed tetanus has a number of aspects. The offending organism must be removed by local debridement, after the patient's spasms are controlled by benzodiazepines. Penicillin or metronidazole is usually administered to kill the bacteria, but may not be a necessary adjunct in therapy. Although, penicillin has been historically considered to be the drug of choice, it has been speculated that penicillin could have an adverse effect by acting synergistically with tetanospasmin. Metronidazole is currently recommended, and there is some evidence that it is associated with an improved prognosis. HTIG is injected intramuscularly: dosage recommendations vary from 500 IU in a single intramuscular injection to 3000-6000 IU injected intramuscularly in several sites. Supportive measures, such as respiratory assistance and intravenous fluids, are often critical to patient survival. Recommended treatment includes benzodiazepines, such as diazepam (Valium). Analgesics that will not cause respiratory depression should be used, and include codeine, meperidine (Demerol), and morphine. Adequate nutritional support should be provided and should consider that the patient's nutritional needs are extraordinarily great.

In cases of clean, minor wounds, tetanus toxoid should be administered only if the patient has not had a booster dose within the past 10 years. For more serious wounds, toxoid should be administered if the patient has not had a booster dose within the past 5 years. All patients who have a reasonable potential for contracting tetanus should receive injections of tetanus immunoglobulin, including those recovering from diagnosed cases of tetanus.

Botulism and *Clostridium Botulinum*

Clinical Manifestations

Botulism is a disease caused by the toxin of a diverse group of clostridia called *C botulinum*. This neurotoxin characteristically causes a symmetrical, descending paralysis (Fig. 18-3). The symptoms of botulism can occur in both the nervous system and the alimentary tract of the patient. Therefore, many diseases enter into the differential diagnosis, including pharyngitis, gastroenteritis, sepsis, intestinal obstruction, myasthenia gravis, encephalitis, muscular dystrophy, electrolyte imbalance, meningitis, poliomyelitis, cerebrovascular accident, Guillain-Barré

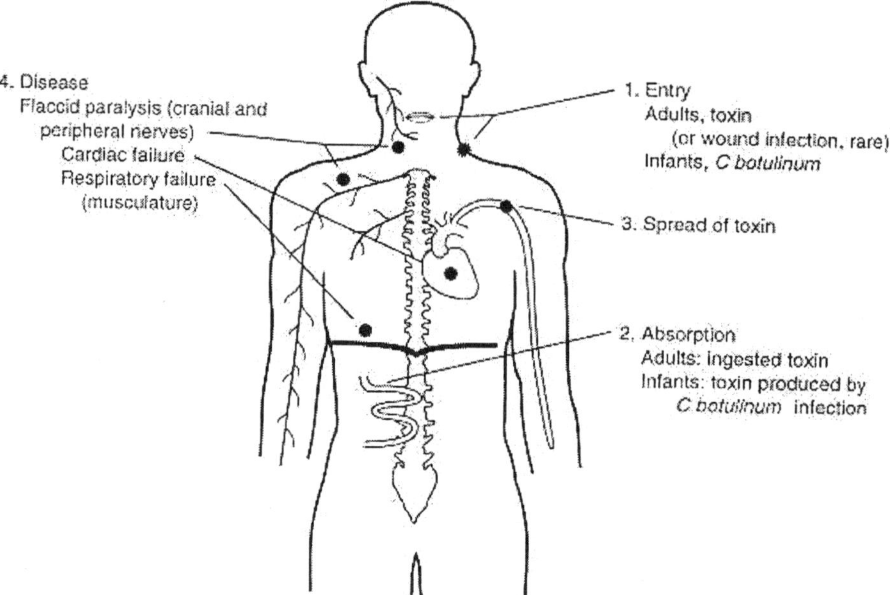

Figure 18-3 Pathogenesis of botulism caused by *C botulinum*.

syndrome, chemical food poisoning, tick paralysis, Reye syndrome, hypothyroidism, heavy metal ingestion, carbon monoxide poisoning, and snake bite. For infant botulism, additional syndromes enter into the differential diagnosis: failure to thrive, acute infantile polyneuropathy, dehydration, and various hereditary and metabolic disorders. Infant botulism often is missed by physicians, but it always should be considered if any of the typical symptoms are present.

Structure, classification and antigenic Types

Unlike most species of bacteria, which comprise strains that have a close genetic relationship and similar cultural characteristics, the *C botulinum* "species" consists of several distinct groups of organisms that have a common name solely because they produce similar toxins. The name *C botulinum* is thus only a convenience that reflects the medical importance of the species. A strain of *C botulinum* usually produces only one of seven toxin types, designated A, B, C, D, E, F, and G. The toxins produced by *C botulinum* types C and D have been shown to be coded by the genetic material contained in bacteriophages that infect the bacteria. *C botulinum* types C and D can be interconverted by use of specific bacteriophage. Types C and D even can be transformed into *C novyi* by curing bacteria of the botulinum phage, and substituting a phage that codes for a *C novyi* toxin. To add to the confusion, a few strains of *C baratii* and *C butyricum* have been reported to secrete botulinum toxin. In addition, certain proteolytic strains of *C botulinum* are indistinguishable from *C sporogenes* except by toxin assay. Thus, the production of a pharmacologically similar neurotoxin is the single distinctive property of a clostridium that places it in the botulinum "species." All the organisms that produce this toxin are anaerobic rods, Gram-positive, and are motile by peritrichous flagella. Oval, subterminal spores are produced in extremely variable numbers, depending on the particular isolate and on the culture medium. Cultural reactions vary greatly, and the species includes highly proteolytic and nonproteolytic strains as well as saccharolytic and nonsaccharolytic strains.

Of the seven serologically distinct neurotoxins produced by *C botulinum* (A, B, C, D, E, F, and G), humans are most susceptible to types A, B, E, and F. Types C and D are most toxic for animals. Type G is rare, with only a few reported human cases. The toxins often are released from the bacteria as inactive proteins that must be cleaved by a protease to expose the active site. These proteases may be produced by the cell itself or may be in the body fluids of the infected host. Type A toxin is the most potent poison known; ingestion of only 10-8 grams of this toxin can kill a human. Put another way, the amount of toxin that could be held on the tip of a dissecting probe could kill 40 medical students.

Pathogenesis

The pathogenicity of *C botulinum* depends entirely on neurotoxin production (Fig. 18-3). In humans, these toxins cause disease in three ways: the well-known form of food poisoning results from ingestion of toxin in improperly preserved food; wound botulism, a rare disease, results from *C botulinum* growing in the necrotic tissue of a wound; and infant botulism is caused when the organism grows and produces toxin in the intestines of infants.

From its site of entry into the body, the toxin travels through the blood and lymphatic systems (and possibly the nervous system). It then becomes fixed to cranial and peripheral nerves, but exerts almost all of its action on the peripheral nervous system. The toxin appears to bind to receptor sites at the neuromuscular junctions of parasympathetic nerves, and inhibits the release of acetylcholine at peripheral cholinergic synapses. The result is flaccid muscular paralysis.

The cranial nerves are affected first, followed by a descending, symmetric paralysis of motor nerves. The early involvement of cranial nerves causes problems with eyesight, hearing, and speech. Double or blurred vision, dilated pupils, and slurred speech are common symptoms. Decreased saliva production causes a dryness of the mouth and throat, and swallowing may be painful. An overall weakness ensues, followed by descending paralysis with critical involvement of the respiratory tree. Death usually is caused by respiratory failure, but cardiac failure also can be the primary cause. Mortality is highest for type A, followed by type E, and then type B, possibly reflecting the affinities of the toxins for neural tissue: type A binds most firmly, followed by type E, then type B. Fatality rates are directly proportional to the infectious dose and inversely proportional to the incubation time of the disease.

Type A toxin is used therapeutically to treat a variety of conditions involving involuntary muscle spasms, including strabismus and certain focal dystonias. This therapy takes advantage the effect of the toxin as a specific muscle relaxant. The therapeutic toxin is a neurotoxin-hemagglutinin complex isolated from *C botulinum* cultures. The extreme potency of type A botulinum toxin requires extreme caution in using this compound as a therapeutic agent.

Food poisoning. In botulism food poisoning, the toxin is produced by the vegetative cells of *C botulinum* in contaminated food, and preformed toxin then is ingested with the contaminated food. The incubation time can vary from a few hours to 10 days, but most commonly is 18-36 hours.Only a small, but effective, percentage of the ingested toxin is absorbed through the intestinal mucosa; the remainder being eliminated in the feces. Gastrointestinal disturbances are early symptoms of the disease in about one-third of the patients with toxin types A or B, and in almost all of the cases involving type E toxin. These symptoms include nausea, vomiting, and abdominal pain. Diarrhea often is present, but constipation also may occur. Symptoms of toxemia then become apparent. No fever occurs in the absence of complicating infections.

Wound botulism. Wound botulism is a rare disease. The initial event is contamination of a wound by *C botulinum.* The organisms are not invasive and are confined to the necrotic tissue, where they replicate and elaborate the lethal neurotoxin. The incubation time varies from a few days to as long as 2 weeks. The only differences in the symptoms of wound botulism and food poisoning (in addition to a possibly longer incubation time) are that wound botulism lacks gastrointestinal symptoms, and a wound exudate or a fever, or both may be present. *C botulinum* may be present in a wound but creates no symptoms of botulism. There have been several recent reports of wound botulism in intravenous drug abusers, who are now considered a population at risk.

Infant botulism. In contrast to food poisoning with toxemia caused by ingestion of preformed toxin, infant botulism results from germination of spores in the gastrointestinal tract. Here vegetative cells replicate and release the botulinum toxin. It is unclear as to why spores can germinate and bacteria replicate in the infant intestine, but phenomenon apears to be related to the compositionof the intestinal flora of infants. Almost all reported cases have occurred in infants between 2 weeks and 6 months of age, with the median age of onset being 2 to 4 months. Toxins A or B are most frequently implicated. In infant botulism, the usual first indication of illness, constipation, is often overlooked. The infant then becomes lethargic and sleeps more than normally. Suck and gag reflexes diminish, and dysphagia often becomes evident as drooling. Later, head control may be lost, and the infant becomes flaccid. In the most severely affected babies, respiratory arrest can occur. Infant botulism can be lethal and is the likely etiologic agent in 4 to 15% of the cases of sudden infant death.

There are scattered reports that *C botulinum* can occasionally multiply and secrete toxin in the intestinal tracts of adults with an altered intestinal flora due to antibiotic therapy or achlorhydria.

Host Defenses

Host defenses against *C botulinum* are undefined. Some people can tolerate ingestion of botulinum toxin better than others. The reason for this phenomenon is obscure, but could be due to differences in the efficiency of uptake of the toxin from the intestine or in transporting the toxin to neural tissue. An attack of botulism does not produce effective immunity. The small amount of toxin in the circulation and its affinity for neural tissue probably prevent adequate amounts of toxin from interacting with the immune system.

Epidemiology

C botulinum spores are found worldwide in the soil (including in sea sediments) and in low numbers in the gastrointestinal tracts of some birds, fish, and mammals. In the United States, the most frequent isolate is type A, followed byB and E, with an occasional isolate of type F. In Europe, B is the most frequent isolate, whereas A is comparatively rare. Despite the worldwide occurrence of *C botulinum* in the environment, wound botulism is a comparatively rare disease.

Originally, botulism food poisoning was thought to be associated only with contaminated meat, especially sausage; however, it is now known that *C botulinum* can grow equally well in many types of food including vegetables, fish, fruits, and condiments. Home canning using inadequate sterilization techniques has been responsible for most cases of botulism during this century. The spores are heat resistant and can survive 100° C for hours, but the toxin is relatively heat labile. The toxin is usually produced at pH 4.8-8.5. However, even acid foods such as canned tomatoes have been responsible for several recent cases of botulism food poisoning. In addition, certain culture conditions have been shown to cause toxin production at pH values lower than 4.6. In general, germination of botulinum spores is favored in food kept at warm temperatures under anaerobic conditions for a long period of time.

C botulinum spores exist throughout the environment; all adults have likely ingested these spores with no ill effects. Because spores can cause poisoning in infants, obvious sources should be eliminated from the infant's environment and especially the infant's diet. Honey is the only dietary ingredient that has been implicated, and honey is no longer recommended for infants under 1 year of age. Most cases are not caused by ingesting honey, however, so this will not eliminate the disease. The other more common environmental sources of spores, such as soil and dust, are not so easily controlled.

Diagnosis

Although, all forms of botulism are difficult to diagnose, prompt diagnosis and treatment are crucial to patient survival. Laboratory tests offer little in establishing an initial diagnosis of botulism, and accordingly, the finding of a normal cerebrospinal fluid can help to eliminate many of the diseases concerned with central nervous system disorders. Differential diagnoses are myriad and include neurological as well as gastrointestinal disorders. The infant with botulism is typically afebrile with generalized weakness, a weak cry, pooling of oral secretions, and poor sucking ability. Constipation may precede the illness by several weeks. An electromyogram pattern of brief, small-amplitude overabundant motor reaction potentials often is seen.

Confirmation of the initial diagnosis rests on demonstrating toxin in the patient's feces, serum, or vomitus. In adult botulism, serum samples rarely yield type A toxin because of the strong affinity of this toxin for neural tissue. In infant botulism, circulating toxin can occasionally be found in the serum. Fecal samples are the best specimens for detecting toxin in botulism food poisoning or infant botulism because only a small percentage of ingested or in situ formed toxin is absorbed through the intestinal mucosa. Toxin may be excreted for days or even weeks following botulism food poisoning. Toxin is usually detected by its lethal effect in mice coupled with neutralization of this effect by specific antisera. In infants, the organism can usually be cultured from the stool.

Control

The best way to control botulism food poisoning is to use adequate food preservation methods and to heat all canned food before eating. Because botulinum toxin is heat-labile, boiling food for a few minutes will eliminate toxin contamination; however, the spores themselves are not destroyed by boiling, and proper canning procedures must be followed to kill clostridial spores.

Once a case of wound botulism or food poisoning has been diagnosed, therapy has four objectives: to eliminate the source of the toxin, to eliminate any unabsorbed toxin, to neutralize any unbound toxin with specific antitoxin, and to provide general supportive care.

Food Poisoning in Adults and Wound Botulism: In food poisoning, the unabsorbed toxin may be eliminated by stomach lavage and high enemas. Although, cathartics may be used to eliminate residual toxin, they may have adverse effects in patients with bowel paralysis. In wound botulism, debridement and antibiotic therapy with penicillin are used to eliminate the offending organism. Antibiotic therapy is of questionable value in food poisoning, but is advocated by those who believe the organism can replicate in the intestinal tract of adults.

For both food poisoning and wound botulism, antitoxin therapy is most effective if administered early; however, clear-cut evidence for the efficacy of antitoxin therapy exists for only type E toxin. Antitoxin is available from the Centers for Disease Control (Atlanta, GA) through State Health Departments; trivalent ABE botulinum antitoxin is currently recommended. Unfortunately, all antitoxins are equine preparations, so a significant percentage ofpatients experience reactions typical of anaphylaxis and serum sickness. Thus, before they receive antitoxin, all patients should be tested for sensitivity to horse serum. The most important aspect of treatment in botulism is close observation of the patient and availability of adequate facilities for immediate respiratory support. Respiratory failure may occur within minutes, and immediate respiratory assistance often saves the lives of patients with botulism toxemia. Due to improvements in supportive care, the mortality rate for botulism has been dramatically reduced from approximately 60% (in the 1940s) to 10%.

All cases of botulism food poisoning should be reported immediately to local, state, or federal authorities, who will then take steps to minimize the chance of an outbreak. All persons suspected of ingesting contaminated food should be closely observed. Antitoxin should be administered both to those with overt symptoms and to those who have definitely ingested contaminated food.

Infant Botulism. The most important aspect of treatment of infant botulism is meticulous supportivecare. Oral antibiotic therapyis not indicated because it may unpredictably alter the intestinal microecology and allow accidental overgrowth of *C botulinum.* Cathartics and enemas are also potentially dangerous. The value of human botulinum antitoxin is disputed, and there is not firm evidence to support its efficacy. The most significant aspect of therapy for infant botulism is supportive care. The infant should be kept under close supervision, with facilities for respiratory support immediately available. The fatality rate for infant botulism is surprisingly less than 5%.

Antibiotic-Associated Diarrhea, Pseudomembranous Colitis, and *Clostridium difficile.*

Clinical Manifestations

Clostridium difficile is a major nosocomial pathogen that causes a spectrum of intestinal disease from uncomplicated antibiotic-associated diarrhea to severe, possibly fatal, antibiotic-associated colitis. Diarrhea has come to be accepted as a natural accompaniment of treatment with many antibiotics. Although, this diarrhea usually causes only minor concern, it can evolve into a life-threatening enterocolitis.

Many antibiotics have been associated with diarrhea and with pseudomembranous colitis, including ampicillin, cephalosporins, clindamycin, and amoxicillin. Patients treated with clindamycin have a higher incidence of *C difficile* disease, but most cases are found in patients treated with other antibiotics because of the more widespread use of these agents. Occasionally, antineoplastic agents that alter the normal intestinal flora may also induce pseudomembranous colitis, with methotrexate most commonly implicated. Chemotherapy-associated *C difficile* disease may not be easily recognized due either to an absence of antibiotic therapy or due to the frequent concomitant use of antibiotics, obscuring true incidence of chemotherapy-associated *C difficile* disease.

Clinical symptoms of *C difficile* disease vary widely from mild diarrhea to severe abdominal pain accompanied by fever (typically >101°F) and severe weakness. Diarrhea is watery and usually nonbloody (Fig. 18-4), but approximately 5 to 10% of patients have bloody diarrhea. Fecal material typically contains excess mucus, and pus or blood may also be noted. Hypoalbuminemia and leukocytosis are common findings. Pathology involves only the colon where there may be disruption of brush border membranes followed by extensive damage to the mucosa. The disease may progress to a pseudomembranous colitis, possibly including intestinal perforation and toxic megacolon. There is a leukocytic infiltrate into the lamina propria accompanied by elaboration of a mixture of fibrin, mucus, and leukocytes, which can form gray, white, or yellow patches on the mucosa. These areas are called pseudomembranes; hence the common term pseudomembranous colitis. Pseudomembranes usually develop after 2-10 days of antibiotic treatment, but they may appear 1-2 weeks after all antibiotic therapy has stopped. Mortality varies, but may be as high as 10% in patients with pseudomembranous colitis. The ultimate cause of death often is difficult to determine, as most patients show a nonspecific deterioration over a period of weeks.

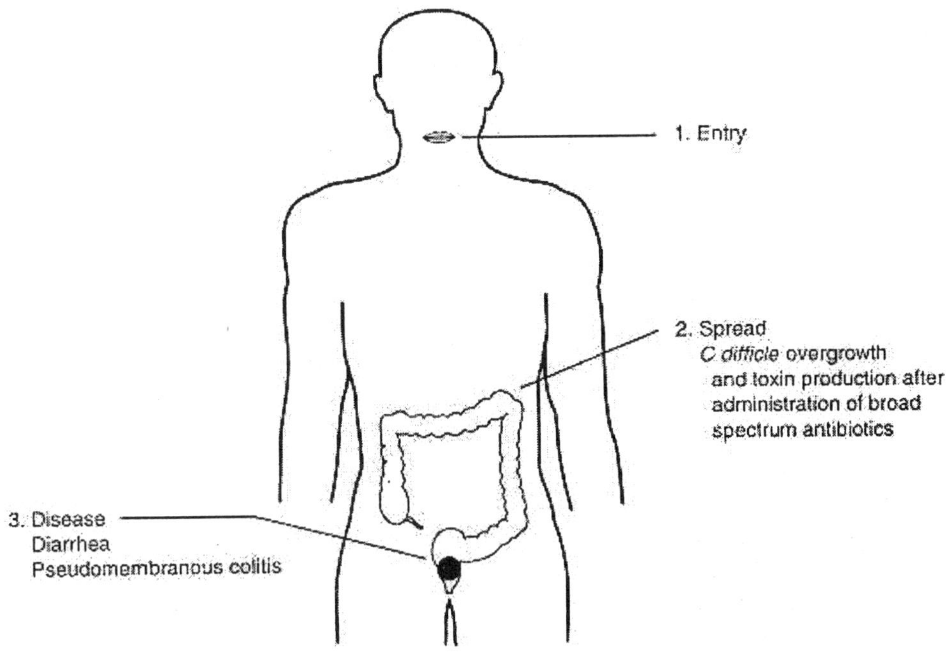

Figure 18-4 Pathogenesis of pseudomembranous colitis caused by *C difficile*.

The incidence of pseudomembranous colitis has been diminishing in recent years, most likely due to early diagnosis of the disease and prompt antimicrobial therapy. However, *C difficile* is now considered a major cause of diarrhea in hospitals and nursing homes. In most instances, once a patient develops antibiotic-associated diarrhea and *C difficile* organisms and/or toxin is detected in the stool, appropriate antimicrobial therapy is begun, and the symptoms are not allowed to progress to the formation of colonic pseudomembranes. Thus, in recent years the terms "*C difficile* diarrhea" and "*C difficile* disease" have come to be associated with a spectrum of diseases, including pseudomembranous colitis as well as diarrhea and colitis in the absence of pseudomembranes. The common factors in all of these diseases are the presence of diarrhea associated with antibiotic therapy and the recovery of *C difficile* organisms and/or toxin from the stool.

Etiologic Agent

C difficile is a slender, gram-positive bacillus that produces large, oval, subterminal spores. It is an anaerobe, and some strains are extremely sensitive to oxygen. *C difficile* is nonhemolytic and does not produce lecithinase or lipase reactions on egg yolk agar. It produces various tissue degradative enzymes, including proteases, collagenases, hyaluronidase, heparinase, and chondroitin-4-sulfatase. The products of fermentation are many and complex and include acetic, butyric, isovaleric, valeric, isobutyric, and isocaproic acids; however, only small amounts of each are produced.

Pathogenesis

C difficile disease is caused by the overgrowth of the organism in the intestinal tract, primarily in the colon (Fig. 18-4). The organism appears unable to compete successfully in the normal intestinal ecosystem, but can compete when normal flora are disturbed by antibiotics, allowing overgrowth of *C difficile.* This organism then replicates and secretes two toxins. Toxin A is an enterotoxin that causes fluid accumulation in the bowel, and it is a weak cytotoxin for most mammalian cells; toxin B is a potent cytotoxin. Nearly all toxigenic strains produce both toxins A and B. Highly toxigenic strains produce high levels of both toxins, while weakly toxigenic strains produce low levels of both toxins. Results from in vitro studies using cultured intestinal epithelial cells have indicated that toxin A causes necrosis, increased intestinal permeability, and inhibition of protein synthesis. Toxin A somehow affects phospholipase A2, resulting in the production of several arachidonic acid metabolites including prostaglandins and leukotrienes. Although, the exact mechanism of endocytosis is unclear, both toxins A and B are internalized by host cells, resulting in alterations in the actin-containing cytoskeleton. Toxin A is a chemotactic factor for granulocytes; both toxins A and B have effects on leukocytes that include alterations in actin cytoskeletal microfilaments, and induction of tumor necrosis factor, interleukin 1, and interleukin 6. These latter effects contribute to the inflammatory response associated with *C difficile* disease.

Both toxins A and B kill experimental animals, and both probably are involved in the pathology of disease. Toxin production causes diarrhea that may progress to pseudomembranous colitis, where the characteristic pseudomembranes are largely limited to the colon. In the intestinal tract, toxin A damages villous tips and brush border membranes, and may result complete in erosion of the mucosa. This tissue damage causes a viscous hemorrhagic fluid response. In contrast, toxin B does not have noticeable enterotoxic activity, but it is lethal when injected into experimental animals. Thus, it seems reasonable to speculate that, in humans, toxin B exerts its pathogenic effect following dissemination through a damaged gut wall to extraintestinal organs. It has been speculated that infants harboring high levels of intestinal toxins A and B are at risk for the systemic toxicity of toxin B if their intestinal barrier is compromised.

Host Defenses

Host defenses for *C difficile* disease are not completely understood, but it seems reasonable to assume that the best host defense against *C difficile* disease is maintenance of the stability of the normal intestinal flora. Certain prostaglandins are known to protect the stomach and small intestine from mucosal necrosis caused by harsh chemicals; and, in experimental animals, prostaglandins have been shown to prevent extensive mucosal damage from *C difficile* toxin. These prostaglandins are also produced by toxin A-induced activation of the arachidonic cascade by phospholipase A2. Production of specific neutralizing antibodies to toxins A and B may participate in host defense, and a specific intestinal secretory IgA response to toxin A is more evident in the colon than the upper intestinal tract, compatible with the colon as the primary site of intestinal disease. The intestinal tract responds to *C difficile* toxins by increased fluid production, by secretory IgA neutralization of toxin, and by mucus production, which may inhibit the attachment of the toxins to their putative receptor sites on intestinal epithelial cells.

Epidemiology

C difficile is a member of the normal intestinal flora of <3% of adults. The organism can be acquired as a nosocomial pathogen and a variable incidence of disease is noted in hospitals and nursing homes. This seems to be due in part to environmental contamination with *C difficile* spores, and in part to different patient populations in various institutions. Patients with *C difficile* diarrhea excrete large numbers of *C difficile* spores, and epidemiological studies have shown that the organism can reside on environmental surfaces as well as on the hands of health care workers. Healthy adults do not carry significant numbers of the organism in their intestinal tracts, but healthy infants may have large numbers of these organisms in their feces. Most studies report a high carriage rate of approximately 50% in neonates, although some studies report a carriage rate of 0 to 6%, likely due to differences in environmental exposure to the organism. The toxins also are present in these infants' stools, and the same amounts of toxins are associated with disease in adults. The toxins typically have no adverse effect in infants, but confound the diagnosis of *C difficile* disease. There is circumstantial evidencesupporting the theory that infants do not develop disease because they lack specific intestinal receptors for *C difficile* toxins. In recent years, *C difficile* as also emerged as one of the causes of chronic diarrhea in AIDS patients.

Diagnosis

It is often difficult to distinguish *C difficile* disease from other intestinal diseases, including ulcerative colitis and Crohn's disease. Diagnosis of *C difficile* disease includes the presence of diarrhea associated with antibiotic therapy in the preceding 4 to 6 weeks, and the recovery of *C difficile* organisms and/or toxin from the stool. However, the isolation of toxigenic *C difficile* from patients is often not a definitive diagnosis because other enteric pathogens are usually not excluded. Many cases of severe diarrhea are caused not by *C difficile*, but are caused by other enteric pathogens such as Campylobacter spp, Salmonella spp, Shigella spp, toxigenic strains of Escherichia coli, etc. Moreover, antimicrobial therapy increases the likelihood of isolating *C difficile* from the fecal flora: *C difficile* can be isolated from the feces of approximately 20 to 40% of asymptomatic hospitalized patients who are receiving antimicrobial therapy. Despite these caveats, *C difficile* is likely responsible for 25% of cases of antibiotic-associated diarrhea and colitis. Diagnosis of pseudomembranous colitis requires demonstration of pseudomembranes by colonoscopy, and *C difficile* can be isolated from the stools of almost all patients with this disease.

A good selective agar medium is used for the isolation of *C difficile* from stool. Toxin detection is also used for diagnosis. Although, the most appropriate test for toxin detection remains controversial, a cellularcytotoxicity test remains the "gold standard"; here, filter sterilized fecal extract (or filter-sterilized broth containing a pure culture of *C difficile*) is added to a monolayer of cultured mammalian cells resulting in a cytopathic effect that is neutralized with specific antiserum. Unfortunately, this test is time-consuming and cumbersome. A rapid latex agglutination test is widely used, but this test is not specific, does not detect toxin A or B, and often results in false negative or false positive reactions. Enzyme-linked immunoassays can be used to detect both toxin A and toxin B, and these tests are useful for diagnosis of *C difficile* disease.

Control

In many cases, symptoms resolve 1-14 days after the offending antibiotic is discontinued, and antibiotic treatment is not needed. Vancomycin or metronidazole are the antibiotics of choice to treat active disease. Oral vancomycin is the "gold standard," and metronidazole is most often used to treat milder infections. Some claim that metronidazole should be considered the drug of choice in all but the most severe cases, based on relative cost of the two drugs and based on prevention of development of vancomycin resistance in enteric bacteria. *C difficile* is susceptible to both of these antimicrobial agents, but relapses occur in 15 to 20% of patients. Some patients have had many repeated relapses. Constipating agents, such as atropine diphenoxylate (Lomotil) or codeine, should not be used. Supportive therapy is needed to compensate for the often severe fluid and electrolyte loss. Health care workers caring for patients infected with C. difficile should wear gloves and strictly adherc to proper hand washing procedures.

Other Pathogenic Clostridia Food Poisoning and *Clostridium perfringens*

C perfringens is a major cause of food poisoning in the United States. The disease results from ingestion of a large number of organisms in contaminated food, usually meat or meat products. Food poisoning usually does not occur unless the food contains at least 106-107 organisms per gram. The spores are ubiquitous and, if present in food, can be triggered to germinate when the food is heated. Some heat-sensitive strains do not need heating to germinate. After germination, the number of organisms quickly increases in warm food because the generation time can be extremely short (minutes) and bacterial multiplication occurs over a wide temperature range. The location of *C perfringens* enterotoxin within the bacterial cell is controversial; some investigators claim that the enterotoxin is localized in the bacterial cytoplasm and others claim that it is associated with the spore coat. Regardless, food poisoning results from the ingestion food contaminated with enterotoxin-producing *C perfringens*. The enterotoxin directly affects the permeability of the plasma membrane of mammalian cells.

C perfringens type A is the usual causative agent, and serotyping is necessary and available for epidemiologic studies. Incubation time is 8-22 hours after ingestion of contaminated food, with a mean of 14 hours. Symptoms include diarrhea, cramps, and abdominal pain. Fever, nausea, and vomiting are rare, and the disease lasts only about 24 hours. The organism and its enterotoxin usually can be isolated from the feces of infected persons. The mortality rate is essentially zero, but elderly and immunologically compromised patients should be closely supervised.

Necrotizing Enteritis and *Clostridium perfringens*

Necrotic enteritis in humans has not been well documented. In adults, the disease appears to result from ingesting large amounts of food contaminated with *C perfringens*, usually type C. It generally follows ingestion of a large meal, implicating bowel distention and bacterial stasis as contributing factors. The intestinal pathology varies considerably, and may include sloughing of intestinal mucosa, submucosa, and mesenteric lymph nodes. Intestinal perforations occur frequently. The best-documented cases of this disease involve the natives of New Guinea, who develop necrotic enteritis ("pig-bel") after eating large quantities of improperly cooked pork that has been contaminated with the bowel contents of the animal. The course of the disease is fulminate, and the mortality rate is high. Scattered cases of necrotizing enteritis with *C perfringens* as the prominent bacterial isolate have been reported in Western countries. In these cases, controversy exists concerning whether *C perfringens* is a primary invader, an accidental contaminant, or an opportunistic pathogen.

Some evidence suggests that acute necrotizing enterocolitis of infants may be caused by a clostridium, but definitive evidence is lacking. The theory is supported by the fact that pneumatosis cystoides intestinalis, a syndrome that can be caused by *C perfringens*, often is present in cases of acute necrotizing enterocolitis of infants. In addition, *C perfringens*, *C butyricum*, *C difficile* and other clostridial species are often isolated in cases of neonatal necrotizing enterocolitis, but a clear pathogenic role for clostridia is yet to be elucidated.

Bacteremia, endometritis and nonbacteremic infections with *Clostridium sordellii*

C sordellii is part of the normal intestinal flora of humans. The organism produces several exotoxins including toxins serologically related to the toxins of *C difficile*. There are scattered reports in the literature of *C sordellii* wound infections, most of which involve significant trauma. *C sordellii* has been occasionally implicated in bone and joint infections, in pulmonary infections, in bacteremia, and in fulminate endometritis. Because many clinical laboratories fail to speciate clostridial pathogens, the pathogenic potential of *C sordellii* is likely underestimated.

Malignancy and *Clostridium septicum*

C septicum is a spindle-shaped rod that is motile in young cultures. The organism produces toxins designated alpha, beta, gamma, and delta; the alpha toxin is necrotizing and lethal for mice. Whether *C septicum* is a member of the host's normal flora or whether it takes advantage of a compromised host is uncertain. The organism is not strongly invasive, but has been associated with gas gangrene. Fewer than 200 cases of invasive disease have been reported, but the majority have a malignancy somewhere in the body. The most frequent association is with colorectal cancer, but other types of malignancies have been noted, including leukemia, lymphoma, and sarcoma. In one survey of *C septicum* bacteremia, 49 of 59 (83%) cases had an underlying malignancy and, in 28 of these cases, the portal of entry appeared to be the distal ileum or the colon. Diabetes mellitus is seen in about 20% of cases. In collective review of 162 cases of nontraumatic *C septicum* infection, 81% of the patients had malignant disease; in contrast, other clostridial species are associated with malignancy in approximately 10% of cases. Thus, in the absence of an overt infection, isolation of *C septicum* should alert the physician to the possible presence of a malignancy, most likely in the ileum or the colon. Immediate antibiotic therapy is indicated because most patients die quickly of the infection if not treated. Penicillin is the antibiotic of choice, but chloramphenicol, carbenicillin, and cephalothin also have been used successfully.

Bacteremia and *Clostridium tertium*

C tertium is an aerotolerant clostridium that is usually considered nonpathogenic. However, there are scattered reports of this organism causing bacteremia. Most cases have involved neutropenic patients, and the gastrointestinal tract appears to be the source of the infection. It is possible that this organism causes many more cases of bacteremia than is currently appreciated. The aerotolerant nature of C *tertium* may result in its misidentification as a Bacillus species.

Other clostridia with potential clinical significance

Clostridium butyricum, C clostrdioforme, C innocuum, and C ramosum are isolated with some frequency from clinical specimens and may have an unrecognized clinical significance. These species are often resistant to clindamycin and cephalosporins. *C ramosum* is usually listed with the ten anaerobic species most frequently isolated from clinical specimens. *C ramosum* frequently is misidentified, as the Gram reaction is lost easily, and spores are difficult to detect. *C clostridioforme* stains Gram-negative and forms characteristic football-shaped cells,rarely sporulates, and may be easily misidentified as Bacteroides sp or Fusobacterium sp.

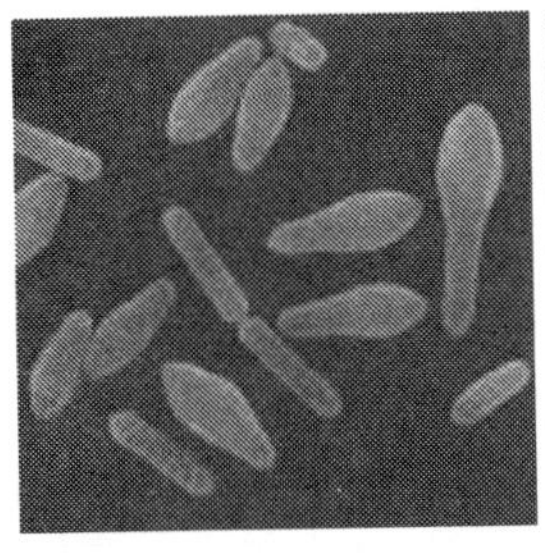
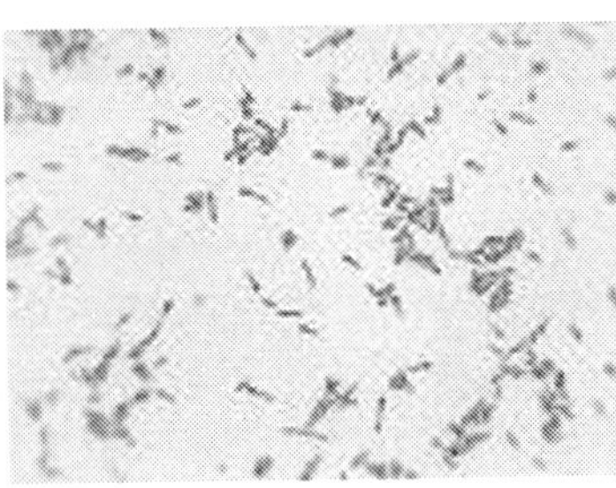
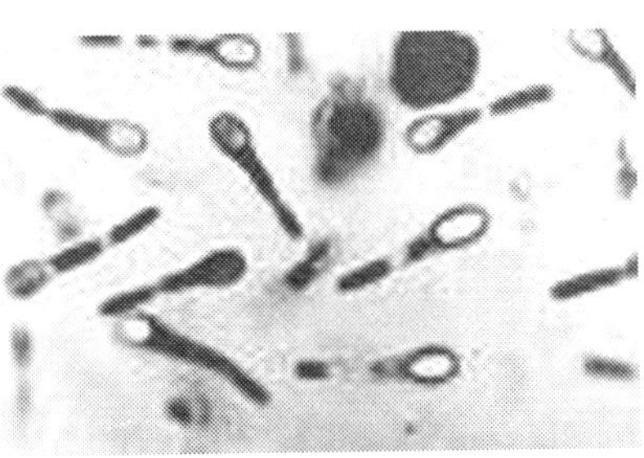

Microscopic and structural observations of ***Clostridium***

REFERENCES

Anand A, Glatt AE. Clostridium difficile infection associated with antineoplastic chemotherapy: a review. Clin Infect Dis 17:109, 1993

Bartlett JG. Clostridium difficile: History of its role as an enteric pathogen and the current state of knowledge about the organism. Clin Infect Dis 18: S265, 1994

Bongaerts GPA, Lyerly DM. Role of toxins A and B in the pathogenesis of Clostridium difficile disease. Microbial Pathogenesis 17: 1, 1994

Bleck TP. Tetanus: pathophysiology, management, and prophylaxis. Disease-A-Month 37: 545, 1991.

Coffield JA, Considine RV, Simpson, LL: Clostridial neurotoxins in the age of molecular medicine. Trends in Microbiology 67: 67, 1994

Finegold SM, George WL (eds): Anaerobic infections in humans. Academic Press, New York, 1989

Hambleton P. Clostridium botulinum toxins: a general review of involvement in disease, structure, mode of action and preparation for clinical use. J Neurol. 239: 16, 1992

Holdeman LV, Cato EP, Moore WEC (ed): Anaerobic laboratory manual, 4th ed. Blacksburg: V.P.I. Anaerobic Laboratory, Virginia Polytechnic Institute and State University, 1977.

Knoop FC, Owens M, Crocker IC. Clostridium difficile: clinical disease and diagnosis. Clin Microbiol Rev 6: 251, 1993

Lorimer JW, Eidus LB. Invasive Clostridium septicum infection in association with colorectal carcinoma. Can J Surg 37: 245, 1994

Lyerly DM, Krivan HC, Wilkins TD: Clostridium difficile: Its disease and toxins. Clin Microbiol Rev 1:1, 1988

Morbidity and Mortality Weekly Report: Tetanus - United States, 1985-1986. 36: 477-481, 1987

Rood JI, Cole ST. Molecular genetics and pathogenesis of Clostridium perfringens. Microbiol Rev 55: 621, 1991

Schofield F: Selective primary health care: strategies for control of disease in a developing world. XXII. Tetanus: a preventable problem. Rev Infect Dis 8: 144, 1986

Speirs G, Warren RE, Rampling A: Clostridium tertium septicemia in patients with neutropenia. J Infect Dis 158: 1336, 1988

Spera RV, Kaplan MH, Allen SL. Clostridium sordellii bacteremia: case report and review. Clin Infect Dis 15: 950, 1992

Stevens DL, Bryant AE. Role of theta toxin, a sulfhydryl-activated cytolysin, in the pathogenesis of clostridial gas gangrene. Clin Infect Dis 16: S195, 1993

Wigginton JM, Thill P. Infant botulism: A review of the literature. Clin Pediatr. 32: 669, 1993

Chapter **18**

Mycobacteria and Nocardia

General Concepts

Mycobacterium tuberculosis Complex:

Mycobacterium tuberculosis, M bovis

Clinical Manifestations

Tuberculosis primarily affects the lower respiratory system and is characterized by a chronic productive cough, low-grade fever, night sweats, and weight loss.

Structure

Mycobacteria are slender, curved rods that are acid fast and resistant to acids, alkalis, and dehydration. The cell wall contains complex waxes and glycolipids. Multiplication on enriched media is very slow, with doubling times of 18 to 24 hours; clinical isolates may require 4 to 6 weeks to grow.

Classification and Antigenic Types

On the basis of growth rate, catalase and niacin production, and pigmentation in light or dark, mycobacteria are classified into members of the *Mycobacterium tuberculosis* complex (*M tuberculosis, M bovis, M africanum, M microtii*) and nontuberculous species. Gene probe technology now facilitates this distinction.

Pathogenesis

Tuberculous mycobacteria enter the alveoli by airborne transmission. They resist destruction by alveolar macrophages and multiply, forming the primary lesion or tubercle; they then spread to regional lymph nodes, enter the circulation, and reseed the lungs. Tissue destruction results from cell-mediated hypersensitivity.

Host Defenses

Susceptibility is influenced by genetic and ethnic factors. Acquired resistance is mediated by T lymphocytes, which lyse infected macrophages directly or activate them via soluble mediators (e.g., gamma interferon) to destroy intracellular bacilli; antibodies play no protective role.

Epidemiology

M tuberculosis is contagious, but only 5-10 percent of infected normal individuals develop active disease. Tuberculosis is most common among the elderly, poor, malnourished, or immunocompromised, especially persons infected with human immunodeficiency virus (HIV). Persistent infection may reactivate after decades owing to deterioration of immune status; exogenous reinfection also occurs.

Diagnosis

Recent infection with *M tuberculosis* results in conversion to a positive Mantoux skin test with purified protein derivative (PPD). A diagnosis of active disease is based on clinical manifestations, an abnormal chest radiograph, acid-fast bacilli in sputum or bronchoscopic specimens and recovery of the organism. Assays based upon amplification of mycobacterial genes in clinical specimens are currently being tested.

Treatment and Control

Therapy consists of a 6 to 9 month course of isoniazid, rifampin, pyrazinamide and ethambutol. Additional drugs may be used if drug resistance is suspected (e.g., infection in Southeast Asian immigrants or in areas where outbreaks of drug-resistant tuberculosis have been documented). If the patient is HIV-positive, treatment for longer periods (9-12 months) is recommended. PPD conversion without other signs or symptoms may warrant prophylactic isoniazid therapy for 6 months. *M bovis* BCG vaccine is used in more than 120 countries, but its efficacy is controversial. Although, BCG has not been used routinely in the U.S., the current epidemic has prompted a reevaluation of its use, especially in high-risk populations.

Nontuberculous Mycobacteria

Clinical Manifestations

Patients exhibit lower respiratory disease similar to tuberculosis (*M kansasii, M avium-intracellulare*), cervical lymphadenitis (*M scrofulaceum*), skin and soft tissue infections (*M ulcerans, M marinum*), or disseminated disease in persons infected with HIV.

Structure

Nontuberculous mycobacteria resemble other mycobacteria. Multiplication is similar to that of other mycobacteria, except for rapidly growing strains (e.g. *M. fortuitum*).

Classification and Antigenic Types

Nontuberculous mycobacteria are classified by pigmentation in the light or dark and by growth rate. Several species contain many serotypes, based upon lipooligosaccharides or peptidoglycolipids.

Pathogenesis

Pathogenesis is similar to that of other mycobacteria. There may be granuloma formation and delayed hypersensitivity.

Host Defenses

Host defenses are similar to those of other mycobacteria. Cell-mediated resistance is important.

Epidemiology

The infection is not transmissible between humans, but is acquired from natural sources (e.g., soil and water). Nontuberculous mycobacteria are important opportunistic pathogens in immunocompromised patients (e.g., patients infected with HIV).

Diagnosis

Diagnosis requires culture and identification.

Treatment

Many of these mycobacteria are resistant to the usual antituberculosis drugs. Therefore, different drugs are often necessary. Surgical resection is sometimes required. No vaccine is available.

Mycobacterium leprae

Clinical Manifestations

Leprosy is an infection of the skin, peripheral nerves, and mucous membranes, leading to lesions, hypopigmentation, and loss of sensation (anesthesia), particularly in the cooler areas of the body.

Structure

Mycobacterium leprae is similar to other mycobacteria; the cell wall contains unique phenolic glycolipids. It cannot be cultivated *in vitro*: it multiplies very slowly *in vivo* (12-day doubling time).

Classification and Antigenic Types

All isolates of *M leprae*, both human and sylvatic, appear to be the same by DNA homology.

Pathogenesis

The spectrum of leprosy (Hansen's disease) ranges from lepromatous (disseminated, multibacillary, with loss of specific cell-mediated immunity) to tuberculoid (localized, paucibacillary, with strong cell-mediated immunity).

Host Defenses

Host defenses are similar to those against other mycobacteria.

Epidemiology

Transmission requires prolonged contact and occurs directly through intact skin, mucous membranes, or penetrating wounds. Armadillos in Louisiana and Texas are naturally infected.

Diagnosis

Diagnosis is based on acid-fast stain and cytologic examination of affected skin and response to the lepromin skin test;

M leprae cannot be cultured.

Treatment

Treatment (including prophylaxis in close contacts) with multi-drug therapy or MDT (dapsone, rifampin,and clofazimine) is performed on anoutpatient basis for 3 to 5 years; vaccination with *M bovis* BCG has been effective in some endemic areas.

Nocardia

Clinical Manifestations

The most common manifestation of nocardial infection is pneumonia: fever, weight loss, cough, pleuritic chest pain, and dyspnea. In about 20 percent of patients there are granulomatous skin lesions and/or central nervous system abnormalities.

Structure

The bacteria are Gram-positive, partially acid-fast rods, which grow slowly in branching chains resembling fungal hyphae.

Classification and Antigenic Types

Three species cause nearly all human infections: *N asteroides*, *N brasiliensis*, and *N caviae*. These are distinguished by proteolytic and fermentation patterns in culture.

Pathogenesis

Infection is by inhalation of airborne bacilli from an environmental source (soil or organic material); the disease is not contagious. Skin lesions caused by *N brasiliensis* often result from direct inoculation. *Nocardia* subverts antimicrobial mechanisms of phagocytes, causing abscess or rarely granuloma formation with hematogenous or lymphatic dissemination to the skin or central nervous system. Mortality is up to 45 percent even with therapy.

Host Defenses

The natural resistance to infection is high in normal individuals,and the disease is usually associated with cellular immune dysfunction, immunoglobulin deficiencies, or leukocyte defects. Acquired resistance is complex and involves activated macrophages, cytotoxic T cells, and neutrophil inhibition.

Epidemiology

Nocardiosis is rare in normal persons. It usually occurs in recipients of organ transplants; in patients with leukemia, lymphoma, humoral, or leukocyte defects; or after prolonged steroid therapy.

Diagnosis

Diagnosis is by Gram stain, modified acid-fast stain, and culturing of organisms from sputum, bronchoscopic specimens (washing, brushing), aspirates of abscesses, or by biopsy.

Treatment

Nocardiosis is treated by prolonged (up to 1 year) therapy with trimethoprim-sulfamethoxazole.

INTRODUCTION

The genera *Mycobacterium* and *Nocardia* have been grouped into the family Mycobacteriaceae within the order Actinomycetales based upon similarities in staining and motility, lack of spore formation, and catalase production. These genera are characterized by the presence of long-chain fatty acids, called mycolic aids, which have the following general structures in their cell walls:

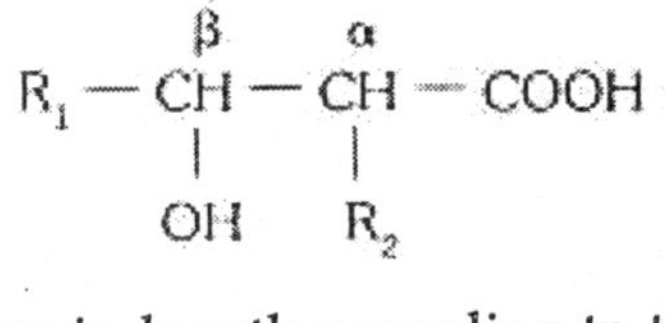

$$\begin{array}{ccccccc} & & \beta & & \alpha & & \\ R_1 & - & CH & - & CH & - & COOH \\ & & | & & | & & \\ & & OH & & R_2 & & \end{array}$$

The side chains (R_1 and R_2) vary in length according to the genus; C60 to C90 in *Mycobacterium*; C40 to C56 in *Nocardia*. Several species produce disease in humans.

Mycobacterium tuberculosis Complex

Clinical Manifestations

The first sign of a new infection is often conversion of the intradermal skin test with purified protein derivative (PPD) to positive or detection of a lesion by chance on a chest x-ray in an otherwise asymptomatic individual (see Diagnosis below). Clinical signs and symptoms develop in only a small proportion (5-10 percent) of infected healthy people. These patients usually present with pulmonary disease; prominent symptoms are chronic, productive cough, low-grade fever, night sweats, easy fatigability, and weight loss. Tuberculosis may present with or also exhibit extrapulmonary manifestations including lymphadenitis; kidney, bone, or joint involvement; meningitis; or disseminated (miliary) disease. Lymphadenitis and meningitis are more common among normal infants with tuberculosis, and all extrapulmonary manifestations are increased in frequency among immunocompromised individuals such as patients on chronic renal dialysis and elderly, malnourished, or HIV-infected individuals.

Structure

Mycobacteria are slender, curved rods in stained clinical specimens. The cell wall is composed of mycolic acids, complex waxes, and unique glycolipids. Figure 33-1 depicts the typical cell wall structure for *M tuberculosis* and other mycobacteria. The mycolic acids containing extremely long (C60 to C90) side chains are joined to the muramic acid moiety of the peptidoglycan by phosphodiester bridges and to arabinogalactan by esterified glycolipid linkages. Species variations are characterized by variation in sugar substitutions in the glycolipids or peptidoglycolipids. The mycobacterial cell wall is acid-fast (i.e., it retains carbolfuchsin dye when decolorized with acid-ethanol). This important property allows differential staining in contaminated clinical specimens such as sputum. Other important wall components are trehalose dimycolate (so-called cord factor, as it is thought to induce growth in serpentine cords on artificial medium) and mycobacterial sulfolipids, which may play a role in virulence. Another unique constituent which may contribute to pathogenesis is lipoarabinomannan (LAM). Purified LAM from virulent and attenuated strains of mycobacteria may differ structurally, and these differences may contribute to their varying abilities to stimulate cytokine production in mononuclear cell cultures.

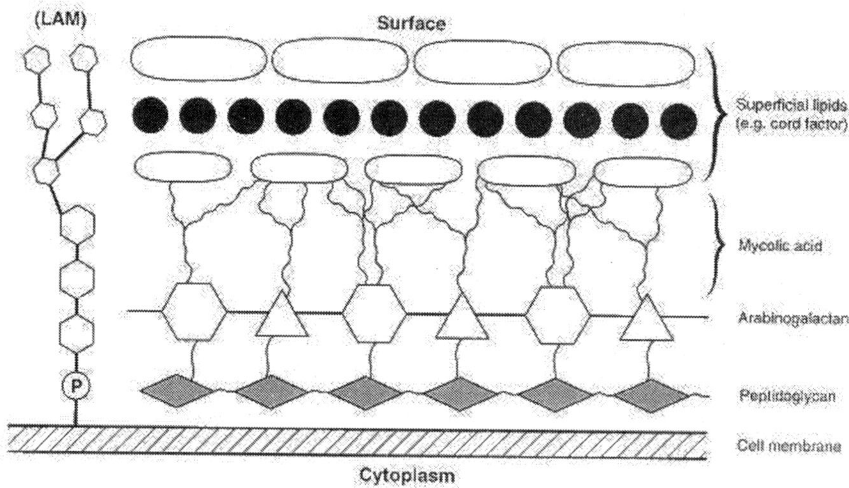

FIGURE 33-1 Complex cell wall structure of mycobacteria.

This unusual cell wall structure endows mycobacteria with resistance to dehydration, acids, and alkalis. The resistance to acids and alkalis is useful in the isolation of mycobacteria from contaminated clinical specimens such as sputum. Treatment of sputum with dilute solutions of sulfuric acid or sodium hydroxide will allow mycobacteria to survive and grow on culture medium in the absence of the members of the respiratory flora.

Another important consequence of the unique cell wall structure of mycobacteria is the adjuvant action of whole cells when mixed with a wetting agent (e.g., Tween) in an oil-water emulsion. Such a mixture is called Freund's complete adjuvant. Researchers have identified muramyl dipeptide as the peptidoglycan component that mediates adjuvanticity (Fig. 33-2).

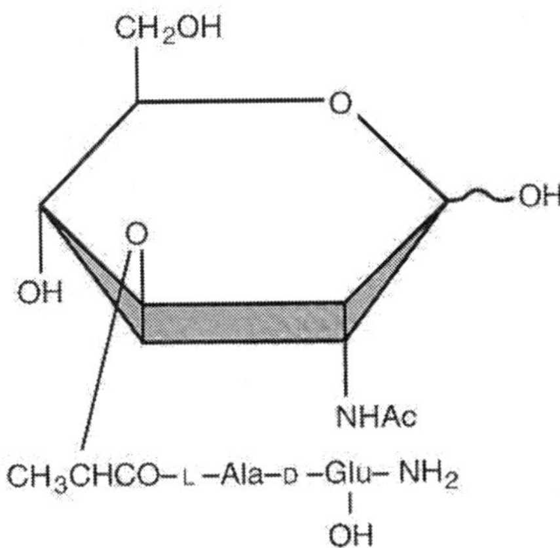

FIGURE 33-2 Structure of muramyl dipeptide from mycobacteria, which mediates adjuvanticity.

Although, mycobacteria are normally cultured from clinical material by inoculation onto enriched agar media containing bovine serum albumin, they can grow on a chemically defined medium containing asparagine, glycerol, and micronutrients. Even under ideal culture conditions *M tuberculosis* and *M bovis* grow very slowly, with doubling times on the order of 18 to 24 hours. This extremely slow growth, even *in vivo,* has two consequences of clinical significance: (1) the infection is an insidious, chronic process, which may take several weeks or months to become clinically patent, and (2) on solid media inoculated with clinical material, identifiable mycobacterial colonies may not appear for 4 to 6 weeks. When they do appear, the colonies are irregular, waxy, and buff colored, with bacteria piled up into clumps or ridges.

Classification and Antigenic Types

With the exception of *M leprae,* the mycobacteria are classified into two broad categories members of the *M tuberculosis* complex *(M tuberculosis, M bovis, M microtii, M africanum)* and nontuberculous mycobacteria (virtually all other species), which often are described based on their growth rate and pigmentation with and without exposure to light. Although, there are antigenic differences among species on the basis of serologic reactions to carbohydrate moieties in the glycolipids, such determinations are not clinically useful. Modern molecular biologic techniques have revealed a remarkable conservation of genes coding for the immunodominant antigens of all mycobacteria, including *M leprae*.

Pathogenesis

In this country, virtually all *M tuberculosis* infections occur by airborne transmission of droplet nuclei containing a few viable, virulent organisms produced by a sputum-positive individual (Fig. 33-3). The bacilli are deposited in the alveolar spaces of the lungs, where they are engulfed by alveolar macrophages. A portion of the infectious inoculum resists intracellular destruction and persists, eventually multiplying and killing the macrophage. The ability of virulent mycobacteria to survive within phagocytes justifies their designation as facultative intracellular pathogens. The mechanisms of intracellular survival are not clear and may vary from species to species. There is some evidence that *M tuberculosis* can prevent phagosome-lysosome fusion. Other studies have demonstrated that virulent mycobacteria can prevent acidification of the phagolysosome, perhaps by modulating the activity of a membrane proton pump. In addition, some of the components of the mycobacterial cell wall (e.g., cord factor) may be directly cytotoxic to macrophages. Although, hemolysins and lipases are produced by *M tuberculosis*, their role in escape of tubercle bacilli from the phagosome, and the importance of extravacuolar organisms in pathogenesis are unknown. Most of the tissue destruction associated with tuberculosis results from cell-mediated hypersensitivity, however, rather than direct microbial aggression.

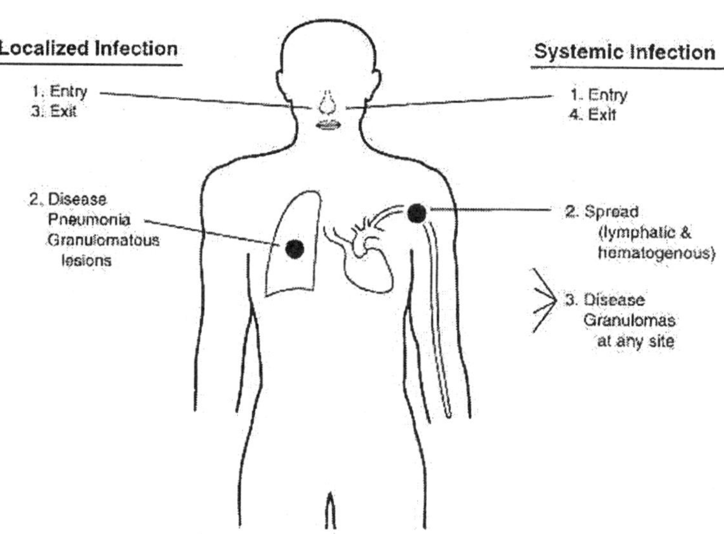

FIGURE 33-3 Pathogenesis of tuberculosis.

Eventually, the accumulating mycobacteria stimulate an inflammatory focus which matures into a granulomatous lesion characterized by a mononuclear cell infiltrate surrounding a core of degenerating epithelioid and multinucleated giant (Langhans) cells. This lesion (called a tubercle) may become enveloped by fibroblasts, and its center often progresses to caseous necrosis. Liquefaction of the caseous material and erosion of the tubercle into an adjacent airway may result in cavitation and the release of massive numbers of bacilli into the sputum. In the resistant host, the tubercle eventually becomes calcified.

Early in infection, mycobacteria may spread distally either indirectly through the lymphatics to the hilar or mediastinal lymph nodes and thence via the thoracic duct into the blood stream, or directly into the circulation by erosion of the developing tubercle into a pulmonary vessel. Extrapulmonary hematogenous dissemination results in the seeding of other organs (e.g., spleen, liver, and kidneys) and, eventually, reinoculation of the lungs. The resulting secondary lung lesions (as opposed to the initial site of implantation) may serve as the origin of reactivation of clinical disease years or decades later owing to the persistence of viable tubercle bacilli. Figure 33-4 illustrates the radiologic differences between primary and secondary tuberculosis. Primary disease is usually characterized by a single lesion in the middle or lower right lobe with enlargement of the draining lymph nodes. Endogenous reactivation is often accompanied by a single (cavitary) lesion in the apical region, with unremarkable lymph nodes and multiple secondary tubercles.

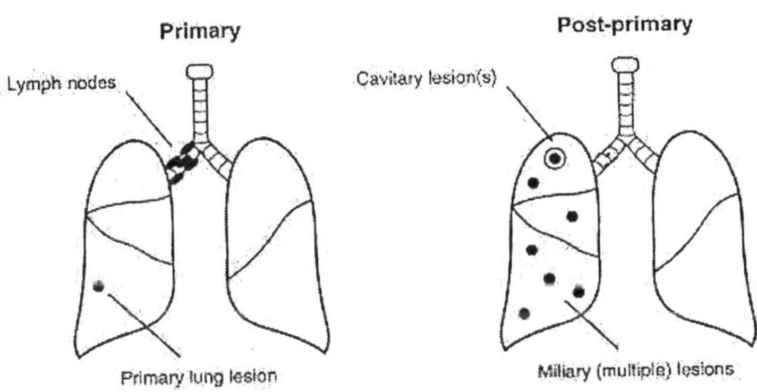

FIGURE 33-4 Radiologic differences between primary and post-primary tuberculosis. Miliary lesions, which are small granulomas, resemble millet seeds spread throughout the lung fields.

In parts of the world where bovine tuberculosis has not been eliminated and where dairy products are not properly treated, direct infection of the gastrointestinal tract may occur by ingestion of virulent *M bovis* organisms. The gut is also exposed occasionally in pulmonary tuberculosis when large numbers of viable *M tuberculosis* cells are coughed up and swallowed.

In either case, the principal site of involvement is the mesenteric lymph nodes, with subsequent dissemination.

Host Defenses

Innate susceptibility to pulmonary infection with *M tuberculosis* is clearly influenced by genetic and/or ethnic variables that have not been defined. Studies of mono- and dizygotic twins and siblings indicate a significant relationship between genotype and resistance to tuberculosis. Studies of inbred experimental animals point to genes both within and outside the major histocompatibility complex that apparently contribute to resistance. In the mouse, at least two genes *(Bcg* and *Tbc)* have been implicated in the expression of innate resistance to mycobacteria. A search for human homologues is currently under way. The mechanism for this genetic effect may reflect the ability of macrophages to process and present mycobacterial antigens to the immune system.

Acquired immunity following mycobacterial infection usually develops within 4

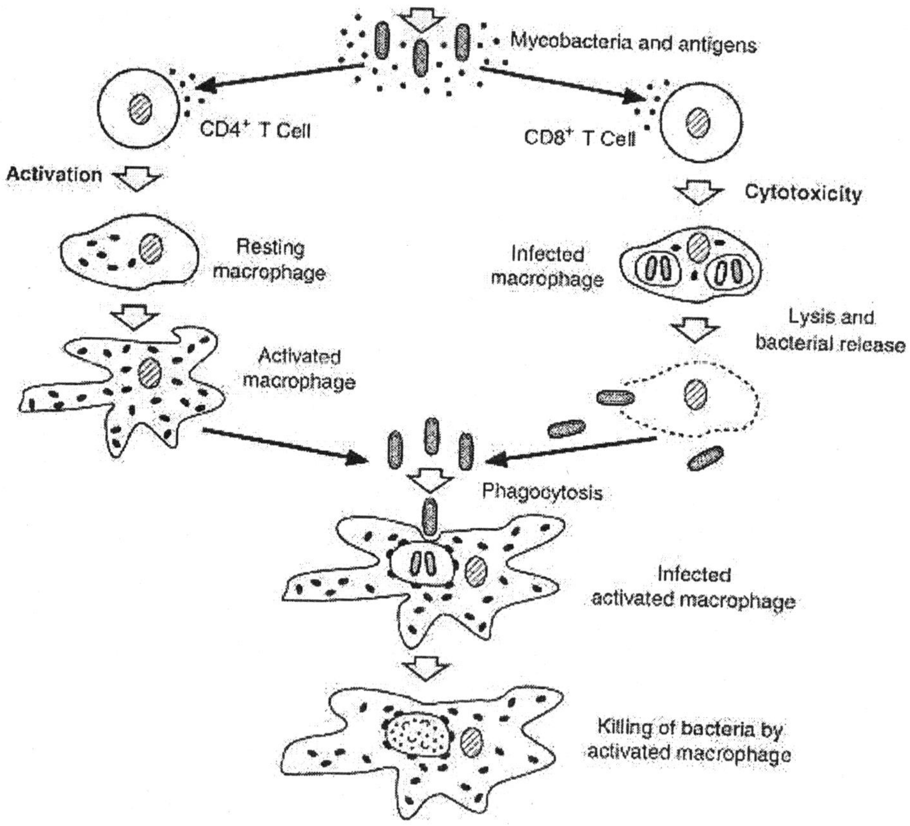

FIGURE 33-5 Principal mechanisms of T-lymphocyte activation or destruction of macrophages following stimulation by mycobacterial antigens. Activation of macrophages can result in bacterial killing while cytotoxicity may release bacteria from phagocytes and allow their engulfment and destruction by activated macrophages.

to 6 weeks and is associated temporarily with the onset of delayed hypersensitivity to mycobacterial antigens such as PPD. Successful acquired resistance is mediated by T lymphocytes. Antimycobacterial antibodies, although present in many patients, do not play a protective role in tuberculosis. Figure 33-5 depicts two of the principal mechanisms by which T lymphocytes activated by specific mycobacterial antigens can limit the replication of tubercle bacilli. Cells of the helper/inducer phenotype ($CD4^+$) up-regulate populations of antigen-specific effector T cells ($CD4^+$) and cytotoxic T cells ($CD8^+$). CD4 cells produce factors (e.g., gamma interferon) that activate macrophages and endow them with enhanced mycobacteriostatic or mycobactericidal capabilities. Unlike normal macrophages, these activated cells can limit the replication of intracellular *M tuberculosis* and may kill tubercle bacilli. CD8 cells attack infected macrophages expressing mycobacterial antigens and lyse the cells, releasing the mycobacteria from their protective niche and exposing them to activated macrophages.

Recently, $CD4^+$ (and perhaps $CD8^+$) T cells have been dichotomized into functional subsets in the murine system based upon their cytokine profiles. So-called "Th1" cells produce interleukin-2 (IL-2) and interferon gamma and promote inflammatory reactions and cell-mediated immunity. "Th2" cells produce interleukin-4, -5, and -10 and drive the immune response toward antibody production. The two subsets apparently cross-regulate principally via opposing actions of interferon gamma and interleukin-4. Although, evidence for this clean functional separation is not as compelling in humans, it is likely that disease resistance in tuberculosis depends upon the predominance of a Th1-like cytokine response to mycobacterial antigens.

Since the vast majority (90-95 percent) of otherwise healthy individuals who are infected with *M tuberculosis* never develop clinically apparent disease, acquired resistance must be quite effective. However, the immune response to mycobacteria is a double-edged sword: the intense cell-mediated hypersensitivity that usually accompanies infection is responsible for much of the pathology associated with clinical tuberculosis. Some clinical studies suggest that inappropriately high levels of circulating cytokines, such as tumor necrosis factor alpha, may be responsible for some of the clinical features of tuberculosis (e.g. fever, weight loss). Therapy with cytokine-blocking drugs, such as pentoxyfilline and thalidomide, may prove to be an important adjunct to standard chemotherapy in some tuberculosis patients. Experimental evidence suggests that protection and hypersensitivity may be mediated by distinct subsets of T lymphocytes and may be directed against different mycobacterial antigens. Much progress has been made in recent years in defining the important antigens of *M tuberculosis* and other mycobacteria by using gene cloning and monoclonal antibody technology. Interestingly, some of these immunodominant mycobacterial antigens show remarkable homology with a family of stress (or heat shock) proteins that are widely conserved in both prokaryotes and eukaryotes, including humans. This observation may have relevance for the autoimmune phenomena (e.g., arthritis) associated with mycobacterial infection in some people. Some highly purified or recombinant antigens are currently being tested for their usefulness in diagnosis or in the development of subunit vaccines to prevent tuberculosis.

Epidemiology

Numerous studies of tuberculosis epidemics in closed populations (e.g., on naval vessels and in nursing homes) document the contagious nature of this infection. Fortunately, overt clinical disease actually develops in only a small percentage of those infected. Identification of recently infected individuals is still important, however, because viable mycobacteria persisting in tissues may lead to endogenous reactivation of tuberculosis later in life (see Treatment and Control, below). Reactivation is usually associated with deterioration of the cell-mediated immune response due to aging or to some associated clinical condition. Exogenous reinfection also has been documented, but most cases of so-called "post-primary" tuberculosis in this country are thought to be the result of endogenous reactivation.

Tuberculosis epidemiology has been clarified significantly by the development of molecular biological techniques which allow the relatively unambiguous identification of a particular clinical isolate. Recent, active transmission of a single strain of *M tuberculosis* would result in clinical isolates from several patients which exhibited identical DNA patterns or "fingerprints." In contrast, endogenous reactivation in a group of patients would likely result in isolates which exhibited unique patterns. The most commonly used procedure for comparing isolates involves the use of molecular probes which bind to segments of DNA called insertion sequences (IS) in the mycobacterial genome. Such a sequence, IS6110, has been employed very successfully to track outbreaks of tuberculosis within institutions (e.g. hospitals) associated with recent transmission. This "fingerprinting" procedure, referred to as restriction fragment length polymorphism (RFLP) analysis, is currently being used to detect previously unsuspected transmission between apparently unrelated patients in the community.

Tuberculosis is particularly common in groups such as the elderly, the chronically malnourished, alcoholics, and the poor. The prevalence of clinical tuberculosis among the homeless in the United States may be up to 300 times higher than the national average rate. In recent years, the incidence of disease in racial minorities in the United States has been more than five times that observed in whites. Of particular concern is the very high incidence of tuberculosis among recent immigrants.

Perhaps the most significant factor influencing the incidence of mycobacterial disease in the United States since 1984 has been the HIV epidemic. HIV-infected individuals have a high incidence of tuberculosis, characterized by frequent extrapulmonary disease. Both primary, pulmonary infection and endogenous reactivation are seen in HIV-positive individuals. Owing to the loss of T-cell function in these patients, the tuberculin skin test may not be reliable and the chest radiograph may not show the classic well-defined primary tubercle. Both of these observations make the diagnosis of tuberculosis in HIV-infected patients more challenging. Furthermore, tuberculosis often occurs early in the course of HIV infection, before a significant decline in $CD4^+$ T cells has occurred. In this sense, tuberculosis in an apparently healthy young adult may "signal" the presence of underlying HIV infection.

Diagnosis

Infection in an asymptomatic individual can be diagnosed with the help of the intradermal PPD skin test. Intradermal introduction of PPD into a previously infected, hypersensitive person results in the delayed (48-72 hr) appearance of an indurated (raised, hard) reaction with or without erythema. It is impossible to distinguish between present and past infection on the basis of a positive tuberculin test. Recent conversion of the reaction from negative to positive warrants clinical attention. Although, multiple-puncture (or tine) tests once were popular for screening for tuberculin hypersensitivity, they are not as accurate as the Mantoux test and should not be used. The Mantoux test requires the intradermal injection of a measured volume (0.1 ml) containing a specified quantity (5 tuberculin units) of PPD. The transverse diameter of induration is measured 48 to 72 hours later. Interpretation varies, as shown in Table 33-1. In a person with symptoms suggestive of tuberculosis, clinical specimens (sputum, bronchial or gastric washings, pleural fluid, urine, or cerebrospinal fluid) should be stained and cultured for acid-fast bacilli. Culture and identification of mycobacteria in such specimens are mandatory for diagnosis. Two types of stains are used specifically for detection of mycobacteria: fluorochrome (recommended) and carbol fuchsin. In smears stained with carbol fuchsin, mycobacteria typically appear as red rods (1-10 μm long and 0.2-0.6 μm wide) and often are beaded or banded, but also may appear coccoid or filamentous. In general, the microscopic appearance of the mycobacteria as slightly curved rods does not provide a species identification but may be suggestive for some species.

TABLE 33-1 Guidelines for Interpretation of the mantoux Test

Diameter of Induration (in mm)	Persons for whom reaction is Considered Positive
≥ 5	Immunosuppressed, a recent close contact of a person with active TB, abnormal chest x-ray consistent with TB
≥ 10	Foreign-born (country with high TB prevalence), low income, injection drug users, residents of correctional facility or nursing home > 70 yr < 18 yr, healthcare workers, mycobacterial lab employee, medical condition associated with increased risk TB (diabetes mellitus, prolonged carticosteroids, gastrectomy, chronic malabsorption, sllicosis, ≥ 10% below ideal body weight)
≥ 15	All other

aPersons infected with human immunodoficiency virus, those on imunosuppressive therapy, these with hemalologic diseases, cancer, or and stage renal disease.

The specificity of stains for AFB typically is $\geq$ 99% and the sensitivity 25% to 75%. A positive stain and negative culture may be caused by nonviable organisms, such as might occur in persons receiving antituberculosis medication. Higher sensitivity of smear results occurs with cavitary lesions, respiratory specimens, increased number of specimens, increased number of mycobacteria present in the sample (above the minimum of 5,000 organisms/ml), the presence of *M tuberculosis* or *M kansasii* species, observer experience, and stain used.

Culture for mycobacteria involves inoculation of solid and broth media. For specimens such as sputum that are contaminated with normal bacterial flora, a selective medium containing antimicrobial agents should be inoculated. Sterile body fluids should be inoculated to solid media and a broth medium. Cultures are incubated at 35° to 37° C in an atmosphere of 5 to 10% CO_2. For specimens from cutaneous sites a second set of cultures should be incubated at 30° C. All cultures should be examined weekly for 8 weeks.

The major advantage of culture on solid media is that it allows visualization of colony morphology and pigmentation, which is useful diagnostically for distinguishing colonies of *M tuberculosis* from those of some nontuberculous mycobacteria. However, they require 3 or 4 weeks. The more rapid broth systems (e.g. Bactec) require only 5 to 12 days, and rely upon the detection of ^{14}C-labeled $C0_2$ produced by growing mycobacteria.

Commercial chemiluminescent DNA probes, gas-liquid chromatography, high-performance liquid chromatography, and thin-layer chromatography allow identification of a few species of mycobacteria within hours after sufficient growth is present on solid or in a liquid medium.

In the future, nucleic acid amplification methods may prove useful for detection of mycobacteria directly in clinical material within 24 hours or less of specimen receipt. Currently, standardized guidelines for susceptibility testing of mycobacteria have been developed only for isolates of *M tuberculosis* a positive BACTEC TB vial (indirect test), or on sputum specimens that are smear-positive (direct test). Using a broth system, results are available 5-7 days after bottles are inoculated.

Treatment and Control

In the United States, tuberculosis in the general population is controlled by intensive case finding (by means of PPD skin testing) and aggressive prophylactic chemotherapy in tuberculin converters. In individuals with clinical disease, short term (6-9 month) ambulatory therapy with so-called first-line anti-mycobacterial drugs, such as isoniazid, rifampin, pyrazinamide, and ethambutol, results in disappearance of viable tubercle bacilli from the sputum, rendering the patient noninfectious. Prompt therapy, even in the absence of other signs or symptoms, is thought to sterilize the tissues and prevent endogenous reactivation of tuberculosis later in life. Patient compliance is probably the single most important variable affecting treatment outcome. Directly observed therapy (DOT) has been instituted in high prevalence areas, especially among recalcitrant patients, as the only reliable means of ensuring that patients complete their treatment successfully.

Multiple-drug resistance (MDR) has become a particularly threatening aspect of the current tuberculosis epidemic in this country. Although, confined at present predominantly to large metropolitan areas (e.g. New York City) MDR strains of *M tuberculosis* have been associated with several outbreaks characterized by rapid progression and high mortality (50-75%). Some progress has been made in the research laboratory in elucidating the mechanisms of drug resistance in mycobacteria. Multiple-drug resistance in mycobacteria is apparently the result of the step-wise accumulation of resistance to individual drugs. For example, mutations in the *catG* and *inhA* genes are associated with isoniazid resistance, while the *rpoB* gene responsible for RNA polymerase is altered in many clinical isolates resistant to rifampin. The threat of drug-resistant tuberculosis has made susceptibility testing mandatory for all initial isolates. Unfortunately, more rapid techniques are needed to provide this information in a timely manner to physicians. Such assays are presently being developed. One of the most exciting involves the use of a luciferase reporter gene which is introduced into the clinical isolate on a mycobacteriophage. Light production in the presence of the drug reveals resistance, and can be detected very quickly. When resistance to two or more of the first line drugs is detected, additional drugs (ethionamide, streptomycin, ciprofloxacin) may be added to the regimen.

A viable, attenuated strain of *M bovis*, called bacille Calmette-Guérin (BCG), after the French microbiologists who developed the strain, has been used in more than 120 countries for many years as a vaccine to prevent clinical tuberculosis. Vaccination is the only feasible approach to controlling this disease in much of the developing world. The efficacy of BCG has varied in field trials from 0 to 85 percent, indicating an influence of unknown local environmental or host factors. BCG is not used in the United States because it results in PPD conversion, thereby interfering with the epidemiological and diagnostic value of the skin test. In the past 5 years, researchers have turned their efforts to the development and animal model testing of new tuberculosis vaccines. Promising results have been obtained in animal experiments with purified subunit vaccines, recombinant BCG strains, and auxotrophic BCG mutants. Excellent animal models for vaccine testing are available using mice, guinea pigs or rabbits infected with *M tuberculosis* by the pulmonary route.

Nontuberculous Mycobacteria

Clinical Manifestations

Nontuberculous mycobacteria, previously referred to as "atypical" mycobacteria, comprise several species, which may produce a wide range of clinical conditions involving several organ systems (Table 33-2). Clinically, pulmonary disease caused by these organisms is virtually indistinguishable from tuberculosis. Disseminated infection is usually limited to immunocompromised patients, particularly HIV-infected individuals, in whom the *M avium-intracellulare* complex is responsible for more than 90 percent of cases. Cervical lymphadenitis due to infection with *M scrofulaceum* is seen especially in children younger than 5 years. Granulomatous skin lesions and soft tissue infections are usually associated with *M marinum* (swimming pool granuloma) or *M ulcerans.*

TABLE 33-2 Clinical Presentation of Nontuberculous mycobacterial infections

Involvement during infection with						
Location	M. avlum complex	M. kansasii	M. scrofulaceum	M. fortultum	M. ulcerans	M. marfnum
Pulmanary	+	+	-		-	-
Cutaneous	-	+	+	+	+	+
Gastrointestinal		+		-		
Genitournary			-	-	-	-
ocular	-		-	+/-	-	-
Lymph node	+	-	+	-	-	-
Dissominated	+	-	-	-	-	-

Classification and Antigenic Types

Table 33-3 lists the nontuberculous mycobacteria associated with human disease and their classification within the Runyon scheme, in which the species in groups I to III are slow growers and those in group IV are rapid growers. The Runyon groups are further characterized by pigment production: nonchromogens (group III) are rarely pigmented; photochromogens (group I) are pigmented only when exposed to light; scotochromogens (group II) form pigment in the dark. Some of these species have been grouped into complexes based on similarities in the clinical condition which they cause. Other associations may be based upon biochemical similarities (e.g., the inclusion of *M scrofulaceum* and *M avium-intracellulare* to form the MAIS complex). Further distinction of multiple serotypes within some species (e.g., the MAIS complex) is based upon variations in the sugar residues on the lipooligosaccharides or the peptidoglycolipids.

TABLE 33-3 Classification Nontuberoulous Mycobacteria

Species	Runyon Group	Pigment Formation	collective Dealgnation of Complex
M kansasll	I		
M marinum	I	Photochromogens	-
M simlae	I		
M Scrofulaceum	II		
M szulgai	II	Scotochromogens	-
M gordenae	II		
M avium	III		
M intracellularae	III	Nonchromogens	MAC*
M ulcerans	III		
M forfultum	IV		
M chelonae	IV	Rapid growers	M forfultum-chelonae complex

* MAC, M avium complex

Epidemiology

A crucial difference between *M tuberculosis* and nontuberculous mycobacteria is the lack of transmission of the latter from patient to patient (Fig. 33-3). There is no evidence that infections caused by nontuberculous mycobacteria are contagious. Rather, the organisms exist saprophytically in the soil or water, occasionally in association with some infected-animal reservoir (e.g., poultry infected with *M avium*). Inhalation or ingestion of viable mycobacteria or introduction of bacilli through skin abrasions initiates the infection. Evidence of geographic concentrations of nontuberculous mycobacteria in the southeastern United States comes from skin test surveys with "tuberculins" made from specific nontuberculous mycobacteria (e.g., PPD-Y for *M kansasii*; PPD-A for *M avium*). In endemic areas many subclinical infections may occur. The ubiquity of nontuberculous mycobacteria makes them ideal opportunists for immunocompromised hosts. Up to 30 percent of patients with acquired immune deficiency syndrome (AIDS) may suffer disseminated mycobacterial infections, most of which are caused by members of the *M avium-intracellulare* complex. Such infections, which are associated with shortened survival, result from environmental exposure. In fact, nontuberculous mycobacteria have been cultured directly from tap water in several hospitals.

Treatment and Control

Many nontuberculous mycobacteria are resistant to the drugs commonly used successfully in the treatment of tuberculosis (e.g., isoniazid, pyrazinamide, and streptomycin). Antibiotic regimens may require several (five or six) drugs including rifampin, which is quite effective against *M kansasii*, or clarithromycin, which has marked activity against the *M avium-intracellulare* complex. Surgical resection is occasionally recommended with or without chemotherapy. In treating disseminated infections in AIDS patients, a regimen of five or six drugs, including clarithromycin, ethambutol and perhaps rifabutin, should be considered.

Mycobacterium leprae

Clinical Manifestations

The clinical spectrum of leprosy (Hansen's disease) reflects variations in three aspects of the illness: bacterial proliferation and accumulation, immunologic responses to the bacillus, and the resulting peripheral neuritis. The disease affects peripheral nerves, skin, and mucous membranes. Skin lesions, areas of anesthesia, and enlarged nerves are the principal signs of leprosy. The disease manifestations fall on a continuum from lepromatous leprosy to tuberculoid leprosy. The polar lepromatous leprosy patient presents with diffuse or nodular lesions (lepromas) containing many acid-fast *M leprae* bacilli cells (multibacillary lesions). These lesions are found predominantly on the cooler surfaces of the body, such as the nasal mucosa and the peripheral nerve trunks at the elbow, wrist, knee, and ankle. Sensory loss results from damage to nerve fibers. On the other hand, polar tuberculoid leprosy consists of a few well-defined anesthetized lesions containing only a few acid-fast bacilli (paucibacillary lesions). Borderline forms of the disease are unstable conditions, presenting with intermediate signs and symptoms.

Pathogenesis

Since *M leprae* has never been cultured *in vitro*, it appears to be an obligate intracellular pathogen that requires the environment of the host macrophage for survival and propagation. Estimates of the replication rate *in vivo* are on the order of 10 to 12 days. The bacilli resist intracellular degradation by macrophages, perhaps by escaping from the phagosome into the cytoplasm, and accumulate to high levels (10^{10} bacilli/g of tissue) in lepromatous leprosy. The peripheral nerve damage appears to be mediated principally by the host immune response to bacillary antigens. Tuberculoid leprosy is characterized by self-healing granulomas containing only a few, if any, acid-fast bacilli.

Host Defenses

The successful host response in tuberculoid leprosy involves macrophage activation and recruitment by T lymphocytes that recognize *M leprae* antigens. Very little circulating antibody against the bacillus is present in tuberculoid leprosy. In contrast, lepromatous leprosy is associated with profound specific anergy (lack of T cell-mediated immunity against *M leprae* antigens) and high levels of circulating antibodies.

These antibodies play no protective role and may actually interfere with effective cell-mediated immunity. There is experimental evidence that components of the leprosy bacillus may induce suppressor T cells or interfere with macrophage function in the lesions. Figure 33-6 summarizes the immunologic and pathologic spectrum of leprosy. Much of the pathology is caused by the host immune response.

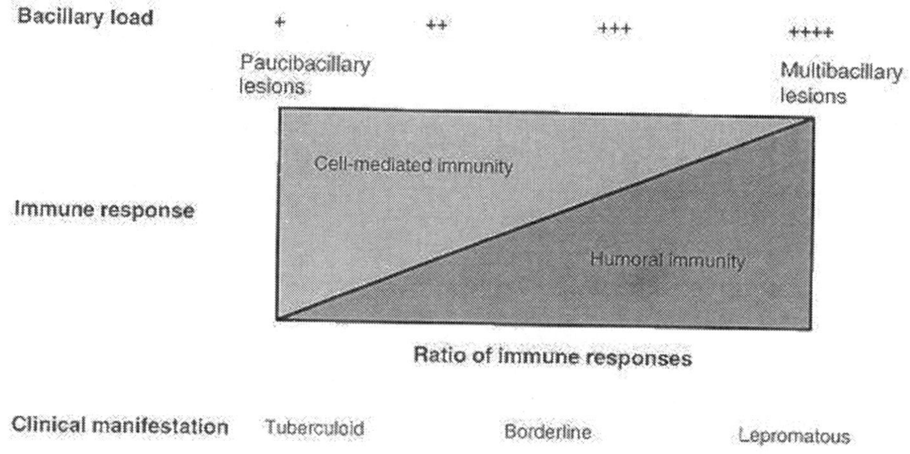

FIGURE 33-6 Pathologic (bacillary load), immunologic, and clinical spectrum of leprosy.

The dichotomy between Th1 and Th2 CD4⁺ T cell cytokine profiles mentioned earlier is even more clear-cut in leprosy. Elegant studies of cytokine production in cells from lepromatous and tuberculoid lesions have revealed a marked propensity for a Th2-like pattern in lepromatous leprosy and a Th1-like pattern in the tuberculoid form. These results imply that the production of IL-2 and IFNg in response to mycobacterial antigens is associated with successful bacillary control. The conclusion has been strengthened by clinical observations of upgrading lesion status in patients receiving IL-2 therapy by direct intralesional injection.

Epidemiology

More than 10 million cases of leprosy are estimated to exist worldwide, predominantly in Asia (two-thirds) and Africa (one-third). Human-to-human transmission requires prolonged contact and is thought to occur via intact skin, penetrating wounds or insect bites, or by inhalation of *M leprae* and deposition on respiratory mucosa (Fig. 33-7). The source of the organism in nature is unknown. Recently, the development of PCR techniques for the detection of *M leprae* DNA in environmental and clinical specimens has allowed investigators to begin to study natural distribution as well as the contribution of asymptomatic human "carriers" to the epidemiology of leprosy. Natural infections have been documented in mangabey monkeys, and in wild armadillos in Texas and Louisiana. Several human infections have been reported following contact with armadillos, but their role in the epidemiology of this disease is controversial. Armadillos experimentally infected with *M leprae* serve as an important source of bacilli for researchers.

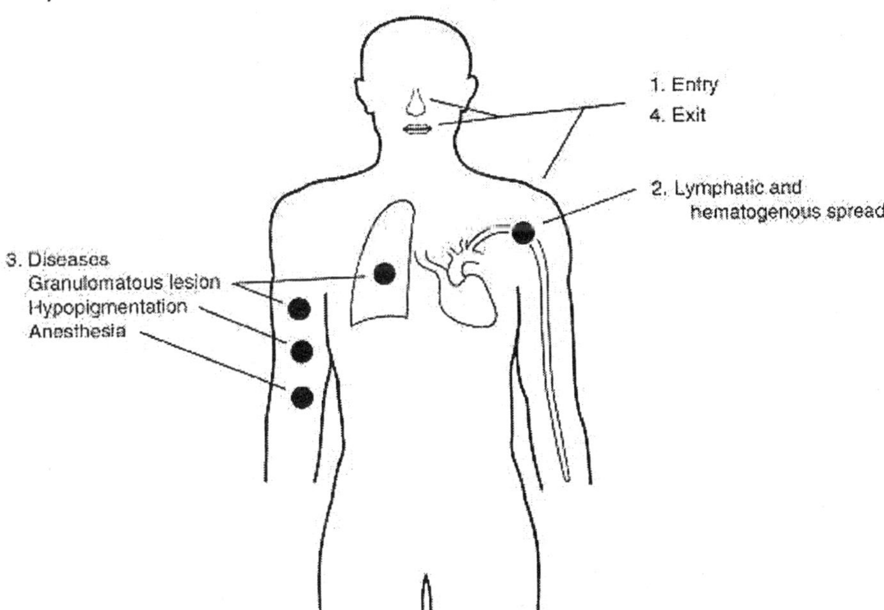

FIGURE 33-7 Pathogenesis of leprosy.

Diagnosis

The diagnosis of leprosy is based on the clinical signs previously discussed, and histologic examination of biopsy specimens taken from lepromas or other skin lesions. A consistent pattern of inflammation plus the presence of acid-fast bacilli is presumptive evidence of infection with *M leprae*. Although, *M leprae* cannot be grown *in vitro*, bacteriologic culturesof clinical material should be done to rule out the presence of other mycobacteria. The lepromin skin test, in which a heat-killed suspension of armadillo-derived *M leprae* is injected into the skin of the patient, has little diagnostic value but will provide information of prognostic importance about the immune status of the individual. The PCR technique mentioned above, by which very small amounts of *M leprae* DNA can be detected directly in clinical specimens, may prove to be a useful diagnostic tool.

Treatment and Control

A variety of combinations of the following drugs (so-called multidrug therapy or MDT) are used to treat leprosy: dapsone, rifampin, clofazimine, and either ethionamide or prothionamide. Paucibacillary cases (tuberculoid and borderline tuberculoid) can be treated in 6 months, although dapsone alone is usually given for up to 3 years after disease inactivity. Therapy for patients with lepromatous or borderline lepromatous leprosy may require primary treatment for 3 years, with dapsone alone continued for the rest of the patient's life. Although, some public health officials believe that MDT alone will result in eradication of leprosy in the near future, other experts are much more cautious. Drug resistance has been documented in *M leprae*. In many cases patient management must include anti-inflammatory therapy to alleviate the immunologic sequelae. Irreversible nerve damage leading to loss of sensation may result in paralysis or occult wounds and deformities. Wound prevention techniques and proper wound care are important.

Nocardia

Clinical Manifestations

Nocardia rarely causes clinical disease except in immunocompromised individuals, especially organ transplant recipients. Ninety percent of such patients present with pulmonary involvement, including cough, pleuritic chest pain, dyspnea, and radiologic abnormalities such as nodules and nodular infiltrates. Other clinical findings include weightloss, malaise, fever, and night sweats. About 20 percent of patients with nocardiosis present with cutaneous lesions, either localized or disseminated, and/or central nervous system involvement. About 50 percent of patients have an associated disease process (another infection or a tumor). Cutaneous infection with *N brasiliensis* results in localized development of granulomata and abscesses with soft tissue and bone involvement (Fig. 33-8).

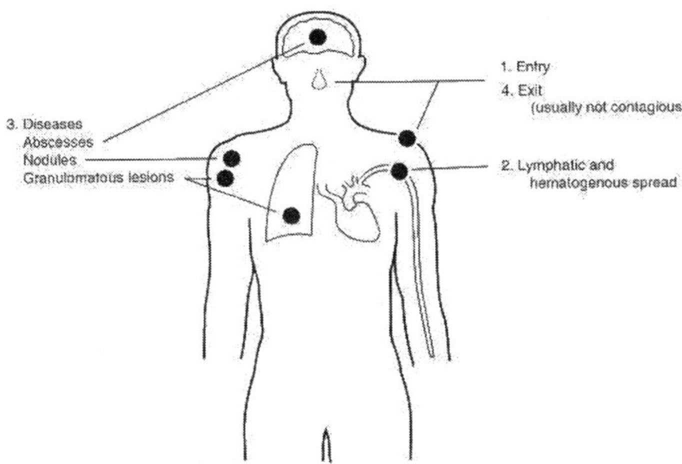

FIGURE 33-8 Pathogenesis of nocardiosis.

Structure

Nocardia organisms are Gram-positive rods, which in old cultures or clinical specimens may appear as branching chains resembling fungal hyphae. Figure 33-9 illustrates a typical colony of *N asteroides* and shows these organisms infecting rabbit alveolar macrophages. The filamentous morphology is evident. Nocardia are weakly acid-fast following staining with the modified Ziehl-Neelsen or Kinyoun stain. Cultures may grow in a few days, but typically require 2 to 3 weeks of incubation. The colony in Figure 33-9 is 3 weeks old.

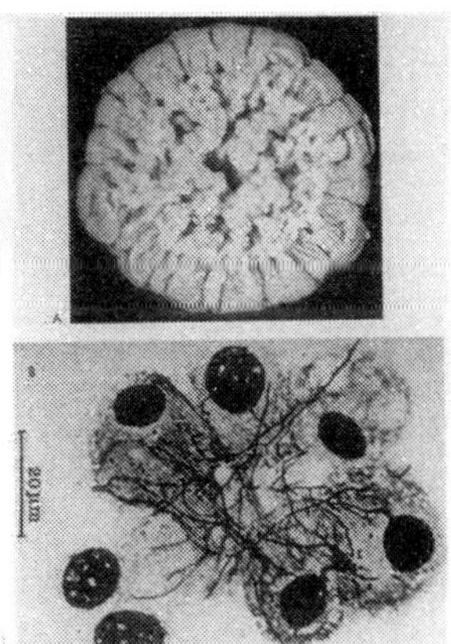

FIGURE 33-9 (A) Colony of *N asteroides* after 3 weeks of growth at 37° C on brain heart infusion agar (X11). (B) Interaction of *N asteroides*, virulent strain 14795, with seven (rabbit) alveolar macrophages (some in the process of fusing) 24 hours post infection. At 3 hours post infection, only rod-shaped, intracellular forms were apparent. At 6 hours post infection, elongation into filaments was evident. Some bacterionemata shown here are intact, and others are fragmenting. Gram-stained cover slip preparation. (Fig. A courtesy of Cynthis Vistica and Blaine L. Beaman. Fig. B from Beaman BL: In vitro response of rabbit alveolar macrophages to infection with *Nocardia asteroides*. Infect Immun 15:934, 1977, with permission.)

Classification and Antigenic Types

Three species of *Nocardia* are responsible for most human infections. *Nocardia asteroides* causes most nocardial pulmonary infections in this country (80 to 90 percent), with *N brasiliensis* (5 to 6 percent) and *N caviae* (3 percent) being recovered from only a few patients with nocardiosis. In the southern United States and in the tropics, *N brasiliensis* is an important agent of cutaneous nocardiosis. The three species can be distinguished by their patterns of proteolytic hydrolysis or of acid fermentation of several substrates.

Pathogenesis

Nocardia cells have been isolated from soil and organic material throughout the world. Natural infections occur in domestic animals. Human infection usually results from the inhalation of airborne bacilli or the traumatic inoculation of organisms into the skin. The infection is not transmissible between individuals. Natural resistance, mediated by intact mucous membranes and alveolar and tissue phagocytes, is quite strong.

In immunocompromised hosts, pulmonary infection results in the formation of abscesses and, rarely, granulomas with hematogenous or lymphatic dissemination to the skin or central nervous system. *Nocardia* can subvert the antimicrobial mechanisms of phagocytes by inhibiting phagosome-lysosome fusion. Owing to the debilitated nature of the infected patients, mortality is high (up to 45 percent), even with appropriate therapy.

Host Defenses

Nocardiosis is usually associated with T-cell dysfunctions, immunoglobulin deficiencies, or leukocyte abnormalities. Acquired resistance to *Nocardia* is complex, involving antibody-dependent phagocytosis by neutrophils, macrophage activation by the products of immune T cells, and the development of cytotoxic T lymphocytes. Neutrophils ingest opsonized bacteria but may not kill them. Macrophage activation is associated with containment and clearance of *Nocardia* organisms from the lungs. In a murine model, resistance to nocardiosis can be transferred with whole spleen cells or splenic T cells from immune mice.

Epidemiology

Although nocardiosis has been diagnosed in individuals with no detectable deficiency of humoral or cell-mediated immunity, it usually occurs in patients whose immune status has been compromised by post-transplant immunosuppressive therapy, leukemia, lymphoma, dysgammaglobulinemia, pancytopenia, humoral defects, chronic granulomatous disease, or steroid therapy. The male/female ratio in nocardiosis is approximately 2:1, and infections occur from infancy to old age. There is no apparent geographic clustering of cases in the United States, except for cutaneous infection with *N brasiliensis*, which is more common in the south.

Diagnosis

Nocardia can be identified presumptively by Gram and acid-fast stains and definitively by culture from appropriate clinical specimens. Sputum culture is useful for patients with a productive cough. The presence of branching, weakly acid-fast organisms in histologic sections, pus, or sputum suggests the clinical diagnosis. In one series, nearly 40 percent of cases required more invasive procedures (e.g., thoracentesis, transtracheal aspiration, bronchial washing, or biopsy) to obtain useful material for stain and culture of *Nocardia*.

Treatment and Control

Antimicrobial therapy with sulfa drugs (e.g., trimethoprim-sulfamethoxazole) is the treatment of choice. The duration of therapy ranges from 2-3 months for minor infections to 1 year for major infections.

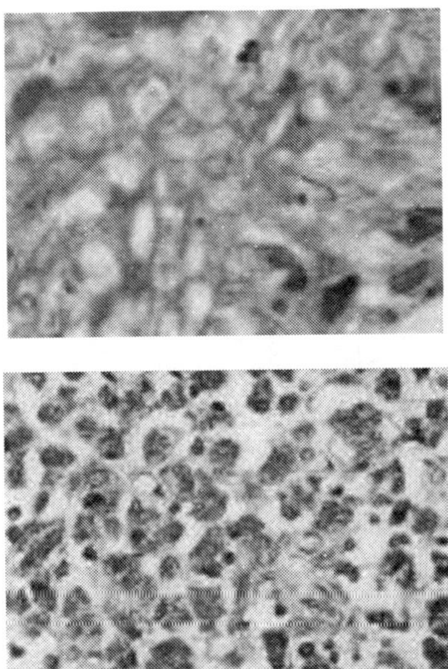

Colony and microscopic appearance of *Mycobacterium*

REFERENCES

Beaman BL, Beaman L: *Nocardia* species: host-parasite relationships. Clin Microbiol Rev 7:213, 1994

Bloom BR (ed): Tuberculosis - Pathogenesis, Protection and Control. ASM Press, Washington, DC, 1994

Johnson JL, Elner JJ, Shiratsuchi H: Monocyte-*Mycobacterium avium* complex interactions: studies of potential virulence factors for humans. Immunol Ser 60:263, 1994

Krahenbuhl JL, Adams, LB: The role of the macrophage in resistance to the leprosy bacillus. Immunol Ser 60:281, 1994

Modlin RL: Th1-Th2 paradigm: insights from leprosy. J Invest Dermatol 102:828, 1994

Reichman LB, Hershfield ES (eds): Tuberculosis - A Comprehensive International Approach. Marcel Dekker, New York, 1993

Rigsby MO, Curis, AM: Pulmonary disease from nontuberculous mycobacteria in patients with human immunodeficiency virus. Chest 106:913, 1994

Rom WN, Garay S (eds): Tuberculosis. Little, Brown and Co, New York, 1995

Shinnick T (ed): Tuberculosis. Curr Top Microbiol Immunol, Springer-Verlag, Heidelberg, 1995

Chapter 19

Mycoplasma

General Concepts

Clinical Manifestations

Mycoplasma pneumoniae infection is a disease of the upper and lower respiratory tracts. Cough, fever, and headache may persist for several weeks. Convalescence is slow. *Ureaplasma urealyticum* infection causes nongonococcal urethritis in men, resulting in dysuria, urgency, and urethral discharge.

Structure, Classification, and Antigenic Types

Mycoplasmas are spherical to filamentous cells with no cell walls. There is an attachment organelle at the tip of filamentous *M pneumoniae, M genitalium,* and several other pathogenic mycoplasmas. Fried-egg-shaped colonies are seen on agar. The mycoplasmas presumably evolved by degenerative evolution from Gram-positive bacteria and are phylogenetically most closely related to some clostridia. Mycoplasmas are the smallest self-replicating organisms with the smallest genomes (a total of about 500 to 1000 genes); they are low in guanine and cytosine. Mycoplasmas are nutritionally very exacting. Many require cholesterol, a unique property among prokaryotes. Ureaplasmas require urea for growth, another unusual property. Mycoplasmas have surface antigens such as membrane proteins, lipoproteins, glycolipids, and lipoglycans. Some of the membrane proteins undergo spontaneous antigenic variation. Antibodies to surface antigens inhibit growth; various serological tests have been developed and are useful in classification.

Pathogenesis

Mycoplasmas are surface parasites of the human respiratory and urogenital tracts. *Mycoplasma pneumoniae* attaches to sialoglycoproteins or sialoglycolipid receptors on the tracheal epithelium via protein adhesins on the attachment organelle. The major adhesin is a 170-kilodalton (kDa) protein, named P1. Hydrogen peroxide and superoxide radicals (O2-) excreted by the attached organisms cause oxidative tissue damage. Pneumonia is induced largely by local immunologic and phagocytic responses to the parasites. Sequelae of *M pneumoniae* infection (mainly hematologic and neurologic) apparently have an autoimmune etiology. Several fastidious mycoplasmas may act as cofactors in activation of the aquired immunodeficiency syndrome (AIDS). Macrophage activation, cytokine induction, and superantigen properties of some mycoplasmal cell components can be considered as pathogenicity factors.

Host Defenses

IgM antibodies, followed by IgG and secretory IgA, are important in host resistance. The importance of cell-mediated immunity is unclear.

Epidemiology

Mycoplasma pneumoniae infection occurs worldwide and is more prevalent in colder months. It affects mainly children ages 5 to 9 years. It is spread by close personal contact and has a long incubation period. *Ureaplasma urealyticum* is spread primarily through sexual contact. Women may be asymptomatic reservoirs.

Diagnosis

Culture of *M pneumoniae* from sputum or a throat swab is possible, but very slow; therefore diagnosis is usually based on serologic tests. Tests using diagnostic DNA probes and amplification of specific genomic mycoplasma sequences by the polymerase-chain reaction (PCR) are being developed.

Control

There is no certified vaccine for *M pneumoniae*. Treatment with erythromycin or tetracyclines is effective in reducing symptoms in both *M pneumoniae* and *U urealyticum* infections.

INTRODUCTION

Mycoplasmas are the smallest and simplest self-replicating bacteria. The mycoplasma cell contains the minimum set of organelles essential for growth and replication: a plasma membrane, ribosomes, and a genome consisting of a double-stranded circular DNA molecule (Fig. 37-1). Unlike all other prokaryotes, the mycoplasmas have no cell walls, and they are consequently placed in a separate class *Mollicutes* (*mollis*, soft; *cutis*, skin). The trivial term mollicutes is frequently used as a general term to describe any member of the class, replacing in this respect the older term mycoplasmas.

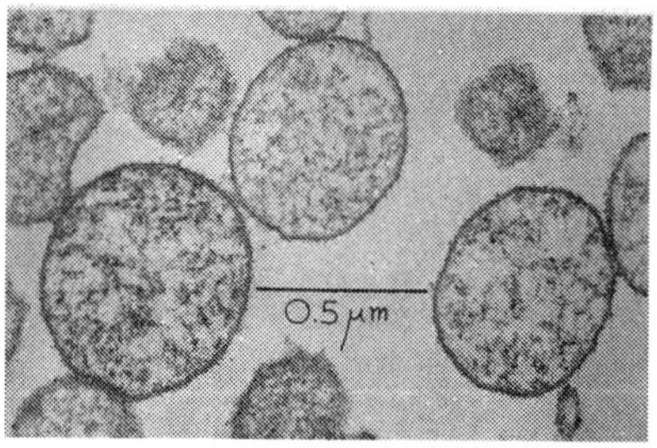

0.5 μm

FIGURE 37-1 Electron micrograph of thin-sectioned mycoplasma cells. Cells are bounded by a single membrane showing in section the characteristic trilaminar shape. The cytoplasm contains thin threads representing sectioned chromosome and dark granules representing ribosomes. (Courtesy of RM Cole, Bethesda, Maryland).

Mycoplasmas have been nicknamed the "crabgrass" of cell cultures because their infections are persistent, frequently difficult to detect and diagnose, and difficult to cure. Contamination of cell cultures by mycoplasmas presents serious problems in research laboratories and in biotechnological industries using cell cultures. The origin of contaminating mycoplasmas is in components of the culture medium, particularly serum, or in the flora of the technician's mouth, spread by droplet infection.

Clinical Presentation

Mycoplasmal pneumonia

The term primary atypical pneumonia was coined in the early 1940s to describe pneumonias different from the typical lobar pneumonia caused by pneumococci. Several common respiratory viruses, including influenza virus and adenovirus, were shown to be responsible for a significant number of these pneumonias. From other cases, many of which developed antibodies agglutinating red blood cells in the cold (cold agglutinins), an unidentified filterable agent was isolated by Eaton and associates and was called *Eaton agent*. This agent was identified as a new *Mycoplasma* species after its successful cultivation on cell-free media in 1962. Named *Mycoplasma pneumoniae*, it was the first clearly documented mycoplasma pathogenic for humans.

The effects of *M pneumoniae* on humans include subclinical infection, upper respiratory disease, and bronchopneumonia. Most human infections do not progress to a clinically evident pneumonia. When pneumonia occurs, the onset generally is gradual and the clinical picture is one of a mild to moderately severe illness, with early complaints referable to the lower respiratory passages. Radiography frequently reveals evidence of pneumonia before physical signs are apparent. Involvement is usually limited to one of the lower lobes of the lungs, and the pneumonia is interstitial or bronchopneumonic. The course of disease varies; remittent fever, cough, and headache persist for several weeks. One of the most consistent clinical features is a long convalescence, which may extend from 4 to 6 weeks. Few fatal cases have been reported. Several unusual complications have been noted, including hemolytic anemia, polyradiculitis, encephalitis, aseptic meningitis, and central nervous system illness such as Guillain-Barré syndrome. In addition, pericarditis and pancreatitis have been observed. These sequelae may be related to the suspected immunopathology of *M pneumoniae* disease (see below).

Nongonococcal Urethritis and Salpingitis

Growing evidence suggests that *Ureaplasma urealyticum* causes nongonococcal urethritis in men free of *Chlamydia trachomatis*, an established agent of nongonococcal urethritis. The wide occurrence of *U urealyticum* in sexually active, symptom-free adults hampers research in this field. Evidence is based primarily on the production of nongonococcal urethritis symptoms in ureaplasma-free and chlamydia-free volunteers by intraurethral inoculation of *U urealyticum* and on a report that this disease could be cured in a chlamydia-free man only when he and his partner were treated simultanously with tetracycline, which eliminated *U urealyticum* from both.

Ureaplasmas have also been associated with chorioamnionitis, habitual spontaneous abortion, and low-weight infants. *Mycoplasma hominis*, a common inhabitant of the vagina of healthy women, becomes pathogenic once it invades the internal genital organs, where it may cause pelvic inflammatory diseases such as tubo-ovarian abscess or salpingitis.

It has been suggested that *Mycoplasma genitalium*, isolated in 1981 from the urethral discharge of two homosexual men, may account for the tetracycline-responsive, nongonococcal urethritis cases in which chlamydias and ureaplasmas cannot be isolated (about 20 percent of all cases). However, *M genitalium* is so fastidious that very few clinical isolates have so far been made on the best mycoplasma medium available. Only the recent application of specific PCR amplification of the organism's DNA in clinical specimens has provided experimental proof for the relative prevalence of *M genitalium* in the human urogenital tract and its apparent role in male urethritis.

Mycoplasmas in AIDS and Immunocompromised Patients

The question of whether mycoplasmas act as co-factors in the development of AIDS has attracted much attention recently. Several mycoplasms have so far been incriminated: *M fermentans*, considered until recently a relatively rare mycoplasma of the human urogenital tract, and *M penetrans* , a newly-discovered human mycoplasma isolated from several AIDS patients. *M pirum*, a mycoplasma of an unknown host, has been recently isolated from the blood of a few AIDS patients. While, *in vitro* studies show that these mycoplasmas may markedly enhance pathogenicity of the human immunodeficiency virus, the possibility that the mycoplasmas may simply represent opportunistic agents found in high frequency in patients with AIDS, cannot be ruled out. Yet on the whole, with the increasing incidence of immunocompromised patients (due to AIDS, organ transplantation, etc.) evidence is accumulating for invasion of tissues and the intracellular location of some mycoplasmas, notably *M fermentans* and *M penetrans*. Extragenital infections by urogenital mycoplasmas are rather common in neonates, immunosuppressed and/or hypogammaglobulinemic patients; clinical symptoms are expressed frequently as arthritis.

Structure, Classification, and Antigenic Types

Distinguishing Properties

The coccus is the basic form of all mycoplasmas in culture. The diameter of the smallest coccus capable of reproduction is about 300 nm. In most mycoplasma cultures, elongated or filamentous forms (up to 100 μm long and about 0.4 μm thick) also occur. The filaments tend to produce truly branched mycelioid structures, hence the name mycoplasma (myces, a fungus; plasma, a form). Mycoplasmas reproduce by binary fission, but cytoplasmic division frequently may lag behind genome replication, resulting in formation of multinuclear filaments (Fig. 37-2).

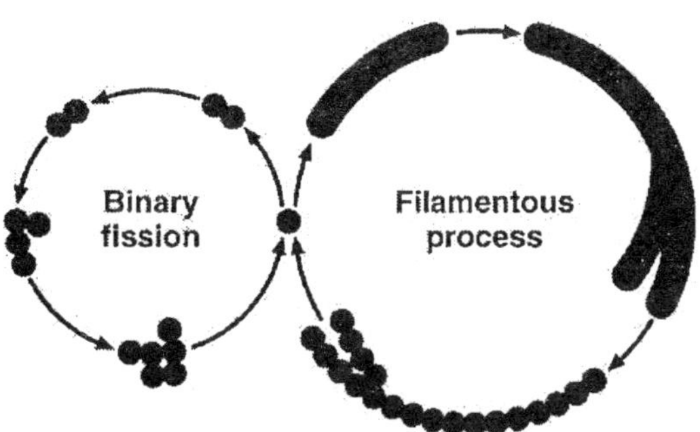

FIGURE 37-2 Schematic presentation of the mode of mycoplasma reproduction. Cells may either divide by binary fission or first elongate to multinucleate filaments, which subsequently breakup to coccoid bodies. (From Razin S: Mycoplasmas: the smallest pathogenic procaryotes. Isr J Med Sci 17:510, 1981, with permission).

Some mycoplasmas possess unique attachment organelles, which are shaped as a tapered tip in *M pneumoniae* and *M genitalium. Mycoplasma pneumoniae* is a pathogen of the respiratory tract, adhering to the respiratory epithelium, primarily through the attachment organelle. Interestingly, these two human mycoplasmas exhibit gliding motility on liquid-covered surfaces. The tip structure always leads, again indicating its importance in attachment. One of the most useful distinguishing features of mycoplasmas is their peculiar fried-egg colony shape, consisting of a central zone of growth embedded in the agar and a peripheral one on the agar surface (Fig. 37-3).

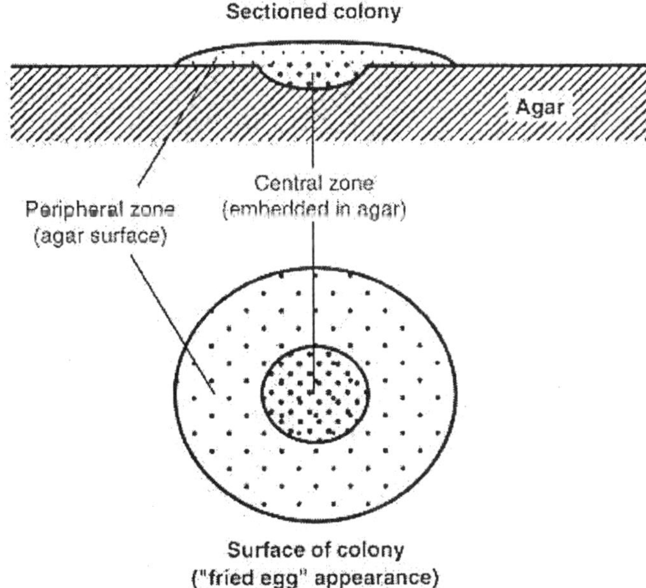

FIGURE 37-3 Morphology of a typical "fried-egg" mycoplasma colony.

The lack of cell walls and intracytoplasmic membranes facilitates isolation of the mycoplasma membrane in a relatively pure form. The isolated mycoplasma membrane resembles that of other prokaryotes in being composed of approximately two-thirds protein and one-third lipid. The mycoplasma lipids resemble those of other bacteria, apart from the large quantities of cholesterol in the sterol-requiring mycoplasmas.

Membrane proteins, glycolipids, and lipoglycans exposed on the cell surface are the major antigenic determinants in mycoplasmas. Antisera containing antibodies to these components inhibit growth and metabolism of the mycoplasmas and, in the presence of complement, cause lysis of the organisms. These properties are used in various serologic tests that differentiate between mycoplasma species and serotypes and detect antibodies to mycoplasmas in sera of patients (see below).

Molecular Biology

The mycoplasma genome is typically prokaryotic, consisting of a circular, double stranded DNA molecule. The *Mycoplasma* and *Ureaplasma* genomes are the smallest recorded for any self-reproducing prokaryote (Table 37-1). Therefore, there are very few genes; in some mycoplasmas the number is estimated at fewer than 500, about one sixth the number of genes in *Escherichia coli*. Mycoplasmas accordingly express a small number of cell proteins and lack many enzymatic activities and metabolic pathways. Their nutritional requirements are correspondingly complex, and they are dependent on a parasitic mode of life.

TABLE 37-1 Taxonomy and Properties of Mycoplasmas Capable of infecting Humans[a]

Genus	No of Established Species	Genome		Cholesterol Requirement	Distinctive Properties	Hosts
		Size (kbp)	G+C Content (mol%)			
Mycoplasma	98	580-1300	23-40	+	None	Humans other animals
Ureaplasma	6	760-1140	27-29	+	Urease positive	Humans other animals

aThe table includes only Mycoplasma and Ureaplasma , the mycoplasma genera species capable of infecting humans. The genera Acholeplama, asteroleplama, Anaeroplasma, Mesoplasma, and Spiroplasama contain species infecting only animals, plants and arthropods.

The dependence of mycoplasmas on their host for many nutrients explains the great difficulty of cultivation in the laboratory. The complex media for mycoplasma culture contain serum, which provides fatty acids and cholesterol for mycoplasma membrane synthesis. The requirement of most mycoplasmas for cholesterol is unique among prokaryotes. The consensus is that only a small fraction of mycoplasmas existing in nature have been cultivated so far. Some of the cultivable mycoplasmas, including the human pathogen *M pneumoniae*, grow very slowly, particularly on primary isolation. *Ureaplasma urealyticum*, a pathogen of the human urogenital tract, grows very poorly in vitro, reaching maximal titers of 107 organisms/ml of culture. *Mycoplasma genitalium*, another human pathogen, grows so poorly in vitro that only a few successful isolations have been achieved.

Glucose and other metabolizable carbohydrates can be used as energy sources by the fermentative mycoplasmas possessing the Embden-Meyerhof-Parnas glycolytic pathway. All mycoplasmas examined thus far possess a truncated, flavin-terminated respiratory system, which rules out oxidative phosphorylation as an ATP-generating mechanism. Breakdown of arginine by the arginine dihydrolase pathway has been proposed as a major source of ATP in nonfermentative mycoplasmas. Ureaplasmas have a requirement, unique among living organisms, for urea. Because they are non-glycolytic and lack the arginine dihydrolase pathway, it has been suggested, and later proven experimentally, that ATP is generated through an electrochemical gradient produced by ammonia liberated during the intracellular hydrolysis of urea by the organism's urease.

The mycoplasma genome is characterized by a low guanine-plus-cytosine content and by a corresponding preferential utilization of codons containing adenine and uracil, particularly in the third position. Most interesting is the use of the universal stop codon UGA as a tryptophan codon in many mycoplasmas, a rare property found so far only in mycoplasmas and in nonplant mitochondria. Resistance of mycoplasmal RNA polymerase to rifampicin is another property distinguishing mycoplasmas from the conventional eubacteria. However, apart from this resistance to rifampicin, the mycoplasmas are susceptible to antibiotics, such as tetracyclines and chloramphenicol, that inhibit protein synthesis on prokaryotic ribosomes.

Phylogeny

As the smallest and simplest self-replicating prokaryotes, the mycoplasmas pose an intriguing question: do they represent the descendents of exceedingly primitive bacteria that existed before the development of a peptidoglycan-based wall, or do they represent evolutionary degenerate eubacterial forms that have lost their cell walls? The balance of the molecular evidence, based largely on comparison of base sequences of the highly conserved ribosomal RNA (rRNA) molecules, particularly of the 16S rRNA type, favors the hypothesis of degenerative evolution. According to Woese and his colleagues, the mycoplasmas evolved as a branch of the low-guanine-plus-cytosine Gram-positive bacteria and are most closely related to two clostridia, *Clostridium innocuum* and *C ramosum*. However, the marked phenotypic and genotypic variability among mycoplasmas has led some workers to conclude that mycoplasmas evolved from a variety of walled bacteria and accordingly have a polyphyletic origin. Woese maintains that the origin of mycoplasmas is monophyletic and explains the great varietyof mycoplasmas by a process of rapid evolution characteristic of the group.

Pathogenesis

All mycoplasmas cultivated and identified thus far are parasites of humans, animals, plants, or arthropods. The primary habitats of human and animal mycoplasmas are the mucous surfaces of the respiratory and urogenital tracts and the joints in some animals. Although, some mycoplasmas belong to the normal flora, many species are pathogens, causing various diseases that tend to run a chronic course (Fig. 37-4).

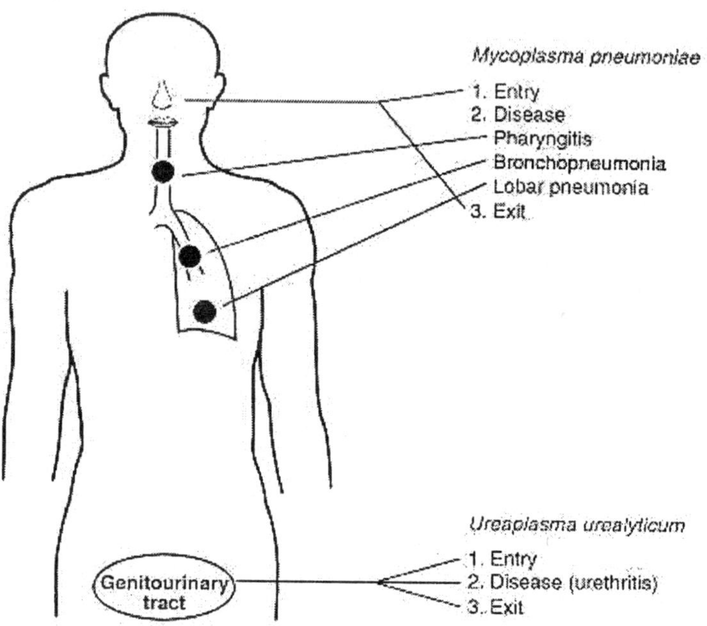

FIGURE 37-4 Pathogenesis and disease sites of infection by *M pneumoniae* and *U urealyticum.*

Most mycoplasmas that infect humans and other animals are surface parasites, adhering to the epithelial linings of the respiratory and urogenital tracts. Adherence is firm enough to prevent the elimination of the parasites by mucous secretions or urine. The intimate association between the adhering mycoplasmas and their host cells provides an environment in which local concentrations of toxic metabolites excreted by the parasite build up and cause tissue damage (Fig. 37-5). Moreover, because mycoplasmas lack cell walls, fusion between the membranes of the parasite and host has been suggested, and some experimental evidence for it has recently been obtained. Membrane fusion would alter the composition and permeability of the host cell membrane and enable the introduction of the parasite's hydrolytic enzymes into the host cell, events expected to cause serious damage. Recent studies have indicated the presence in mycoplasmas of antigenic variability systems. These systems, some of which are already defined in molecular genetic terms, are responsible for rapid changes in major surface protein antigens. The change in the antigenic coat of the parasite helps it to escape recognition by the immune mechanisms of the host.

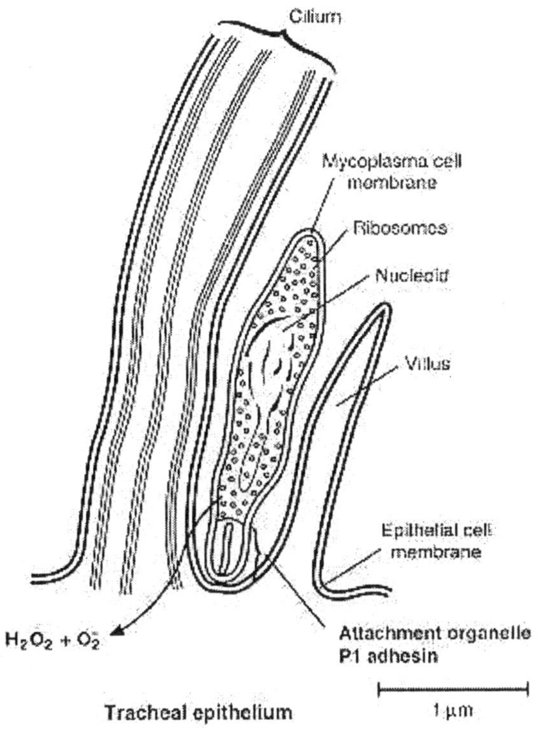

FIGURE 37-5 Schematic presentation of a *M pneumoniae* organism attaching to the surface of the ciliary tracheal epithelium, as seen by electron microscopy of a thin section. The clustering of the P1 adhesin on the surface of the attachment organelle at the tip of the mycoplasma is depicted. The H_2O_2 and O_2 excreted by the mycoplasma penetrate into the host cell and cause oxidative damage.

Because attachment of *M pneumoniae* and *M genitalium* is affected by pretreatment of the host cells with neuraminidase, sialoglycoproteins and/orsialoglycolipids of the host cell membrane appear to be receptor sites for these mycoplasmas. There is evidence that several *M pneumoniae* membrane proteins act as adhesins and that they have high affinity for the specific receptors for *M pneumoniae* on host cells. Monoclonal antibodies to one of these proteins, protein P1 (molecular weight, 170,000 daltons), inhibit attachment of the parasite. Ferritin labeling of the antibodies has shown that P1 concentrates on the tip structure of the mycoplasma, a finding that further supports the notion that the tip serves as an attachment organelle.

The results obtained with *M pneumoniae* were essentially duplicated recently with *M genitalium* and showed that in this organism, which closely resembles *M pneumoniae* morphologically and physiologically, a major adhesin protein, named MgPa, is clustered at the tip organelle. The genes of the major adhesins of *M pneumoniae* (P1) and of *M genitalium* (MgPa) were cloned and sequenced, allowing the characterization of these proteins. The two adhesins are alike in many respects and in fact contain extensive areas of homology, as expressed also by shared epitopes. These two proteins may be the product of an ancestral gene that underwent a horizontal gene transfer event.

The nature of the toxic factors that damage the mucosal surfaces infected by mycoplasmas is still unclear. Toxins are rarely found in mycoplasmas. Consequently, researchers considered whether the end products of mycoplasma metabolism were responsible for tissue damage. Hydrogen peroxide (H_20_2), the end product of respiration in mycoplasmas, has been implicated as a major pathogenic factor ever since it was shown to be responsible for the lysis of erythrocytes by mycoplasmas in vitro; however, the production of H_2O_2 alone does not determine pathogenicity, as the loss of virulence in *M pneumoniae* is not accompanied by a decrease in H_2O_2 production. For the H_2O_2 to exert its toxic effect, the mycoplasmas must adhere closely enough to the host cell surface to maintain a toxic, steady-state concentration of H_2O_2 sufficient to cause direct damage, such as lipid peroxidation, to the cell membrane. The accumulation of malonyldialdehyde, an oxidation product of membrane lipids, in cells exposed to *M pneumoniae* supports this notion. Moreover, *M pneumoniae* inhibits host cell catalase by excreting superoxide radicals (O_2-). This would be expected to further increase the accumulation of H_2O_2 at the site of parasite-host cell contact (Fig. 37-6).

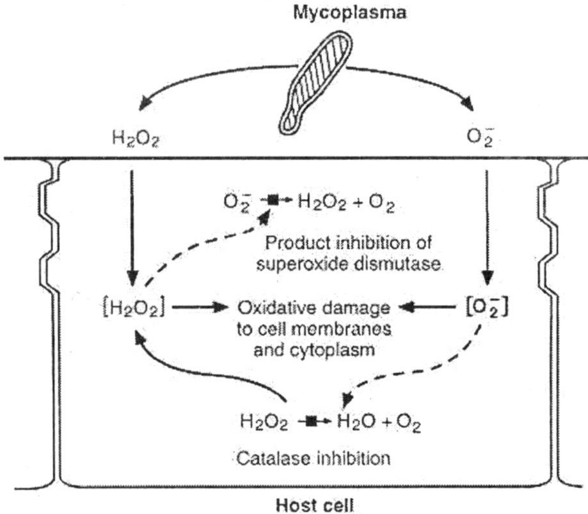

FIGURE 37-6 Proposed mechanism of oxidative damage to host cells by adhering *M pneumoniae* by increasing concentrations of H_2O_2 and O_{2-}. (Modified from Almagor M, Kahane I, Yatziv S: Role of superoxide anion in host cell injury induced by *Mycoplasma pneumoniae* infection. J Clin Invest 73:842, 1984, with permission).

There is evidence that both organism-related and host-related factors are involved in the pathogenesis of mycoplasma infections. Mycoplasmas activate macrophages, and induce cytokine production and lymphocyte proliferation; the rat pathogen, *Mycoplasma arthritidis,* produces a potent superantigen. Thus, in the case of *M pneumoniae,* the host may be largely responsible for the pneumonia by mounting a local immune response to the parasite. Syrian hamsters inoculated intranasally with *M pneumoniae* show patchy bronchopneumonic lesions consisting of infiltration of mononuclear cells. The ablation of thymic function before the experimental infection prevents development of the characteristic pulmonary infiltration, but lengthens the period during which the organisms may be isolated from the lungs. When thymic animals are allowed to recover and then reinfected, an exaggerated and accelerated pneumonic process occurs. Epidemiologic data also suggest that repeated infections in humans are required before symptomatic disease occurs: serum antibodies to *M pneumoniae* can be found in most children 2 to 5 years of age, although the illness occurs with greatest frequency in individuals 5 to 15 years of age.

An immunopathologic mechanism also may explain the complications affecting organs distant from the respiratory tract in some patients infected with *M pneumonia.* Various autoantibodies have been detected in the sera of many of these patients, including cold agglutinins reacting with the erythrocyte I antigen, and antibodies reacting with lymphocytes, smooth muscle cells, and brain and lung antigens. Serologic cross-reactions between *M pneumoniae* and brain and lung antigens have been demonstrated, and these antigens are probably related to the glycolipids of *M pneumoniae* membranes, which are also found in most plants and in many bacteria. Clearly, host reaction varies markedly, as only about half of the patients develop cold agglutinins and complications are rare, even among individuals with anti-tissue globulins.

Host Defenses

Infection with *M pneumoniae* induces the development of serum antibodies that fix complement, inhibit growth of the organism and lyse the organism in the presence of complement. Generally, the first antibodies produced are of the IgM class, whereas later in convalescence the predominant antibody is IgG. Secretory IgA antibodies also develop and appear to be important in host resistance. The first infection in infancy usually is asymptomatic and generates a brief serum antibody response. Recurrent infections generate a more prolonged systemic antibody response and increasing numbers of circulating antigen-responsive lymphocytes. By late childhood, clinically apparent lower respiratory disease, including pneumonia, becomes more common. Therefore, mycoplasma respiratory disease manifestations appear to vary, depending on the state of local and systemic immunity at the time of reinfection. One hypothesis is that local immunity mediates resistance to infection and that systemic immunity contributes substantially to the pulmonary and systemic reaction characteristic of *M pneumoniae* pneumonia.

The relative importance of humoral and cell-mediated immunity in resistance to respiratory mycoplasma infections is still unclear. For many mycoplasma infections, such as bovine pleuropneumonia, resistance can be transferred with convalescent-phase serum, but this may not be true for all mycoplasma respiratory diseases. For example, resistance of rats to pulmonary disease induced by *M pulmonis* can be transferred only with spleen cells obtained from previously infected animals. Although, IgA antibody may be important in resistance to mycoplasmas, other factors seem to be involved in resistance to pulmonary disease, and these factors may not be the same for all mycoplasma infections.

Epidemiology

One of the most puzzling features of *M pneumoniae* pneumonia is the age distribution of patients. In a survey conducted between 1964 and 1975 of more than 100,000 individuals in the Seattle area, the age-specificattack rate was highest among 5- to 9-year-old children. Rates of *M pneumoniae* pneumonia in the youngest age group, 0 to 4 years old, were about one-half those in school-age children, but considerably higher than in adults. *Mycoplasma pneumoniae* pneumonia was rarely observed in infants younger than 6 months, suggesting maternally conferred immunity (Fig. 37-7). *Mycoplasma pneumoniae* accounts for 8 to 15 percent of all pneumonias in young school-age children. In older children and in young adults, the organism is responsible for approximately 15 to 50 percent of all pneumonias. Infection with *M pneumoniae* occurs worldwide all year round but shows a predilection for the colder months, apparently because of the greater opportunityfor transmission by droplet infection. *Mycoplasma pneumoniae* appears to require close personal contact to spread; successful spreading usually occurs in families, schools, and institutions. The incubation period ranges from 2 to 3 weeks.

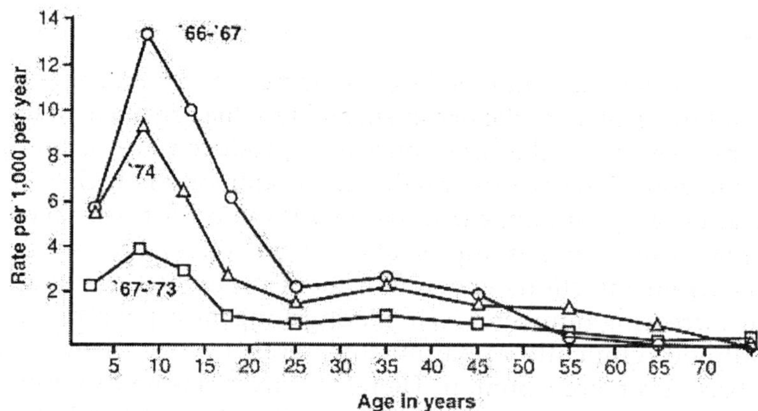

FIGURE 37-7 Incidence of *M pneumoniae* pneumonia in Seattle by age, for two epidemics (1966-67 and 1974) and the endemic periods (1967-73). (From Foy HM, Kenny GE, Cooney MK, Allen ID: Long-term epidemiology of infections with Mycoplasma pneumoniae pneumonia. J Infect Dis 139:681, 1979, with permission).

Ureaplasma urealyticum is spread primarily through sexual contact. Colonization has been linked to the frequency of sexual intercourse and the number of sexual partners. Women may be asymptomatic reservoirs of infection.

Diagnosis

Culture is essential for definitive diagnosis (See below).

Culture

A routine mycoplasma medium consists of heart infusion, peptone, yeast extract, salts, glucose or arginine, and horse serum (5 to 20 percent). Fetal or newborn calf serum is preferable to horse serum. To prevent the overgrowth of the fast-growing bacteria that usually accompany mycoplasmas in clinical materials, penicillin, thallium acetate or both are added as selective agents. For *Ureaplasma* culture, the medium is supplemented with urea and its pH is brought to 6.0. *Ureaplasm a* and *M genitalium* are relatively sensitive to thallium, which is, therefore, omitted from their culture media. For *M pneumoniae* isolation, nasopharyngeal secretions are inoculated into a selective diphasic medium (pH 7.8) made of mycoplasma broth and agar and supplemented with glucose and phenol red. When *M pneumoniae* grows in this medium, it produces acid, causing the color of the medium to change from purple to yellow. Broth from the diphasic medium is subcultured to mycoplasma agar when a color change occurs, or at weekly intervals for a minimum of 8 weeks.

Identification

Colonies appearing on the plates can be identified as *M pneumoniae* by staining directly on agar with homologous fluorescein-conjugated antibody or by demonstrating that a specific antiserum to *M pneumoniae* inhibits their growth on agar. Colonies of ureaplasmas are usually minute (less than 100 μm in diameter); because of urea hydrolysis and ammonia liberation, the medium becomes alkaline. When manganous sulfate is added to the medium, the ureaplasma colonies stain dark brown. Isolates can be characterized in more detail by a variety of biochemical and serologic tests. More sophisticated tests, including electrophoretic analysis of cell proteins, DNA-DNA hybridization tests, mycoplasmal DNA cleavage patterns by restriction endonucleases, and PCR tests employing species-specific primers for amplification, may be performed in a research laboratory.

Serodiagnosis and Molecular Probes

Serodiagnosis consists of examining serum samples for antibodies that inhibit the growth and metabolism of the organism or fix complement with mycoplasmal antigens. A fourfold or greater rise in IgG titer is considered indicative of recent infection, whereas a sustained high antibody titer may not be significant, because a relatively high level of antibody may persist for at least 1 year after infection. A variety of rapid tests based on indirect hemagglutination of erythrocytes or latex particles coated with *M pneumoniae* antigens have been developed, and some are commercially available.The cold agglutinin test is less useful because only about one-half of patients develop cold agglutinins and because these antibodies also are induced by a great many other conditions.

Present techniques for laboratory diagnosis of *M pneumoniae* infections are of little use to the clinician because recovery by culture and identification of the mycoplasmas take at least 1 to 2 weeks. Methods for rapid laboratory diagnosis, such as direct demonstration of organisms in the respiratory specimens by nucleic acid amplification techniques, have promise but diagnostic kits are not yet commercially available.

Control

Prevention

Chemoprophylaxis of mycoplasma infections is not recommended, and no vaccine is available. Prior natural infection appears to provide the most effective resistance; however, evidence shows that *M pneumoniae* infections recur at intervals of several years. These observations suggest that immunity to a single natural infection is relatively short-term.

Treatment

The mycoplasmas are sensitive to tetracyclines, macrolides, and the newer quinolones, but are resistant to antibiotics that specifically inhibit bacterial cell wall synthesis. Tetracycline or erythromycin is recommended for treatment of *M pneumoniae* pneumonia, although effective treatment of the symptoms usually is not accompanied by eradication of the organism from the infected host. To prevent recurrence of nongonococcal urethritis caused by *U urealyticum*, sexual partners should be treated simultaneously with tetracycline. The incidence of tetracycline-resistant strains of *U urealyticum* and *M hominis* is on the rise.

REFERENCES

Blanchard A, Montagnier L: AIDS-associated mycoplasmas. Annu Rev Microbiol 48:687,1994

Cassell GH (ed) : The changing role of mycoplasmas in respiratory disease and AIDS. Clin Infect Dis 17 (Suppl 1), 1993

Cole BC, Griffiths MM: Triggering and exacerbation of autoimmune arthritis by the *Mycoplasma arthritidis* superantigen MAM. Arthritis Rheum 36:994, 1993

Kenny GE, Kaiser GG, Cooney MK, et al: Diagnosis of *Mycoplasma pneumoniae* pneumonia: sensitivities and specificities of serology with lipid antigen and isolation of the organism on soy peptone medium for identification of infections. J Clin Microbiol 28:2087, 1990

Maniloff J, McElhaney RN, Finch LR, Baseman JB (eds): Mycoplasmas, Molecular Biology and Pathogenesis. American Society for Microbiology, Washington, 1992

Pietsch K, Ehlers S, Jacobs E: Cytokine gene expression in the lungs of BALB/c mice during primary and secondary intranasal infection with *Mycoplasma pneumoniae*. Microbiology 140:2043, 1994

Razin S: Molecular biology and genetics of mycoplasmas (*Mollicutes*). Microbiol Rev 49:419, 1985

Razin S: DNA probes and PCR in diagnosis of mycoplasma infections. Molec Cellular Probes 8:497, 1994

Razin S, Barile MF (eds): The Mycoplasmas, Vol. 4, Mycoplasma Pathogenicity Academic Press, Orlando, 1985

Razin S, Jacobs E: Mycoplasma adhesion. J Gen Microbiol 138:407, 1992

Razin S, Tully JG (eds): Molecular and Diagnostic Procedures in Mycoplasmology. Academic Press, Orlando, 1995

Rottem S, Kahane I (eds): Mycoplasma Cell Membranes. Plenum Press, New York, 1993

Wise KS: Adaptive surface variation in mycoplasmas. Trends Microbiol 1:59, 1993

Chapter 20

Chlamydia

General Concepts

Clinical Manifestations

Ocular Infections: *Chlamydia trachomatis* causes trachoma and inclusion conjunctivitis. Trachoma is characterized by the development of follicles and inflamed conjunctivae. The cornea may become cloudy and vascularized; repeated infections are a common cause of blindness. Inclusion conjunctivitis is a milder inflammatory conjunctival infection with purulent discharge.

Genital Infections: Some *C trachomatis* strains cause genital infections, including nongonococcal urethritis in men and acute salpingitis and cervicitis in women. Other strains cause lymphogranuloma venereum, a venereal disease with genital lesions and regional lymph node involvement (buboes).

Respiratory Infections: *Chlamydia psittaci* usually causes an influenza like illness called psittacosis. *Chlamydia pneumoniae* (TWAR organism) causes atypical pneumonitis in humans.

Structure, Classification, and Antigenic Types

Chlamydiae are obligate intracellular bacteria. They lack several metabolic and biosynthetic pathways and depend on the host cell for intermediates, including ATP. Chlamydiae exist as two stages: (1) infectious particles called elementary bodies and (2) intracytoplasmic, reproductive forms called reticulate bodies. The chlamydiae consist of three species, *C trachomatis, C psittaci,* and *C pneumoniae.* The first two contain many serovars based on differences in cell wall and outer membrane proteins. *Chlamydia pneumoniae* contains one serovar the TWAR organism.

Pathogenesis

Chlamydiae have a hemagglutinin that may facilitate attachment to cells. The cell-mediated immune response is largely responsible for tissue damage during inflammation, although an endotoxin-like toxin has been described.

Host Defenses

Antibodies develop during infection, but they do not prevent reinfection. The precise role of cell-mediated immunity is not known.

Epidemiology

Trachoma occurs worldwide and is prevalent in Africa and Asia. *Chlamydia trachomatis* usually is inoculated into the eye by contaminated fingers or fomites or, in neonates, by passage through an infected birth canal. Genital infections are spread venereally, and respiratory infections usually by inhalation. Psittacosis is acquired from infected birds.

Diagnosis

The clinical presentation is often diagnostic; the diagnosis may be confirmed by serology (complement fixation or microimmunofluorescence tests) on sera and/or tears.

Control

Tetracycline and erythromycin are the drugs of choice. Penicillin is not effective.

INTRODUCTION

The chlamydiae are a small group of nonmotile coccoid bacteria that are obligate intracellular parasites of eukaryotic cells. Chlamydial cells are unable to carry out energy metabolism and lack many biosynthetic pathways; therefore they are entirely dependent on the host cell to supply them with ATP and other intermediates. Because of their dependence on host biosynthetic machinery, the chlamydiae were originally thought to be viruses; however, they have a cell wall and contain DNA, RNA, and ribosomes and therefore are now classified as bacteria. The group consists of a single genus, *Chlamydia* (order Chlamydiales, class Chlamydiaceae). This genus contains the species *C trachomatis* and *C psittaci*, as well as a new organism, the TWAR organism, which has recently been proposed as a third species (*C pneumoniae*). All three species cause disease in humans. *Chlamydia psittaci* infects a wide variety of birds and a number of mammals, whereas *C trachomatis* is limited largely to humans. *Chlamydia pneumoniae* (TWAR organism) has been found only in humans.

Clinical Manifestations

Chlamydia trachomatis Infections

The diseases caused by chlamydiae are summarized in Table 39-1 and Figure 39-1.

TABLE 39-1 Human Diseases Caused by Chlamydia

Species	Serotypes	Disease(s)
C trachomaatis	A, B, Ba, C	Trachoma (typerendemic, a leading cause of blindness in humans); sexually trasmitted infections of the genitals
	D, E, F, G, H, I, J, K	Inclusion conjunctivitis (adult and newborn): cervicitis: salpingitis; proctitis epidolymitis, and pneumonia of newborns lymphogranulonia venreurn
C psiftacl	L-1, L-2, L-3, many unidentilied	Pneumonia (psillacosis)
C pneumoniae	TWAR*	Pneumonia

*The name is derived from the isolates TW-183 and AR-39.
(Modifed from Schachter J: Chlamydial infection. N Engl J Med 298 428, 1978, with permission.)

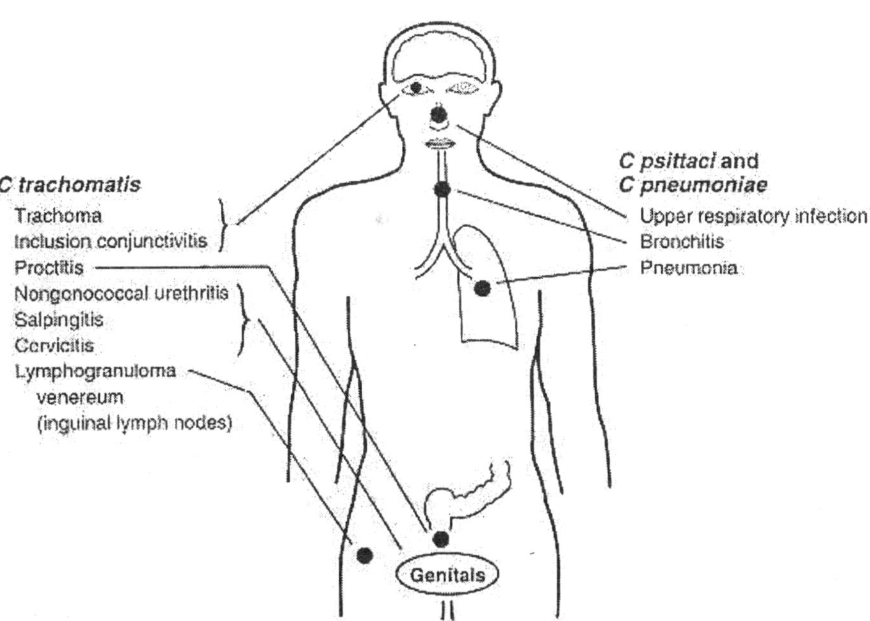

FIGURE 39-1 Clinical manifestations of chlamydial infections.

Trachoma, a *C trachomatis* infection of the conjunctival epithelial cells, results in subepithelial infiltration of lymphocytes, leading to the development of follicles. The infected epithelial cells contain cytoplasmic inclusion bodies. As a result of damage to the epithelial cells, fibroblasts and blood vessels invade the infected area, a pannus forms, and the cornea becomes vascularized and clouded. The eyelids become scarred and malformed, causing trichiasis, an abnormal inward growth of the eyelashes. Continual scraping of the cornea by the eyelashes leads to corneal opacification and blindness.

Chlamydia trachomatis also causes inclusion conjunctivitis, an eye disease of children and adults that is milder than trachoma. It consists of purulent conjunctivitis that heals spontaneously without scarring.

Chlamydia trachomatis also causes sexually transmitted genital and rectal infections. The frequency of *C trachomatis* infections in men may equal or exceed the frequency of gonorrhea. Nongonoccocal urethritis, epididymitis, and proctitis in men can result from infection with *C trachomatis*. Superinfection of gonorrhea patients with *C trachomatis* also occurs. Acute salpingitis and cervicitis in young women can be caused by a *C trachomatis* infection ascending from the cervix. A high rate genital tract coinfection by *C trachomatis* in women with gonorrhea has been reported. *Chlamydia trachomatis* was isolated from the fallopian tubes of infected women. In one report *C trachomatis* elementary bodies attached to spermatozoa were recovered from the peritoneal cavity of patients with salpingitis.

Neonates exposed to *C trachomatis* in an infected birth canal may develop acute conjunctivitis within 5 to 14 days. The disease is characterized by marked conjunctival erythema, lymphoreticular proliferation, and purulent discharge. Untreated infections can develop into pneumonitis; this type of pneumonitis occurs only during the first 4 to 6 months of life.

Recently, *C trachomatis* has been suspected of causing lower respiratory tract infections in adults, and several cases of *C trachomatis* pneumonia have been reported in immunocompromised patients from whom the pathogen was isolated. Evidence also indicates that *C trachomatis* may cause pneumonia or bronchopulmonary infections in immunocompetent persons.

Polyarthritis in lambs, calves, and possibly humans also may be caused by *C trachomatis.*

Lymphogranuloma venereum is a human venereal disease caused by *C trachomatis* strains different from the strains that cause trachoma (Table 39-1). The disease usually occurs in men and involves inguinal lymphadenopathy. Signs of lymphogranuloma venereum appear a few days after venereal exposure. The initial lesions, or vesicles, appear in the urogenital tract in men and women. If the disease does not heal spontaneously, regional lymph nodes become involved.

Chlamydia psittaci Infections

Chlamydia psittaci infects birds through the respiratory tract. Humans exposed to dead or living infected birds may develop fever, a mild influenza like disease, or toxic fulminating pneumonitis after an incubation period of 2 to 4 weeks. *Chlamydia psittaci* can cause pneumonia in cats and sheep as well as in humans. Other strains of *C psittaci* can cause abortions in animals.

TWAR Organism Infections

Recently, a new *Chlamydia* strain (designated *C pneumoniae* serovar TWAR organism) that spreads from person to person in human populations was reported to cause outbreaks of respiratory tract infections in immunocompetent persons.

Latent *Chlamydia* Infections

Latent and inapparent infections of humans, other mammals, and birds are sometimes caused by chlamydiae. The agents of lymphogranuloma venereum, for example, may persist in infected humans for years before the disease becomes apparent. Individuals may develop acute trachoma years after leaving areas endemic for trachoma.

Structure

The chlamydiae exist in nature in two forms: (1) a nonreplicating, infectious particle called the elementary body (EB), 0.25 to 0.3 μm in diameter, that is released from ruptured infected cells and can be transmitted from one individual to another (C *trachomatis, C pneumoniae*) or from infected birds to humans (*C psittaci*), and (2) an intracytoplasmic form called the reticulate body (RB), 0.5 to 0.6 μm in diameter, that engages in replication and growth (Fig. 39-2 and 39-3). The elementary body, which is covered by a rigid cell wall, contains a DNA genome with a molecular weight of 66 X 107 (about 600 genes, one-quarter of the genetic information present in the DNA of *Escherichia coli*). A cryptic DNA plasmid (7,498 base pairs) is also found. It contains an open reading frame for a gene involved in DNA replication. In addition, the elementary body contains an RNA polymerase responsible for the transcription of the DNA genome after entry into the host cell cytoplasm and the initiation of the growth cycle. Ribosomes and ribosomal subunits are present in the elementary bodies. Throughout the developmental cycle, the DNA genome, proteins, and ribosomes are retained in the membrane-bound prokaryotic cell (reticulate body).

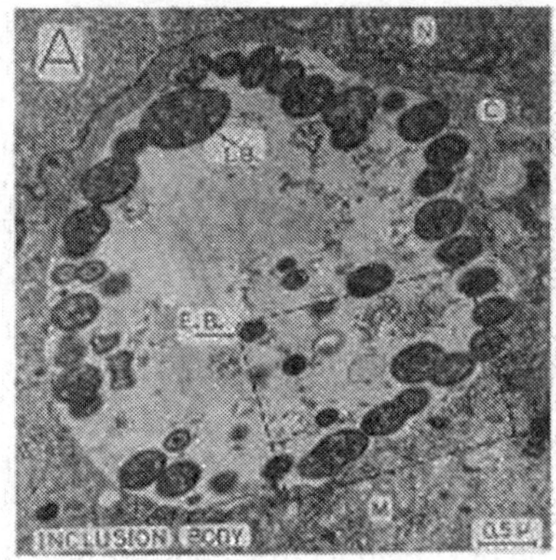

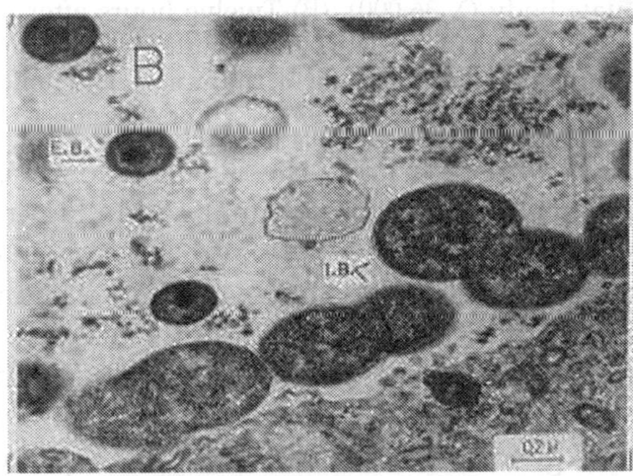

FIGURE 39-2 (A) Electron micrograph of C trachomatis inclusion body in cytoplasm (C) of infected cell. Part of the nucleus (N) and mitochondria (M) can also be seen. (B) Enlarged view of inclusion body showing elementary bodies (E.B.) and reticulate (initial) bodies (I.B.). (Courtesy of Y. Becker, Jerusalem, Israel.)

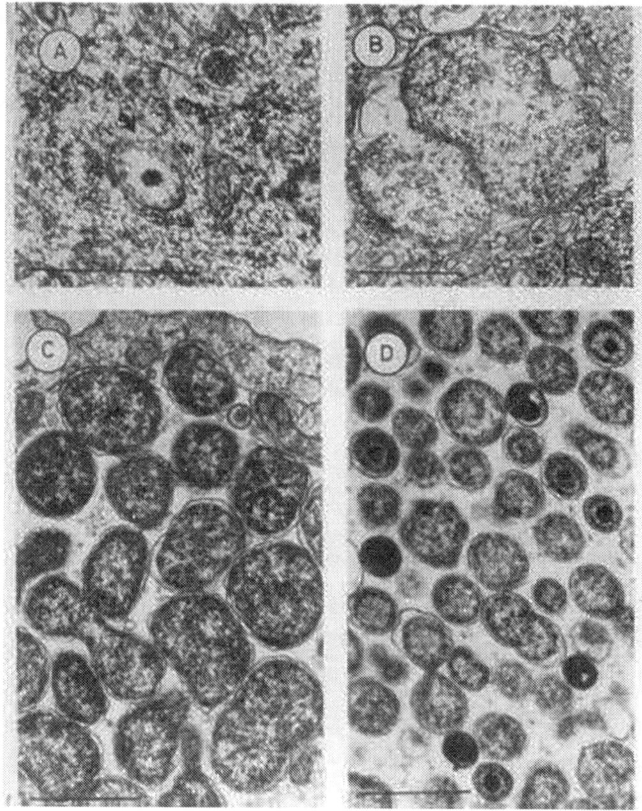

FIGURE 39-3 Developmental cycle of C psittaci in L cells (mouse fibroblasts). (A) Bar, 1 μm. At 2.5 hours after infection, the figure shows an elementary body (arrow) that has just begun to differentiate into a reticulate body (X 36,000). (B) Twelve hours after infection (X 23,000). (C) Twenty hours after infection (X 23,000) (D) Thirty hours after infection (X 23,000). (From Tribby II E, Friis RR, Moulder JW: Effect of chloramphenicol, rifampicin, and nalidixic acid on Chlamydia psittaci growing in L cells. J Infect Dis 127158, 1973, with permission.)

A complex series of events occurs during the developmental cycle of chlamydiae. These and the effects on the host cell are summarized in Figure 39-4 and Table 39-2. Studies on the growth cycle of *C trachomatis* and *C psittaci* in cell cultures in vitro revealed that the infectious elementary body develops into a noninfectious reticulate body (RB) within a cytoplasmic vacuole in the infected cell. There is an eclipse phase of about 20 hours after entry of the elementary body into the infected cell, during which the infectious particle develops into a reticulate body. In these structures the chlamydial genome is transcribed into RNA, proteins are synthesized, and the DNA is replicated. The reticulate body divides by binary fission to form particles which, after synthesis of the outer cell wall, develop into new infectious elementary body progeny. The yield of chlamydial elementary bodies is maximal 36 to 50 hours after infection.

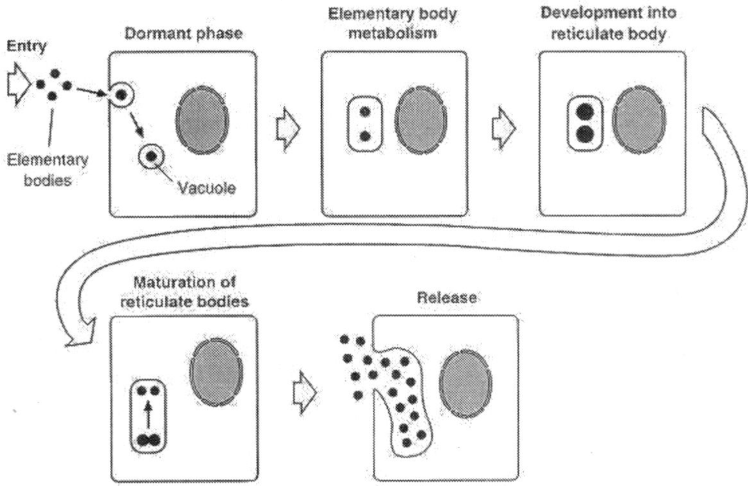

FIGURE 39-4 Developmental cycle of the chlamydiae.

TABLE 39-2 Stage in the Developmental Cycle of Chalmydiae

Agent	Host Cell
Stage 1. Dormant phage. Elementary bodies (EDs) have no or little metabolic activity, contain a DNA genome (ca. 600 genes), plasmid DNA (ca. 8 kbp), ribosomal subunits, cytoplasm, cell membrane, and cell wall containing a major outer membrane protein (MOM), three cysteine rich outer membrane proteins (60 to 62, 15 and 74 kDa), two cell binding proteins (31 and 18 kDa),	Host cell can be in any phase of its growth cycle at infection.
Stage 2. Initiation of EB metabolism. EBs adsorb to host cell membrane. The organization of the EB cell wall changes. (deficiency of the cystline rich proteins; no change in MOMP). EBs adsorb to hot cells and utilize glucose-6 phosphate as substrate.	Duration: 0-12 h after infection. Cell phagocylizes EDs. Cell borns vacuole around EDs. Mitochondria support EB development. Cellular enzyme in glucose metabolic pailway is utilized by EB.
EBs utilize mitochondrial from compact organization to loose arrangement, possibly for transcriptional processes.	Host cell nucleic acid metabolism continues. Nucleolides in host cell pool are available for the prokaryotic parasite.
DNA-dependent RNA polymeraes motecules attached to DNA genome, are activated, and transcribe the genome. At this stage. EB development becomes sensitive to rifampin, Protein synthesis with existing ribosomes begins in EBs.	

Low level of DNA synthesis in EB is detected since development is sensitive to 5 X 10^{-4} M hydroxyurea, EB mass increases because of macromolecule synthesis.	
Stage 3. Development of EB into reticulate body. Rate of DNA biosynthesis increases. DNA is replicated by a DNA-development DNA polymerase encoded by chlamydial and plasmid DNAs. Replication of DNA is semiconservative, with synthesis of shown Oxazaki-like fragments on the DNA template	Duration: 12-35 h after infection. Cell respired to enlargement of EB by increasing the size of vacuole in which the agent develops.
DNA synthesis is accompanied by endogenous synthesis of thymidine, probably from deoxycytidine precursor.	Host cell can supply some of the nucleotide precursors.
DNA synthesis is inhibited by folic acid analogs and hydroxyurea.	
DNA-development RNA polymerase transcribes ribosomal genes in chlamydial DNA. ribosomal of rRNA are synthesized and processed to yield two RNA species, of 1.1 and	
O.55 MDa rRNA species assemble into 50S and 30S ribosomal subunits by combining with ribosomal proteins systhesized according to genetic information in the chlamydial DNA.	
Glucose 6-phosphate is catabolized by enzymes in the agent. In C Itachorrnalis only, glycogene is synthesized and deposited in vacuole, where RBs develop.	
Synthesis of DNA, ribosmes, and proteins is coupled with formation of new RBs by a binary Bssion-like process. At this stage the cytoplasmic vacuoles are linked with RBs. Chlamydial enzymes are involved in the biosynthesis of the RB-limiting membranes. Chlamydial enzymes synthesize RB cell walls	Duration: 20-48 in after infection. Inhibition of protein synthesis of cytoplasmic polyribosomes b chloramphenlcol does not affect chlamydial life cycle. The cytoplasmic vacuole containing the RBS is markedly enlarged. In C particle the vacuole this most of the cell cytoplasm. In C trachomatis the vacuole is more rigid and defined. DNA synthesis in the nucleus is affected.
Stage 4. Maturation of RBs, formation of EBs.	
DNA synthesis in RBs is coupled with division of particles into two smaller daughter cells (pre-EBs). Internal organization in pre-EB articles the DNA genome is condensed in the center of the particle, with cell wall loosely arranged around the particle	

RBs start the synthesis of cysterine rich proteins, Cell wall synthesis loads to rigid EB particles. At this stage, EBs are formed and the life cycle is completed, EBs are released from the ruptured cells.	Vacuole is disrupted, and host cell death is inevitable.

(Modified form Becker Y: chlamydia: Molecular biology of precaryotic obligate parasites of eucaryocytes Microbiol Rev 42 299 1978, with permission.)

Classification, and Antigenic Types

Several distinct antigenic components have been recognized in *C trachomatis* and *C psittaci*, some group specific and others species specific. Detergents have been used to extract antigens from elementary bodies and reticulate bodies. *Chlamydia pneumoniae* (TWAR organism) is serologically unique and differs from *C trachomatis* species and all *C psittaci* strains tested.

The outer chlamydial cell wall contains several antigenic proteins, including a 40-kilodalton (kDa) major outer membrane protein (MOMP), a 60- to 62-kDa and 15-kDa, cysteine-rich proteins, a 74 kDa species-specific protein, and 31- and 18-kDa eukaryotic cell-binding proteins, which share the same primary sequence.

Hyperimmune mouse antiserum against the 40-kDa MOMP protein from serotype L2 reacted with elementary bodies of *C trachomatis* serotypes Ba, E, D, K, L1, L2, and L3 during indirect immunofluorescence but failed to react with serotypes A, B, C, F, G, H, I, and J or with *C psittaci*. Indeed, cloning and sequencing of the *C trachomatis* MOMP gene revealed the same number of amino acids for serovars L2 and B, while the MOMP gene of serovar C contained codons for three additional amino acids. The antigenic diversity of the chlamydial MOMP was reflected in four sequence-variable domains, two of which are candidates for the putative type-specific antigenic determinants. The basis for MOMP differences among *C trachomatis* serovars were clustered nucleotide substitutions for closely related serovars and insertions and deletions for distantly related serovars. When MOMP is inserted into the outer elementary body envelope, exposed domains of MOMP serve as both serotyping and protective antigenic determinants. Predominantly conserved regions of C and B serotypes are interspersed with short variable domains.

Three monoclonal antibodies that recognize epitopes on cysteine-rich membrane proteins interact with all 15 human *C trachomatis* serotypes, establishing the species specificity of this antigen. Monoclonal antibodies to the 15-kDa cysteine-rich protein showed biovar specificity and species specificity. The 60- to 62-kDa and 15-kDa cysteine-rich proteins are highly immunogenic in the natural infection, but the antibodies do not neutralize the infectivity of *C trachomatis* elementary bodies.

Pathogenesis

Spread of Agents

Human diseases caused by chlamydiae can be divided into two types: (1) chlamydial agents transmitted by direct contact (*C trachomatis* genital and ocular infections, *C pneumoniae* ocular infection) and (2) chlamydial agents that are transmitted by the respiratory route (*C psittaci* and *C pneumoniae.*)

The spread of *C trachomatis* from person to person may cause trachoma, inclusion conjunctivitis, or lymphogranuloma venereum. Transmission of *C trachomatis* from the urogenital tract to the eyes and vice versa occurs via contaminated fingers, towels, or other fomites and, in neonates, by passage through an infected birth canal. These diseases appear in an epidemic form in populations with low standards of hygiene. *Chlamydia trachomatis* genital infections are sexually transmitted. *Chlamydia psittaci* is transmitted from infected birds or animals to humans through the respiratory tract. *Chlamydia pneumoniae* spreads from infected individuals by respiratory tract infections hut is not sexually transmitted.

Chlamydial Diseases

Chlamydial agents are intracytoplasmic obligate parasites of mammalian cells and can damage infected cells in tissues. The elementary bodies are infectious particles that can be transmitted from the infected tissues to uninfected tissues in the same person (transfer of *C trachomatis* elementary bodies from an infected genital tract to the eyes and vice versa) or from a person with atypical pneumonia (caused by *C psittaci* or *C pneumoniae*) to healthy individuals (respiratory release of elementary bodies). In the infected individuals the chlamydial agent causes tissue damage and induction of interleukin-1a, interleukin-1ß, and tumor necrosis factor alpha, which are cytokines involved in the inflammation process. Ocular infections by *C trachomatis* and sometimes *C pneumoniae* strains cause acute purulent conjunctivitis either due to infection of the neonate during passage through the birth canal or due to subsequent infections leading to scarring of the conjunctiva and to blindness subsequent to mucopurulent follicular conjunctivitis. *Chlamydia trachomatis* infection also spreads through sexual contact when urethritis or cervicitis is present. The genital tract infection serves as a source of infectious elementary bodies for the eyes.

The recently recognized *C pneumoniae* isolates cause mild to severe pneumonia, prolonged bronchitis, pharyngitis, sinusitis, and a febrile illness in humans. The agent does not cause death in patients without complications.

Host Defenses

Nonspecific Responses

Infections with chlamydial agents evoke responses from the blood vessels (ocular trachoma), connective tissue (scars in *C trachomatis* infections), and lymphocyte infiltration (pannus). Chlamydial infections are characterized by chronic inflammation. The mechanisms that trigger migration of lymphocytes or connective tissue to the site of *C trachomatis* infection in the eyes are not known. However, coculture of *C trachomatis* (serovar L2) with human blood monocytes induced the production of interleukin-1, an important mediator of inflammation and scarring. Interleukin-1a and interleukin-1ß can be induced in human monocytes by *C trachomatis* lipopolysaccharide. Release of angiogenesis factors from infected cells may cause proliferation of blood vessels in the infected eye. Fever accompanies *C psittaci* pneumonitis. It was reported that tumor necrosis factor is induced by *C trachomatis* infection in athymic nude mice.

Cultured chlamydiae are sensitive to interferon, which is produced by cultured cells infected with chlamydiae.

Immune Response in Humans

All chlamydial infections induce IgM, IgG, IgA, and IgE antibodies, but these antibodies do not prevent reinfection. Although, secretions from trachomatous eyes contain specific antitrachoma IgG and IgA antibodies, these antibodies do not impede the infection. Moreover, antibodies that bind to *C trachomatis* elementary bodies do not impair their infectivity in cell cultures. However, the addition of anti-gamma globulin to antibody-treated elementary bodies neutralizes their infectivity. Monoclonal antibodies to proteins in the outer elementary body envelope were reported to neutralize elementary body infectivity. Most patients with *C trachomatis* infections have antibodies that react with the *C trachomatis* cell wall proteins. Sera from individuals with genital infections caused by *C trachomatis* also reacted with the 60- to 62-kDa cysteine-rich proteins of all the *C trachomatis* serotypes. The precise role of cell-mediated immunity is not known.

Epidemiology

Trachoma is still prevalent in Africa and Asia (more than 500 million people are estimated to have the disease), and sporadic cases occur all over the world. The disease flourishes in hot, dry areas where there is a shortage of water and where standards of hygiene are low. The agent isspread to the eyes by flies, dirty towels, fingers, or cosmetic eye pencils. The initial infection usually occurs in childhood, and the active disease eventually appears (mostly by 10 to 15 years of age). Trachoma may leave a residuum of permanent lesions that can lead to blindness. *Chlamydia trachomatis* also resides in the genital tract, cervix, and urethra of adults, and genital infection is spread sexually. Lymphogranuloma venereum persists in the genital tract of infected persons. Because *C trachomatis* is able to infect both the eyes and the urogenital tract, antitrachoma campaigns involving only ocular treatments are futile.

Chlamydia psittaci, the cause of psittacosis in birds and occasionally in humans, is carried by wild and domestic birds, including poultry. The severity of psittacosis in humans has been considerably reduced by the susceptibility of *C psittaci* to antibiotics.

Chlamydia pneumoniae spreads in human populations by respiratory tract infections. It is the agent of atypical pneumonia in hospitalized patients as well as in young individuals with an acute respiratory disease. It has caused epidemics in Scandinavia. Studies of the prevalence of antibodies to *C pneumoniae* in humans around the world showed that it also prevails in Japan, Panama, and North America.

Diagnosis

Most diseases caused by the chlamydiae are diagnosed on the basis of their clinical manifestations. Eye damage caused by *C trachomatis* is typical, as are the vesicles in the infected urogenital tract. Diagnosis of pneumonitis requires laboratory testing.

Chlamydia trachomatis can be identified microscopically in scrapings from the eyes or the urogenital tract. Inclusion bodies in scraped tissue cells are identified by iodine staining of glycogen present in the cytoplasmic vacuoles in infected cells. To isolate the agent, cell homogenates that contain the chlamydial elementary bodies are centrifuged onto the cultured cells (e.g., irradiated McCoy cells). After incubation, typical cytoplasmic inclusions are seen in the cells stained with Giemsa stain or iodine. Staining with iodine can distinguish between inclusion bodies of *C trachomatis* and *C psittaci*, as only the former contain glycogen. Each chlamydial agent can also be identified by using specific immunofluorescent antibodies prepared against either *C trachomatis* or *C psittaci*. Homogenates or exudates of infected tissues also have been used to isolate the agent in the yolk sac of embryonated eggs.

Sera and tears from infected humans are used to detect anti-*Chlamydia* antibodies by the complement fixation or microimmunofluorescence tests. The latter is useful for identifying specific serotypes of *C trachomatis*; however, even detection of anti-*C trachomatis* antibodies of the IgM class cannot be used diagnostically for genital infections, because similar antibodies are found in *Chlamydia*-negative patients. Fluorescent monoclonal antibodies are used to stain *C trachomatis* elementary bodies in urethral and cervical exudates.

It is possible to diagnose *C trachomatis* in tissue biopsy specimens by *in situ* DNA hybridization with cloned *C trachomatis* DNA probes. DNA from *C trachomatis* isolates can be examined by restriction endonuclease analysis. The DNA cleavage pattern of *C trachomatis* isolates differs greatly from that of DNA from *C psittaci* isolates. DNAs of the agents of trachoma and lymphogranuloma venereum differ in their cleavage patterns, and this allows identification of the biovars.

Chlamydia pneumoniae DNA has 10 percent homology with *C trachomatis* or *C psittaci*; *C pneumoniae* isolates have 100 percent homology. *Chlamydia pneumoniae* isolates can be diagnosed by hybridization with a specific DNA probe that does not hybridize to other chlamydiae. Two additional serologic tests are in use: the microimmunofluorescence test with *C pneumoniae*-specific elementary body antigen, and the complement fixation test, which measures *Chlamydia* antibodies.

Control

Attempts to use *C trachomatis* vaccines for prophylaxis and treatment of trachoma have failed. The course of trachoma is more severe in immunized than in nonimmunized individuals. Specific anti-*Chlamydia* antibodies fail to neutralize chlamydial elementary bodies *in vivo*.

Tetracycline and erythromycin are the antibiotics commonly used to treat chlamydial infections in humans. Penicillin is not effective. Patients with trachoma have been treated effectively with erythromycin, rifampin, sulfonamides, chloramphenicol, and tetracyclines. Repeated treatment cycles of long-acting sulfonamides also have been used in local or systemic treatment of trachoma infections. In trachoma patients with trichiasis, corrective surgery is necessary. Patients with inclusion conjunctivitis usually are not treated, because the infection is self-limiting and relatively mild.

Tetracyclines or sulfonamides sometimes are effective in patients with lymphogranuloma venereum, but treatment does not always improve the condition. Tetracycline treatment of gonorrhea in patients infected with gonococci or *Chlamydia* is more effective against postgonococcal urethritis than is treatment with penicillin.

REFERENCES

Baehr W, Zhang YX, Joseph T, et al: Mapping antigenic domains expressed by *Chlamydia trachomatis* major outer membrane protein genes. Proc Natl Acad Sci USA 85:4000, 1988

Beatty WL, Morrison RP, Byrne GI. Persistent chlamydiae: from cell culture to a paradigm for chlamydial pathogenesis. Microbiol Rev 58:686, 1994

Becker Y: The agent of trachoma. Monogr Virol 7,1974

Becker Y: *Chlamydia*: molecular biology of procaryotic obligate parasites of eucaryotes. Microbiol Rev 42:274, 1978

Cook PJ, Honeybourne D: Chlamydia pneumoniae (Review) J Antimicrobial Chemotherapy. 34(6):859, 1994

Grayston JT: *Chlamydia pneumoniae*, strain TWAR. Chest 95:664, 1989

GU L, Wenman WM, Remacha M, et al: Chlamydia trachomatis RNA polymerase alpha subunit: sequence and structural analysis. J Bact 177:2594, 1995

Hatch TP, Miceli M, Sublett JE: Synthesis of disulfide-bonded outer membrane proteins during the developmental cycle of *Chlamydia psittaci* and *Chlamydia trachomatis*. J Bacteriol 165:379, 1986

Hatt C, Ward ME, Clarke IN: Analysis of the entire nucleotide sequence of the cryptic plasmid of *Chlamydia trachomatis* serovar L1. Evidence for involvement in DNA replication. Nucleic Acids Res 16:4053, 1988

Magee DM, Williams DM, Smith JG, et al: Role of CD8 T cells in primary Chlamydia infection. Infection and Immunity 63:516, 1995

Moulder JW: Order II Chlamydiales Storz and Page 1971,334AL. p. 729. In Krieg NR, Holt JG (eds), Bergey's Manual of Systematic Bacteriology. Vol 1. Williams & Wilkins, Baltimore, 1984

Munday PE, Thomas BJ, Gilroy CB, et al: Infrequent detection of Chlamydia trachomatis in a longitudinal study of women with treated cervical infection. Genitourin Med 71:24, 1995.

Peterson EM, de la Maza LM: Restriction endonuclease analysis of DNA from *Chlamydia trachomatis* biovars. J Clin Microbiol 26:635, 1988

Rothermel CD, Schachter J, Lavrich P, et al: *Chlamydia trachomatis*-induced production of interleukin-1 by human monocytes. Infect Immun 57:2705, 1989

Schachter J: Chlamydiae: exotic and ubiquitous. West J Med 132:238, 1980

Schachter J, Crossman M: Chlamydial infections. Annu Rev Med 32:45, 1981

Stephens RS, Tam MR, Kuo CC, Nowinski RC: Monoclonal antibodies to *C.trachomatis*: antibody specificity and antigen characterizations. J Immunol 128:1083, 1982

Tribby II E, Friis RR, Moulder JW: Effect of chloramphenicol, rifampicin and nalidixic acid on *Chlamydia psittaci* growing in L cells. J Infect Dis 127:155, 1973

Weinstock H, Dean D, Bolan G: Chlamydia trachomatis infections. (Rev) Infectious Disease Clinics of North Americia 8(4):797, 1994

Wenman WM, Lovett MA: Expression in *E coli* of *C trachomatis* antigen recognized during human infection. Nature (London) 296:68, 1982

Williams DM, Bonewald LF, Roodman GD, et al: Tumor necrosis factor alpha is a cytotoxin induced by murine *Chlamydia trachomatis* infection. Infect Immun 57:1351, 1989

Wong KC, Ho BS, Egglestone SI, et al: Duplex PCR system for simultaneous detection of *Neisseria gonorrhoeae* and *Chlamydia trachomatis* in clinical specimens. J Clin Pathol 48:101, 1995.

Yong EC, Chinn GS, Caldwell HD, Kuo CC: Reticulate bodies as single antigen in *Chlamydia trachomatis* serology with microimmunofluorescence. J Clin Microbiol 10:351, 1979

Zhang YX, Morrison SG, Caldwell HD, Baehr W: Cloning and sequence analysis of the major outer membrane protein genes of two *Chlamydia psittaci* strains. Infect Immun 57:1621, 1989

Chapter 21

Rickettsiae

General Concepts

Rickettsiae

The rickettsiae are a diverse collection of obligately intracellular Gram-negative bacteria found in ticks, lice, fleas, mites, chiggers, and mammals. They include the genera *Rickettsiae, Ehrlichia, Orientia,* and *Coxiella.* These zoonotic pathogens cause infections that disseminate in the blood to many organs.

Rickettsia

Clinical Manifestations

Rickettsia species cause Rocky Mountain spotted fever, rickettsialpox, other spotted fevers, epidemic typhus, and murine typhus. *Orientia* (formerly *Rickettsia*) *tsutsugamushi* causes scrub typhus. Patients present with febrile exanthems and visceral involvement; symptoms may include nausea, vomiting, abdominal pain, encephalitis, hypotension, acute renal failure, and respiratory distress.

Structure, Classification, and Antigenic Types

Rickettsia species are small, Gram-negative bacilli that are obligate intracellular parasites of eukaryotic cells. This genus consists of two antigenically defined groups: spotted fever group and typhus group, which are related; scrub typhus rickettsiae differ in lacking lipopolysaccharide, peptidoglycan, and a slime layer, and belong in the separate, although related, genus *Orientia.*

Pathogenesis

Rickettsia and *Orientia* species are transmitted by the bite of infected ticks or mites or by the feces of infected lice or fleas. From the portal of entry in the skin, rickettsiae spread via the bloodstream to infect the endothelium and sometimes the vascular smooth muscle cells. *Rickettsia* species enter their target cells, multiply by binary fission in the cytosol, and damage heavily parasitized cells directly.

Host Defenses

T-lymphocyte-mediated immune mechanisms and cytokines, including gamma interferon and tumor necrosis factor alpha, play a more important role than antibodies.

Epidemiology

The geographic distribution of these zoonoses is determined by that of the infected arthropod, which for most rickettsial species is the reservoir host.

Diagnosis

Rickettsioses are difficult to diagnose both clinically and in the laboratory. Cultivation requires viable eukaryotic host cells, such as antibiotic-free cell cultures, embryonated eggs, and susceptible animals. Confirmation of the diagnosis requires comparison of acute- and convalescent-phase serum antibody titers.

Control

Rickettsia species are susceptible to the broad-spectrum antibiotics, doxycycline, tetracycline, and chloramphenicol. Prevention of exposure to infected arthropods offers some protection. A vaccine exists for epidemic typhus but is not readily available.

Ehrlichia

Clinical Manifestations

Ehrlichia species cause ehrlichioses that vary in severity from a life-threatening febrile disease that resembles Rocky Mountain spotted fever, except for less frequent rash, to an infectious mononucleosis-like syndrome.

Classification and Antigenic Types

Ehrlichia sennetsu, E chaffeensis, and the human granulocytic ehrlichia arc genetically distinct and are easily distinguished antigenically.

Pathogenesis

A reservoir of *E chaffeensis* is deer, and for both human monocytic and granulocytic ehrlichiosis are transmitted when ticks bite human skin and inoculate organisms, which then spread by the bloodstream. Macrophages or neutrophils have cytoplasmic vacuoles that contain ehrlichiae dividing by binary fission in each of these ehrlichioses.

Host Defenses

Host defenses against *E chaffeensis* include cytokine-mediated restriction of iron supplies to the ehrlichiae.

Epidemiology

Sennetsu ehrlichiosis has been documented in Japan and Malaysia. Human infections with *E chaffeensis*- and *E phagocytophila*-like organisms have been found recently. Human monocytic ehrlichiosis originates in most of the Atlantic, southeastern, and south central states from New Jersey to Texas. Human granulocytic ehrlichiosis has been identified in the upper midwest and New England thus far.

Diagnosis

Clinical and laboratory clues must be confirmed serologically or by polymerase chain reaction detection of specific ehrlichial DNA.

Coxiella

Clinical Manifestations

Coxiella burnetii causes Q fever, which may present as an acute febrile illness with pneumonia or as a chronic infection with endocarditis.

Structure, Classification, and Antigenic Types

Coxiella burnetii varies in size and has an endospore-like form. This species has lipopolysaccharide and phage type diversity.

Pathogenesis

Coxiella burnetii organisms are transmitted to the human lungs by aerosol from heavily infected placentas of sheep and other mammals and disseminate in the bloodstream to the liver and bone marrow, where they are phagocytosed by macrophages. Growth within phagolysosomes is followed by formation of T-lymphocyte-mediated granulomas. In the few patients who develop serious chronic Q fever, heart valves contain organisms within macrophages.

Host Defenses

Host defense depends on T lymphocytes and gamma interferon.

Epidemiology

Q fever is found worldwide. It is associated mainly with exposure to infected placentas and birth fluids of sheep and other mammals.

Diagnosis

The disease is difficult to diagnose clinically, and cultivation poses a biohazard. Therefore, serology is the mainstay of laboratory diagnosis.

Control

Antibiotics are effective against acute Q fever. A vaccine containing killed phase I organism shows promise in protecting against infection.

Bartonella

Bartonella (Rochalimaea) quintana, the agent of trench fever, was formerly considered as a rickettsial agent. It can be cultured outside of eukaryotic cells and is transmitted to humans via lice. Trench fever was a significant medical problem during World

War I and has reappeared among homeless and alcoholic persons. Recently, cat scratch disease and bacillary angiomatosis and peliosis were discovered to be caused in most cases by a related organism, *B henselae*. *Bartonella bacilliformis* has long been known to cause a sand fly-transmitted acute infection in South America that destroys the red blood cells and a chronic infection that causes a vascular tumor-like lesion similar to those of *B henselae*.

INTRODUCTION

Rickettsiae are small, Gram-negative bacilli that have evolved in such close association with arthropod hosts that they are adapted to survive within the host cells. They represent a rather diverse collection of bacteria, and therefore listing characteristics that apply to the entire group is difficult. The common threads that hold the rickettsiae into a groupare their epidemiology, their obligate intracellular lifestyle, and the laboratory technology required to work with them. In the laboratory, rickettsiae cannot be cultivated on agar plates or in broth, but only in viable eukaryotic host cells (e.g., in cell culture, embryonated eggs, or susceptible animals). The exception, which shows the artificial nature of using obligate intracellular parasitism as a defining phenotypic characteristic, is *Bartonella (Rochalimaea) quintana*, which is cultivable axenically, but was traditionally considered as a rickettsia. The diversity of rickettsiae is demonstrated in the variety of specific intracellular locations where they live and the remarkable differences in their major outer membrane proteins and genetic relatedness (Table 38-1). An example of extreme adaptation is that the metabolic activity of *Coxiella burnetii* is greatly increased in the acidic environment of the phagolysosome, which is a harsh location for survival for most other organisms. Obligate intracellular parasitism among bacteria is not unique to rickettsiae. Chlamydiae also have evolved to occupy an intracellular niche, and numerous bacteria (e.g., *Mycobacteria, Legionella, Salmonella, Shigella, Francisella,* and *Brucella*) are facultative intracellular parasites. In contrast with chlamydiae, all rickettsiae can synthesize ATP. *Coxiella burnetii* is the only rickettsia that appears to have a developmental cycle.

TABLE 38-1 Properties of Selected Rickettsial Organisms

Species	Cellular Location	Delevopmental Cycle	Axenic Cultivation	Lipopolysaccaharide	Major Outer Membrane proteins (KDa) Antigens	Percent G+C Content
Rickettsia rickettsII	Endothelial and smooth muscle cytosol and nucleus	No	No	Yes	190, 135	33

Species	Cellular Location	Delevop mental Cycle	Axenic Cultivation	Lipopolysa ccaharide	Major Outer Membrane proteins (KDa) Antigens	Percent G+C Content
Rickettsia Prowazekll	Endothelial cytosol	No	No	Yes	135	29
Orientla (Rickersia) Isutsugam oshi	Cytosol rarely nuacleus	No	No	No	70, 54-56, 46-47	28.5-30
Ehrlichia chaffeensis	Cytoplasic endosomal membrane-bound vacuole	No	No	Unknown	30, 29, 28	Unknown
Coxiella bumell	Macrophage phagoysoso me	Appare ntly	No	Yes	28	43
Bartonella quintane	Extracellular, attached to touse gut epithelium	No	Yes	Yes	100, 75, 60, 35, 17	39-41

Some organisms in the family *Rickettsiaceae* are closely related genetically (e.g., *Rickettsia rickettsii, R akari, R prowazekii*, and *R typhi*); others are related less closely to *Rickettsia* species (e.g., *Ehrlichia* and *Bartonella*); and others not related to *Rickettsia species* (e.g., *C burnetii*). The phenotypic traits of the medically important organism *Orientia* (*Rickettsia*) *tsutsugamushi* suggest that the species may be an example of convergent evolution in a similar ecologic niche.

Rickettsioses are zoonoses that, except for Q fever, are usually transmitted to humans by arthropods (tick, mite, flea, louse, or chigger) (Table 38-2). Therefore, their geographic distribution is determined by that of the infected arthropod, which for most rickettsial species is the reservoir host. Rickettsiae are important causes of human diseases in the United States (Rocky Mountain spotted fever, Q fever, murine typhus, sylvatic typhus, human monocytic ehrlichiosis, human granulocytic ehrlichiosis, and rickettsialpox) and around the world (Q fever, murine typhus, scrub typhus, epidemic typhus, boutonneuse fever, and other spotted fevers) (Table 38-2).

TABLE 38-2 Distinguishing Characteristics of Rickettsial Diseases

Disease	Organism	Geographic Distribution	Ecologic Niche	Transmission to Human	Pathologic Basis (Injury)	Rash	Eschar	Serologic Diagnosis
			Ticks	Tik bite	Microvascular	98%	Rare	IFA, LA, IHA, FIA, CF
			Ticks	Tik bite	Microvascular	97%	50%	IFALA, CF
			Micro	Mle bite	Microvascular	100%	92%	IFA, CF
			Ticks	Tik bite	Microvascular	100%	77%	IFA, CF
			Ticks	Tik bite	Microvascular	92%	75%	CF
			Unknown	Arthropod bite	Microvascular	100%	48%	IFA, CF
			Humans, flying squires	Louso focus	Microvascular	100%	None	IFA, IHA, CIA, CF
			faos, rats	Flea feces	Microvascular	80%	None	IFA, IHA, EIA, LA, CF
			Cnggors	Chigger bite	Microvascular	50%	35%	IFAM EIA
			Unknown	Unknown	lymphoc hyperplasia	Very rare	None	IFAM CF
			Other	Tik bite	Penvaseultis, granulomas	40%	None	IFA
			Unknown	Tik bite	Unknown	Rare	None	IFA
			Ticks engulates	Aerosol from infected pacenata of sheep, goats, cattle	Poeurerio, grnebmas to fever and bone marow, chronic endocarets	Rare	None	IFA, EIA, CF
			Humans	Louse bite and toes	Perivasultis	Yes	None	IHA, EIA, CF
			Cats	Cat seraph or bite	Graruorras: vascular Protteration	Rare	None	IFA, EIA
			Humans	Sandfy bite	Acute hemolysis chone vascular perforation	yes chronic phase	None	EIA

Rickettsiae of the Spotted Fever and Typhus Groups

The rickettsial diseases are arranged into several major categories (Table 38-2), the first two of which are the spotted fever and typhus fever groups.

Clinical Manifestations

Rocky Mountain Spotted Fever

Rocky Mountain spotted fever is among the most severe of human infectious diseases, with a mortality of 20 to 25 percent unless treated with an appropriate antibiotic. The severity and mortality are greater for men, elderly persons, and black men with glucose-6-phosphate dehydrogenase deficiency. Although, in theory, the disease is always curable by early, appropriate treatment, the case fatality rate is still 4 percent. The incidence of disease parallels the geographic distribution of infected *Dermacentor variabilis* ticks in the eastern United States and *D andersoni* in the Rocky Mountain states, where the infection was first recognized. Rocky Mountain spotted fever was subsequently recognized in the eastern United States. The incidence has declined in the Rocky Mountain states and increased dramatically in the southeastern United States and Oklahoma. Currently most cases actually occur in the Atlantic states from Maryland to Georgia, as well as in Oklahoma, Missouri, Kansas, Ohio, Tennessee, Arkansas, and Texas, although cases are reported in nearly every state. In the southeastern states, the disease occurs during the seasonal activity of *D variabilis* ticks (April through September) and affects children more frequently than adults. Significant changes in incidence do occur. From a low of 199 cases reported in 1959, the annual number of cases rose steadily to a peak of 1,192 cases in 1981, with a subsequent decline and plateau of approximately 700 cases since 1985. The reasons for these fluctuations are unclear.

The rickettsiae are maintained in nature principally by transovarial transmission from infected female ticks to infected ova that hatch into infected larval offspring (Fig. 38-1). A low rate of acquisition of rickettsiae by uninfected ticks occurs when the ticks feed upon small mammals with enough rickettsiae in their blood to establish tick infection. This effect replenishes lines of infected ticks that are occasionally killed by massive rickettsial overgrowth. A recently observed factor of potential importance in this balance of nature is the interference phenomenon, by which infection of ticks with nonpathogenic spotted fever group rickettsiae prevents the establishment of infection by *R rickettsii*.

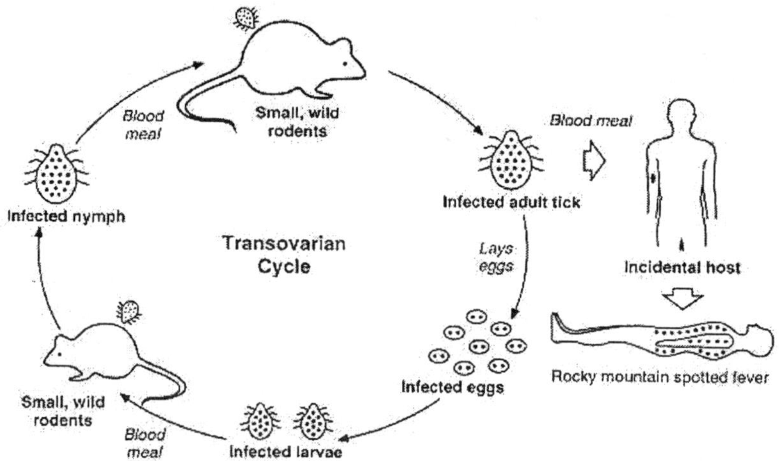

FIGURE 38-1 Transovarian passage of *R rickettsii* in the tick vector is an important cycle in maintaining the infection in nature from one generation of tick to another. Horizontal transmission (i.e., acquisition of the bacteria by uninfected ticks feeding on infected animals) occurs less often and is not shown. Humans become incidental hosts after being bitten by an infected adult tick.

The clinical gravity of Rocky Mountain spotted fever is due to severe damage to blood vessels by *R rickettsii*. This organism is unusual among rickettsiae in its ability to spread and invade vascular smooth muscle cells as well as endothelium. Damage to the blood vessels in the skin in locations of the rash leads to visible hemorrhages in one-half of all infected persons (Fig. 38-2). Attempted plugging of vascular wall destruction consumes platelets, with consequent thrombocytopenia also affecting approximately one-half of the patients.

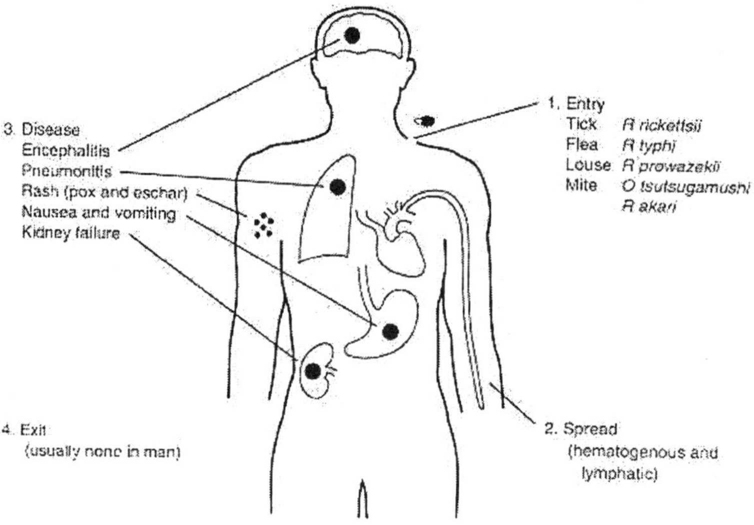

FIGURE 38-2 Common clinical manifestations of the rickettsial diseases.

Rickettsialpox and Other Spotted Fevers

In the 1940s an epidemic of disease characterized by fever, rash, and cutaneous necrosis appeared in one area of New York City. The etiology was traced to *R akari* transmitted by the bite of mites (*Liponyssoides sanguineus*) that infested the numerous mice in an apartment house in this area. The disease was named rickettsialpox because many patients had blister-like rashes resembling those of chickenpox. Epidemics were diagnosed in other cities, and *R akari* has been isolated in other countries (e.g., the Ukraine). Perhaps because this nonfatal disease is seldom considered by physicians, or its incidence is truly low, the diagnosis is rarely made. Transovarial transmission in the mite and periodic documentation of cases assure us that the etiologic agent is still with us.

Boutonneuse fever, so called because of the papular rash in some cases, has many synonyms, reflecting different geographic regions of occurrence (e.g., Mediterranean spotted fever, Kenya tick typhus, and South African tick bite fever). Cases are observed in the United States in travelers returning from endemic areas. The agent, *R conorii*, is closely related to *R. rickettsii*. Severe disease resembling Rocky Mountain spotted fever can cause death in high-risk groups (e.g., elderly, alcoholic, and glucose-6-phosphate dehydrogenase-deficient patients). Cutaneous necrosis caused by rickettsial vascular infection at the tick bite site of inoculation, known as an eschar or tache noire, is observed in only half the patients with boutonneuse fever. The curiously high prevalence of antibodies reactive with *R conorii* in healthy populations in endemic regions might be explained by missed diagnosis of prior illness, subclinical infection, infection with an antigenically related but less pathogenic rickettsia, or nonspecificity of the laboratory test.

Other spotted fevers occur in geographic distributions of little concern to many physicians in the United States. North Asian tick typhus caused by *R sibirica*, Queensland tick typhus caused by *R australis*, and the recently discovered oriental spotted fever caused by *R japonica* demonstrate that spotted fever group rickettsiae occur worldwide.

Epidemic Typhus and Brill-Zinsser Disease

Epidemics of louse-borne typhus fever have had important effects on the course of history; for example, typhus in one army but not in the opposing force has determined the outcome of wars. Populations have been decimated by epidemic typhus. During and immediately after World War I, 30 million cases occurred, with 3 million deaths. Unsanitary, crowded conditions in the wake of war, famine, flood, and other disasters and in poor countries today encourage human louse infestation and transmission of *R prowazekii*. Epidemics usually occur in cold months in poor highland areas, such as the Andes, Himalayas, Mexico, Central America, and Africa. Lice live in clothing, attach to the human host several times daily to take a blood meal, and become infected with *R prowazekii* if the host has rickettsiae circulating in the blood. If the infected louse infests another person, rickettsiae are deposited on the skin via the louse feces or in the crushed body of a louse. Scratching inoculates rickettsiae into the skin.

Between epidemics *R prowazekii* persists as a latent human infection. Years later, when immunity is diminished, some persons suffer recrudescent typhus fever (Brill-Zinsser disease). These milder sporadic cases can ignite further epidemics in a susceptible louse-infested population. In the United States Brill-Zinsser disease is seen in immigrants who suffered typhus fever before entering the country. In the eastern United States, sporadic human cases of *R prowazekii* infection have been traced to a zoonotic cycle involving flying squirrels and their own species of lice and fleas.

Murine Typhus

Murine typhus is prevalent throughout the world, particularly in ports, countries with warm climates, and other locations where rat populations are high. *Rickettsia typhi* is associated with rats and fleas, particularly the oriental rat flea, although other ecologic cycles (e.g., opossums and cat fleas) have been implicated. Fleas are infected by transovarian transmission or by feeding on an animal with rickettsiae circulating in the blood. Rickettsiae are shed from fleas in the feces, from which humans acquire the infection through the skin, respiratory tract, or conjunctiva. During the 1940s more than 4,000 cases of murine typhus occurred annually in the United States. The incidence declined coincident with increased utilization of the insecticide DDT. Although, the infection and clinical involvement affects the brain, lungs, and other visceral organs in addition to the skin, mortality in humans is less than 1 percent.

Structure, Classification, and Antigenic Types

Rickettsia species include two antigenically defined groups that are closely related genetically but differ in their surface-exposed protein and lipopolysaccharide antigens. These are the spotted fever and typhus groups. The organisms in these groups are smaller (0.3 µm by 1.0 µm) than most Gram-negative bacilli that live in the extracellular environment (Fig. 38-3). They are surrounded by a poorly characterized structure that is observed as an electron-lucent zone by transmission electron microscopy and is considered to represent a polysaccharide-rich slime layer or capsule. The cell wall contains lipopolysaccharides, a major component that differs antigenically between the typhus group and the spotted fever group. These rickettsiae also contain major outer membrane proteins with both cross-reactive antigens and surface-exposed epitopes that are species specific and easily denatured by temperatures above 54°C. The major outer membrane protein of typhus group rickettsiae has an apparent molecular mass of 120,000 Da. Spotted fever group rickettsiae generally have a pair of analogous proteins with some diversity of their molecular masses. *Rickettsia prowazekii* has a transport mechanism that exchanges ATP for ADP in its intracellular environment, thus providing a means to usurp host cell energy sources under favorable circumstances. Rickettsiae also are able to synthesize ATP via metabolism of glutamate. Adaptation to the intracellular environment is further evidenced in a variety of transport mechanisms to obtain crucial substances such as particular amino acids from cytoplasmic pools in the host cell. These adaptations and the presence of numerous independent metabolic activities demonstrate that rickettsiae are not degenerate forms of bacteria, but rather have evolved successfully for survival with an intracellular life-style.

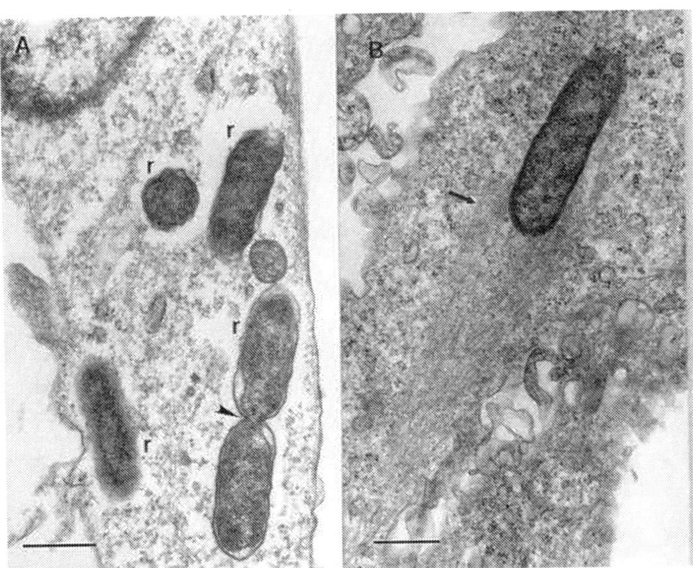

FIGURE 38-3 (A) Organisms of *Rickettsia conorii* (*r*) in a cultured human endothelial cell are located free in the cytosol. One rickettsia is dividing by binary fission (arrowhead). (B) These rickettsiae can move inside the cytoplasm of the host cell because of the propulsive force created by the "tail" of host cell actin filaments (arrow). Bars = 0.5 µm.

Pathogenesis

Rickettsiae are transmitted to humans by the bite of infected ticks and mites and by the feces of infected lice and fleas. They enter via the skin and spread through the bloodstream to infect vascular endothelium in the skin, brain, lungs, heart, kidneys, liver, gastrointestinal tract, and other organs (Fig. 38-1). Rickettsial attachment to the endothelial cell membrane induces phagocytosis, soon followed by escape from the phagosome into the cytosol (Fig. 38-4). Rickettsiae divide inside the cell. *Rickettsia prowazekii* remains inside the apparently healthy host cell until massive quantities of intracellular rickettsiae accumulate and the host cell bursts, releasing the organisms. In contrast, *R rickettsii* leaves the host cell via long, thin cell projections (filopodia) after a few cycles of binary fission. Hence, relatively few *R rickettsii* organisms accumulate inside any particular cell, and rickettsial infection spreads rapidly to involve many other cells. Perhaps because of the numerous times the host cell membrane is traversed, there is an influx of water that is initially sequestered in cisternae of cytopathically dilated rough endoplasmic reticulum in the cells more heavily infected with *R rickettsii*.

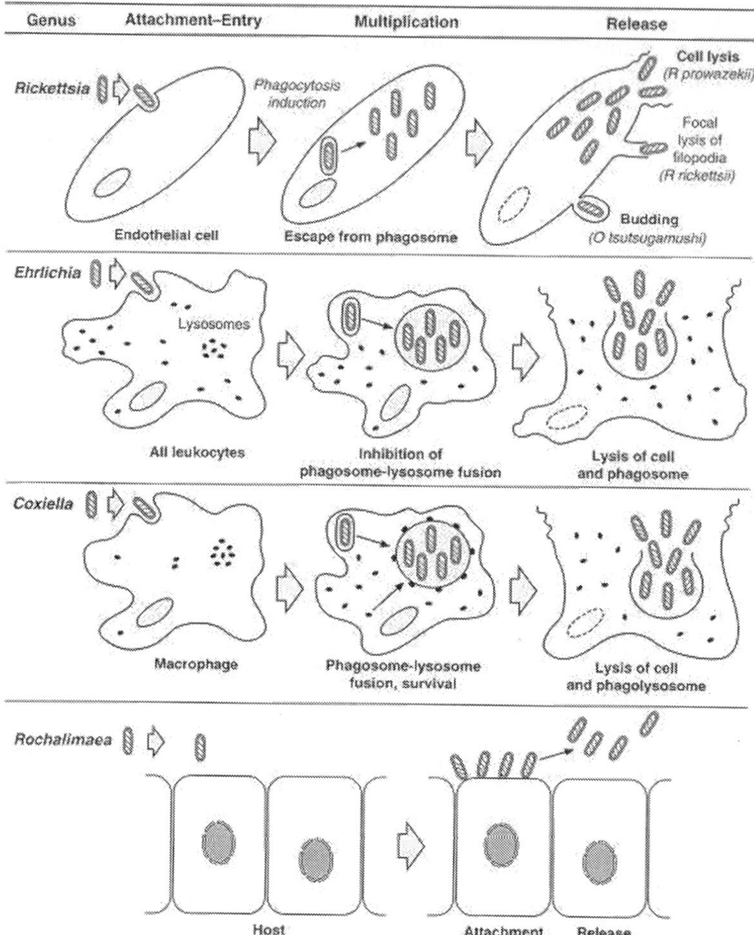

FIGURE 38-4 Pathogenesis of the rickettsial agents illustrating unique aspects of their interactions with eukaryotic cells.

The bursting of endothelial cells infected with *R prowazekii* is a dramatic pathologic event. The mechanism is unclear, although phospholipase activity, possibly of rickettsial origin, has been suggested. Injury to endothelium and vascular smooth muscle cells infected by *R rickettsii* seems to be caused directly by the rickettsiae, possibly through the activity of a rickettsial phospholipase or rickettsial protease or through free-radical peroxidation of host cell membranes. Host immune, inflammatory, and coagulation systems are activated and appear to benefit the patient. Cytokines and inflammatory mediators account for an undefined part of the clinical signs. Rickettsial lipopolysaccharide is biologically relatively nontoxic and does not appear to cause the pathogenic effects of these rickettsial diseases.

The pathologic effects of these rickettsial diseases originate from the multifocal areas of endothelial injury with loss of intravascular fluid into tissue spaces (edema), resultant low blood volume, reduced perfusion of the organs, and disordered function of the tissues with damaged blood vessels (e.g., encephalitis, pneumonitis, and hemorrhagic rash).

Diagnosis

Diagnosis of rickettsial infections is often difficult. The clinical signs and symptoms (e.g., fever, headache, nausea, vomiting, and muscle aches) resemble many other diseases during the early stages when antibiotic treatment is most effective. A history of exposure to the appropriate vector tick, louse, flea, or mite is helpful but cannot be relied upon. Observation of a rash, which usually appears on or after day 3 of illness, should suggest the possibility of a rickettsial infection but, of course, may occur in many other diseases also. Knowledge of the seasonal and geographic epidemiology of rickettsioses is useful, but is inconclusive for the individual patient. Except for epidemic louse-borne typhus, rickettsial diseases strike mostly as isolated single cases in any particular neighborhood. Therefore, clinico-epidemiologic diagnosis is ultimately a matter of suspicion, empirical treatment, and later laboratory confirmation of the specific diagnosis.

Because rickettsiae are both fastidious and hazardous, few laboratories undertake their isolation and diagnostic identification (Fig. 38-5). Some laboratories are able to identify rickettsiae by immunohistology in skin biopsies as a timely, acute diagnostic procedure, but to establish the diagnosis physicians usually rely on serologic demonstration of the development of antibodies to rickettsial antigens in serum collected after the patient has recovered. Currently, assays that demonstrate antibodies to rickettsial antigens themselves (e.g., the indirect fluorescence antibody test or latex agglutination) are preferable to the nonspecific, insensitive Weil-Felix test that is based on the cross-reactive antigens of OX-19 and OX-2 strains of *Proteus vulgaris*.

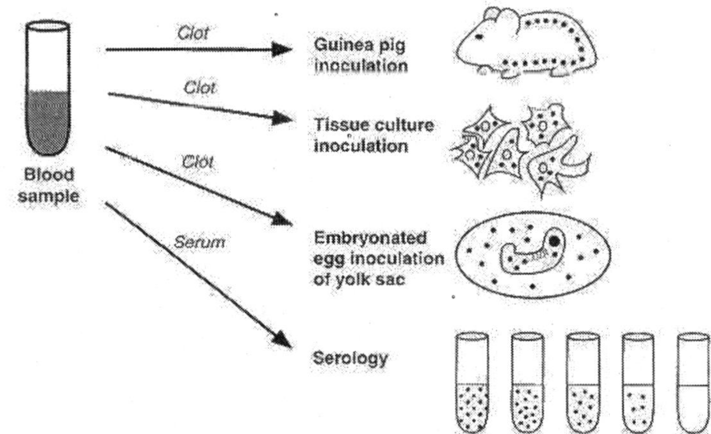

FIGURE 38-5 Laboratory methods used in confirming a diagnosis of rickettsial infection. These bacteria can be cultivated as indicated, but use of serology is more common.

Control

Although, early treatment with doxycycline, tetracycline, or chloramphenicol is effective in controlling the infection in the individual patient, this action has no effect on rickettsiae in their natural ecologic niches (e.g., ticks). Human infections are prevented by control of thevector and reservoir hosts. Massive delousing with insecticide can abort an epidemic of typhus fever. Prevention of attachment of ticks and their removal before they have injected rickettsiae into the skin reduces the likelihood of a tick-borne spotted fever. Control of rodent populations and of the access of rats and mice to homes and other buildings may reduce human exposure to *R typhi* and *R akari*.

Vaccines against spotted fever and typhus group rickettsiae have been developed empirically by propagation of rickettsiae in ticks, lice, embryonated hen eggs, and cell culture. Vaccines containing killed organisms have provided incomplete protection. A live attenuated vaccine against epidemic typhus has proved successful, but is accompanied by a substantial incidence of side effects, including a mild form of typhus fever in some persons. The presence of strong immunity in convalescent subjects indicates that vaccine development is feasible, but it requires further study of rickettsial antigens and the effective anti-rickettsial immune response. T-lymphocyte-mediated immune mechanisms, includingeffects of the lymphokines, gamma interferon tumor necrosis factor, and interleukin-1, seem most important.

Orientia (Rickettsia) tsutsugmushi and Scrub Typhus

Although, the agents of scrub typhus bear a single taxonomic name, *Orientia (Rickettsia) tsutsugamushi*, these interrelated organisms are somewhat heterogeneous and differ strikingly from *Rickettsia* species of the spotted fever and typhus groups.

Clinical Manifestations

Patients with scrub typhus often have only fever, headache, and swollen lymph nodes and in some cases myalgia, gastrointestinal complains, or cough beginning 6 to 21 days following exposure to the vector. Fewer than half of the patients have an eschar at the site where the larval mite fed and the classic rash. The mortality varies but averages 7 percent without anti-rickettsial treatment.

Structure, Classification, and Antigenic Types

Orientia (Rickettsia) tsutsugamushi is a very labile rickettsia that is particularly difficult to propagate and separate from the host cells in which it grows. In contrast with spotted fever group and typhus group rickettsiae, *O tsutsugamushi* does not seem to possess lipopolysaccharides, peptidoglycan, a slime layer, or other T-independent antigens. The rickettsial cell wall consists of proteins linked by disulfide bonds. Antigenically distinguishable strains represent only part of what seems to be a great antigenic mosaic. Immunity to infection with the homologous strain wanes within a few years; cross-protective immunity to heterologous strains disappears within a few months. The reasons for this lack of long-term immunity are unclear.

Pathogenesis

Orientia (*Rickettsia*) *tsutsugamushi* is injected into the skin during feeding by a larval trombiculid mite (chigger). An eschar often forms at this location. Rickettsiae spread via the bloodstream and damage the microcirculation of the skin (rash), lungs (pneumonitis), brain (encephalitis), and other organs. The generalized enlargement of lymph nodes is unique among rickettsial diseases. *Orientia* (*Rickettsia*) *tsutsugamushi* is phagocytosed by the host cell, escapes from the phagosome into the cytosol, divides by binary fission, and is released from projections of the cell membrane (Fig. 38-4). The pathogenic mechanism of *O tsutsugamushi* is not known.

Epidemiology

Scrub typhus occurs where chiggers infected with virulent rickettsial strains feed upon humans. *Leptotrombidium deliense* and other mites are found particularly in areas where regrowth of scrub vegetation harbors the *Rattus* species that are hosts for the mites. Some of these foci are quite small and have been referred to as mite islands. Because *O tsutsugamushi* is transmitted transovarially from one generation of mites to the next, these dangerous areas tend to persist for as long as the ecologic conditions, including scrub vegetation, persist. Truly one of the neglected diseases, scrub typhus occurs over a vast area, including Japan, China, the Philippines, New Guinea, Indonesia, other islands of the southwest Pacific Ocean, southeastern Asia, northern Australia, India, Sri Lanka, Pakistan, Russia, and Korea. Recognized in western countries mainly because of large numbers of infections of military personnel during World War II and the Vietnam War, scrub typhus perennially affects native populations. Reinfection and undiagnosed infections are highly prevalent. Mortality ranges from 0 to 35 percent and has not been correlated with any specific factor.

Diagnosis

Classic textbook cases with fever, headache, eschar, and rash are far outnumbered by cases that lack rash or eschar. Such cases are usually misdiagnosed. Laboratory diagnosis is unavailable in many areas where scrub typhus occurs. Isolation of rickettsiae requires inoculation of mice or cell culture. Serologic diagnosis is made by specific methods (indirect fluorescence antibody test or enzyme immunoassay) or by the older method of demonstrating cross-reactive antibodies that agglutinate the OXK strain of *P mirabilis*.

Control

Scrub typhus can be treated with doxycycline, tetracycline, or chloramphenicol. Chigger repellents may prevent exposure. Prophylaxis with weekly doses of doxycycline during and for 6 weeks after exposure protects against scrub typhus. Attempts to develop a safe, effective vaccine have failed.

Ehrlichia

According to the evolutionary scheme suggested by 16S rRNA sequence homology, ehrlichiae are genetically related to *Rickettsia* species. The genus *Ehrlichia* contains Gram-negative bacteria that reside in a cluster (morula) within membrane-bound cytoplasmic vacuoles of monocytes and macrophages, or polymorphonuclear leukocytes. Ehrlichiae have been implicated as the agents of diseases of horses (*E ristícii* and *E equi*), dogs (*E canis, E ewingii* and *E platys*, a platelet pathogen), and other animals. *Ehrlichia sennetsu* causes a human disease in Japan resembling infectious mononucleosis. Ehrlichiae are unusual in their cell wall structure and they can establish persistent infections.

In 1987 the first case of human ehrlichiosis was reported in the United States. A severely ill man with multiorgan system involvement had morula inclusions demonstrated in peripheral blood leukocytes. Subsequently, cases of human monocytic ehrlichiosis have been documented mainly in eastern and southern states between New Jersey and Texas. The infection has varied from severe and sometimes fatal, mimicking Rocky Mountain spotted fever, to oligosymptomatic and asymptomatic forms. A history of tick bite and the seasonal and geographic occurrence correlate with the predominant tick vector, *Amblyomma americanum*. Illness is often accompanied by leukopenia, thrombocytopenia, and damage to the liver. Lesions include perivasculitis in the central nervous system, kidney, heart, and lungs and granulomas in the bone marrow and liver. Clinical diagnosis is difficult. Laboratory diagnosis by indirect fluorescence antibody assay or polymerase chain reaction is not widely available. *Ehrlichia chaffeensis* morulae are difficult to detect in peripheral blood leukocytes.

In 1994 another serious new infectious disease, human granulocytic ehrlichiosis, was reported. Ehrlichiae seen within morulae in neutrophils in smears of peripheral blood were identified as very closely related to *E phagocytophila* (a European tick-transmitted infection of sheep, cattle, goats, and deer) and *E equi*. The causative organism, like other granulocytic ehrlichiae, has never been cultivated. Human granulocytic ehrlichiosis has been associated with the deer tick, *Ixodes scapularis*, and thus is found as far north as Minnesota, Wisconsin, and New England. Laboratory diagnosis is practically achieved by visualizing morulae in neutrophils, as serology and polymerase chain reaction for the agent are presently research procedures. Sometimes fatal, human granulocytic ehrlichiosis, like *E chaffeensis* infection, can be treated effectively with doxycycline.

Coxiella burnetii and Q Fever

Coxiella burnetii is sufficiently different genetically from the other rickettsial agents that it is placed in a separate group. Unlike the other agents, it is very resistant to chemicals and dehydration. Additionally, its transmission to humans is by the aerosol route, although a tick vector is involved in spread of the bacteria among the reservoir animal hosts.

Clinical Manifestations

Q fever is a highly variable disease, ranging from asymptomatic infection to fatal chronic infective endocarditis (Fig. 38-6). Some patients develop an acute febrile disease that is a nonspecific influenza-like illness or an atypical pneumonia. Other patients are diagnosed after identification of granulomas in their liver or bone marrow. The most serious clinical conditions are chronic *C burnetii* infections, which may involve cardiac valves, the central nervous system, and bone.

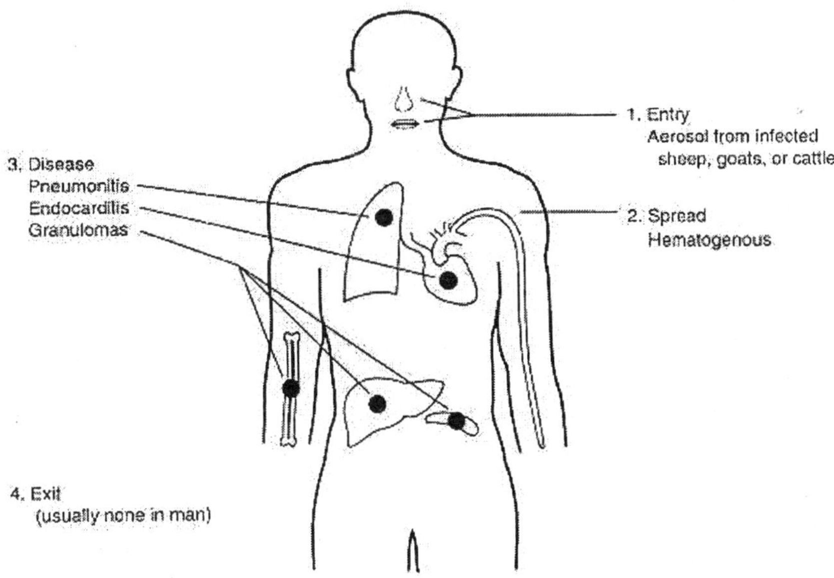

FIGURE 38-6 Clinical manifestations of Q fever.

Structure, Classification, and Antigenic Type

Coxiella burnetii is an obligately intracellular bacterium with some peculiar characteristics. It is small, generally 0.25 µm by 0.5 to 1.25 µm. However, there is considerable ultrastructural pleomorphism, including small- and large-cell variants and possible endospore-like forms, suggesting a hypothetical developmental cycle. Among rickettsiae, *C burnetii* is the most resistant to environmental conditions, is the only species that resides in the phagolysosome, is activated metabolically by low pH, and has a plasmid. The extensive metabolic capacity of *C burnetii* suggests that its obligate intracellular parasitism is a highly evolved state rather than a degenerate condition. The cell wall is typical of Gram-negative bacteria and contains peptidoglycan, proteins, and lipopolysaccharide. When propagated under laboratory conditions in embryonated eggs or cell culture, *C burnetii* undergoes phase variation analogous to the smooth to rough lipopolysaccharide variation of members of the Enterobacteriaceae. Phase I is the form found in nature and in human infections. The phase II variant contains truncated lipopolysaccharide, is avirulent, and is a poor vaccine.

Pathogenesis

Human Q fever follows inhalation of aerosol particles derived from heavily infected placentas of sheep, goats, cattle, and other mammals. *Coxiella burnetii* proliferates in the lungs, causing atypical pneumonia in some patients. Hematogenous spread occurs, particularly to the liver, bone marrow, and spleen. The disease varies widely in severity, including asymptomatic, acute, subacute, or chronic febrile disease, granulomatous liver disease, and chronic infection of the heart valves. The target cells are macrophages in the lungs, liver, bone marrow, spleen, heart valves, and other organs. *Coxiella burnetii* is phagocytosed by Kupffer cells and other macrophages and divides by binary fission within phagolysosomes (Fig. 38-3). Apparently it is minimally harmful to the infected macrophages. Different strains have genetic and phenotypic diversity. The lipopolysaccharides are relatively nonendotoxic. Host-mediated pathogenic mechanisms appear to be important, especially immune and inflammatory reactions, such as T-lymphocyte-mediated granuloma formation.

Epidemiology

Coxiella burnetii infects a wide variety of ticks, domestic livestock, and other wild and domestic mammals and birds throughout the world. Most human infections follow exposure to heavily infected birth products of sheep, goats, and cattle, as occurs on farms, in research laboratories, and in abattoirs. *Coxiella burnetii* is also shed in milk, urine, and feces of infected animals. Animals probably become infected by aerosol and by the bite of any of the 40 species of ticks that carry the organisms.

Diagnosis

Clinical diagnosis depends upon a high index of suspicion, careful evaluation of epidemiologic factors, and ultimately, confirmation by serologic testing. Although, *C burnetii* can be isolated by inoculation of animals, embryonated hen eggs, and cell culture, very few laboratories undertake this biohazardous approach. Likewise, the diagnosis is seldom made by visualizationof the organisms in infected tissues. Acute Q fever is diagnosed by demonstration of the development of antibodies to protein antigens of *C burnetii* phase II organisms. Chronic Q fever endocarditis is diagnosed by demonstration of a high titer of antibodies, particularly IgG and IgA, against the lipopolysaccharide antigens of *C burnetii* phase I organisms in patients with signs of endocarditis whose routine blood cultures contain no organisms.

Control

Antibiotic treatment is more successful in ameliorating acute, self-limited Q fever than in curing life-threatening chronic endocarditis. Reduction in exposure to these widespread organisms is difficult because some serologically screened animals that have no detectable antibodies to *C burnetii* still shed organisms at parturition. Persons with known occupational hazards (e.g., Australian abattoir workers) have benefitted from a vaccine composed of killed phase I organisms. This vaccine is not readily available, but offers promise for development of safe, effective immunization.

Bartonella

It has been recognized recently that organisms thought to be closely related to rickettsiae such as the louse-borne causative agent of trench fever, *Bartonella* (formerly *Rochalimaea*) *quintana*, in fact, belong in the genus *Bartonella*. These bacteria can be cultivated in cell-free medium and hence do not fit the criterion of definition of rickettsiae as obligately intracellular bacteria. *Bartonella quintana* infections were a serious medical problem during World War I. Soldiers in the trenches were infested with body lice that passed *B quintana* in their feces onto the skin. Individuals who have recovered from trench fever continue to have *R quintana* circulating in this stage of infection and may serve as sources of infection for lice, which can transmit the infection to others.

In association with the AIDS epidemic, another species *B henselae* (in addition to *B quintana)* has been discovered to be the cause of opportunistic infections often masquerading as hemangioma-like lesions of skin and visceral organs, bacillary angiomatosis. *Bartonella henselae* was recognized subsequently to be the long sought after cause of cat scratch disease, which usually manifested as a self-limited enlargement and inflammation of lymph nodes of several months duration in the regional drainage of a cat scratch or bite.

Bartonella bacilliformis transmitted by the sandfly in certain regions of Western South America invades human red blood cells, causing acute, often severe, hemolytic anemia. In chronic infections, there are skin lesions known as *verruga peruana* (Peruvian warts) that are similar to those of bacillary angiomatosis. A Peruvian medical student, Daniel Carrion, proved these lesions to be caused by an infectious agent in 1885 when he fatally inoculated himself with material from a *verruga peruana*. He died of the acute infectious hemolytic anemia known today as Oroya fever or, in his memory, Carrion's disease.

REFERENCES

Adal KA, Cockerell CJ, Petri WA: Cat scratch disease, bacillary angiomatosis, and other infections due to *Rochalimaea*. N Engl J Med 330:1509-1515, 1994

Audy JR (ed): Red Mites and Typhus. University Press, New York, 1968

Bakken JS, Dumler JS, Chen S-M, et al: Human granulocytic ehrlichiosis in the upper Midwest United States. A new species emerging? JAMA 272:212-218, 1994

Brouqui P, Dumler JS, Raoult D: Immunohistologic demonstration of *Coxiella burnetii* in the valves of patients with Q fever endocarditis. Am J Med 97:451-453, 1994

Drancourt M, Mainardi JL, Brouqui P, et al: *Bartonella* (*Rochalimaea*) *quintana* endocarditis in three homeless men. N Engl J Med 332:419-423, 1995

Fishbein DB, Dawson JE, Robinson LE: Human ehrlichiosis in the United States, 1985 to 1990. Ann Int Med 120:736-743, 1994

Hechemy KE, Paretsky D, Walker DH, Mallavia (eds): Rickettsiology: Current Issues and Perspectives. Vol 590. NY Acad Sci, New York, 1990

Helmick CG, Bernard KW, D'Angelo LJ: Rocky Mountain spotted fever: clinical, laboratory and epidemiological features of 262 cases. J Infect Dis 150:480, 1984

Kaplowitz LG, Fischer JJ, Sparling PF: Rocky Mountain spotted fever: a clinical dilemma. Curr Clin Top Infect Dis 2:89, 1981

Marrie TJ: Q Fever. Vol I. The Disease. CRC Press, Boca Raton, FL. 1990

McDade JE. Newhouse VF: Natural history of *Rickettsia rickettsii*. Annu Rev Microbiol 40:287, 1986

McDade JE, Shepard CC, Redus MA, et al: Evidence of *Rickettsia prowazekii* infections in the United States. Am J Trop Med Hyg 29:277, 1980

Moulder JW (ed): Intracellular Parasitism. CRC Press, Boca Raton, FL. 1989

Walker DH (ed): Biology of Rickettsial Disease. Vols I and II. CRC Press, Boca Raton, FL, 1988

Williams WJ, Radulovic S, Dasch GA, et al: Identification of *Rickettsia conorii* infection by polymerase chain reaction in a soldier returning from Somalia. Clin Infec Dis 19:93-99, 1994

Wolbach SB, Todd JL, Palfrey FW: Etiology and Pathology of Typhus. The League of Red Cross Societies Harvard Press, Cambridge, Mass, 1922

Zinsser H (ed): Rats, Lice, and History: Little, Brown, New York, 1935

Chapter 22

Legionella

General Concepts

Clinical Manifestations

The most common presentation of *Legionella pneumophila* is acute pneumonia (legionellosis); potentially any species of *Legionella* may cause the disease. Extrapulmonary disease (e.g., pericarditis and endocarditis) is rare. Less often, disease presents as a nonpneumonic epidemic, influenzalike illness called Pontiac fever.

Structure, Classification, and Antigenic Types

Legionella species are Gram-negative bacilli. There are currently 39 species and 60 distinct antigenic types of *Legionella.*

Pathogenesis

Legionella bacilli reside in surface and drinking water and are usually transmitted to humans in aerosols. The bacteria multiply intracellularly in alveolar macrophages. Recruited neutrophils and monocytes, as well as bacterial enzymes, produce destructive alveolar inflammation. Direct inoculation of surgical wounds by contaminated tap water has been described.

Host Defenses

Nonspecific physical and inflammatory pulmonary defenses are important, but cell-mediated immunity is critical. Immunologically activated monocytes and macrophages restrict intracellular bacterial growth. The role of humoral immunity is unclear.

Epidemiology

Legionella species are widespread in nature. Interactions with other environmental organisms may facilitate growth. Disease may be sporadic or epidemic and may occur in the community or in hospitals. People with compromised host defenses are at increased risk.

Diagnosis

Legionellosis can be suspected clinically, but diagnosis can be confirmed only by laboratory testing. The preferred method is culturing on special charcoal-containing agar.

Control

Decontamination of identified environmental sources is of primary importance for prevention. The drug of choice is erythromycin. Immunization works in experimental animals but has not been attempted in humans.

INTRODUCTION

Legionella was first recovered from the blood of a soldier more than 50 years ago, but its importance as a human pathogen was not recognized until 1976, when a mysterious epidemic of pneumonia struck members of the Pennsylvania American Legion. The disease was dubbed Legionnaire's disease by the press. Within 6 months a bacterium, subsequently named *Legionella pneumophila,* had been isolated and definitively established as the agent, thanks to the efforts of many investigators from Pennsylvania and the Centers for Disease Control and Prevention in Atlanta. A general term for disease produced by *Legionella* species is legionellosis.

Clinical Manifestations

The clinical manifestations of *Legionella* infections are primarily respiratory (Fig. 40-1). Two very different kinds of respiratory illness may result from infection; the reasons for this dichotomy are not understood (Table 40-1). The most common presentation is acute pneumonia, which varies in severity from mild illness that does not require hospitalization (walking pneumonia) to fatal multilobar pneumonia. Typically, patients have high, unremitting fever and cough but do not produce much sputum. Extrapulmonary symptoms, such as headache, confusion, muscle aches, and gastrointestinal disturbances, are common. Most patients respond promptly to appropriate antimicrobial therapy, but convalescence is often prolonged (lasting many weeks or even months).

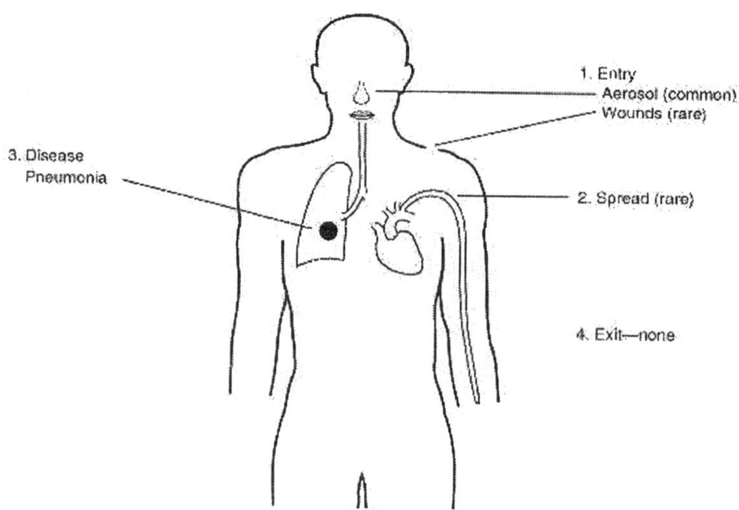

FIGURE 40-1 Pathogenesis of legionellosis.

TABLE 40-1 Clinical manifestations of Legionella Infections

Disease	pneumonia	Occurrence	Attack Rate	Incubation Period	Species Implicated
Legionnaire's disease	Almost always	Epidermic, sporadic, nosocomial community	Low	Long (days)	All, especially L pneumophila and L modedel
Pontiac fever	Never	Epidemic community	High	Short (hours)	L pneumophila, L ansia L micdadel, L leolel
Disseminated infection	Usually	Rare			L pnedmophila
Primary Wound infection	Rarely	Sporadic, nosocomial	Rare		L pnedmophila L dumollii

The second form of respiratory illness is called Pontiac fever after the city in Michigan where the first epidemic was recognized. This uncommon manifestation of infection resembles acute influenza, including fever, headache, and severe muscle aches. It is self-limited, and convalescence is uneventful.

Bacteremia occurs during *Legionella* pneumonia, and symptomatic infection outside the lungs occasionally develops. Under special conditions, bacteria introduced through portals other than the lungs, such as surgical wounds, may cause disease.

Structure

Legionella cells are thin, somewhat pleomorphic Gram-negative bacilli that measure 2 to 20 μm (Fig. 40-2). Long, filamentous forms may develop, particularly after growth on the surface of agar. Ultrastructurally, *Legionella* has the inner and outer membranes typical of Gram-negative bacteria. It possesses pili (fimbriae), and most species are motile by means of a single polar flagellum.

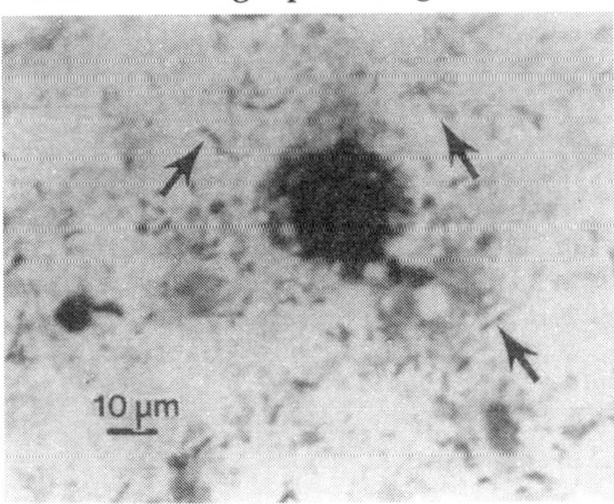

FIGURE 40-2 Impression smear from the lung of a patient fatally infected with *L pneumophila* serogroup 1, demonstrating many thin, Gram-negative bacilli (arrows). These bacteria stain less intensely with safranin than do enteric bacilli.

It is ironic that *Legionella* species are sometimes referred to as fastidious bacteria, because they may grow luxuriantly in tap water and can multiply in the usually hostile environment of phagocytic cells. They are fastidious only in regard to the media commonly used in laboratories. The primary growth factor required is L-cysteine, a nutrient that is also essential for *Francisella tularensis.* Ferric iron is also essential, and other compounds are necessary for optimal growth. Energy is derived from amino acids rather than carbohydrates.

Classification and Antigenic Types

Molecular characterization of strains isolated from patients in Pennsylvania led to the creation of a new family of bacteria, Legionellaceae, as well as a new genus, *Legionella.* The genus now has 39 species, defined by studies of DNA homology. Only one genus within the family has been recognized. Immunologic diversity within species is reflected in the creation of serogroups (Table 40-2). *Legionella pneumophila* holds the record, with 14 distinct serologic types. Important antigens include outer membrane proteins, some of which are species specific antigens, and the lipopolysaccharide thatis the major serogroup specific antigen.Strains may be divided into subtypes by antigenic analysis, using panels of monoclonal antibodies, or by characterization of bacterial enzyme systems. These increasingly fine distinctions can be very valuable for epidemiologic study but do not affect clinical decisions.

TABLE 40-2 Pathogenic Legionella Species Isolated from Humans*

Species	Number of Serogroups
L pnedmophila	14
L micdadel	1
L dumollii	1
L bozemanil	2
L gormanil	1
L leelei	2
L hackeliae	2
L israelensis	1
L jordanis	1
L sainfhelarsf	2
L longbeahae	2
L maceachemll	1
L oakridgensis	1
L wadsworthil	1
L biminghamersis	1
L clnoclnnatiensis	1
L anisa	1
L tussonensis	1
L lansingensis	1

*Species isolated from the environment only: Leglonelia cherrll, L erythra, l jamestownensis, L paristensis, L shakespearel, L santhcrucis, L steigerwaltil, L adelaidensis, L fairfleldensis, L brunensis, L moravioa, L guinlivianil, L gratlane, L quatelrensis, L nautarum, L worslelensis, L londinlensis, Lgeesliana, L rubnilvens, L spirtensis.

Pathogenesis

The pathogenesis of *Legionella* infections begins with a supply of water containing virulent bacteria and with a means for dissemination to humans (Fig. 40-1). Person-to-person transmission has never been demonstrated, and *Legionella* is not a member of the bacterial flora of humans.

Infection begins in the lower respiratory tract. Alveolar macrophages, which are the primary defense against bacterial infection of the lungs, engulf the bacteria; however, *Legionella* is a facultative intracellular parasite and multiplies freely in macrophages (Fig. 40-3). The bacteria bind to alveolar macrophages via the complement receptors and are engulfed into a phagosomal vacuole. However, by an unknown mechanism, the bacteria block the fusion of lysosomes with the phagosome, preventing the normal acidification of the phagolysosome and keeping the toxic myeloperoxidase system segregated from the susceptible bacteria. The bacilli multiply within the phagosome. Thus, a cellular compartment that should be a death trap instead becomes a nursery. Eventually, the cell is destroyed, releasing a new generation of microbes to infect other cells.

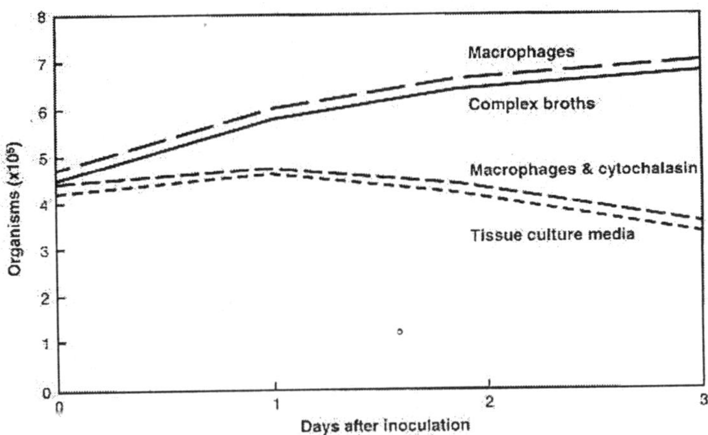

FIGURE 40-3 Growth of *Legionella* cells in vitro. This facultative intracellular pathogen grows well in complex broths that provide all necessary nutrients. The usual tissue culture media, which are adequate to support the growth of human and animal cells, cannot support the growth of Legionella cells. The bacteria also grow well within alveolar macrophages that have been maintained in cell culture. If phagocytosis is prevented by treatment with cytochalasin, however, the bacteria are denied access to the intracellular environment and growth does not occur.

Bacterial growth, activation of the complement system, and/or the death of alveolar macrophages produce powerful chemotactic factors that elicit an influx of monocytes and polymorphonuclear neutrophils (Fig. 40-4). Leaky capillaries allow the transudation of serum and deposition of fibrin in the alveoli. The result is a destructive pneumonia that obliterates the air spaces and compromises respiratory function (Fig. 40-5). Dissemination of bacteria to sites outside the lung occurs at least partially via macrophages, but only rarely does an inflammatory response develop.

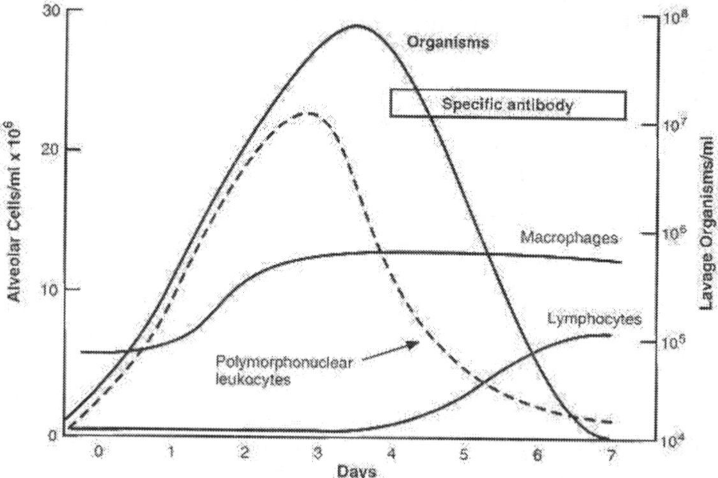

FIGURE 40-4 Inflammatory response in experimental *Legionella* pneumonia. Alveolar macrophages are the only resident cells in the air spaces of the lungs. Exponential bacterial growth begins soon after infection at a time when only macrophages are present; bacteria are most numerous 3 to 4 days later, but have virtually disappeared by the end of the first week. Infection elicits a large influx of polymorphonuclear leukocytes, followed by monocytes from the peripheral blood. Fluid exudation into the alveoli follows the pattern of the polymorphonuclear leukocytes. Specific immunologic responses (i.e., antibody and lymphocyte influx) are detectable 4 to 5 days after infection.

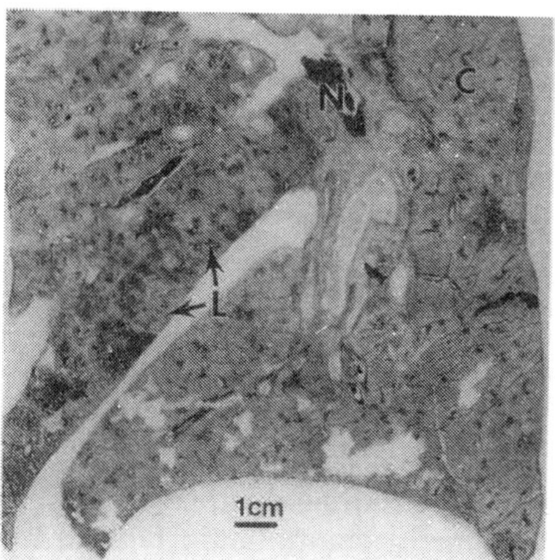

FIGURE 40-5 Acute *L pneumophila* pneumonia. Papermounted whole-lung section, unstained. The air spaces are filled with fibrin and inflammatory cells. The consistency of the completely consolidated lower lobe (C) resemble that of an adjacent hilar lymph node (N). The lobular nature of the process is accentuated by gray carbon accumulation around the terminal bronchioles in the centers of the lobules (L).

The symptoms of *Legionella* infection undoubtedly result from a combination of physical interference with oxygenation of blood, ventilation-perfusion imbalance in the remaining lung tissue, and release of toxic products from bacteria and inflammatory cells. Bacterial factors include a protease that may be responsible for tissue damage. Cellular factors include interleukin-1, which produces fever after it is released from monocytes, and tumor necrosis factor, which may be responsible for some of the systemic symptoms.

Virulence appears to be multifactorial. An outer membrane protein that functions as a metalloprotease and a cytoplasmic membrane heat-shock protein elicit protective immune responses, but are not essential for expression of virulence. A gene that encodes a 29 Kd protein and plays a role in cellular infection has been identified. Mutations of the gene are associated with decreased virulence.

Host Defenses

Risk factors for legionnaire's disease include conditions that compromise both the specific and non-specific defenses. The fact that patients with chronic heart and lung disease are at increased risk of developing serious *Legionella* pneumonia suggests that the integrity of physical clearance mechanisms, such as the mucociliary escalator of the tracheobronchial tree, is an important element of the defenses. Nonimmunologic antibacterial factors normally found in respiratory secretions, such as lactoferrin or lysozyme, may also play a role.

Inflammatory cell defenses play both positive and negative roles. The human alveolar macrophage and its relative, the recruited blood monocyte, abdicate their normal roles as primary antibacterial defenses in *Legionella* infections. The major reason why mice are very resistant to experimental *Legionella* pneumonia is probably that their alveolar macrophages do not support intracellular bacterial growth.

The role of polymorphonuclear leukocytes is less clear. These cells do not support bacterial growth in vitro and are only minimally bactericidal. Treatment with cytokines such as gamma interferon marginally increases their bactericidal activity. Neutropenia is not an important risk factor.

The most impressive risk factors for human disease are various types of immunosuppression. In a small outbreak of disease caused by contaminated nebulizers, pneumonia developed most often in patients being treated with corticosteroids.

The critical component of the immune system in resistance to legionellosis has not yet been pinpointed. Attention has focused on cell-mediated immunity because *Legionella* is a facultative intracellular pathogen. Infected patients produce a cell-mediated immune response that can be detected by measuring lymphocyte blastogenesis after exposure to *Legionella* antigen. Lymphocytes appear in the air spaces of experimentally infected animals about 5 days after an acute infection (Fig. 40-4). In contrast to naive alveolar macrophages, which are permissive for intracellular bacterial growth, activated alveolar macrophages or peripheral blood monocytes restrict bacterial multiplication in vitro (Fig.40-6). The macrophages can be activated by treatment with lymphokines produced by specifically stimulated lymphocytes. Gamma interferon, which can substitute for the lymphokines, is an important mediator. It has been suggested that restriction of entry of iron (an important growth factor for *Legionella*) into the phagosome inhibits intracellular growth.

The role of humoral immunity is less clear. Antibody in all immunoglobulinclasses is made after human or experimental infection. This antibody serves an opsonizing function in vitro, facilitating the phagocytosis of bacteria by polymorphonuclear leukocytes, macrophages, and monocytes. Antibody does not kill most strains of *Legionella*; however, so that the outcome of the interaction depends on the capa bilities of the phagocytic cell (Fig. 40-6). The classic pathway of the complement system is activated by *L pneumophila*, enhancing phagocytosis still further. *Legionella micdadei* activates the alternative complement pathway as well, so that opsonization of this species can occur even before an immunologically specific antibody response is mounted. One can construct scenarios from in vitro data in which antibody is deleterious as well as helpful. Experimental studies with animals support a protective role for antibody.

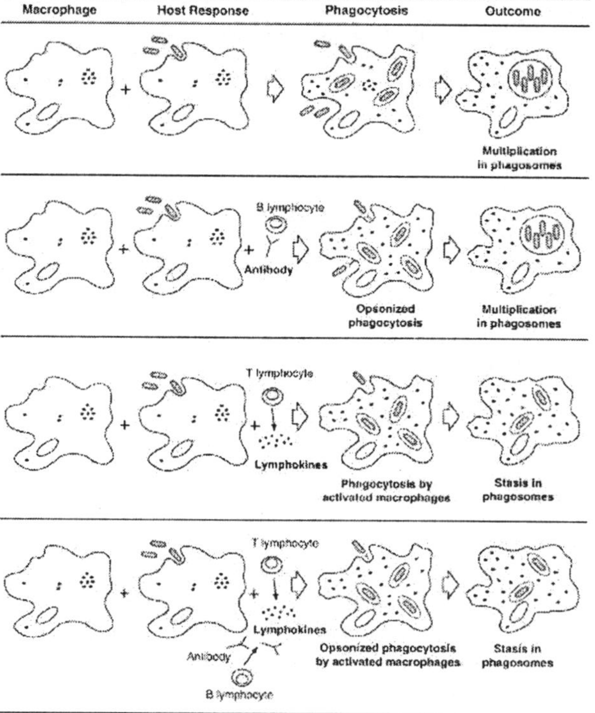

FIGURE 40-6 Multiplication of *Legionella* is inhibited in activated macrophages, whereas growth in normal macrophages provides a preferred environment, enhancing bacterial growth.

Epidemiology

The epidemiology of *Legionella* infections is a complex equation that is composed of the aquatic environment (including representatives of multiple microbial phyla, humans, mechanical devices and medical facilities), dissemination from the environment to the host, and host susceptibility. The complexity of the environmental interactions rivals those of viral and parasitic infections.

The only documented source of *Legionella* species is water, particularly the surface waters of rivers and lakes and drinking water. *Legionella* does not multiply in sterile tap water, but the addition of free-living amoebae results in growth of *Legionella* in vitro. The relationship with a variety of amoebae is particularly interesting. These single-cell animals, which can be viewed as nature's macrophages, occur in the same aquatic environment as do *Legionella* organisms (Fig. 40-7) and support the intracellular growth of *Legionella* in much the same way. Another factor that favors the survival of *Legionella* in natural or treated waters is its relative resistance to the effects of chlorine and heat; *Legionella* can find refuge in relatively inhospitable environments such as hot-water tanks.

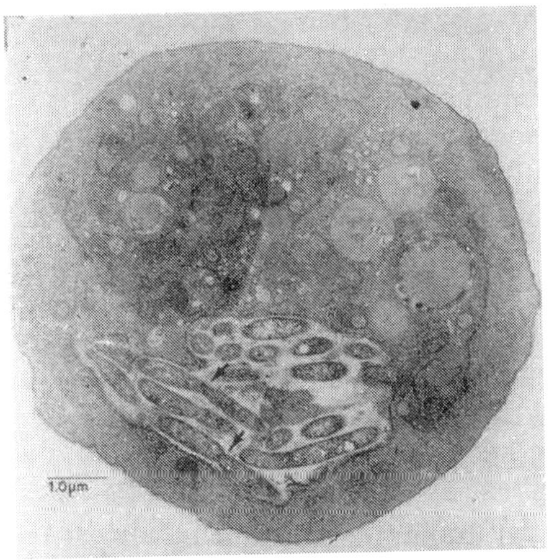

FIGURE 40-7 Electron micrograph showing *L pneumophila* serogroup 1 in the process of dividing (arrows) within a vesicle of an amoeba (Hartmanella veriformis) cell. (X 18,500.) (Courtesy of Barry S. Fields, Centers for Disease Control.)

The second factor in the epidemiology equation is dissemination of bacteria from the environment to the host. In most cases the link is an aerosol of water contaminated with the organisms. Evaporative condensers and cooling towers are proven sources of outdoor infection. Indoors, nebulizers and humidifiers filled with contaminated drinking water have disseminated *Legionella* to susceptible patients. The automatic misting devices that keep supermarket produce fresh have even been fingered as culprits in outbreaks of pneumonia. Aerosols are produced in numerous ways in our environment, from taking a shower to flushing the toilet.

An epidemic of Pontiac fever caused by *Legionella anisa* was associated with an ornamental fountain in a public place. Clusters of *Legionella* pneumonia have occurred after exposure to whirlpool spas in hotels or cruise ships. In most cases we do not know the source of the infection. Direct infection of surgical wounds has been linked to washing of patients with tap water that harbored pathogenic *Legionella* organisms.

The final factor in the equation is the susceptibility of patients. The diseases and conditions that serve as risk factors are concentrated in health care facilities, so it should be no surprise that many of the epidemics of disease have been nosocomial.

Legionella infections may be sporadic or epidemic, community acquired or nosocomial. There is great geographic variation in the frequency of infection even within communities, presumably reflecting the presence of suitable aquatic environments and susceptible subjects. Both sporadic and epidemic cases are more common during summer than winter months, perhaps because ofincreased use of air-cooling equipment that generates aerosols.

Diagnosis

There are no reliable distinguishing clinical features of *Legionella* pneumonia, so the diagnosis must come from the laboratory. Some clinical features suggest legionnaire's disease; however, and should prompt the selection of appropriate laboratory tests (Table 40-3). The diagnosis is confirmed in the laboratory by culture, demonstration of bacterial antigen in body fluids, or detection or a serologic response.

The preferred diagnostic method is culturing, because it is both sensitive and specific; however, appropriate specimens are not always available. The laboratory must be alerted to the possibility of legionellosis, because specially designed media must be used. The medium of choice is buffered charcoal-yeast extract - a-ketoglutarate medium. This medium contains yeast extract, iron, L-cysteine, and a-ketoglutarate for bacterial growth; activated charcoal to inactivate toxic peroxides that develop in the media; and buffer with a pK at pH 6.9, the optimum for growth of *Legionella* organisms. Addition of albumin to the media may further facilitate growth of species other than *L pneumophila.* For contaminated specimens such as sputum, antibiotics should be added. Morphologically distinctive bacterial colonies can usually be detected within 3 to 5 days and identified presumptively as *Legionella* species if the isolated bacteria depend on cysteine for growth. The identification can be confirmed by specific immunologic typing of the isolated bacteria or, in problematic cases, by molecular analysis.

Direct detection of bacterial antigen in clinical specimens is potentially much faster than culturing. Unfortunately, direct immunofluorescence detection of *Legionella* antigen in respiratory specimens is neither sensitive nor specific enough to warrant general use. A commercially available radioimmunoassay for bacterial antigen in urine is satisfactory, but is available only for serogroup 1 of *L pneumophila* and requires the use and disposal of radiochemicals.

Serologic diagnosis is moderately sensitive and reasonably specific. It should be considered as an adjunct to diagnosis by culture. Indirect immunofluorescence has been used most frequently.

It is important to use an assay that detects IgM and IgG. The advantages of serologic diagnosis are that it is performed on easily obtained blood specimens and can detect mild or even asymptomatic infection. The major disadvantage of the technique is that paired acute and convalescent-phase sera are essential. The convalescent-phase specimen must be obtained at least 6 weeks after onset of the infection, by which time the physician and patient have often lost interest in the enterprise. Furthermore, in non-epidemic situations when the prevalence of disease is low, cross-reactions with other bacteria may create an unacceptable low predictive value for a positive test. Cross-reactions among *Legionella* species and serogroups make assignment of a species-specific diagnosis impossible on serologic grounds.

Control

Because the interactions between *Legionella* organisms, the environment, and the host are so complex, the incidence of disease may be controlled in several ways. If an aquatic source of infection can be found, elimination of *Legionella* from the source is an effective control mechanism. This genus is so common in water systems that molecular analysis of environmental and clinical stains is often helpful in pinpointing the source. Unfortunately, decontamination can be expensive. The two most common means of eradicating *Legionella* are periodic superheating of water with attendant dangers of scalding, and continuous chlorination, which accelerates deterioration of plumbing systems unless carefully monitored. Even "chlorinated" drinking water must be treated because the levels of chlorine decrease with increasing distance from the distribution center, particularly in hot water. Constant vigilance must be maintained to prevent return of the unwanted pathogens. Elimination of *Legionella* spp. from all environmental sites is unlikely ever to be accomplished.

Immunization has been proposed as a means of preventing *Legionella* infection in susceptible populations. This approach works in experimental animals but has not been attempted in humans.

Pontiac fever does not require antimicrobial therapy. The preferred drug for symptomatic *Legionella* infections is erythromycin. If the patient is seriously ill, it is important to deliver the antibiotic intravenously at first; subsequently, oral therapy may be used. Rifampin is sometimes added as a second antibiotic in seriously ill patients. Retrospective analysis of the antibiotics used to treat the Pennsylvania legionnaires suggests that erythromycin was the most effective agent. If these patients had been hospitalized in major medical centers, they would undoubtedly have received the latest antimicrobial agents and it might have taken us years to determine that erythromycin is the drug of choice. Experimentally, both erythromycin and rifampin inhibit the growth of *Legionella* organisms in infected macrophages, but do not kill the bacteria. Prophylactic antibiotic therapy may be useful for patients at high risk of serious disease, such as transplant recipients, when a documented epidemic is occurring.

It may appear that we are defenseless against *Legionella* infection, because the most effective type of host defense shows only very modest bactericidal abilities in vitro. In fact most infections are subclinical, and mortality is low in patients who are not immunocompromised. Similarly, even susceptible experimental animals survive infection unless moderately large doses of bacteria are given. The defense mechanisms probably function better in vivo than in vitro. The action of host defenses may also be additive in vivo. One can construct a scenario by which bacteria are increasingly phagocytosed by cells that do not permit bacterial growth. The net result is a decreasing number of extracellular bacteria and hence a decreased source of infection for a decreasing population of permissive cells. Obsolescent inflammatory cells in the lungs are removed by the mucociliary escalator and expectorated as sputum. Therefore, the infection may begin with a bang, but it ends in most cases with a whimper.

REFERENCES

Barbaree JM, Breiman RF, Dufour AP: Legionella: Current Status and Emerging Perspectives. American Society for Microbiology. Washington, D.C. 1993

Bhopal R: Source of infection for sporadic *Legionnaires;* disease: a review. J Infect Rev 30(1):9, 1995.

Dowling JN, Saha AK, Glew RH: Virulence factors of the family *Legionellaceae*. Microbiol Rev 56:32, 1992

Fang GD, Yu VL, Vickers RM: Disease due to the *Legionellaceae* (other than *Legionella pneumophila*). Historical, microbiological, clinical, and epidemiological review. Medicine (Baltimore) 68:116, 1989

Fraser DW, Tsai TR, Orenstein W, et al: Legionnaires' disease: description of an epidemic of pneumonia. N Engl J Med 297:1189, 1977

McDade JE, Shepard CC, Fraser DW, et al: Legionnaires' disease: isolation of a bacterium and demonstration of its role in other respiratory disease. N Engl J Med 297:1197, 1977.

Ott M: Genetic approaches to study *Legionella* pneumophila pathogenicity. FEMS Microbiology Rev 14(2):161, 1994.

Winn WC Jr., Myerowitz RL: The pathology of the *Legionella* pneumonias. A review of 74 cases and the literature. Hum Pathol 12:401, 1981

Winn WC Jr.: *Legionella*: historical perspective. Clin Microbiol Rev 1:60, 1988

Chapter **23**

Leptospira

General Concepts

Clinical Manifestations

Leptospira interrogans causes leptospirosis, a usually mild febrile illness that may result in liver or kidney failure.

Structure, Classification, and Antigenic Types

Leptospira is a flexible, spiral-shaped, Gram-negative spirochete with internal flagella. *Leptospira interrogans* has many serovars based on cell surface antigens.

Pathogenesis

Leptospira enters the host through mucosa and broken skin, resulting in bacteremia. The spirochetes multiply in organs, most commonly the central nervous system, kidneys, and liver. They are cleared by the immune response from the blood and most tissues but persist and multiply for some time in the kidney tubules. Infective bacteria are shed in the urine. The mechanism of tissue damage is not known.

Host Defenses

Serum antibodies are responsible for host resistance.

Epidemiology

Leptospirosis is a worldwide zoonosis affecting many wild and domestic animals. Humans acquire the infection by contact with the urine of infected animals. Human-to-human transmission is extremely rare.

Diagnosis

Clinical diagnosis is usually confirmed by serology. Isolation of spirochetes is possible, but it is time-consuming and requires special media.

Control

Animal vaccination and eradication of rodents are important. Treatment with tetracycline and penicillin G is effective. No human vaccine is available.

Borrelia

Clinical Manifestations

Borrelia recurrentis (louse borne) and *B hermsii and B turicatae* (tick borne) cause relapsing fevers: influenza-like febrile diseases that follow a relapsing and remitting course. Myocarditis is a rare sequela. *Borrelia burgdorferi* causes Lyme disease, a multisystem, relapsing febrile disease with a rash and manifestations such as arthritis, carditis, and neuritis.

Structure, Classification, and Antigenic Types

Like *Leptospira, Borrelia* is a flexible, spiral-shaped, Gram-negative spirochete with internal flagella. *Borrelia* species are differentiated primarilyon the basis of vectors and DNA homology.

Pathogenesis

Borrelia is transmitted by tick or louse bites. The relapsing-fever borreliae cause recurrent febrile bacteremias separated by remissions during which the borreliae are sequestered in tissues; each resurgence involves a change in cell surface antigens. Lyme disease may have different manifestations at different times; recurrences and late sequelae may appear for many years. The pathogenesis of borrelial diseases is not understood.

Host Defenses

Serum antibodies are responsible for host resistance.

Epidemiology

The tick-borne relapsing fevers and Lyme disease are zoonoses with rodents as the major reservoir; incidence and distribution depend mainly on the biology of the tick vectors. Louseborne relapsing fever has no animal reservoir and causes epidemics in crowded, unsanitary populations.

Diagnosis

The clinical diagnosis is confirmed by serology and also by microscopic visualization of the organism in blood of relapsing fever patients.

Control

Areas known to harbor infected ticks and lice should be avoided. Tetracycline is an effective treatment. No vaccines are available.

Spirillum

Clinical Manifestations

Spirillum causes rat bite fever, with ulceration at the site of the bite, lymphadenopathy, rash, and a relapsing fever.

Structure, Classification, and Antigenic Types

Spirillum is Gram negative but, unlike *Leptospira* and *Borrelia,* has a rigid cell wall and external flagella. *Spirillum* species are differentiated on the basis of cell morphology.

Pathogenesis

Spirillum is transmitted by the bite of an infected rat. The mechanism of pathogenesis is not understood.

Host Defenses

Serum antibodies are responsible for host resistance.

Epidemiology

Rat bite fever occurs worldwide but is most common in Asia.

Diagnosis

The clinical diagnosis is confirmed by microscopic visualization of the organism in exudates from the initial lesion, aspirates of involved lymph nodes, or blood. No specific serologic test is available.

Control

Eradication of rodents is the main means of control. Treatment with penicillin is effective. No vaccine is available.

INTRODUCTION

Leptospira, Borrelia, and *Spirillum* cause disease characterized by clinical stages with remissions and exacerbations. *Leptospira* organisms are very thin, tightly coiled, obligate aerobic spirochetes characterized by a unique flexuous type of motility. The genus is divided into two species: the pathogenic leptospires *L interrogans* and the free-living leptospire *L biflexa.* Serotypes of *L interrogans* are the agents of leptospirosis, a zoonotic disease. The primary hosts for this disease are wild and domestic animals, and the disease is a major cause of economic loss in the meat and dairy industry. Humans are accidental hosts in whom this disseminated disease varies in severity from subclinical to fatal. The first human case of leptospirosis was described in 1886 as a severe icteric illness and was referred to as Weil's disease; however, most human cases of leptospirosis are nonicteric and are not life-threatening. Recovery usually follows the appearance of a specific antibody.

In contrast to the pathogenic leptospires, serotypes of *L biflexa* exist in water and soil as free-living organisms. Although, *L biflexa* has been isolated from mammalian hosts on occasion, no pathology has been found, and it does not infect experimental animals. Because of the widespread distribution of *L biflexa* in fresh water and the capability of leptospires to pass through 0.45 to 0.22-µm-pore-size sterilizing filters, they have been found as contaminants of filter-sterilized media.

Borrelia species are responsible for the relapsing fevers and Lyme disease. The organisms are transmitted to humans primarily by lice or ticks. Relapsing fevers are acute recurrent illnesses characterized by febrile episodes that recede spontaneously but generally reappear with decreasing intensity and duration. *Borrelia recurrentis* is responsible for the louse-borne or epidemic type of relapsing fever with humans serving as the reservoir host. The disease does not occur in the United States. In the western United States and Canada *B hermsii* and *B turicatae* are the most frequent causes of tick-borne or endemic type of relapsing fever, with *B hermsii* responsible for most human cases. Rodents are the primary reservoir for these borreliae. Lyme disease is another tick-borne illness and is caused by *B burgdorferi*. The disease occurs in the north temperate zone. The majority of cases in the United States occur in the north central and northeastern states and California. Rodents are the major reservoir for this spirochete. Antibodies play an important role in immunity to borrelial infections.

A single member of the genus *Spirillum, S minum,* is pathogenic for humans. *Spirillum minum* causes one type of rat bite fever, which is characterized by recurrent fever. The pathogenesis of the organism is obscure, but the host can produce a spirillicidal antibody.

Leptospira Leptospirosis

Clinical Manifestations

Clinical manifestations of leptospirosis are associated with a general febrile disease and are not sufficiently characteristic for diagnosis. As a result, leptospirosis often is initially misdiagnosed as meningitis or hepatitis. Typically, the disease is biphasic, which an acute leptospiremic phase followed by the immune leptospiruric phase. The three organ systems most frequently involved are the central nervous system, kidneys, and liver (Fig. 35-1). After an average incubation period of 7 to 14 days, the leptospiremic acute phase is evidenced by abrupt onset of fever, severe headache, muscle pain, and nausea; these symptoms persist for approximately 7 days. Jaundice occurs during this phase in more severe infections. With the appearance of antileptospiral antibodies, the acute phase of the disease subsides and leptospires can no longer be isolated from the blood. The immune leptospiruric phase occurs after an asymptomatic period of several days. It is manifested by a fever of shorter duration and central nervous system involvement (meningitis). Leptospires appear in the urine during this phase and are shed for various periods depending on the host. The more severe form of leptospirosis is frequently associated with infections having the serotype icterohaemorrhagiae and is often referred to as Weil's disease.

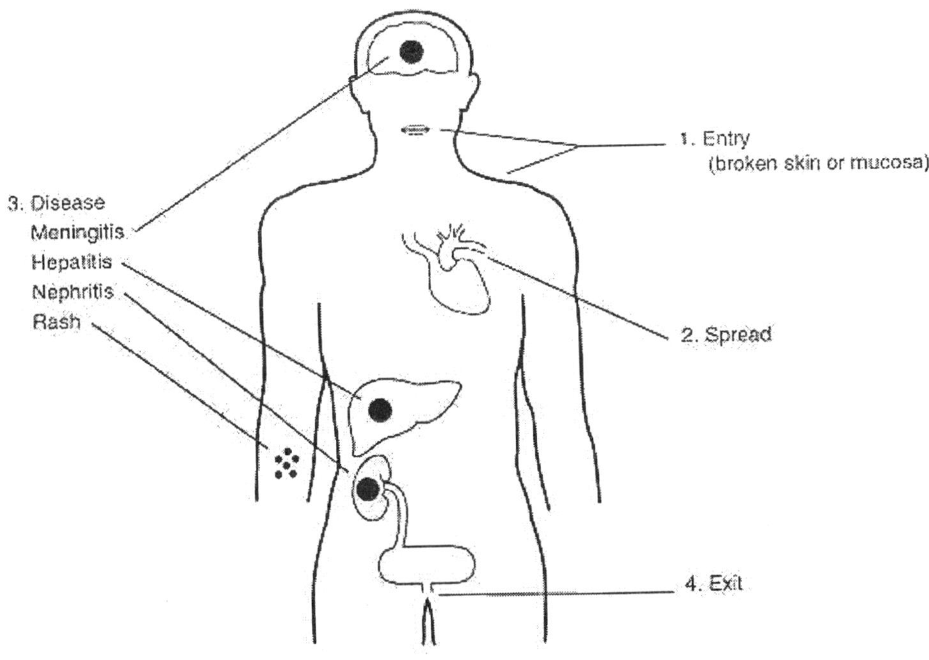

FIGURE 35-1 Clinical manifestations of leptospirosis.

Structure, Classification, and Antigenic Types

Leptospira has the general structural characteristics that distinguish spirochetes from other bacteria (Fig. 35-2). The cell is encased in a three- to five-layer outer membrane or envelope. Beneath this outer membrane are the flexible, helical peptidoglycan layer and the cytoplasmic membrane; these encompass the cytoplasmic contents of the cell. The structures surrounded by the outer membrane are collectively called the protoplasmic cylinder An unusual feature of the spirochetes is the location of the flagella, which lie between the outer membrane and the peptidoglycan layer. They are referred to as periplasmic flagella. The periplasmic flagella are attached to the protoplasmic cylinder subterminally at each end and extend toward the center of the cell. The number of periplasmic flagella per cell varies among the spirochetes. The motility of bacteria with externalflagella is impeded in viscous environments, but that of spirochetes is enhanced. The slender (0. 1 µm by 8 to 20 µm) leptospires are tightly coiled, flexible cells (Fig. 35-3). In liquid media, one or both ends are usually hooked. Leptospires are too slender to be visualized with the bright-field microscope but are clearly seen by dark-field or phase microscopy. They do not stain well with aniline dyes.

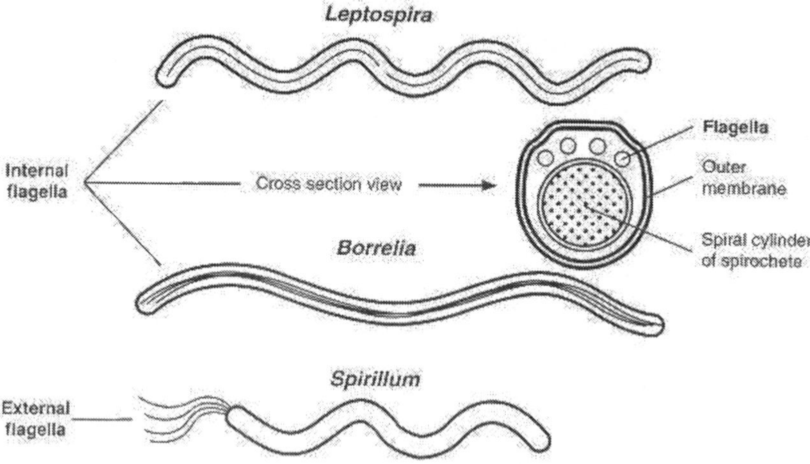

FIGURE 35-2 Morphological comparison of Leptospira, Borrelia, and Spirillum.

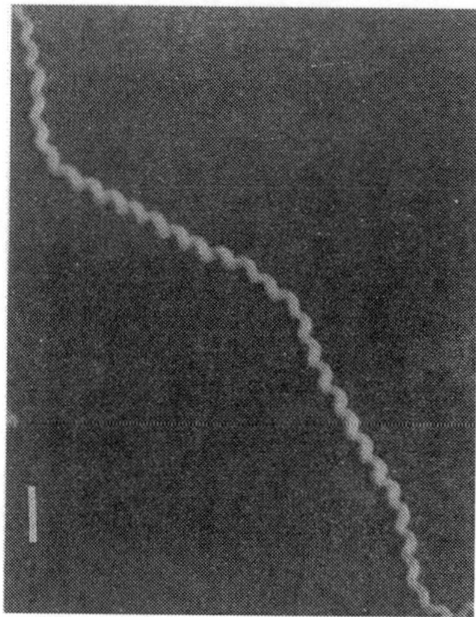

FIGURE 35-3 Electron micrograph of Leptospira interrogans serovar icterohaemorrhagiae. Bar equals 0.5 µm.

The leptospires have two periplasmic flagella, one originating at each end of the cell. The free ends of the periplasmic flagella extend toward the center of the cell, but do not overlap as they do in other spirochetes. The basal bodies of *Leptospira* periplasmic flagella resemble those of Gram-negative bacteria, whereas those of other spirochetes are similar to the basal bodies of Gram-positive bacteria. *Leptospira* differs from other spirochetes in lacking glycolipids and having diaminopimelic acid rather than ornithine in its peptidoglycan.

The leptospires are the most readily cultivated of the pathogenic spirochetes. They have relatively simple nutritional requirements; long-chain fatty acids and vitamins B1 and B12 are the only organic compounds known to be necessary for growth.

When cultivated in media of pH 7.4 at 30°C, their average generation time is about 12 hours. Aeration is required for maximal growth. They can be cultivated in plates containing soft (1 percent) agar medium, in which they form primarily subsurface colonies.

The two species, *L interrogans* and *L biflexa*, are further divided into serotypes based on their antigenic composition. More than 200 serotypes have been identified in *L interrogans*. The most prevalent serotypes in the United States are canicola, grippotyphosa, hardjo, icterohaemorrhagiae, and pomona. Genetic studies have demonstrated that serologically diverse serotypes may be present in the same genetic group. At least seven species of pathogenic leptospires have been identified by nucleotide analysis.

Pathogenesis

The mucosa and broken skin are the most likely sites of entry for the pathogenic leptospires (Fig. 35-1). A generalized infection ensues, but no lesion develops at the site of entry. Bacteremia occurs during the acute, leptospiremic phase of the disease. The host responds by producing antibodies that, in combination with complement, are leptospiricidal. The leptospires are rapidly eliminated from all host tissues except the brain, eyes, and kidneys. Leptospires surviving in the brain and eyes multiply slowly if at all; however, in the kidneys they multiply in the convoluted tubules and are shed in the urine (the leptospiruric phase). The leptospires may persist in the host for weeks to months; in rodents they may be shed in the urine for the lifetime of the animal. Leptospiruric urine is the vehicle of transmission of this disease.

The mechanism by which leptospires cause disease remains unresolved, as neither endotoxin nor exotoxins have been associated with them. The marked contrast between the extent of functional impairment in leptospirosis and the scarcity of histologic lesions suggests that most damage occurs at the subcellular level. Damage to the endothelial lining of the capillaries and subsequent interference with blood flow appear responsible for the lesions associated with leptospirosis. The most notable feature of severe leptospirosis is the progressive impairment of hepatic and renal function. Renal failure is the most common cause of death. The lack of substantial cell destruction in leptospirosis is reflected in the complete recovery of hepatic and renal function in survivors. Although, spontaneous abortion is common in infected cattle and swine, only recently has a human case of fatal congenital leptospirosis been documented.

The host's immunologic response to leptospirosis is thought to be responsible for lesions associated with the late phase of this disease; this helps to explain the ineffectiveness of antibiotics once symptoms of the disease have been present for 4 days or more.

Host Defenses

Nonspecific host defenses appear ineffective against the virulent leptospires, which are rapidly killed in vitro by the antibody-complement system; virulentstrains are more resistant to this leptospiricidal activity than are avirulent strains. Immunity to leptospirosis is primarily humoral; cell-mediated immunity does not appear to be important, but may be responsible for some of the late manifestations of the disease. Immunity to leptospirosis is serotype specific and may persist for years. Immune serum has been used to treat human leptospirosis and passively protects experimental animals from the disease. The survival of leptospires within the convoluted tubules of the kidneys may be related to the ineffectiveness of the antibody-complement system at this site. Previously infected animals can become seronegative and continue to shed leptospires in their urine, possibly because of the lack of antigenic stimulation by leptospires in the kidneys.

Epidemiology

Leptospirosis is a worldwide zoonosis with a broad spectrum of animal hosts. The primary reservoir hosts are wild animals such as rodents, which can shed leptospires throughout their lifetimes. Domestic animals are also an important source of human infections. Leptospires have been isolated from approximately 160 mammalian species in the temperate zone. The disease is more widespread in tropical countries, where the infectious agent may be one of many serotypes carried by a large variety of hosts.

Direct or indirect contact with urine containing virulent leptospires is the major means by which leptospirosis is transmitted. As mentioned above, leptospires from urine-contaminated environments, such as water and soil, enter the host through the mucous membranes and through small breaks in the skin. Moist environments with a neutral pH provide suitable conditions for survival of leptospires outside the host. Urine-contaminated soil can remain infective for as long as 14 days. In humans, leptospirosis has occurred in an infant being breast-fed by a mother with the disease. The cellular structure of leptospires causes them to be susceptible to killing by adverse conditions such as dehydration, exposure to detergents, and temperatures above 50°C. Most cases of leptospirosis occur during summer and fall.

Diagnosis

Because clinical manifestations of leptospirosis are too variable and nonspecific to be diagnostically useful, microscopic demonstration of the organisms, serologic tests, or both are used in diagnosis. The microscopic agglutination test is most frequently used for serodiagnosis. The organisms can be isolated from blood or urine on commercially available media, but the test must be requested specifically because special media are needed. Isolation of the organisms confirms the diagnosis.

Control

Human leptospirosis can be controlled by reducing its prevalence in wild and domestic animals. Although, little can be done about controlling the disease in wild animals, leptospirosis in domestic animals can be controlled through vaccination with inactivated whole cells or an outer membrane preparation. If vaccines do not contain a sufficient immunogenic mass, the resulting immune response protects the host against clinical disease but not against development of the renal shedder state. Because a multiplicity of serotypes may exist in a given geographic region and the protection afforded by the inactivated vaccines is serotype specific, the use of polyvalent vaccines is recommended. Vaccines for human use are not available in the United States.

Although, the leptospires are susceptible to antibiotics such as penicillin and tetracycline in vitro, use of these drugs in the treatment of leptospirosis is somewhat controversial. Treatment is most effective if initiated within a week of disease onset. At later times, immunologic damage may already have begun, rendering antimicrobial therapy less effective. Doxycycline has been used successfully as a chemoprophylactic agent for military personnel training in tropical areas.

Borrelia

Clinical Manifestations

Once the relapsing-fever borreliae have entered the host, they cause a generalized infection, apparent after an incubation period of approximately 1 week (Fig. 35-4). The onset of the disease, which is associated with numerous spirochetes in the blood is abrupt, with fever, headache, and muscle pain that persists for 4 to 10 days, followed by an afebrile period of 5 to 6 days correlated with the absence of spirochetemia. Usually, a single relapse occurs in louse-borne relapsing fever.

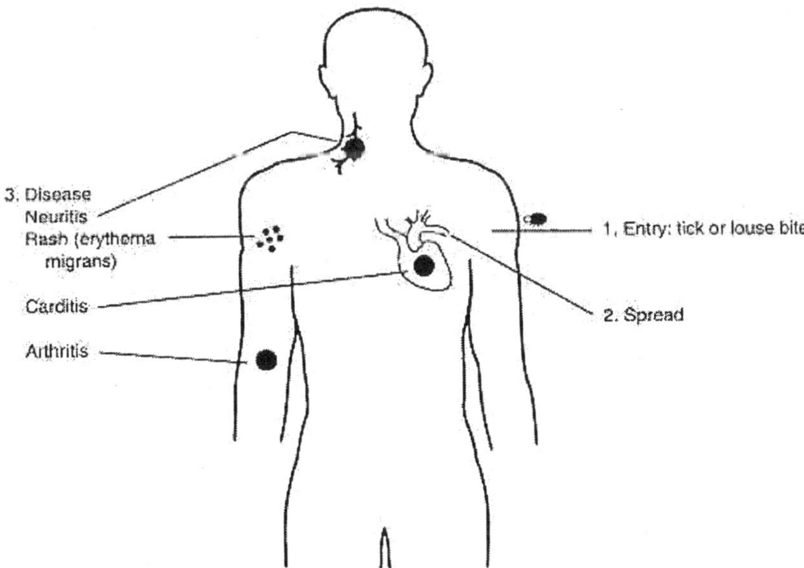

FIGURE 35-4 Pathogenesis of Borrelia infection.

The clinical features of relapsing fever, other than its recurrent pattern, are not diagnostic. The mortality in untreated epidemic relapsing fever can be higher than 40 percent, and myocarditis probably is the most common cause of death. The tick-borne relapsing fever is similar to the louse-borne disease, but is less severe (mortality, 0 to 8 percent), and several relapses of decreasing intensity are commonly experienced.

Structure, Classification, and Antigenic Types

Borrelia has morphologic characteristics similar to those of *Leptospira,* except that cells average 0.2 to 0.5 µm by 4 to 18 µm and have fewer coils (Fig. 35-5). Seven to twenty periplasmic flagella originate at each end and overlap at the center of the cell. In contrast to the *Leptospira* peptidoglycan, that of *Borrelia* contains ornithine rather than diaminopimelic acid. Basal bodies of periplasmic flagella of borreliae resemble those in Gram-positive bacteria. Because of their larger diameter, borreliae are more readily stained with aniline dyes than are other spirochetes. Their lipid components are unusual in that they include cholesterol; this substance has been found in only one other bacterial genus, *Mycoplasma.* The nutritional requirements of the borreliae are more complex than those of leptospires. Glucose, amino acids, long-chain fatty acids, N-acetylglucosamine, and several vitamins are some of their required organic nutrients. The borreliae are microaerophilic organisms. *Borrelia hermsii* has a generation time of 12 hours when cultivated in artificial media at 35°C compared with only 6 to 10 hours in the mouse.

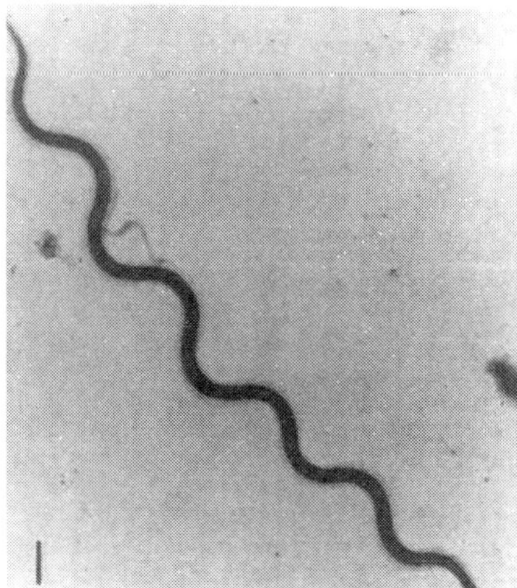

FIGURE 35-5 Electron micrograph of Borrelia hispanica. **Bar equals 1 µm.**

Pathogenesis

In most cases, borreliae must rely on an insect vector to transmit the organisms through the epidermis (Fig. 35-4). The site of entry is usually not prominent as the organisms are not clinically recognized until they enter the blood. The mechanisms by which they reach the bloodstream are unknown. The relapses are due to the ability of borreliae to undergo multiple cyclic antigenic variations (Fig. 35-6). As antibodies for the predominant antigenic type multiplying within the host appear, these organisms "disappear" from the peripheral blood and are replaced by a different antigenic variant within a few days. This process may occur several times in an untreated host, depending on the infecting *Borrelia* strain.

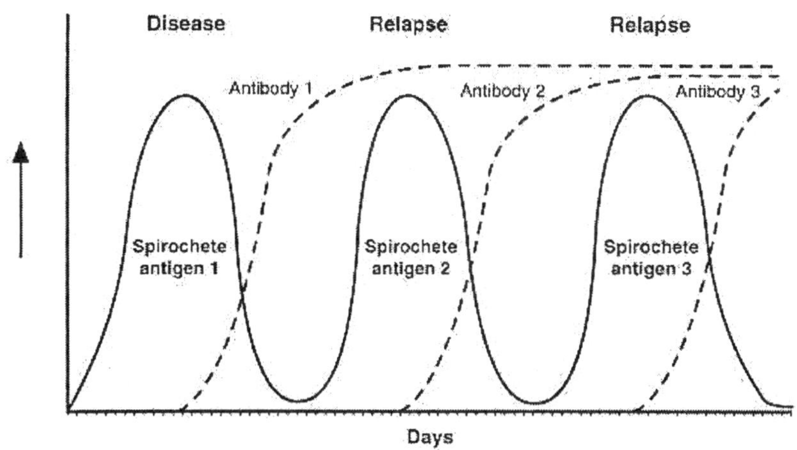

FIGURE 35-6 Immunoavoidance mechanism of Borrelia illustrating the emergence of antigenic variants during infection.

The mechanism by which borreliae cause Lyme disease has not been elucidated. Early Lyme disease is characterized by an expanding annular red rash, erythema migrains, in approximately 70 percent of patients, which is frequently accompanied by fever, fatigue, headache, and muscle and joint pain. Arthritis, neuritis, and carditis may also be present during early Lyme disease. Persistent neurologic and arthritic infections (lasting for months to years) may occur in some patients. In contrast to the relapsing fevers, there are very few spirochetes in Lyme disease, but viable *B burgdorferi* organisms are necessary for the disease to manifest itself. Although, the disease has the same general features in the United States and Europe, arthritis occurs more frequently in the former and neurologic disease in the latter.

Host Defenses

Borrelia appears to be resistant to nonspecific host defense mechanisms and elicits an inflammatory response consisting of mononuclear cells. These spirochetes are rapidly killed in vitro by the antibody-complement system. Immunity to the borrelioses is primarily humoral, and immune serum passively protects experimental animals from infection. Several genospecies of *B burgdorferi* have been reported.

Epidemiology

Borrelia is transmitted to humans by the body louse or ticks. *Borrelia recurrentis,* the cause of louse-borne (epidemic) relapsing fever, is carried by the human louse *Pediculus humanus.* The louse ingests the bacterium while feeding on a borrelemic host. The organisms multiply in the hemolymph and central ganglion of the louse. Because other organs are not invaded, transovarial transmission does not occur in the louse. In addition, *Borrelia* organisms are released when the louse is injured by host activities such as scratching. Accordingly, one louse can infect only a single host and is infective only for its life-span of around 1 month. Therefore, humans rather than the louse are the reservoir of this disease. The louse-borne disease is called epidemic relapsing fever because it can be rapidly disseminated under conditions of overcrowding and poor personal hygiene, such as during wars and natural disasters. Relapsing fever, which depends on the aforementioned conditions favoring the multiplicationand transfer of the human louse, has disappeared from the United States except for imported cases. Ethiopia appears to have the highest incidence of this disease.

The tick-borne relapsing fever is called endemic relapsing fever because it occurs whenever humans are exposed to infected ticks. The soft ticks *Ornithodoros hermsi* and *O turicata* most frequently transmit the disease in the United States. These ticks often obtain their blood meal at night, and because the tick bite is usually painless and feedings are short (5 to 20 minutes), people may not be aware of having been bitten. The designation of species of these borreliae is based on their vector (e.g., *B hermsii* is associated with *O hermsi).* Genetic studies have shown that this basis is incorrect. The two North American species, *B hermsii* and *B turicatae,* actually represent a single species. The ticks usually become infected by feeding on borrelemic rodents. In contrast to the louse, all tissues of the tick are invaded, resulting in transovarial transmission andthe presence of borreliae in salivary and coxal (basal segment of appendage) secretions. These spirochetes in the salivary and coxal secretions enter the host through the bite wound while the tick is feeding (less than 1 minute may be required for transmission). The largest outbreak of tick-borne relapsing fever in the western hemisphere occurred in 1973 on the north rim of the Grand Canyon, where 62 persons staying in log cabins developed the disease.

Borrelia burgdorferi, the agent of Lyme disease, is transmitted by members of the *Ixodes ricinus* complex (hard ticks). In the northeastern and midwestern United States the spirochete is transmitted by the deer tick, *I scapularis,* whereas the western black-legged tick, *I pacificus,* is the primary vector in the western United States. However, only one *Borrelia* species, *B burgdorferi,* appears responsible for the disease in North America, and it does not undergo the antigenic changes of the relapsing-fever borreliae. Transovarial transmission of the spirochete is infrequent, with fewer than 1 percent of the larvae infected. The pinhead-sized nymph form of the tick is the primary source ofhuman infections, but the disease canalso be transmitted by the large adult female tick. Feedings are lengthy and require several days. The spirochetes are present in the midgut of the tick, and 12 to 24 hours is required before the spirochetes are transmitted. Within endemic areas of Lyme disease, 20 to 60 percent of *I scapularis* may be carriers of *B burgdorferi* and a similar percentage of white-footed mice, a major reservoir host, are infected. The white-tailed deer, although not a reservoir host for the spirochete, plays a critical role in the life cycle of the tick, and Lyme disease occurs in areas in which deer are present. Dogs and birds are important in the dispersal of infected ticks to new locations. The number of cases of Lyme disease reported in the United States is about 10,000 per year.

Diagnosis

Clinical features of the relapsing fevers other than their recurring pattern are not diagnostic. Diagnosis is based primarily on demonstration of the spirochetes in blood during febrile episodes by dark-field examination, use of stained blood smears, or mouse inoculation. Antibody detection by indirect immunofluorescence assay is available. There is strong cross-reactivity with antibodies to *B burgdorferi* and weaker reactivity with those to *Treponema pallidum.*

The characteristic expanding red skin lesion, erythema migrans, is diagnostic for Lyme disease. However, 30 percent of patients do not develop this rash. The usual symptoms of early disease (fever, fatigue, headache, and muscle and joint pain) are too nonspecific to be diagnostic. Although, *B burgdorferi* has been isolated from blood, skin, and cerebrospinal fluid, this is a low yield procedure and is not recommended. Serologic tests are used most commonly for diagnosis of Lyme disease.

Unfortunately, most of the currently available tests are unable to detect early antibody responses to *B burgdoferi* (which may take up to two months to develop), thus the diagnosis of a patient without erythema migrans who is suspected of having early Lyme disease can be difficult. Moreover, false positive results may occur due to cross-reacting antibodies against other bacteria, especially *T pallidum* or other spirochetes, and patients with connective tissue disorders may demonstrate false serologic results. Despite these limitations, however, if patients being tested are carefully selected for symptoms compatible with Lyme disease and epidemiologic factors are included in consideration of the diagnosis, the positive predictive value of serologic tests can be quite acceptable, especially if positive serologic results are confirmed by a Western blotting procedure.

Many of the problems with serological testing for Lyme disease are based on interlaboratory differences in the substrate antigens, in antigen preparation, and in interpretive criteria. To improve the sensitivity and specificity of serodiagnosis of Lyme disease, evaluating serum samples from patients with symptoms of Lyme disease by a two-step process which includes a sensitive screening test such as immunofluorescence assay or enzyme immunoassay is recommended. Samples judged equivocal or positive by the screening should be further tested by Western blot. Both IgM and IgG Western blot should be performed for persons with suspected Lyme disease who present within the first four weeks of disease onset.

Only IgG Western should be performed for patients presenting later because interpretation of IgM band patterns after that time is less reliable. Serologic testing of a convalescent serum sample two to four weeks after the first sample is recommended for patients who have symptoms of early Lyme disease and negative screening test results.

Control

Relapsing fevers and Lyme disease are prevented by avoiding the vectors. It is important to be aware of endemic areas and to take proper precautions. When in potential tick habitats, one should wear clothingthat covers as much of the skin as possible and use tick repellents. Periodic skin inspection and tick removal prevent Lyme disease. A Lyme disease vaccine may be available in the near future. The relapsing-fever and Lyme disease borreliae have similar antibiotic susceptibilities. Early Lyme disease can be effectively treated with oral tetracyclines and semisynthetic penicillins. Arthritic and neurologic disorders are treated with high-dose intravenous penicillin G or ceftriaxone. Patients who have failed to respond to penicillin or tetracycline therapy have been effectively treated with ceftriaxone.

Spirillum

Spirillum is one cause and one form of rat bite fever. Spirillar rat bite fever is characterized by ulceration at the site of the bite, lymphadenopathy, rash, and a relapsing fever. The spirilla differ from the spirochetes in that the former possess the rigid cell wall and external flagella typical of Gram-negative bacterial morphology. A single species, *S minus,* is pathogenic for humans and is the agent of one type of rat bite fever, referred to as spirillar fever to distinguish it from that caused *by Streptobacillus moniliformis* (streptobacillary fever). *Spirillum minus* is an aerobic, short, spiral-shaped, Gram-negative rod (0.2 to 0.5 µm by 3 to 5 µm) with two to three coils and bipolar tufts of flagella. It has not been cultivated successfully on an artificial medium.

Rat bite fever is primarily a disease of wild rodents. Human infection follows the bite of a rodent or rodent-ingesting animal (Fig. 35-7). After an incubation period of about 2 weeks, the site of the bite becomes inflamed and painful, and a chancre-like ulceration may occur. Associated with this eruption are inflammation and enlargement of the adjacent lymphatics and lymph nodes, fever, headache, and a rash radiating from the wound site and lasting about 48 hours. In untreated patients the symptoms subside, only to reappear in 3 to 9 days. This relapsing fever may last for weeks to months. Because of the low incidence of this disease, little is know of its pathogenesis. Diagnosis is established by demonstrating *S minus* in dark-field preparations of lesions and adjacent lymph node exudates or blood. If this fails, laboratory animals free of spirilla are inoculated and examined for the development of a spirillemia. The disease has been successfully treated with penicillin. Improvement of sanitary conditions to minimize rodent contact with humans is the best preventive measure for rat bite fever.

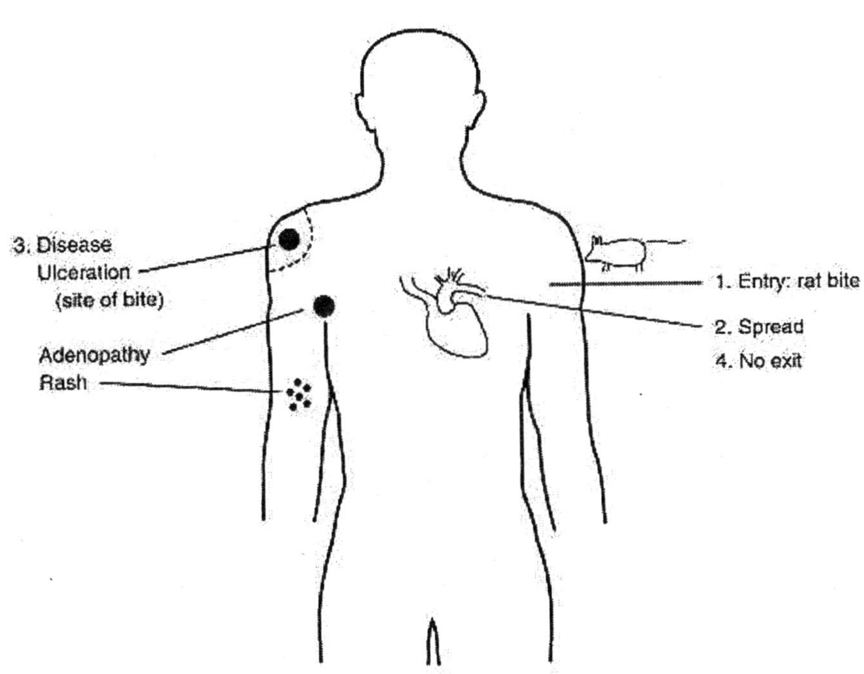

FIGURE 35-7 Pathogenesis of Spirillum infection.

REFERENCES

Baranton G, Postic D, Saint Girons I et al.:Delineation of *Borrelia burgdorferi* sensu stricto, *Borrelia garinii* sp nov and group VS 461 associated with Lyme borreliosis. Int J Syst Bacteriol 42:378, 1992

Burgdorfer W, Barbour AG, Hayes SF et al: Lyme disease: a tick-borne spirochetosis. Science 216:1317, 1982

Felsenfeld 0: *Borrelia*. Strains, Vectors, Human and Animal Borreliosis. Warren Green, St. Louis, 1971

Johnson RC (ed): The Biology of Parasitic Spirochetes. Academic Press, San Diego, 1976

Johnson RC (ed): Lyme disease. Rheum Dis Clin North Am 15:627, 1989

Kaufmann AE, Weyant RS: *Leptospiraceae*. In Murray PR, Baron EJ, Pfaller MA, Tenover FC, Yolken RH (eds): Manual of Clinical Microbiology. 6th Ed. American Society for Microbiology, Washington DC, 1995

Rahn DW, Malawista SE: Lyme disease: recommendations for diagnosis and treatment. Ann Int Med 114:472,1991

Schwan TG, Burgdorfer W, Rosa PA: *Borrelia*. In Murray PR, Baron EJ, Pfaller MA, Tenover FC, Yoken RH (eds): Manual of Clinical Microbiology. 6th Ed. American Society for Microbiology, Washington DC, 1995

Steere AC, Grodzicki RL, Kornblatt AN et al: The spirochetal etiology of Lyme disease. N Engl J Med 308:733, 1983

Chapter **24**

Actinomyces, Propionibacterium propionicus, and Streptomyces

General Concepts

Actinomyces and Propionibacterium

Clinical Manifestations

Typical actinomycosis in humans is a chronic disease caused by *Actinomyces israelii, A gerencseriae* and *Propionibacterium propionicus* (previously *Arachnia propionica*). The infection is characterized by persisting swelling, suppuration, and formation of abscesses with draining sinuses. Major types are cervicofacial, thoracic, and abdominal. In addition to the pathogens in actinomycosis, some *Actinomyces* species have also been isolated from other mixed anaerobic infections, eye infections, blood and the urinary tract.

Structure

Gram-positive rods grow as filaments, branching rods, and diphtheroidal rods.

Classification, Antigenic and DNA Types

There are 19 species of *Actinomyces* associated with humans and animals. Most species represent a distinct serologic group, often with subtypes.*P. propionicus* has two serotypes. Recently, strains of some species have been typed by DNA fingerprinting and ribotyping. Classification is based on analysis of DNA and rRNA, cell wall structure, metabolic end products, biochemical reactions, and serology.

Pathogenesis

In actinomycosis, endogenous oral bacteria are introduced into tissue. Abscesses with fibrous walls and pus with sulfur granules develop. Lesions spread by direct extension. A similar etiology may apply to other types of infection, including those with the more recently described *Actinomyces*, although this is not known with certainty.

Host Defenses

An intact mucosa is the first line of defense. Another possible defense is cell-mediated and humoral immunity.

Epidemiology

Actinomycosis is endemic; sporadic cases occur worldwide. Generally the infection is unrelated to age, sex, season, or occupation. Some *Actinomyces* spp are associated with other mixed anaerobic infections and infections in severely compromised patients. These can be significant opportunist pathogens.

Diagnosis

Actinomycosis is suggested by a suppurative lesion with Gram-positive filaments in the exudate. Sulfur granules may be present. The diagnosis of actinomycosis and other infections involving *Actinomyces*, including those in compromised patients is confirmed by isolation and identification of the bacteria.

Control

Good oral hygiene may help in the prevention of actinomycosis. Actinomycosis is treated with antibiotics and surgical drainage of lesions. In all *Actinomyces* infections, penicillin is the drug of choice.

Streptomyces

Clinical Manifestations

Symptoms of mycetoma include actinomycotic mycetoma. Lesions are swollen and localized at the site of the trauma. There are multiple abscesses, draining sinuses, and pus with granules.

Structure

Gram-positive bacteria form filaments and branching filaments, a nonfragmenting substrate mycelium, and an aerial mycelium with spores.

Classification and Antigenic Types

Over 400 species have been described based on morphology, pigmentation, cell wall type, lipid composition, and numerical taxonomy. As a result of numerical taxonomy using well-defined stable characters for discrimination, it seems likely that this number of species will be reduced.

Pathogenesis

Most species are not pathogenic. *Streptomyces somaliensis* causes mycetoma. Some species cause plant diseases.

Host Defenses

Resistance depends on a well-developed cell-mediated immune system.

Epidemiology

Streptomyces species are found worldwide in soil. They are important in soil ecology and as antibiotic producers. Mycetoma occurs mainly in tropical and subtropical areas.

Diagnosis

The diagnosis is suggested by clinical features of mycetoma and characteristic pus granules; it is confirmed by isolation and identification of microorganisms.

Control

Treatment is with long-term antibiotics and surgery.

INTRODUCTION

Actinomyces spp and *Propionibacterium propionicus* (previously *Arachnia propionica*) are members of a large group of pleomorphic Gram-positive bacteria, many of which have some tendency toward mycelial growth. Both are members of the oral flora of humans or animals. *Actinomyces* species, in particular, are major components of dental plaque. *A. israelii*, *A gerencseriae* (previously *A israelii* serotype II), and *P propionicus* cause actinomycosis in humans and animals. Other species of *Actinomyces* can be involved in mixed anaerobic and other infections, where they may not always play an obviously pathogenic role. In addition, some coryneform bacteria (diphtheroids) isolated from clinical samples, which had been placed into the Centers for Disease Control Coryneform groups 1, 2 and E, have been identified as new species of *Actinomyces.*

Actinomycosis

Clinical Manifestations

Actinomycosis is a chronic disease characterized by the production of suppurative abscesses or granulomas that eventually develop draining sinuses (Fig. 34-1). These lesions discharge pus containing the organisms (Fig. 34-2). In long-standing cases, the organisms are found in firm, yellowish granules called sulfur granules (Fig 34-3). The disease is usually divided into three major clinical types, cervicofacial, thoracic, and abdominal, but primary infections may involve almost any organ. Secondary spread of the disease is by direct extension of an existing lesion without regard to anatomic barriers. Hematogenous secondary spread of the organisms, except from thoracic lesions, is not common.

Actinomycosis is almost always a mixed infection; a variety of other oral bacteria can be found in the lesion with *Actinomyces* or *P propionicus*. The role of these associated bacteria is not clear. It has been shown that succinic acid produced by *Actinomyces* can promote the growth of some Gram -ve anaerobes. Also, in experimental infections, *Eikenella corrodens* and *Actinobacillus actinomycetemcomitans* enhance the virulence of *Actinomyces*. The mycelial masses of *Actinomyces* reduce the rate of penetration of antibiotics and may physically protect the associated bacteria. In addition, associated organisms which produce ß-lactamases can complicate treatment. Therefore, the dense granules of *Actinomyces* and the presence

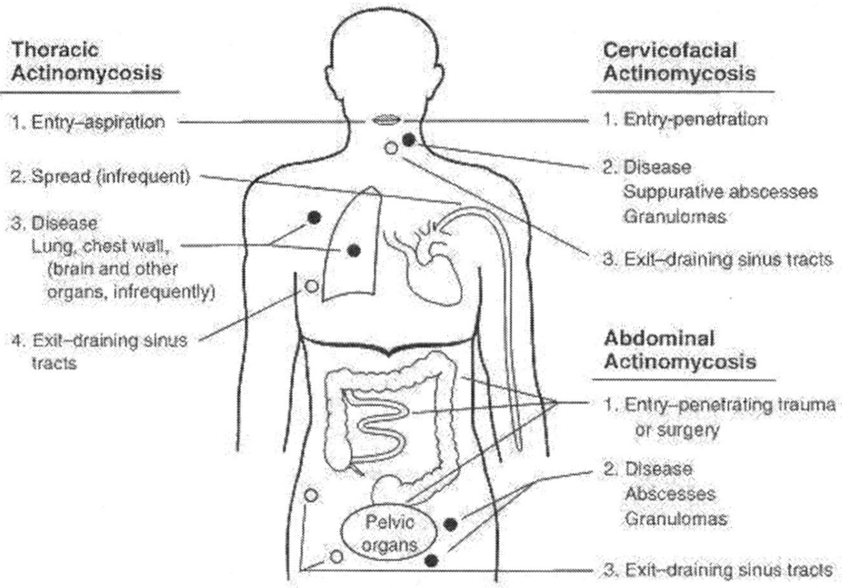

FIGURE 34-1 Pathogenesis and disease sites of three major forms of actinomycosis.

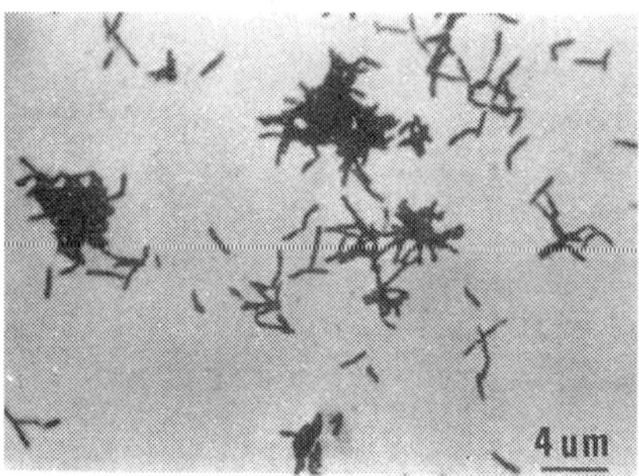

FIGURE 34-2 Gram stain of *A israelii* showing diphtheroidal rods and short branching filaments.

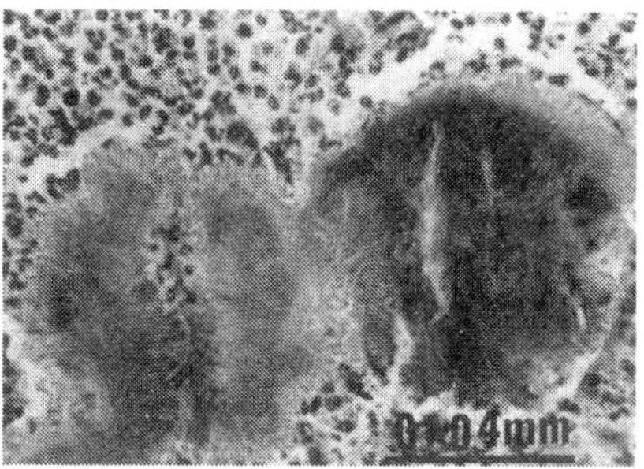

FIGURE 34-3 Sulfur granule from human actinomycosis tissue section (hematoxylin and eosin stain). (From Slack JM, Gerencser MA: *Actinomyces*, Filamentous Bacteria: Biology and Pathogenicity. p. 95. Burgess, Minneapolis, 1975, with permission.)

of associated bacteria can enhance the virulence of the infection and influence the mode of use of antibiotics, thereby adding to the difficulty of treating the disease.

Cervicofacial infections involve the face, neck, jaw, or tongue and usually occur following an injury to the mouth or jaw or a dental manipulation such as extraction. The disease begins with pain and firm swelling along the jaw and slowly progresses until draining sinuses are produced.

Thoracic actinomycosis results from aspiration of pieces of infectious material from the teeth and may involve the chest wall, the lungs, or both. The symptoms are similar to those of other chronic pulmonary diseases, and the disease is often difficult to diagnose. Thoracic disease may spread extensively to adjacent tissues or organs and often disseminates through the bloodstream, resulting in abscesses in distant sites such as the brain.

Abdominal actinomycosis is often associated with abdominal surgery, accidental trauma, or acute perforative gastrointestinal disease. Persistent purulent drainage after surgery or abdominal masses resembling tumors may be the first sign of infection.

Actinomycosis may affect almost any organ. For example, *Actinomyces* and *P propionicus* cause a lacrimal canaliculitis with concretions in the canaliculi that are persistent. *A naeslundii* and *A odontolyticus* can also infect the eyes. In recent years, *Actinomyces* and *P propionicus* have been isolated with increasing frequency from female pelvic infections associated with wearing an intrauterine contraceptive device. There has been considerable interest in the possible role of *Actinomyces* in human periodontal disease and root surface caries. Evidence suggests that these bacteria are not involved in the more destructive forms of periodontal disease. *Actinomyces naeslundii* and possibly other species may be involved in gingivitis and mild forms of periodontitis. They may also facilitate colonization of the gingivae through coaggregation with Gram -ve anaerobes. *Actinomyces naeslundii* has also been associated with root surface caries and can be the predominant organism in some lesions.

Structure

Despite their name, which means "ray fungus," *Actinomyces* are typical bacteria. Both *Actinomyces* and *P propionicus* are Gram-positive filamentous rods that are not acid fast and are nonmotile (Fig. 34-4A). As in other Gram-positive bacteria, the cell wall peptidoglycan contains muramic acid, *N*-acetylglucosamine, glutamic acid, and one or two additional amino acids. *Actinomyces* species have lysine or lysine plus ornithine in the peptidoglycan, whereas *P propionicus* contains L-diaminopimelic acid and glycine. Most strains of *A viscosus* and *A naeslundii* bear well-developed long, thin surface fibrils. Pili (fimbriae) on *A viscosus* and *A naeslundii* are of two types. Type 1 pili are involved in attachment of the bacteria to hard surfaces in the mouth, whereas type 2 pili are involved in coaggregation reactions with other bacteria.

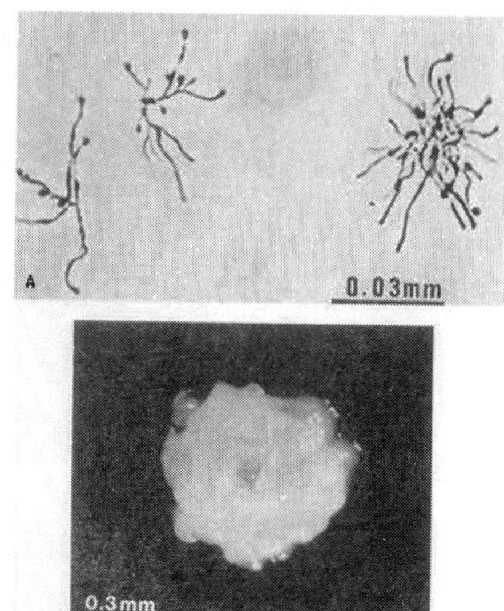

FIGURE 34-4 Characteristic colonies of *A israelii*. **(A) Microcolony at 24 hrs. Brain heart infusion agar shows branching filaments with no distinct center: spider colony. (B) Mature colony at 14 days. Brian heart infusion agar, heaped colony with central depression. (From Slack, JM, Landfried S, Gerencser MA: Morphological, biochemical, and serological studies on 64 studies on 64 strains of *Actinomyces israeli*. J Bacteriol 97:873, 1969, with permission.)**

Classification and Antigenic Types

Actinomyces and *P. propionicum* are irregular, nonspore-forming, Gram-positive rods. *Actinomyces* contains 19 species (Table 34-1), of which *A israelii* and *A gerencseriae* are the most common human pathogens. The genera *Actinomyces* and *Propionibacterium* are defined by chemotaxonomic tests, e.g., cell wall composition, fermentation products, cellular fatty acids; DNA homology; and analysis of their ribosomal RNA (rRNA). Using these methods some of the CDC Cornyeform groups (see above) have been classified as *Actinomyces*.

Actinomyces and *P propionicus* grow well on most rich culture media. They are best described as aerotolerant anaerobes. The species vary in oxygen requirements: *A viscosus* and *A naeslundii* for example, grow best in an aerobic environment with carbon dioxide, whereas *A israelii* requires anaerobic conditions for growth. Both *Actinomyces* and *P propionicus* obtain energy from the fermentation of carbohydrates. The major end products of glucose fermentation by *Actinomyces* are acetic, lactic, formic, and succinic acids, whereas *Propionibacterium* produces propionic, acetic, and formic acids with a trace of succinic acid.

All the *Actinomyces* species that have been examined serologically can be separated from the other species and from *P propionicus*. *A naeslundii, A viscosus, A odontolyticus*, and *A bovis* each have at least two serotypes. *P propionicum* strains can also be separated into two serotypes.

Table 34-1 Species of *Actinomyces*

Isolated from	Human Pathogen	Isolated from Animals	Pathogen
A israelil	++++[a]	*A bovis*	+++
A gerencseriae[a]	+++	*A pyogenes*	+++
A naeslundil[a]	++	*A hordeovuineris*	+
A odontolyticus	++	*A hyovaginalis*	+
A meyeri	+	*A suis*	+
A neuii	+	*A viscosus*	+
A bernardiae	+	*A denticolens*	?
A radingae	+	*A howellii*	?
A turicensis	+	*A slackii*	?
A pyogenes	+	*A georgiae*	?

[a] ++++ to + indication of incidence of species in infections, pathogenicity not known.
[b] Previously *A. israelii* serotype II
[c] Now includes human strains of *A. viscosus* (Johnson et al 1990).
[d] *A. viscosus* is now limited to animals (Johnson et al 1990).
Data from : Dent and Williams 1986; Schaal 1986; Johnson et al. 1990; Ludwig et al, 1992; Collins et al, 1993; Funke et al, 1994, 1995; Wust et al, 1995.

The chemical composition and cellular location of some *Actinomyces* antigens are known. One group of carbohydrate antigens is cell-wall associated, protease resistant, and heat stable. *Actinomyces viscosus* also has an amphipathic antigen that is a fatty acid-substituted heteropolysaccharide and different from the teichoic and lipoteichoic acids found in most Gram-positive bacteria. The pili of *A viscosus* and *A naeslundii* are of two antigenic types, which correlate with the different functions of type 1 and 2 pili (see above).

Recently, *Actinomyces* strains have been characterized on the basis of their DNA fingerprints in attempts to show the transmission of specific oral strains among persons. Also, strains have been ribotyped, by identifying rRNA gene sequences in the DNA fingerprint. Ribotyping of oral *A naeslundii* strains has shown that there are up to 5 ribotypes in an individual and that it is uncommon for individuals to share ribotypes. These newer genetic typing methods are likely to improve our knowledge of the biology of *Actinomyces*, their epidemiology and their role in disease.

Pathogenesis

Actinomycosis results when bacteria resident in the mouth are introduced into the tissues. The mechanisms by which *Actinomyces* and *P propionicum* produce disease are not clear. Pathogenesis may involve the ability of these organisms to suppress some of the immune functions of the host. Studies of oral disease have shown that *Actinomyces* are chemotactic, activate lymphocyte blastogenesis, and stimulate the release of lysosomal enzymes from polymorphonuclear leukocytes and macrophages.

With the exception of *A. pyogenes,* which produces a soluble toxin and a hemolysin that can be neutralized by antiserum, *Actinomyces* and *P. propionicus* do not produce exotoxins or significant amounts of other toxic substances. Factors that would aid in tissue invasion and abscess formation have not been demonstrated.

Host Defenses

Antibodies to *Actinomyces* circulate in some healthy individuals and in individuals with gingivitis and periodontitis, as well as in those with clinical actinomycosis. This humoral response probably does not play a major role in defense against actinomycosis. An intact mucosa is the first line of defense, because *Actinomyces* and *P propionicus,* like other anaerobes in the normal flora, must gain access to tissue with an impaired blood supply to establish an infection. Once the organisms have gained access to tissues, the cell-mediated immune response of the host may limit the extent of the infection, but may also contribute to tissue damage. Humoral responses could play a role in infections with some of the other *Actinomyces* sp, such as *A bernardiae* and *A neuii,* which have been isolated from blood cultures.

Epidemiology

Actinomyces israelii, A gerencseriae, A georgiae, A naeslundii, A odontolyticus, A meyeri, possibly *A pyogenes* and *P propionicus* are normal inhabitants of the human mouth and are found in saliva, on the tongue, in gingival crevice debris, and frequently in tonsils in the absence of clinical disease. Some data suggest that *A israelii* may also be a common inhabitant of the female genital tract. The natural habitats are not known for the more recently described *Actinomyces* from human infections *A bernardiae, A neuii, A radingae,* and *A turicensis* are not known. *Actinomyces bovis* is not found in humans and, to date, the other species described from animals, with the exception of *A pyogenes,* have not been found in human specimens. Among the animal pathogens, *A bovis* causes lumpy jaw in cattle; *A viscosus* and *A hordeovulneris* infect dogs; *A pyogenes* causes infections in a range of domestic animals, including cattle, sheep and goats; and *A suis* and *A hyovaginalis* produce infections in swine. Other species appear to be relatively non-pathogenic, *A denticolens, A howellii* and *A slackii* in cattle; *A viscosus* in rodents and both *A naeslundii* and *A viscosus* in zoo animals, including primates. *Actinomyces* and *P propionicus* are not found in the soil or on vegetation.

Actinomycosis occurs worldwide. No relationship to race, age, or occupation has been noted, but the disease appears more often in men than in women. Except for human bite wounds, no evidence exists to support person-to-person or animal-to-human transmission of *Actinomyces.* Compromised patients may be infected by the more recently described *Actinomyces.*

Diagnosis

A search for *Actinomyces* sp. and *P propionicucus* usually is based on a tentative clinical diagnosis of actinomycosis, but these bacteria should be considered whenever a direct Gram stain of pus or suppurative exudate shows Gram-positive, non-acid-fast rods in diphtheroidal arrangements with or without branching. Specimens are first examined for the presence of granules. If present, the granules are crushed, Gram stained, and examined for Gram-positive rods or branching filaments. Washed, crushed granules or well-mixed pus in the absence of granules is cultured on a rich medium, such as brain heart infusion blood agar and incubated anaerobically and aerobically with added carbon dioxide. Plates are examined after 24 hours and after 5 to 7 days for the characteristic colonies of *A israelii*, *A gerencseriae* and *P propionicus*, which resemble those of *A israelii* (Fig. 34-4B). Colonies of other *Actinomyces* spp. can take a variety of forms, including smooth domed or flat colonies of various sizes. Isolates morphologically resembling *Actinomyces* and *P propionicus* are identified by determining the metabolic end products by gas-liquid chromatography and by performing a series of biochemical tests. Immunofluorescence tests are useful for serologic identification of isolates and the direct demonstration of the organisms in clinical samples.

Control

Due to the mixed nature of the infection and the presence of granules (see above) successful treatment of actinomycosis requires long-term antibiotic therapy combined with surgical drainage of the lesions and excision of damaged tissue. *Actinomyces* spp and *P propionicus* are susceptible to penicillins, the cephalosporins, tetracycline, chloramphenicol, and a variety of other antibiotics. Penicillin is the drug of choice for infections with all species of *Actinomyces*; significant drug resistance is unknown.

Streptomyces

Like *Actinomyces* and *Propionibacterium*, *Streptomyces* belongs to the large group of filamentous bacteria known as actinomycetes, but *Streptomyces* species have a well-developed substrate mycelium, produce an aerial mycelium with chains of spores, and are strict aerobes. *Streptomyces* are common in soil and give it its characteristic earthy odor. They seldom produce human infections, but are important as producers of antibiotics.

Clinical Manifestations

Some *Streptomyces* species, principally *S somaliensis*, cause actinomycotic mycetoma, which is indistinguishable from that caused by *Nocardia* species (see Ch. 33), other actinomycetes, and some fungi. Lesions usually occur on the extremities, most often on the feet (Fig. 34-5). They appear as localized swollen nodules that slowly enlarge. Multiple abscesses form, and draining sinuses open to the surface and discharge pus and granules.

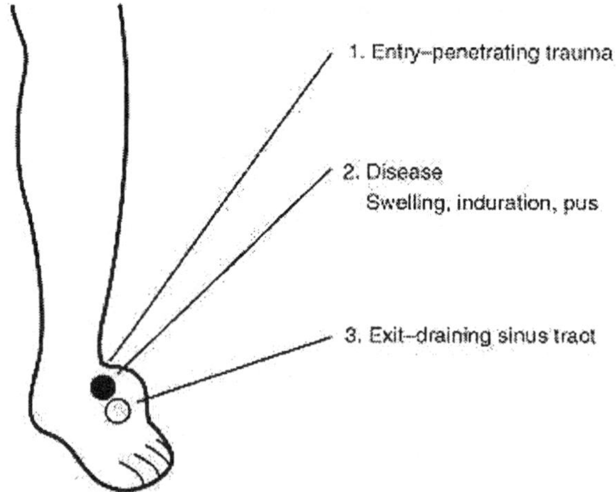

FIGURE 34-5 Pathogenesis of actinomycotic mycetoma.

Structure

Streptomyces species are Gram-positive, aerobic, filamentous bacteria with an extensive substrate mycelium that does not fragment and aerial hyphae with chains of spores produced by hyphal segmentation. Colonies are initially smooth but become powdery or cottony as the aerial mycelium and spores develop. The colonies grow slowly, requiring 7 to 10 days to develop aerial hyphae.

Classification and Antigenic Types

Species identification within the genus *Streptomyces* is complicated. More than 400 species have been described on the basis of morphology, pigmentation, cell wall structure, chemical composition, some biochemical tests, and serologic relationships. *Streptomyces* species have been divided into seven groups on the basis of the color of the mature aerial mycelium. Further subdivision is based on the morphology of the spore chains. In addition, cell wall peptidoglycan structure and the types of sugars in whole-cell hydrolysates are used extensively to characterize aerobic actinomycetes. *Streptomyces* has a type 1 cell wall containing L-diaminopimelic acid and glycine and no characteristic cell wall sugar. Chemical analyses and numerical taxonomy have identified 77 groups (clusters) of *Streptomyces*, some of which comprise a single species, while others include several species (species groups). Although, the results of these studies suggest that the numbers of species of *Streptomyces* should be reduced, the genus will remain complex.

Pathogenesis

Soil organisms are introduced into the tissue by trauma, most often minor trauma caused by thorns, splinters, or an abrasion (Fig. 34-5). The initial lesions spread to subcutaneous tissue and then into bone. Radiographs show multiple small granulomas with pus in the center and cavities in the bone. Continued spread of the cutaneous lesions leads to the formation of draining sinuses. The discharged pus contains granules, whose size and color provide a clue to the etiologic agent. For example, *S somaliensis* granules are large and yellow to brown.

Host Defenses

Mycetoma develops in only some of the many individuals exposed to the organisms. Development of disease seems to be dependent on a deficiency in cell-mediated immunity. A relationship between susceptibility to mycetoma and deficiency in cell-mediated immunity has been shown in animal studies.

Epidemiology

Streptomyces spp occur worldwide in the soil, in water, and on organic debris. They play an important role in soil ecology by decomposing organic matter and contributing to soil fertility. The major medical significance of this genus is as a producer of antibiotics. About 85 percent of the known antibiotics, including streptomycin, chloramphenicol, and tetracycline, are produced by streptomycetes.

Mycetoma occurs mainly in tropical and subtropical areas. The dominant agent in a particular area depends on the prevalence of these agents in the soil.

Diagnosis

Diagnosis is usually based on clinical grounds when there is a well-established lesion. Determination of antibodies in serum, usually by counter-immunoelectrophoresis, is useful. Pus should be examined for the color, size, consistency, and microscopic appearance of granules, which will usually identify the agent of mycetoma. Definitive diagnosis depends on isolation and identification of the bacteria.

Control

Long-term antibiotic treatment and surgical management are necessary. *In vitro, S somaliensis* is sensitive to rifampicin, erythromycin, tobramycin, fusidic acid and streptomycin. Strains tested were resistant to trimethoprim. For *S somaliensis* infection, treatment with streptomycin and either co-trimoxazole or dapsone is recommended. The average duration of treatment is about 10 months.

REFERENCES

Barsotti O, Morrier JJ, Decoret D, et al: An investigation into the use of restriction endonuclease analysis for the study of transmission of *Actinomyces*. J Clin Periodontol. 20:436, 1993

Bowden GH, Johnson J, Schachtele C: Characterization of *Actinomyces* with genomic DNA fingerprints and rRNA gene probes. J Dent Res 72:1171, 1993

Brown DA, Fischlschweiger S, Birdsell DC: Morphological, chemical and antigenic characterization of cell walls of the oral pathogenic strains *Actinomyces viscosus* T14V and T14AV. Arch Oral Biol 25:451, 1980

Brown JR: Human actinomycosis: a study of 181 subjects. Hum Pathol 4:319, 1973

Causey WA: Actinomycosis. p.383. In Vinken PV, Bruyn GW (eds): Handbook of clinical Neurology. Vol. 35. North Holland, Amsterdam, 1978

Funke G, Ramos PC, Fernández-Garayzábal JF, et al: Description of human-derived Centers for Disease Control Coryneform group 2 bacteria as *Actinomyces bernardiae.* I J Syst Bacteriol 45:57, 1995

Funke G, Stubbs S, von Graevenitz A, et al: Assignment of human-derived CDC Group 1 corynebacterium bacteria and CDC group 1-like coryneform bacteria to the genus *Actinomyces* as *Actinomyces neuii* subsp. *neuii* sp nov, subsp. nov, and *Actinomyces neuii* subsp *anitratus* subsp nov. Int J Syst Bacterio 144:167, 1994

Gerencser MA: Actinomycosis. p. 551. In Balows A, Hausler WJ, Jr, Ohashi M, (eds): Laboratory Diagnosis of Infectious Diseases: Principles and Practice. Vol. 1. Bacterial, Mycotic and Parasitic Diseases. Springer-Verlag, New York, 1988

Goodfellow M, Mordarski M, William ST (eds): The Biology of the Actinomycetes. Academic Press (London), London, 1984

Johnson JL, Moore LVH, Kaneko B, et al: *Actinomyces georgiae* sp nov, *Actinomyces gerencseriae* sp nov, Designation of two genospecies of *Actinomyces naeslundii,* and inclusion of *A naeslundii* serotypes II and III and *Actinomyces viscosus* serotype II in *A naeslundii* genospecies 2. Int J Syst Bacteriol 40:273, 1990

Jordan HV, Kelley DM, Heeley JD: Enhancement of experimental actinomycosis in mice by Eikenella corrodens. Infect Immun 46:367, 1984

Lerner PI: Pneumonia due to Actinomyces, Arachnia, and Nocardia. p. 514. In

Pennington JE (ed): Respiratory Infections: Diagnosis and Management. 2nd ed. Raven Press, New York, 1988

Lopatin DE, Peebles FL, Woods RW, et al: *In vitro* evaluation in man of immuno-stimulation by subfractions of Actinomyces viscosus. Arch Oral Biol 25:23, 1980

Ludwig W, Kirchhof G, Weizenegger M, et al: Phylogenetic evidence for the transfer of *Eubacterium suis* to the genus *Actinomyces* as *Actinomyces suis* comb. nov. Int J Syst Bacteriol 42:161, 1992

Mahgoub ES: Mycetoma. p. 633. In Balows A, Hausler WJ, Jr, Ohashi M, et al(eds): Laboratory Diagnosis of Infectious Diseases: Principles and Practice. Vol. 1. Bacterial, Mycotic and Parasitic Diseases. Springer-Verlag, New York, 1988

Perrson E, Holmberg K: A longitudinal study of *Actinomyces israelii* in the female genital tract. Acta Obstet Gynecol Scand 63:207, 1984.

Schaal KP: Genus Arachnia, p. 1332, and Genus Actinomyces, p. 1383: In Sneath PH, Mair NS, Sharpe ME, Holt JG (eds): Bergey's Manual of Systematic Bacteriology. Vol. 2. Williams & Wilkins, Baltimore, 1986

Schaal Kp, Hee-Joo L: Actinomycete infections in humans - a review. Gene 115:201, 1992

Slack JM, Gerencser MA: Actinomyces. Filamentous Bacteria: Biology and Pathogenicity. Burgess, Minneapolis, 1975

Wüst J, Stubbs S, Weiss N, et al: Assignment of Actinomyces-like (CDC coryneform group E) bacteria to the genus *Actinomyces* as *Actinomyces radingae* sp nov and *Actinomyces turicensis* sp nov. Let Appl Microbiol 20:76, 1995

Chapter 25

Pasteurella, Yersinia, and Francisella

General Concepts

Pasteurella

Clinical Manifestations

In cattle, sheep and birds Pasteurella causes a life-threatening pneumonia. Pasteurella is non-pathogenic for cats and dogs and is part of their normal nasopharyngeal flora. In humans, Pasteurella causes chronic abscesses on the extremities or face following cat or dog bites.

Structure, Classification, and Antigenic Types

Pasteurellae are small, nonmotile, Gram-negative coccobacilli often exhibiting bipolar staining. Pasteurella multocida occurs as four capsular types (A, B, D, and E), and 15 somatic antigens can be recognized on cells stripped of capsular polysaccharides by acid or hyaluronidase treatment. Pasteurella haemolytica infects cattle and horses.

Pathogenesis

Human abscesses are characterized by extensive edema and fibrosis. Encapsulated organisms resist phagocytosis. Endotoxin contributes to tissue damage.

Host Defenses

Encapsulated bacteria are not phagocytosed by polymorphs unless specific opsonins are present. Acquired resistance is humoral.

Epidemiology

Pasteurella species are primarily pathogens of cattle, sheep, fowl, and rabbits. Humans become infected by handling infected animals.

Diagnosis

Diagnosis depends on clinical appearance, history of animal contact, and results of culture on blood agar. Colonies are small, nonhemolytic, and iridescent. The organisms are identified by biochemical and serologic methods.

Control

Several vaccines are available for animal use, but their effectiveness is controversial. No vaccines are available for human use. Treatment requires drainage of the lesion and prolonged multidrug therapy. Pasteurella multocida is susceptible to sulfadiazine, ampicillin, chloramphenicol, and tetracycline.

Yersinia

Clinical Manifestations

Yersinia pestis causes bubonic and pneumonic plague. Bubonic plague is transmitted by the bite of infected rat fleas. Swollen, blackened lymph nodes (buboes) develop, followed by septicemia and hemorrhagic pneumonia and death. The pneumonic form spreads directly from human to human via respiratory droplets. Outbreaks are explosive in nature, and invariably lethal. Yersinia enterocolitica causes severe diarrhea and local abscesses, and Y pseudoterberculosis causes severe enterocolitis.

Structure, Classification, and Antigenic Types

Yersinia are small, Gram-negative coccobacilli showing bipolar staining. The capsular or envelope antigen is heat labile. Somatic antigens V and W are associated with virulence.

Pathogenesis

In bubonic plague, the bacilli spread from a local abscess at the flea bite site to draining lymph nodes; followed rapidly by septicemia and hemorrhagic pneumonia. Yersinia enterocolitica enters via the Peyer's patches following ingestion of contaminated water or food and cause severe liver and splenic abscesses. Yersinia pseudotuberculosis causes enlarged, caseous nodules in the Peyer's patches and mesenteric lymph nodes.

Host Defenses

Specific anti-envelope antibodies are opsonic and protective. Cell-mediated resistance may also be involved.

Epidemiology

Yersinia pestis is primarily a rat pathogen. Human infections are initially transmitted by rat fleas, but later the disease may shift into the pneumonic form and continue by direct person- to-person spread. Yersinia enterocolitica, a pathogen of deer and cattle spreads to humans via contaminated drinking water.

Diagnosis

Early clinical diagnosis is essential in plague. Blood cultures are positive for Y pestis. Sputum may show large numbers of small bacilli when stained with fluorescent antibody. Yersinia pestis is an extremely infectious hazard for nursing and laboratory personnel.

Control

Control of rats and rat fleas is crucial. Laboratory personnel should be vaccinated. Yersinia pestis is susceptible to sulfadiazine, streptomycin, tetracycline, and chloramphenicol. Yersinia enterocolitica is best controlled by purifying drinking water and pasteurizing dairy products. Yersinia pseudotuberculosis disease requires aggressive treatment with ampicillin and tetracycline.

Franciscella

Clinical Manifestations

Francisella tularensis causes tularemia, with high fever, acute septicemia and toxemia. Oral infection causes typhoid-like disease.

Structure, Classification, and Antigenic Types

The organisms are small, nonmotile, Gram-negative coccobacilli. Franciscella is nutritionally demanding. It is biochemically similar to the brucellae, but antigenically distinct.

Pathogenesis

A local abscess at the site of infection is followed by septicemia with rapid spread to the liver and spleen; 30 percent of untreated patients die.

Host Defenses

Cell-mediated immunity is protective and long lasting.

Epidemiology

Franciscella is primarily a pathogen of squirrels and rabbits; humans are infected by the bite of an infected deerfly or tick or by handling infected rabbit carcasses or eating undercooked meat.

Diagnosis

Cultivation from blood or biopsy material is difficult and slow. Blood smears can be stained with specific fluorescent antibody. Hemagglutinins appear in 10 to 12 days; a rising titer is diagnostic.

Control

A live attenuated vaccine is available for laboratory personnel. Goggles should be worn in the laboratory to preventconjunctival infection. Franciscella is susceptible to streptomycin, tetracycline, and chloramphenicol.

INTRODUCTION

The genus Pasteurella was originally proposed and described by Trevisan in 1887. It consisted of a group of nonmotile, small (0. 7 µm by 0. 5 µm), Gram-negative coccobacilli often exhibiting a characteristic type of bipolar staining (Fig. 29-1). Most members of this genus are associated with severe, life-threatening systemic diseases involving both hemorrhagic pneumonia and septicemia. The first pathogen to be studied (called at that time Pasteurella septica) was shown to be responsible for hemorrhagic septicemia in cattle and sheep, and fowl cholera in chickens. This organism, used by Pasteur for his milestone vaccination studies in 1880, is now called Pasteurella multocida. Adult animals may carry this organism as part of their normal nasopharyngeal or gingival microflora and may infect young, susceptible animals which develop a fulminating, rapidly lethal hemorrhagic pneumonia. The incubation period for this disease may be as short as 12 hours with very high mortality rates (80 to 100 percent). The disease can spread explosively through an apparently normal herd or flock.

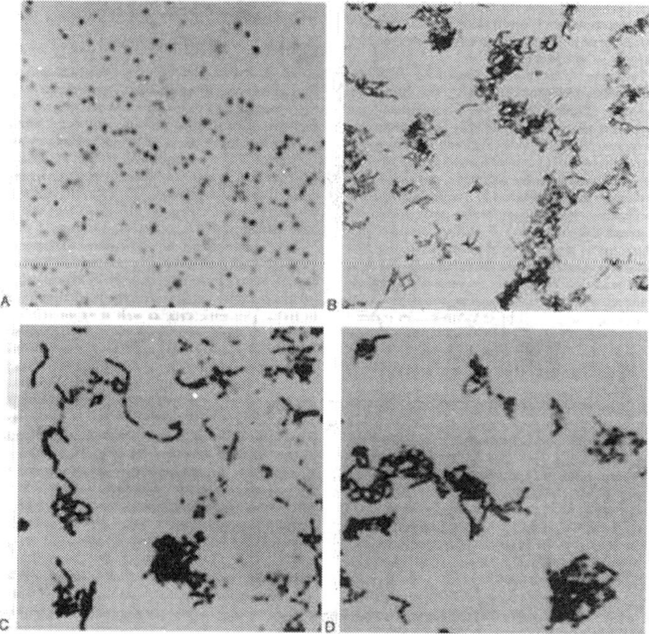

Figure 29-1 (A) Pasteurella multocida growth on brain-heart infusion agar. (B) Francisella tularensis growth on chocolate blood agar. (C) Yersinia pestis growth on tryptosis soy agar. (D) Yersinia enterocolitica growth on tryptose soy agar. (Gram stain, X1,200.)

Two other important pathogens were initially included in this genus. The first was Pasteurella pestis (the plague bacillus), which was isolated and described almost simultaneously by Kitasato and by Yersin in 1894. This organism is primarily a pathogen of the rat (one of a select group of acute bacterial pathogens for this host).

For taxonomic reasons, it was decided in 1971 to place the plague bacillus in a new genus as Yersinia pestis, together with Y pseudotuberculosis and Y enterocolitica. The latter infect a variety of rodent species but can cause severe intestinal disease in humans. Finally, a third genus was created for an organism originally grouped with P pestis, but now known as Francisella tularensis, the agent of tularemia in rodents and humans. The various diseases caused by these three genera, together with the vectors responsible for their spread to humans, are summarized in Table 29-1.

TABLE 29-1 Disease and Vector Comparisons for Pasteurella, Yersinia, and Francisella

Organism	Host Species	Vector	Disease
P mullocida	Cattle birds, Humans	Aerosols Cat and dog bites	Pneurnonia and septicemia Localized abscess
Y pestis	Rats, humans	Rat flea bite Aerosol	Bubonic plague Pneumonic plague
Y enterocolitica	Cattle, birds, humans	Infected water and foods	enterocolitis, systemic abscesses
Y pseudotuberculosis	Rodents, humans	Infected water	Enterocolitis
F lufarensis	Rodents, squirrels, humans	Ticks, deerflies, frauma, infected rabbit meat	Tularemia, septicemia, typhoid like enteritis

Metabolically, these organisms are facultative anaerobes which grow best on nutrient media enriched with blood, hematin, or catalase. Most members show a restricted fermentative capacity. Although, they grow well when incubated at 37°C, they can also multiply at room temperature, when some species produce putative virulence factors that help to establish the pathogen within the tissues. Vaccines are of limited value and aggressive treatment with broad-spectrum antibiotics is required to control human infections.

Pasteurella

Clinical Manifestations

Although, P multocida is an awesome pathogen for young cattle and birds, it may occur as a relatively benign member of the nasopharyngeal microflora of adult cattle, rabbits, cats, and dogs which do not usually develop severe pulmonary infections. Humans do not usually develop acute pulmonary disease, although there have been reports of a mild pasteurellosis in some cattle handlers. It seems likely that this organism cancolonize normal human nasopharyngeal membranes.

Most human infections with P multocida occur as localized abscesses of the extremities or face as a result of cat or dog bites (Fig. 29-2). Historically, such infections were first associated with tiger bites in India. Domestic cats, as well as exotic felines, carry P multocida as part of their normal gingival microflora; organisms introduced into human tissues as a result of bites and scratches enter the subcutaneous tissues, quickly producing severe local abscesses, which eventually spread to the draining lymph nodes (Fig. 29-2). These abscesses require surgical drainage due to their extensive edema and fibrosis (which also reduces the effectiveness of chemotherapeutic intervention, although this organism is susceptible to most antibiotics in vitro).

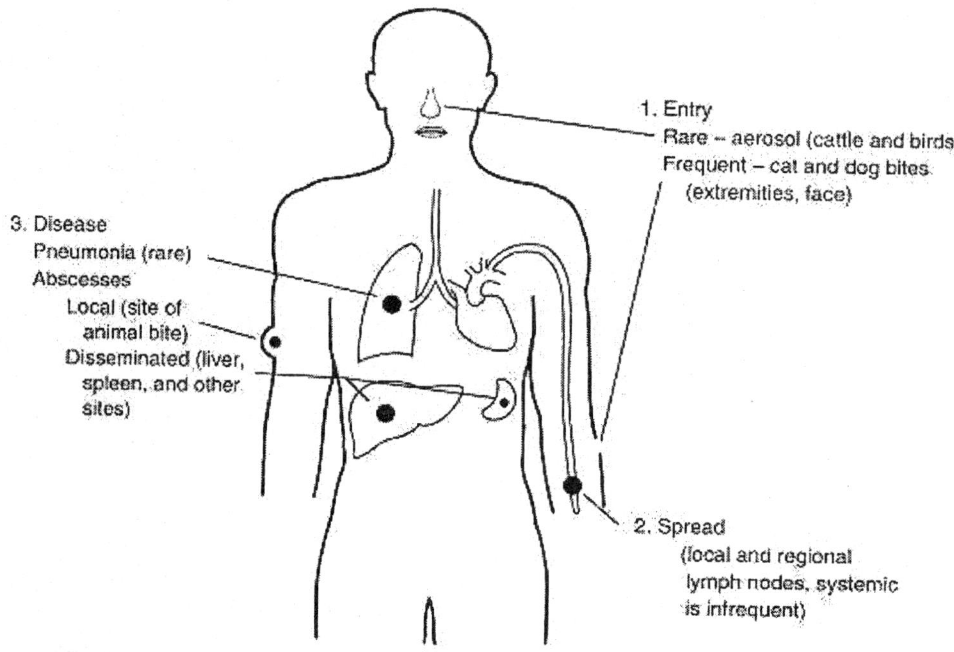

Figure 29-2 Pathogenesis of P multocida

Adult cattle may carry virulent strains of P multocida as part of their normal nasopharyngeal flora with no obvious sign of infection until they are stressed by dehydration, poor nutrition, overcrowding, or an intercurrent viral infection. The dynamics of the resulting pulmonary infection are complex, and the controlling factors are still poorly understood. Young animals are infected by older carriers shortly after birth, while still immunologically immature, and develop an acute pneumonia and septicemia, especially when herded into crowded transports or feedlots (hence the name shipping fever). Death can occur 12 to 18 hours after the onset of symptoms. Similarly, chicken and turkey flocks can be decimated by an overwhelming pneumonia that sweeps through a previously healthy flock, apparently as a result of waterborne spread from a single infected carrier.

Rabbits are highly susceptible to chronic nasopharyngeal infections caused by P multocida, developing a characteristic "snuffles", often associated with a purulent otitis media. This infection may progress to a life-threatening hemorrhagic pneumonia when the animal is stressed as a result of the hyperimmunization procedures used during antibody production. Attempts to develop pasteurella-free rabbit breeding stocks have had only limited success.

Mice are extraordinarily susceptible to parenteral and aerogenic challenge with P multocida, especially by strains obtained from cattle and fowl sources. Introduction of fewer than 10 viable P multocida serotype A organisms into the lungs of a normal mouse is followed by logarithmic growth in which the systemic disease overwhelms the host defenses in a matter of hours. The virulent encapsulated organism resists phagocytosis and multiplies freely within the extracellular spaces and fluids of the lung (Fig. 29-3). The host dies with no sign of any humoral or cellular immune response. In fact, the rate of growth by the pathogen in vivo is little different from that observed in laboratory media. Large numbers of viable bacilli appear within the bloodstream in a matter of hours and can be recovered from virtually every organ of the moribund host. Curiously, the same organism shows little ability to cross the intact intestinal mucosa, and large numbers of viable bacilli can be introduced intragastrically with no harm to the host, provided that appropriate precautions are taken to prevent accidental inoculation into the lung.

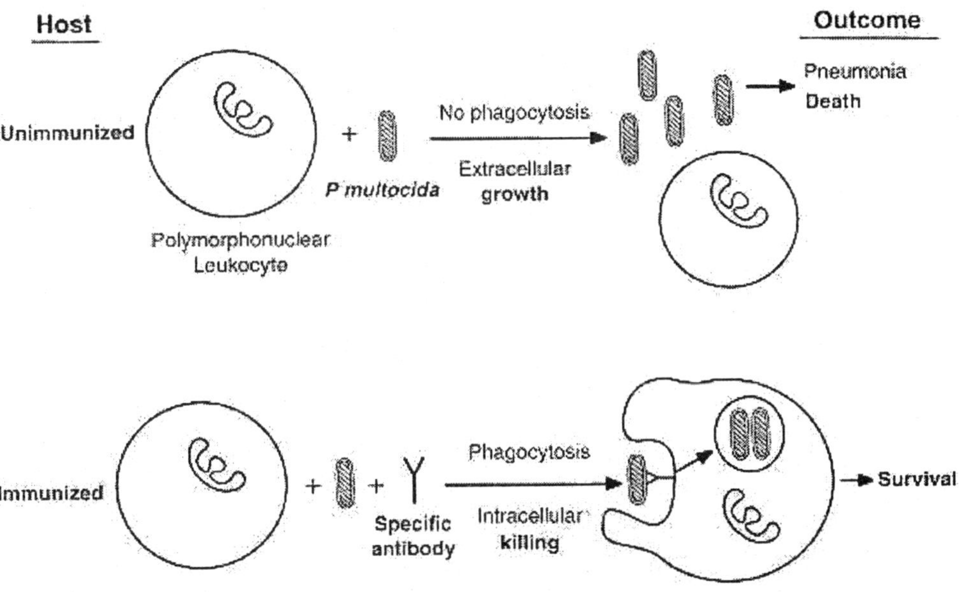

Figure 29-3 Protection against P multocida is mediated by opsonic antibodies.

Pasteurella hemolytica causes a life-threatening hemorrhagic pneumonia in horses and cattle, but does not appear to infect humans.

Structure, Classification, and Antigenic Types

Pasteurella multocida is a small, nonmotile Gram-negative coccobacillus, which often exhibits bipolar staining, in which the ends of the bacilli stain more intensely than the middle. These bacteria possess both capsular and somatic antigens, and isolates can be divided into four distinct capsular types (A, B, D, and E) on the basis of an indirect hemagglutination test. As many as 15 somatic antigens can be recognized once the capsular layer has been removed by acid or hyaluronidase treatment. Most fowl and human isolates are type A, whereas most cattle strains are type B. Type E strains are associated mostly with bovine hemorrhagic septicemia in central Africa. Freshly isolated strains may produce large mucoid colonies rich in hyaluronic acid. However, after several transfers on solid media, most virulent strains produce smooth, translucent colonies that become rough (untypable) following prolonged cultivation on laboratory media.

Pathogenesis

Pasteurella species are primarily pathogens of cattle, sheep, fowl, and rabbits. Humans may become infected while handling infected animals. Human abscesses are characterized by extensive edema and fibrosis. Encapsulated organisms resist phagocytosis. Endotoxin contributes to tissue damage.

Host Defenses

Pasteurella multocida is an extracellular parasite that multiplies freely at the site of implantation despite the influx of polymorphonuculear leukocytes (PMNs) into the lesion. In the absence of specific opsonins (immune antibodies), the organisms are not phagocytosed but multiply rapidly within the tissue fluids. Acquired resistance to pasteurellosis is humorally mediated and protection can be passively transferred to naive recipients by means of hyperimmune serumbut not by spleen cells harvested from the same donor. Once opsonized, the pasteurellae are rapidly phagocytosed and inactivated, so that the number of viable bacilli within the lesion declines sharply instead of multiplying. More importantly, hematogenous spread (a feature of the normal infection pattern) is completely ablated by specific antibodies, preventing the establishment of a fatal pneumonia. This protective effect occurs long before any mononuclear cell response has time to develop. Killed whole-cell vaccines (bacterins) induce a substantial B-cell response, leading to copious plasma cell production and the release of specific anticapsular antibodies into the bloodstream (Fig. 29-3). These opsonins can passively transfer high levels of specific antibacterial immunity to naive recipients, at least in the laboratory. The situation is more complicated in practice. Field trials with a number of multivalent P multocida vaccines containing adjuvant have yielded generally disappointing and inconsistent results, possibly owing to the large number of different serotypes present in the environment of the test population under most field test conditions. Recently, several live attenuated vaccines (usually presented in drinking water) have been claimed to be effective against fowl cholera outbreaks in turkey flocks.

Diagnosis

Pasteurella multocida grows readily on nutrient blood agar to produce small, nonhemolytic, iridescent colonies. Highly mucoid colonies are occasionally seen on primary isolation. There is no growth on MacConkey agar, and most strains ferment only glucose and sucrose, with no gas production.

Control

Pasteurella multocida is susceptible to sulfadiazine, ampicillin, chloramphenicol, and tetracycline in vitro. Because of the acute nature of most animal infections, chemotherapy is of limited value. In human cases, localized abscesses resolve quite slowly and may require prolonged therapy with multiple antibiotics. The organism is susceptible to mild heat (55°C), as well as to exposure to most hospital disinfectants. Organisms in dried blood may remain viable at room temperature for several weeks. Vaccines are not available for human use.

Yersinia

The genus Yersinia contains three species of medical importance: Y pestis, the agent of bubonic and pneumonic plague, and Y pseudotuberculosis and Y enterocolitica, both of which can result in severe gastroenteritis, with local abscess formation and death as a result of peritonitis.

Clinical Manifestations

Yersinia pestis is primarily a rodent pathogen, with humans being an accidental host when bitten by an infected rat flea (Fig. 29-4). The flea draws viable Y pestis organisms into its intestinal tract with its blood meal. These organisms multiply in situ sufficiently to block the proventriculus, and some organisms are regurgitated into the next bite wound, transferring the infection to a new host. While growing in the invertebrate host, Y pestis loses its capsular layer, and most of the organisms are phagocytosed and killed by the polymorphonuclear leukocytes which enter the infection site in large numbers. However, a few bacilli are taken up by tissue macrophages, which are unable to kill them but provide a protected environment for the organisms to resynthesize their capsular and other virulence antigens. The re-encapsulated organisms kill the macrophage and are released into the extracellular environment, where they resist phagocytosis by the polymorphs. The resulting infection quickly spreads to the draining lymph nodes, which become hot, swollen, tender, and hemorrhagic, giving rise to the characteristic black buboes responsible for the name of this disease (Fig. 29-4). Within hours of the initial flea bite, the infection spills out into the bloodstream, leading to substantial involvement of the liver, spleen, and lungs. As a result, the patient develops a severe bacterial pneumonia, exhaling large numbers of viable organisms into the air during coughing fits. Up to 90 percent of untreated patients will die representing a highly contagious health hazard to nursing staff. As the epidemic of bubonic plague develops (especially under conditions of severe overcrowding, malnutrition, and heavy ectoparasite infestation), it eventually shifts into a predominately pneumonic form (Fig. 29-4), which is far more difficult to control and which has 100 percent mortality. Experimentally, a conjunctival infection route has been demonstrated in monkeys and guinea pigs, and it is likely that many laboratory-derived infections occur via this route.

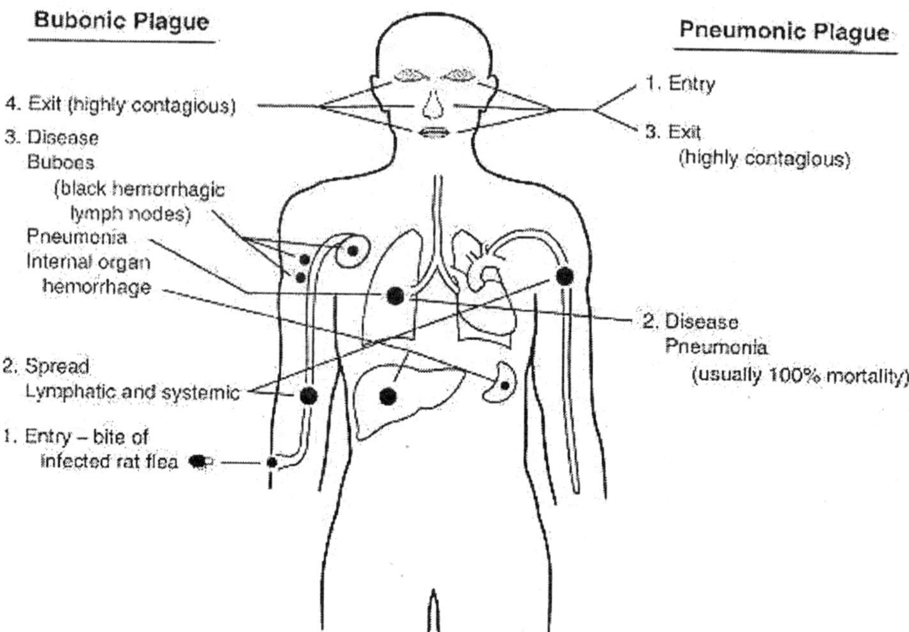

Figure 29-4 Pathogenesis of Y pestis in plague patients.

Structure, Classification, and Antigenic Types

Yersinia pestis is a small, Gram-negative coccobacillus, which frequently shows strong bipolar staining. However, pleomorphic and club-shaped forms are not unusual. Freshly isolated cultures often exhibit substantial slime production, due to a so-called capsular or envelope antigen which is heat labile and is readily lost when the organism is growing in vitro or in the insect vector (Table 29-2). Fully virulent strains possess V and W (virulence) antigens, which are highly toxic for the mouse and, to a lesser extent, for guinea pigs (Table 29-2).

TABLE 29-2 Anitgenic Makeup of Y Pestis

Antigen	Composition	Function	Protection
Envelope (F1)			
A	Soluble polysaccharide-protein	Immunogen	+
B	Soluble polysaccharide	Species-specific antigen	-
C	Insoluble polysaccharide	Nonimmunogen	-

Somatic (O)			
1	Unknown	Virlence antigen	+
3	Corresponds to F1	species-specific antigen	+
4	Heat-stable protein	Nonimmunogen	-
5	Heat-stable protein	Shared with Y psoudotuberculosis	-
8	Heat-labile polypeptide	Toxin	-
V	protein	Associated with virutencenhibitis	+
W	protein	Share with Y pseudoluberculosis	+
Rough	Heat-stable polysaccharide	Share with Y pseudoluberculosis	-

Pathogenesis

The virulence of Y pestis strains can be equated to the rate of growth (or elimination) of the organisms in the spleen following intravenous inoculation (Fig. 29-5). The most virulent strains multiply logarithmically with no initial lag phase, reaching lethal proportions within 2 or 3 days. Infected animals exhibit a progressive septicemia and die as a result of a hemorrhagic pneumonia. Less virulent strains begin to multiply in vivo only after an initial lag period and this slowed early growth allows the host defenses time to mount an effective immune response (Fig. 29-5).

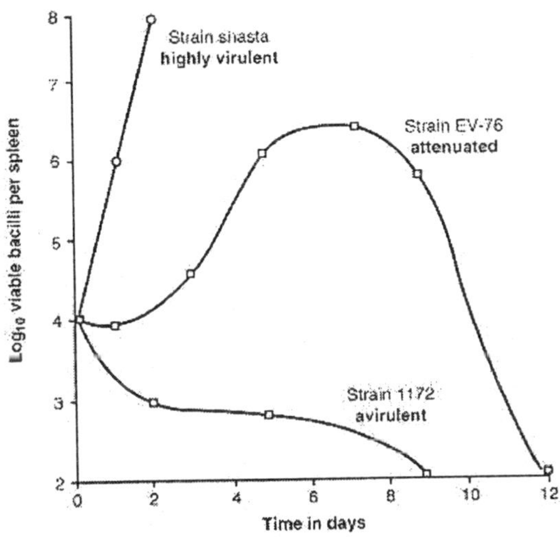

Figure 29-5 Growth of Y pestis in intravenously infected mice showing combined viable counts for spleen and liver homogenates. The virulence of Y pestis correlates with rate of growth in the mouse. The highly virulent Shasta strain killed 100 percent of infected animals within 3 days. The attenuated strain EV-76 gave rise to a self-limiting infection that induced an excellent immune response. The avirulent variant 1122 did not induce a protective immune response. (Data from Walker DL, Foster LE, Chen TH et al: Studies on immunization against plague. V. Multiplication and persistence of virulent and avirulent P pestis in mice and guinea pigs. J Immunol 70:245, 1953.)

Host Defenses

The major defense against Y pestis infection is the development of specific anti-envelope (F1) antibodies, which serve as opsonins for the virulent organisms, allowing their rapid phagocytosis and destruction while still within the initial infectious locus (Fig. 29-6). Although, the V and W antigens are associated with virulence, a number of avirulent strains may also possess them, and some individuals possessing high anti-VW antibody titers will nevertheless undergo a second attack of this disease. Therefore, the immune mechanism(s) against this disease is extremely complex and involves a combination of humoral and cellular factors. The convalescent host is solidly immune (at least for a time) to virulent rechallenge, the inoculum being eliminated as though the organisms were completely avirulent. Killed Y pestis vaccines (especially when given with a suitable adjuvant) induce some measure of host protection, although this will be less effective than that afforded by the live infection.

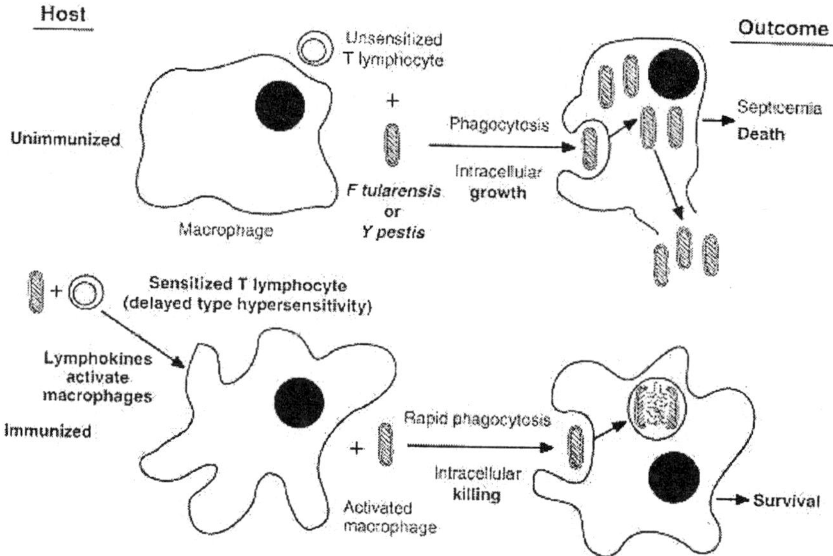

Figure 29-6 Protection against F tularensis or Y pestis is cell-mediated.

Epidemiology

Bubonic plague the Black Death one of the great epidemic scourges of mankind, swept across Europe and Asia in a series of devastating pandemics during the Middle Ages. This disease may have been responsible for the death of one-third of the world's population at that time. Then, for largely unknown reasons, bubonic plague suddenly ceased to be an important pandemic disease, and no major epidemics have occurred in Europe or North America in more than a century. Sporadic outbreaks of sylvatic plague still occur in wild rats, squirrels and prairie dogs in the western United States, and endemic plague has been reported in parts of Southeast Asia. Although, no recent deaths due to plague have been reported in the United States, occasional isolates of Y pestis appear to be fully virulent for experimental animals. The striking change in the epidemiology of this disease is probably due to such nonspecific factors as improved rodent control and the widespread use of insecticides against the insect vector. Recent advances in our understanding of the molecular biology of microbial virulence factors associated with this pathogen offer the promise of improved subunitimmunogens capable of inducing a fully effective acquired resistance.

Diagnosis

Yersinia infections must be diagnosed quickly due to the extraordinary virulence of these organisms. Death from pneumonic plague can occur in as little as 24 hours after the first appearance of clinical symptoms. Sputum specimens from these patients contain large numbers of Gram-negative coccobacilli. Blood cultures are positive, and lymph node biopsy material shows a massive inflammatory cell infiltrate, together with numerous cell-free coccobacilli. The organisms can be identified using a fluorescent antibody staining technique, and the epidemiology of the outbreak can be traced by bacteriophage typing.

Yersinia pestis poses a serious infectious hazard for nursing and laboratory personnel. Protective clothing and a full face respirator should always be worn when working with this organism. Cultivation and virulence testing of this organism should be attempted only in P-3 containment facilities by staff who have been immunized recently with live attenuated vaccine. Animals should be checked to ensure that they are free of ectoparasites.

Control

Yersinia can be killed by mild heat (55°C) and by treatment with 0. 5 percent phenol for 15 minutes. It is susceptible to sulfadiazine, streptomycin, tetracycline, and chloramphenicol in vitro. Thus far, few drug-resistant strains have emerged. Control measures against plague center largely on rat flea eradication programs, which have been credited with preventing epidemics of plague in Europe in 1945 and in Southeast Asia during the Vietnam war. Attempts to eradicate the rodent reservoir have been unsuccessful, and it seems unlikely that rat plague will ever be completely eliminated world-wide.

Yersinia Pseudotuberculosis

Yersinia pseudotuberculosis is a natural pathogen of rodents and birds but can infect humans, causing a severe enterocolitis with enlarged caseousnodules in the Peyer's patches and the mesenteric lymph nodes. These lesions often resemble those seen during intestinal tuberculosis. The organism is highly infectious for guinea pigs and can result in devastating outbreaks of pseudotuberculosis in breeding colonies, with very high mortality rates. This infection is virtually impossible to eliminate once established. Yersinia pseudotuberculosis can be readily distinguished from other Yersinia species because of its motility when grown at 25°C.

In humans, Y pseudotuberculosis causes severe intestinal abscesses that require aggressive chemotherapy with ampicillin and tetracycline. No vaccine is available.

Yersinia Enterocolitica

Yersinia enterocolitica is a natural pathogen of cattle, deer, pigs, and birds. Most infected animals recover from their primary disease, remaining healthy carriers indefinitely. The organism is excreted in large numbers in the feces by infected carriers and can contaminate drinking water and dairy products. Oral infection results in a severe diarrhea in humans, together with necrosis of the Peyer's patches, chronic lymphadenopathy, and hepatic and splenic abscesses (Fig. 29-7). An increasing number of human outbreaks have been reported in recent years, mostly in colder climates. This may reflect a greater awareness of the disease, (together with improved isolation and diagnostic procedures) rather than an actual increase in the overall incidence of this disease in humans.

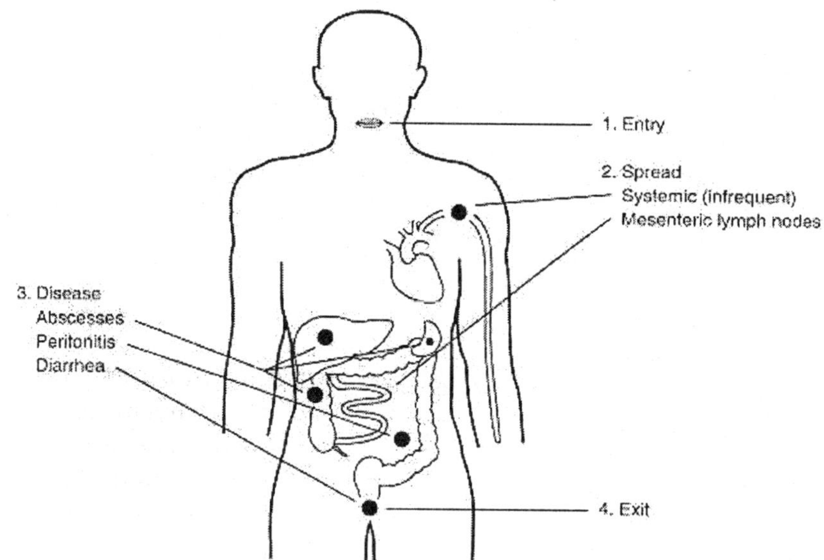

Figure 29-7 Pathogenesis of Y enterocolitica.

Most Y enterocolitica isolates are avirulent for laboratory rodents. Recently, several mouse virulent strains have been isolated from human pathologic material. Orally infected mice develop progressive involvement of the ileal Peyer's patches, with abscess formation in the mesenteric lymph nodes, liver, and spleen. Eventually, most of the animals die when these intestinal abscesses undergo perforation and peritonitis. The lesions typically contain large numbers of PMNs but there is also a strong mononuclear cell response, which may be important to the successful control of this infection.

The best prevention methods for *Y enterocolitica* infections are adequate water purification and milk pasteurization. Once the infection becomes established within the gut-associated lymphoid tissues, it produces chronic abscesses, which require aggressive chemotherapy involving a combination ofampicillin, chloramphenicol, and polymyxin. No vaccine is available for this infection.

Francisella

Clinical Manifestations

Francisella tularensis causes tularemia, which is spread naturally to humans directly by ticks and deerflies (Fig. 29-8). Most strains that infect rabbits are highly infectious and virulent for humans. The subcutaneous infectious dose may be as low as 10 viable bacilli, with a mortality as high as 30 percent in untreated patients. Infections may result from local trauma incurred while skinning anddressing infected rabbit carcasses. Hence, protective gloves and goggles should always be worn while performing this chore in an endemic area. Humans can also contract the disease by eating inadequately cooked, infected rabbit meat. The resulting tularemia is a severe typhoid like intestinal disease, with local abscess formation in the Peyer's patches and the mesenteric lymph nodes (Fig. 29-7). It is associated with high fever and a severe toxemia (septicemia). Francisella tularensis is a facultative intracellular parasite, which induces a strong mononuclear cell immune response on the part of the host defenses (Fig. 29-3). A humoral response also develops, although the precise nature of the relationship between the specific antibodies and resistance to the naturally acquired disease is still not altogether clear. Actively infected mice develop a strong, delayed-type skin hypersensitivity to sensitins produced by this organism.

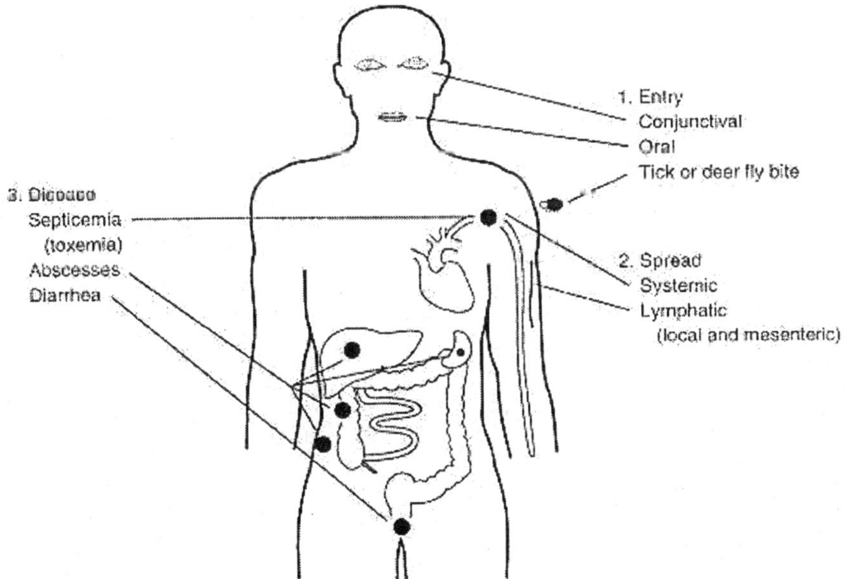

FIGURE 29-8 Pathogenesis of F tularensis.

Laboratory infections may occur via the conjunctival route; this probably explains the high infection rate seen in laboratory personnel working with this pathogen. Goggles and a face mask should always be worn when working with virulent strains of this organism. Staff should be immunized with the live attenuated vaccine. Animal infection studies must be performed under P-3 containment conditions, and, whenever possible, experimental studies should use the vaccine strain.

Structure, Classification, and Antigenic Types

Francisella tularensis is a nonmotile, Gram-negative coccobacillus, which forms small translucent colonies on glucose blood agar or on Dorset egg slants. The organism grows readily in developing chicken embryos. Nutritionally and biochemically it bears a close resemblance to the Brucellae, but it can be differentiated from members of this genus on the basis of DNA homology tests. It is a natural pathogen of rodents (squirrels and rabbits mainly), but can be carried by birds, which usually develop latent infections.

Pathogenesis

Mice, rats, guinea pigs, and rabbits are readily infected with F tularensis via the subcutaneous, nasal, or conjunctival routes. Virulent strains multiply logarithmically within the liver and spleen (but not the lungs), and death usually occurs 5 to 8 days later. In sublethally challenged animals, the systemic infection peaks and declines rapidly, a response associated with cell-mediated immunity, which can be transferred adoptively to naiverecipients by splenic immuneT cells, but not by hyperimmune serum. Most clinical isolates lose virulence when maintained for long periods on laboratory media and eventually cannot produce progressive disease in susceptible animals.

Host Defenses

Acquired resistance following recovery from tularemia is cellular, long lasting and highly protective.

Diagnosis

Isolation of F tularensis from pathologic material can be difficult and slow. Best growth occurs on cysteine-glucose-blood agar, but plates should be incubated at 37°C for at least 3 weeks before being discarded as negative. Smears of pathologic material or blood cultures may be stained using fluorescent antibodies directed against specific surface antigens of the organism. Hemagglutinins appear in serum samples some 10 to 12 days after infection and slowly increase in titer for up to 8 weeks. A rising titer is always diagnostic of active disease.

Control

Francisella tularensis is suseptible to inactivation by mild heat (55°C for 10 minutes) and disinfectants. It is susceptible to streptomycin, tetracycline, and chloramphenicol in vitro. Relapses are not uncommon if treatment is stopped before all the viable bacilli have been eliminated from the tissues. Infection control measures usually entail the elimination of the insect vectors.

Killed F tularensis vaccines are not very effective, even when presented in adjuvant. A live attenuated vaccine has been developed and should be used to immunize laboratory staff working with this organism.

REFERENCES

Butler T: Yersinia infections: centennial of the discovery of the plague bacillus. Clin Inf Dis 19:655, 1994

Collins FM: Mechanisms of resistance to P multocida infection. A review. Cornell Vet 67:103, 1977

Cover TL, Aber RC: Yersinia enterocolitica. N Eng. J Med. 321:16, 1989

Crook TL, Tempest B: Plague. A clinical review of 27 cases. Arch Intern Med 152:1253, 1992

Falkow S: Molecular Koch's postulates applied to microbial pathogenicity. Rev Infect Dis 10:S274, 1988

Koskela P, Salminen A: Humoral immunity against Francisella tularensis after natural infection. J Clin Microbial 22:973, 1985

Sanford JP: Landmark perspective: tularemia. J Am Med Assoc 250:3225, 1988

Tarnvik A, Löfgren ML, Löfgren S et al: Long-lasting cell-mediated immunity induced by live Francisella tularensis vaccine. J Clin Microbiol 22:527, 1985

Weber DJ, Wolfson JS, Swartz MN, Hooper DC: Pasteurella multocida infections. Reports of 34 cases and a review of the literature. Medicine 63:133, 1984

Chapter 26

Treponema

General Concepts

Clinical Manifestations

Treponemes cause diverse clinical manifestations. In patients with acquired venereal syphilis, there is an initial genital tract lesion (primary stage) followed by disseminated lesions (secondary stage) and, in approximately one-third of untreated individuals, cardiovascular and neurologic problems (tertiary stage). Infection during pregnancy (congenital syphilis) may result in fetal death or birth defects. Yaws, pinta, and endemic syphilis, the nonvenereal treponematoses, are usually present as skin or mucous membrane lesions. Soft tissue and bone lesions also can occur with yaws and endemic syphilis.

Structure and Biology

Treponemes are helically coiled, corkscrew-shaped cells, 6 to 15 µm long and 0.1 to 0.2 µm wide. They have an outer membrane which surrounds the periplasmic flagella, a peptidoglycan-cytoplasmic membrane complex, and a protoplasmic cylinder. Multiplication is by binary transverse fission. Treponemes have not yet been cultured in vitro.

Classification and Antigenic Types

Classification of the pathogenic treponemes is based primarily upon the clinical manifestations of the respective diseases they cause. *Treponema pallidum* subsp *pallidum* causes venereal syphilis; *T pallidum* subsp *pertenue* causes yaws; *T pallidum* subsp *endemicum* causes endemic syphilis; and *T carateum* causes pinta. Venereal syphilis is transmitted by sexual contact; the other diseases are transmitted by close nonvenereal contact.

Pathogenesis

Treponemes are highly invasive pathogens which often disseminate relatively soon after inoculation. Evasion of host immune responses appears to be, at least in part, due to the unique structure of the treponemal outer membrane (i.e., its extremely low content of surface-exposed proteins). Although, treponemes lack classical lipopolysaccharide (endotoxin), they possess abundant lipoproteins which induce inflammatory processes.

Host Defenses

Various studies suggest that both cellular and humoral processes contribute to host defenses against treponemal infection. Clearance of treponemes from local sites appears to be due to phagocytosis by macrophages.

Epidemiology

Humans are the only source of treponemal infection; there are no known nonhuman reservoirs. Venereal syphilis is distributed worldwide, and over the past several decades has become a significant public health problem in many underdeveloped countries. Infectivity rates correspond to the most sexually active age groups. Following the adoption of penicillin as the mainstay of syphilotherapy, the number of new syphilis cases progressively decreased until 1958, after which the trend reversed and a steady increase has occurred. The late 1980's experienced a major increase in the incidence of early syphilis cases which was largely related to crack cocaine usage among inner city minorities. Improved surveillance methods have helped to control this syphilis epidemic. Despite extensive eradication campaigns, yaws remains widespread in the tropics. Pinta remains endemic in Central and South America, and endemic syphilis is present in certain regions of the Middle East. The pathogenic treponemes have many cross-reacting antigens, and untreated infection is believed to confer partial protection against the other treponemal diseases.

Diagnosis

Diagnosis relies heavily on clinical manifestations. In addition, the finding of treponemes within exudative lesions and positive serology aids the diagnosis.

Control

Control of venereal and nonvenereal treponematoses is based upon active surveillance and treatment of contacts. Penicillin treatment eradicates all stages, including congenital infection in pregnancy.

INTRODUCTION

The genus *Treponema* contains both pathogenic and nonpathogenic species. Human pathogens cause four treponematoses: syphilis (*T pallidum* subsp *pallidum*), yaws (*T pallidum* subsp *pertenue*), endemic syphilis (*T pallidum* subsp *endemicum*), and pinta (*T carateum*). Nonpathogenic treponemes may be part of the normal flora of the intestinal tract, the oral cavity, or the genital tract. Some of the oral treponemes have been associated with gingivitis and periodontal disease.

Clinical Manifestations

Because syphilis exhibits diverse clinical manifestations that mimic many other infectious and noninfectious disorders (Fig. 36-1), it has earned a reputation as "the great imitator." Yaws, pinta, and endemic syphilis also have highly variable manifestations. Treponemal infections are unique in that they are characterized by distinct clinical stages. Multiplication of the organisms at the initial site of entry produces the primary stage. The dissemination of treponemes to other tissues results in the secondary stage. After a relatively prolonged period, in some cases 20 to 30 years, the tertiary or late stage evolves. *Treponema pallidum* subsp *pallidum*, the most invasive of the pathogenic treponemes, produces highly destructive lesions in almost any tissue of the body, including the central nervous system. *Treponema carateum* is the least invasive and causes only cutaneous disease. *Treponema pallidum* subspp *pertenue* and *endemicum* are intermediate in invasiveness and cause destructive lesions in bones and soft tissues.

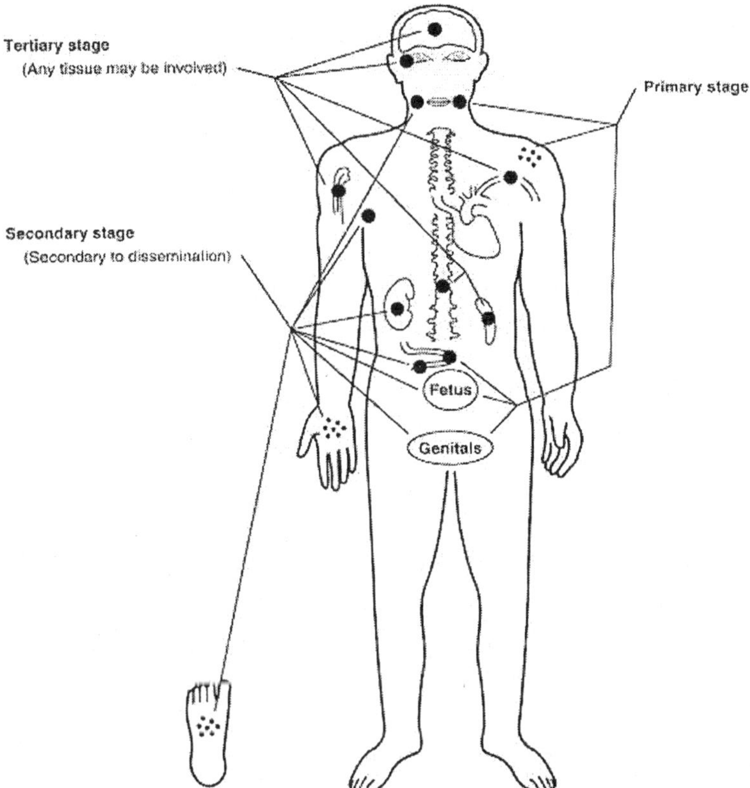

FIGURE 36-1 Clinical manifestations of syphilis.

Venereal syphilis is the prototype treponemal disease and the only treponematosis of significance in developed countries. Clinical manifestations of syphilis are complex and the periods associated with each stage vary greatly. (Fig. 36-2 depicts averages.) After an incubation period of 10 to 90 days, extensive multiplication of treponemes at the site of entry produces erythema and induration. The resultant papule eventually progresses to a superficial ulcer with a firm base called a hard chancre. (*Haemophilus ducreyi* causes soft chancre, which differs in that it is flat, centrally umbilicate, tender, and painful.) Numerous treponemes are present in this highly contagious, open lesion. Regional lymph nodes enlarge, causing regional lymphadenopathy. After 2 to 6 weeks of symptoms, this primary lesion heals, leaving only remnants of scar tissue.

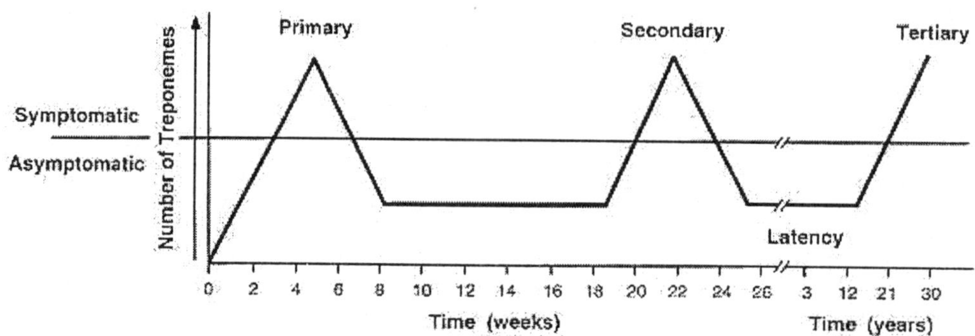

FIGURE 36-2 Development of the clinical stages of syphilis over time.

After an asymptomatic period of 2 to 24 weeks, the secondary or disseminated stage begins. Organisms multiply in many different tissues. Clinical manifestations include slight fever, generalized lymphadenopathy, malaise, and a mucocutaneous rash. The rash initially appears on the palms and soles and eventually spreads to other areas. The rash may be macular, papular, follicular, papulosquamous, or pustular. Superficial sores (mucous patches) may occur on mucous membranes of the mouth, vagina, or anus, while wart-like lesions called condylomata lata may form in moist intertriginous areas. All of these lesions teem with treponemes and are highly contagious. Deposition of immune complexes consisting of treponemal antigens and host antibodies in glomerular basement membranes may produce nephrotic syndrome. Two to six weeks after the onset of secondary syphilis, host defenses bring about healing. About 25 percent of untreated patients experience recurrences of this secondary stage in the first several years following infection.

The period between secondary and tertiary syphilis, termed latency, can last for many years. Early latency refers to the first 4 years when secondary relapses may occur; late latency is the asymptomatic period beyond 4 years. During this latter period, the patient harbors infectious organisms, especially in the spleen and lymph nodes and blood serology remains positive.

Tertiary syphilis can affect almost any tissue. Approximately 80 percent of fatalities are caused by cardiovascular involvement, while most of the remaining 20 percent are from neurologic involvement. Cardiovascular problems are usually attributed to local inflammation induced by the multiplication of treponemeswithin the wall of the thoracic aorta. The subsequent aortitis produces complications such as aneurysms and coronary artery stenosis. Neurologic syphilis may be meningeal, meningovascular, parenchymatous, or various combinations thereof. If the parenchymatous form involves the brain, it is called generalized paresis; if it involves the spinal column, it is called tabes dorsalis. Complications of neurosyphilis include dementia, loss of proprioception, strokes, and blindness. For unclear reasons, cardiovascular syphilis is much less common than during the pre-antibiotic era. Gummas are highly destructive tertiary syphilitic lesions that usually occur in skin and bones but may also occur in other tissues. They are necrotizing granulomas with numerous lymphocytes, giant cells, and epithelioid cells, but few treponemes. A delayed hypersensitivity response to the small numbers of treponemes in the lesions may be responsible for the development of gummatous disease. Gummas also have become rare in the post-antibiotic era.

Besides the three stages of disease in adults, *T pallidum* subsp *pallidum* also damages fetuses. If a woman is pregnant and has symptomatic or asymptomatic early syphilis, hematogenously disseminating organisms may pass through the placenta to infect the fetus. Approximately 50 percent of fetuses are aborted or stillborn; the rest exhibit diverse syphilitic stigmata. In early congenital syphilis, signs are apparent before the age of two years. These include mucocutaneous lesions, osteochondritis (especially within the long bones), anemia, and hepatosplenomegaly. In late congenital syphilis, an infected child appears normal past two years of age and then exhibits syphilitic manifestations, such as interstitial keratitis and blindness, tooth deformation (notched incisors and moon molars), eighth-nerve deafness, neurosyphilis, rhagades (fissures at mucocutaneous junctions), cardiovascular lesions, Clutton's joints (fluid accumulation on knee), and bone deformation of the legs, nasal septum, and hard palate. Combinations of these stigmata often occur. In late congenital syphilis, three commonly observed manifestations, called Hutchinson's triad, are interstitial keratitis, notched incisors, and eighth-nerve deafness.

Structure and Biology

Treponemes are helically coiled, corkscrew-shaped organisms 6 to 15 µm long and 0.1 to 0.2 µm wide (Fig. 36-3). The organisms stain poorly with aniline dyes. Treponemes in tissues can be visualized by silver impregnation methods. Live treponemes, which are too slender to be seen by conventional light microscopy, can be visualized by using dark-field microscopy. *Treponema pallidum* subsp *pallidum* exhibits characteristic motility that consists of rapid rotation about its longitudinal axis and bending, flexing, and snapping about its full length.

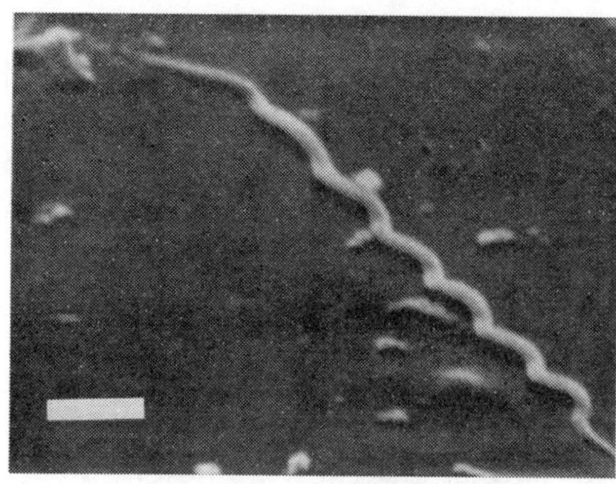

FIGURE 36-3 Scanning electron micrograph of *T pallidum*. (From Fitzgerald TJ, Cleveland P, Johnson RC et al: Scanning electron microscopy of Treponema pallidum (Nichols strain) attached to cultured mammalian cells. J Bacteriol 130:1333, 1977, with permission.)

Treponema pallidum subsp *pallidum* is a fastidious organism that exhibits narrow optimal ranges of pH (7.2 to 7.4), Eh (230 to240 mV), and temperature (30 to 37°C). It is rapidly inactivated by mild heat, cold, desiccation, and most disinfectants. Traditionally this organism has been considered a strict anaerobe, but it is now known to be microaerophilic. Treponemes multiply by binary transverse fission. The in vivo generation time is relatively long (30 hours). Despite intense efforts over the past 75 years, *T pallidum* subsp *pallidum* has not been successfully cultured in vitro. Viable organisms can be maintained for 18 to 21 days in complex media, while limited replication has been obtained by co-cultivation with tissue culture cells. The other three pathogenic treponemes also have not been successfully grown in vitro.

The composition of *T pallidum* subsp *pallidum* (dry weight) is approximately 70 percent proteins, 20 percent lipids, and 5 percent carbohydrates. This lipid content is relatively high for bacteria. The lipid composition of *T pallidum* is complex, consisting of several phospholipids, including cardiolipin, and a poorly characterized glycolipid which is biochemically and immunologically distinct from lipopolysaccharide. While antigenic analysis of *T pallidum* subsp *pallidum* has been hampered by the inability to grow this organism in vitro, this situation has largely been circumvented through the use of modern molecular techniques, including monoclonal antibodies and recombinant DNA. During the course of infection, antibodies develop to a number of treponemal proteins, most notably the lipoproteins and flagella.

Although, treponemes possess both outer and cytoplasmic membranes (Fig. 36-4), they differ considerably in structure from enteric Gram-negative bacteria. Figure 36-5 is a diagrammatic cross-section of *T pallidum* subsp *pallidum*. The organism has an outer membrane containing an extremely low density of surface-exposed transmembrane proteins. Typically, three flagella originate from each end of the bacterium, and, winding about the bacterium within the periplasmic space, overlap at the midpoint. The presence of peptidoglycan in the cell wall, originally surmised on the basis of the bacterium's exquisite sensitivity to penicillin, has been confirmed by biochemical analysis. Unlike Gram-negative bacteria in which the peptidoglycan underlies the outer membrane, in treponemes the murein layer overlies the cytoplasmic membrane. The cytoplasmic membrane covers the protoplasmic cylinder; this membrane contains the majority of the bacterium's integral membrane proteins and is particularly abundant in lipid-modified polypeptides (lipoproteins).

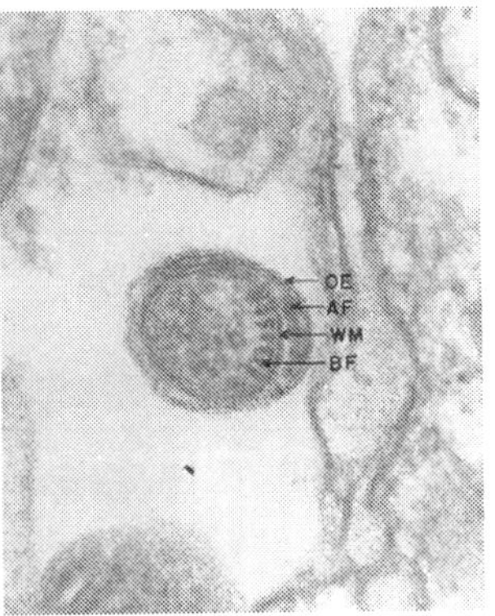

FIGURE 36-4 Transmission electron micrograph of cross-section of *T pallidum*. Abbreviations: OE, outer envelope (membrane); AF, axial filament; WM, cell wall membrane; BF, body fibrils. (From Johnson RC, Ritzi DM, Levermore BP et al: Outer envelope of virulent Treponema pallidum. Infect Immun 8:294, 1973, with permission.)

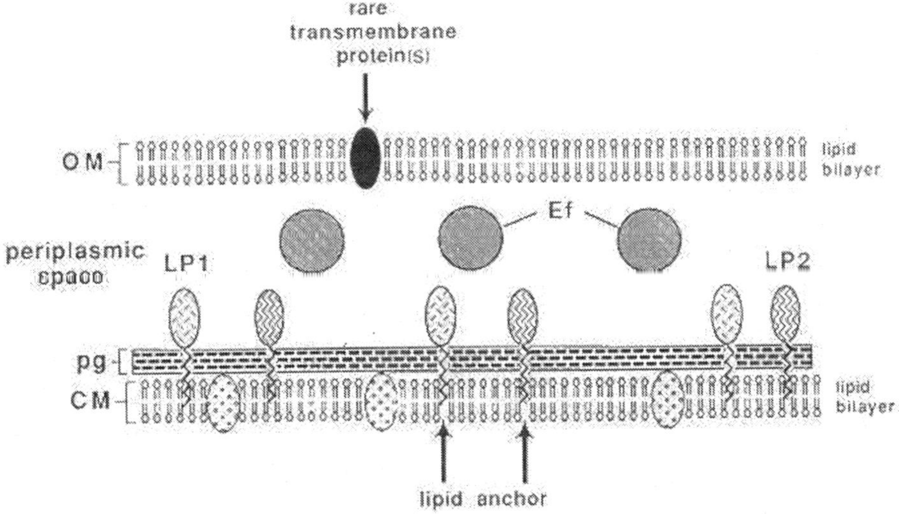

FIGURE 36-5 Proposed structure of treponemal outer and cytoplasmic membranes. Abbreviations: OM, outer membrane; Ef (endoflagella or periplasmic flagella); LP1, 2, lipoproteins; pg, peptidoglycan; CM, cytoplasmic membrane. (From Cox DL, Chang P, McDowall AW, and Radolf JD: The outer membrane, not a coat of host proteins, limits the antigenicity of virulent Treponema pallidum. Infect Immun 60:1076, with permission.

Classification and Antigenic Types

The pathogenic treponemes cannot be distinguished by morphologic, antigenic, biochemical, or genetic criteria. Differentiation of the treponematoses is based on geographic location, modes of transmission, and clinical manifestations (Table 36-1). Similarities in treponemal infections include their generalized nature, regional and general lymphadenopathy, chronicity, spontaneous healing, asymptomatic periods, and relatively painless symptoms. Although, specific strain differentiation is not available, different human isolates have been characterized. These isolates exhibit various degrees of virulence as determined by animal inoculation studies.

TABLE 36-1 Characteristics of the Four Treponematoses

Characteristic	Syphitis (subsppallidum)	Yaws (subsppertenue)	Pinta (subspcarateum)	Endemic syphilis (subspendemicum)
Epidemiology				
Other names	Venereal syphilis	Frambesia plan	Carate, cute	Bejel, dichuchwa
Prevalence	Worldwide areas	Hot, humid areas	Hot, humid	Hot, dry areas
Locations	Worldwide	Tropics	Central and south America	Deserts
Age group	Adults	Children adolescents	Children	Children, aduits
Spread	Venereal	Skin	Skin	Mucous membranes
Congenital infection	Yes	No	No	Rarely
Disease characteristics				
Incubation period	10-90 days	14-28 days	2-6 months	?
Invasiveness	High	Intermediate	Low	Intermediate
Perivascular (culfing)	Yes	No	Yes	Yes
Tissues	All	Skin, bones, soft tissues	Skin	Mucous membranes
Predominant cellular infiltrate	Lymphocytes, plasma cells	Mostly plasma cells	Mostly lymphocytes	Skin, muscles, bones
Destructive lesions	Yes	Yes	No	Lymphocytes, plasma cells
Granulomas	Yes	Yes	No	Yes
Gumas	Yes	Yes	No	Yes
Condylomata lata	Yes	Yes	No	Yes

Pathogenesis

Untreated syphilis is a slowly evolving chronic disease that transpires in stages separated by asymptomatic intervals. Humans are the only natural host for *T pallidum* subsp *pallidum*, and infection occurs through sexual contact. The organisms penetrate mucous membranes or enter minuscule breaks in the skin. Experiments in both humans and rabbits using standardized inocula indicated that less than 10 organisms are capable of producing infection. In women the initial lesion is usually on the labia, the walls of the vagina, or the cervix; in men it is on the shaft or glans of the penis. A chancre also may occur on lips, tongue, tonsils, anus, or other skin areas. The observation, made in a number of in vitro studies, that*T pallidum* subsp *pallidum* and subsp *pertenue* specifically attach to numerous cell types is believed to reflect the ability of these bacteria to infect diverse tissues and organs. To disseminate away from the site of initial entry, organisms must traverse the viscous ground substance between tissue cells. There is evidence that *Treponema pallidum* subsp *pallidum* elaborates an enzyme capable of degrading hyaluronic acid within the ground substance, thereby potentially facilitating hematogenous dissemination of organisms.

Blood from a patient with incubating syphilis may be infectious long before the appearance of a chancre, as transfused blood from patients with incubating syphilis has transmitted the infection. Although, neurological symptoms may not become apparent until years after infection, invasion of the central nervous system by *T pallidum* occurs relatively early. A high proportion of patients with early syphilis have CSF abnormalities, despite a lack of neurological symptoms and treponemes often can be recovered by rabbit inoculation with either normal or abnormal spinal fluids during this period. Conversely, a normal spinal fluid examination two or more years after infection indicates a low risk of subsequent development neurosyphilis. Other target tissues include lymph nodes, skin, mucous membranes, liver, spleen, kidneys, heart, bones, joints, larynx, and eyes. It is generally believed that the motility of the organisms contributes to the invasiveness of these bacteria. Studies with cultured human endothelial cells have demonstrated that *T pallidum* rapidly penetrates between tight endothelial cell junctions, and this may be the mode of invasion at the various sites of dissemination.

The most prominent and widespread histopathologic feature of syphilitic infection is perivascular inflammation. This typically consists of proliferation of adventitial cells; perivascular cuffing with lymphocytes, monocytes and plasma cells; and swelling and proliferation of endothelial cells (Fig. 36-6). Occasionally these changes progress to frank vasculitis with ischemic necrosis of tissues. Granulomatous changes are characteristic of tertiary disease but may be seen in secondary syphilis as well. *Treponema pallidum* subsp *pallidum* appears to lack potent toxins, and it is believed that tissue damage results from the host's cellular inflammatory response. Treponemal lipoproteins have been shown to be capable of activating relevant immune effector cells, most notably macrophages and endothelial cells. While evasion of host cellular and humoral immune responses is presumably prerequisite to establishment of persistent infection, the mechanisms responsible for this are poorly understood. It is known, however, that the surface of the organism reacts poorly with specific antibodies; the paucity of surface-exposed proteins is believed to explain this phenomenon (Fig. 36-5). Circulating immune complexes containing treponemal antigens are regularly present in secondary syphilis and may contribute to clinical manifestations, most notably the nephrotic syndrome occasionally seen in the secondary stage of the disease.

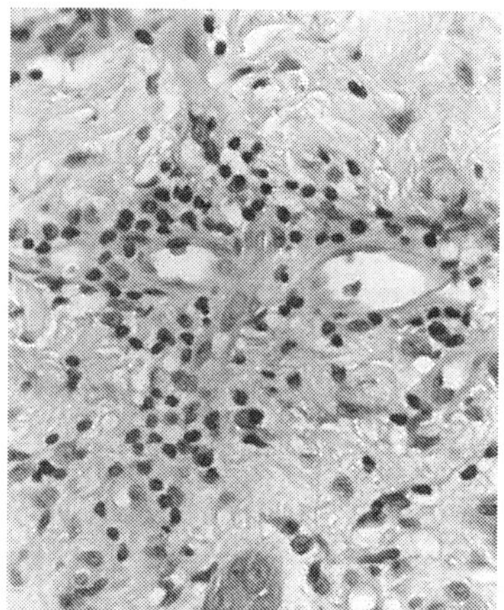

FIGURE 36-6 Perivascular dermal infiltrate containing mononuclear and plasma cells in secondary syphilis.

Host Defenses

Several clinical studies support the notion that humans develop some degree of resistance to treponemal infection. The Oslo study, begun in the early 1900s, involved over 2,000 untreated primary and secondary syphilis patients observed for 30 to 50 years. Approximately 25 percent of these patients developed secondary syphilis and 13 percent developed tertiary syphilis. The fact that 75 percent of the patients did not progress beyond primary syphilis has been taken to indicate the development of some degree of immunity. The Sing-Sing study, in which human volunteers were intradermally inoculated with virulent *T pallidum* subsp *pallidum* demonstrated that resistance to infection occurs in late syphilis.

The precise contribution of humoral and cellular immune mechanisms to host defenses in syphilitic infection is not clear. Treponemes are capable of disseminating hematogenously despite high titers of circulating antibodies directed against numerous treponemal proteins. On the other hand, the presence of both organisms and antibody-synthesizing host cells (i.e., plasma cells) in the same tissue areas suggests some role for antibodies. Passive transfer of immune serum does not protect experimental syphilitic rabbits against challenge inoculation, although, some modification of treponemal lesions occurs. Injection of immune serum before *T pallidum* subsp *pallidum* challenge lengthens the incubation period, reduces the severity of lesions, and speeds healing. Thus, antibodies appear to be partially but not solely responsible for healing and immunity. A mononuclear and plasma cell infiltrate occurs early in infection and is especially prominent in perivascular areas. Immunohistochemical studies indicate that a large proportion of the mononuclear cells are macrophages, while the lymphocytes are a mixture of $CD4^+$ and $CD8^+$ T cells. The roles of these immune cell types are still being clarified. In vitro, macrophages readily phagocytose and process treponemal organisms and histopathological studies suggest a similar process occurs in vivo. Activated macrophages, in turn, amplify T cell stimulation.

Epidemiology

Humans are the only source of treponemal infection; there are no known nonhuman reservoirs. Venereal syphilis is distributed worldwide, and over the past several decades has become a significant public health problem in many underdeveloped countries. Infectivity rates correspond to the most sexually active age groups, being highest in the 20-to 24-year age group, slightly lower in the 15- to 19- year age group, and lower still in the 25- to 29- year age group. The peak incidence of syphilis was observed in 1946 to 1947 (Fig. 36-7). In the late 1940s, it was discovered that *T pallidum* is exquisitely sensitive to penicillin G, and penicillin was found to be effective in eradicating syphilis of all clinical stages as well as the congenital infection. Following the adoption of penicillin as the mainstay of syphilotherapy, the number of new syphilis cases progressively decreased until 1958, after which the trend reversed and a steady increase has occurred. The late 1980's experienced a major increase in the incidence of early syphilis cases which was largely related to crack cocaine usage among inner city minorities. Improved surveillance methods have helped to control this syphilis epidemic. Approximately one to 10 percent of persons with gonorrhea may have concurrent syphilitic infection. Because gonorrhea has a shorter incubation period (two to eight days) and has painful symptoms, the patient seeks treatment before syphilitic lesions develop. Most regimens used to treat gonorrhea also eradicate incubating syphilis.

FIGURE 36-7 Incidence of new cases of primary and secondary syphilis in the United States from 1955 through 1993.

Despite extensive eradication campaigns, yaws remains widespread in the tropics. Pinta remains endemic in Central and South America, and endemic syphilis is present in certain regions of the Middle East. The pathogenic treponemes have many cross-reacting antigens, and untreated infection is believed to confer partial protection against the other treponemal diseases. In areas in which yaws is endemic, the incidence of syphilis is low. After effective campaigns to eradicate yaws, the incidence of syphilis eventually increases. Similar epidemiologic observations have been made after the local eradication of pinta and endemic syphilis.

Diagnosis

Definitive diagnosis of syphilis is complicated by the inability to cultivate *T pallidum* subsp *pallidum* in vitro. Clinical manifestations, demonstration of treponemes in lesion material, and serologic reactions are used for diagnosis. In many cases, clinical manifestations are highly characteristic. If manifestations include one or more cutaneous exudative lesions, motile treponemes can often be visualized within lesion exudate by dark-field microscopy.

Serologic tests are a mainstay of syphilis diagnosis. They are the only means of identifying asymptomatically infected individuals. More than 200 serologic tests have been developed over the years and fall into two general categories: (1) "nontreponemal" tests, which measure antibodies directed against lipid antigens, principally cardiolipin, thought to be derived from host tissues and (2) "treponemal," which detect antibodies directed against protein constituents of *T pallidum* subsp *pallidum*. Examples of the former are the Venereal Disease Research Laboratory (VDRL) and Rapid Plasma Reagin (RPR) tests; examples of the latter are the Fluorescent *T pallidum* Antibody-Absorption (FTA-ABS) and Microhemagglutination for *T pallidum* (MHA-Tp) tests. Both treponemal and nontreponemal serological tests have been highly standardized by the Centers for Disease Control and Prevention (CDC). The sensitivity of the nontreponemal and treponemal tests varies with the stage of the disease. The results of nontreponemal tests usually parallel the extent of infection; titers tend to be highest during secondary syphilis and subside during subclinical infection (latency) or following antibiotic therapy. The treponemal tests often remain reactive for life.

Two terms relevant to syphilis serodiagnostic testing are sensitivity and specificity. The perfect test, not yet developed, would detect 100 percent of the treponemal infections and would be nonreactive in all other diseases. Sensitivity refers to the ability to detect the tested variable, in this case syphilis. A false-negative occurs when serum from a syphilitic patient fails to react. Specificity refers to the ability to recognize when the variable is not present (i.e., to exclude syphilis in nonsyphilitic patients). A false-positive occurs when serum from a nonsyphilitic patient reacts positively. In general, treponemal tests are more sensitive and more specific than the nontreponemal tests. However, mathematical models have shown that maximal sensitivity and specificity are achieved if patients are screened with a nontreponemal test and positive sera confirmed by a treponemal test. A number of clinical conditions may cause false-positive nontreponemal tests. These include leprosy, tuberculosis, malaria, infectious mononucleosis, collagen disorders, systemic lupus erythematosus, rheumatoid arthritis, pregnancy, and drug abuse. Individuals with false-positive nontreponemal tests are identified by virtue of the fact that their treponemal tests are nonreactive.

Congenital syphilis is difficult to diagnose in asymptomatically infected neonates because maternal antibodies (IgG) which pass through the placenta and enter the fetal circulation cause reactivity in both nontreponemal and treponemal tests. In uninfected infants, such maternal antibodies disappear by 3 months. Because of the presence of maternal antibodies in the newborn, quantitative VDRL or RPR tests should be performed monthly over the first 6 months. If the titer increases or stabilizes and does not decrease, congenital syphilis is indicated and the baby should be treated accordingly.

Control

The current worldwide prevalence of syphilis emphasizes the need for continued preventive measures and strategies. Unfortunately, effective measures are limited (Table 36-2). Short of abstinence, the condom remains the method of choice for prevention of sexual transmission. Topical application of antibiotics, chemicals, creams, or lotions and thorough washing with soap and water after sexual contact are highly ineffective. A vaccine appears to be the only hope for future control of syphilis. Despite intense research in this area, only limited progress has been made. Two other control measures are important. The first is to educate people about the early clinical manifestations of primary syphilis, so that they can seek treatment before infecting others. The second, for which epidemiology programs have been established, is to trace contacts of syphilitic patients; these contacts are then treated prophylactically before onset of clinical manifestations.

TABLE 36-2 Effective and Ineffective Preventive Measures against Syphilis

Very Effective	Totally Ineffective
Abstinence	Topical antibiotics
Condoms	Topical chemicals
	Topical creams, including spermicides
	Thorough washing with soap and water

Penicillin remains the drug of choice for treating syphilis. Penicillin resistance has not yet emerged, unlike the situation for gonorrhea. In non-penicillin-allergic patients without central nervous system involvement, infection is usually treated with benzathine penicillin G, a long-acting penicillin preparation which produces treponemicidal levels in serum for up to ten days. Patients with central nervous system involvement (neurosyphilis) should receive high dose intravenous penicillin for 10 to 14 days. Penicillin-allergic, nonpregnant patients with early syphilis can be treated with tetracycline. Penicillin-allergic, pregnant patients and patients with neurosyphilis must be desensitized to penicillin because of the lack of effective alternative therapies. A Jarisch-Herxheimer reaction occasionally follows treatment of secondary syphilis. This systemic reaction is associated with the rapid death of treponemes. Between 2 and 12 hours after antibiotic therapy, headache, malaise, slight fever, chills, muscle aches, and intensification of syphilitic lesions occur. These manifestations resolve in fewer than 12 hours. This benign reaction requires no prophylactic measures and, importantly, indicates effective therapy.

Both nontreponemal and treponemal antibodies remain detectable for long periods, even after effective treatment. The recommended procedure for verifying a cure involves tracking the nontreponemal antibodies until they are undetectable or reach a stable, low titer. Nontreponemal tests in patients with primary and secondary syphilis should be nonreactive within 6 to 12 months and 12 to 18 months after treatment, respectively. A cure may be difficult to verify in patients with long standing infection; in these patients, nontreponemal and treponemal antibodies may be detected years after treatment.

Other Treponematoses

Yaws

Yaws, caused by *T pallidum* subsp *pertenue,* predominates in the tropical areas of Africa, South America, India, Indonesia, and the Pacific Islands. Its highly contagious nature is indicated by an estimated 50 million cases worldwide. Transmission occurs through nonsexual human-to-human contact. Most cases are in children and adolescents. In endemic areas, 75 percent of the population contract yaws before reaching 20 years of age.

The primary lesion, or mother yaw, develops within 2 to 4 weeks at the site of skin entry as a painless erythematous papule or group of papules. Lesions enlarge and ulcerate, exuding a serous fluid with a bloody tinge that is swarming with organisms. These lesions heal within one to several months, leaving an atrophic, depressed scar. The treponemes disseminate, and, within 1 to 12 months, secondary lesions evolve that are quite similar to the mother yaw. Crops of these lesions develop initially on the face and moist areas of the body and then spread to the trunk and arms. Infection of the soles and palms is characteristic, as it is in syphilis. Elevated granulomatous papules may enlarge to a diameter of 5 cm and then heal, leaving areas of depigmentation. Successive crops of these lesions occur for many months. Histopathology is similar to that observed in syphilis, with minimal vascular changes and no endothelial cell proliferation. The late destructive stage, sometimes called tertiary yaws, involves treponemal infection of the bones and periosteum, especially the long bones of legs and forearms, and the bones of the feet and hands. Pathologic findings are similar to those seen in the tertiary stage of syphilis. Highly destructive gummas also may occur within the bones and soft tissues. Diagnosis depends on geographic location, clinical manifestations, demonstration of treponemes within exudates, and positive serology. In areas in which syphilis and yaws coexist, definitive diagnosis is unnecessary since both can be readily eradicated by penicillin.

Pinta

Pinta, caused by *T carateum*, is endemic in the tropical areas of Central and South America. Recently, the total number of cases has been estimated at 500,000. Transmission occurs through human-to-human nonsexual contact. Most cases initially occur in children and adolescents.

The primary lesion develops within 2 to 6 months at the site of skin entry as a flat, erythematous papule or group of papules. These lesions and occasional satellite lesions enlarge over several months and produce plaques with scaly surfaces. Secondary lesions occur after 2 to 18 months or longer and involve ulceration and hyperchromic patches. Typically the hands, feet, and scalp are infected. Late stages of pinta involve patches of hyperchromia and achromia, irregular acanthosis, and epidermal atrophy. Lesions heal initially with hyperpigmentation but, over time, become depigmented and hyperkeratotic due to scarring. The treponemes disturb normal melanin pigmentation and produce the characteristic skin manifestations within 2 to 5 years.

The different stages of this disease are not clearly separated, and overlap of manifestations is common. Diagnosis relies on geographic location, clinical manifestations, demonstration of organisms in exudates, and positive serology. Penicillin is the antibiotic of choice. Contrary to syphilis and yaws, in which the lesions heal rapidly following antibiotic treatment, pinta lesions may require 1 year to fully resolve. After primary or early secondary manifestations, skin pigmentation returns to normal. In later manifestations, however, pigmentation remains altered permanently.

Endemic Syphilis

Endemic syphilis, caused by *T pallidum* subsp *endemicum*, is found in the desert areas of the Middle East and Central and South Africa. Transmission is through human-to-human nonsexual contact. Most cases are contracted by children past the age of two years. Transmission of endemic syphilis, like that of yaws and pinta, is associated with poor hygiene.

Clinical manifestations can be quite similar to those of syphilis and yaws. The site of entry is usually the mucous membranes of the eyes and mouth. The primary lesion, a small papule, is detectable in only one percent of cases. After two to three months, secondary lesions or plaques develop in mucous membranes, skin, muscles, and bone. These oozing papules erode, harden, become condylomatous, and eventually heal. Clinical manifestations are then not apparent for 5 to 15 years (latency). Late endemic syphilis develops in the skin and skeletal system. Skin lesions may be superficial, nodular, or tuberous, or they may be highly destructive, deep gummas. Destructive bone lesions frequently localize in the tibia. Diagnosis depends on geographic location, clinical manifestations, treponemes in the exudate, and positive serology. Penicillin eradicates endemic syphilis.

REFERENCES

Crissey JT, Denenholz DA: Syphilis. Clin Dermatol 2:1, 1984

Lukehart SA, Holmes KK: Syphilis. In: Harrison's Principles of Internal Medicine (E. Braunwald et al., eds.) McGraw-Hill Book Company, New York, 1994

Musher DM: Biology of *Treponema pallidum*. In: Sexually Transmitted Diseases (Holmes KK et al, eds.), Second Edition, McGraw-Hill Book Company, New York, 1990

Dunn RA, Rolfs RT: The resurgence of syphilis in the United States. Curr Opin Infect Dis 4:3, 1991

Koff AB, Rosen T: Nonvenereal Treponematoses: yaws, endemic syphilis, and pinta. J Am Acad Dermatol 29:519, 1993

Jackman JD, Radolf JD: Cardiovascular syphilis. Am J Med 87:425, 1989.

Larsen SA, Steiner BM, Rudolph AH: Laboratory diagnosis and interpretation of tests for syphilis. Clin Microbiol Rev 8:1, 1995

Miller JN: Value and limitations of non-treponemal and treponemal tests in the laboratory diagnosis of syphilis. Clin Obstet Gynecol 18:191, 1975

Chiu MJ, Cockerell CJ, Houpt KR et al: Spirochetal infections of the skin. In: Atlas of Infectious Diseases. Mandell GE and Stevens DL (eds) Churchill Livingstone, Philadelphia, 1994

Chapter 27

Escherichia Coli in Diarrheal Disease

General Concepts

Clinical Manifestations

Depending on the virulence factors they possess, virulent Escherichia coli strains cause either noninflammatory diarrhea (watery diarrhea) or inflammatory diarrhea (dysentery with stools usually containing blood, mucus, and leukocytes).

Structure, Classification, and Antigenic Types

These are Gram-negative bacilli of the family Enterobacteriaceae. Virulent strains differ from nonvirulent E coli only in possessing genetic elements for virulence factors. Strains producing enterotoxins are enterotoxigenic E coli (ETEC).

Pathogenesis

Transmission is by the fecal-oral route. Pili (fimbriae) allow the bacteria to colonize the ileal mucosa. Cytotonic enterotoxins (encoded on plasmid or bacteriophage DNA) induce watery diarrhea. Plasmid-encoded invasion factors permit invasion of the mucosa, and plasmid- or bacteriophage-encoded cytotoxic enterotoxins induce tissue damage; the presence of either of these factors induces a host inflammatory reaction with an influx of lymphocytes and resulting dysentery.

Host Defenses

Gastric acid and intestinal transit time are important defenses. Specific intestinal immunoglobulin A (IgA) develops and appears to be protective.

Epidemiology

Infection is common where sanitation is poor; both infants and susceptible travelers to developing countries are particularly at risk. The disease is most serious in infants.

Diagnosis

The diagnosis is suggested by the clinical picture and confirmed by stool culture. Serotyping and tests for virulence factors are occasionally performed for outbreaks.

Control

Prevention depends on sanitary measures to prevent fecal-oral transmission; handwashing and proper preparation of food; chlorination of water supplies; and sewage treatment and disposal. Parenteral or oral fluid and electrolyte replacement is used to prevent dehydration. Broad-spectrum antibiotics are used in chronic or life-threatening cases.

INTRODUCTION

Escherichia coli is a common member of the normal flora of the large intestine. As long as these bacteria do not acquire genetic elements encoding for virulence factors, they remain benign commensals. Strains that acquire bacteriophage or plasmid DNA encoding enterotoxins or invasion factors become virulent and can cause either a plain, watery diarrhea or an inflammatory dysentery. These diseases are most familiar to Westerners as traveler's diarrhea, but they are also major health problems in endemic countries, particularly among infants. Three groups of E coli are associated with diarrheal diseases. Escherichia coli strains that produce enterotoxins are called enterotoxigenic E coli (ETEC). There arenumerous types of enterotoxin. Some of these toxins are cytotoxic, damaging the mucosal cells, whereas others are merely cytotonic, inducing only the secretion of water and electrolytes. A second group of E coli strains have invasion factors and cause tissue destruction and inflammation resembling the effects of Shigella (EIEC). A third group of serotypes, called enteropathogenic E coli (EPEC), are associated with outbreaks of diarrhea in newborn nurseries, but produce no recognizable toxins or invasion factors. Figure 25-1 presents a summary of the diseases caused by virulent E coli.

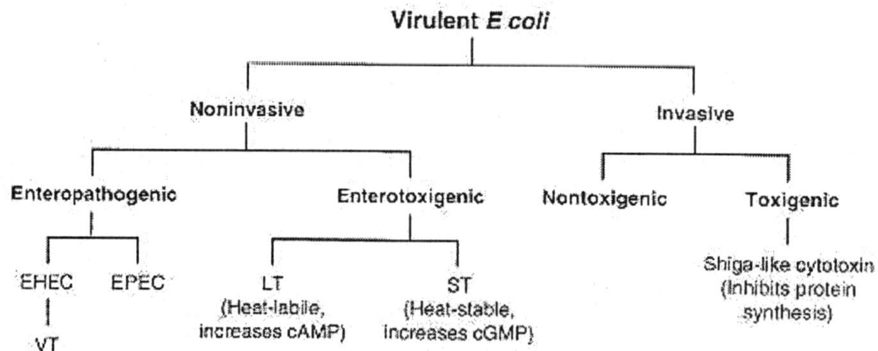

FIGURE 25-1 Virulence mechanisms of E coli.

Noninflammatory Diarrheas Caused by Enterotoxigenic *Escherichia Coli*

Clinical Manifestations

The diarrheal disease caused by ETEC is characterized by a rapid onset of watery, nonbloody diarrhea of considerable volume, accompanied by little or no fever (Fig. 25-2). Other common symptoms are abdominal pain, malaise, nausea, and vomiting. Diarrhea and other symptoms cease spontaneously after 24 to 72 hours.

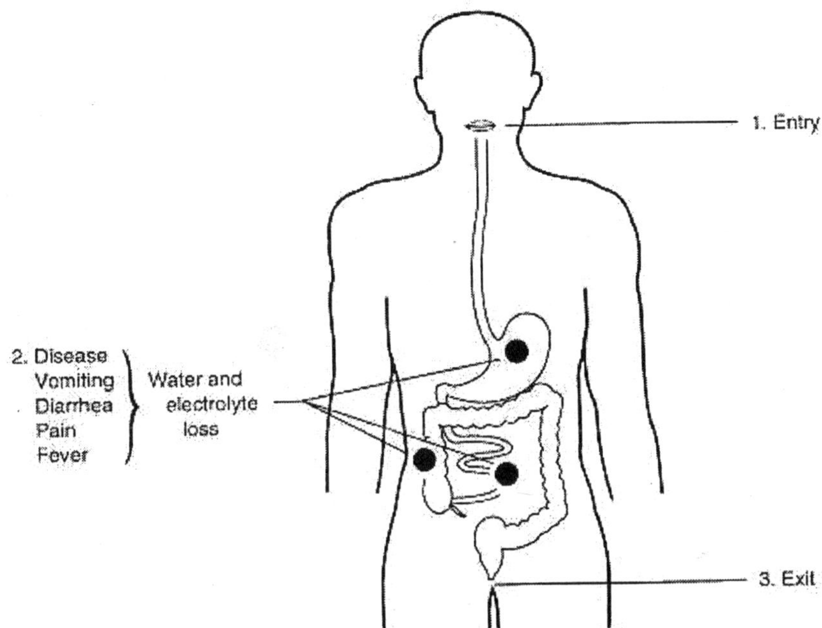

FIGURE 25-2 Pathogenesis of E coli diarrheal disease.

Structure, Classification, and Antigenic Types

ETEC organisms are Gram-negative, short rods not visibly different from E coli found in the normal flora of the human large intestine. Virulence-associated fimbriae are too small to be seen by light microscopy. All ETEC contain plasmids, but this is also not a distinguishing feature unless gene probe techniques are used to detect specific virulence-associated genes on these plasmids.

E coli organisms are serogrouped according to the presence or absence of specific heat-stable somatic antigens (O antigens) composed of polysaccharide chains linked to the core lipopolysaccharide (LPS) complex common to all Gram-negative bacteria. O specificity is determined by sugar or amino-sugar composition and by the sequence of these outer polysaccharide chains. More than 170 different O-specific antigens have been defined since Kauffmann began this method of typing E coli in 1943. In normal smooth strains, which are typable, the core LPS is buried beneath the O antigen. Also occurring are untypable O-minus mutants in which the core LPS is exposed; these are called rough strains. There is considerable cross-reactivity among E coli O antigens; also, many O groups of E coli are cross-reactive or identical with specific O groups of Shigella, Salmonella, or Klebsiella.

Escherichia coli serotypes are specific O-group/H-antigen combinations. The H antigens are the flagellar antigens, of which there are at least 56 types. Escherichia coli isolates may be nonmotile and nonflagellated and hence H negative (H). H typing is important for E coli associated with diarrheal disease for two reasons.

First, a strain causing an outbreak or epidemic can be differentiated from the normal stool flora by its unique O:H antigenic makeup. Second, most ETEC belong to specific serotypes (Table 25-1); this relationship facilitates their identification even in isolated cases. The reason for the close association between specific serotypes and the production of plasmid-determined virulence factors remains a mystery.

TABLE 25-1 Major Enterotoxigenic E Coli Serogroups and Serotypes Grouped According to Their Colonization Factor Antigens[a]

CFA/I	CFA/II	CFA/IV	Other
O15:H11, O15: H	O6:H16, O6:H	O25:H42	O159:H4
O25:H12, O63:H	O8:H9, O8:H	O29:H21	O9
O78:H11, O78:H12	O8O, O85	O115:H4O	
O78:, O128ac	O115:51	O148:H48	
O167:H21	O139		
O153:H12, O153:H45			

a No particular ETEC serotype has been associated with CFA/III or longus.

One plasmid-encoded but enterotoxin- and serotype-independent pilus (a long polar structure termed longus) has been reported recently in ETEC; like the CFAs, production of this pilus is restricted to E coli isolated from human sources.

Most E coli isolates also produce heat-labile, surface-associated proteins antigenically unrelated to O and H. These antigens can be seen in electron micrographs as filamentous structures called pili (fimbriae), which are much thinner and usually more rigid than flagella. Commensal E coli strains usually produce so-called common pili, which are defined as a specific set of an antigen. When E coli possessing common pili are mixed with erythrocytes (the standard test uses guinea pig erythrocytes), rapid hemagglutination occurs. This hemagglutination is blocked and also reversed by millimolar concentrations of the carbohydrate mannose.

ETEC possess specialized pili, antigenically unrelated to common pili, which act as ligands to bind the bacterial cells to specific complex carbohydrate receptors on the epithelial cell surfaces of the small intestine. Since this interaction results in colonization of the intestine by ETEC, with subsequent multiplication on the gut surface, these pili are termed colonization-factor antigens (CFAs). Most ETEC isolates produce either CFA/I, CFA/II or CFA/IV, whereas CFA/III and an undetermined number of other CFAs occur on other particular serotypes (Table 25-1). CFA-type pili play a major role in host specificity; for instance, different CFAs (e.g., K88, K99, and 987P) are produced by E coli that cause acute diarrhea in domestic animals.

A simple presumptive assay for CFAs on E coli is a test for mannose-resistant (non-common pili) hemagglutination reaction with either human or bovine erythrocytes. However, identification must be confirmed by reaction of the bacteria with antibody directed against a specific CFA or polymerase chain reaction (PCR) assay for specific CFA genes.

Genes coding for the production of CFAs reside on the ETEC virulence plasmids, usually on the same plasmids that carry the genes for one or both of the two types of E coli enterotoxin, heat-labile enterotoxin (LT) and heat-stable enterotoxin (ST). Most cases of ETEC diarrhea are caused by E coli possessing a CFA and both LT and ST; fewer are caused by those possessing a CFA and only one toxin (usually LT); and the fewest are caused by E coli that lack a CFA and possess only ST.

Pathogenesis

Escherichia coli diarrheal disease is contracted orally by ingestion of food or water contaminated with a pathogenic strain shed by an infected person. ETEC diarrhea occurs in all age groups, but mortality is most common in infants, particularly in the most undernourished or malnourished infants in developing nations.

The pathogenesis of ETEC diarrhea involves two steps: intestinal colonization, followed by elaboration of diarrheagenic enterotoxin(s) (Fig 25-3). ST is actually a family of toxic peptides ranging from 18 to 50 amino acid residues in length. Those termed STa can stimulate intestinal guanylate cyclase, the enzyme that converts guanosine 5'-triphosphate (GTP) to cyclic guanosine 5'-monophosphate (cGMP). Increased intracellular cGMP inhibits intestinal fluid uptake, resulting in net fluid secretion. Those termed STb do not seem to cause diarrhea by the same mechanism. One method for testing suspect E coli isolates for ST production involves injection of culture supernatant fluids into the stomach of infant mice and seeing whether diarrhea ensues. Specific DNA gene probes and PCR assays have been developed to test isolated colonies for the presence of genes encoding ST and LT.

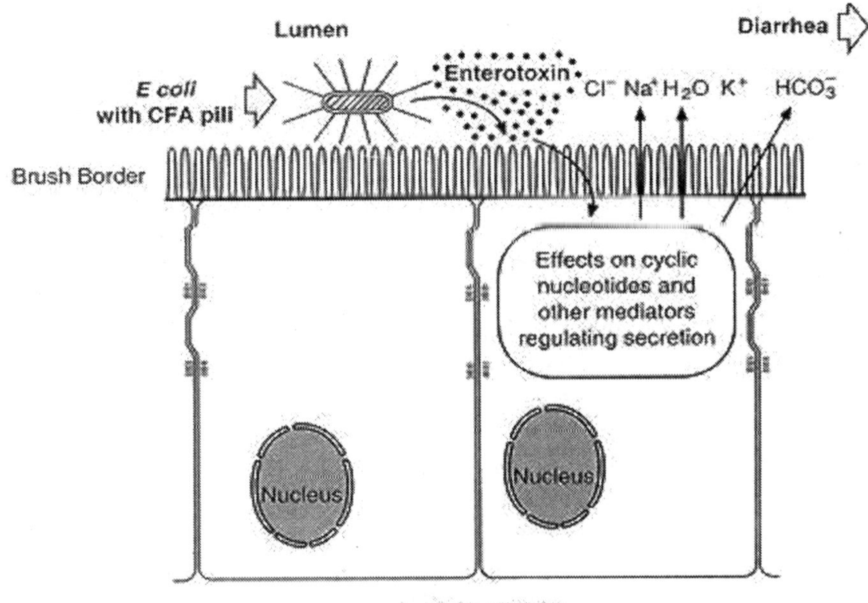

FIGURE 25-3 Cellular pathogenesis of E coli having CFA pili.

The E coli LTs are antigenic proteins whose mechanism of action is similar to that of Vibrio cholerae enterotoxin. LT shares antigenic determinants with cholera toxin, and its primary amino acid sequence is similar.

LT is composed of two types of subunits. One type of subunit (the B subunit) binds the toxin to the target cells via a specific receptor that has been identified as Gm1 ganglioside. The other type of subunit (the A subunit) is then activated by cleavage of a peptide bond and internalized. It then catalyzes the ADP-ribosylation (transfer of ADP-ribose from nicotinamide adenine dinucleotide [NAD]) of a regulatory subunit of membrane-bound adenylate cyclase, the enzyme that converts ATP to cAMP. This activates the adenylate cyclase, which produces excess intracellular cAMP, which leads to hypersecretion of water and electrolytes into the bowel lumen.

LT production is demonstrable by serologic methods, testing for diarrheagenic activity in ligated rabbit intestine, and by testing for specific cAMP-mediated morphological changes in cultured Y-1 adrenal tumor cells or Chinese hamster ovary (CHO) cells.

Host Defenses

As in any orally transmitted disease, the first line of defense against ETEC diarrhea is gastric acidity. Other nonspecific defenses are small-intestinal motility and a large population of normal flora in the large intestine.

Information about intestinal immunity against diarrheal disease is still somewhat superficial. However, intestinal secretory immunoglobulin (IgA) directed against surface antigens such as the CFAs and against LT appears to be the key to immunity from ETEC diarrhea. Passive immune protection of infants by colostral antibody is important. Human breast milk also contains nonimmunoglobulin factors (receptor-containing molecules) that can neutralize E coli toxins and CFAs.

Epidemiology

Escherichia coli diarrheal disease of all types is transmitted from person to person with no known important animal vectors. The incidence of E coli diarrhea is clearly related to hygiene, food processing sophistication, general sanitation, and the opportunity for contact. The geographic frequency of ETEC diarrhea is inversely proportional to the sanitation standards. Single-source outbreaks of ETEC diarrhea involving contaminated water supplies or food have been found in adults in the United States and Japan. Adults traveling from temperate climates to more tropical areas typically experience traveler's diarrhea caused by ETEC. This phenomenon is not readily explained, but contributing factors are low levels of immunity and an increased opportunity for infection.

Diagnosis

ETEC diarrhea is characterized by copious watery diarrhea with little or no fever. The diarrheal stool yields a virtually pure culture of E coli. Since the disease is self-limiting, virulence testing of isolates and serotyping is impractical except in an outbreak situation. Confirmation is achieved by serotyping, serologic identification of a specific CFA on isolates, demonstration of LT or ST production, and identification of genes encoding these virulence factors (Fig. 25-4).

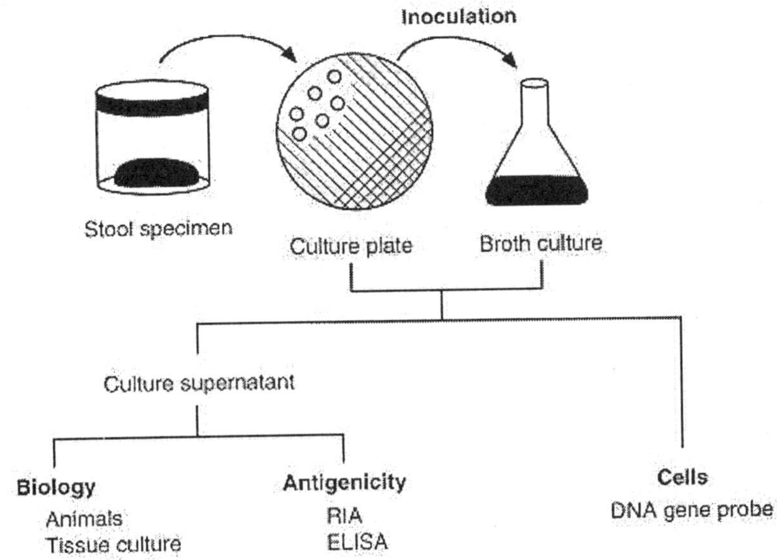

FIGURE 25-4 Laboratory methods for isolation and identification of ETEC.

Control

Escherichia coli diarrheal disease is best controlled by preventing transmission and by stressing the importance of breast-feeding of infants, especially where ETEC is endemic. The best treatment is oral fluid and electrolyte replacement (intravenous in severe cases). Antibiotics are not recommended because this practice leads to an increased burden of antibiotic-resistant pathogenic E coli and of more life-threatening enteropathogens.

Inflammatory Diarrheas Caused by Enteroinvasive, Cytotoxic, and Enteropathogenic Escherichia Coli

Clinical Manifestation

Diarrhea caused by the enteroinvasive, cytotoxic, and enteropathogenic (EPEC) strains of E coli ranges from very mild to severe. Illness is usually protracted and accompanied by fever. Infection with a few serogroups (O157, O26) is characterized by bloody diarrhea (hemorrhagic colitis). Infection with the Shigella-like serogroups presents as bacillary dysentery (i.e., abdominal pain and scanty stool containing blood and mucus).

Structure, Classification, and Antigenic Types

As in the case with ETEC, these strains of E coli are not detectably different in structure from E coli of the normal flora. The EPEC serogroups listed in Table 25-2 were the first E coli groups to be recognized as causative agents of diarrhea in infants. Their status as pathogens remained controversial for decades, mainly because the same O groups can be isolated from healthy contacts in outbreaks and from healthy adults. Recent work has proven that these E coli serogroups possess an antigenic adherence factor (termed bundle-forming pilus, or BFP); the gene for BFP is carried by a plasmid termed EAE (enteroaggregative Escherichia coli) plasmid. BFP is responsible for the initial attachment of EPEC to intestinal target cells.

TABLE 25-2 Common Serologic Types of Non-0ETEC E Coli Grouped According to Their Mechanism of Pathogenesis[a]

Classic EPEC Serogroups	Enteroivasive Shigella-Like Serogroups	Enterohecoorrhageic Serogroups
O16,O26,O44	I28acm,O29,O124	O26:H11,O111,O55o
O66, O111	O138,O143, O144	O157:H7
O112, O114, O119	O162,O154	
O125,O126,O1127		
O128B,O142		

aOccasional isolates fall to show the expected correlation between serogroup and disease type: this is due to the genetics of virulence factors (i.e. transmissibility of virulence-associated plasmids and bacteriophage between E coli strains in vivo as well as spontaneous loss of these factors in vitro)

bUsually EAF positive other EPEC serogroups usually EAF negative.

A small but important group of EPEC includes serotypes O157:H7, O26:H11, and some O111 isolates. These cause epidemic hemorrhagic colitis. Serotype O157:H7 is often associated with food-borne outbreaks.

Table 25-2 lists the Shigella-like enteroinvasive E coli serotypes (i.e., those with somatic antigens reactive with specific anti-Shigella typing serum). Also, like Shigella, these E coli strains are non-motile and therefore H negative. These serogroups usually do not harbor ETEC virulence plasmids and therefore are usually CFA negative.

Pathogenesis

Escherichia coli strains belonging to the classic EPEC serogroups (Table 25-2) bind intimately to the epithelial surface of the intestine, usually the colon, via the adhesive BFP. The lesion caused by EPEC consists mainly of destruction of microvilli. There is no evidence of tissue invasion. Cell damage occurs in two steps, collectively termed attaching and effacing; first is intimate contact, sometimes characterized as pedestal formation; second is loss of microvilli which is the result of rearrangement of the host cell cytoskeleton. Loss of microvilli leads to malabsorption and osmotic diarrhea. Diarrhea is persistent, often chronic, and accompanied by fever. EPEC are negative for ST and LT, but most strains produce relatively small amounts of a potent Shiga-like toxin that has both enterotoxin and cytotoxin activity.

The E coli strains associated with hemorrhagic colitis (enterohemorrhagic E coli, or EHEC) most notably O157:H7, produce relatively large amounts of the bacteriophage-mediated Shiga-like toxin. This toxin is called Vero toxin (VT), or Vero cytotoxin after its cytotoxic effect on cultured Vero cells. Many strains of O157:H7 also produce a second cytotoxin (Shiga-like toxin 2, or Vero toxin 2), which is similar in effect but antigenically different. The Shigella-like E coli strains are highly virulent; oral exposure to a very small number of these invasive bacteria causes severe illness. The site of the infection is the colon, where adherence is rapidly followed by invasion of the intestinal epithelial cells (Fig. 25-5). An acute inflammatory response and tissue destruction produce diarrhea with little fluid, much blood, and sheets of mucus containing polymorphonuclear cells. Invasive E coli, like Shigella, causes a rapid keratoconjunctivitis when placed on the conjunctiva of the guinea pig eye (Sereny test). Virulent Sereny test-positive isolates carry a large (usually 140-megadalton) plasmid responsible for this property.

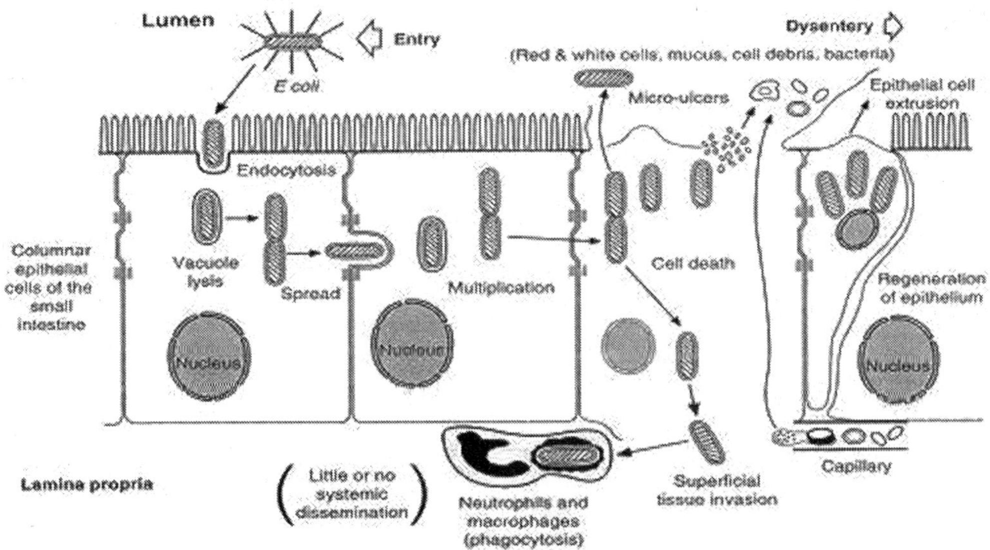

FIGURE 25-5 Cellular pathogenesis of invasive E coli

Host Defenses

Host defenses against EPEC are the same as those for ETEC. These defenses are frequently deficient or lacking in the infant and the elderly, which is consistent with the epidemiology of EPEC illness. An important example is the role of the immune system. Passive immune protection of infants by colostral antibody is important; breast-feeding is especially relevant where crowding and poor economic conditions prevail. Infection with these pathogens often excites an inflammatory cell response in the intestine, as is frequently reflected in the diarrheal symptoms.

Epidemiology

The geographic distribution of all EPEC is generally the same as for the ETEC, with a more severe disease in infants and young children, and so EPEC are much less important in traveler's diarrhea. Common-source community outbreaks are rare in geographic areas with satisfactory sanitation. However, sporadic cases are seen in the United States, Canada, and Europe, and outbreaks occur in these areas, but most commonly in close-contact institutions such as hospital nurseries, day-care centers, and nursing homes.

Diagnosis

Diagnosis is usually based on the symptomatology described above. Enterohemorrhagic E coli, such as serotype O157:H7, is suspected in the setting of copious bloody diarrhea without fecal leukocytes or fever, especially when symptoms include hemolytic uremic syndrome, or HUS. Escherichia coli serotyping is useful in chronic cases and in outbreaks, because identification of the agent and its antibiotic sensitivity pattern are valuable in these situations. Testing for specific EPEC virulence factors is usually impractical because it can be done only in reference and specialized research laboratories.

Control

Prevention and control are generally the same as for ETEC. Intervention of the fecal-oral transmission cycle is most effective in institutional situations. Broad-spectrum antibiotics are recommended in chronic and/or life-threatening cases.

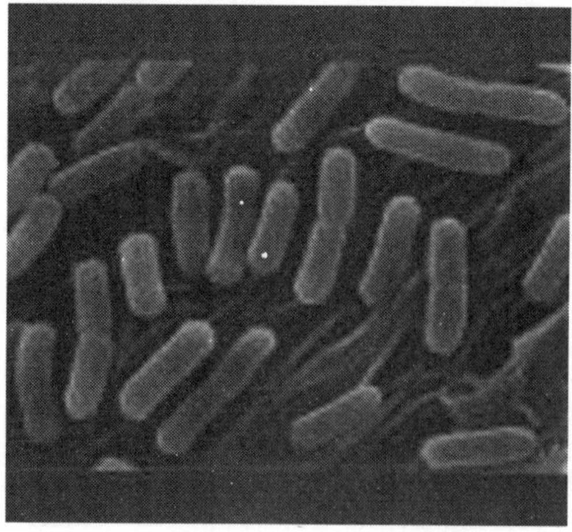

Microscopic appearance of *E.coli*

REFERENCES

Blanco J, Blanco M, Gonzalez EA et al: Serotypes and colonization factors of enterotoxigenic Escherichia coli isolated in various countries. Eur J Epidemiol 9:489, 1993

Chapman PA, Siddons CA, Wright DJ et al: Cattle as a possible source of verocytotoxin-producing Escherichia coli O157 infections in man. Epidemiol Infect 111:439, 1993

Donnenberg MS, Tacket CO, James SP et al: Role of the eaeA gene in experimental enteropathogenic Escherichia coli infection. J Clin Invest 92:141, 1993

Dytoc M, Sone R, Cockerill, F, III et al: Multiple determinants of verotoxin-producing Escherichia coli O157:H7 attachment-effacement. Infect Immun 61:3382, 1993

Evans DJ, Jr., Evans DG: Colonization factor antigens of human pathogens. Current Topics Microbiol Immunol 151:129, 1990

Giron JA, Ho ASY, Schoolnik GK: An inducible bundle-forming pilus of enteropathogenic Escherichia coli. Science 254:710, 1991

Giron JA, Levine MM, Kaper JB: Longus: a long pilus ultrastructure produced by human enterotoxigenic Escherichia coli. Molec Microbiol 12:71, 1994

Spangler BD: Structure and function of cholera toxin and the related Escherichia coli heat-labile enterotoxin. Microbiol Rev 56:622, 1992

Tesh VL, O'Brien AD. Adherence and colonization mechanisms of enteropathogenic and enterohemorrhagic Escherichia coli. Microb Pathogenesis 12:245, 1992

Wenneras C, Svennerholm AM, Ahren C, Czerkinsky C: Antibody-secreting cells in Human peripheral blood after oral immunization with an inactivated enterotoxigenic Escherichia coli vaccine. Infect Immun 60:2605, 1992

Chapter **28**

Escherichia, Klebsiella, Enterobacter, Serratia, Citrobacter, and Proteus

General Concepts

Clinical Manifestations

The genera Escherichia, Klebsiella, Enterobacter, Serratia, and Citrobacter (collectively called the coliform bacilli) and Proteus include overt and opportunistic pathogens responsible for a wide range of infections. Many species are members of the normal intestinal flora. *Escherichia coli* (E coli) is the most commonly isolated organism in the clinical laboratory.

Enteric Infections: E coli is a major enteric pathogen, particularly in developing countries. The principal groups of this organism responsible for enteric disease include the classical enteropathogenic serotypes (EPEC), enterotoxigenic (ETEC), enteroinvasive (EIEC), enterohemorrhagic (EHEC), and enteroggregative (EAEC) strains which are described in detail in Chapter 25.

Nosocomial Infections: Coliform and Proteus bacilli currently cause 29 percent of nosocomial (hospital-acquired) infections in the United States. In order of decreasing frequency, the major sites of nosocomial infection are the urinary tract, surgical sites, bloodstream, and pneumonias. This group of nosocomial pathogens are responsible for 46% of urinary tract and 24% of surgical site infections, 17% of the bacteremias, and 30% of the pneumonias. E coli is the premier nosocomial pathogen.

Community-Acquired Infections: E coli is the major cause of urinary tract infections, including prostatitis and pyelonephritis; Proteus, Klebsiella, and Enterobacter species are also common urinary tract pathogens. Proteus mirabilis is the most frequent cause of infection-related kidney stones. Klebsiella pneumoniae causes a severe pneumonia; K rhinoscleromatis causes rhinoscleroma; and K ozaenae is associated with ozena, an atrophic disease of the nasal mucosa.

Structure, Classification, and Antigenic Types

The coliforms and Proteus are Gram negative bacilli. All genera except Klebsiella are flagellated. Some strains produce capsules. Virulence often depends on the presence of attachment pili (which can be characterized by specific hemagglutinating reactions). Sex pili also may be present. The major classes of antigens used in defining strains are H (flagellar), O (somatic), and K (capsular).

Pathogenesis

E coli enteropathogens have diverse mechanisms for disease production which include different toxins and colonization factors (see Ch. 25). Specific serotypes of coliforms and Proteus with particular virulence factors often preferentially infect specific extraintestinal sites. E coli bacilli in extraintestinal infections have soluble and cell-bound hemolysins, siderophores, capsules, and adherence pili.

Host Defenses

Coliforms and Proteus species rarely cause extraintestinal disease unless host defenses are compromised. Disruption of the normal intestinal flora by antibiotic therapy may allow resistant nosocomial strains to colonize or overgrow. The skin and mucosae may be breached by disease, trauma, operation, venous catheterization, tracheal intubation, etc. Immunosuppressive therapy also increases the risk of infection.

Epidemiology

The epidemiology of coliform and Proteus infections involves many reservoirs and modes of transmission. The infecting organism may be endogenous or exogenous. Transmission may be direct or indirect; vehicles include hospital food and equipment, intravenous solutions, and the hands of hospital personnel. Nosocomial strains progressively colonize the intestine and pharynx with increasing length of hospital stay, resulting in an increased risk of infection.

Diagnosis

The clinical picture depends on the site of infection; diagnosis relies on culturing the organism and on biochemical and/or serologic identification. A variety of phenotypic (i.e., biotyping, serotyping, antibiograms, bacteriocin and phage typing) and genotypic (i.e., plasmid analysis, RFLP, ribotyping, and PCR) methods are used for epidemiological investigations.

Control

The most effective way to reduce transmission of nosocomial organisms is for all hospital personnel to wash hands meticulously after attending to each patient. Vaccines and hyperimmune sera are not currently available. Various antibiotics are the backbone of treatment; drug resistance (often multiple) due to conjugative plasmids is a major problem.

INTRODUCTION

The Gram-negative bacilli of the genera Escherichia, Klebsiella, Enterobacter, Serratia, Citrobacter, and Proteus (Table 26- 1) are members of the normal intestinal flora of humans and animals and may be isolated from a variety of environmental sources. With the exception of Proteus, they are sometimes collectively referred to as the coliform bacilli because of shared properties, particularly the ability of most species to ferment the sugar lactose.

Many of these microorganisms used to be dismissed as harmless commensals. Today, they are known to be responsible for major health problems worldwide. A limited number of species, including E coli, K pneumoniae, Enterōbacter aerogenes, Enterobacter cloacae, S marcescens, and P mirabilis, are responsible for most infections produced by this group of organisms. The increasing incidence of the coliforms, Proteus, and other Gram-negative organisms in diseases reflects in part a better understanding of their pathogenic potential but more importantly the changing ecology of bacterial disease. The widespread and often indiscriminate use of antibiotics has created drug-resistant Gram-negative bacilli that readily acquire multiple resistance through transmission of drug resistance plasmids (R factors). Also, development of new surgical procedures, health support technology, and therapeutic regimens has provided new portals of entry and compromised many host defenses.

Clinical Manifestations

As opportunistic pathogens, the coliforms and Proteus take advantage of weakened host defenses to colonize and elicit a variety of disease states (Fig. 26-1). Together, the many disease syndromes produced by these organisms are among the most common infections in humans requiring medical intervention.

TABLE 26-1 Taxnomy of Selected Coliform Bacilli and Proteus in Human Clinical Specimens

Organism	Other and older Designations
Esherichia (6 species in Bergeys marual)[a]	
E coli	
E coli inactive	E coli (locose negative, nonmotio, anaorogenic)
E forgusonll	
E hermarill	
E vulneris	
Klebsiella (7 species)	
K pneumniao[b]	Friedlander's bacillus
K oxyloca	Indole positive K pnoumoniae
K ozaenae	K pneumorrise subsp ozanae
K rhinoscleromatis	K pneurmoniae subsp (hinosferomatis)
K planticola	Omithine-positve K Oxytoca
Enterobacter (13 species)	
E aerogenes[b]	
E cloacae	
Eagglomerans	Pantoea agglomerans, Enrinia Herbicola
E sakazakil	E cloacae (yellow pigmentec)
E gergoviae	
Serraitia (11 species)	

S marcescens	
S rubdase	
S liquefaciens group	S liquefacierns, S grmesll, S proteamaculans
S ficeria	
S fonticota	
S odorifera	
S plymollrica	
Citrobacter (13 species)	
C lreundil	
C diversus	C koseri
C amalonaficus	C intermedius biotype
proteus (4 species)	
P mlrobitis[c]	
P vulgans	
P peruneri	

aHoll, J,G, of al, 1994, Bergey's manual of determinative bacteriology, 9th ed. The William Co, Baltimore.

bMajor coliforms in noscomial and/or commonly-acquired human diseases (prevalent in published reports).

Crganism previously disignated as profeus morganll and Proteus rellgerll are now classified n the genera Morganella and Providencia, respectively.

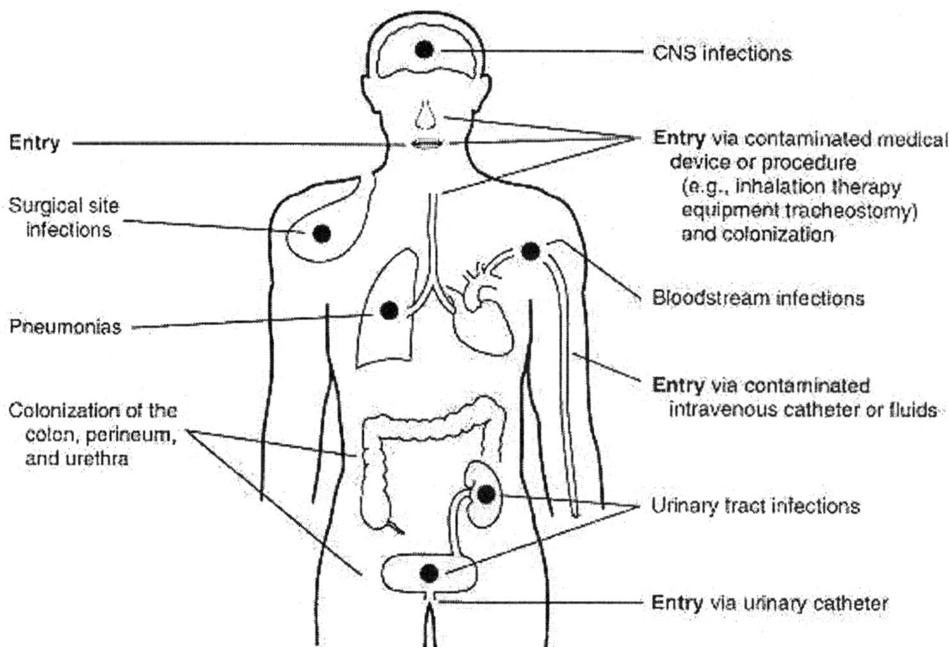

FIGURE 26-1 Sites of colonization and extraintestinal disease production by the coliforms and Proteus.

Enteric Infections

The role of E coli as a major enteric pathogen, particularly in developing countries, is discussed in detail in Ch. 25. However, the different types of E coli associated with enteric infections and which are classified into five groups according to their virulence properties are briefly described here: Enteropathogenic (EPEC) serotypes in the past were associated with serious outbreaks of diarrhea in newborn nurseries in the US. They remain an important cause of acute infantile diarrhea in developing countries. Disease is rare in adults. Enteroinvasive (EIEC) types produce disease resembling shigellosis in adults and children. Enterotoxigenic (ETEC) types are a major cause of traveler's diarrhea, and of infantile diarrhea in developing countries. Enterohemorrhagic E. coli (EHEC) occur largely as a single serotype (O157:H7) causing sporadic cases and outbreaks of hemorrhagic colitis characterized by bloody diarrhea. EHEC also may cause hemolytic uremic syndrome (HUS), an association of hemolytic anemia, thrombocytopenia, and acute renal failure. Enteroaggregative (EAEC) types exhibit a characteristic aggregative pattern of adherence and produce persistent gastroenteritis and diarrhea in infants and children in developing countries.

Nosocomial Infections

The etiology of nosocomial infections has markedly changed during past decades. Streptococci were the major nosocomial pathogens in the preantibiotic era. However, following the introduction and useof sulfonamides and penicillin, Staphylococcus aureus became the predominant pathogen in the 1950's. Aerobic gram negative rods gained prominence as nosocomial pathogens with widespread use of aminoglycosides and first generation cephalosporins through the early 1970's. Subsequent widespread use of broad spectrum cephalosporins was associated with changes in the frequency and etiology of nosocomial infections into the 1980's with the trend towards certain gram-positive pathogens. For example, in nosocomial bloodstream infections from 1980 to 1989 marked increases in the incidence of coagulase-negative staphylococci, S. aureus, enterococci, and Candida albicans infections occurred.

The coliforms and Proteus were responsible for 29 percent of nosocomial (hospital-acquired) infections in the United States from 1990 through 1992 based on data from hospitals participating in the National Nosocomial Infections Survey (NNIS) (Table 26-2). Estimates of nosocomial infections in US hospitals suggest that about 5 percent of the estimated 40 million annual admissions, or 2 million patients, had at least one nosocomial infection. Thus, the coliforms and Proteus probably are responsible for hospital-acquired infections in approximately 600,000 patients each year. Aside from the enormous cost measured in human life, nosocomial infections prolong the duration of hospitalization by an average of 4 days and increase the cost of medical care by $4.5 billion a year in 1992 dollars.

TABLE 26-2 Frequency of Selected Pathogens Causing Nosocomial infections[a]

	PERCENTAGE OF INFECTIONS AT SITE					
Organism	**Urimary Tract**	**Surgical Site**	**Primary Bloodstroom**	**Pneumonias**	**Other**	**All Sites**
E coli	25	8	5	4	4	12
Klebsiella	8	4	5	9	4	6
Enterobacter	5	7	4	11	4	6
Serratia marcescens	1	1	1	3	1	1
Cilrobacter	2	1	1	1	1	1
Proleus mirabitis	5 [46]	3 [24]	1 [17]	2 [30]	2 [16]	3 [29]
Pseudomonas aerugnoas	11	8	3	16	6	9
Stophyloccocus aureus	2	19	18	20	17	12
Coegulase-negatice Staph	4	14	31	2	14	11
enterococcus	16	12	9	2	5	10
Candida albicaris	6	3	6	5	5	5
Other pathogens	13	20	19	25	37	24
% of All Isolates	36.0	16.3	13.4	12.6	21.3	100

aData from NNIS system for 1990 through 1992. Emori. T.G, ad R.P Gaynes, 1993. An overview of nosccomila infections, including the role of the microbioligy laboratory. Clin. Microbiol. Rov.6:428-442.

The highest numbers of nosocomial infections in the NNIS occur in surgical and medicine services. Among surgical patients, highest rates of nosocomial infections occur with surgery on the stomach (21%) and bowel (19%), craniotomies (18%), coronary artery bypass graft procedures (11%) and other cardiac surgery (10%). High rates also are observed with burn (15%) and high-risk nursery patients (14%). In order of decreasing frequency, the major sites of nosocomial infection are the urinary tract, surgical sites, bloodstream, and lower respiratory tracts. The coliforms and Proteus were responsible for 46% of urinary tract and 24% of surgical site infections, 17% of the bacteremias, and 30% of the pneumonias from 1990 through 1992. Escherichia coli, the predominant nosocomial pathogen, is the major cause of infection in the urinary tract and is common in other body sites. Staphylococcus aureus and Pseudomonas aeruginosa are currently the most common pathogens in nosocomial pneumonias, followed by Enterobacter and Klebsiella. Coagulase-negative staphylococci have replaced E coli as the predominant pathogen in primary bloodstream infections. The major causes of surgical site infections are S aureus, coagulase- negative staphylococci, and enterococci.

Other coliform bacilli and Proteus have been incriminated in various hospital-acquired infections. Klebsiella, Enterobacter, and Serratia species are frequent causes of bacteremia at some medical centers and also are frequently involved in infections associated with respiratory tract manipulations, such as tracheostomy and procedures using contaminated inhalation therapy equipment. Klebsiella and Serratia species commonly cause infections following intravenous and urinary catheterization and infections complicating burns. Proteus species frequently cause nosocomial infections of the urinary tract, surgical wounds, and lower respiratory tract. Less frequently, Proteus species cause bacteremia, most often in elderly patients. A series of nationwide outbreaks of bacteremia (1970 to 1971 and 1973), caused by contaminated commercial fluids for intravenous injections, involved Enterobacter cloacae, Enterobacter agglomerans, and C freundii.

The role of Citrobacter species in human disease is not as great as that of the other coliforms and Proteus. Citrobacter freundii and C diversus (C koseri) have been isolated predominantly as superinfecting agents from urinary and respiratory tract infections. Citrobacter septicemia may occur in patients with multiple predisposing factors; Citrobacter species also cause meningitis, septicemia, and pulmonary infections in neonates and young children. Neonatal meningitis produced by C diversus, while uncommon, is associated with a very high frequency of brain abscesses, death, and mental retardation in survivors. Although, E coli and group B streptococci cause most cases of neonatal meningitis, the most common cause of brain abscesses in neonatal meningitis is P mirabilis.

Immunocompromised patients often develop non-hospital-acquired infections with coliforms. For example, group B streptococci and E coli are responsible for most cases of neonatal meningitis, with the latter accounting for about 40 percent of cases. Infections seen in cancer patients with solid tumors or malignant blood diseases frequently are caused by E coli, Klebsiella, Serratia, and Enterobacter species. Such infections often have lethal course. Individuals who are immunosuppressed by therapy (e.g., cancer patients or transplant recipients) or by congenital defects of the immune system may develop Klebsiella, Enterobacter, and Serratia infections. Many additional factors such as diabetes, trauma, and chronic lung disease may predispose to infection by coliforms and other microbes.

Community-Acquired Infections

The coliform organisms and Proteus species are major causes of diseases acquired outside the hospital; many of these diseases eventually require hospitalization. Escherichia coli causes approximately 85 percent of cases of urethrocystitis (infection of the urethra and bladder), about 80 percent of cases of chronic bacterial prostatitis, and up to 90 percent of cases of acute pyelonephritis (inflammation of the renal pelvis and parenchyma). Approximately one half of females have had a urinary tract infection by their late twenties due to E coli from their fecal flora. Proteus, Klebsiella, and Enterobacter species are among the other organisms most frequently involved in urinary tract infections. Proteus, particularly P mirabilis, is believed to be the most common cause of infection-related kidney stones, one of the most serious complications of unresolved or recurrent bacteriuria.

Klebsiella was first recognized clinically as an agent of pneumonia. Klebsiella pneumoniae accounts for a small percentage of pneumonia cases; however, extensive damage produced by the organism results in high case fatality rates (up to 90 percent in untreated patients). Klebsiella rhinoscleromatis is the agent of rhinoscleroma, a chronic destructive granulomatous disease of the respiratory tract that is endemic in Eastern Europe and Central America. Klebsiella ozaenae, a rare cause of serious infection, is classically associated only with ozena, an atrophy of nasal mucosal membranes with a mucopurulent discharge that tends to dry into crusts; however, recent studies indicate that the organism may cause various other diseases including infections of the urinary tract, soft tissue, middle ear, and blood.

Distinctive Properties

Structure and Antigens

The generalized structure and antigenic composition of coliform bacilli, as well as of Proteus and other members of the family Enterobacteriaceae, are depicted schematically in Figure 26-2. A more detailed figure of the structure is presented in Chapter 2. The major antigens of coliforms are referred to as H,K, and O antigens. The coliforms and Proteus are divided into serotypes on the basis of combinations of these antigens; different serotypes may have different virulence properties or may preferentially colonize and produce disease in particular body habitats. The H antigen determinants are flagellar proteins. Escherichia coli, Enterobacter, Serratia, Citrobacter, and Proteus organisms are peritrichous (i.e., they have flagella that grow from many places on the cell surface). Klebsiella species are nonmotile and nonflagellated and thus have no H antigens.

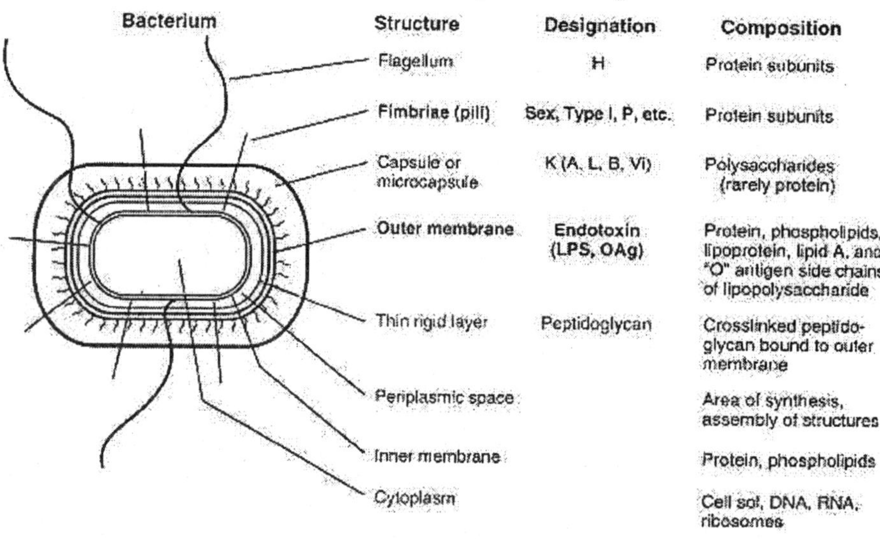

FIGURE 26-2 Structure and antigenic composition of coliforms and Proteus species.

Some strains of coliform and Proteus species have pili (fimbriae). Pili are associated with adhesive properties and, in some cases, are correlated with virulence. Different pilial colonization factors generally are detectable as hemagglutinins that can be distinguished by the type of erythrocyte agglutinated and by the susceptibility of the hemagglutination to inhibition by the sugar mannose. Sex pili, which have receptors for "male" specific bacterial viruses and are genetically determined by extrachromosomal plasmids, are important in coliform ecology and in the epidemiology of diseases produced by coliforms and Proteus species in that sex pili are involved in genetic transfer by conjugation (e.g., chromosome-mediated and plasmid-mediated drug resistances or virulence factors).

Major Surface Antigens

K antigens (capsule antigens) are components of the polysaccharide capsules. Certain K antigens (e.g., K88 and K99 of E coli) are pilus-like proteins. The K antigens often block agglutination by specific O antisera. In the past, K antigens routinely were differentiated into A, L, and B groups on the basis of differences in their lability to heat; however, these criteria are subject to difficulties that make the distinction tenuous. Some Citrobacter serotypes produce Vi (virulence) antigen, a K antigen also found in Salmonella typhi. Species of Proteus, Enterobacter, and Serratia apparently have no regular K antigens. However, the K antigens are important in the pathogenesis of some coliforms. A diffuse slime layer of variable thickness (the M antigen) also may be produced but, unlike the K antigens, it is nonspecific and is serologically cross-reactive among different organisms.

The outer membrane of the bacterial cell wall of these species contains receptors for bacterial viruses and bacteriocins (plasmid-encoded, antibioticlike bactericidal proteins called colicins in E coli that are active against the same or closely related species). The outer membrane also contains lipopolysaccharide (LPS), of which the lipid A portion is endotoxic and the O (somatic) antigen is serotype specific. The serologic specificity of the O antigens is based on differences in sugar components, their linkages, and the presence or absence of substituted acetyl groups. Loss of the O antigen by mutation results in a smooth-to-rough transformation, which often involves changes in colony type and saline agglutination, as well as loss of virulence. Certain strains of P vulgaris (OX-19, OX-2,and OX-K) produce O antigens that are shared by some rickettsiae. These Proteus strains are used in an agglutination test (the Weil-Felix test) for serum antibodies produced against rickettsiae of the typhus and spotted fever groups (see Ch. 38).

Toxins

Enterotoxigenic strains of Klebsiella, Enterobacter, Serratia,Citrobacter, and Proteus also have been isolated from infants and children with acute gastroenteritis. The enterotoxins of at least some of these organisms are of the heat-labile and heat-stable types and have other properties in common with the E coli toxins (see Ch.

25). However, the importance of the coliforms and Proteus, other than E coli, in enteric infections is questionable.

Pathogenesis

The process of disease production by coliforms is, in many cases, poorly understood. Production of disease by coliforms or Proteus species in extraintestinal sites often involves specific serotypes of the organisms and special virulence factors. For example, respiratory tract infections by K pneumoniae predominantly involve capsular types 1 and 2, whereas urinary tract infections often involve types 8, 9, 10, and 24. Similarly, only a few polysaccharide K antigens (types 1, 2, 3, 5, 12, and 13) of E coli are found with high frequency in urinary tract and other extraintestinal infections. These observations suggest that different serotypes may have specific pathogenicities. An alternative explanation is that such strains may simply be the most prevalent types in the normal gut flora.

There is good evidence for specific pathogenicity in E coli strains that cause extraintestinal infections (Table 26-3). Approximately 80 percent of E coli isolates involved in neonatal meningitis carry the K1 antigen, a fact attributable, at least in part, to the higher resistance to phagocytosis of K1-positive strains. Certain O antigens (O7 and O18) are found in combination with K1, usually in strains that are isolated from cases of neonatal bacteremia and meningitis and that show increased resistance to the bactericidal effects of serum complement. Interestingly, the E coli K1 antigen, composed of neuraminic acid, shows immune cross-reactivity with the group B meningococcal polysaccharide capsule.

TABLE 26-3 Virulence Factors of E coli Isolates from Extrantestinal Infections

Virulence Factor	proposed Role(s) in Pathogenesis
Col V plasmid	Codes for a siderophore (aerobaclin) for Fe chelation Increases bacterial resistance to serum
Hamolysin	Damages host cells Releases Fe from red blood cells
Enterochelin	Chalates Fe for bacterial uptake
K1 antigen	Impedes phagocytosis Blocks binding of C3b opsonin
P-pill	Allow bacteria to bind to P blood group antigens on urinary tract cells (especially in kidneys)
Type 1 pill	Allow bacteria to bind to (1) bladder epithelium (2) Tamm-Horstall glycoprotein and (3) D-mannose residues on a variety of cells.

Escherichia coli strains isolated from extraintestinal infections often possess a number of properties not usually found in random fecal isolates. These include production of soluble and cell-bound hemolysins, the colicin V plasmid, production of the siderophores aerobactin and enterochelin, and special pilial antigens for adherence to target cells. The hemolysin kills host cells and makes iron more available by releasing hemoglobin-bound iron from lysed red cells. To strip iron from the host iron-binding proteins (transferrin and lactoferrin), E coli produces siderophores of both the hydroxamate (aerobactin) and phenolate (enterochelin) types. Common or type 1 pili may mediate adherence to bladder cells; P-pili are virulence factors for strains causing pyelonephritis; S-pili, which recognize O-linked sialo-oligosaccharidesof glycophorin A, are associated with meningitis and urinary tract infections. Certain afimbrial adhesions and outer membrane proteins also have been associated with urinary tract infections.

The enzyme urease, produced by Proteus, and to a lesser extent by Klebsiella species, is thought to play a major role in the production of infection-induced urinary stones. Urease hydrolyzes urea to ammonia and carbon dioxide. Alkalinization of the urine by ammonia can cause magnesium phosphate and calcium phosphate to become supersaturated and crystallize out of solution to form, respectively, struvite and apatite stones. Bacteria within the stones may be refractory to antimicrobial therapy. Large stones may interfere with renal function. The ammonia produced by urease activity may also damage the pithelium of the urinary tract.

Except in cases of bacteremia and other systemic infection, there is little evidence that endotoxin plays a role in most coliform and Proteus diseases. Humans with coliform bacteremia show many of the typical effects of endotoxin, including fever, depletion of complement, release of inflammatory mediators, lactic acidosis, hypotension, vital organ hypoperfusion, irreversible shock, and death.

Host Defenses

It cannot be overemphasized that coliforms (except for E coli in enteric diseases) and Proteus species are unlikely to cause disease unless the local or generalized host defenses fail in some way. The normal gastrointestinal flora, which includes E coli and, frequently, other coliforms and Proteus species in small numbers, is important in preventing disease through bacterial competition. Prolonged antibiotic therapy compromises this defense mechanism by reducing susceptible components of the normal flora, permitting nosocomial coliform strains or other bacteria to colonize or overgrow.

The organisms may breach anatomic barriers through third-degree burns, ulcers associated with solid tumors of the skin and mucous membranes, intravenous catheters, and surgical or instrumental procedures on the biliary, gastrointestinal, and genitourinary tracts. The lungs may be violated by instrumentation, as in tracheal intubation, or even by aerosols from contaminated nebulizers or humidifiers, which carry organisms to the terminal alveoli.

Corticosteroid administration, radiotherapy, and the increased steroid levels associated with pregnancy tend to decrease host control over infections (e.g., by depressing the immune response). Cytotoxic drugs also are immunosuppressive. Cancer-or drug-induced neutropenia is an important predisposing factor in bacteremia. Devitalized tissue or foreign bodies may be a source of organisms and may also shelter the organisms from phagocytes and antimicrobial factors.

The interaction of multiple predisposing factors often determines the clinical course and outcome of coliform or Proteus infection. For example, the mortality of bacteremia increases progressively when the underlying disease (e.g., cancer or diabetes) is rated as nonfatal, ultimately fatal (death within 5 years), or rapidly fatal (death within 1 year). Similarly, coliform and Proteus infections commonly are more severe in the very old and very young.

Epidemiology

The epidemiology of coliform and Proteus infections is complex and involves multiple reservoirs and modes of transmission. Klebsiella, Enterobacter, Serratia, Citrobacter, and Proteus species live in water, soil, and occasionally food and, in many cases, form part of the intestinal flora of humans and animals. Escherichia coli is believed not to be free living, and its presence in environmental samples is taken as indicating recent fecal contamination. In fact, water quality is determined by the presence of the rapid lactose fermenting E coli, Klebsiella, and Enterobacter (coliform counts) and E coli (fecal coliform counts) using special selective media.

Coliform and Proteus organisms causing infection may be exogenous or endogenous. While most nosocomial infections appear to arise from endogenous flora, studies of hospitalized adults and infants have shown that the intestinal tract is progressively colonized by nosocomial coliforms with increasing length of hospitalization. Patients being treated with antibiotics, severely ill patients, and (probably) infants are more likely to be colonized, and other sites of colonization such as the nose and throat may be important in such patients. Colonized patients have a higher risk of nosocomial infection than patients who are not colonized.

The bacteria may be acquired indirectly via various vehicles or by direct contact. A variety of vehicles have been implicated in the spread of nosocomial pathogens. For example, Klebsiella, Enterobacter, and Serratia species have all been recovered in large numbers from hospital food, particularly salads, with the hospital kitchen being a primary source . An outbreak of urinary tract infections due to multiply drug-resistant S marcescens was associated with contaminated urine- measuring containers and urinometers. Serious outbreaks or individual cases of bacteremia due to coliforms have been associated with extrinsic contamination of intravenous fluids or caps during manufacture and with extrinsic contamination of intravenous fluids and administration sets in the hospital environment. Other medical devices and medications have served as vehicles for the spread of nosocomial pathogens. Occasionally, transmission may be via members of the hospital staff who are colonized with nosocomial pathogens in the rectum or vagina or on the hands; however, passive carriage on the hands of medical personnel constitutes the major mode of transmission.

Certain properties of the coliforms may be important in the epidemiology of hospital-acquired infections. Coliform bacteria other than E coli frequently are found in tap water or even distilled or deionized water. They may persist or actively multiply in water associated with respiratory therapy or hemodialysis equipment. Klebsiella, Enterobacter, and Serratia species, like Pseudomonas species, may exhibit increased resistance to antiseptics and disinfectants. The same group of coliforms has a selective ability over other common nosocomial pathogens (including E coli, Proteus species, Pseudomonas aeruginosa, and staphylococci) to proliferate rapidly at room temperature in commercial parenteral fluids containing glucose.

Diagnosis

Because the coliforms and Proteus can cause many types of infection, the clinical symptoms rarely permit a diagnosis. Culturing and laboratory identification are usually required. Selected characteristics that are useful in the differentiation of coliform bacilla and Proteus species foundin human clinical specimens are shown in Table 26-4. The organisms have simple nutritional requirements and grow well on mildly selective media commonly used for members of the Enterobacteriaceae, but not on some moderately and highly selective enteric plating media (Salmonella-Shigella, bismuth sulfite, and brilliant green agar). Extraintestinal specimens such as urine, purulent material from wounds or abscesses, sputum, and sediment from cerebrospinal fluid should be plated for isolation on blood agar and a differential medium such as MacConkey or eosin-methylene blue agar. The finding of more than 105 organisms/ml in clean voided midstream urine is often taken as "significant bacteriuria." However, in acutely symptomatic females and with other types of specimens (i.e., those obtained by catheterization or suprapubic aspiration) from either sex, a more appropriate threshold, particularly in the presence of pus cells and the absence of epithelial cells, might be more than 102 colonies of a known uropathogen/ml. Because urine is a good growth medium for many microbes, specimens should be refrigerated (4°C) if transport to the laboratory is delayed longer than 30 minutes, unless a urine transport container with preservative is used.

TABLE 26-4 Differentiation of Coliform Bacilli and Proteus found in Human Clinical Specimens

Organism	Motility	Lactose	Indole	Urease	H_2S	Other
Eschenchis						
E coli	+	+	+	-	-	
Klebsiella						
K pneumonlae	-	+	-	+	-	large mucold colonles
K oxytoca	-	+	+	+	-	large mucold colonles
Enterobacter						
E aerogenas	+	+	-	-	-	some strains mucold, LD+, AD-
E cloacae	+	+	-	d	-	LD-, AD+
E sakazkll	+	+	[-]	-	-	yellow pigment, LD-, AD+
E gegoviae	+	d	-	+	-	LD+, AD-

Pantoea						
P agglomorans	+	d	[-]	[-]	-	some strains yellow pigment, LD-, AD-
Setratia						
S marcascens[a]	+	-	-	-	-	some strains red pigment,
S rubldeae	+	+	-	-	-	
Citrobacter						
C fround	+	d	-	d	+	
C koseri	+	d	+	d	-	
Proteus						
P miratitis	+	-	-	+	+	*swarming motility
pvulgaris	+	-	+	+	+	*swarming motility

+(90% strains positive), d (26-75% strains positive), [-] (11.25% strains positive), - (0-10% strains positive) LD (lysline decarboxylase), AD (arginline dihydrolase)

* S liquifaciens group and S licaria have same live reactions shown.

S lonticola, S odorlera, and S plymuthica have same live reactions shown.

* C amelonaticus and other Cilrobacter have same live reactions shown.

Isolation of certain coliforms or Proteus species from fecal specimens may be facilitated by adding a moderately selective medium such as xylose-lysine-desoxycholate (XLD) or Hektoen enteric agar. Use of tetrathionate or selenite broth for enrichment of enterotoxigenic strains from feces is not recommended because both media inhibit various genera of coliforms. The strong (E coli, K pneumoniae, Enterobacter aerogenes) and occasionally the slow or weak (Serratia, Citrobacter) lactose-fermenting coliforms produce characteristic pigmented colonies on the enteric plating media. A striking characteristic of Proteus species is their propensity to swarm over the surface of most plating media, making the isolation of other organisms in mixed cultures difficult. The swarming growth appears as a rapidly spreading thin film, sometimes with changing patterns of whirls and bands. Sorbitol MacConkey agar is useful for screening EHEC (commonly E. coli O157:H7) on which sorbitol-negative colonies are nonpigmented and considered suspicious for the organism. Unless the physician specifically requests that the laboratory look for the possibility of E coli as an enteropathogen, tests for pathogenic strains, including toxin assays, serotyping, and serogrouping, will not be done.

In cases of suspected bacteremia, replicate bottles (one cultured aerobically, the other anaerobically) containing 25 to 100 ml of appropriate medium with anticoagulant (e.g., sodium polyanetholesulfonate) are inoculated with 10-ml portions of blood. It is usually necessary to take multiple specimens, both before and after antibiotic therapy is started. It is important to take specimens after antibiotic treatment is started so that therapeutic failure can be recognized while the bacteremia may still be amendable to more aggressive medical or surgical treatment.

All of the coliforms and Proteus species are Gram negative, facultative anaerobic, non-spore- forming rods that are typically motile, except for Klebsiella, which is nonmotile. The oxidase test is negative, and nitrates are reduced to nitrites. Proteus species and all coliforms ferment glucose, but fermentationof other carbohydrates varies. Lactose usually is fermented rapidly by Escherichia, Klebsiella and some Enterobacter species and more slowly by Citrobacter and some Serratia species. Proteus, unlike the coliforms, deaminates phenylalanine to phenylpyruvic acid, and it does not ferment lactose. Typically, Proteus is rapidlyurease positive. Some species of Klebsiella, Enterobacter, and Serratia produces a positive urease reaction, but they do so more slowly. A battery of tests for biochemical properties is required to identify the coliforms and Proteus to the species level. Commercial identification systems are now widely used by most US clinical laboratories and consist of "kits" or miniaturized biochemical tests which areread manually (e.g., API-20E and BBL Crystal) or automatically (e.g., Vitek or MicroSCAN).

The coliforms are characterized by great antigenic diversity caused by various combinations of specific H, K, and O antigens. For example, approximately 50 H, 90 K, and 160 O antigens have been identified among various strains of E coli. In contrast, Klebsiella, with no H antigens, has 10 O antigens and approximately 80 K antigens. Serologic identification of the coliforms and Proteus species, commonly by reference laboratories, is an extremely important epidemiologic tool. Similarly, other phenotying methods including biotyping (biochemical profiles), antibiograms (patterns of resistance to antimicrobal agents), and bacteriocin and phage typing have been widely used in epidemiologic studies, particularly of multiresistant isolates of coliforms and Proteus. Recently, genotyping methods such as plasmid profiles (determined by agarose gel electrophoresis), RFLP (restriction fragment link polymorphism) of total DNA, pulsed-field gel electrophoresis, targeted analysis of DNA polymorphism, ribotype, and arbitrarily primed PCR (polymerase chain reaction) have been used in epidemiological studies. In hospital-acquired infections, for example, the same or a small number of serologic or plasmid types suggests single sources of infection. The finding of multiple serotypes or plasmid profiles suggests multiple sources of infection or endogenous infections.

Control

Prevention of coliform and Proteus infections, particularly those that are hospital acquired, is difficult and perhaps impossible. Sewage treatment, water purification, proper hygiene, and other control methods for enteric pathogens will reduce the incidence of E coli enteropathogens. However, these control measures are rarely available in less developed regions of the world. Breast-feeding is an effective means of limiting outbreaks of enteropahogens in infants. Aggressive infection control committees in hospitals can do much to reduce nosocomial infections through identification and control of predisposing factors, education and training of hospital personnel, and limited microbial surveillance. Except for investigations of potential outbreaks, routine culturing of personnel, patients, and the environment is not warranted. Selective decontamination of the digestive tract with a suitable nonabsorbable antimicrobial regimen may be useful during outbreaks caused by nosocomial coliforms and Proteus. Meticulous hand washing after each patient contact a highly effective means of reducing the transmission of nosocomial pathogens (Fig. 26-3)is infrequently or poorly performed by some hospital personnel. In a study conducted in an intensive care unitfollowing an educational campaign on the importance of hand washing, the compliance was 17 percent for physicians, 100 percent for nurses, 82 percent for respiratory technicians, and 88 percent for diagnostic services personnel.

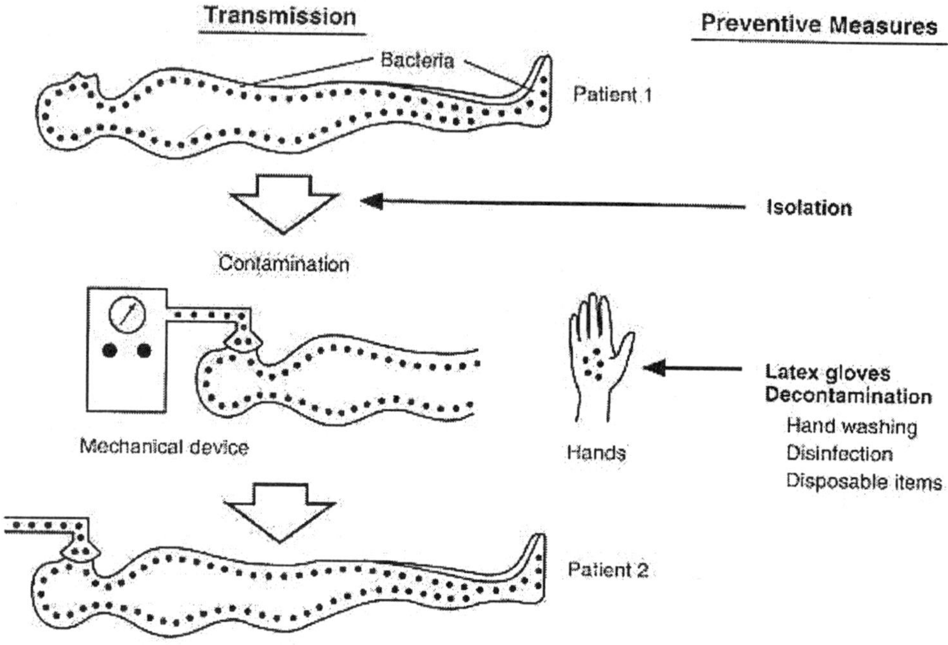

FIGURE 26-3 Major routes of transmission and prevention of spread of nosocomial pathogens.

Active or passive immunization against coliforms and Proteus species is not practiced. However, vaccines or hyperimmune sera for the six common Gram negative pathogens (E coli, Klebsiella, Enterobacter, Serratia, Pseudomonas aeruginosa, and Proteus) probably would have a major impact on morbidity and mortality from nosocomial infections. In a trial, the mortality was reduced markedly in a group of patients with Gram-negative bacteremia who had been given antiserum against a mutant E coli with an exposed lipopolysaccharide core.

Ampicillin, sulfonamides, cephalosporins, tetracycline, trimethoprim-sulfamethoxazole, nalidixic acid, ciprofloxacin, and nitrofurantoin have been useful in treating urinary tract infections by coliforms and Proteus species. Gentamicin, amikacin, tobramycin, ticarcillin/clavulate, imipenem, aztreonam, and a variety of third-generation cephalosporins may be effective for systemic infections; however, laboratory tests for drug susceptibility are essential. For example, resistence of E coli to ampicillin, and first generation cephalosporins is increasing rapidly to the extent that they can no longer be considered primary drugs of choice in empirical treatment of urinary tract infections. Likewise, emergence of coliforms with chromosomal or plasmid-encoded extended spectrum B-lactamase activity is causing global problems with resistance to third generation cephalosporins. Some coliforms have multiple resistance due to the presence of R plasmids transmissible by conjugation. Conjugative resistance plasmids allow the transfer of resistance genes among species and genera that normally do not exchange chromosomal DNA (Ch. 5). In some cases, resolution of the infection may require drainage of abscesses or other surgical intervention.

Measures commonly used to control epidemics of antibiotic resistant Gram-negative bacilli have included: (1) closing the unit to new admissions until control of the outbreak is underway; (2) reinforcing hand-washing practices; (3) gown and glove isolation, often combined with isolation of patients in separate quarters; and (4) restricting the use of the antibiotic to which the offending clone is resistant.

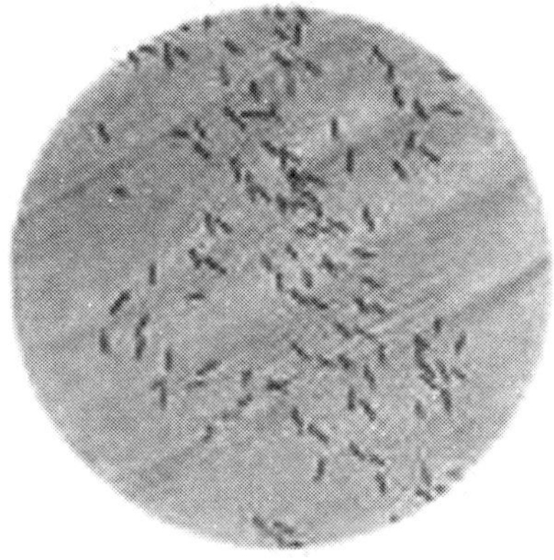

Cultural morphology of *Enterobacteriacae*

REFERENCES

Beck-Sague C, Villarino E, Giuliano D, et al: Infectious diseases and death among nursing home residents: results of surveillance in 13 nursing homes. Infect Control Hosp Epidemiol 15:494, 1994

Bergeron MG: Treatment of pyelonephritis in adults. Med Clin North Amer 79:619, 1995

Bingen, E: Applications of molecular methods to epidemiologic investigations of nosocomial infections in a pediatric hospital. Infect Control Hosp Epidemiol 15, 1994

Bodey, GP, Elting LS, Rodriguez S, Hernandez M: Klebsiella bacteremia: a 10-year review in a cancer institution. Cancer 64:2368, 1989

Brun-Buisson C, Legrand, P: Can topical and nonabsorbable antimicrobials prevent cross transmission of resistant strains in ICUs? Infect Control Hosp Epdemiol 15:447, 1994

Conley JM, Hill S, Ross J et al: Handwashing practices in an intensive care unit: the effects of an educational program and its relationship to infection rates. Am J Infect Control 17:330, 1989

Emori GT, Gaynes RP: An overview of nosocomial infections, including the role of the microbiology laboratory. Clin Microbiol Rev 6: 428, 1993

Foxman, B, Zhang L, Palin K, et al: Bacterial virulence characteristics of Escherichia coli isolates from first-time urinary tract infection. Infect Dis 171:1514, 1995

Gordon MC, Hankins GDV: Urinary tract infections and pregnancy. Comp Ther 15:52, 1989

Johnson JR, Stamm WE: Urinary tract infections in women: diagnosis and treatment. Ann Intern Med 111:906, 1989

Horan TC, Culver DH et al: Nosocomial infections in surgical patients in the United States, January 1986-June 1992. Infect Control Hosp Epidemiol 14:73, 1993

Lipsky BA: Urinary tract infections in men: epidemiology, pathophysiology, diagnosis, and treatment. Ann Intern Med 110:138, 1989

Marshall JC, Christou NV, Horn R, Meakins JL: The microbiology of multiple organ failure: the proximal gastrointestinal tract as an occult reservoir of pathogens. Arch Surg 123:309, 1988

Mobley HLT, Chippendale GR: Hemagglutinin, urease, and hemolysin production by Proteus mirabilis. J Infect Dis 161:525, 1990

Morris JG, Lin FYC, Morrison CB et al: Molecular epidemiology of neonatal meningitis due to Citrobacter diversus: a study of isolates from hospitals in Maryland, J Infect Dis 154:409, 1986

Nyström B: Impact of handwashing on mortality in intensive care: examination of the evidence. Infect Control Hosp Epidemiol 15:435, 1994

Saito H, Elting L, Bodey GP, Berky P: Serratia bacteremia: review of 118 cases. Rev Infect Dis 11:912,1989

Schaberg, DR, Culver, DH, Gaynes, RP: Major trends in the microbial etiology of nosocomial infection. Amer Med 91(suppl 3B): 72S, 1991

Stamm, WE: Catheter-associated urinary tract infections: epidemiology, pathogenesis, and prevention. Amer Med 91(suppl 3B):65S, 1991

Swartz MN: Hospital-acquired infections: diseases with increasingly limited therapies. Proc. Natl. Acad. Sci. USA 91:2420, 1994

Toltzis P, Blumer JL: Antibiotic-resistant gram-negative bacteria in the critical care setting. Pediatric Clin North Amer 42:687, 1995

van Saene HKF, Nunn AJ et al: Viewpoint: survival benefit by selective decontamination of the digestive tract (SDD). Infect Control Hosp Epidemiol 15:443, 1994.

Chapter **29**

Shigella

General Concepts

Clinical Manifestations

Symptoms of shigellosis include abdominal pain, tenesmus, watery diarrhea, and/ or dysentery (multiple scanty, bloody, mucoid stools). Other signs may include abdominal tenderness, fever, vomiting, dehydration, and convulsions.

Structure, Classification, and Antigenic Types

Shigellae are Gram-negative, nonmotile, facultatively anaerobic, non-spore-forming rods. Shigella are differentiated from the closely related Escherichia coli on the basis of pathogenicity, physiology (failure to ferment lactose or decarboxylate lysine) and serology. The genus is divided into four serogroups with multiple serotypes: A (S dysenteriae, 12 serotypes); B (S flexneri, 6 serotypes); C (S boydii, 18 serotypes); and D (S sonnei, 1 serotype).

Pathogenesis

Infection is initiated by ingestion of shigellae (usually via fecal-oral contamination). An early symptom, diarrhea (possibly elicited by enterotoxins and/or cytotoxin), may occur as the organisms pass through the small intestine. The hallmarks of shigellosis are bacterial invasion of the colonic epithelium and inflammatory colitis. These are interdependent processes amplified by local release of cytokines and by the infiltration of inflammatory elements.Colitis in the rectosigmoid mucosa, with concomitant malabsorption, results in the characteristic sign of bacillary dysentery: scanty,. unformed stools tinged with blood and mucus.

Host Defenses

Inflammation, copious mucus secretion, and regeneration of the damaged colonic epithelium limit the spread of colitis and promote spontaneous recovery. Serotype-specific immunity is induced by a primary infection, suggesting a protective role of antibody recognizing the lipopolysaccharide (LPS) somatic antigen. Other Shigella antigens include enterotoxins, cytotoxin, and plasmid-encoded proteins that induce bacterial invasion of the epithelium. The protective role of immune responses against these antigens is unclear.

Epidemiology

Shigellosis is endemic in developing countries were sanitation is poor. Typically 10 to 20 percent of enteric disease, and 50% of the bloody diarrhea or dysentery of young children, can be characterized as shigellosis, and the prevalence of these infections decreases significantly after five years of life. In developed countries, single-source, food or water-borne outbreaks occur sporadically, and pockets of endemic shigellosis can be found in institutions and in remote areas with substandard sanitary facilities.

Diagnosis

Shigellosis can be correctly diagnosed in most patients on the basis of fresh blood in the stool. Neutrophils in fecal smears is also a strongly suggestive sign. Nonetheless, watery, mucoid diarrhea may be the only symptom of many S sonnei infections, and any clinical diagnosis should be confirmed by cultivation of the etiologic agent from stools.

Control

Prevention of fecal-oral transmission is the most effective control strategy. Severe dysentery is treated with ampicillin, trimethoprim-sulfamethoxazole, or, in patients over 17 years old, a 4-fluorquinolone such as ciprofloxacin. Vaccines are not currently available, but some promising candidates are being developed.

INTRODUCTION

Gram-negative, facultative anaerobes of the genus Shigella are the principal agents of bacillary dysentery. This disease differs from profuse watery diarrhea, as is commonly seen in choleraic diarrhea or in enterotoxigenic Escherichia coli diarrhea, in that the dysenteric stool is scant and contains blood, mucus, and inflammatory cells. In some individuals suffering from shigellosis, however, moderate volume diarrhea is a prodrome or the sole manifestation of the infection. Bacillary dysentery constitutes a significant proportion of acute intestinal disease in the children of developing countries, and this infection is a major contributor to stunted growth of these children. Shigellosis also presents a significant risk to travelers from developed countries when visiting in endemic areas, and sporadic food or water-borne outbreaks occur in developed countries.

The pathogenic mechanism of shigellosis is complex, involving a possible enterotoxic and/or cytotoxic diarrheal prodrome, cytokine-mediated inflammation of the colon, and necrosis of the colonic epithelium. The underlying physiological insult that initiates this inflammatory cascade is the invasion of Shigella into the colonic epithelium and the lamina propria. The resulting colitis and ulceration of the mucosa result in bloody, mucoid stools, and/or febrile diarrhea.

Clinical Presentation

Shigellosis has two basic clinical presentations: (1) watery diarrhea associated with vomiting and mild to moderate dehydration, and (2) dysentery characterized by a small volume of bloody, mucoid stools, and abdominal pain (cramps and tenesmus) (Table 22-1). Volunteer challenge studies show that shigellosis can be evoked by an extremely small inoculum (10-100 organisms), and the time of onset of symptoms is somewhat influenced by the size of the challenge. The salient point is that shigellosis is an acute infection with onset of symptoms usually occurring within 24-48 hours of ingestion of the etiologic agent. The average duration of symptoms in untreated adults is 7 days, and the organism may be cultivated from stools for 30 days or longer.

TABLE 22-1 Clinical characteristics of shigellosis

Symptom				**Approximate Percentage of patients**[a]		
		S sonnel		S flexneri		S dysenteriae
Wartery diarrhea	75		30		30	
Stool mucus		50		75		95
Stool blood		10		50		80
Abdominal pain	50		70		85	
Vomiting		60		30		40
Fever		5		10		10

a Based on data from Dacca Hospital, Dacca, Bangladesh

The clinical features of shigellosis aresummarized in Figure 22-1. Watery diarrhea occurs as a prodrome, or as the sole clinical manifestation, in a majority of patients infected with S sonnei. Diarrhea is often a prodome of the dysentery characterizing infection with other species of Shigella. Recently discovered enterotoxins secreted by S flexneri may contribute to the diarrheal phase as the etiologic agents traverse the small intestine. However, diarrhea is most common in patients who have colitis involving the transverse colon or cecum. These patients evidence net water secretion and impaired absorption in the inflamed colon. In patients experiencing dysentery, involvement is most severe in the distal colon, and the resulting inflammatory colitis is evidenced in frequent scanty stools reflecting the ileocecal fluid flow. Dysentery is also characterized by the daily loss of 200-300 ml of serum protein into the feces. This loss of serum proteins results in depletion of nitrogen stores that exacerbates malnutrition and growth stunting. Depletion of immune factors also increases the risk of concurrent, unrelated infectious disease and contributes to substantial mortality.

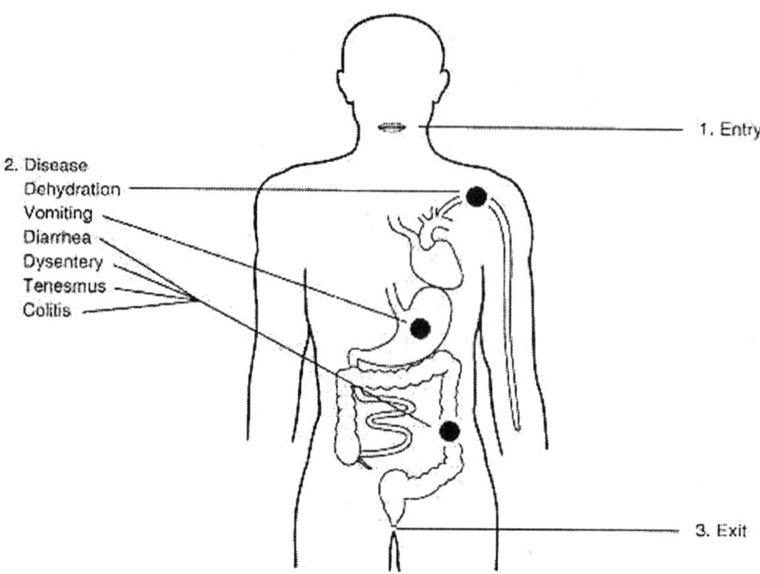

FIGURE 22-1 Pathogenesis of shigellosis in humans.

Possible complications of shigellosis include bacteremia, convulsions and other neurological complications, reactive arthritis, and hemolytic-uremic syndrome. Bacteremia occasionally accompanies S dysenteriae serotype 1 infections in malnourished infants, but this complication is uncommon in otherwise healthy individuals. Convulsions have been reported in up to 25% of Shigella infections involving children under the age of 4 years. Both high fever and a family history of seizures are risk factors for a convulsive episode. Ekiri syndrome, an extremely rare, fatal encephalopathy has also been described in Japanese children with S sonnei or S flexneri infections. Reactive arthritis, a self-limiting sequela of S flexneri infection, occurs in an incidence as high as 2% in individuals expressing the HLA-B27 histocompatibility antigen. Hemolytic-uremic syndrome, characterized by a triad of microangiopathic hemolytic anemia, thrombocytopenia, and acute renal failure, is a rare complication in children infected with S dysenteriae serotype 1.

Structure, Classification, and Antigenic Types

Organisms of the genus Shigella belong to the tribe Escherichia in the family Enterobacteriaceae. In DNA hybridization studies, Escherichia coli and Shigella species cannot be differentiated on the polynucleotide level; however, the virulence phenotype of the latter species is a distinctive distinguishing feature. Enteroinvasive E coli (EIEC), are very similar to shigellae biochemically and they also evoke diarrhea and/or dysentery. Some EIEC are also serologically related to shigellae. For example, EIEC serotype O124 agglutinates in S dysenteriae serotype 3 antiserum.

The genus Shigella is differentiated into four species: S dysenteriae (serogroup A, consisting of 12 serotypes); S flexneri (serogroup B, consisting of 6 serotypes); S. boydii (serogroup C, consisting of 18 serotypes); and S sonnei (serogroup D, consisting of a single serotype). Serogoups A, B, and C are very similar physiologically while S. sonnei can be differentiated from the other serogroups by positive b-D-galactosidase and ornithine decarboxylase biochemical reactions. The identification of shigellae by species in the clinical laboratory is usually accomplished by slide agglutination using commercially available, absorbed rabbit antisera.

Pathogenesis

Pathology

The rectosigmoidal lesions of shigellosis resemble those of ulcerative colitis. With frequencies indicated in Figure 22-2, there is proximal extension of erythema, edema, loss of vascular pattern, focal hemorrhage, and adherent layers of purulent exudate. Biopsy specimens from affected areas are typically edematous, with capillary congestion, focal hemorrhage, crypt hyperplasia, goblet cell depletion, mononuclear and polymorphonuclear (PMN) cell infiltration, shedding of epithelial cells and erythrocytes, and microulcerations.

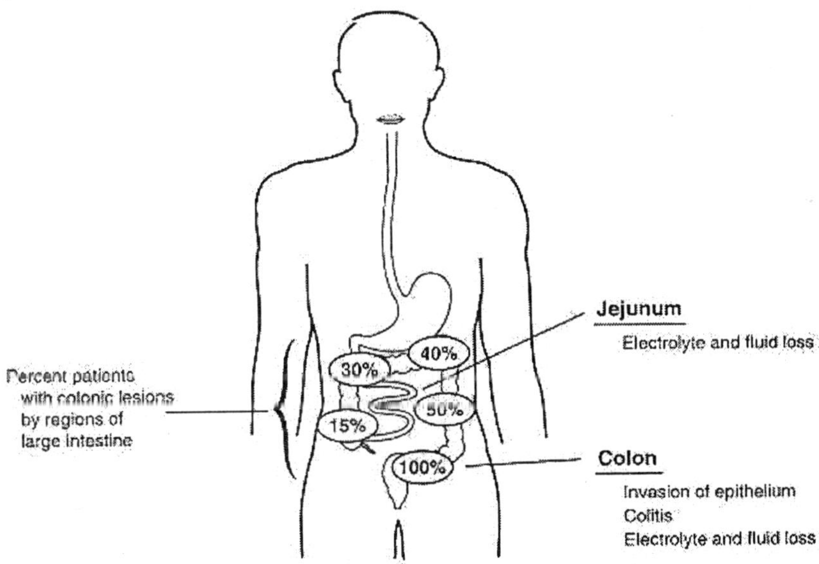

FIGURE 22-2 Gross pathology of shigellosis.

The pathogenic mechanism that underlies these pathological manifestations is diagrammed in Figure 22-3. This cartoon incorporates experimental observations from tissue cultures and from animal models of shigellosis such as rabbit ligated ileal loops injected with virulent organisms. In the latter model, Shigella infection is initiated at the membranous (M) cells that are associated with macroscopic lymphoid follicles (Peyer's patches). Biopsy studies in rhesus monkeys suggest that shigellae also infect microscopic lymphoid follicles of the primate colon. During the early stages of infection, bacteria are transcytosed through the M cells into the subepithelial space. In the subepithelial space, the organisms are phagocytosed by resident macrophages. However, virulent shigellae are not killed and digested in the macrophage phagolysome. The bacteria lyse the phagosome and initiate apoptosis (programmed cell death). During this process, the infected macrophage releases the inflammatory cytokine IL-1, which elicits infiltration of PMN.

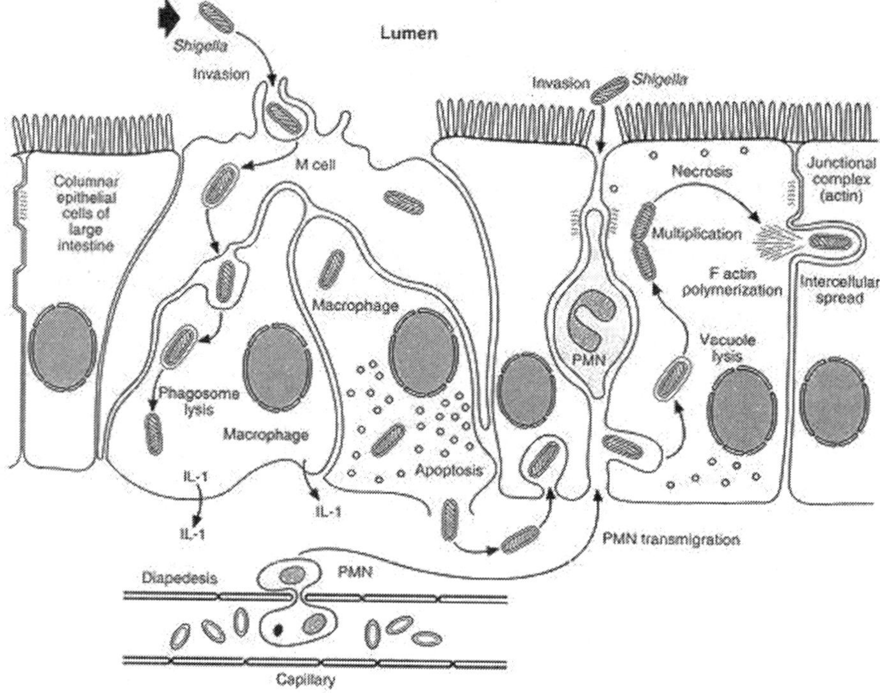

FIGURE 22-3 Histopathology of acute colitis following peroral infection with shigellae. The organisms are initially ingested by membranous (M) cells that are associated with lymphoid microfollicles in the colon. After transcytosis through the M cell, the bacteria are deposited into the subepithelial space where they are phagocytosed by macrophages. The macrophage phagosome is subsequently degraded, and the intracellular shigellae cause release of IL-1 that evokes an influx of polymorphonuclear leukocytes (PMN). Eventually the infected macrophages undergo apoptosis (programmed cell death), and the bacteria are released onto the basolateral surface of adjacent colonic enterocytes. In addition, PMN transmigration through the epithelium disrupts tight junctions, allowing shigellae to migrate into the subepithelial space. The bacteria infect enterocytes by induced endocytosis, and the endocytic vacuoles are subsequently degraded. The intercellular shigellae attach to actin in the enterocyte junctional complex, multiply, and spread to contiguous enterocytes by induced actin polymerization. Ultimately, the infected enterocytes die, and the resulting necrosis of the epithelium, in conjunction with the continuing inflammatory response, constitutes the lesions of shigellosis.

Transmigration of infiltrating PMNs through the tight junctions of local epithelial cells and into the intestinal lumen allows the reverse migration of shigellae from the lumen into the subepithelial spaces. These organisms then infect the columnar epithelial cells by inducing endocytic uptake at the basolateral surface. Immediately after infection of enterocytes, intracellular shigellae lyse endocytic vacuoles and attach to the actin cytoskeleton in the area of the junctional complex. As these organisms multiply within the enterocyte cytoplasm, occasional daughter cells induce polar nucleation of filamentous actin resulting in a "tail" that propels the shigellae into protrusions impinging on contiguous enterocytes. Plasma membranes enveloping the organisms are again lysed, and the organisms are deposited within the contiguous host cell resulting in intercellular bacterial spread.

In summary, shigellosis can be characterized as an acute inflammatory bowel disease initiated by the uptake of only a few organisms into lymphoid follicles. Intracellular replication and intercellular spread leads to an amplified inflammatory cascade at the initial site of entry, and as this inflammation persists and expands, the infiltration of PMN facilitates the entry of additional bacteria into the epithelium. The inflammatory infiltrate can also cause detachment of sheets of epithelial cells in areas devoid of lymphoid structures or bacterial cells.

Genetics of Virulence

Shigella are exquisitely adapted for reproduction within the colonic epithelium of the human host. Many of the bacterial virulence determinants that mediate the complex interactions between these bacteria and mammalian host cells have been identified by genetic and immunological means. These virulence determinants are encoded by large extra-chromosomal elements (plasmids) that are functionally identical in all Shigella species and in EIEC. A complex of two plasmid-encoded determinants, designated Invasion Plasmid Antigens (Ipa) B and C, is recognized by antibody in the sera of convalescent patients. Ipa proteins are maximally expressed in conditions approximating the intestinal lumen (e.g., bile salts, high osmolarity, and human body temperature), and release of the IpaBC complex is triggered by contact with the mammalian host cell. This complex induces the endocytic uptake of shigellae by M cells, epithelial cells, and macrophages. IpaB also mediates lysis of endocytic vacuoles in epithelial cells or macrophages. In the latter case, Ipa proteins also cause release of the IL-1 cytokine and macrophage apoptosis. Another plasmid-encoded virulence determinant is secreted at the poles of Shigella daughter cells as these organisms multiply within the cytoplasm of infected host cells. This InterCellular Spread (IcsA) protein elicits polymerization of filamentous actin. Formation of this actin tail provides a motive force for shigellae impinging on the plasma membrane of the infected cell. The resulting protrusions deform the plasma membrane of contiguous cells. The IcsB plasmid-encoded protein then lyses the plasma membranes, resulting in intercellular bacterial spread. Biochemical characterization of the interaction between these Shigella virulence determinants and host cell components is a remaining research challenge.

Characterizing and enhancing the neutralizing potential of antibody recognizing these protein virulence determinants is also an important research goal.

Toxins

Spent medium from S flexneri or EIEC cultures elicits fluid accumulation in rabbit ligated ileal loops and ion secretion in isolated ileal tissue. Using these assays, enterotoxins designated ShET1 and ShET2 have been identified, and the genetic loci encoding these toxins have been localized to the chromosome and plasmid, respectively. ShET1 is neutralized by convalescent sera from volunteers challenged with S flexneri 2a, suggesting that this toxic moiety is expressed by shigellae growing in the human intestine. The ShET1 locus is present on the chromosome of S flexneri 2a, but it is only occasionally found in other serotypes. In contrast, ShET2 is more widespread and detectable in 80% of shigellae representing all four species. These enterotoxins may elicit the diarrheal prodrome that often precedes bacillary dysentery; however, their role in the disease process remains to be defined by controlled challenge studies using toxin-negative mutants.

S dysenteriae serotype 1 expresses Shiga toxin, an extremely potent, ricin-like, cytotoxin that inhibits protein synthesis in susceptible mammalian cells. This toxin also has enterotoxic activity in rabbit ileal loops, but its role in human diarrhea is unclear, since shigellae apparently express a number of enterotoxins. Experimental infection of rhesus monkeys with S dysenteriae 1, and with a Shiga toxin-negative mutant, suggests that this cytotoxin causes capillary destruction and focal hemorrhage that exacerbates dysentery (see Table 22-1). More importantly, Shiga toxin is associated with the hemolytic-uremic syndrome, a complication of infections with S dysenteriae 1. Closely related toxins are expressed by enterohemorrhagic E coli (EHEC) including the potentially lethal, food-borne O157-H7 serotype.

Host Defense

Shigellae are remarkably infectious enteric pathogenes that can cause disease after the ingestion of as few as 10 organisms. Nonetheless, shigellosis is normally an acute, self-limiting disease that exemplifies the regenerative capacity of the intestinal epithelium. Shigella virulence probably reflects both the efficient uptake by the follicle associated epithelium (M cells) and the amplifying effect of the inflammatory cascade generated by apoptic macrophages. Tenesmus and evacuation of mucus by intestinal goblet cells may effectively eliminate both extracellular shigellae and infected enterocytes from the intestinal lumen, but this defensive response, in conjunction with PMN infiltration, also constitutes the definitive sign of bacillary dysentery.

In endemic areas, shigellosis is essentially a childhood disease, and the incidence decreases drastically in the indigenous population over 5 years of age. Controlled volunteer challenge studies in North American adults also indicate that prior infection with S flexneri protects against reinfection with the homologous serotype (70% efficacy). Serotype-specific immune protection against shigellosis suggests that antibody recognizing the O-polysaccharide of LPS protects against clinical symptoms. Ingested bovine colostrum containing antibody recognizing the O-polysaccharide of S flexneri 2a passively protects volunteers challenged with the homologous Shigella serotype. These observations have encouraged the development of a number of parenteral and mucosally administered O-polysaccharide vaccines that are currently in safety and/or efficacy trials. These vaccines offer the possibility of effective control of shigellosis independent of the needed improvements in the public health infrastructure of developing countries, but licensure and delivery of practical Shigella vaccines remains a distant prospect.

Epidemiology

Humans are the primary reservoir of Shigella species, with captive subhuman primates as accidental hosts. In developing countries with prevailing conditions of inadequate sanitation and overcrowded housing, the infection is transmitted most often by the excreta of infected individuals via direct fecal-oral contamination. Flies may contribute to spread from feces to food. The most common species, S dysenteriae and S flexneri, are also the most virulent. In developed countries, sporadic common-source outbreaks, predominantly involving S sonnei, are transmitted by uncooked food or contaminated water. The latter outbreaks usually involve semipublic water systems such as those found in camps, trailer parks, and Indian reservations. Direct fecal-oral spread can also occur in institutional environments such as child day-care centers. mental hospitals, and nursing homes. Homosexual men are also at increased risk for direct transmission of Shigella flexneri infections, and chronic, recrudescent illness complicating HIV infection has been reported.

Diagnosis

Clinical

Patients presenting with watery diarrhea and fever should be suspected of having shigellosis. The diarrheal stage of the infection cannot be distinguished clinically from other bacterial, viral, and protozoan infections. Nausea and vomiting can accompany Shigella diarrhea, but these symptoms are also observed during infections with nontyphoidal salmonellae and enterotoxigenic E coli. Bloody, mucoid stools are highly indicative of shigellosis, but the differential diagnosis should include EIEC, Salmonella enteritidis, Yersinia enterocolitica, Campylobacter species, and Entamoeba histolytica. Although, blood is common in the stools of patients with amebiasis, it is usually dark brown rather than bright red, as in Shigella infections. Microscopic examination of stool smears from patients with amebiasis should reveal erythrophagocytic trophozoites in the absence of PMN, whereas bacillary dysentery is characterized by sheets of PMN. Sigmoidoscopic examination of a shigellosis patient reveals a diffusely erythematous mucosal surface with small ulcers, whereas amebiasis is characterized by discrete ulcers in the absence of generalized inflammation.

Laboratory

Although, clinical signs may evoke the suspicion of shigellosis, diagnosis is dependent upon the isolation and identification of Shigella from the feces. Positive cultures are most often obtained from blood-tinged plugs of mucus in freshly passed stool specimens obtained during the acute phase of disease. Rectal swabs may also be used to culture shigellae if the specimen is processed rapidly or is deposited in a buffered glycerol saline holding solution. Isolation of shigellae in the clinical laboratory typically involves an initial streaking for isolation on differential/ selective media with aerobic incubation to inhibit the growth of the anaerobic normal flora. Commonly used primary isolation media include MacConkey, Hektoen Enteric Agar, and Salmonella-Shigella (SS) Agar. These media contain bile salts to inhibit the growth of other Gram-negative bacteria and pH indicators to differentiate lactose fermenters (Coliforms) from non-lactose fermenters such as shigellae. A liquid enrichment medium (Hajna Gram-negative broth) may also be inoculated with the stool specimen and subcultured onto the selective/differential agarose media after a short growth period. Following overnight incubation of primary isolation media at 37° C, colorless, non-lactose-fermenting colonies are streaked and stabbed into tubed slants of Kligler's Iron Agar or Triple Sugar Iron Agar. In these differential media, Shigella species produce an alkaline slant and an acid butt with no bubbles of gas in the agar. This reaction gives a presumptive identification, and slide agglutination tests with antisera for serogroup and serotype confirm the identification.

Some E coli biotypes of the normal intestinal flora closely resemble Shigella species (i.e. they are nonmotile, delayed lactose fermenters). These coliforms can usually be differentiated from shigellae by the ability to decarboxylate lysine. However, some coliforms cause enteroinvasive disease because they carry the Shigella-like virulence plasmid, and these pathogens are conventionally identified by laborious serological screening for EIEC serotypes. Sensitive and rapid methodology for identification of both EIEC and Shigella species utilizes DNA probes that hybridize with common virulence plasmid genes or DNA primers that amplify plasmid genes by polymerase chain reaction (PCR). Enzyme-linked immunosorbentassay (ELISA) using antiserum or monoclonal antibody recognizing Ipa proteins can also be used to screen stools for enteroinvasive pathogens. These experimental diagnostic techniques are useful for epidemiological studies of enteroinvasive infections, but they are probably too specialized for routine use in the clinical laboratory.

Treatment

Although, severe dehydration is uncommon in shigellosis, the first consideration in treating any diarrheal disease is correction of abnormalities that result from isotonic dehydration, metabolic acidosis, and significant potassium loss. The oral rehydration treatment developed by the World Health Organization has proven effective and safe in the treatment of acute diarrhea, provided that the patient is not vomiting or in shock from severe dehydration. In the latter case, intravenous fluid replacement is required until initial fluid and electrolyte losses are corrected. With proper hydration, shigellosis is generally a self-limiting disease, and the decision to prescribe antibiotics is predicated on the severity of disease, the age of the patient, and the likelihood of further transmission of the infection. Effective antibiotic treatment reduces the average duration of illness from approximately 5-7 days to approximately 3 days and also reduces the period of Shigella excretion after symptoms subside. Absorbable drugs such as ampicillin (2 g/day for 5 days) are likely to be effective when the isolate is sensitive. Trimethoprim (8 mg/kg/day) and sulfamethoxazole (40 mg/kg/day) will eradicate sensitive organisms quickly from the intestine, but resistance to this agent is increasing. Ciprofloxacin (1 g/day for 3 days) is effective against multiple drug resistant strains, but this antibiotic is not approved by the United States Food and Drug Administration for use in children less than 17 years of age because there is a theoretical risk of cartilage damage. Opiates, such as paregoric, induce intestinal stasis and may promote bacterial invasion, prolonging the febrile state.

Control

As is the case with other intestinal infections, the most effective methods for controlling shigellosis are provision of safe and abundant water and effective feces disposal. These public health measures are, at best, long range strategies for control of enteric infections in developing countries. The estimated five million deaths annually attributed to diarrheal disease in these countries, in addition to the malabsorption and growth stunting among survivors, require more immediate and practical approaches. The most effective intervention strategy to minimize morbidity and mortality would involve comprehensive media and personal outreach programs consisting of the following components: (1) education of all residents to actively avoid fecal contamination of food and water and to encourage hand washing after defecation; (2) encourage mothers to breast-feed infants; (3) promote the use of oral rehydration therapy to offset the effects of acute diarrhea; (4) encourage mothers to provide convalescent nutritional care in the form of extra food for children recovering from diarrhea or dysentery.

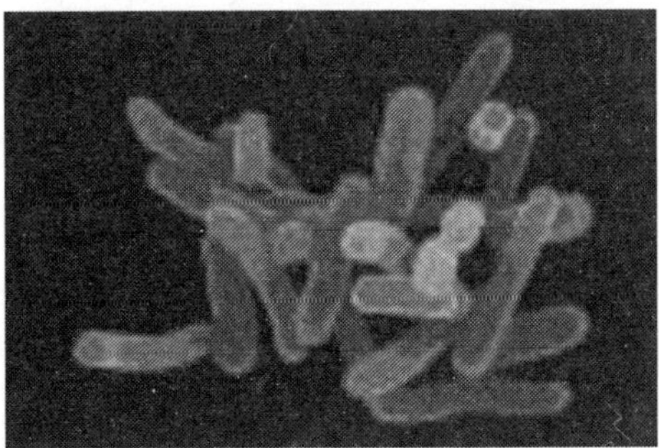

Colony Morphology of *Shigella*

REFERENCES

Bennish ML, Salam, MA, Khan WA et al: Treatment of shigellosis. 3. Comparison of one-dose or 2-dose ciprofloxacin with standard 5-day therapy - a randomized, blinded trial. Ann Int Med 117:727, 1992

Butler T, Speelman P, Kabir I et al: Colonic dysfunction during shigellosis. J Infect Dis 154:817, 1986

Davis H, Taylor JP, Perdue JN, et al: A shigellosis outbreak traced to commercially distributed shredded lettuce. Am J Epidemiol 128:1312, 1988

Fasano A, Noriega FR, Maneval DR, et al: Shigella enterotoxin 1: an enterotoxin of Shigella flexneri 2a active in rabbit small intestine in vivo and in vitro. J Clin Invest 95:000, 1995. (in press)

Goldberg MB, Bârzu O, Parsot C: Unipolar localization and ATPase activity of IncA, a Shigella flexneri protein involved in intracellular movement. J Bacteriol 175:2189, 1993

Hale TL: Genetic basis of virulence in Shigella species. Micro Rev 55:206, 1991

High N, Mounier J, Prévost MC, Sansonetti PJ: IpaB of Shigella flexneri causes entry into epithelial cells and escape from the phagocytic vacuole. EMBO J 11:1991, 1992

Keusch, GT, Bennish ML: Shigellosis. p. 593. In Evans A S, Brachman PS (ed): Bacterial Infections of Humans, Epidemiology and Control. Plenum, New York, 1991

Mathan M, Mathan VI: Morphology of rectal mucosa of patients with shigellosis. Rev Infect Dis 13 (Suppl 4): S314, 1991

Perdomo OJJ, Cavaillon JM, Huerre M et al: Acute inflammation causes epithelial invasion and mucosal destruction in experimental shigellosis. J Exp Med 180:1307, 1994

Perdomo OJJ, Gounon P, Sansonetti: Polymorphonuclear leukocyte transmigration promotes invasion of colonic epithelial monolayer by Shigella flexneri. J Clin Invest 93:633, 1994

Vasselon T, Mounier J, Hellio R, et al: Movement along actin filaments of the perijunctional area and de novo polymerization of cellular actin are required for Shigella flexneri colonization of epithelial Caco-2 cell monolayers. Infect Immun 60:1031, 1992

Wharton M, Spiegel RA, Horan JM et al: A large outbreak of antibiotic-resistant shigellosis at a mass gathering. J Infect Dis 162:1324, 1990

Zychlinsky A, Fitting C, Cavaillon J-M, et al: Interleukin 1 is released by murine macrophages during apoptosis induced by Shigella flexneri. J Clin Invest 94:1328, 1994

Chapter **30**

Salmonella

General Concepts

Clinical Manifestations

Salmonellosis ranges clinically from the common Salmonella gastroenteritis (diarrhea, abdominal cramps, and fever) to enteric fevers (including typhoid fever), which are life threatening febrile systemic illness requiring prompt antibiotic therapy. Focal infections and an asymptomatic carrier state occur. The most common form of salmonellosis is a self-limited, uncomplicated gastroenteritis.

Structure, Classification, and Antigenic Types

Salmonella species are Gram-negative, flagellated facultatively anaerobic bacilli characterized by O, H, and Vi antigens. There are over 1800 known serovars which current classification considers to be separate species.

Pathogenesis

Pathogenic salmonellae ingested in food survive passage through the gastric acid barrier and invade the mucosa of the small and large intestine and produce toxins. Invasion of epithelial cells stimulates the release of proinflammatory cytokines, which induce an inflammatory reaction. The acute inflammatory response causes diarrhea and may lead to ulceration and destruction of the mucosa. The bacteria can disseminate from the intestines to cause systemic disease.

Host Defenses

Both nonspecific and specific host defenses are active. Non-specific defenses consist of gastric acidity, intestinal mucus, intestinal motility (peristalsis), lactoferrin, and lysozyme. Specific defenses consist of mucosal and systemic antibodies and genetic resistance to invasion. Various factors affect susceptibility.

Epidemiology

Non-typhoidal salmonellosis is a worldwide disease of humans and animals. Animals are the main reservoir, and the disease is usually food borne, although it can be spread from person to person. The salmonellae that cause Typhoid fever and other enteric fevers spread mainly from person-to-person via the fecal-oral route and have no significant animal reservoirs. Asymptomatic human carriers ("typhoid Marys") may spread the disease.

Diagnosis

Salmonellosis should be considered in any acute diarrheal or febrile illness without obvious cause. The diagnosis is confirmed by isolating the organisms from clinical specimens (stool or blood).

Control

Effective vaccines exist for typhoid fever but not for non-typhoidal salmonellosis. Those diseases are controlled by hygienic slaughtering practices and thorough cooking and refrigeration of food.

INTRODUCTION

Salmonellae are ubiquitous human and animal pathogens, and salmonellosis, a disease that affects an estimated 2 million Americans each year, is common throughout the world. Salmonellosis in humans usually takes the form of a self-limiting food poisoning (gastroenteritis), but occasionally manifests as a serious systemic infection (enteric fever) which requires prompt antibiotic treatment. In addition, salmonellosis causes substantial losses of livestock.

Clinical Manifestations

Some infectious disease texts recognize three clinical forms of salmonellosis: (1) gastroenteritis, (2) septicemia, and (3) enteric fevers. This chapter focuses on the two extremes of the clinical spectrumgastroenteritis and enteric fever. The septicemic form of salmonella infection can be an intermediate stage of infection in which the patient is not experiencing intestinal symptoms and the bacteria cannot be isolated from fecal specimens. The severity of the infection and whether it remains localized in the intestine or disseminates to the bloodstream may depend on the resistance of the patient and the virulence of the Salmonella isolate.

The incubation period for Salmonella gastroenteritis (food poisoning) depends on the dose of bacteria. Symptoms usually begin 6 to 48 hours after ingestion of contaminated food or water and usually take the form of nausea, vomiting, diarrhea, and abdominal pain. Myalgia and headache are common; however, the cardinal manifestation is diarrhea. Fever (38°C to 39°C) and chills are also common. At least two-thirds of patients complain of abdominal cramps. The duration of fever and diarrhea varies, but is usually 2 to 7 days.

Enteric fevers are severe systemic forms of salmonellosis. The best studied enteric fever is typhoid fever, the form caused by S typhi, but any species of Salmonella may cause this type of disease. The symptoms begin after an incubation period of 10 to 14 days. Enteric fevers may be preceded by gastroenteritis, which usually resolves before the onset of systemic disease. The symptoms of enteric fevers are nonspecific and include fever, anorexia, headache, myalgias, and constipation. Enteric fevers are severe infections and may be fatal if antibiotics are not promptly administered.

Structure, Classification, and Antigenic Types

Salmonellae are Gram-negative, flagellated, facultatively anaerobic bacilli possessing three major antigens: H or flagellar antigen; O or somatic antigen; and Vi antigen (possessed by only a few serovars). H antigen may occur in either or both of two forms, called phase 1 and phase 2. The organisms tend to change from one phase to the other. O antigens occur on the surface of the outer membrane and are determined by specific sugar sequences on the cell surface. Vi antigen is a superficial antigen overlying the O antigen; it is present in a few serovars, the most important being S typhi.

Antigenic analysis of salmonellae by using specific antisera offers clinical and epidemiological advantages. Determination of antigenic structure permits one to identify the organisms clinically and assign them to one of nine serogroups (A-I), each containing many serovars (Table 1). H antigen also provides a useful epidemiologic tool with which to determine the source of infection and its mode of spread.

TABLE 21-1 Ecologic Classification of Salmonellae

Species	Represontative Serovar(s)[a]	Reservoir (Host preferences)
S choferaesuls	One only	Animals (swine)
S typh	One only	Humans
S enteritidis	Paratyphl-A	Humans
	Schollmuellenri	
	Puflorum	Animals (fowl)
	Dublin	Animals (cattle)
	Typhimurium	
	Derby	
	Enteritidis	Humans and many animals
	Heidelberg and hundreds of related serovars	

a It is now accepted practice to refer to the 1,800 serovars of samonella as though they constituted separate species (e.g. S pullorum).

(Adapted from Grady FG, Keusch GT, N Engl J Med 285:831, 1972, with permission.)

As with other Gram-negative bacilli, the cell envelope of salmonellae contains a complex lipopolysaccharide (LPS) structure that is liberated on lysis of the cell and, to some extent, during culture. The lipopolysaccharide moiety may function as an endotoxin, and may be important in determining virulence of the organisms. This macromolecular endotoxin complex consists of three components, an outer O-polysaccharide coat, a middle portion (the R core), and an inner lipid A coat. Lipopolysaccharide structure is important for several reasons. First, the nature of the repeating sugar units in the outer O-polysaccharide chains is responsible for O antigen specificity; it may also help determine the virulence of the organism.

Salmonellae lacking the complete sequence of O-sugar repeat units are called rough because of the rough appearance of the colonies; they are usually avirulent or less virulent than the smooth strains which possess a full complement of O-sugar repeat units. Second, antibodies directed against the R core (common enterobacterial antigen) may protect against infection by a wide variety of Gram-negative bacteria sharing a common core structure or may moderate their lethal effects. Third, the endotoxin component of the cell wall may play an important role in the pathogenesis of many clinical manifestations of Gram-negative infections. Endotoxins evoke fever, activate the serum complement, kinin, and clotting systems, depress myocardial function, and alter lymphocyte function. Circulating endotoxin may be responsible in part for many of the manifestations of septic shock that can occur in systemic infections.

Pathogenesis

Salmonellosis includes several syndromes (gastroenteritis, enteric fevers, septicemia, focal infections, and an asymptomatic carrier state) (Fig. 1). Particular serovars show a strong propensity to produce a particular syndrome (*S typhi, S paratyphi-A, and S schottmuelleri* produce enteric fever; *S choleraesuis* produces septicemia or focal infections; *S typhimurium* and *S enteritidis* produce gastroenteritis); however, on occasion, any serotype can produce any of the syndromes. In general, more serious infections occur in infants, in adults over the age of 50, and in subjects with debilitating illnesses.

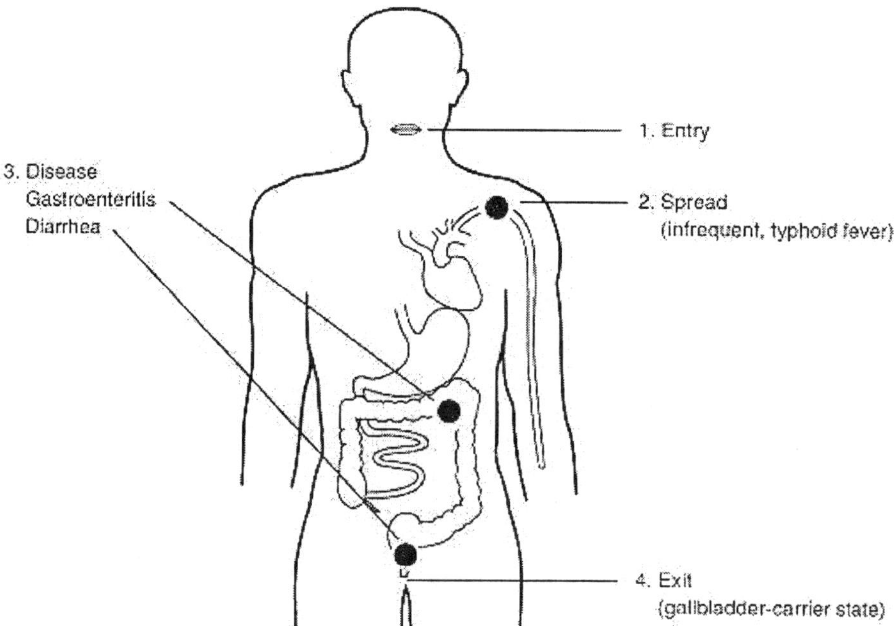

Figure 21-1 Pathogenesis of Salmonellosis.

Most non-typhoidal salmonellae enter the body when contaminated food is ingested (Fig. 2). Person-to-person spread of salmonellae also occurs. To be fully pathogenic, salmonellae must possess a variety of attributes called virulence factors. These include (1) the ability to invade cells, (2) a complete lipopolysaccharide coat, (3) the ability to replicate intracellularly, and (4) possibly the elaboration of toxin(s). After ingestion, the organisms colonize the ileum and colon, invade the intestinal epithelium, and proliferate within the epithelium and lymphoid follicles. The mechanism by which salmonellae invade the epithelium is partially understood and involves an initial binding to specific receptors on the epithelial cell surface followed by invasion. Invasion occurs by the organism inducing the enterocyte membrane to undergo "ruffling" and thereby to stimulate pinocytosis of the organisms (Fig. 3). Invasion is dependent on rearrangement of the cell cytoskeleton and probably involves increases in cellular inositol phosphate and calcium. Attachment and invasion are under distinct genetic control and involve multiple genes in both chromosomes and plasmids.

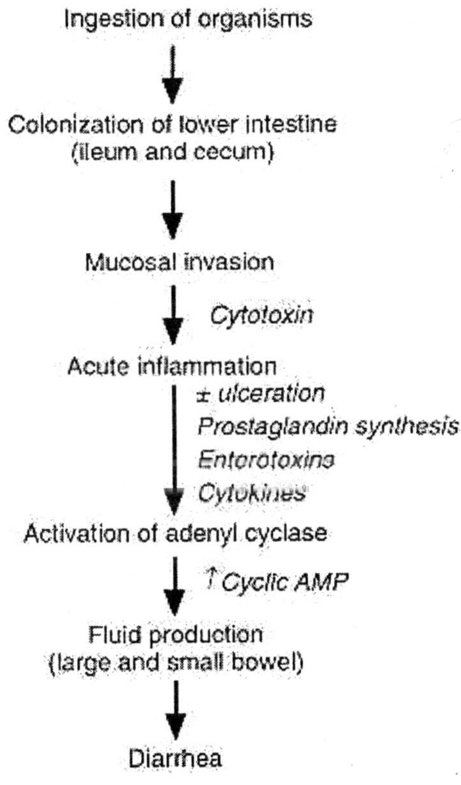

Figure 21-2 Scheme of the Pathogenesis of *Salmonella enterocolitis* and diarrhea.

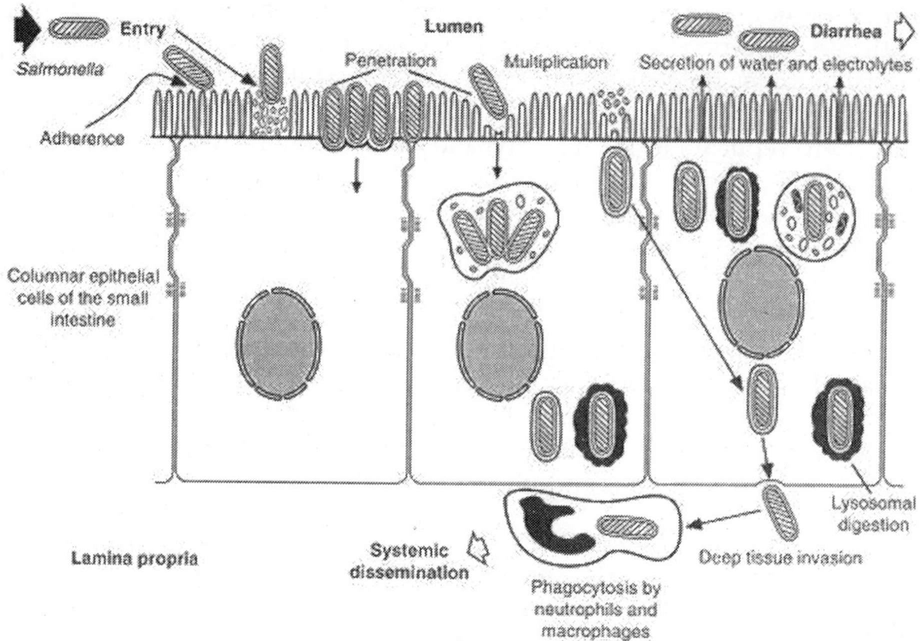

Figure 21-3 Invasion of intestinal mucosa by Salmonella.

After invading the epithelium, the organisms multiply intracellularly and then spread to mesenteric lymph nodes and throughout the body via the systemic circulation; they are taken up by the reticuloendothelial cells. The reticuloendothelial system confines and controls spread of the organism. However, depending on the serotype and the effectiveness of the host defenses against that serotype, some organisms may infect the liver, spleen, gallbladder, bones, meninges, and other organs (Fig. 1). Fortunately, most serovars are killed promptly in extraintestinal sites, and the most common human Salmonella infection, gastroenteritis, remains confined to the intestine.

After invading the intestine, most salmonellae induce an acute inflammatory response, which can cause ulceration. They may elaborate cytotoxins that inhibit protein synthesis. Whether these cytotoxins contribute to the inflammatory response or to ulceration is not known. However, invasion of the mucosa causes the epithelial cells to synthesize and release various proinflammatory cytokines, including: IL-1, IL-6, IL-8, TNF-2, IFN-U, MCP-1, and GM-CSF. These evoke an acute inflammatory response and may also be responsible for damage to the intestine. Because of the intestinal inflammatory reaction, symptoms of inflammation such as fever, chills, abdominal pain, leukocytosis, and diarrhea are common. The stools may contain polymorphonuclear leukocytes, blood, and mucus.

Much is now known about the mechanisms of Salmonella gastroenteritis and diarrhea. Figures 2 and 3 summarize the pathogenesis of Salmonella enterocolitis and diarrhea. Only strains that penetrate the intestinal mucosa are associated with the appearance of an acute inflammatory reaction and diarrhea (Fig. 4); the diarrhea is due to secretion of fluid and electrolytes by the small and large intestines. The mechanisms of secretion are unclear, but the secretion is not merely a manifestation of tissue destruction and ulceration. Salmonella penetrate the intestinal epithelial cells but, unlike Shigella and invasive E. coli, do not escape the phagosome. Thus, the extent of intercellular spread and ulceration of the epithelium is minimal. Salmonella escape from the basal side of epithelial cells into the lamina propria. Systemic spread of the organisms can occur, giving rise to enteric fever. Invasion of the intestinal mucosa is followed by activation of mucosal adenylate cyclase; the resultant increase in cyclic AMP induces secretion. The mechanism by which adenylate cyclase is stimulated is not understood; it may involve local production of prostaglandins or other components of the inflammatory reaction. In addition, Salmonella strains elaborate one or more enterotoxin-like substances which may stimulate intestinal secretion. However, the precise role of these toxins in the pathogenesis of Salmonella enterocolitis and diarrhea has not been established.

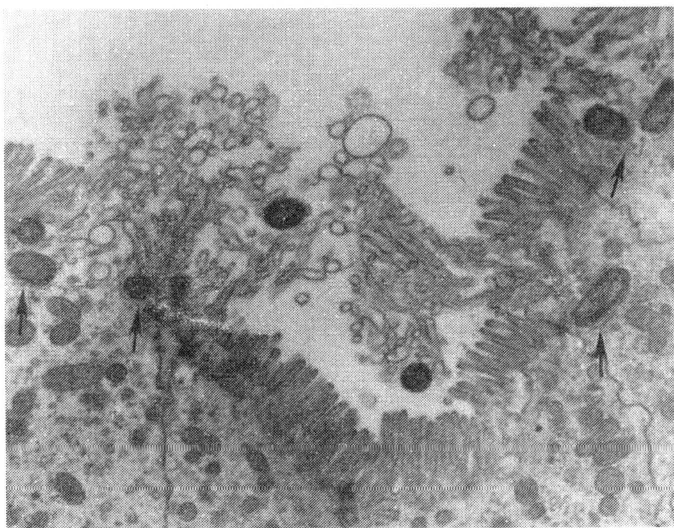

Figure 21-4 Electron photomicrograph demonstrating invasion of guinea pig ileal epithelial cells by Salmonella *typhimurium*. Arrows point to invading Salmonella organisms. (Courtesy Akio Takeuchi, Walter Reed Army Institute of Research, Washington, D.C.).

Host Defenses

Various host defenses are important in resisting intestinal colonization and invasion by Salmonella (Table 2). Normal gastric acidity (pH < 3.5) is lethal to salmonellae. In healthy individuals, the number of ingested salmonellae is reduced in the stomach, so that fewer or no organisms enter the intestine. Normal small intestinal motility also protects the bowel by sweeping ingested salmonellae through quickly. The normal intestinal microflora protects against salmonellae, probably through anaerobes, which liberate short-chain fatty acids that are thought to be toxic to salmonellae. Alteration of the anaerobic intestinal flora by antibiotics renders the host more susceptible to salmonellosis. Secretory or mucosal antibodies also protect the intestine against salmonellae. Animal strains genetically resistant to intestinal invasion by salmonellae have been described. When these host defenses are absent or blunted, the host becomes more susceptible to salmonellosis;factors that render the host more susceptible to salmonellosis are listed in Table 3. For example, in AIDS, Salmonella infection is common, frequently persistent and bacteremic, and often resistant to even prolonged antibiotic treatment. Relapses are common. The role of host defenses in salmonellosis is extremely important, and much remains to be learned.

TABLE 21-2 Host Defenses Against salmonellae

Host Defense	Examples of Factors
Gastric Factors	Gastric acidity Rate of gastric emptying
Intestinal Factors	Intestinal motility Normal intestinal flora Mucus Secretory antibodies Genetic resistance to invasion
Nonspecific and Other Possible Factors	Nutritinal state lactolerrin Gut reticuloendothelial cells Lysozyme

TABLE 21-3 Factors Increasing Susceptibility to Salmonellosis

Location of Factor	specific Condition
Stomach	Achlorhydria Gastric Surgery
Intesline	Antiboitic administration Gastrointestinal surgery Idiopathic inflammatoy bowel disease
Hemolytic Anemias	Espcially sickle cell anemia and other hemoglobinopathies
Impaired Systemic	Carcinomatosis, leukemias, lymhomas Diabetes mellitus, Immunosupperssive drugs, acquired Immunodeficiency syndrome (AIDS), etc.

Epidemiology

Contaminated food is the major mode of transmission for non-typhoidal salmonellae because salmonellosis is a zoonosis and has an enormous animal reservoir. The most common animal reservoirs are chickens, turkeys, pigs, and cows; dozens of other domestic and wild animals also harbor these organisms. Because of the ability of salmonellae to survive in meats and animal products that are not thoroughly cooked, animal products are the main vehicle of transmission. The magnitude of the problem is demonstrated by the following recent yields of salmonellae: 41% of turkeys examined in California, 50% of chickens cultured in Massachusetts, and 21% of commercial frozen egg whites examined in Spokane, WA.

The epidemiology of typhoid fever and other enteric fevers primarily involves person-to-person spread because these organisms lack a significant animal reservoir. Contamination with human feces is the major mode of spread, and the usual vehicle is contaminated water. Occasionally, contaminated food (usually handled by an individual who harbors S typhi) may be the vehicle. Plasmid DNA fingerprinting and bacteria phage lysotyping of Salmonella isolates are powerful epidemiologic tools for studying outbreaks of salmonellosis and tracing the spread of the organisms in the environment.

In typhoid fever and non-typhoidal salmonellosis, two other factors have epidemiologic significance. First, an asymptomatic human carrier state exists for the agents of either form of the disease. Approximately 3% of persons infected with S typhi and 0.1% of those infected with non-typhoidal salmonellae become chronic carriers. The carrier state may last from many weeks to years. Thus, human as well as animal reservoirs exist. Interestingly, children rarely become chronic typhoid carriers. Second, use of antibiotics in animal feeds and indiscriminant use of antibiotics in humans increase antibiotic resistance in salmonellae by promoting transfer of R factors.

Salmonellosis is a major public health problem because of its large and varied animal reservoir, the existence of human and animal carrier states, and the lack of a concerted nationwide program to control salmonellae.

Diagnosis

The diagnosis of salmonellosis requires bacteriologic isolation of the organisms from appropriate clinical specimens. Laboratory identification of the genus Salmonella is done by biochemical tests; the serologic type is confirmed by serologic testing. Feces, blood, or other specimens should be plated on several nonselective and selective agar media (blood, MacConkey, eosin-methylene blue, bismuth sulfite, Salmonella-Shigella, and brilliant green agars) as well as intoenrichment broth such as selenite or tetrathionate. Any growth in enrichment broth is subsequently subcultured onto the various agars. The biochemical reactions of suspicious colonies are then determined on triple sugar iron agar and lysine-iron agar, and a presumptive identification is made. Biochemical identification of salmonellae has been simplified by systems that permit the rapid testing of 10-20 different biochemical parameters simultaneously. The presumptive biochemical identification of Salmonella then can be confirmed by antigenic analysis of O and

H antigens using polyvalent and specific antisera. Fortunately, approximately 95% of all clinical isolates can be identified with the available group A-E typing antisera. Salmonella isolates then should be sent to a central or reference laboratory for more comprehensive serologic testing and confirmation.

Control

Salmonellae are difficult to eradicate from the environment. However, because the major reservoir for human infection is poultry and livestock, reducing the number of salmonellae harbored in these animals would significantly reduce human exposure. In Denmark, for example, all animal feeds are treated to kill salmonellae before distribution, resulting in a marked reduction in salmonellosis. Other helpful measures include changing animal slaughtering practices to reduce cross-contamination of animal carcasses; protecting processed foods from contamination; providing training in hygienic practices for all food-handling personnel in slaughterhouses, food processing plants, and restaurants; cooking and refrigerating foods adequately in food processing plants, restaurants, and homes; and expanding of governmental enteric disease surveillance programs.

Recently, The U.S. Department of Agriculture has approved the radiation of poultry to reduce contamination by pathogenic bacteria, e.g. salmonella and campylobacter. Unfortunately, radiation pasteurization has not yet been widely accepted in the U.S. Adoption and implementation of this technology would greatly reduce the magnitude of the salmonella problem.

Vaccines are available for typhoid fever and are partially effective, especially in children. No vaccines are available for non-typhoidal salmonellosis. Continued research in this area and increased understanding of the mechanisms of immunity to enteric infections are of great importance.

General salmonellosis treatment measures include replacing fluid loss by oral and intravenous routes, and controlling pain, nausea, and vomiting. Specific therapy consists of antibiotic administration. Typhoid fever and enteric fevers should be treated with antibiotics. Antibiotic therapy of non-typhoidal salmonellosis should

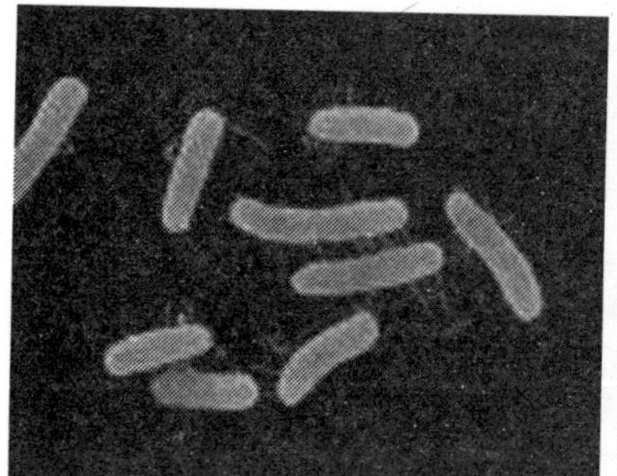

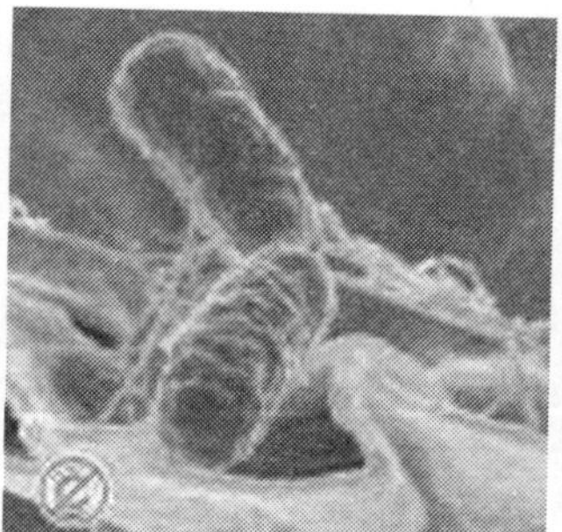

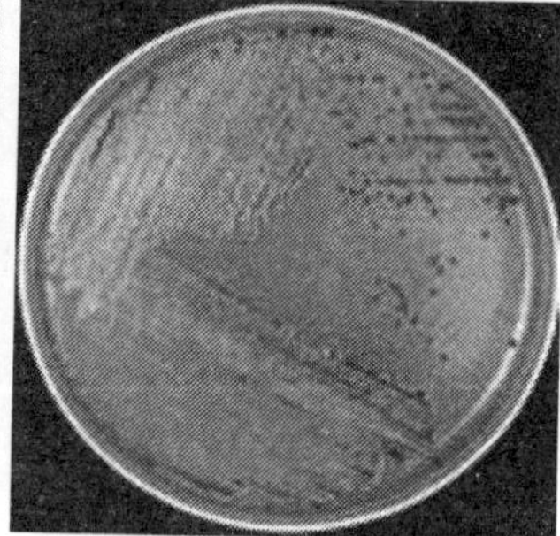

Colony and cultural morphology of *Salmonella*

be reserved for the septicemic, enteric fever, and focal infection syndromes. Antibiotics are not recommended for uncomplicated Salmonella gastroenteritis because they do not shorten the illness and they significantly prolong the fecal excretion of the organisms and increase the number of antibiotic-resistant strains.

REFERENCES

Black PH, Kunz LJ, Swartz MN: Salmonellosisa review of some unusual aspects. N Engl J Med 262:811, 864, 921, 1960.

Chopra AK, Peterson JW, Chary P et al: molecular characterization of an enterotoxin from Salmonella typhimurium. Microb Pathogen 16:85, 1994.

Finlay RB, Heffron F, Falkow S: Epithelial cell surfaces induce Salmonella proteins required for bacterial adherence and invasion. Science 243: 940, 1989.

Finlay BB, Leung KY, Rosenshine I et al: Salmonella interactions with the epithelial cell. A model to study the biology of intracellular parasitism. ASM News 58:486, 1992

Galan JE, Curtiss R: Cloning and molecular characterization of genes whose products allow Salmonella typhimurium to penetrate tissue culture cells. Proc Natl Acad Sci USA 86:6383,1989

Giannella RA: Importance of the intestinal inflammatory reaction in Salmonella-mediated intestinal secretion. Infect Immune 23:140, 1979

Giannella RA, Broitman SA, Zamcheck N: Influence of gastric acidity on bacterial and parasitic enteric infections: A perspective. Ann Intern Med 78:271,1973

Giannella RA, Formal SB, Dammin GJ et al: Pathogenesis of salmonellosis. Studies of fluid secretion, mucosal invasion, and morphologic reaction in the rabbit ileum. J Clin Invest 52:441, 1973

Giannella RA, Gots RE, Charney AN et al: Pathogenesis of Salmonella-mediated intestinal fluid secretion: activation of adenylate cyclase and inhibition by indomethacin. Gastroenterology 69:1238, 1975

Mishu B, Koehler J, Lee LA et al: Outbreaks of Salmonella enteritidis infections in the United States, 1985-1991. J Infect Dis 169:547, 1994.

Rubin RH, Weinstein L: Salmonellosis: Microbiologic, Pathologic and Clinical Features. Stratton Intercontinental Medical Book Corp, New York, 1977

Stephen J, Wallis TS, Starkey WG, et al: Salmonellosis: in retrospect and prospect. p. 175. In Evered D, Whelan J (eds): Microbial Toxins and Diarrhea Disease. Ciba Foundation Symposium 112. Pitman Press, London, 1985

Chapter **31**

Campylobacter and Helicobacter

General Concepts

Campylobacter Jejuni and other *Enteric Campylobacters*

Clinical Manifestations

Campylobacter species cause acute gastroenteritis with diarrhea, abdominal pain, fever, nausea, and vomiting. Recently, Campylobacter infections have been identified as the most common antecedent to an acute neurological disease, the Guillain-Barré syndrome.

Structure

Campylobacter species are Gram-negative, microaerophilic, non-fermenting, motile rods with a single polar flagellum; they are oxidase-positive and grow optimally at 37° or 42°C.

Classification and Antigenic Types

Campylobacter species have many serogroups, based on lipopolysaccharide (O) and protein (H) antigens. However, only a few serogroups account for most human isolates in a given geographic region. *C jejuni* possesses several common surface-exposed antigens, including porin protein and flagellin.

Pathogenesis

The bacteria colonize the small and large intestines, causing inflammatory diarrhea with fever. Stools contain leukocytes and blood. The role of toxins in pathogenesis is unclear. *C jejuni* antigens that cross-react with one or more neural structures may be responsible for triggering the Guillian-Barre syndrome.

Host Defenses

Nonspecific defenses such as gastric acidity and intestinal transit time are important. Specific immunity, involving intestinal immunoglobulin (IgA) and systemic antibodies, develops. Persons deficient in humoral immunity develop severe and prolonged illnesses.

Epidemiology

C jejuni and *C coli* infections are endemic worldwide and hyperendemic in developing countries. Infants and young adults are most often infected. Disease incidence peaks in the summer. Domestic and wild animals are the reservoirs for the organisms. Outbreaks are associated with contaminated animal products or water.

Diagnosis

Observation of darting motility in fresh fecal specimens or of vibrio forms on Gram stain permit presumptive diagnosis; definitive diagnosis is established by stool culture, and occasionally by blood culture.

Control

Control depends on measures to prevent transmission from animal reservoirs to humans.

Helicobacter Pylori *and other Gastric* Helicobacter-*like Organisms*

Clinical Manifestations

Helicobacter pylori is associated with chronic superficial gastritis (stomach inflammation) and plays a role in the pathogenesis of peptic ulcer disease. Increasing evidence indicates that H pylori infection is important in causing gastric carcinoma and lymphoma. Acute infection may cause vomiting and upper gastrointestinal pain; hypochlorhydria and intense gastritis develop. Chronic infection usually is asymptomatic.

Structure

This Gram-negative curved or spiral rod is distinguished by multiple, sheathed flagellae and abundant urease.

Classification and Antigenic Types

The antigenic structures are not completely defined and no universal typing scheme has been developed; strains may be differentiated by genotypic methods including restriction endonuclease analysis, and polymerase chain reaction (PCR).

Pathogenesis

Helicobacter pylori is sheltered from gastric acidity in the mucus layer and a small proportion of cells adheres to the gastric epithelium. The microorganism does not appear to invade tissue. Production of urease, a vacuolating cytotoxin, and the cagA-encoded protein is associated with injury to the gastric epithelium.

Host Defenses

Local and systemic humoral immune responses are essentially universal, but are not able to clear infection.

Epidemiology

H pylori infection has a worldwide distribution; about 1/3 of the world's population is infected. The prevalence of infection increases with age. The major, if not exclusive, reservoir is humans but the exact modes of transmission are not known. H pylori has now been isolated from feces and dental plaque.

Diagnosis

Examination of gastric biopsy or stained smears allows presumptive diagnosis; definitive diagnosis is made by culture. Recently, non-invasive techniques such as the urea breath test and serologic tests have been developed to diagnose H pylori infection, with accuracy exceeding 95 percent.

Control

Several indications have emerged for the use of antimicrobial therapies that eradicate H pylori infection. No vaccine is yet available.

Other Pathogenic Camplyobacter and Helicobacter Species

Campylobacter fetus causes bacteremia in compromised hosts and self-limited diarrhea in previously healthy individuals. Helicobacter cinaedi and H fennelliae cause enteric and extraintestinal diseases and are more common in homosexual men and in travelers.

INTRODUCTION

Campylobacter and Helicobacter are Gram-negative microaerophilic bacteria that are widely distributed in the animal kingdom. They have been known as animal pathogens for nearly 100 years. However, because they are fastidious and slow-growing in culture, they have been recognized as human gastrointestinal pathogens only during the last 20 years. They can cause diarrheal illnesses, systemic infection.

TABLE 23-1 Campylobacter, Helicobacter and related Species Associated With Clinical Manifestations of human Infection

Intestinal Group	Gastric Group
C jejunl	*H pylori*
C coli	*H heflmannll*
C lari	*H rappinl*
C fetus	
C upsatiensis	
C hyointestinalis	
C hyolnteslinalis	
C sputorium	
Arcobacler cryaerophila	
A bulzlerl	
A skltrowll	
H cinaedl	
H lennelliae	

chronic superficial gastritis, peptic ulcer disease, and can lead to gastric carcinoma.

Table 23-1 lists the Campylobacter species known to be pathogenic for humans. Campylobacter jejuni, and, less often, C coli and C lari are the most common bacterial causes of acute diarrheal illnesses in developed countries. Helicobacter pylori (formerly known as Campylobacter pylori), which was first cultured from gastric biopsy tissues in 1982, causes chronic superficial gastritis and is associated with the pathogenesis of peptic ulcer disease and gastric cancer. Campylobacter fetus subspecies fetus occasionally causes systemic illnesses in compromised hosts.

Campylobacter Jejuni and other Enteric Campylobacters

Clinical Manifestations

The symptoms and signs of Campylobacter enteritis are not distinctive enough to differentiate it from illness caused by many other enteric pathogens. Symptoms range from mild gastrointestinal distress lasting 24 hours to a fulminating or relapsing colitis that mimics ulcerative colitis or Crohn's disease (Figure 23-1). The predominant symptoms experienced by individuals in developed countries are diarrhea, abdominal pain, fever, nausea, and vomiting. A history of grossly bloody stools is common, and many patients have at least one day with eight or more bowel movements; fecal leukocytes are usually present. A cholera-like illness with massive watery diarrhea may also occur. Campylobacter enteritis usually is self-limiting with gradual improvement in symptoms over several days. Most patients recover within a week, but 10%-20% experience relapse or a prolonged severe illness. Toxic megacolon, pseudomembranous colitis, and massive lower gastrointestinal hemorrhage also have been described. Mesenteric adenitis and appendicitis have been reported in children and young adults. Bacteremia is uncommon (<1%) in immunocompetent patients with *C jejuni* infection.

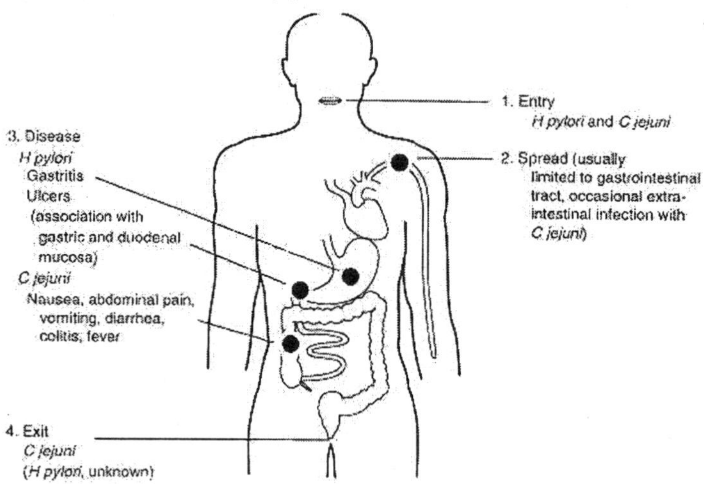

Figure 23-1 Pathogenesis of Campylobacter and Helicobacter infection in humans.

Among populations in developing countries, infection by *C jejuni* and closely related organisms is associated with much milder illness, without bloody diarrhea, fever or fecal leukocytes. Asymptomatic infection is much more common than in the developed countries, especially in older children and adults. However, when travelers from developed countries acquire *C jejuni* infections in developing countries, the symptoms are those associated with an inflammatory process. This indicates that organisms in the developing countries are fully pathogenic. Guillain-Barré syndrome is an uncommon consequence of *C jejuni* infection that is present 2-3 weeks after the diarrheal illness. However, because of the high incidence of Campylobacter infection, it has been estimated to be the trigger of 20 to 40 percent of all cases of Guillain-Barré syndrome.

Structure

Campylobacter jejuni, like all Campylobacter species, is a microaerophilic, non-fermentative Gram-negative organism. The name Campylobacter, meaning "curved rod," describes the appearance of the organisms (Figure 23-2). In young cultures, organisms are comma shaped, spiral, S shaped, or gull-winged shaped; as cultures age or are subjected to atmospheric or temperature stresses, round or coccoid forms appear.

C jejuni, which is structurally similar to other Gram-negative bacilli, is motile, with a single flagellum at one or both poles of the cell. The cell envelope has an inner bipolar lipid cell membrane, a thin peptidoglycan layer, an outer bipolar lipid layer with the lipid moiety of a lipopolysaccharide layer embedded in it, and the carbohydrate portion extending to the surface of the cell. Interspersed in the outer membrane layer are membrane proteins, some of which are exposed to the surface and are antigenic for infected hosts. Many Campylobacter species contain surface proteins that are external to the outer membrane. Campylobacter lipopolysaccharide has endotoxin activity similar to that of other Gram-negative bacteria.

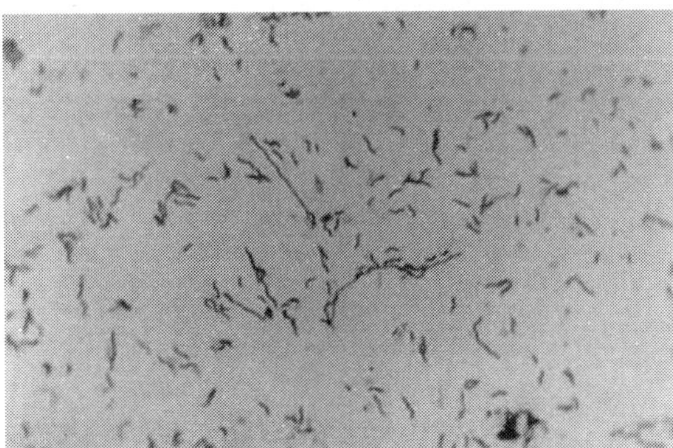

Figure 23-2 Forty-eight-hour culture of *C jejuni* (originally King's "related vibrios"), showing typical thin, comma-, S-, or gull-winged shaped forms. In broth cultures, chained organisms may appear as elongated forms. All forms are Gram negative and motile (X1000). (Courtesy of Robert Weaver, Ph.D.)

All Campylobacter species except *H pylori* are similar in structure and appearance. The Campylobacter and Helicobacter species and subspecies may be differentiated by biochemical markers (Table 23-2).

TABLE 23-2 Differentiation of Campylobacter and Helicobacter Species Related to Human disease

Test									
								Growth	
								Susceptiblity	
Species	Catalase	H_2S	Hippurate Test	Urease activity	25°C	37°C	42°C	Cph[a]	Na[b]
C jejuni	+	+	+	-	-	+	+	R	S
C coli	+	+	-	-	-	+	+	R	S
C lari	+	+	-	-	-	+	+	R	R
C upellensis	-/W*	+	-	-	-	+	+	S	S
C fetus (subsp fetus)	+	V	-	-	+	+	V	S	S
H cinaedi	W	+	-	-	-	+	-	S	S
H fannelliae	W	+	-	-	-	+	-	S	S
H pyfori	+	+	-	+	-	+	-	S	R

* weakly positive

V variable

a Cph: Copilotin

b NA: Nalidixic acid

Classification and Antigenic Types

Based on heat-labile antigens, at least 108 serogroups of both *C jejuni* and C coli have been described. In addition, 47 and 18 different heat-stable somatic (O) antigens have been described among isolates of *C jejuni* and C coli organisms, respectively. Although, geographic differences in the prevalence of serogroups exist, 10 O-groups account for about 70% of human infections. Similarly, only a few serogroups account for most human isolates in any geographic region. Serotyping has been of value in numerous epidemiologic investigations.

Despite the antigenic diversity of these organisms, *C jejuni* possesses several common surface-exposed antigens which have been used for development of serological tests. Major antigens include the porin protein (Mr 45,000), flagellin (Mr 63,000), and a group of proteins around Mr 30,000 that appear important in adhesion. The flagellar proteins undergo antigenic phase variation.

Pathogenesis

As with other enteric pathogens, the attack rate of *C jejuni* varies with the ingested dose. In outbreaks of Campylobacter enteritis, the incubation period has ranged from 1-7 days, with most illness developing 2-4 days after infection. Infection leads to multiplication of organisms in the intestines. Patients shed 106 to 109 Campylobacter per gram of feces, concentrations similar to those shed in Salmonella and Shigella enteric infections. The sites of tissue injury include the small and large intestines, and the lesions show an acute exudative and hemorrhagic inflammation. Patients frequently have colonic involvement consisting of inflammation of the lamina propria with neutrophils, eosinophils, and mononuclear cells. Destruction of epithelial glands with crypt abscess formation occurs in severe cases (Figure 23-3). The pathologic lesions seen in Campylobacter colitis are difficult to distinguish from those in ulcerative colitis. Therefore, before ulcerative colitis can be diagnosed, infection by Campylobacter and related organisms should be ruled out.

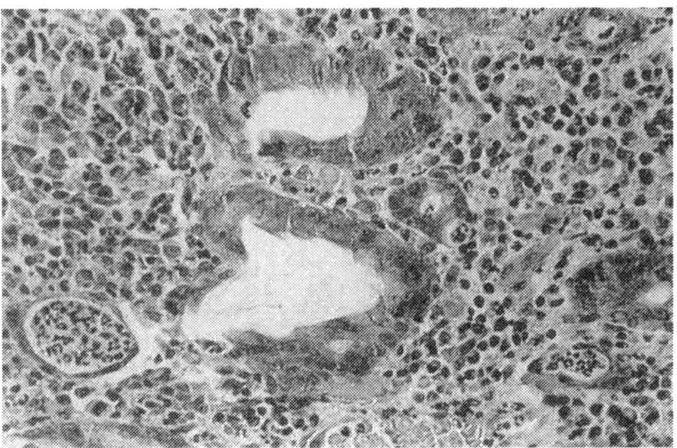

Figure 23-3 Rectal biopsy from a patient with Campylobacter colitis.There is increased cellularity of the lamina propria with neutrophils, plasma cells, and eosinophils. Glandular epithelial cells are degenerated and thinned, with loss of goblet cells. A crypt abscess is present. (Hematoxylin and eosin stain; X 250).

The mechanisms by which *C jejuni* causes illness are uncertain. Cellular infiltration in colonic biopsy specimens of patients with Campylobacter infections and the occasional presence of bacteremia suggest that these organisms may be invasive. That most Campylobacter enteritis in developed countries is associated with fever and the presence of fecal leukocytes and blood in the stool also is consistent with the invasive characteristics of the organisms. *C jejuni* is invasive in vitro in chicken embryo cells and causes bacteremia in experimentally infected mice, rabbits, calves, chickens and monkeys.

Some *C jejuni* isolates elaborate very low levels of cytotoxins similar to Shiga toxin. Some isolates have been reported to elaborate an enterotoxin similar to cholera toxin. Enterotoxin production has been more frequently observed in isolates from developing countries, where infection by *C jejuni* has been associated with watery diarrhea. However, the clinical significance of the toxigenicity of these organisms is still unclear. Strains lacking detectable enterotoxin production and with low level in vitro cytotoxin production were fully virulent in volunteers. *Campylobacter jejuni* may adhere in vitro in several tissue culture lines. This may be important in intestinal colonization or may enhance tissue invasion. A superficial antigen (PEB1) that appears to be the major adhesin is conserved among*C jejuni* strains. However, the actual in vivo significance of adherence remains undefined.

Host Defenses

C jejuni and related organisms are capable of infecting healthy persons as well as immunocompromised patients. The minimal infection-causing dose of *C jejuni* is not known, although volunteers who have ingested as few as 800 organisms have become ill. *Campylobacter jejuni* can be killed by hydrochloric acid, suggesting that normal gastric acidity may be an important barrier against infection.

Neutrophils often are observed in the feces of patients infected with *C jejuni,* and colonic biopsy specimens from patients with *Campylobacter colitis* have shown marked infiltration with neutrophils, suggesting that these cells may be important in host defense. In mice, macrophages are important for clearance of bacteremia, and, in vitro, *C jejuni* antigens stimulate a T-cell response.

Acutely infected persons frequently develop elevated specific serum ImmunoglobulinA (IgA), IgG, and IgM titers, which may persist for several weeks. Experimentally infected animals and humans manifest specific intestinal IgA production. Whether the antibody response eliminates the infection or protects against reinfection is not known. However, upon challenge with *C jejuni,* human volunteers with elevated specific serum IgA levels were likely to develop asymptomatic infection with only a brief duration of pathogen excretion. In contrast, hypogammaglobulinemic persons and those with acquired immune deficiency syndrome (AIDS) are at increased risk for severe, recurrent or bacteremic infections. *C jejuni* isolates are usually susceptible to complement-mediatedkilling by normal serum. Regardless of the exact host defense mechanisms involved, most *C jejuni* infections resolve spontaneously.

Epidemiology

In developed countries, *C jejuni* is an important cause of diarrhea, particularly in children and young adults (Figure 23-4). Between 3 and 14 percent of patients with diarrhea who seek medical attention are infected with *C jejuni*. Prolonged asymptomatic carriage is rare. The attack rate is highest in children less than 1 year old, and gradually decreases throughoutchildhood. A second peak occurs in young adults (18 to 29 years old). Although, *C jejuni* enteritis occurs throughout the year, the highest isolation rates occur in summer, as with other enteric pathogens. In contrast, up to 40 percent of healthy children in developing countries may carry the organism at any time. This is an age-related phenomenon, with the highest excretion rates in very young children. Case-to-infection ratios decline with age, which probably is indicative of acquisition of immunity due to recurrent exposure.

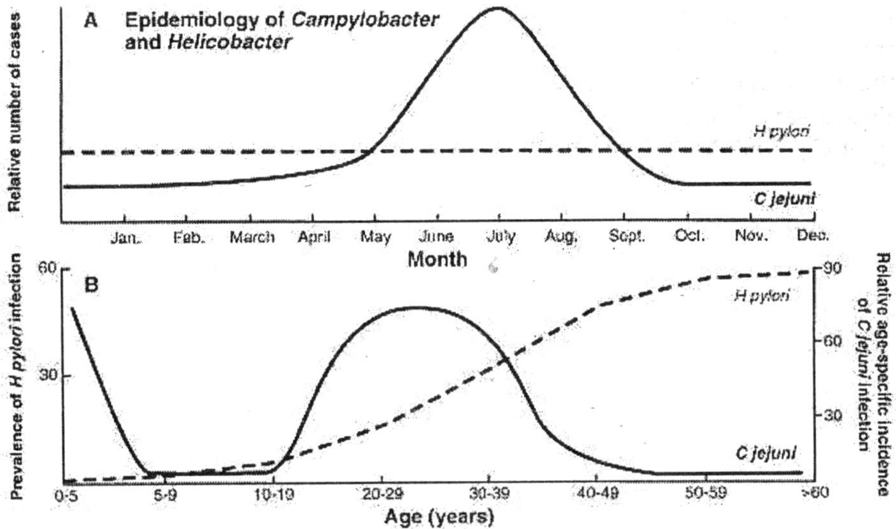

Figure 23-4 Comparison of the epidemiology of *C jejuni* (___) and H pylori (— —) by seasonal distribution by month (panel A) and by age (panel B) in the United States.

The ultimate reservoir for *C jejuni* is gastrointestinal tract of many wild animals, and a variety of domestic animals, including food animals (cattle, sheep, poultry, swine, and goats). More than 50 percent of poultry sold in the United States is contaminated with *C jejuni*. Transmission from food sources accounts for most human infections. Rodents and pets including dogs, cats, and birds also may transmit infection to humans, and excreta from wild animals may contaminate water supplies. Therefore, *C jejuni* infection may be transmitted via food, water, or direct contact with infected animals; in rare cases it may be transmitted from person-to-person.

Diagnosis

Campylobacter enteritis is hard to distinguish from enteritis caused by other pathogens. The presence of neutrophils or bloodin the feces of patients with acute diarrheal illnesses is an important clue to Campylobacter infection. Darting motility in a fresh fecal specimen observed by dark-field or phase-contrast microscopy or characteristic vibrio forms visible after Gram staining permit a presumptive diagnosis. The diagnosis is confirmed by isolating the organism from a fecal culture or, rarely, from a blood culture (Figure 23-5). Because of its growth requirement for microaerobic atmosphere, special laboratory methods are needed to isolate *C jejuni*. Plating methods must be selective to inhibit the growth of competing microorganisms in the fecal flora. The traditional approach to isolating *C jejuni* has been to use media that contain antibiotics to which *C jejuni* is resistant but most members of the usual flora are susceptible. However, owing to their motility and small diameter, Campylobacter organisms have been isolated by filtration methods that do not use antibiotic-containing media. Use of filters (pore size 0.6μm) in conjunction with non-selective media improves stool culture yields of both *C jejuni* and the atypical enteric Campylobacters. Polymerase chain reaction (PCR)-based techniques have been developed for rapid detection, culture confirmation and for typing of *C jejuni* strains.

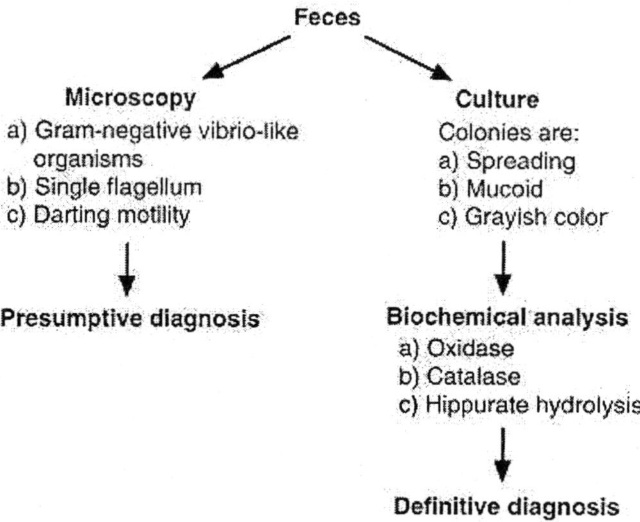

Figure 23-5 Detection of *C jejuni* and related enteric bacteria.

Because Campylobacter is microaerophilic, cultures must be incubated in an environment with reduced oxygen, optimally between 5 and 10 percent. The optimal temperature for growth is 42°C for *C jejuni*, and 37°C for many of the other enteric Campylobacters. When selective methods are used, suspicious colonies can be readily identified by their spreading character, mucoid appearance, and grayish color. The series of biochemical reactions outlined in Table 23-2 can differentiate the Campylobacter species. Serologic methods for diagnosis are only research tools at the present. A non-radioactive gene probe is available for rapid identification of *C jejuni* and C coli from isolated colonies.

Control

Control of Campylobacter enteritis depends largely on interrupting the transmission of the organism to humans from farm and domestic animals, food of animal origin, or contaminated water. Individuals can reduce the risk of Campylobacter infection by properly cooking and storing meat and dairy products, avoiding contaminated drinking water and unpasteurized milk, and washing their hands after contact with animals or animal products.

Fluid and electrolyte replacement are the cornerstones for treatment. Specific treatment with antimicrobial agents indicated for persons with severe or prolonged symptoms. However, for mild infections, the efficacy of treatment with antimicrobial agents has not yet been demonstrated. When treatment is required, erythromycin or ciprofloxacin appear to be the agents of choice. The presence of several surface-exposed, broadly specific proteins may permit vaccine development.

Other Pathogenic Campylobacter Species

Campylobacter fetus subsp fetus, well known as an animal pathogen, may cause bacteremia and other extraintestinal infections in compromised hosts, as well as an uncommon self-limited diarrheal illness in previously healthy persons. Recognized complications of C fetus infection include meningitis, endocarditis, pneumonia, thrombophlebitis, septicemia, arthritis, and peritonitis.

Virtually all C fetus isolates from humans possess lipopolysaccharide molecules with long polysaccharide side chains. Two major serogroups, A and B, have been identified. A microcapsule of high molecular-weight, antigenically related surface-array proteins has been associated with serum and phagocytosis resistance. These proteins apparently mediate serum resistance by inhibiting the binding of complement component C3b, thereby conferring to the organism a significant survival advantage. C fetus can undergo antigenic variation, switching the particular S-layer protein expressed.

Helicobacter Pylori and other Gastric Helicobacter-Like Organisms

Clinical Manifestations

Helicobacter pylori has repeatedly been shown to be associated with chronic superficial gastritis (CSG), which involves the antrum and the fundus of the stomach (Figure 23-1). Essentially all infected persons develop chronic superficial gastritis and it has clearly been shown worldwide that H pylori is the major cause of this lesion. Most of the patients with H pylori-induced chronic superficial gastritis are asymptomatic. The organisms are present on the luminal surface of mucus-secreting cells and within gastric pits, but do not invade tissue. Colonization of the affected areas of the gastric mucosa may be patchy (heavily colonized areas may be adjacent to those with no colonization). Organisms are generally not present over areas of intestinal metaplasia in the gastric mucosa. This CSG is nearly always present in patients with either gastric or duodenal ulcers. Essentially all patients with duodenal ulcers harbor H pylori in the duodenum. In duodenal ulcer disease, H pylori is associated with gastric metaplasia, but not with normal duodenal mucosa. The association of H pylori infection and gastric metaplasia is highly associated with active duodenitis.

H. pylori causes the most common form of chronic gastritis (CSG), and chronic gastritis is a well known risk factor for the development of gastric carcinoma. The epidemiologic characteristics of H pylori infection are similar to those observed in the epidemiology of adenocarcinoma of the stomach. In addition, the development of intestinal metaplasia and atrophic gastritis, two risk factors for gastric cancer, are associated with H pylori infection. All these data and prospective epidemiologic studies indicate that infection of humans with *H. pylori* is causally associated with the risk of developing gastric cancer. *H. pylori* infection also is associated with risk of developing gastric lymphoma.

Structure

H pylori differs genetically from members of the genus *Campylobacter*, and has been reclassified from *Campylobacter* (where it was initially placed) to the separate genus *Helicobacter*. *H pylori* organisms are microaerophilic, nonsporulating, Gram-negative curved rods, 3.5 µm long and 0.5 to 1 µm wide, with a spiral periodicity in fresh cultures and spherical (coccoid) forms present in older cultures. *H pylori* further differs from Campylobacter species in having multiple polar sheathed flagellae (Figure 23-6), a unique composition of cell wall fatty acids, and a smooth surface. Most *Campylobacter* species contain either unipolar or bipolar single unsheathed flagellae and have a wrinkled surface. Unlike most *Campylobacters*, *H pylori* produces urease and does not grow when incubated below 30°C. Growth is best on chocolate or blood agar plates after incubation for 2 to 5 days; for liquid media, either a blood or a hemin source appears essential.

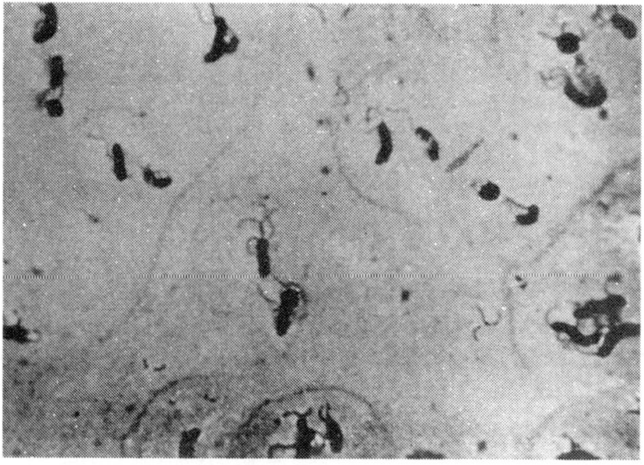

Figure 23-6 Seventy-two hour culture of *H pylori* showing typical thin, comma- or S-shaped forms. All forms are Gram-negative and motile with multiple sheathed flagella. Old cultures may presented coccoid forms (X 1000) (Courtesy of Donna R. Murray, PhD.)

Classification and Antigenic Types

The antigenic nature of *H. pylori* has not been completely defined. The whole-cell and outer-membrane profiles of all H pylori isolates have major similarities and are substantially different from those of *C jejuni* and *C fetus*. However, *H pylori* have strain-specific protein and lipopolysaccharide antigens, so it may be possible to type the organism. Simple systems for biotyping and serotyping *H pylori* are not yet available, but strains can be differentiated by genotypic-based techniques such as restriction endonuclease analysis, polymerase chain reaction, and restriction fragment length polymorphism (RFLP).

Pathogenesis

Helicobacter pylori are readily killed by brief exposure to hydrochloric acid solutions with pH below 4.0. This is paradoxical for an organism whose primary residence is the gastric lumen. However, several factors may explain this phenomenon. Firstly, *H. pylori* lives in the mucus layer overlaying the gastric mucosa, a niche protected against gastric acid. This mucus is relatively thick and viscous and maintains a pH gradient from approximately pH 2 adjacent to the gastric lumen to pH 7.4 immediately above the epithelial cells. Secondly, *H pylori* is among the most efficient producers of urease. An important effect of this metabolic activity is the release of ammonia, which buffers acidity. Third, *H pylori* is highly motile even in very viscous mucus. This motility may allow organisms to migrate to the most favorable pH gradient. Finally, acute H pylori infection is associated with hypochlorhydria. *H pylori* only overlays gastric-type but not intestinal-type epithelial cells; a proportion of the bacterial cells are adherent.

Inflammatory infiltrates with polymorphonuclear leukocytes, eosinophils, and an increased number of lymphocytes are observed in the epithelium and lamina propria (Figure 23-7). The exact mechanism by which *H pylori* causes tissue injury is unknown. At present there is little evidence for direct tissue invasion by H pylori. For pathogenic organisms that do not invade tissue, the lesions are likely to reflect a response to extracellular products such as exotoxins. An 87kDa cytotoxin that induces vacuolation of eukaryotic cells is expressed in vitro by about 50% of strains.

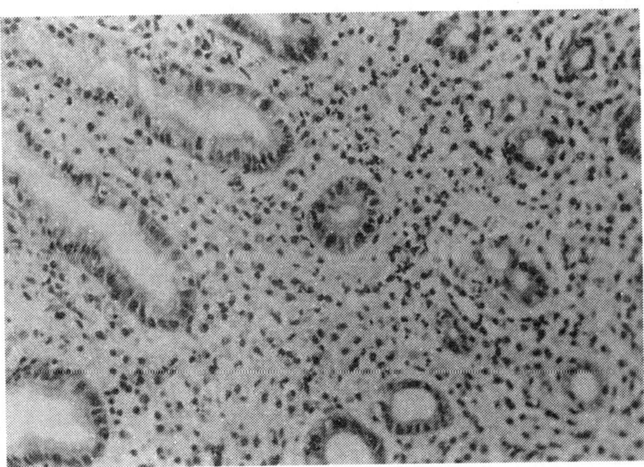

Figure 23-7 Antral gastric biopsy from a patient with H pylori gastritis. There is increased cellularity of the epithelium and lamina propria with neutrophils, eosinophils, and lymphocytes (Hematoxylin and eosin stain, x 100.) (Courtesy of Donna R. Murray, PhD.)

However, vacA, the gene encoding this toxin, is present in all strains but has substantial variability. Strains from patients with ulcers are more often toxin-producing than are strains from patients with gastritis only. *H pylori* also appears to affect the gastric mucus layer in which it resides. Isolates cultured in vitro produce an extracellular protease. This proteolytic activity affects the ability of mucus to retard diffusion of hydrogen ions. Mucus depletion over inflamed tissues is characteristic of H pylori-associated gastritis. Ammonia, produced by urease, is known to be toxic to eukaryotic cells and may potentiate mucosal injury. *H pylori* strains from patients with duodenal ulceration more frequently express a highly antigenic protein of 120-128kDa than *H pylori* strains from patients with gastritis. The gene encoding this protein, termed cagA, only is present in about 60% of *H pylori* strains.

Host Defenses

Although, gastric acid plays an important role in protection against many enteric organisms, it is not a sufficient barrier to prevent colonization of the gastric mucosa by *H pylori.* Infected persons develop high titers of serum and local IgA and IgG antibodies to *H pylori.* Longitudinal serologic studies show that *H pylori* can persist for years or longer despite these high antibody levels. It is not known whether these specific antibodies play any protective role, such as inhibiting adherence or promoting opsonophagocytosis. The role of cell-mediated immunity to these persistent pathogens has only been explored in recent years. Activation of mononuclear cells by *H pylori* induces production of tumor necrosis factor a, Interleukin-1 and other cytokines. Differences among infected hosts in cell-mediated immunity are possible determinants of outcome variability.

Epidemiology

H pylori is found worldwide and affects persons from diverse socioeconomic strata. The prevalence of these infections, as documented by both histologic and serologic studies, rises with age, as does gastritis. Person-to-person transmission is the major, if not exclusive, source of infection. *H pylori* has been isolated from dental plaque, and DNA products may be detected in saliva by PCR. *H pylori* has been isolated from feces. These data indicate potential routes of transmission of *H pylori. H pylori* is frequently isolated from asymptomatic persons who have no dyspeptic or ulcer-related symptoms. H pylori infection is more common among populations from developing areas than in more industrialized countries. Moreover, high prevalence of infection has been observed among persons in settings where sanitary conditions are suboptimal suggesting that fecal-oral transmission occurs. Infection, defined by seropositivity, persists for years and possibly for life. The annual incidence of infection in adult populations in developed countries is approximately 1 percent. On occasion, transmission occurs from person to person via contaminated endoscopes.

Other gastric Helicobacter-like organisms have now been observed in a variety of animals, including rodents, primates, swine, and ferrets, but with the exception of primates and possibly cats, these isolates are clearly different from human isolates.

Human exposure to non-human primates is not sufficiently frequent to explain the wide prevalence of *H pylori* infection in humans. Food-borne transmission would not be unusual for an enteric pathogen, but no other environmental reservoirs of H pylori have been identified.

Diagnosis

H pylori can be presumptively identified in freshly prepared gastric biopsy smears by phase-contrast microscopy, based on the characteristic motility of the microorganisms, and by staining histologic sections from gastric biopsies with Gram (carbol fuchsin counterstain), Warthin-Starry silver, Giemsa, or acridine orange stains. Organisms also can be seen directly in fixed tissue stained with hematoxylin and eosin. *H pylori* may be isolated from gastric tissue or from biopsies of esoageal or duodenal tissue containing gastric metaplasia (Figure 23-8) using nonselective media, such as chocolate agar, or antibiotic-containing selective media, such as those of Skirrow or Goodwin. Spiral organisms that are oxidase-, catalase-, and urease-positive can be identified as *H. pylori*. Culture allows determination of antimicrobial susceptibilities. In gastric biopsies, *H. pylori* also can be diagnosed presumptively, on the basis of the presence of preformed urease. DNA probe and PCR methodologies have been developed as well.

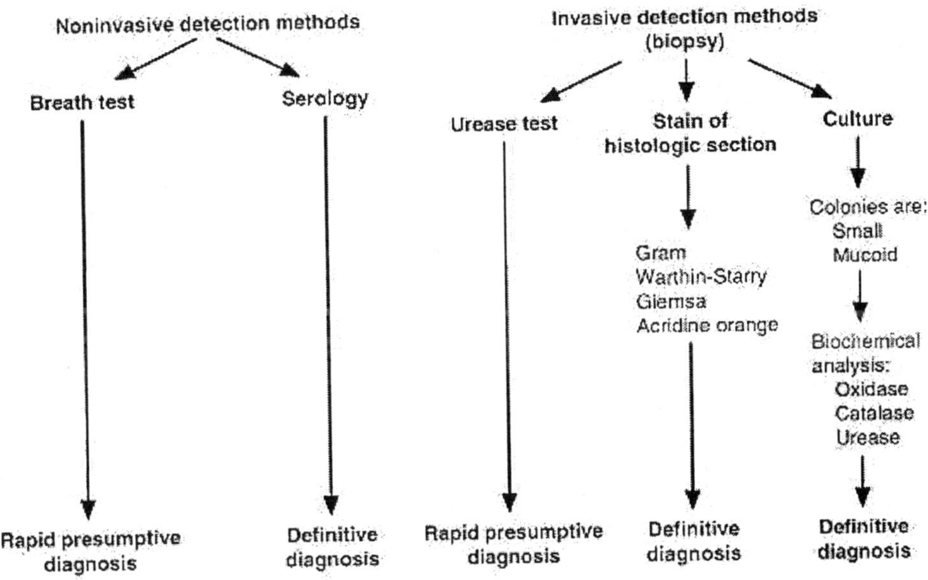

Figure 23-8 Detection methods for *H pylori*

All of the above tests require endoscopy and biopsy. A non-invasive technique known as the urea breath test has been developed to diagnose *H pylori* infection. Infection can also be diagnosed accurately by detecting serum antibodies to *H pylori* antigens. These methods may be more sensitive than diagnostic methods involving biopsies. These non-invasive methods will greatly facilitate diagnosis in individual patients, aid studies of the epidemiology of this infection, and be useful for evaluation of the efficacy of antimicrobial therapy. A number of kits now are commercially available.

Control

Antimicrphobial therapy for treatment of this infection has emerged as the most important means to resolve *H pylori* infection. Antimicrobial therapy is now one of the primary therapies for duodenal ulceration. Studies to identify the best combinations of antibiotics are being done. However, for most cases of *H pylori*-associated non-ulcer dyspepsia, data related to efficacy of antimicrobial therapy are not clear.

Other Pathogenic Helicobacter Species

Helicobacter cinaedi and *Helicobacter fennelliae* are two newly recognized *Helicobacter* species, formerly identified as Campylobacter species which have been associated with enteric and extraintestinal diseases; they are more common in homosexual men, and in travelers to developing countries.

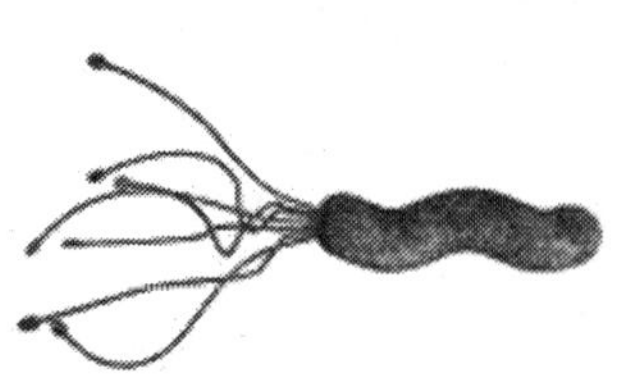

Helicobacter pylori

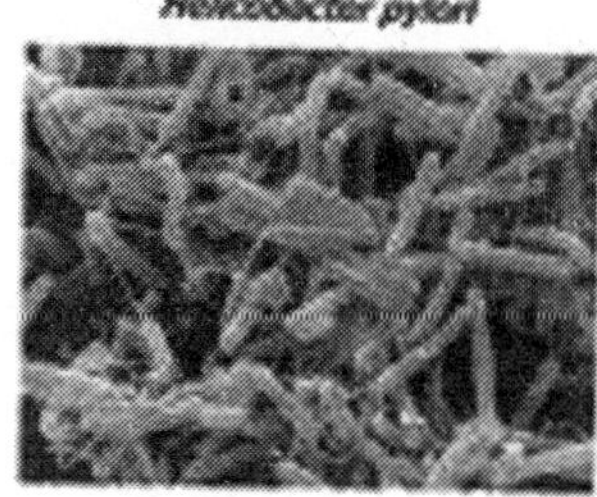

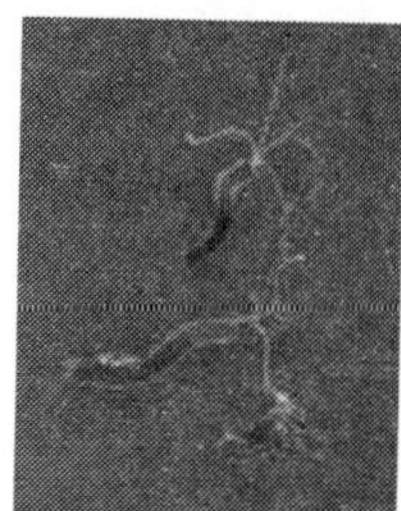

Structural and microscopic view of *H. pylori*

REFERENCES

Blaser MJ, Parsonnet J: Parasitism by the "slow" bacterium Helicobacter pylori leads to altered gastric homeostasis and neoplasia. J Clin Invest 94: 4-8, 1994

Dooley CP, Cohen H, Fitzgibbons PL, et al: Prevalence of Helicobacter pylori infection and histologic gastritis in asymptomatic persons. N Engl J Med 321: 1562, 1989

Goodwin CS, Armstrong JA, Chilvers T, et al.: Transfer of Campylobacter pylori and Campylobacter mustelae to Helicobacter mustelae comb. nov. respectively. Int J Syst Bacteriol 39: 397, 1989

Graham DY, Klein PD, Evans DJ Jr, et al.: Campylobacter pylori detected non-invasively by the 13C-urea breath test. Lancet i: 1174, 1988

Allos-Mishu B, Blaser MJ: Campylobacter jejuni and the expanding spectrum of related infections. Clin Infect Dis 20: in press, 1995

Penner JL: The genus Campylobacter: A decade of progress. Clin Microbiol Rev 1: 157, 1988

Perez-Perez GI, Dworkin BM, Chodos JE, et al: Campylobacter pylori antibodies in humans. Ann Intern Med 109: 11, 1988

Nachamkin I, Blaser MJ, Tompkins LS, editors. Campylobacter jejuni - current strategy and future trends. Washington D.C.: American Society for Microbiology, 1992, p.3-296.

Chapter **32**

Cholera, Vibrio cholerae O1 and O139, and Other Pathogenic Vibrios

General Concepts

Cholera and *Vibrio cholerae*

Clinical Manifestations

Cholera is a potentially epidemic and life-threatening secretory diarrhea characterized by numerous, voluminous watery stools, often accompanied by vomiting, and resulting in hypovolemic shock and acidosis. It is caused by certain members of the species Vibrio cholerae which can also cause mild or inapparent infections. Other members of the species may occasionally cause isolated outbreaks of milder diarrhea whereas others the vast majority are free-living and not associated with disease.

Structure, Classification, and Antigenic Types

Vibrios are Gram-negative, highly motile curved rods with a single polar flagellum. They tolerate alkaline media that kill most intestinal commensals, but they are sensitive to acid. Numerous free-living vibrios are known, some potentially pathogenic. Until 1992, cholera was caused by only two serotypes, Inaba (AC) and Ogawa (AB), and two biotypes, classical and El Tor, of toxigenic O group 1 V cholerae. These organisms may be identified by agglutination in O group 1-specific antiserum directed against the lipopolysaccharide component of the cell wall and by demonstration of their enterotoxigenicity. In 1992, cholera caused by serogroup O139 (synonym "Bengal"; the 139th and latest serogroup of V cholerae to be identified) emerged in epidemic proportions in India and Bangladesh. This serovar is identified by 1) absence of agglutination in O group 1 specific antiserum; 2) by agglutination in O group 139 specific antiserum; and 3) by the presence of a capsule.

Pathogenesis

Cholera is transmitted by the fecal-oral route. Vibrios are sensitive to acid, and most die in the stomach. Surviving virulent organisms may adhere to and colonize the small bowel, where they secrete the potent cholera enterotoxin (CT, also called "choleragen"). This toxin binds to the plasma membrane of intestinal epithelial cells and releases an enzymatically active subunit that causes a rise in cyclic adenosine 51-monophosphate (cAMP) production. The resulting high intracellular cAMP level causes massive secretion of electrolytes and water into the intestinal lumen.

Host Defenses

Gastric acid, mucus secretion, and intestinal motility are the prime nonspecific defenses against V cholerae. Breastfeeding in endemic areas is important in protecting infants from disease. Disease results in effective specific immunity, involving primarily secretory immunoglobulin (IgA), as well as IgG antibodies, against vibrios, somatic antigen, outer membrane protein, and/or the enterotoxin and other products.

Epidemiology

Cholera is endemic or epidemic in areas with poor sanitation; it occurs sporadically or as limited outbreaks in developed countries. In coastal regions,it may persist in shellfish and plankton. Long-term convalescent carriers are rare. Enteritis caused by the halophile V parahaemolyticus is associated with raw or improperly cooked seafood.

Diagnosis

The diagnosis is suggested by strikingly severe, watery diarrhea. For rapid diagnosis, a wet mount of liquid stool is examined microscopically. The characteristic motility of vibrios is stopped by specific antisomatic antibody. Other methods are culture of stool or rectal swab samples on TCBS agar and other selective and nonselective media; the slide agglutination test of colonies with specific antiserum; fermentation tests (oxidase positive); and enrichment in peptone broth followed by fluorescent antibody tests, culture, or retrospective serologic diagnosis. More recently the polymerase chain reaction (PCR) and additional genetically-based rapid techniques have been recommended for use in specialized laboratories.

Control

Control by sanitation is effective but not feasible in endemic areas. A good vaccine has not yet been developed. A parenteral vaccine of whole killed bacteria has been used widely, but is relatively ineffective and is not generally recommended. An experimental oral vaccine of killed whole cells and toxin B-subunit protein is less than ideal. Living attenuated genetically engineered mutants are promising, but such strains can cause limited diarrhea as a side effect. Antibiotic prophylaxis is feasible for small groups over short periods.

Other Vibrio Infections

Other serogroups of V cholerae may cause diarrheal disease and other infections but are not associated with epidemic cholera. Vibrio parahaemolyticus is an important cause of enteritis associated with the ingestion of raw or improperly prepared seafood. Other Vibrio species, including *V vulnificus*, can cause infections of humans and other animals including fish. Campylobacter species (formerly included with vibrios) can cause enteritis. C pylori, now known as Helicobacter pylori, is associated with gastric and duodenal ulcers (see Ch. 23).

INTRODUCTION

Vibrios are highly motile, gram-negative, curved or comma-shaped rods with a single polar flagellum. Of the vibrios that are clinically significant to humans, Vibrio cholerae O group 1, the agent of cholera, is the most important. Vibrio cholerae was first isolated in pure culture by Robert Koch in 1883, although it had been seen by other investigators, including Pacini, who is credited with describing it first in Florence, Italy, in 1854.

Cholera is a life-threatening secretory diarrhea induced by an enterotoxin secreted by *V. cholerae*. Cholera and the cholera enterotoxin are increasingly recognized as the prototypes for a wide variety of non-invasive diarrheal diseases, collectively known as the enterotoxic enteropathies; of these, diarrhea due to enterotoxigenic strains of *Escherichia coli* (see Ch. 26) is the most important. Cholera remains a major epidemic disease. There have been seven great pandemics. The latest, which started in 1961, invaded the Western Hemisphere (for the first time this century) with a massive outbreak in Peru in 1991. There have since been more than a million cases in Central and South America as well as a few imported cases in the U.S. and Canada. *V. cholerae* serogroup O139, which arose in October of 1992 in India and Bangladesh, may become the cause of the 8th great pandemic of cholera.

Other vibrios may also be clinically significant in humans, and some are known to cause diseases in domestic animals. Nonpathogenic vibrios are widely distributed in the environment, particularly in estuarine waters and seafoods. For this reason, isolation of a vibrio from a patient with diarrheal disease does not necessarily indicate an etiologic relationship.

Vibrio Cholerae

Clinical Manifestations

Following an incubation period of 6 to 48 hours, cholera begins with the abrupt onset of watery diarrhea (Fig. 24-1). The initial stool may exceed 1 L, and several liters of fluid may be secreted within hours, leading to hypovolemic shock. Vomiting usually accompanies the diarrheal episodes. Muscle cramps may occur as water and electrolytes are lost from body tissues. Loss of skin turgor, scaphoid abdomen, and weak pulse are characteristic of cholera. Various degrees of fluid and electrolyte loss are observed, including mild and subclinical cases. The disease runs its course in 2 to 7 days; the outcome depends upon the extent of water and electrolyte loss and the adequacy of water and electrolyte repletion therapy. Death can occur from hypovolemic shock, metabolic acidosis, and uremia resulting from acute tubular necrosis.

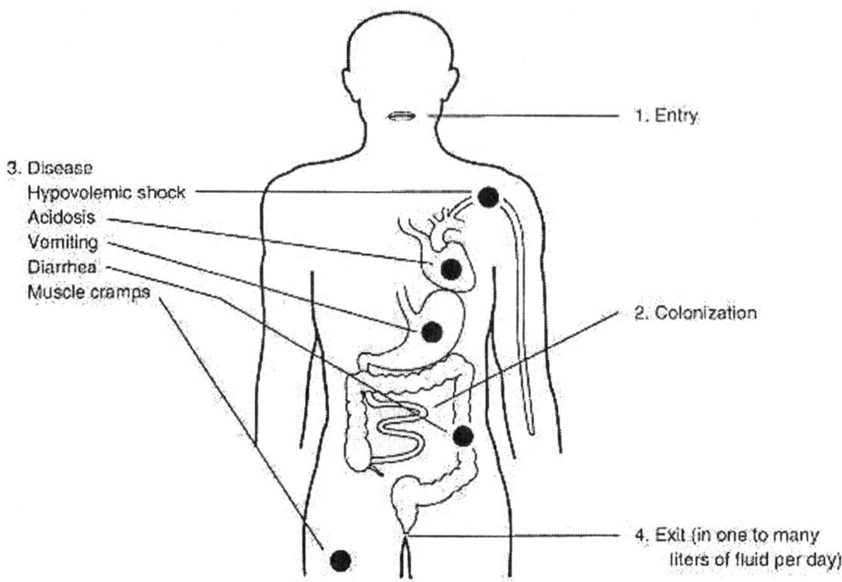

FIGURE 24-1 Pathophysiology of cholera.

Structure, Classification, and Antigenic Types

The cholera vibrios are Gram-negative, slightly curved rods whose motility depends on a single polar flagellum. Their nutritional requirements are simple. Fresh isolates are prototrophic (i.e., they grow in media containing an inorganic nitrogen source, a utilizable carbohydrate, and appropriate minerals). In adequate media, they grow rapidly with a generation time of less than 30 minutes. Although, they reach higher population densities when grown with vigorous aeration, they can also grow anaerobically. Vibrios are sensitive to low pH and die rapidly in solutions below pH 6; however, they are quite tolerant of alkaline conditions. This tolerance has been exploited in the choice of media used for their isolation and diagnosis.

Until 1992, the vibrios that caused epidemic cholera were subdivided into two biotypes: classical and El Tor. Classical *V. cholerae* was first isolated by Koch in 1883. Subsequently, in the early 1900s, some vibrios resembling *V. cholerae* were isolated from Mecca-bound pilgrims at the quarantine station at *El Tor*, in the Sinai peninsula, that had been established to try to control cholera associated with pilgrimages to Mecca. These vibrios resembled classical *V cholerae* in many ways but caused lysis of goat or sheep erythrocytes in a test known as the Greig test. Because the pilgrims from whom they were isolated did not have cholera, these hemolytic El Tor vibrios were regarded as relatively insignificant except for the possibility of confusion with true cholera vibrios. In the 1930s, similar hemolytic vibrios were associated with relatively restricted outbreaks of diarrheal disease, called paracholera, in the Celebes. In 1961, cholera caused by El Tor vibrios erupted in Hong Kong and spread virtually worldwide. Although, in the course of this pandemic most *V. cholerae* biotype El Tor strains lost their hemolytic activity, a number of ancillary tests differentiate them from vibrios of the classical biotype.

The operational serology of the cholera vibrios which belong in O antigen group 1 is relatively simple. Both biotypes (El Tor and classical) contain two major serotypes, Inaba and Ogawa (Fig. 24-2). These serotypes are differentiated in agglutination and vibriocidal antibody tests on the basis of their dominant heat-stable lipopolysaccharide somaticantigens. The cholera group has a common antigen, A, and the serotypes are differentiated by the type-specific antigens, B (Ogawa) and C (Inaba). An additional serotype, Hikojima, which has both specific antigens, is rare. V. cholerae O139 appears to have been derived from the pandemic El Tor biotype but has lost the characteristic O1 somatic antigen; it has gained the ability to produce a polysaccharide capsule; it produces the same cholera enterotoxin; and it seems to have retained the epidemic potential of O1 strains.

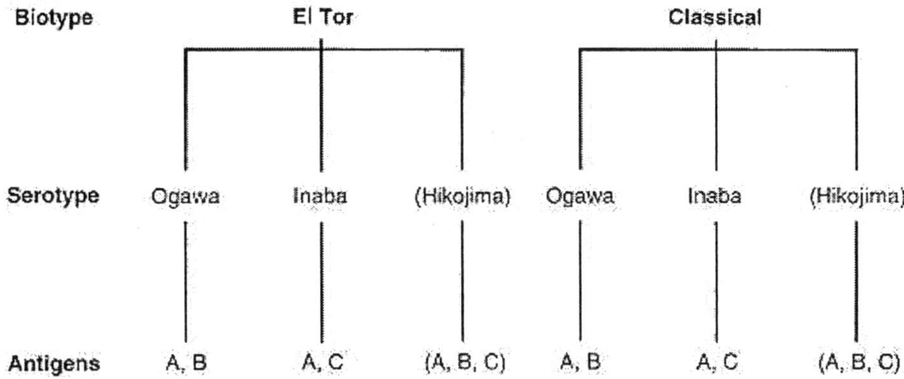

FIGURE 24-2 Vibrio cholerae (O group 1 antigen).

Other antigenic components of the vibrios, such as outer membrane protein antigens, have not been extensively studied. The cholera vibrios also have common flagellar antigens. Cross-reactionswith Brucella and Citrobacterspecies have been reported. Because of DNA relatedness and other similarities, other vibrios formerly called "nonagglutinable" are now classified as *V. cholerae*. The term nonagglutinable is a misnomer because it implies that these vibrios are not agglutinable; in fact, they are not agglutinable in antisera againstthe O antigen group 1 cholera vibrios, but they are agglutinable in their own specific antisera. More than 139 serotypes are now recognized. Some strains of non-O group 1 V cholerae cause diarrheal disease by means of an enterotoxin related to the cholera enterotoxin and, perhaps, by other mechanisms, but these strains have not been associated with devastating outbreaks like those caused by the true cholera vibrios. Recently, vibrio strains that agglutinate in some O group 1 cholera diagnostic antisera but not in others have been isolated from environmental sources. Volunteer feeding experiments have shown that these atypical O group 1 vibrios are not enteropathogenic in humans. Recent studies using specific toxin gene probes indicate that these environmental isolates not only are nontoxigenic, but also do not possess any of the genetic information encoding cholera toxin, although some isolates from diarrheal stools do.

The cholera vibrios cause many distinctive reactions. They are oxidase positive. The O group 1 cholera vibrios almost always fall into the Heiberg I fermentation pattern; that is, they ferment sucrose and mannose but not arabinose, and they produce acid but not gas. *Vibrio cholerae* also possesses lysine and ornithine decarboxylase, but not arginine dihydrolase. Freshly isolated agar-grown vibrios of the El Tor biotype, in contrast to classical *V. cholerae*, produce a cell-associated mannose-sensitive hemagglutinin active on chicken erythrocytes. This activity is readily detected in a rapid slide test. In addition to hemagglutination, numerous tests have been proposed to differentiate the classical and El Tor biotypes, including production of a hemolysin, sensitivity to selected bacteriophages, sensitivity to polymyxin, and the Voges-Proskauer test for acetoin. El Tor vibrios originally were defined as hemolytic. They differed in this characteristic from classical cholera vibrios; however, during the most recent pandemic, most El Tor vibrios (except for the recent isolates from Texas and Louisiana) had lost the capacity to express the hemolysin. Most El Tor vibrios are Voges-Proskauer positive and resistant to polymyxin and to bacteriophage IV, whereas classical vibrios are sensitive to them. As both biotypes cause the same disease, these characteristics have only epidemiologic significance. Strains of the El Tor biotype, however, produce less cholera enterotoxin, but appear to colonize intestinal epithelium better than vibrios of the classical variety. Also, they seem some what more resistant to environmental factors. Thus, El Tor strains have a higher tendency to become endemic and exhibit a higher infection-to-case ratio than the classical biotype.

Pathogenesis

Recent studies with laboratory animal models and human volunteers have provided a detailed understanding of the pathogenesis of cholera. Initial attempts to infect healthy American volunteers with cholera vibrios revealed that the oral administration of up to 1011 living cholera vibrios rarely had an effect; in fact, the organisms usually could not be recovered from stools of the volunteers. After the administration of bicarbonate to neutralize gastric acidity, however, cholera diarrhea developed in most volunteers given 104 cholera vibrios. Therefore, gastric acidity itself is a powerful natural resistance mechanism. It also has been demonstrated that vibrios administered with food are much more likely to cause infection.

Cholera is exclusively a disease of the small bowel. To establish residence and multiply in the human small bowel (normally relatively free of bacteria because of the effective clearance mechanisms of peristalsis and mucus secretion), the cholera vibrios have one or more adherence factors that enable them to adhere to the microvilli (Fig. 24-3). Several hemagglutinins and the toxin-coregulated pili have been suggested to be involved in adherence but the actual mechanism has not been defined. In fact, there may be multiple mechanisms. The motility of the vibrios may affect virulence by enabling them to penetrate the mucus layer. They also produce mucinolytic enzymes, neuraminidase, and proteases. The growing cholera vibrios elaborate the cholera enterotoxin (CT or choleragen), a polymeric protein (Mr 84,000) consisting of two major domains or regions. The A region (Mr 28,000), responsible for biologic activity of the enterotoxin, is linked by noncovalent interactions with the B region (Mr 56,000), which is composed of five identical noncovalently associated peptide chains of Mr 11,500. The B region, also known as choleragenoid, binds the toxin to its receptors on host cell membranes. It is also the immunologically dominant portion of the holotoxin. The structural genes that encode the synthesis of CT reside on a transposon-like element in the V cholerae chromosome, in contrast to those for the heat-labile enterotoxins (LTs) of E coli (Ch. 25), which are encoded by plasmids. The amino acid sequences of these structurally, functionally, and immunologically related enterotoxins are very similar. Their differences account for the differences in physicochemical behavior and the antigenic distinctions that have been noted. There are at least two antigenically related but distinct forms of cholera enterotoxin, called CT-1 and CT-2. Classical O1 V cholerae and the Gulf Coast El Tor strains produce CT-1 whereas most other El Tor strains and O139 produce CT-2. Vibrio cholerae exports its enterotoxin, whereas the E coli LTs occur primarily in the periplasmic space. This may account for the reported differences in severity of the diarrhea caused by these organisms.

Studies in adult American volunteers have shown that 5μ g of CT, administered orally with bicarbonate, causes 1 to 6 L of diarrhea; 25μg causes more than 20 L.

Synthesis of CT and other virulence-associated factors such as toxin-coregulated pili are believed to be regulated by a transcriptional activator, Tox R, a transmembrane DNA-binding protein.

The molecular events in these diarrheal diseases involve an interaction between the enterotoxins and intestinal epithelial cell membranes (Fig. 24-4). The toxins bind through region B to a glycolipid, the GM1 ganglioside, which is practically ubiquitous in eukaryotic cell membranes. Following this binding, the A region, or a major portion of it known as the A1 peptide (Mr 21,000), penetrates the host cell and enzymatically transfers ADP-ribose from nicotinamide adenine dinucleotide (NAD) to a target protein, the guanosine 5'-triphosphate (GTP)-binding regulatory protein associated with membrane-bound adenylate cyclase. Thus, CT (and LT) resembles diphtheria toxin in causing transfer of ADP-ribose to a substrate. With diphtheria toxin, however, the substrate is elongation factor 2 and the result is cessation of host cell protein synthesis. With CT, the ADP-ribosylation reaction essentially locks adenylate cyclase in its "on mode" and leads to excessive

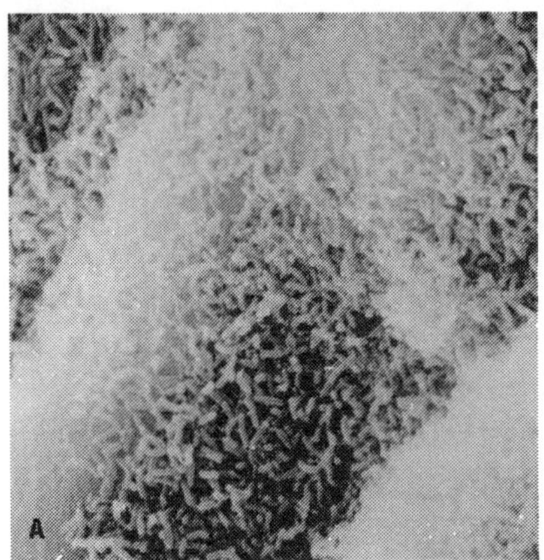

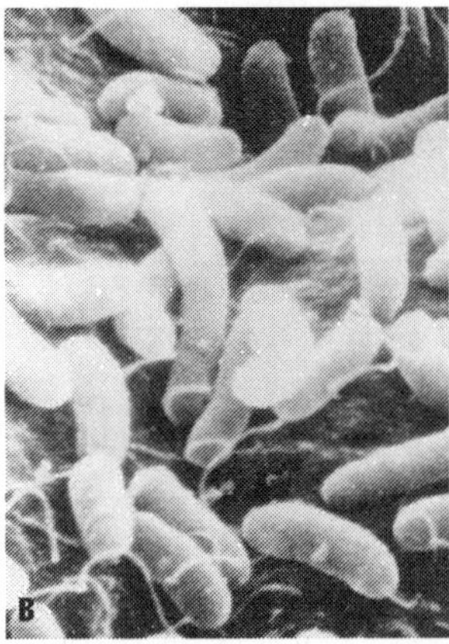

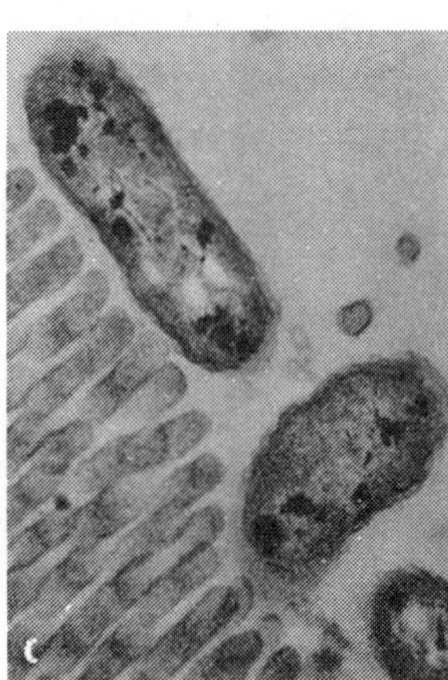

FIGURE 24-3 Vibrio cholerae attachment and colonization in experimental rabbits. The events are assumed to be similar in human cholera. (A) Scanning electron microscopy during early infection. Curved vibrios adhering to epithelial surface. (Approximately X 4,000.) (B) Higher power scanning electron micrograph showing single polar flagellum of the cholera vibrios. (C) Transmission electron microscopy of vibrios in both end-on and horizontal modes close to tips of microvilli. (From Nelson ET, Clements JD, Finkelstein RA: Vibrio cholerae adherence and colonization in experimental cholera: electron microscopic studies. Infect Immun 14:527, 1976, with permission.)

production of cyclic adenosine 51-monophosphate (cAMP). Pertussis toxin, another ADP-ribosyl transferase, also increases cAMP levels, but by its effect on another G-protein, Gi (Fig. 24-5). The subsequent cAMP-mediated cascade of events has not yet been delineated, but the final effect is hypersecretion of chloride and bicarbonate followed by water, resulting in the characteristic isotonic voluminous cholera stool. In hospitalized patients, this can result in losses of 20 L or more of fluid per day. The stool of an actively purging, severely ill cholera patient can resemble rice water the supernatant of boiled rice. Because the stool can contain 108 viable vibrios per ml, such a patient could shed 2 X 1012 cholera vibrios per day into the environment. Perhaps by production of CT, the cholera vibrios thus ensure their survival by increasing the likelihood of finding another human host. Recent evidence suggests that prostaglandins may also play a role in the secretory effects of cholera enterotoxin. Recent studies in volunteers using genetically-engineered Tox- strains of V cholerae have revealed that the vibrios have putative mechanisms in addition to CT for causing (milder) diarrheal disease. These include Zot (for Zonula occludens toxin) and Ace (for accessory cholera enterotoxin), and perhaps others, but their role has not been established conclusively. Certainly CT is the major virulence factor and the act of colonization of the small bowel may itself elicit an altered host response (e.g., mild diarrhea), perhaps by a trans-membrane signaling mechanism.

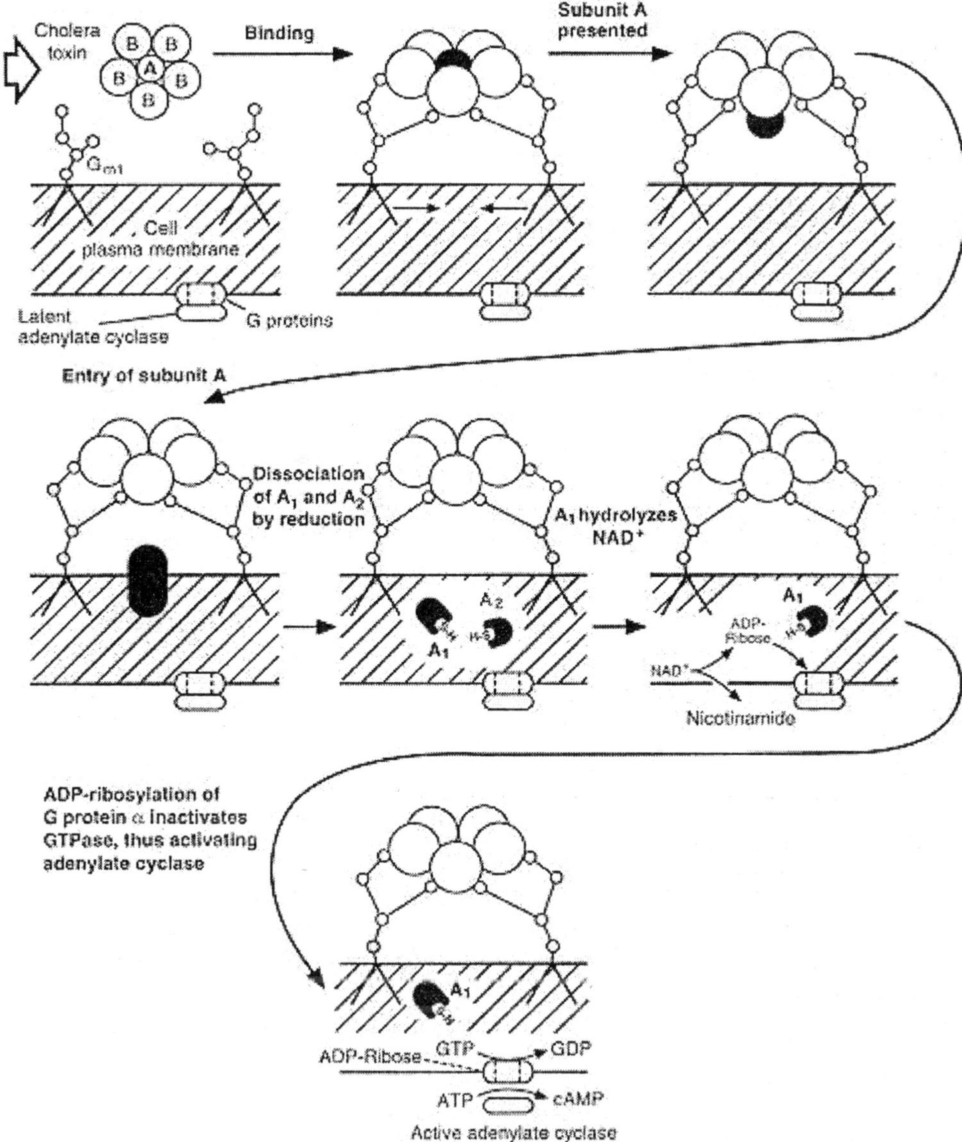

FIGURE 24-4 Mechanism of action of cholera enterotoxin. Cholera toxin approaches target cell surface. B subunits bind to oligosaccharide of GM1 ganglioside. Conformational alteration of holotoxin occurs, allowing the presentation of the A subunit to cell surface. The A subunit enters the cell. The disulfide bond of the A subunit is reduced by intracellular glutathione, freeing A1 and A2. NAD is hydrolyzed by A1, yielding ADP-ribose and nicotinamide. One of the G proteins of adenylate cyclase is ADP-ribosylated, inhibiting the action of GTPase and locking adenylate cyclase in the "on" mode (Modified from Fishman PH: Mechanism of action of cholera toxin: events on the cell surface. p. 85. In Field M, Fordtran JS, Schultz SG (eds): Secretory Diarrhea. Waverly Press, Baltimore, 1980, with permission.)

Various animal models have been used to investigate pathogenic mechanisms, virulence, and immunity. Ten-day-old suckling rabbits develop a fulminating diarrheal disease after intraintestinal inoculation with virulent V cholerae or CT. Adult rabbits are relatively resistant to colonization by cholera vibrios; however, they do respond, with characteristic out pouring of fluid, to the intraluminal inoculation of live vibrios or enterotoxin in surgically isolated ileal loops. Suckling mice are susceptible to intragastric inoculation of vibrios and to orally administered toxin. Adult conventional mice are also susceptible to orally administered toxin, but resist colonization except in isolated intestinal loops. Interestingly, however, germ-free mice can be colonized for months with cholera vibrios. They rarely show adverse effects, although they are susceptible to cholera enterotoxin. Dogs have been used experimentally, although they are relatively refractory and require enormous inocula to elicit choleraic manifestations. Chinchillas also are susceptible to diarrhea following intraintestinal inoculation with moderate numbers of cholera vibrios. Infections initiated by extraintestinal routes of inoculation (e.g., intraperitoneal) largely reflect the toxicity of the lipopolysaccharide endotoxin. The intraperitoneal infection in mice has been used to assay the protective effect of conventional killed vibrio vaccines (no longer widely used).

Various animals, including humans, rabbits, and guinea pigs, also respond to intradermal inoculation of relatively minute amounts of CT with a characteristic delayed (maximum response at 24 hours), sustained (visible up to 1 week or more),

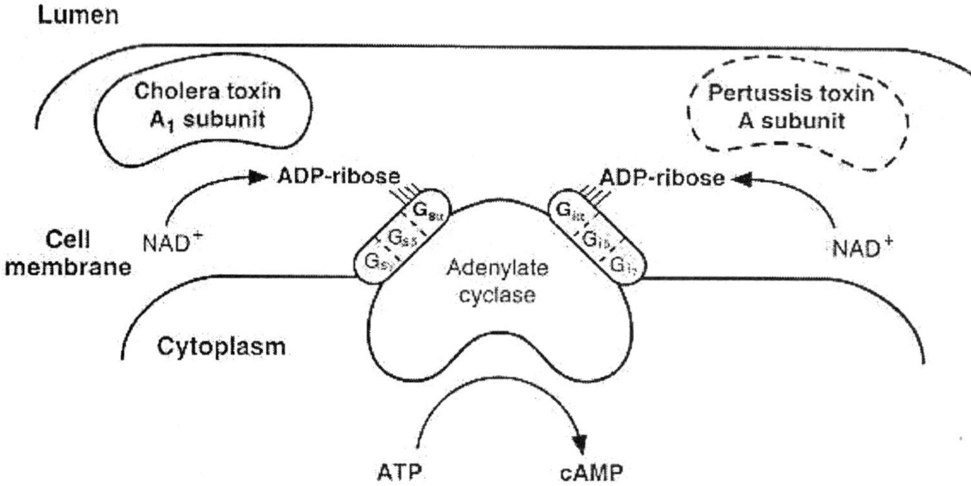

FIGURE 24-5 Comparison of activities of cholera enterotoxin (CT) with pertussis toxin (PT). The a-subunits of Gs and Gi, with GTP-binding sites, are ADP-ribosylated, respectively, by A1 peptide of CT or by the A subunit of PT, preventing, respectively, the hydrolysis of Gs-GTP to GDP or the responsiveness of Gi to inhibitory hormones, both effectively producing increases in adenylate cyclase activity. (Modified from Gill DM, Woolkalis M: Toxins which activate adenylate cyclase. CIBA Found Symp 112:57, 1985, with permission.)

erythematous, edematous induration associated with a localized alteration of vascular permeability. In laboratory animals, this response can be measured after injecting a protein-binding dye, such as trypan blue, that extravasates to produce a zone of bluing at the site of intracutaneous inoculation of toxin. This observation has been exploited in the assay of CT and its antibody and in the detection of other enterotoxins.

In addition, because of the broad spectrum of activity of CT on cells and tissues that it never contacts in nature, various in vitro systems can be used to assay the enterotoxin and its antibody. In each, the toxin causes a characteristically delayed, but sustained, activation of adenylate cyclase and increased production of cAMP, and it may cause additional, readily recognizable, morphologic alterations of certain cultured cell lines. The cells most widely used for this purpose are Chinese hamster ovary (CHO) cells, which elongate in response to pictogram doses of the toxin, and mouse Y-l adrenal tumor cells, which round up. Cholera toxin has become an extremely valuable experimental probe to identify other cAMP-mediated responses. It also activates adenylate cyclase in pigeon erythrocytes, a procedure that was used by D. Michael Gill to define its mode of action.

These assays and models also have been applied in the study of an expanding number of CT-related and unrelated enterotoxins. These include the LTs of E coli, which are structurally and immunologically similar to it and are effective in any model that is responsive to CT. The family of small molecular weight heat-stable enterotoxins (ST) of E coli, which activate guanylate cyclase, and which are rapidly active in the infant mouse and certain other intestinal models, are clearly unrelated to CT. CT-related enterotoxins have been reported from certain nonagglutinable (non-O group I) Vibrio strains and a Salmonella enterotoxin was shown to be related immunologically to CT. CT-like factors from Shigella and V parahaemolyticus have thus far been demonstrated only in sensitive cell culture systems. Other enterotoxins and enterocytotoxins, which elicit cytotoxic effects on intestinal epithelial cells, also have been described from Escherichia, Klebsiella, Enterobacter, Citrobacter, Aeromonas, Pseudomonas, Shigella, V parahaemolyticus, Campylobacter, Yersinia enterocolitica, Bacillus cereus, Clostridium perfringens, C difficile, and staphylococci. Escherichia coli, some vibrio strains, and some other enteric bacteria produce cytotoxins that, like Shiga toxin of Shigella dysenteriae, act on Vero (African green monkey kidney) cells in vitro. These toxins have been called Shiga-like toxins, Shiga toxin-like toxins, Vero toxins, and Vero cytotoxins. The classic staphylococcal enterotoxins perhaps should more properly be called neurotoxins, as they seem to affect the central nervous system rather than the gut directly to cause fluid secretion or histopathologic effects.

Host Defenses

Infection with cholera vibrios results in a spectrum of responses. These range from no observed manifestations except perhaps a serologic response (the most common) to acute purging, which must be treated by hospitalization and fluid replacement therapy; this is the classic response. The reasons for these differences are not entirely clear, although it is known that individuals differ in gastric acidity and that hypochlorhydric individuals are most prone to cholera. Whether individuals differ in the availability of intestinal receptors for cholera vibrios or for their toxin has not been established. Prior immunologic experience of subjects at risk is certainly a major factor. For example, in heavily endemic regions such as Bangladesh, the attack rate is relatively low among adults in comparison with children. In neoepidemic areas, cholera is more frequent among the working adult population. Resistance is related to the presence of circulating antibody and, perhaps more importantly, local immunoglobulin A (IgA) antibody against the cholera bacteria or the cholera enterotoxin or both. Intestinal IgA antibody can prevent attachment of the vibrios to the mucosal surface and neutralize or prevent binding of the cholera enterotoxin. For reasons that are not clear, individuals of blood group O are slightly more susceptible to cholera. Breastfeeding is highly recommended as a means of increasing immunity of infants to this and other diarrheal disease agents.

Recovery from cholera probably depends on two factors: elimination of the vibrios by antibiotics or the patient's own immune response, and regeneration of the poisoned intestinal epithelial cells. Treatment with a single 200-mg dose of doxycycline has been recommended. As studies in volunteers demonstrated conclusively, the disease is an immunizing process. Patients who have recovered from cholera are solidly immune for at least 3 years.

Cholera vaccines consisting of killed cholera bacteria administered parenterally have been used since the turn of the century. However, recent controlled field studies indicate that little, if any, effective immunity is induced in immunologically virgin populations by such vaccines, although they do stimulate preexisting immunity in the adult population in heavily endemic regions. Controlled studies have likewise shown that a cholera toxoid administered parenterally was ineffective in preventing cholera. Probably the natural disease should be simulated to induce truly effective immunity although a parenterally administered conjugate vaccine consisting of the polysaccharide of the vibrio LPS covalently linked to cholera toxin has given promising results in preliminary studies. Studies in volunteers have shown that orally administered, chemically mutagenized or genetically engineered mutants which do not produce CT or produce only its B subunit protein can induce immunity against subsequent challenge. However, most of these candidate vaccines also produce unacceptable side effects primarily mild to moderate diarrhea. An exception is strain CVD103-HgR (a mercury resistant A-B+ derivative of classical biotype Inaba serotype strain 569B). This strain has minimal reactogenicity but does not colonize well and therefore has to be given in higher doses. Field studies with this strain are in progress. Combined preparations of bacterial somatic antigen and toxin antigen have been reported to act synergistically in stimulating immunity in laboratory animals; that is, the combined protective effect is closer to the product than to the sum of the individual protective effects. However, a large field study evaluating such nonviable oral vaccines in Bangladesh revealed that neither the whole-cell bacterin nor the killed vibrios supplemented with the B-subunit protein of the cholera enterotoxin induced sufficient long term protection, especially in children, to justify their recommendation for public health use. No clear-cut advantage of the inclusion of the B-subunit was demonstrated.

In any case, even if these vaccines were effective, the requirement for large and repeated doses would make them too expensive for use in the developing areas that are usually afflicted with epidemic cholera. Moreover, they were clearly less effective in children the primary target population in heavily endemic areas. Neither the killed whole cell vaccine nor strain CVD103-HgR could be expected to protect against the new O139 serovar.

Epidemiology

Humans apparently are the only natural host for the cholera vibrios. Cholera is acquired by the ingestion of water or food contaminated with the feces of an infected individual. Previously, the disease swept the world in six great pandemics and later receded into its ancestral home in the Indo-Pakistani subcontinent. In 1961, the El Tor biotype (a subset distinguished by physiologic characteristics) of V cholerae, not previously implicated in widespread epidemics, emerged from the Celebes (now Sulawesi), causing the seventh great cholera pandemic. In the course of their migration, the El Tor biotype cholera vibrios virtually replaced V cholerae of the classic biotype that formerly was responsible for the annual cholera epidemics in India and East Pakistan (now Bangladesh). The pandemic that began in 1961 is now heavily seeded in Southeast Asia and in Africa. It has also invaded Europe, North America, and Japan, where the outbreaks have been relatively restricted and self-limited because of more highly developed sanitation. Several new cases were reported in Texas in 1981 and sporadic cases have since been reported in Louisiana and other Gulf Coast areas. This now endemic focus appears to be due to a clone which is unique from the pandemic strain. In 1991, the pandemic strain hit Peru with massive force and has since spread through most of the Western Hemisphere, causing more than a million cases. Fortunately, mortality has been less than 1 percent because of the effectiveness of oral rehydration therapy. The vibrios surprised us again, in 1992, with the emergence of O139 in India and Bangladesh. For a while it appeared that O139 would replace O1 (both classical and El Tor) but it has exhibited quiescent periods when O1 reemerges.

Cholera appears to exhibit three major epidemiologic patterns: heavily endemic, neoepidemic (newly invaded, cholera-receptive areas), and, in developed countries with good sanitation, occasional limited outbreaks. These patterns probably depend largely on environmental factors (including sanitary and cultural aspects), the prior immune status or antigenic experience of the population at risk, and the inherent properties of the vibrios themselves, such as their resistance to gastric acidity, ability to colonize, and toxigenicity. In the heavily endemic region of the Indian subcontinent, cholera exhibits some periodicity; this may vary from year to year and seasonally, depending partly on the amount of rain and degree of flooding. Because humans are the only reservoirs, survival of the cholera vibrios during interepidemic periods probably depends on a relatively constant availability of low-level undiagnosed cases and transiently infected, asymptomatic individuals. Long-term carriers have been reported but are extremely rare. The classic case occurred in the Philippines, where "cholera Dolores" harbored cholera vibrios in her gallbladder for 12 years after her initial attack in 1962. Her carrier state resolved spontaneously in 1973; no secondary cases had been associated with her well-marked strain. Recent studies, however, have suggested that cholera vibrios can persist for some time in shellfish, algae or plankton in coastal regions of infected areas and it has been claimed that they can exist in "a viable but nonculturable state."

During epidemic periods, the incidence of infection in communities with poor sanitation is high enough to frustrate the most vigorous epidemiologic control efforts. Although, transmission occurs primarily through water contaminated with human feces, infection also may be spread within households and by contaminated foods. Thus, in heavily endemic regions, adequate supplies of pure water may reduce but not eliminate the threat of cholera.

In neoepidemic cholera-receptive areas, vigorous epidemiologic measures, including rapid identification and treatment of symptomatic cases and asymptomatically infected individuals, education in sanitary practices, and interruption of vehicles of transmission (e.g., by water chlorination), may be most effective in containing the disease. In such situations, spread of cholera usually depends on traffic of infected human beings, although spread between adjacent communities can occur through bodies of water contaminated by human feces. John Snow was credited with stopping an epidemic in London, England, by the simple expedient of removing the handle of the "Broad Street pump" (a contaminated water supply) in 1854, before acceptance of the "germ theory" and before the first isolation of the "Kommabacillus" by Robert Koch.

In such developed areas as Japan, Northern Europe, and North America, cholera has been introduced repeatedly in recent years, but has not caused devastating outbreaks; however, Japan has reported secondary cases and, in 1978, the United State experienced an outbreak of about 12 cases in Louisiana. In that outbreak, sewage was infected, and infected shellfish apparently were involved. Interestingly, the hemolytic vibrio strain implicated was identical to one that caused an unexplained isolated case in Texas in 1973.

Diagnosis

Rapid bacteriologic diagnosis offers relatively little clinical advantage to the patient with secretory diarrhea, because essentially the same treatment (fluid and electrolyte replacement) is employed regardless of etiology. Nevertheless, rapid identification of the agent can profoundly affect the subsequent course of a potential epidemic outbreak. Because of their rapid growth and characteristic colonial morphology, V cholerae can be easily isolated and identified in the bacteriology laboratory, provided, first, that the presence of cholera is suspected and, second, that suitable specific diagnostic antisera are available. The vibrios are completely inhibited or grow somewhat poorly on usual enteric diagnostic media (MacConkey agar or eosin-methylene blue agar). An effective selective medium is thiosulfate-citrate-bile salts-sucrose (TCBS) agar, on which the sucrose-fermenting cholera vibrios produce a distinctive yellow colony. However, the usefulness of this medium is limited because serologic testing of colonies grown on it occasionally proves difficult, and different lots vary in their productivity. This medium is also useful in isolating V parahaemolyticus. They can also be isolated from stool samples or rectal swabs from cholera cases on simple meat extract (nutrient) agar or bile salts agar at slightly alkaline pH values. Following observation of characteristic colonial morphology with a stereoscopic microscope using transmitted oblique illumination, microorganisms can be confirmed as cholera vibrios by a rapid slide agglutination test with specific antiserum. Classic and El Tor biotypes can be differentiated at the same time by performing a direct slide hemagglutination test with chicken erythrocytes: all freshly isolated agar-grown El Tor vibrios exhibit hemagglutination; all freshly isolated classic vibrios do not. In practice, this can be accomplished with material from patients as early as 6 hours after streaking the specimen in which the cholera vibrios usually predominate. However, to detect carriers (asymptomatically infected individuals) and to isolate cholera vibrios from food and water, enrichment procedures and selective media are recommended. Enrichment can be accomplished by inoculating alkaline (pH 8.5) peptone broth with the specimen and then streaking for isolation after an approximate 6-hour incubation period; this process both enables the rapidly growing vibrios to multiply and suppresses much of the commensal microflora.

The classic case of cholera, which includes profound secretory diarrhea and should evoke clinical suspicion, can be diagnosed within a few minutes in the prepared laboratory by finding rapidly motile bacteria on direct, bright-field, or dark-field microscopic examination of the liquid stool. The technician can then make a second preparation to which a droplet of specific anti-V cholerae O group 1 antiserum is added. This quickly stops vibrio motility. Another rapid technique is the use of fluorescein isothiocyanate-labeled specific antiserum (fluorescent antibody technique) directly on the stool or rectal swab smear or on the culture after enrichment in alkaline peptone broth. For cultural diagnosis, both nonselective and selective (TCBS) media may be used. Although, demonstration of typical agglutination essentially confirms the diagnosis, additional conventional tests such as oxidase reaction, indole reaction, sugar fermentation reactions, gelatinase, lysine, arginine, and ornithine decarboxylase reactions may be helpful. Tests for chicken cell hemagglutination, hemolysis, polymyxin sensitivity, and susceptibility to phage IV are useful in differentiating the El Tor biotype from classic V cholerae. Tests for toxigenesis may be indicated.

Diagnosis can be made retrospectively by confirming significant rises in specific serum antibody titers in convalescents. For this purpose, conventional agglutination tests, tests for rises in complement-dependent vibriocidal antibody, or tests for rises in antitoxic antibody can be employed. Convenient microversions of these tests have been developed. Passive hemagglutination tests and enzyme-linked immunosorption assays (ELISAs) have also been proposed.

Cultures that resemble V cholerae but fail to agglutinate in diagnostic antisera (nonagglutinable or non-O group 1 vibrios) present more of a problem and require additional tests such as oxidase, decarboxylases, inhibition by the vibriostatic pteridine compound 0/129, and the "string test." The string test demonstrates the property, shared by most vibrios and relatively few other genera, of forming a mucus-like string when colony material is emulsified in 0.5 percent aqueous sodium deoxycholate solution. Additional tests for enteropathogenicity and toxigenesis may be useful. Genetically based tests such as PCR are increasingly being used in specialized laboratories.

Control

Treatment of cholera consists essentially of replacing fluid and electrolytes. Formerly, this was accomplished intravenously, using costly sterile pyrogen-free intravenous solutions. The patient's fluid losses were conveniently measured by the use of buckets, graduated in half-liter volumes, kept underneath an appropriate hole in an army-type cot on which the patient was resting. Antibiotics such as tetracycline, to which the vibrios are generally sensitive, are useful adjuncts in treatment. They shorten the period of infection with the cholera vibrios, thus reducing the continuous source of cholera enterotoxin; this results in a substantial saving of replacement fluids and a markedly briefer hospitalization. Note, however, that fluid and electrolyte replacement is all-important; patients who are adequately rehydrated and maintained will virtually always survive, and antibiotic treatment alone is not sufficient.

Recently it has been recognized that almost all cholera patients and others with similar severe secretory diarrheal disease can be maintained by fluids given orally if the solutions contain a usable energy source such as glucose. Because of this discovery, packets containing appropriate salts are distributed by such organizations as WHO and UNICEF to cholera-afflicted areas, where they are dissolved in water as needed. One such formulation, called ORS for oral rehydration salts, contains NaCl, 3.5 g; KCl,1.5 g; $NaHCO_3$, 2.5 g (or trisodium citrate, 2.9 g); and glucose, 20.0 g. This mixture is dissolved in 1 L of water and taken orally in increments. Flavoring may be added. Improved versions of ORS, including rice-based formulations that reduce stool output and can be made at home, have been recommended. Unfortunately, this technique, which will save countless millions of lives in developing countries, has not yet been widely accepted by practicing physicians in developed countries.

The possibility of pharmacologic intervention (e.g., a pill that will stop choleraic diarrhea after it has started), has been considered. Two drugs, chlorpromazine and nicotinic acid, have been effective in experimental animals, although the precise mechanism of action has yet to be defined.

Like smallpox and typhoid, cholera under natural circumstances appears to affect only humans; therefore, V cholerae as an etiologic entity could conceivably disappear with the last human infection. Nevertheless, the spectrum of cholera-like diarrheal diseases probably will persist for some time.

Cholera is essentially a disease associated with poor sanitation. The simple application of sanitary principles protecting drinking water and food from contamination with human feces would go a long way toward controlling the disease. However, at present, this is not feasible in the underdeveloped areas that are afflicted with epidemic cholera or are considered to be cholera receptive. Meanwhile, development of a vaccine that would effectively prevent colonization and manifestations of cholera would be extremely helpful. As indicated above, such vaccines are presently being tested. Antibiotic or chemotherapeutic prophylaxis is feasible and may be indicated under certain circumstances. It also should be mentioned that the incidence of cholera is significantly higher in formula-fed than in breast-fed babies.

Present information indicates that V parahaemolyticus enteritis could be almost completely prevented by applying appropriate procedures to prevent multiplication of the organisms in contaminated seafood, such as keeping it refrigerated continually.

Other Vibrio Infections

Other vibrios may be clinically significant also. These include non-O group 1 V cholerae. Vibrio parahaemolyticus,a halophilic (salt-loving) vibrio associated with enteritis is acquired by ingestion of raw or improperly cooked seafoods. Another halophilic vibrio, which ferments lactose and for this reason was called the L + vibrio, has recently been identified as V vulnificus. It has been associated with wound infections as well as fatal septicemias. Other groups of vibrios, previously referred to as group F and EF-6, have recently been classified into species: V fluvialis, V hollisae, V furnissia, and V damsela. Vibrio mimicus is a recently described sucrose-negative species. Vibrio fetus, a group of anaerobic to microaerophilic spirally curved rods associated with venereally transmitted infertility and abortion in domestic animals, is now called Campylobacter jejuni and is considered to belong in the family Spirillaceae rather than in the family Vibrionaceae. Campylobacter jejuni has been associated with dysentery-like gastroenteritis, duodenal and gastric ulcers, as well as with other types of infection, including bacteremic and central nervous system infections in humans (see Ch. 23). Another vibrio-like organism, Helicobacter pylori (formerly known as C pylori) causes gastritis and predisposes to duodenal ulcers and gastric cancer. Although, some similarities in habitat and other properties occur, members of the family Vibrionaceae are separated taxonomically from members of the family Enterobacteriaceae. The oxidase test (vibrios are usually oxidase positive) is particularly useful. Other vibrios exist, and some of these may be responsible for diseases in fish and other lower animals. As vibrios are widely distributed in the environment, particularly in estuarine waters and in seafoods, reports of their isolation from patients with diarrheal disease do not necessarily always imply an etiologic relationship.

Cholera-like vibrios have been reported in Maryland's Chesapeake Bay but have not been associated with any human cases despite more than 15 years of extensive surveillance. These vibrios are probably nonpathogenic nonagglutinable (non-O group 1) vibrios, or the atypical O group 1 vibrios mentioned above, which do not contain the genes for toxin production, do not colonize, and are avirulent.

Relatively little is known about the epidemiology of nonagglutinable vibrios. When sought, these vibrios have been found widely in brackish surface waters (sewers, marshes, bogs, and coastal areas), and are generally more numerous in warmer months. They appear to be free-living aquatic organisms; whether particular subsets are potential pathogens is not yet clear. Strains isolated from humans with diarrheal disease more frequently give positive responses in assays for enterotoxins or enteropathogenicity, but the pathogenic mechanism of other isolates associated with shellfish remains undefined. An epidemiologic pattern is more evident with V parahaemolyticus, which is clearly part of the normal flora of coastal and estuarine waters throughout the world. Although, originally recognized in Japan, V parahaemolyticus enteritis has been reported virtually worldwide within the last decade. Its reported frequency varies widely, partly because of inherent differences in distribution and partly because many laboratories do not use the appropriate culture medium (TCBS) to isolate these organisms. Two types of clinical syndromes, both usually self-limited, have been observed. The most common is a watery diarrhea, perhaps with associated abdominal cramps, nausea, vomiting, and fever, with a modal incubation period of 15 hours. A dysenteric syndrome with a short incubation period of 2 1/2 hours also has been described. In Japan, about 24 percent of reported cases of food poisoning are attributed to V parahaemolyticus. The disease occurs primarily during summer, possibly reflecting the increased presence of the organism in the marine environment during those months, as well as the enhanced opportunity for it to multiply in unrefrigerated foods. It appears to be transmitted exclusively by food, primarily raw or improperly prepared seafood. As growth of this organism is inhibited at temperatures below 15° C, rapid cooling and refrigeration of seafoods that areeaten raw would vastly reduce the incidence of disease. The organisms are killed by heating to 65° C for 10 minutes; therefore, properly handled cooked seafood should present no problem. The role played in virulence and pathogenesis by the thermostable direct hemolysin, which is responsible for the positive Kanagawa phenomenon (a hemolytic reaction around colonies growing on a particular blood agar medium), is not yet fully defined. This hemolysin is clearly associated with pathogenicity, but whether it is merely an associated marker or intimately involved in the disease process awaits further research. Be this as it may, only strains that possess the Kanagawa hemolysin are considered pathogenic. In laboratory studies, the isolated hemolysin has been reported to be cytotoxic, cardiotoxic, and lethal.

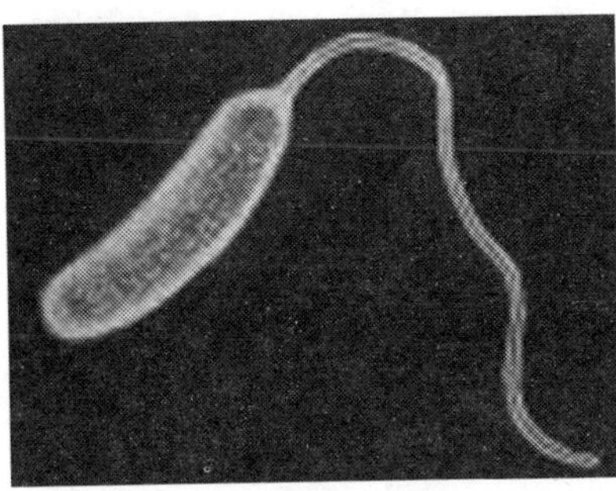

Structural appearance of *Vibrio*

REFERENCES

Albert MJ: Vibrio cholerae O139 Bengal. J Clin Microbiol 32:2345, 1994

Barua D, Greenough III WB: Cholera. Plenum Book Company, New York and London, 1992

Blake JD, Weaver RE, Hollis DG: Diseases of humans (other than cholera) caused by vibrios. Annu Rev Microbiol 34:341, 1980

Clemens JD et al: Field trial of cholera vaccines in Bangladesh: results from three-year follow-up. Lancet 335:270, 1990

Finkelstein RA: Cholera. Crit Rev Microbiol 2:553, 1973

Finkelstein RA: Cholera. In Germanier R (ed): Bacterial

vaccines. Academic Press, San Diego, 1984

Finkelstein RA: Cholera, the cholera enterotoxins, and the cholera enterotoxin-related enterotoxin family. p. 85. In Owen P, Foster TS (eds): Immuno-chemical and Molecular Genetic Analysis of Bacterial Pathogens. Elsevier, Amsterdam, 1988

Finkelstein RA, Burks MF, Zupan A et al: Epitopes of the cholera family of enterotoxins. Rev Infect Dis 9:544, 1987

Gill DM: Seven toxic peptides that cross cell membranes. p. 291. In Jeljaszewicz I,

Wadstrom T (eds): Bacterial Toxins and Cell Membranes. Academic Press, San Diego, 1978

Hoge CW, Watsky D, Peeler RN et al: Epidemiology and spectrum of Vibrio infections in a Chesapeake Bay community. J Infect Dis 160:985, 1989

Kaper JB, Morris Jr JG, Levine MM: Cholera. Clin Microbiol Rev 8:48, 1995

Kaper JB, Moseley SL, Falkow S: Molecular characterization of environmental and nontoxigenic strains of Vibrio cholerae. Infect Immun 32:661, 1981

Levine MM, Kaper JBV, Black RE, Clements ML: New knowledge on pathogenesis of bacterial enteric infections as applied to vaccine development. Microbiol Rev 47:510, 1983

Levine MM, Kaper JP, Herrington D et al: Volunteer studies of deletion mutants of Vibrio cholerae O1 prepared by recombinant techniques. Infect Immun 56:161, 1988

Marchlewicz BA, Finkelstein RA: Immunologic differences among the cholera/coli family of enterotoxins. Diagn Microbiol Infect Dis 1: 129, 1983

Mekalanos JJ, Swartz DJ, Pearson GDN et al: Cholera toxin Miller VL, Taylor RK, Mekalanos JJ: Cholera toxin transcriptional activator Tox R is a transmembrane DNA binding protein. Cell 48:271, 1987

Morris JG, Jr, Black RE: Cholera and other vibrioses in the United States. N Engl J Med 312:343, 1985

Moss J, Vaughn M: Activation of adenylate cyclase by choleragen. Annu Rev Biochem 48:581, 1979

Ouchterlony O, Holmgren J (eds): Cholera and related diarrheas; molecular aspects of a global health problem. 43rd Nobel Symposium, co-sponsored by the World Health Organization. S Karger, Basel, 1980

Peterson JW, Ochoa LG: Role of prostaglandins and cAMP in the secretory effects of cholera toxin. Science 245:857, 1989

Wachsmuth IK, Blake PA, Olsvik O. Vibrio cholerae and Cholera: Molecular to Global Perspectives. ASM Press, Washington, DC,1994

World Health Organization: Diarrheal diseases control programme. Report of the tenth meeting of the technical advisory group (Geneva, March 1317, 1989). WHO/D/89 32:1, 1989

Van Heyningen WE, Seal JR: Cholera: The American Scientific Experience, 1947-1980. Westview Press, Boulder CO, 1983

Chapter **33**

Bordetella

General Concepts

Clinical Manifestations

Bordetella pertussis causes whooping cough (pertussis), an acute respiratory infection marked by severe, spasmodic coughing episodes during the paroxysmal phase. Leukocytosis with lymphocytosis is also common during this phase of the illness. Dangerous complications are bronchopneumonia and acute encephalopathy. *Bordetella parapertussis* can cause a milder form of pertussis.

Structure

The bordetellae are small, Gram-negative, aerobic coccobacilli. *Bordetella pertussis* produces a number of virulence factors, including pertussis toxin, adenylate cyclase toxin, filamentous hemagglutinin, and hemolysin. Agglutinogens and other outer membrane proteins are important antigens.

Classification and Antigenic Types

The genus *Bordetella* contains the species *B pertussis* and *B parapertussis*, which cause pertussis in humans. Other members of the genus are *B bronchiseptica*, causing respiratory disease in various animals and occasionally in humans, and *B avium* as well as *B hinzii*, which cause respiratory disease in poultry and are very rarely found in humans.

Bordetellae are characterized by culture characteristics, biochemical tests, and nucleic acid analysis. Some of them show reversible antigenic modulation under certain culture conditions, and they mutate through several antigenically distinct phases when grown on agar.

Pathogenesis

Transmission is by droplets. The bacteria colonize only ciliated cells of the respiratory mucosa, and they multiply rapidly.

Bacteremia does not occur. The roles of the various toxins in pathogenesis are unclear.

Host Defenses

Infection induces substantial immunity, although the protective antigens have not been identified conclusively. Both nonspecific and specific defenses participate in the response to disease.

Epidemiology

The human respiratory mucosa is the natural habitat for *B pertussis* and *B parapertussis*. Transmission is almost always directly from person to person. Patients are most infectious during the early, catarrhal phase of the disease and remain infectious for about 5 weeks. Pertussis is a common and dangerous childhood disease in unvaccinated populations.

Diagnosis

B pertussis can be cultured on modified Bordet-Gengou medium, charcoal-horseblood agar (Regan-Lowe) or grown in supplement Stainer-Scholte broth. Bordetella DNA can also be detected by PCR. Circulating antibodies appear in week 3 of illness and peak in the eighth to tenth week. Antibodies can be demonstrated by an enzyme-linked immunosorbent assay. Detection of specific IgA provides evidence of natural infection.

Control

Treatment with erythromycin does not alter the course of disease, but reduces the infectious period to 5 to 10 days. Inactivated whole-cell vaccines are highly effective, but occasionally cause toxic side effects. Acellular vaccines with fewer side effects have been licensed for booster vaccination and will possibly be also licensed for primary vaccination.

INTRODUCTION

Bordetella pertussis was first isolated in pure culture in 1906 by Bordet and Gengou. Today, *B pertussis* belongs to the genus *Bordetella* in the family Alcaligenaceae, which contains several species of closely related bacteria with similar morphology. *B pertussis* and *B parapertussis* cause whooping cough (pertussis) in humans. Other members of the genus are *B bronchiseptica*, which causes respiratory disease in various animals and is only occasionally found in humans.Recent additions to the genus are *B avium* and *B hinzii*, which both cause respiratory disease in poultry and are very rarely found in humans.

Clinical Manifestations

After an incubation period of 1 to 2 weeks, whooping cough begins with the catarrhal phase (Fig. 31-1). This phase lasts 1 to 2 weeks and is usually characterized by low-grade fever, rhinorrhea, and progressive cough; the patient is highly infectious. The subsequent paroxysmal phase, lasting 2 to 4 weeks, is characterized by severe and spasmodic cough episodes. At the end of the catarrhal phase, a leukocytosis with an absolute and relative lymphocytosis frequently begins, reaching its peak at the height of the paroxysmal stage. At this time, the total blood leukocyte levels may resemble those of leukemia ($\geq$ 100,000/mm^3), with 60 to 80 percent being lymphocytes. The convalescent phase, lasting 1 to 3 weeks, is characterized by a continuous decline of the cough before the patient returns to normal. Serious complications, sometimes fatal, are bronchopneumonia and acute encephalopathy, the latter being characterized primarily by convulsions and frequently resulting in death or lifelong brain damage.

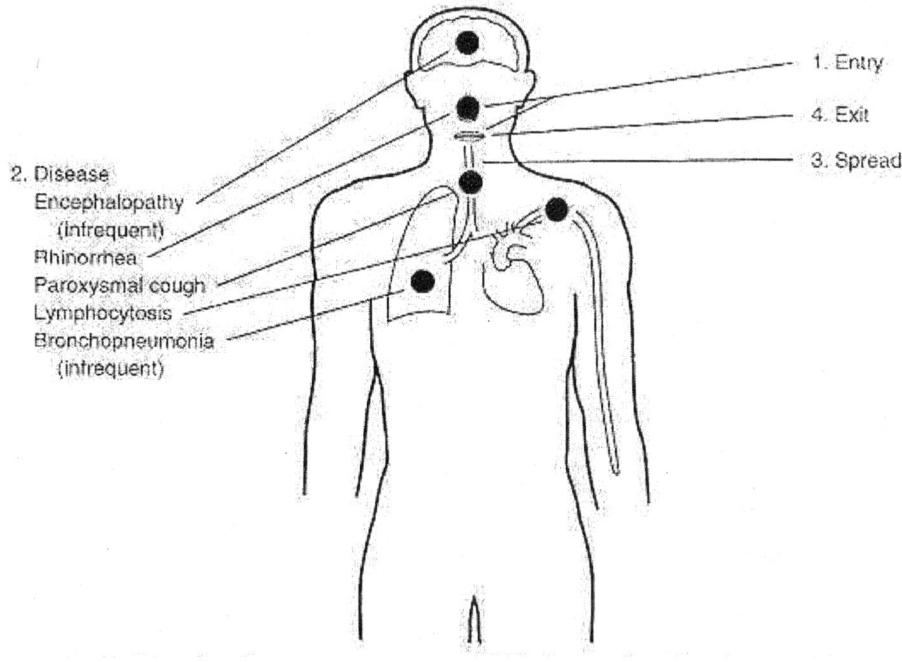

FIGURE 31-1 Pathogenesis of whooping cough.

Structure

Bordetella pertussis is a small (approximately 0.8 µm by 0.4 µm), rod-shaped, coccoid, or ovoid Gram-negative bacterium that is encapsulated and does not produce spores. It is a strict aerobe. It is arranged singly or in small groups and is not easily distinguished from Haemophilus species. *B pertussis* and *B parapertussis* are nonmotile. Numerous antigens and biologically active structural components have been demonstrated in *B pertussis* (Fig. 31-2), although their exact chemical structure and location in the bacterial cell are known only in part.

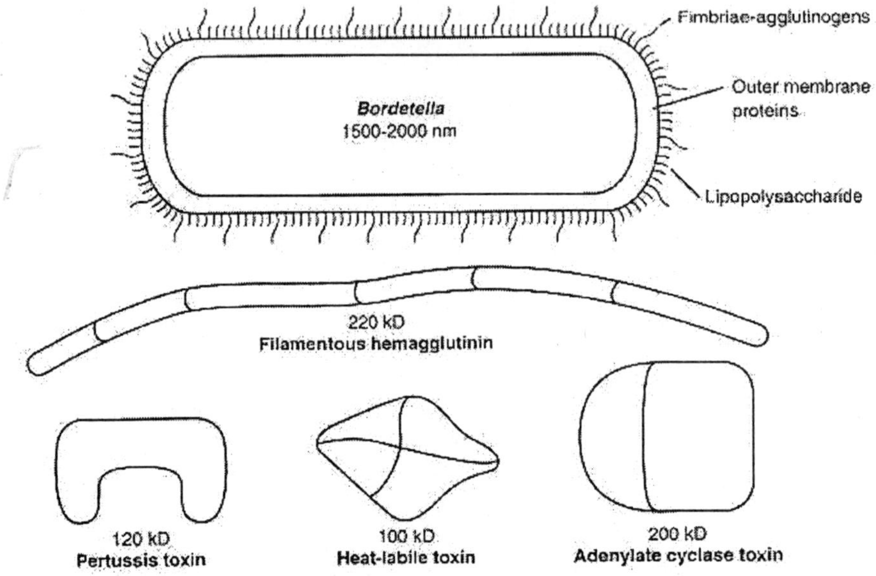

FIGURE 31-2 Virulence factors of *B pertussis.*

Pertussis Toxin

Various immunologic, physiologic, and pharmacologic effects are induced by killed *B pertussis* cells in experimental animals (e.g., increased sensitivity to histamine and serotonin and active and passive anaphylaxis). Adjuvant activity, leukocytosis, splenomegaly, cell proliferation, hypoglycemia, and hypoproteinemia also occur. Many additional features have been described, including increased sensitivity to factors such as endotoxins, X irradiation, infection, cold stress, pollen extracts, peptone shock, and methacholine; increased resistance to infection; increased capillary permeability; and accelerated production of experimental "allergic" encephalomyelitis.

It is now generally accepted that all those biologic activities are caused by a single active protein produced by *B pertussis.* To avoid confusion caused by the many different names for this protein, the uniform term pertussis toxin was proposed by Pittman. Pertussis toxin is a protein exotoxin, secreted during in vivo and in vitro growth; it consists of five different subunits, designated S1, S2, S3, S4, and S5. Since the toxin molecule contains two S4 subunits, it is a hexamer. Like many other protein toxins, pertussis toxin consists of an A subunit that carries the biologic activity and a B subunit that binds the complex to the cell membrane. In pertussis toxin, the S 1 subunit constitutes the A protomer and the B oligomer is formed by the remaining five subunits (Fig. 31-3). The toxin binds to cell receptors by two dimers, one consisting of S2 and S4 and the other of S3 and S4. Since glutaraldehyde-inactivated pertussis toxin is capable of adherence, this binding activity evidently has little to do with the various toxic activities of pertussis toxin. The toxin reacts with different cell types, including T lymphocytes, and acts on different cellular regulatory processes. Pertussis toxin is a member of the family of ADP-ribosylating bacterial toxins. The S1 subunit of pertussis toxin ADP-ribosylates the Cys352 of protein Gi (GTP-binding protein), as well as the corresponding cysteine of protein Ga and of transducin. Although, pertussis toxin is synthesized solely by *B pertussis*, both *B parapertussis* and *B bronchiseptica* possess genes for pertussis toxin without expressing them. *Bordetella parapertussis* expresses pertussis toxin when the toxin gene from the *B pertussis* chromosome is introduced into *B parapertussis*.

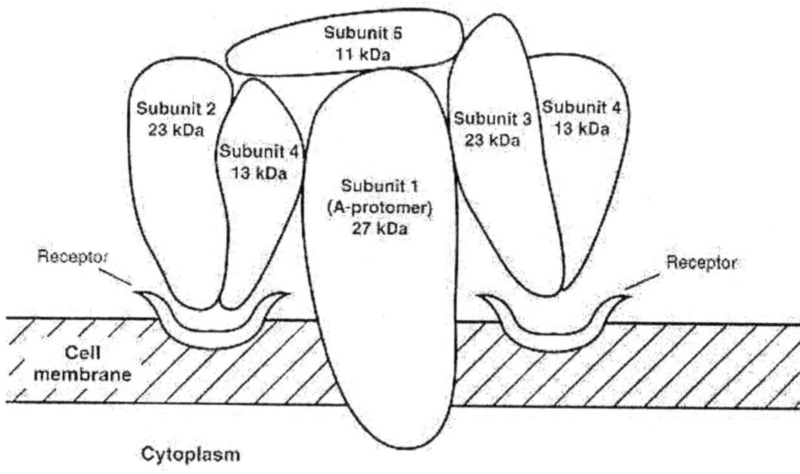

FIGURE 31-3 Binding of pertussis toxin to cell membranes.

Like many other bacteria, *B pertussis* possesses hemagglutinating activity, expressed as its capacity to agglutinate red cells from geese, chickens, and other animals. Pertussis toxin is one of the hemagglutinins, whereas another component with hemagglutinating activity is called filamentous hemagglutinin. This component appears as fine filaments, about 2 nm in diameter and 40 to 100 nm in length. Like pertussis toxin, it has hemagglutinating activity as well as the ability to effect the adherence of *B pertussis* to cilia by its lectin-like binding to lactose-containing moieties.

Heat-Labile Toxin

The heat-labile toxin of *Bordetella* is a proteinaceous dermonecrotic toxin with a molecular weight of about 100,000, localized in the protoplasm. This toxin produces strong vasoconstrictive effects, which are probably important during the initial phase of pertussis by their action on the respiratory tract. Thus, heat-labile toxin, in association with tracheal cytotoxin and lipopolysaccharide, possibly causes tissue damage in the respiratory tract.

Adenylate Cyclase Toxin

Adenylate cyclase toxin is a protein toxin that penetrates the host cell, is activated by host cell calmodulin, and increases intracellular CAMPmassively. The increase of CAMP, which is rather short-lived as in contrast to the action of pertussis toxin, is associated with the inhibition ofphagocytic cell oxidative responses and natural killer cell (NK) activity.

Tracheal Cytotoxin

Tracheal cytotoxin, which is chemically related to peptidoglycan, destroys the ciliated cell population of a hamster trachea in 60 to 96 hours.

Lipopolysaccharide

The heat-stable *Bordetella* lipooligosaccharide (LOS) endotoxin is similar in structure, chemical composition, and biologic activity to other endotoxins produced by Gram-negative bacteria. Endotoxin from *B pertussis*, of which two types can be distinguished, is serologically different from corresponding preparations from *B parapertussis* and *B bronchiseptica.* It is remarkable that heat-labile toxin, adenylate cyclase toxin, tracheal cytotoxin, and LPS are formed by the three *Bordetella* species, whereas pertussis toxin is produced solely by *B pertussis.*

Agglutinogens

The agglutinogens are surface antigens responsible for agglutination of the bacterial cells in the presence of their corresponding antibodies. To date, 14 different agglutinogens (AGG 1 through AGG 14) have been distinguished. AGG1 is specific for *B pertussis*, and is associated with lipooligosaccharide. AGG14 is thought to be specific for *B parapertussis*. The AGGs 2 and 3 (formerly 2, 3, 4, 6) are associated with different types of fimbriae of *B pertussis*, which may also be true for the AGGs 8, 9 and 10 of *B parapertussis*.

Outer Membrane Proteins

At least four different outer membrane protein structures are distinguished on *B pertussis*; they are designated OMP 15, OMP 18, OMP 69, and OMP 91. They are believed to be protective antigens.

Classification and Antigenic Types

The genus *Bordetella* contains species of serologically related bacteria with similar morphology, size, and staining reactions. *B pertussis* and *B parapertussis* are genomically extremely closely related. Other members of the genus are *B bronchiseptica*, which by DNA-DNA and DNA-rRNA hybridization is also closely related. Recent additions tothe genus are *B avium* (formerly designated *Alcaligenes faecalis*) and *B hinzii* (formerly designated *A faecalis* type II), which cause respiratory disease in poultry and are very rarely found in humans.

Bordetella pertussis was first isolated in pure culture in 1906 and was long considered the sole agent of whooping cough. Later studies revealed that this disease also can be caused in a mild form by *B parapertussis* and occasionally by *B bronchiseptica*. A phenomenon of *B pertussis* organisms is their variation during growth on agar plates: the antigenically competent, smooth, virulent form (phase I) can mutate to the antigenically incomplete, nonvirulent, rough form (phase IV). This change is associated with a loss of capacity to synthesize pertussis toxin, filamentous hemagglutinin, heat-labile toxin, adenylate cyclase toxin, agglutinogens, and certain outer membrane proteins. There are also two intermediate forms, called phases II and III.

In addition to this spontaneous phase variation, *B pertussis* undergoes antigenic modulation in response to changes in environmental conditions, such as growth at low temperatures or on agar plates with high concentrations of $MgSO_4$ or nicotinic acid. *Bordetella pertussis* organisms grown under such conditions are avirulent and are characterized by the loss of the capacity to synthesize the numerous toxic factors and other structural components. Both phase variation and antigenic modulation are reversible and also occur in *B parapertussis* and *B bronchiseptica*. Both phenomena are under the control of a single genetic locus. The virulent strains are therefore designated Bvg+, and the avirulent strains Bvg-. Phase variation has been observed in vivo. Another type of serotype variation in *B pertussis* the loss of one or more agglutinogensoccurs independently of phase variation.

Pathogenesis

The agent of whooping cough is transmitted primarily via droplets. Infection results in colonization and rapid multiplication of the bacteria on the mucous membranes of the respiratory tract. Bacteremia does not occur. Electron microscopic studies have demonstrated that phase I strains of *B pertussis* adhere only to the tuft of ciliated cells in the mucosa of the human respiratory tract; no attachment to nonciliated cells was observed. Convincing experimental data indicate that the adherence of *B pertussis* to human cilia is effected by a synergistic action of pertussis toxin and filamentous hemagglutinin, each acting as a bivalent bridge between the bacterium and the ciliary receptor (Fig. 31-4).

Studies of numerous *B pertussis* toxins and their corresponding biologic activities have yielded plausible explanations for many of the symptoms of whooping cough. These include, for example, the frequent occurrence of absolute lymphocytosis (an unusual phenomenon in bacterial infections), hypoglycemia, and the adjuvant effect of pertussis toxin on the immune response to unrelated antigens. The finding that phase I isolates of *B bronchiseptica* produce almost complete ciliostasis within 3 hours in ciliated epithelial cell outgrowths from canine tracheal explants may be explained by the action of adenylate cyclase toxin and tracheal cytotoxin. The same toxins evidently inhibit the phagocytic activities of the host. In humans, an initial local peribronchial lymphoid hyperplasia occurs, accompanied or followed by

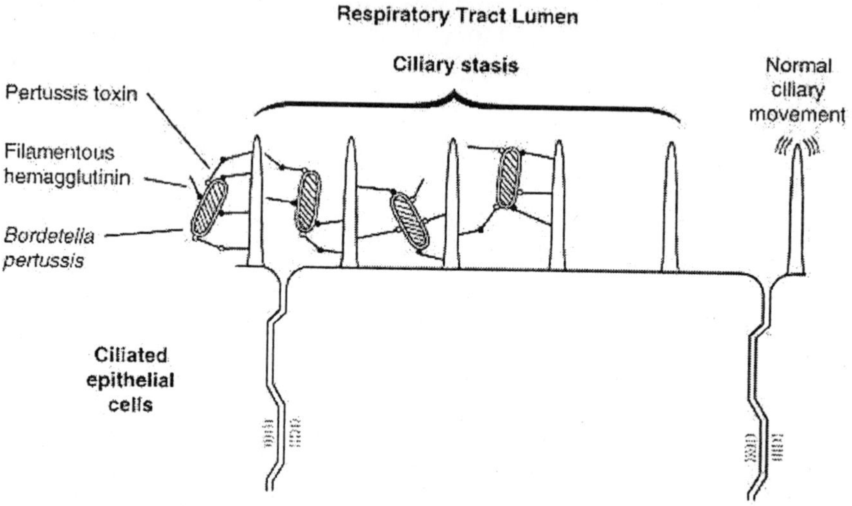

Respiratory Tract Lumen

Ciliary stasis

Normal ciliary movement

Pertussis toxin

Filamentous hemagglutinin

Bordetella pertussis

Ciliated epithelial cells

FIGURE 31-4 Synergy between pertussis toxin and the filamentous hemagglutinin in binding to ciliated respiratory epithelial cells.

necrotizing inflammation and leukocyte infiltration in parts of the larynx, trachea, and bronchi. Usually, peribronchiolitis and variable patterns of atelectasis and emphysema also develop.

To date, there is no plausible explanation for the development of the paroxysmal coughing syndrome characteristic of pertussis. According to Pittman, pertussis is mediated by pertussis toxin and is characterized by a two-stage process infection (colonization) and disease thus resembling other bacterial toxicoses such as diphtheria, tetanus, and cholera. This fascinating idea can be accepted only if it is clearly demonstrated that pertussis toxin causes paroxysmal coughing. Such a demonstration is lacking. Moreover, paroxysmal coughing occurs in infections with *B parapertussis*, which does not synthesize pertussis toxin. On the other hand, an additional infection with *B pertussis* cannot be excluded in such cases. There is no convincing explanation for the acute encephalopathy sometimes observed in pertussis. Research into the pathogenetic mechanisms of pertussis are hampered by the lack of an adequate animal model showing the characteristic paroxysmal coughing syndrome and by the limited opportunity to perform direct studies of the respiratory tract of babies and children.

Host Defenses

A case of whooping cough confers substantial immunity, which usually lasts for many years. Second infections of adults, usually with a typical symptoms and thus not regularly diagnosed as pertussis, may be more frequent than previously assumed. Immunity acquired after infection with *B pertussis* does not protect against the other *Bordetella* species. Pertussis toxin is assumed to be one essential protective immunogen, but numerous findings indicate that other components, such as filamentous hemagglutinin, heat-labile toxin, agglutinogens, outer membrane proteins, and adenylate cyclase toxin, may also contribute to immunity after infection or vaccination. The immunogenicity of the substances may be significantly increased by the presence of pertussis toxin. This synergism indicates that pertussis toxin could function as an adjuvant to a variety of protective antigens of *B pertussis*. The defense mechanisms are both nonspecific (local inflammation, increase in macrophage activity, and production of interferon) and specific (proliferation of B and T cells). The basis of immunity in whooping cough is, however, incompletely understood. A role of circulating antibody in immunity is indicated by the correlation between protection of human vaccinees and their serum agglutinin titers. However, effective immunity does not necessarily depend on the presence of serum agglutinins, and immunity to whooping cough may therefore be mediated essentially by cellular mechanisms. This cell-mediated immunity may be considered the crucial carrier of long-term immunity, and titers of specific humoral antibodies may diminish over the years. This may be the reason why infants usually do not benefit significantly from maternal antibody.

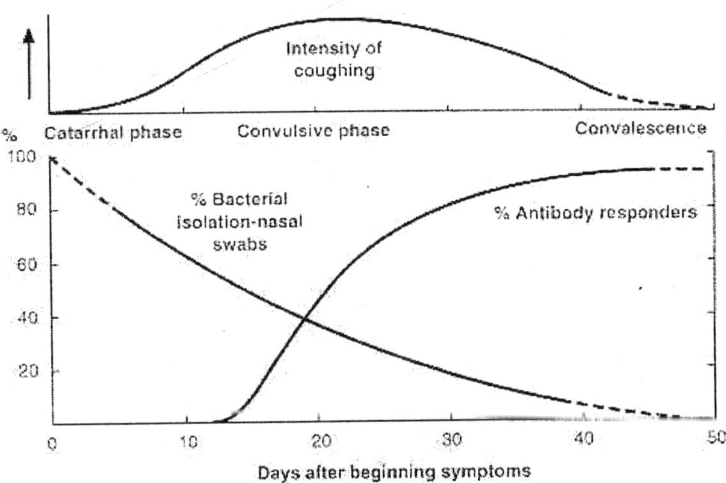

FIGURE 31-5 Relationship of *B pertussis* to the developing antibody response during whooping cough.

Epidemiology

The mucous membranes of the human respiratory tract are the natural habitat for *B pertussis* and *B parapertussis*. Although, *B pertussis* can survive outside the body for a few days and so may be transmitted by contaminated objects, most infections occur after direct contact with diseased persons specifically, by inhalation of bacteria-bearing droplets expelled in cough spray. The patient is most infectious during the early catarrhal phase, when clinical symptoms are relatively mild and noncharacteristic (Fig. 31-5). Subclinical cases may have similar epidemiologic significance. Healthy carriers of *B pertussis* or *B parapertussis* are assumed to play no significant epidemiologic role. The natural habitat of *B bronchiseptica* is the respiratory tract of smaller animals such as rabbits, cats, and dogs. Therefore, human infections with *B bronchiseptica* are extremely rare and occur only after close contact with carrier animals.

Whooping cough, a highly communicable, worldwide infection, was once common and dangerous, killing many thousands of children per year. Widespread vaccination has caused a continuous decrease in incidence and mortality over the years, but large numbers of patients still die in countries where vaccination is inadequate. Whooping cough is mainly an infection of infants and children, although susceptibility is general. The disease is especially dangerous in the first 6 months of life. Neither season nor climate seems to affect the morbidity rate.

Diagnosis

Bordetellae can be cultured from nasopharyngeal swabs or nasopharyngeal secretions. The sensitivity of the method depends mainly on the technique of taking the swabs or secretions. Swabs (one for each nostril) should be introduced deeply into the nose as to reach the nasopharynx. Swabs should be made of dacron or calcium alginate, and they should be transported in half strength charcoal blood agar. Secretions should be from the nasopharynx using a suction device with a mucus trap. Nasopharyngeal secretions should be immediately plated onto Regan-Lowe medium, which has replaced Bordet-Gengou medium as the medium of choice. The transportation time for both materials should be kept as short as possible. For culture isolation, Bordet-Gengou agar containing blood, potato extract, and glycerol remains one of the effective means, although minor modifications regarding blood concentrations and addition of penicillin and nicotinamide have been recommended. For routine use, charcoal-blood agar (REGAN-LOWE medium) is most widely used. A (2,6-O-dimethyl)-b-cyclodextrin supplemented STAINER-SCHOLTE broth can be used as an enrichment medium. The *Bordetella* species do not need factors X and V (NAD+ and hemin).

Bordetella pertussis usually grows after 3 to 4 days of incubation at 37° C. The small, transparent colonies are indistinguishable from those of *B bronchiseptica,* but usually are smaller than those of *B parapertussis*. All three species produce hemolysis. Biochemically they are relatively inert and do not ferment carbohydrates or produce H_2S and indole. An important characteristic of *B parapertussis* is its capacity to produce brown pigmentation on blood-free peptone agar. *B pertussis* and *B parapertussis* can be distinguished by certain biochemical and culture characteristics (Table 31-1) in addition to slide agglutination with specific antisera. *B bronchiseptica* as well as *B avium* and *B hinzii* can be differentiated by conventional methods for typing gram-negative nonfermenting rods (such as API-NE).

TABLE 31-1 Differential Characteristics of B pertussis and B parapertussis

Characteristic	B pertussis	B parapertussis
Molility	-[a]	-
Growth on blood-free peptone agar	-	+
pigment production	-	+
Nitrate reduction	-	-
Urea hydrolysis	-	+
Oxidase reaction	+	-

a+, Present; - absent.

Detection of *B pertussis* and *B parapertussis* DNA by PCR has been described, and various primers and detection methods were applied. Ongoing studies will define the role of a standardized PCR method in pertussis diagnostics.

Circulating antibodies, appearing as late as week 3 of illness and reaching their maximum at weeks 8 to 10, have been demonstrated by agglutination and complement fixation tests. The agglutination test is applied mainly in epidemiologic studies. Although, no direct relationship has been shown between the agglutinin concentration and the degree of protection, high agglutinin titers (>1:320) are assumed to correlate with protection from disease. Modern serologic techniques, such as enzyme-linked immunosorbent assay (ELISA), have been used to detect IgG, IgM, IgA, and IgE antibodies to both whole *B pertussis* cells and certain isolated components. In accordance with other serologic methods, seroconversion could be observed only 2 to 4 weeks after the onset of the disease (Fig. 31-5). The detection of specific IgA and IgM antibodies, however, is indicative of recent infection and is useful for the differential diagnosis of pertussiform syndromes of longer duration. IgA antibodies to pertussis toxin and FHA (filamentous hemagglutinin are found mainly after natural infection; the same is true of secretory IgA in nasopharyngeal secretions, which usually appears during week 2 or 3 of illness. Unfortunately, infants do not regularly produce IgA antibody before 6 months of age. Specific IgM antibodies may be used in infants as an indicator of acute infection.

Control

Although, *B pertussis* is susceptible in vitro to several antibiotics, such as tetracycline, erythromycin, and chloramphenicol, the efficacy of these drugs in patients during the paroxysmal phase is not convincing. Treatment with erythromycin, which is usually considered the antibiotic of choice, will eliminate viable *B pertussis* organisms from the respiratory tract within a few days. The treatment, however, has no influence on the course of the disease. Human hyperimmune pertussis globulin is still used occasionally, but no reliable data support its efficacy. Further treatment is symptomatic.

Susceptible children (unimmunized children without a history of whooping cough) should have no contact with pertussis patients during the first 4 weeks of illness, although such isolation is often difficult. A patient treated with erythromycin may be contagious for only 5 to 10 days. Exposed immunized children younger than 4 years are given booster doses of pertussis vaccine. Exposed unimmunized children are given erythromycin for 10 days after contact is discontinued or after the patient ceases to be contagious.

Pertussis vaccine is produced from smooth forms (phase I) of the bacteria as a killed whole-cell vaccine. In the United States, pertussis vaccination of infants and children is recommended. Owing to a relatively mild course of disease and to occasional neurologic complications after vaccination, it has been argued by numerous pediatricians that general vaccination with the whole-cell vaccine is no longer justified.

Acellular pertussis vaccines have been developed, and were licensed in Japan since 1981 for children (older than two years), and also have been used in infants since 1990. These vaccines are composed very differently and contain various amounts of structural components from the bacteria. Components available for vaccine production include pertussis toxin (which is detoxified), filamentous hemagglutinin, a 69 kDa outer-membrane protein called pertactin, and fimbrial antigens 2 and 3. Some of these vaccines have been licensed in the U.S. for booster vaccinations since 1991. Recent data suggest that after primary vaccinations of infants these vaccines can convey similar levels of protection as the whole-cell vaccine. Thus, acellular vaccines have also been licensed in some European countries for primary vaccination.

REFERENCES

Alouf JE, Freer JH: Sourcebook of bacterial protein toxins. Academic Press, London, 1991

Arai H, Munoz J: Crystallization of pertussigen from *Bordetella pertussis*. Infect Immun 31:495, 1981

Bemis DA, Kennedy JR: An improved system for studying of *Bordetella bronchiseptica* on the ciliary activity of canine tracheal epithelial cells. J Infect Dis 144:349, 1981

Bordet J, Gengou U: Le microbe de la coqueluche. Ann Inst Pasteur 20:731, 1906

Finger H, Heymer B, Hof H et al: Ueber Struktur und biologische Aktivitat von *Bordetella pertussis* Endotoxin. Zentralbl Bakteriol Mikrobiol Hyg 1 Abt OrigA 235:56, 1976

Finger H, Wirsing von Koenig CH: Enhancement and suppression of immune responsiveness by bacteria, bacterial products and extracts. p. 16. In Zschiesche W (ed): Immune Modulation by Infectious Agents. Gustav Fischer Verlag, Jena, 1987

Finger H, Wirsing von Koenig CH: Serological diagnosis of whooping cough. Dev Biol Stand 61:331, 1985

Goldman WE, Klapper DG, Basemann JB: Detection, isolation and analysis of a released *Bordetella pertussis* product toxic to cultured tracheal cells. Infect Immun 36:782, 1982

Goodman YE, Wort AJ, Jackson FL: Enzyme-linked immunosorbent assay for detection of pertussis immunoglobulin A in nasopharyngeal secretions as an indicator of recent infection. J Clin Microbiol 13:286, 1981

Kersters K, Hinz K-H, Hertle A et al: *Bordetella avium* sp. nov., isolated from the respiratory tract of turkeys and other birds. Int J Syst Bacteriol 34:56, 1984

Meade BD, Mink CM, Manclark CR: Serodiagnosis of pertussis, p. 322. In: Manclark CR: Proc. 6th Intl Symp Pertussis, DHHS (FDA) Publication No. 90-1164; Bethesda, MD, 1990

Meade BD, Bollen A: Recommendations for use of the polymerase chain reaction in the diagnosis of *Bordetella pertussis* infections. J Med Microbiol 41:51-55, 1994

Monack D, Munoz JJ, Peacock MG et al: Expression of pertussis toxin correlates with pathogenesis in *Bordetella* species. J Infect Dis 159:205, 1989

Munoz JJ, Bergman RK: *Bordetella pertussis*. Vol. 4. Marcel Dekker, New York, 1977

Pittman M: Pertussis toxin: the cause of the harmful effects and prolonged immunity in whooping cough. A hypothesis. Rev Infect Dis 1:402, 1979

Pittman M: The concept of pertussis as a toxin-mediated disease. Pediatr Infect Dis J 3:467, 1984

Robinson A, Duggleby CJ, Gorringe AR et al: Antigenic variation in *Bordetella pertussis*. p. 147. In Birbeck TH, Penn CW (eds): Antigenic Variation and Infectious Diseases. Society for General Microbiology, IRL Press, Oxford, 1986

Vandamme P, Hommez J, Vancanneyt M, Monsieurs M, Hoste B, Cookson BT, Wirsing von Konig CH, Kersters K, Blackall PJ: *Bordetella hinzii* sp. nov. isolated from poultry and humans. Int J Syst Bact 45:37-45, 1995

Wardlaw AC, Parton R: Pathogenesis and Immunity in Pertussis. John Wiley & Sons, New York, 1988

Weiss AA, Falkow S: Genetic analysis of phase change in *Bordetella pertussis*. Infect Immun 43:263, 1984

Wirsing von Koenig CH, Tacken A, Finger H: Use of supplemented Stainer-Scholte broth for the isolation of *Bordetella pertussis* from clinical material. J Clin Microbiol 26:2558, 1988

Chapter 34

Brucella

General Concepts

Clinical Manifestations

Brucellosis is a severe acute febrile disease caused by bacteria of the genus Brucella. Relapses are not uncommon; focal lesions may occur in bones, joints, genitourinary tract, and other sites. Hypersensitivity reactionscan follow occupational exposure. Infection may be subclinical. Chronic infections may occur.

Structure

Brucellae are Gram-negative coccobacilli; non-spore-forming and non-motile; aerobic, but may need added CO_2.

Classification and Antigenic Types

Three species (B melitensis, B abortus, B suis) are important human pathogens; B canis is of lesser importance. Species are differentiated by production of urease and H_2S, dye sensitivity, cell wall antigens and phage sensitivity. The major species are divided into multiple biovars.

Pathogenesis

Portals of entry are the mouth, conjunctivae, respiratory tract and abraded skin. Organisms spread, possibly in mononuclear phagocytes, to reticuloendothelial sites. Small granulomas reveal a mononuclear response; hypersensitivity is a major factor.

Host Defenses

Effective host defense depends mainly upon cell-mediated immunity.

Epidemiology

Brucellosis is a zoonosis, acquired from handling of infected animals or consuming contaminated milk or milk products. Exposure is frequently occupational. The disease is now uncommon in the United States and Britain but common in the Mediterranean and Arabian Gulf regions, Latin America, Africa, and parts of Asia.

Diagnosis

Diagnosis can be made clinically if there is a history of exposure. Blood cultures may be positive in early disease but serology is mainstay of diagnosis. Interpretation is complicated by subclinical infections and persistent levels of antibody.

Control

Brucellosis is prevented by pasteurizing milk, eradicating infection from herds and flocks, and observing safety precautions (protective clothing and laboratory containment). The disease is treated with doxycycline, streptomycin and rifampin.

INTRODUCTION

Bacteria of the genus Brucella cause disease primarily in domestic, feral and some wild animals and most are also pathogenic for humans. In animals, brucellae typically affect the reproductive organs, and abortion is often the only sign of the disorder. Human brucellosis is either an acute febrile disease or a persistent disease with a wide variety of symptoms. It is a true zoonosis in that virtually all human infections are acquired from animals. The disease is controlled by the routine practice of pasteurizing milk and milk products, as well as by comprehensive campaigns to eradicate the disease by destroying domestic animals which exhibit positive serologic reactions to brucellae. Vaccines providing some protection to cattle, sheep and goats are available.

Clinical Manifestations

The presentation of brucellosis ischaracteristically variable. The incubation period is often difficult to determine but is usually from 2 to 4 weeks. The onset may be insidious or abrupt. Subclinical infection is common.

In the simplest case, the onset is influenza like with fever reaching 38 to 40°C. Limb and back pains are unusually severe, however, and sweating and fatigue are marked. The leukocyte count tends to be normal or reduced, with a relative lymphocytosis. On physical examination, splenomegaly may be the only finding. If the disease is not treated, the symptoms may continue for 2 to 4 weeks. Many patients will then recover spontaneously but others may suffer a series of exacerbations. These may produce an undulant fever in which the intensity of fever and symptoms recur and recede at about 10 day intervals. Anemia is often a feature. True relapses may occur months after the initial episode, even after apparently successful treatment.

Most affected persons recover entirely within 3 to 12 months but some will develop complications marked by involvement of various organs, and a few may enter an ill-defined chronic syndrome. Complications include arthritis, often sacroiliitis, and spondylitis (in about 10 percent of cases), central nervous system effects including meningitis (in about 5%), uveitis and, occasionally, epididymoorchitis. In contrast to animals, abortion is not a feature of brucellosis in pregnant women. Hypersensitivity reactions, which may mimic the symptoms of an infection, may occur in individuals who are exposed to infective material after previous, even subclinical, infection.

Structure

Brucellae are Gram-negative coccobacilli (short rods) measuring about 0.6 to 1.5 µm by 0.5-0.7 µm. They are non-sporing and lack capsules or flagella and, therefore, are non-motile. The outer cell membrane closely resembles that of other Gram-negative bacilli with a dominant lipopolysaccharide (LPS) component and three main groups of proteins. The guanine-plus-cytosine content of the DNA is 55-58 moles/cm. No Brucella species has been found to harbor plasmids naturally although they readily accept broad-host-range plasmids.

The metabolism of the brucellae is mainly oxidative and they show little action on carbohydrates in conventional media. They are aerobes but some species require an atmosphere with added CO_2 (5-10 percent). Multiplication is slow at the optimum temperature of 37°C and enriched medium is needed to support adequate growth.

Brucella colonies become visible on suitable solid media in 2-3 days. The colonies of smooth strains are small, round and convex but dissociation, with loss of the O chains of the LPS, occurs readily to form rough or mucoid variants. These latter forms are natural in B canis and B ovis as the LPS of these lack O chains.

Classification and Antigenic Types

Distinguishing features of the six species of Brucella and their preferred hosts are shown in Table 28-1. B abortus, B melitensis and B suis are serious pathogens in humans, B canis causes mild disease and the other two species have not affected humans.

TABLE 28-1 Differential characteristics of Brucella species

			Lysis by phages					
			To		R/C			
Species	Colony Morphology	Serum Requirement	RTD[a]	10 x RTD	RTD	Oxidase	Urease	Preferred Host
B melliensis	Smooth	-	-	-	-	+	+	Sheep, goats
B nbortus	Smooth	-[b]	+	+	-	+	+	Cattle
B suis	Smooth	-	-	+	-	+	+	Biovar1:pigs Biovar2: hares pigs Biovar3:pigs Biovar4:reindoor Biovar5: wild rodents
B neotomae	Smooth	-	-	+	-	-	+	Desert wood rats
B ovis	Rough	+	-	-	+	-	-	Rams
B canis	Rough	-	-	-	+	+	+	Dogs

a Intermediate rate, some strains rapid.
b Except B abortus biovar 2, which generally requires secure for growth on primary isolation.
c Except B abortus biovar 3 strains isolated in Senegal and Guinea Bissau, which are negative.
d Intermediate rate except for reference strain 544 and occasional field strains, which are negative.
e High rate.
f Minute plaques.
g RTD, routine test dilution.

A culture can be identified as belonging to the genus Brucella on the basis of colonial morphology, staining and slide agglutination with anti-Brucella serum, smooth or rough. Further classification is best done in a specialized laboratory. Identification to species level may be done by the procedures shown in Table 28-1. Further differentiation to biovars may be useful and is illustrated in Table 28-2. As a further refinement, tests for the oxidative metabolism of certain aminoacids and carbohydrates have been devised. Modern DNA hybridization tests, however, show that the currently named species show a high degree of homology and suggest that the genus could be appropriately reclassified as having a single species.

TABLE 28-2 Biovar Differentiation of Brucella Species

				Growth on Dyes[a]		Aglutination in Serum[b]		
Species	Biovar	CO_2 Requirement	H_2S production	Thionine	Basic fuchsine	A	M	R
B melitensis	1	-	-	+	+	-	+	-
	2	-	-	+	+	+	-	-
	3	-	-	+	+	+	+	-
B abortus	1	+	+	-	+	+	-	-
	2	+	+	-	-	+	-	-
	3	+	+	+	+	+	-	-
	4	+	+	-	+	-	+	-
	5	-	-	+	+	-	+	-
	6	-	-	+	+	+	-	-
	9	±	+	+	+	-	+	-
B suis	1	-	+	+	-	+	-	-
	2	-	-	+	-	+	-	-
	3	-	-	+	+	+	-	-
	4	-	-	+	-	+	+	-
	5	-	-	+	-	-	+	-
B neoromae		-	+	-	-	+	-	-
B ovis		+	-	+	-	-	-	+
B canis		-	-	+	-	-	-	+

a Dye concentration, 20mg/ml in serum dextrose medium (1:50,000)
b A, A monospecific antiserum; M, M monospecific antiserum; R, rough Brucells antiserum
c usually positive on primary isolation.
d Some strains isolated in Canada, Britain, and the United States do not grow on dyes.
e Some basic fuchsine resistant strains have been isolated in South America and South East Asia.
f Negative for most strains.
g Growth will accrual 10 mg of (1: 100,000) thioninemt.

The application of techniques of molecular biology have allowed the cloning and characterization of several genes coding for outer membrane proteins, the use of PCR to identify the presence of brucellar DNA at genus and species level and the demonstration of species specific patterns of restriction fragment length polymorphism. It is predictable that this work will be extended to improve diagnostic tests and even vaccine development.

Two different O chains in brucellae occur in the LPS of the brucellae with smooth colonies. These are called A and M, nominally indicating abortus and melitensis antigens. ('Nominally', because some abortus biovars carry the M antigen and some common melitensis biovars the A antigen.) Both O chains have been shown to be homopolymers of 4,6-dideoxy-4-formamido-d-mannopyranose; theydiffer only in that in the A chain the sugar molecules are always linked 2-1, whereas the M chain has every fifth junction a 3-1 linkage. In routine serology, smooth species of brucellae cross-react almost completely with each other, but not with rough species and vice versa. Monospecific polyclonal sera reacting only to A or M antigens are prepared by cross absorption and monoclonal antibodies specific for A and M antigens are now available, indicating that there is at least one unique epitope on each type of chain.

Pathogenesis

Brucellae are facultative intracellular parasites, multiplying mainly in monocyte-macrophage cells. This characteristic dominates the pathology, clinical manifestations and therapy of the disease.

The organisms may gain entry into the body through a variety of portals (Fig. 28-1). Because the infection is systemic, it is often not possible to determine which

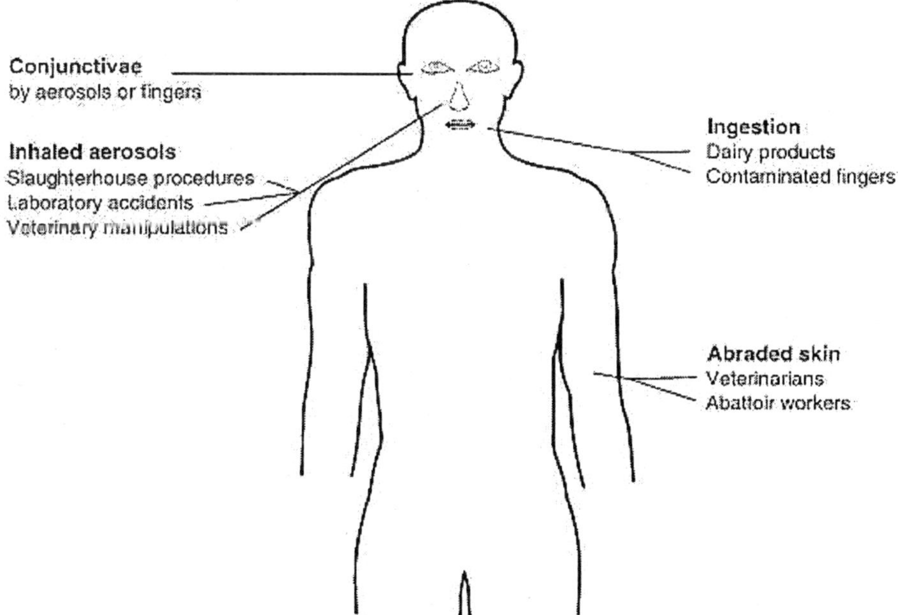

Figure 28-1 Portals of entry for Brucella species.

portal was involved in a particular case. Oral entry, by ingestion of contaminated animal products (often raw milk or its derivatives) or by contact with contaminated fingers, probably represents the most common route of infection even though this portal may not be the most vulnerable one. Inhalation of aerosols containing the bacteria, or aerosol contamination of the conjunctivae, is another route. Inhalation probably underlies some industrial outbreaks. Percutaneous infection through skin abrasions or by accidental inoculation has frequently been demonstrated.

Brucella species differ markedly in their capacity to cause invasive human disease. Brucella melitensis is the most pathogenic; B abortus is associated with less frequent infection and a greater proportion of subclinical cases. The virulence of B suis strains for humans varies but is generally intermediate.

Animal studies suggest that invading brucellae are rapidly phagocytosed by polymorphonuclear leukocytes. Brucellae are frequently able to survive and multiply in these cells because they inhibit the bactericidal myeloperoxidase-peroxide-halide system by releasing 5'-guanosine and adenine. Early in infection, macrophages are also relatively ineffectual in killing the intracellular brucellae (Fig. 28-2). In systemic spread, it is not clear whether the bacteria are transported within neutrophils and macrophages or in the blood stream outside cells but organisms may disseminate widely from regional lymphoid tissue appropriate to the portal of entry and may localize in certain target organs such as lymph nodes, spleen,

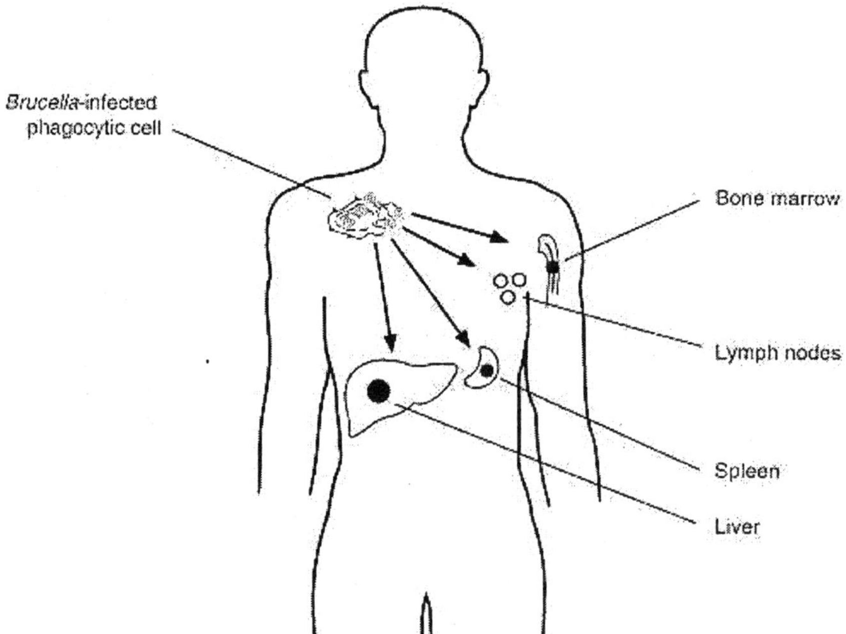

Figure 28-2 Spread of Brucella in the body.

liver, bone marrow, and (especially in animals) the reproductive organs. The presence of meso-erythritol in the testicles and seminal vesicles of bulls, rams, goats, and boars and in the products of conception in pregnant ruminants and pigs stimulates enormous multiplication of brucellae. Erythritol represents a potent localizing factor in the relevant species, but is absent in humans.

In humans, the tissue lesions produced by Brucella species consist of minute granulomas that are composed of epithelioid cells, polymorphonuclear leukocytes, lymphocytes and some giant cells. In cases of infection with B melitensis these granulomas are particularly small although the toxemia associated with this organism is great. Necrosis is not common, and abscesses do not form, except in B suis infection. The fact that humans rapidly develop hypersensitivity to brucellar antigens suggests that many of the symptoms of human brucellosis result from the reaction of the host defenses.

Host Defenses

The specific host defenses against brucellae resemble those against other intracellular bacteria and are both humoral (antibody-mediated) and cell-mediated. Passively administered monoclonal antibody directed against LPS has been shown to reduce the numbers of brucellae surviving in the spleens and livers of experimental mice, indicating a role for antibody in protection. However, the principal component in defense against brucellae is cell-mediated. Macrophages have been shown to process brucellar antigen and present this to T lymphocytes which produce lymphokines. These agents, of which interferon is the most active in this context, activate the formerly ineffective macrophages to greater bactericidal potency. Depletion of gamma interferon makes experimental animals vulnerable to infection. T cell-derived lymphokines are also involved in attracting cells to the foci of infection. This leads to granuloma formation. While this contributes to the pathology, it also delivers the activated macrophages to the site where they are needed. This inflammatory response is enhanced by cytokines, such as the colony-stimulating factors, tumor necrosis factor and interleukin-1, produced by a number of cell types.

Mice which have survived brucellosis are protected against further challenge, and there is clinical evidence that complete recovery from a natural infection is associated with at least a degree of residual resistance in humans.

Epidemiology

The reservoirs of brucellosis are various wild, feral and (particularly) domestic animals (Fig. 28-3). In ruminants, enormous numbers of bacteria are shed widely from infected products of conception, whether aborted or born at term. Brucellae frequently invade the mammary gland of infected ruminants. This organ can even be directly infected by any of the major species of Brucella, for example by contaminated hands. This may allow milk cows, for example, to excrete large numbers of organisms not only B abortus, but B melitensis or B suis and such milk spread has resulted in extensive outbreaks of brucellosis.

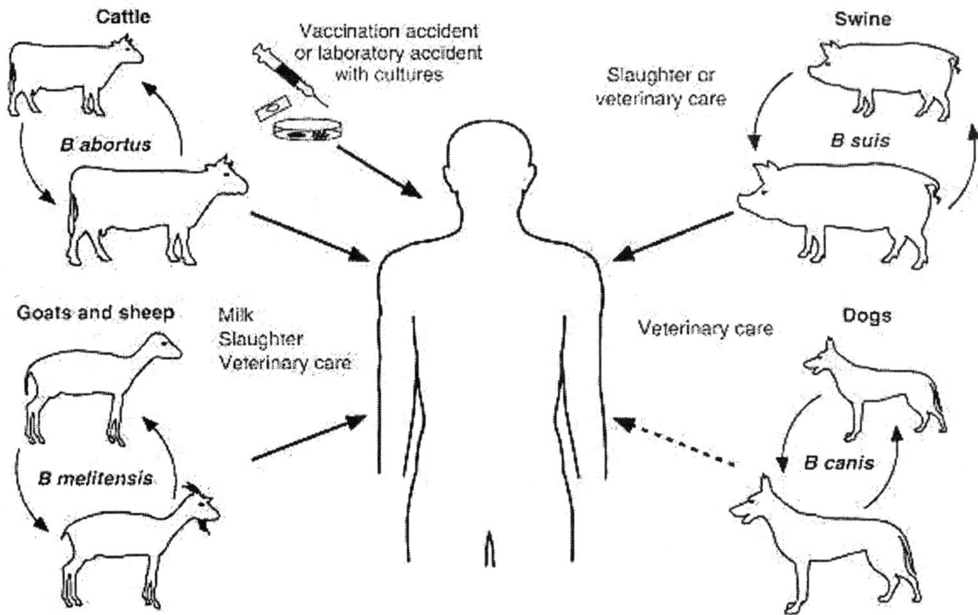

Figure 28-3 Sources of Brucella infection.

Brucella melitensis in sheep and goats represents, by far, the most important source of brucellosis in humans. This species of Brucella is not enzootic in the United States, Canada, northern Europe, Australasia or South East Asia. It is prevalent in Latin America, the Mediterranean area, Central Asia and, especially, in the countries around the Arabian Gulf. Humans are principally infected by the handling of parturient animals and the consumption of raw milk and milk products, especially fresh soft cheeses. In many cases, the vehicle of infection is uncertain and bacteria-laden dust is suspected.

Brucella suis occurs in most areas in which pigs are kept. It affects both sexes of swine causing infertility, abortion, orchitis and lesions of bones and joints. The prevalence is generally low except in parts of South America and South East Asia. In both south eastern United States and in Australia, particularly Queensland, populations of feral swine are heavily infected. Apart from the rare cases in which cows' milk is infected, infections with B suis in humans occur in people handling pigs on farms and during slaughtering and processing - including the hunting of feral swine. While an eradication campaign in the United States has made much headway, abattoir outbreaks still have occurred recently.

Bovine brucellosis, caused by B abortus, has been eradicated from Canada, Japan, northern Europe and Australasia. Cases in humans tend to be sporadic and often stem from occupational exposure. Infection can be acquired by drinking unpasteurized milk, but this a relatively inefficient mode of spread, while abattoir workers can be significantly exposed and veterinarians are doubly at risk from attending parturient cattle and from accidental inoculation with live vaccine.

Cases of B canis infection in man have tended to occur only in dog handlers. Close, frequent contact seems to be necessary for transmission.

Cases of brucellosis have continued to present in some regions from which brucellosis has been effectively eradicated. These are caused, usually by B melitensis, in travelers to popular tourist destinations, like Mexico and the Mediterranean region, in which this organism is highly prevalent, and by the importation of infected dairy products. Although, person to person spread is quite exceptional, the species of principal importance are all potent causes of laboratory infections. Hence stringent safety measures, including adequate containment to reduce the hazard of aerosol spread, are essential. The situation has been aggravated by the failure of certain popular laboratory kits to identify Brucella spp appropriately.

Diagnosis

The diagnosis of brucellosis is primarily dependent on clinical suspicion allied with the taking of an adequate history of possible exposure - including during travel. Presentation can, however, be highly atypical and focal lesions may present decades after exposure.

Unequivocal diagnosis requires isolation of the organism. Blood culture is the method of choice but specimens need to be obtained early in the disease and cultures may need to be incubated for up to four weeks. Even so, failure to grow the organism is common, especially in cases of B abortus infection, and isolation rates of only 20-50% are reported even from experienced laboratories. Modern commercial systems are hampered by the small amount of CO_2 produced during growth. Culture from bone marrow and from presenting foci may be successful. Presumptive identification of cultures from morphology and slide agglutination with specific antiserum should be followed by further work in a reference facility. Molecular techniques for typing are being developed.

Serology remains the mainstay of laboratory diagnosis, but the interpretation of results is fraught with difficulties. The large number of techniques in use is evidence of the problems. The standard serum agglutination test (SAT) has been augmented by the modified Coombs' (antiglobulin) technique and the use of 2-mercaptoethanol to separate the actions of specific IgG and IgM. These classical methods may, in time, be supplanted by EIA (enzyme immunoassay) tests, designed to differentiate between specific IgM and IgG antibodies. While the SAT titers commonly decline after recovery from infection and antiglobulin test levels are maintained much longer, the IgM antibody that is ccmmonly measured by the SAT does not fall away as regularly as in some infections. Nevertheless, persisting levels of antibody may indicate a remaining focus of infection and specific IgG levels rise again with a true relapse.

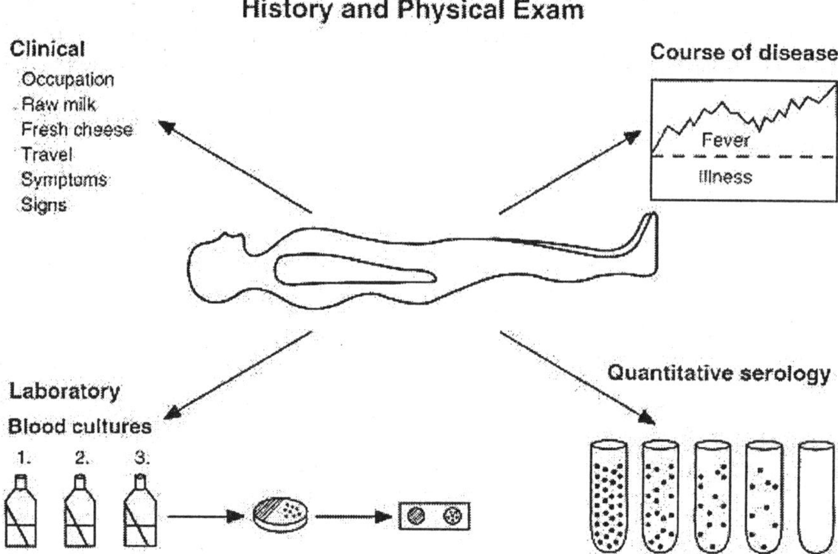

Figure 28-4 Diagnosis of brucellosis.

Further, because cases often are investigated late in their course, rising titers are frequently missed; the variability of individual responses and the frequency of subclinical infections make the interpretation of single high titers subject to error. All serologic tests have to be interpreted with caution in the light of clinical data and in the context of the local prevalence of brucellosis. Moreover, serum from persons with tularemia may show cross-reactions with Brucella antigen.

The diagnosis of the chronic brucellosis syndrome, without specific localization, is often very unsatisfactory. When cultures are negative and the results of serologic tests are equivocal a confident diagnosis is often impossible.

Control

Individuals who are occupationally exposed can be protected to some extent by wearing impermeable clothing, rubber boots, gloves and face masks and by practicing good personal hygiene. Pasteurization of milk for drinking and for incorporation into other dairy products is effective in protecting consumers. No widely accepted vaccines for humans have been developed but progress in the understanding of brucellar epitopes and of immunology could change this.

However, eradication of brucellosis from domestic animals reduces dramatically the threat to humans and has been successful in several countries. In eradication campaigns, the level of enzootic disease can first be reduced by intensive use of live, attenuated vaccines (B abortus strain 19 in cattle, B melitensis strain Rev. 1 for sheep and goats) particularly inimmature animals. Thereafter, theemphasis shifts to the detection of infected herds (by skin tests in sheep; serologic tests on milk or blood samples taken at sale or slaughter in cattle) and individual animals (by serologic tests) and to the elimination of the latter by slaughter.

Humans are treated with combinations of antibiotics for from 4 to 6 weeks. Doxycycline and rifampin form the basis, with cotrimoxazole replacing doxycycline in children, but fewer relapses are reported with regimens including two weeks of daily streptomycin. Azithromycin has shown promising results in experimental models.

REFERENCES

Allardet-Servent A, Bourg G, Ramuz M et al: DNA polymorphism in strains of the genus Brucella. J Bacteriol 170: 4603, 1988

Alton GG, Jones LM, Angus RD et al: Techniques for the brucellosis laboratory. INRA, Paris, 1988

Bricker BJ, Halling SM: Differentiation of Brucella abortus bv. 1,2, and 4, Brucella melitensis, Brucella ovis and Brucella suis bv. 1 by PCR. J Clin Microbiol 32: 2660, 1994

Chomel BB, DeBess EE, Mangiamele DM et al. Changing trends in the epidemiology of human brucellosis in California from 173 to 1992: a shift towards foodborne transmission. J Infect Dis 170: 1216, 1994

Joint FAO/WHO Expert Committee on Brucellosis: Sixth Report, Technical Report Series 740, World Health Organization, Geneva, 1986

Lang R, Shasha B, Ifrach N et al: Therapeutic effects of roxithromycin and azithromycin in experimental murine brucellosis. Chemotherapy 40: 252, 1994

Montejo JM, Alberda I, Glez-Zarate P et al: Open, randomized therapeutic trial of six antimicrobial regimens in the treatment of human brucellosis. Clin Infect Dis 16: 671, 1993

Spink, WW: The Nature of Brucellosis. Minneapolis: University of Minnesota Press, 1956

Young, EJ, Corbel, MJ (eds): Brucellosis: Clinical and Laboratory aspects. CRC Press, Boca Raton, 1989

Chapter 35

Haemophilus Species

General Concepts

Clinical Manifestations

Type b Haemophilus influenzae can cause meningitis, epiglottitis, bacteremia, and cellulitis. Nontypable H influenzae can cause otitias media, sinusitis, tracheobronchitis, and pneumonia. Other Haemophilus species and the syndromes they cause include H parainfluenzae (pneumonia and endocarditis), H ducreyi (genital chancre), and H aegyptius (conjunctivitis or Brazilian purpuric fever).

Structure, Classification, and Antigenic Types

Haemophilus species are Gram-negative coccobacilli similar in ultrastructural features to other pathogenic bacilli. Haemophilus influenzae requires hemin (factor X) and NAD+ (factor V) for growth. Other Haemophilus species require only NAD+ and therefore grow on blood agar. Typable H influenzae isolates are classified on the basis of seven antigenically distinct capsular polysaccharides; isolates lacking these polysaccharides are called nontypable.

Pathogenesis

Type b H influenzae colonizes the nasopharynx, and may penetrate the epithelium and capillary endothelium to cause bacteremia. Meningitis may result from direct spread via lymphatic drainage or from hematogenous spread. Nontypable H influenzae colonizes the nasopharynx and, to a lesser extent, the trachea and bronchi and may infect mucosa damaged by viral disease or cigarette smoking. Lipooligosaccharide is largely responsible for inflammation; exotoxins do not play a role.

Host Defenses

Serum antibody to the capsule (in the case of typable H influenzae) or to somatic antigens is bactericidal and promotes phagocytosis.

Epidemiology

Haemophilus influenzae colonizes healthy children and adults (although the rate of colonization is far greater for nontypable than for type b H influenzae) and is spread by direct contact, secretions, and/or aerosol. Haemophilus ducreyi is spread by venereal contact. There is no animal reservoir for these organisms.

Diagnosis

Respiratory secretions and cerebrospinal fluid must be culturedon chocolate agar. Blood cultures are positive in meningitis. Capsular antigen may be detected in cerebrospinal fluid for early identification if Gram stain is unsuccessful. Haemophilus ducreyi grows on Mueller-Hinton agar with 5 percent sheep blood in a CO_2 enriched atmosphere.

Control

Recommended treatment includes ampicillin for strains of H influenzae that do not make ß-lactamase and a third generation cephalosporin or chloramphenicol for strains that do. Ampicillin or amoxicillin together with a substance, such as clavulanic acid, that blocks the activity of ß-lactamase is also effective, but does not reliably treat meningitis. Tetracyclines remain effective in treating sinusitis or respiratory infection proven to be due to nontypable H influenzae. Use of polyribosylribitol phosphate (PRP) vaccine and, more recently, protein-conjugated PRP has vastly reduced the frequency of infection due to type b H influenzae.

INTRODUCTION

The genus Haemophilus includes a number of species that cause a wide variety of infections but share a common morphology and a requirement for blood-derived factors during growth that has given the genus its name. Haemophilus influenzae, the major pathogen, can be separated into encapsulated or typable strains, of which there are seven types (a through f including e') based on the antigenic structure of the capsular polysaccharide, and unencapsulated or nontypable strains. Type b H influenzae is by far the most virulent organism in this group, commonly causing bloodstream invasion and meningitis in children younger than 2 years. Nontypable strains are frequent causes of respiratory tract disease in infants, children, and adults.

Other Haemophilus species cause disease less frequently. Haemophilus parainfluenzae sometimes causes pneumonia or bacterial endocarditis. Haemophilus ducreyi causes chancroid. Haemophilusaphrophilus is a member of the normal flora of the mouth and occasionally causes bacterial endocarditis. Haemophilus aegyptius, which causes conjunctivitis and Brazilian purpuric fever, and Haemophilus haemolyticus used to be separated on the basis of their ability to agglutinate or lyse red blood cells, but both are now included among the nontypable H influenzae strains.

Clinical Manifestations

Haemophilus species cause a variety of clinical syndromes (Fig. 30-1). Until the implementation of widespread vaccination programs, type b H influenzae was the most common cause of meningitis in children between the ages of 6 months and 2 years. In this situation, headache is followed rapidly by development of a stiff neck, with progression to coma and, in the absence of treatment, death. Emergent treatment reduces the incidence of, but does not eliminate, sequelae such as deafness and learning disabilities. Type b H influenzae also causes cellulitis and epiglottitis, a condition in which the epiglottitis becomes inflamed and swells, closing off the upper airway. Suffocation can be prevented in some cases only by performing a tracheostomy. Nontypable H influenzae strains commonly cause infection of the middle ear (otitis media), which manifests as an earache with fever in babies and young children. In adults, these organisms cause bronchitis and pneumonia, especially if some underlying disease of the bronchi and lungs is present. Nontypable H influenzae strains also commonly cause acute or chronic sinusitis in patients of all ages.

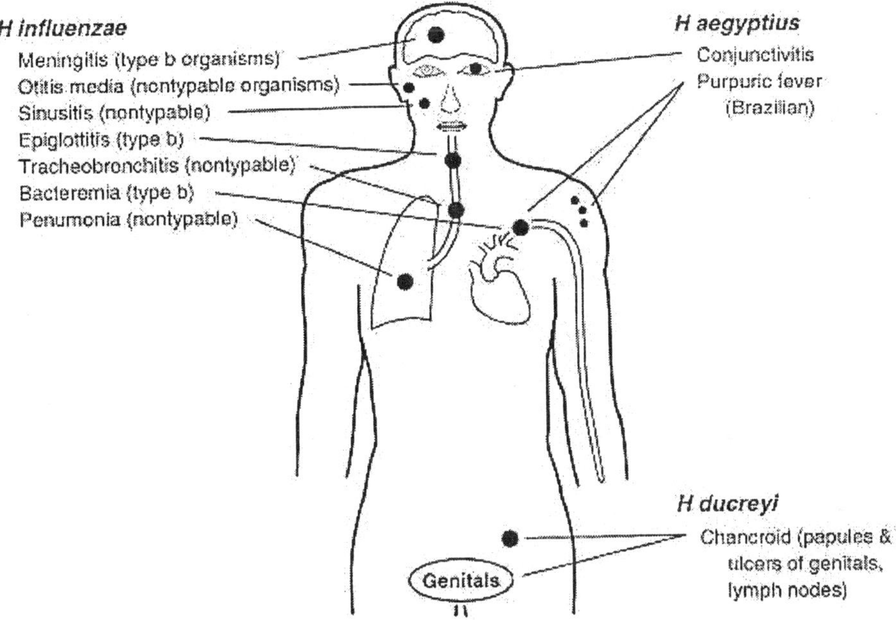

FIGURE 30-1 Clinical presentation of Haemophilus infections.

Chancroid is a venereal disease caused by H ducreyi. Lesions that resemble a syphilitic chancre result from sexual contact with an infected individual; they are usually found on the genitals. Unlike syphilitic chancres, the lesions are painful and are associated with a remarkable degree of swelling of lymph nodes in the inguinal area.

Structure, Classification, and Antigenic Types

Haemophilus species are Gram-negative coccobacilli that share common ultrastructural features with other Gram-negative bacilli. Their cell walls contain lipooligosaccharide, which resembles the lipopolysaccharide of Gram-negative bacilli but has shorter side chains (hence the designation oligosaccharide rather than polysaccharide). Haemophilus species have generally been thought not to make toxins or other extracellular products that account for their ability to produce infection. These organisms require hemin (factor X) and/or nicotinamide adenine dinucleotide (NAD+) (factor V) for growth. Whereas NAD+ is released into the medium by red blood cells and is available to the bacteria in blood agar, hemin is bound to red blood cells and is not released into the medium unless the cells are broken up, as in chocolate agar. Haemophilus influenzae requires both factors X and V; accordingly, it grows on chocolate agar but not on blood agar (Fig. 30-2), although, it may appear on a blood agar plate as tiny satellite colonies around the colonies of other bacteria that have lysed red blood cells. Haemophilus parainfluenzae requires only factor V and therefore is able to grow on blood agar (however, recent reports suggest that many isolates identified as H parainfluenzae actually are H paraphrophilus). The long-prevailing notion that H ducreyi grows only in clotted rabbit blood has been dispelled by recent studies that show slow growth of this organism in Mueller-Hinton agar containing 5 percent sheep blood. All Haemophilus species grow more readily in an atmosphere enriched with CO_2; H ducreyi and some nontypable H influenzae strains will not form visible colonies on culture plates unless grown in CO_2-enriched atmosphere.

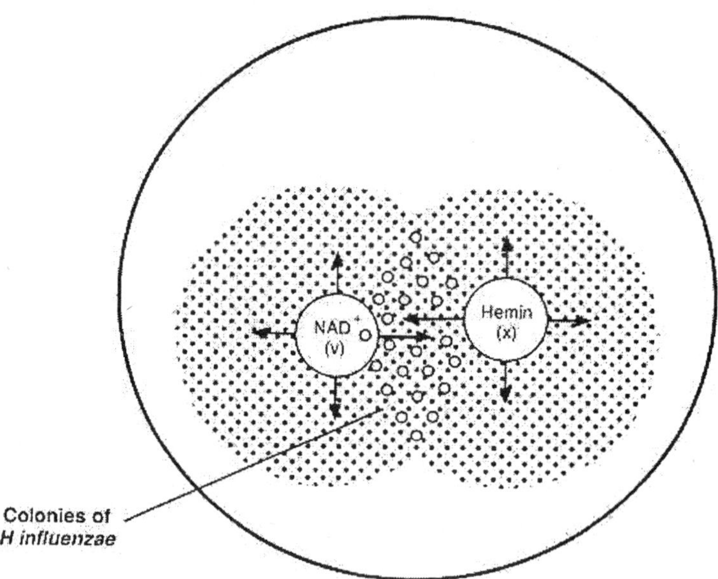

FIGURE 30-2 Growth of H influenzae requires both NAD+ (factor V) and hemin (factor X). Bacterial colonies occur only where both substances have diffused. (The hemin within intact erythrocytes is not accessible to the bacteria unless the erythrocytes are lysed by other bacteria or by heating the medium to make chocolate agar.)

Haemophilus influenzae strains are classifiedas either serotypable (if they display a capsular polysaccharide antigen) or nontypable (if they lack a capsule). The word "type" as applied to H influenzae refers to this serotyping scheme. There are six generally recognized types: a, b, c, d, e, and f. A seventh type has been designated e′ because its polysaccharide is closely related to that of type e. Antiserum to type e′ H influenzae is not routinely available. These types may be identified by an agglutination reaction that uses antisera raised in rabbits; with this method, however, cross-reactions with somatic antigens may cause nontypable strains to be designated erroneously as typable. This kind of error is eliminated by using counterimmunoelectrophoresis, in which migration under an electric current removes somatic (protein) antigens from the reaction, leaving only capsular polysaccharides to react with antibody.

The presence of a polyribosyl ribitol phosphate (PRP) capsule is an important virulence factor: it renders type b H influenzae resistant to phagocytosis by polymorphonuclear leukocytes in the absence of specific anticapsular antibody. Susceptibility to the bactericidal effect of serum depends on the presence of antibodies to a number of antigenic sites, including the lipooligosaccharideor outer membrane proteins designated as P1 and P2. Type b H influenzae is plainly the most virulent of the Haemophilus species; 95 percent of bloodstream and meningeal Haemophilus infections in children are due to this organisms. In contrast, in adults, nontypable strains of H influenzae are the most common cause of Haemophilus infection, presumably because most adults have acquired antibody to PRP.

The relative place of H influenzae biogroup aegyptius, the cause of Brazilian purpuric fever, remains to be determined. Most, but not all, strains that cause this syndrome contain a unique 24 megadalton plasmid and a 79 kilodalton outer membrane protein, either or both of which may mediate virulence. These organisms also have at least some of the genetic material that codes for encapsulation of type b H influenzae.

By using a series of biochemical reactions, H influenzae alsomay be classified into six biotypes designated I through VI. Most type b H influenzae strains fall into biotypes I or II, whereas most non typableH influenzae strains fall into biotypes II through VI. Several interesting clinical correlations have recently emerged. Biotype I isolates appear to have a predilection for causing pneumonia. Nearly all genital isolates, as well as bloodstream isolates from infected neonates or from women with puerperal sepsis, are biotype IV. In addition, biotype III, which agglutinates red blood cells in vitro and includes H influenzae biogroup aegyptius, has been implicated as a common cause of conjunctivitis. There is no explanation for these clinically observed associations.

Pathogenesis

The pathogenesis of H influenzae infections is not completely understood, but the presence of the type b polysaccharide capsule is a major factor in virulence. Encapsulated organisms can penetrate the epithelium of the nasopharynx and invade blood capillaries directly. Nontypable strains are less invasive, but they, as well as typable strains, induce an inflammatory response that causes disease; production of exotoxins is not thought to play a role in pathogenicity. Nontypable

H influenzae strains colonize the nasopharynx of most normal individuals, but type b H influenzae strains are found in only 1 to 2 percent of normal children. Outbreaks of type b infection occur, especially in nurseries and child care centers; prophylactic administration of antibiotics may be used. Vaccination with type b polysaccharide appears to be effective in preventing infection, and vaccines are now available for routine use.

Meningitis

The pathogenesis of meningitis due to type b H influenzae has been well studied. These organisms colonize the nasopharynx and spread from one human to another by direct contact or via secretions and/or aerosol. They penetrate epithelial layers and capillary endothelium by unknown mechanisms, reaching the meninges either directly via lymphatic drainage from the nasopharynx or indirectly by causing bacteremia with subsequent seeding of the highly vascular choroid plexus.

Most cases of H influenzae meningitis in adults are due to nontypable strains. The pathogenesis of these infections differs from that of type b H influenzae. Nontypable strains are unencapsulated and therefore less virulent, and they are unable to penetrate directly into capillaries. Rather, they gain entry to the central nervous system by direct extension, often associated with infection of the sinuses or middle ear and/or with trauma involving the sinuses or skull. Thus, about 50 percent of adults with H influenzae meningitis have a history of prior head trauma with or without a documented cerebrospinal fluid leak, and another 25 percent have chronic otitis media. Also, H influenzae is second only to Streptococcus pneumoniae as a cause of recurring meningitis, an unusual syndrome attributed to a connection between the sinuses and the subarachnoid space, usually via a tear in the dura. The clinical picture of meningitis caused by typable or nontypable H influenzae is similar to that caused by other bacteria, such as S pneumoniae. Since Haemophilus species do not produce substances that obviously damage mammalian tissues, bacterial replication is probably the usual pathway for disease production, with triggering of the complement cascade by classic and alternative pathways, followed by accumulation of inflammatory cells.

Cellulitis and Epiglottitis

Cellulitis and epiglottitis are discussed together because their pathogenesis is probably quite similar. Both are due to type b H influenzae, are likely to cause associated bacteremia, and occur more frequently in children than adults. Epiglottitis can be regarded as a cellulitis of the relatively loose submucosal connective tissues of the epiglottis. In this syndrome, a sore throat rapidly progresses to difficulty in breathing, stridor, obstruction of the air ways, and respiratory arrest. Local extension from the colonized nasopharynx through soft tissues is probably responsible for epiglottitis. Cellulitis often involves the face or neck. It sometimes seems to start at the buccal mucosa and extend outward, supporting the idea that it also results from local extension. The often-repeated teaching that facial cellulitis due to Haemophilus causes a distinctive bluish tinge enabling it to be distinguished from cellulitis caused by other bacteria, defies reason and is best ignored.

Respiratory Disease

Nontypable H influenzae is a major pathogen that colonizes the human respiratory tract. Adherence of bacteria to mammalian tissues, which is mediated by pili (fimbriae), is thought to be an important precursor to colonization, and infection of the upper airways is associated with the presence of pili. Respiratory infections caused by these organisms include sinusitis, otitis media, acute tracheobronchitis, and pneumonia.

Purulent material aspired from acutely infected paranasal sinuses in children or adults or from behind an infected tympanic membrane in babies and young children commonly contains nontypable H influenzae. Studies of outer membrane protein profiles have shown that middle ear and nasopharyngeal isolates are identical, supporting the notion that colonization of the eustachian tube, followed by obstruction and infection, is probably responsible. Repeated bouts of otitis media are thought to be due to different strains; each infection may be associated with emergence of antibody to distinctive surface proteins. The decreasing frequency of otitis media with age is due in part to anatomic changes and in part to immunity to H influenzae.

The situation is somewhat more complex for bronchopulmonary disease. Nontypable H influenzae is found in the nasopharynx and in sputum cultures of nearly one-half of adults with chronic bronchitis. Not surprisingly, this organism is also recovered from the large airways via bronchoscopy, since upper-airway bacteria are carried along by the bronchoscope.On the basis of these observations, as well as of other observations on the presence of antibody in serum samples of patients with chronic bronchitis, some British investigators concluded that H influenzae actually plays a causative role in what they call chronic bronchitisour chronic obstructive pulmonary disease. With the use of semiquantitative techniques, H influenzae have been shown to be very scarce in the sputum of patients with chronic bronchitis who are in a stable clinical state (and perhaps results from oropharyngeal contamination of the specimen), but that, at least in some patients, large numbers are present during an exacerbation. This observation fits with studies in which transtracheal aspiration has revealed the trachea to be free of Haemophilus isolates in stable patients with emphysema, but to contain these organisms in some patients who have purulent sputum.

In any case, nontypable H influenzae is certainly a prominent cause of acute tracheobronchitis or pneumonia in patients who have underlying chronic bronchitis, emphysema, or obstructive pulmonary disease. Other debilitating diseases such as malnutrition, lung cancer, and alcoholism are also often present. Symptoms of acute bronchitis include increased shortness of breath, cough, and production of purulent sputum; in more severe cases, fever and an increased white blood cell count may also be present. Pneumonia may result, with patchy or segmental pulmonary infiltrates detectable by radiography. Gram stain of sputum reveals a profusion of Gram-negative coccobacilli (Fig. 30-3), but no other pathogenic bacteria. A properly plated specimen yields a nearly pure growth of Haemophilus. When S pneumoniae organisms are present in small numbers (for example, 1 S pneumoniae colony to 100 Haemophilus colonies), it is less certain that Haemophilus is the sole pathogen. Blood cultures are positive in 10 to 15 percent of patients with Haemophilus pneumonia. Nontypable H influenzae is second only to S pneumoniae as the cause of bacterial pneumonia in middle-aged men. Although, type b H influenzae is more virulent, pneumonia due to this organism is much less common, probably because of its vastly lower incidence of colonization. Compared with nontypable H influenzae pneumonia, the underlying pulmonary disease may not be so prominent. In type b H influenzae disease, the onset is more acute and blood cultures are more likely to be positive. Haemophilus influenzae received its name because it was first isolated from the lungs of individuals who died during an epidemic of influenza virus infection in 1890. It is not possible to determine whether bacterial infection was due to typable or nontypable isolates.

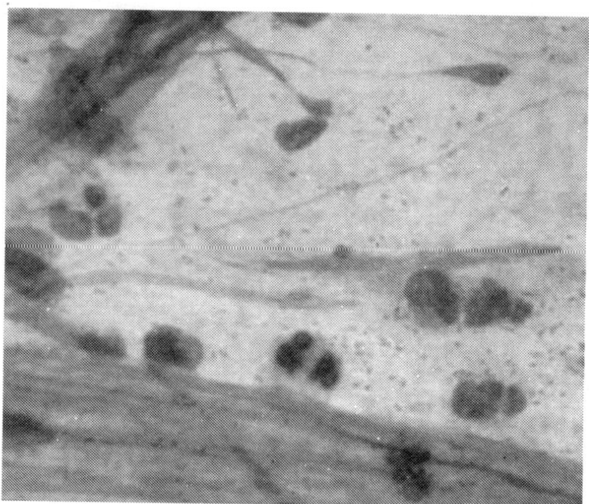

FIGURE 30-3 Gram-stained sputum showing profuse Gram-negative coccobacilli with no other bacterial forms present (original magnification X 440). Cultures showed nontypable H influenzae as the overwhelmingly predominant isolate. Quantitative culture showed 7 X 108 CFU/ml of sputum.

Brazilian Purpuric Fever

In the past few years, a syndrome of fulminating illness with substantial mortality characterized by nausea, vomiting, hemorrhagic skin lesions, fever, prostration, and shock has been recognized under the name Brazilian purpuric fever. Haemophilus influenzae biogroup aegyptius can be cultured from the blood of affected patients. Many have had a history of conjunctivitis in the weeks preceding onset of the disease. As noted above, most of the responsible organisms differ from other H influenzae biogroup aegyptius strains that only cause conjunctivitis in having a unique plasmid and a 79 kD outer membrane protein, as well as containing genetic material that hybridizes with the capsular locus of H influenzae.

Miscellaneous

Type b H influenzae use to be a relatively common cause of septic arthritis in children and results from hematogenous dissemination. Interestingly, this organism only rarely caused osteomyelitis; the reasons for this discrepancy are unknown.

Nontypable H influenzae biotype IV tends to colonize the female genital tract and may cause puerperal fever and/or neonatal sepsis. Infection in the mother is relatively mild, but it may be fulminating in the newborn infant. It is unknown whether nontypable H influenzae biotype IV has any special virulence beyond the factors that promote adherence to vaginal epithelial cells.

Some patients have bacteremia due to type b H influenzae without an apparent focus of infection. There is good evidence that if this condition is left untreated in children, some source for the infection (e.g., meningitis) will become apparent within 24 to 48 hours. However, in this situation one cannot be certain that the meninges were not seeded secondarily to the bacteremia. In adults, there appears to be a syndrome of bacteremia due to nontypable H influenzae for which no focus ever becomes apparent.

Host Defenses

For many years it was believed that bactericidal antibody directed against PRP capsule of type b H influenzae was entirely responsible for host resistance to infection. However, more recent studies have stressed a role for antibody to somatic antigens as well. For example, antibody to PRP can often be detected in the sera of children on admission to the hospital with sepsis due to type b H influenzae. In addition, adsorption of immune serum with PRP alone does not remove its protective capabilities, whereas adsorption with whole organisms does. Finally, immunization with ribosomes is protective in animal models of infection. Separation of the outer membrane of type b H influenzae into its many protein constituents by polyacrylamide gel electrophoresis (PAGE) combined with analysis of antibody responses during infection has suggested that antibody to any of a number of individual membrane proteins may be associated with immunity. Bactericidal antibodies that react with individual outer membrane proteins or with lipooligosaccharide constituents have been identified. These findings support, on a molecular basis, the potential importance of antibody to noncapsular antigens in immunity to type b H influenzae infection. Opsonizing antibody may also play a role in protection and may be directed against PRP or somatic constituents (Fig. 30-4).

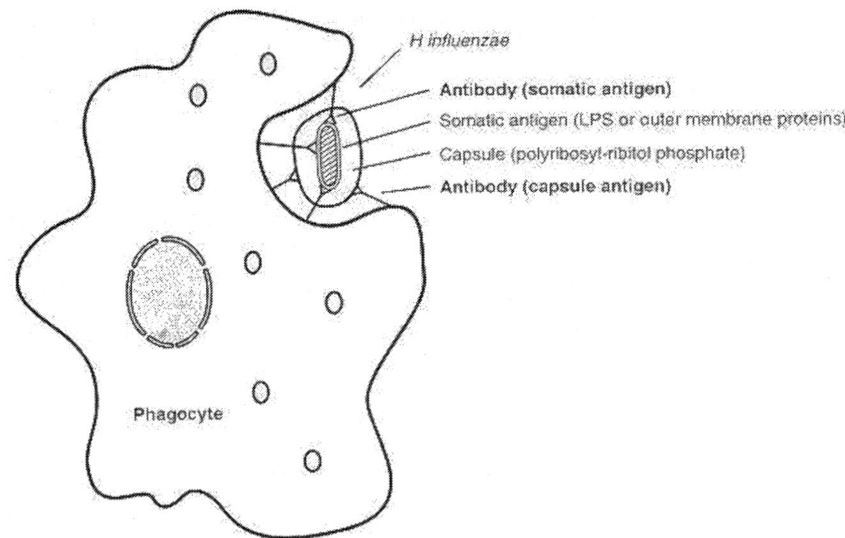

FIGURE 30-4 Macrophage or polymorphonuclear leukocyte phagocytosing H influenzae coated with antibodies specific for the capsule and somatic antigen.

Recent studies of nontypable H influenzae strains have shown that bactericidal antibody to outer membrane proteins develops in infants in response to otitis media caused by these organisms. Normal adults generally have both bactericidal and opsonizing antibodies directed against nontypable H influenzae. Although, levels of opsonizing antibody may be low in adults who develop acute nontypable H influenzae infection, substantial levels of bactericidal activity are present in serum at the time infection is diagnosed. It is not clear why this should occur. In some instances a blocking effect by secretory IgA in bronchial secretions might be responsible. Alternatively, the extensive structural damage to the bronchi and lungs that predisposes to serious nontypable H influenzae infection may allow proliferation of the bacteria unchecked by normal serum defense mechanisms.

Epidemiology

Haemophilus organisms spread directly among individuals without a known contribution from environmental sources or animal reservoirs. Nontypable H influenzae strains are found in the nasopharynx of many healthy subjects, depending upon the frequency and intensity with which they are sought. By contrast, type b H influenzae is found only in 1 to 2 percent of healthy children, and its spread to previously uncolonized children in the early years is associated with a substantially increased risk of infection. Families and day care centers are important sources for dissemination of these organisms.

Diagnosis

Haemophilus influenzae meningitis cannot be distinguished on the basis of clinical presentation, physical examination, or cerebrospinal fluid abnormalities from meningitis due to other common bacterial pathogens. The cerebrospinal fluid in untreated patients contains an average of 2 X 107 bacteria/ml, so that microscopic examination, especially in the absence of prior antibiotic therapy, should reveal the infecting organisms. Detection of capsular material in the cerebrospinal fluid by counter immunoelectrophoresis is helpful in cases in which the Gram stain is not conclusive; this technique is especially important in patients who have received enough antibiotic to suppress the growth of organisms in cultures of cerebrospinal fluid, but not enough to be curative.

The bacteriologic diagnosis of pneumonia or acute febrile purulent tracheobronchitis due to H influenzae is made by finding myriad small, somewhat pleomorphic, Gram-negative coccobacilli in Gram-stained sputum (Fig. 30-3) and by culturing H influenzae as the overwhelmingly predominant isolate; the mean number of viable organisms per milliliter of infected sputum is about 5 X 108. Blood cultures may be positive in 10 to 15 percent of patients with pneumonia and are negative in those with acute febrile tracheobronchitis.

Endocarditis due to H parainfluenzae tends to be associated with large vegetations that embolize to large arteries such as femoral or carotid, causing a limb to turn blue and cool, or producing a stroke. The etiologic diagnosis of endocarditis is, of course, established by blood culture. Recent studies have suggested that many isolates previously identified as H parainfluenzae are, in fact, H paraphrophilus. Chancres due to H ducreyi are tender, somewhat irregular, and slightly indurated; they may be confused with primary syphilitic chancres, traumatic lesions of the penis (especially with bacterial superinfection), fixed drug eruptions, or ulcerated herpetic lesions. The diagnosis is established by culturing the causative organism on Mueller-Hinton agar supplemented with 5 percent sheep blood and incubating it for 96 hours in a CO_2-enriched atmosphere.

Control

Outbreaks of serious infection due to type b H influenzae can be prevented by vaccination or prophylactic therapy. Initial trials of vaccination with type b H influenzae PRP were disappointing, because this polysaccharide in its pure form is not immunogenic in infants, the group most at risk of infection. Later studies showed that injection of PRP conjugated to a protein, such as diphtheria toxoid, that serves as an adjuvant results in good antibody responses in infants. Clinical trials with these vaccines have been successful, and preparation of PRP linked to outer membrane proteins or ribosomes are currently in widespread use.

Once an outbreak of type b H influenzae infection has been documented, infants and toddlers who are in intimate contact with colonized or infected individuals have a greatly increased, albeit still small, likelihood of developing serious infection. The use of rifampin prophylaxis to prevent or eradicate nasopharyngeal colonization has been recommended. This measure is controversial, however, because if widely applied it might encourage the emergence of rifampin-resistant organisms, and also because the cost to prevent each potential case of meningitis is high.

The mainstay of therapy for H influenzae infection used to be ampicillin, since isolates were uniformly susceptible to 0.5 µg/ml. In the last few years an increasing proportion of H influenzae isolates have produced ß-lactamase. In most medical centers, 25 to 30 percent of type b isolates and a somewhat smaller percentage of nontypable isolates are now resistant to penicillin or ampicillin; in some centers, 50 to 60 percent of type b H influenzae isolates are ampicillin resistant. Very rarely, an isolate resists ampicillin but does not produce ß-lactamase; decreased penetration into the bacterium is thought to be responsible. Treatment with a combination of amoxicillin and clavulanic acid (a substance that covalently binds ß-lactamase) is effective against ß-lactamase-producing strains, but has not been recommended for treating meningitis. Chloramphenicol was long considered the drug of choice for meningitis caused by a penicillin-resistant H influenzae strain, and it is still highly effective. Third-generation cephalosporins, such as ceftriaxone or cefotaxime, are effective against H influenzae and penetrate the meninges well; these drugs are useful in treating H influenzae meningitis. The addition of corticosteroids may reduce the incidence of complications such as deafness.

In addition to the above-named drugs, tetracyclineand sulfa drugs are effective in treating upper and lower respiratory infections caused by Haemophilus. Erythromycin should not generally be used to treat H influenzae infections; many isolates are resistant, and documentation of susceptibility in routine clinical laboratories is subject to error.

The spread of soft chancre due to H ducreyi is best prevented by use of a condom during sexual intercourse. Two-thirds of H ducreyi isolates produce ß-lactamase. All isolates are susceptible in vitro to erythromycin, and excellent clinical results have been obtained.

REFERENCES

Doern GV, Jones RN: Antimicrobial susceptibility testing of Haemophilus influenzae,

Branhamella catarrhalis, and Neisseria gonorrhoeae. Antimicrob Agents Chemother 32: 1747, 1988

Eskola J, Peltola H, Takala AK et al: Efficacy of Haemophilus influenzae type b polysaccharide-diphtheria toxoid conjugate vaccine in infancy. N Engl J Med 317:717, 1987

Groeneveld K, van Alphen L, Eijk PP, et al: Endogenous and exogenous reinfections by Haemophilus influenzae in patients with chronic obstructive pulmonary disease: The effect of antibiotic treatment on persistence. J Infect Dis 161:512, 1990

Hammond GW, Slutchuk M, Scatliff J et al: Epidemiologic, clinical, laboratory, and therapeutic features of an urban outbreak of chancroid in North America. Rev Infect Dis 2:867. 1980

Harabuchi Y, Faden H, Yamanaka N, et al: Nasopharyngeal colonization with nontypable Haemophilus influenzae and recurrent otitis media. J Infect Dis 170:862, 1994

Harrison LH, de Silva GA, Pittman M et al: Epidemiology and clinical spectrum of Brazilian purpuric fever. J Clin Microbiol 27:599, 1989

Jorgensen JH, Doern GV, Maher LA, et al: Antimicrobial resistance among respiratory isolates of Haemophilus influenzae, Moraxella catarrhalis, and Streptococcus pneumoniae in the United States. Antimicrob Agents and Chemother 34(11):2075, 1990

Lesse AJ, Gheesling LL, Bittner WE, et al: Stable, conserved outer membrane epitope of strains of Haemophilus influenzae biogroup aegyptius associated with Brazilian Purpuric Fever. Infect and Immun, p. 1351, 1992

Mason EO, Jr, Kaplan SL, Lamberth LB et al: Serotype and ampicillin susceptibility of Haemophilus influenzae causing systemic infections in children: 3 years of experience. J Clin Microbiol 15:543, 1982

Murphy TF, Apicella MA: Nontypable Haemophilus influenzae: a review of clinical aspects, surface antigens, and the human immune response to infection. Rev Infect Dis 9:1, 1987

Murphy TF, Berstein JM, Dryja DM et al: Outer membrane protein and lipooligosaccharide analysis of paired nasopharyngeal and middle ear isolates in otitis media due to nontypable Haemophilus influenzae: pathogenic and epidemiological observations. J Infect Dis 156:723, 1987

Musher D, Goree A, Murphy T et al: Immunity to Haemophilus influenzae type b in young adults: correlation of bactericidal and opsonizing activity of serum with antibody to polyribosylribitol phosphate and lipooligosaccharide before and after vaccination. J Infect Dis 154:935, 1986

Musher DM, Kubitschek KR, Crennan J et al: Pneumonia and acute febrile tracheobronchitis due to Haemophilus influenzae. Ann Intern Med 99:444, 1983

Osterholm MT, Pierson LM, White KE et al: The risk of subsequent transmission of Haemophilus influenzae type b disease among children in day care. N Engl J Med 316:1, 1987

Sell SH, Wright PF (eds): Haemophilus influenzae: Epidemiology, Immunology, and Prevention of Disease. Elsevier Biomedical, New York, 1982 St. Geme JW,III, Falkow S: Infect and Immun, p. 4036, 1990

Wallace RJ, Jr, Baker CJ, Quinones FJ et al: Nontypable Haemophilus influenzae (biotype 4) as a neonatal, maternal, and genital pathogen. Rev Infect Dis 5:123, 1983.

Chapter **36**

Corynebacterium Diphtheriae

General Concepts

Clinical Manifestations

Corynebacterium diphtheriae infects the nasopharynx or skin. Toxigenic strains secrete a potent exotoxin which may cause diphtheria. The symptoms of diphtheria include pharyngitis, fever, swelling of the neck or area surrounding the skin lesion. Diphtheritic lesions are covered by a pseudomembrane. The toxin is distributed to distant organs by the circulatory system and may cause paralysis and congestive heart failure.

Structure, Classification, and Antigenic Types

Corynebacterium diphtheriae is a nonmotile, noncapsulated, club-shaped, Gram-positive bacillus. Toxigenic strains are lysogenic for one of a family of corynebacteriophages that carry the structural gene for diphtheria toxin, *tox*. *Corynebacterium diphtheriae* is classified into biotypes (mitis, intermedius, and gravis) according to colony morphology, as well as into lysotypes based upon corynebacteriophage sensitivity. Most strains require nicotinic and pantothenic acids for growth; some also require thiamine, biotin, or pimelic acid. For optimal production of diphtheria toxin, the medium should be supplemented with amino acids and must be deferrated.

Pathogenesis

Asymptomatic nasopharyngeal carriage is commonin regions where diphtheria is endemic. In susceptible individuals, toxigenicstrains cause disease by multiplying and secreting diphtheria toxin in either nasopharyngeal or skin lesions. The diphtheritic lesion is often covered by a pseudomembrane composed of fibrin, bacteria, and inflammatory cells. Diphtheria toxin can be proteolytically cleaved into two fragments: an N-terminal fragment A (catalytic domain), and fragment B (transmembrane and receptor binding domains).Fragment A catalyzes the NAD^+-dependent ADP-ribosylation of elongation factor 2, thereby inhibiting protein synthesis in eukaryotic cells. Fragment B binds to the cell surface receptor and facilitates the delivery of fragment A to the cytosol.

Host Defenses

Protective immunity involves an antibody response to diphtheria toxin following clinical disease or to diphtheria toxoid (formaldehyde-inactivated toxin)following immunization.

Epidemiology

Corynebacterium diphtheriae is spread by droplets, secretions, or direct contact. *In situ* lysogenic conversion of nontoxigenic strains to a toxigenic phenotype has been documented. Infection is spread solely among humans, although toxigenic strains have been isolated from horses. In regions where immunization programs are maintained, isolated outbreaks of disease are often associated with a carrier who has recently visited a subtropical region where diphtheria is endemic. Large-scale outbreaks of disease may occur in populations where active immunization programs are not maintained.

Diagnosis

Clinical diagnosis depends upon culture-proven toxigenic *C diphtheriae* infection of the skin, nose, or throat combined with clinical signs of nasopharyngeal diphtheria (e.g., sore throat, dysphagia, bloody nasal discharge, pseudomembrane). Toxigenicity is identified by a variety of *in vitro* (e.g., gel immunodiffusion, tissue culture) or *in vivo* (e.g., rabbit skin test, guinea pig challenge) methods.

Control

Immunization with diphtheria toxoid is extraordinarily effective. Diphtheria patients must be promptly treated with antitoxin to neutralize circulating diphtheria toxin.

INTRODUCTION

Diphtheria is a paradigm of the toxigenic infectious diseases. In 1883, Klebs demonstrated that *Corynebacterium diphtheriae* was the agent of diphtheria. One year later, Loeffler found that the organism could only be cultured from the nasopharyngeal cavity, and postulated that the damage to internal organs resulted from a soluble toxin. By 1888, Roux and Yersin showed that animals injected with sterile filtrates of *C diphtheriae* developed organ pathology indistinguishable from that of human diphtheria; this demonstrated that a potent exotoxin was the major virulence factor.

Diphtheria is most commonly an infection of the upper respiratory tract and causes fever, sore throat, and malaise. A thick, gray-green fibrin membrane, the pseudomembrane, often forms over the site(s) of infection as a result of the combined effects of bacterial growth, toxin production, necrosis of underlying tissue, and the host immune response. Recognition that the systemic organ damage was due to the action of diphtheria toxin led to the development of both an effective antitoxin-based therapy for acute infection and a highly successful toxoid vaccine.

Although, toxoid immunization has made diphtheria a rare disease in those regions where public health standards mandate vaccination, outbreaks of diphtheria still occur in nonimmunized and immunocompromised groups. In marked contrast, widespread outbreaks of diphtheria reaching epidemic proportions have been observed in those regions where active immunization programs have been halted.

Clinical Manifestations

There are two types of clinical diphtheria: nasopharyngeal and cutaneous. Symptoms of pharyngeal diphtheria vary from mild pharyngitis tohypoxia due to airway obstruction by the pseudomembrane (Fig. 32-1). The involvement of cervical lymph nodes may cause profound swelling of the neck (bull neck diphtheria), and the patient may have a fever (103 °F). The skin lesions in cutaneous diphtheria are usually covered by a gray-brown pseudomembrane. Life-threatening systemic complications, principally loss of motor function (e.g., difficulty in swallowing) and congestive heart failure, may develop as a result of the action of diphtheria toxin on peripheral motor neurons and the myocardium.

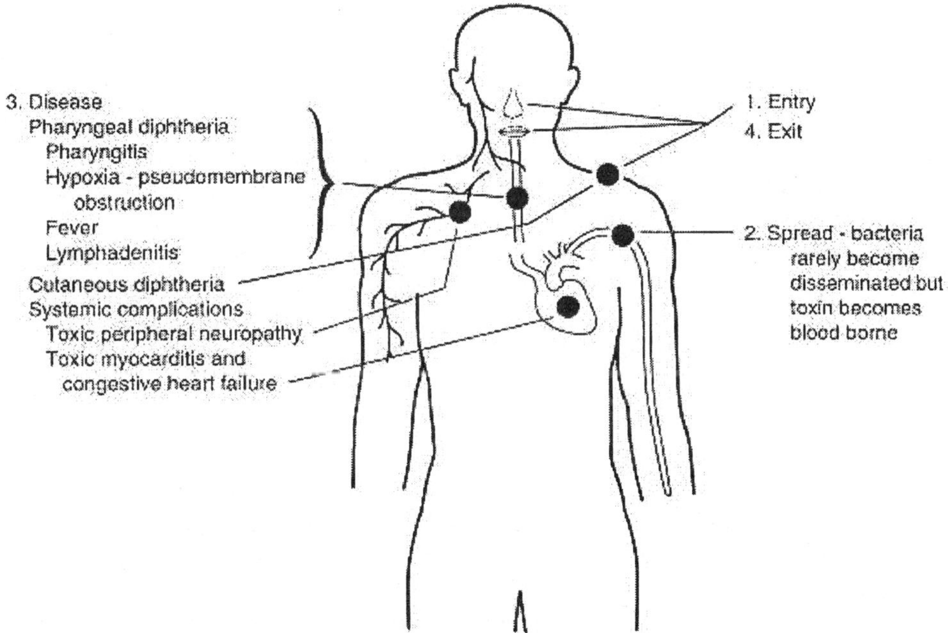

Figure 32-1 Pathogenesis of diphtheria.

Structure, Classification, and Antigenic Types

Corynebacterium diphtheriae is a Gram-positive nonmotile, club-shaped bacillus. Strains growing in tissue, or older cultures *in vitro*, contain thin spots in their cell walls that allow decolorization during the Gram stain and result in a Gram-variable reaction. Older cultures often contain metachromatic granules (polymetaphosphate) which stain bluish-purple with methylene blue. The cell wall sugars include arabinose, galactose, and mannose. In addition, a toxic 6,6'-diester of trehalose containing corynemycolic and corynemycolenic acids in equimolar concentrations may be isolated. Three distinct cultural types, mitis, intermedius, and gravis have been recognized (Table 1).

Most strains require nicotinic and pantothenic acids for growth; some also require thiamine, biotin, or pimelic acid. For the optimal production of diphtheria toxin the medium should be supplemented with amino acids and must be deferrated.

As early as 1887, Loeffler described the isolation from healthy individuals of avirulent (nontoxigenic) *C diphtheriae* that were indistinguishable from the virulent (toxigenic) strains isolated from patients. It is now recognized that avirulent strains of *C diphtheriae* may be converted to the virulent phenotype following infection and lysogenization by one of a number of distinct corynebacteriophages that carry the structural gene for diphtheria toxin, *tox*. Lysogenic conversion from the avirulent to virulent phenotype may occur *in situ*, as well as *in vitro*. The diphtheria toxin structural gene is not essential for either corynebacteriophage or *C diphtheriae*. Despite this observation, genetic drift of diphtheria toxin has not been observed.

Pathogenesis

The pathogenesis of diphtheria is based upon two primary determinants: (1) the ability of a given strain of *C diphtheriae* to colonize in the nasopharyngeal cavity and/or on the skin, and (2) its ability to produce diphtheria toxin. Since those determinants involved in colonizationof the host are encoded by the bacteria, and the toxin is encoded by the corynebacteriophage, the molecular basis of virulence in *C diphtheriae* results from the combined effects of determinants carried on two genomes. Nontoxigenic strains of *C diphtheriae* are rarely associated with clinical disease; however, they may become highly virulent following lysogenic conversion to toxigenicity.

Colonization

Little is known of the colonization factors of *C diphtheriae*. However, it is apparent that factors other than the production of diphtheria toxin contribute to virulence. Epidemiologic studies have demonstrated that a given lysotype may persist in the population for extended periods. It may later be supplanted by another lysotype. The emergence and subsequent predominance of a new lysotype in the population are presumably due to its ability to colonize and effectively compete in their segment of the nasopharyngeal ecologic niche. *Corynebacterium diphtheriae* may produce a neuraminidase that cleaves sialic acid from the cell surface into its pyruvate and N-acetylneuraminic acid components. Cord factor (6,6'-di-O-mycoloyl-a,a'-D-trehalose) is a surface component of *C diphtheriae*, but its role in colonization of the human host is unclear.

TABLE 32-1 Biochemical Properties useful in Distinguishing Corynebacterium species isolated from the Human Oropharynx and Nasopharynx[a]

	Production of					Fermentation of				
Strain	Metachromatic Granules	Catalase	Pyreznamidase	Gelatinase	Urease	Lactose	Maltose	Trehalose	Starch[b]	Glucose
C diptheriae										
var mitis	+	+	-	-	-	-	+	-	-	+
var graivs	+	+	-	-	-	-	+	-	+	+
var intermedius	+	+	-	—	—	-	+	—	-	+
C vlcerans	+	+	-	⊕	⊖	-	+	⊕	+	+
C psedotuberculosis	+	+	-	-	+	-	+	-	-	+
C pseudodiphthenticum	+	+	+	-	+	-	-	-	-	-
C xerosis	+	+	+	-	-	-	⊖	-	-	+

[a]+ All strains tested positive rare negative strains may be found; -, all strains tested negative: , rare positive strains may be found.

[b]Because soluble starch contains some glucose, Laundry starch fermentation tests.

[c]C pseudotberculosis os found only rarely in the human throat.

(Data from Barksdle L of al: phospholipase D acitivity of corynsbacterium pseudotuberculosis (corynetacterium ovis) and Corynebacterium ulcerans, a distinctive marker within the genus Corynebacterium J Clin Microbial 13:335, 1981)

Diphtheria Toxin Production

The structural gene for diphtheria toxin, *tox*, is carried by a family of closely related corynebacteriophages of which the b-phage is the most extensively studied (Fig. 32-2). The regulation of diphtheria *tox* expression is mediated by an iron-activated repressor, DtxR, which is encoded on the *C diphtheriae* genome. The expression of *tox* depends on the physiologic state of *C diphtheriae*. Under conditions in which iron becomes the growth-rate limiting substrate, iron dissociates from DtxR, the *tox* gene becomes derepressed, and diphtheria toxin is synthesized and secreted into the culture medium at maximal rates (Fig. 32-3).

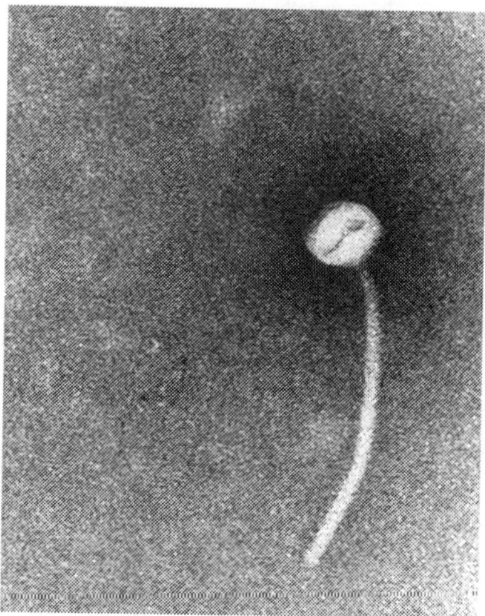

Figure 32-2 Electron micrograph of corynebacteriophage ß, which carries *tox*. Following lysogenic conversion with corynebacteriophage ß, or closely related corynebacteriophages, nontoxigenic strains of *C diphtheriae* become toxigenic.

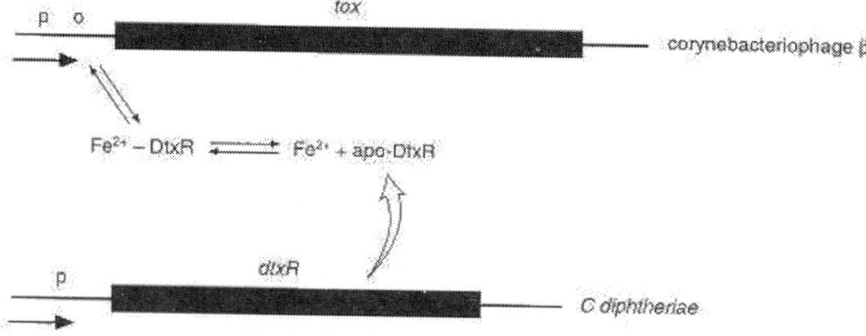

Figure 32-3 Model of the regulation of diphtheria *tox* expression by iron-activated DtxR. *C diphtheriae* encodes the structural gene for the *tox* repressor *dtxR*. In the presence of iron, apo-DtxR becomes activated and binds to the *tox* operator, thereby preventing transcription. When iron becomes the growth rate-limiting substrate, the iron-repressor complex disociates and the diphtheria *tox* gene becomes derepressed.

Diphtheria toxin is extraordinarily potent; in sensitive species (e.g., humans, monkeys, rabbits, guinea pigs) as little as 100 to 150 ng/kg of body weight is lethal. Diphtheria toxin is composed of a single polypeptide chain of 535 amino acids. Biochemical genetic and X-ray crystallographic analysis show that the toxin is composed of three structural/functional domains: an N-terminal ADP-ribosyltransferase (catalytic domain); (2) a region which facilitates the delivery of the catalytic domain across the cell membrane (transmembrane domain); and (3) the eukaryotic cell receptor binding domain (Fig. 32-4). Following mild digestion with trypsin and reduction under denaturing conditions, diphtheria toxin may be specifically cleaved in its protease-sensitive loop into two polypeptide fragments (A and B). Fragment A is the N-terminal 21 kDa component of the toxin and contains the catalytic center for the ADP-ribosylation of elongation factor 2 (EF-2) according to the following reaction:

The C-terminal fragment, fragment B, carries the transmembrane and receptor binding domains of the toxin.

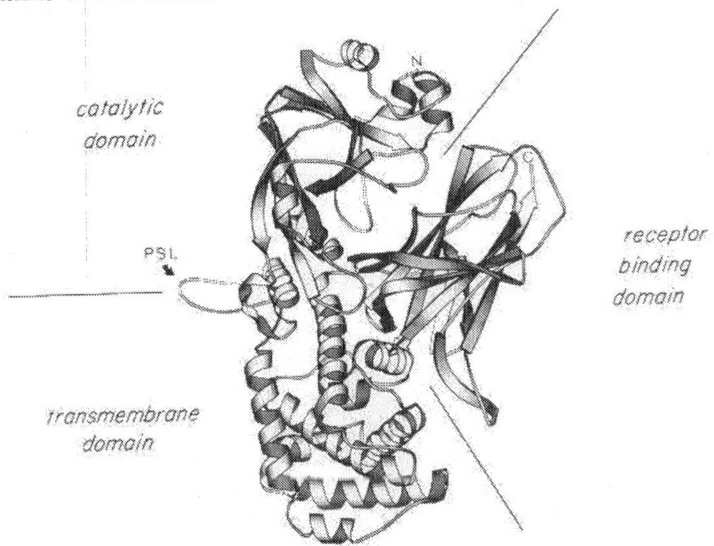

Figure 32-4 Ribbon diagram of the X-ray crystal structure of monomeric native diphtheria toxin (modified from Bennett MJ, Choe S, Eisenberg D: Domain swapping: Entangling alliances between proteins. Proc Natl Acad Sci, USA, 91:3127, 1994). The relative positions of the catalytic, transmembrane, and receptor-binding domains are shown. Intact toxin may be cleaved by trypsin-like proteases at Arg190, Arg192, and/or Arg193, which are positioned in the protease-sensitive loop (PSL). Following reduction of the disulfide bridge between Cys186 and Cys201, the toxin may be separated into fragments A and B. The amino terminus (N) and carboxy terminus (C) of the intact toxin are shown. The ribbon diagram was generated using the software program MOLESCRIPT.

The intoxication of a single eukaryotic cell by diphtheria toxin involves at least four distinct steps (Fig. 32-5): (1) the binding of the toxin to its cell surface receptor; (2) clustering of charged receptors into coated pits and internalization of the toxin by receptor-mediated endocytosis; following acidification of the endocytic vesicle by a membrane-associated, ATP-driven proton pump, (3) the insertion of the transmembrane domain into the membrane and the facilitated delivery of the catalytic domain to the cytosol, and (4) the ADP-ribosylation of EF-2, which results in the irreversible inhibition of protein synthesis. It has been shown that a single molecule of the catalytic domain delivered to the cytosol is sufficient to be lethal for the cell.

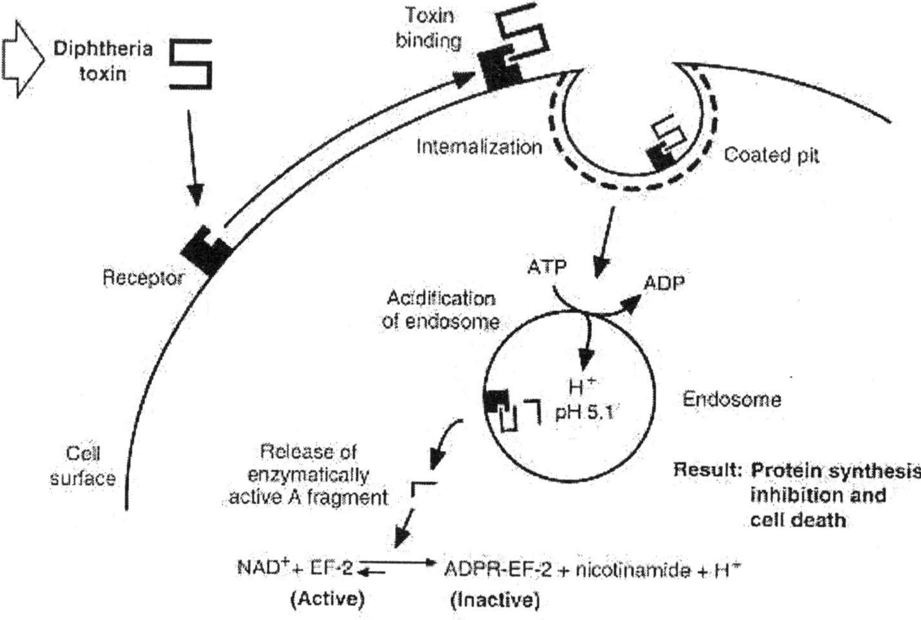

Figure 32-5 Schematic diagram of the diphtherial intoxication of a sensitive eukaryotic cell. The toxin binds to its cell surface receptor and is internalized by receptor-mediated endocytosis; upon acidification of the endosome the transmembrane domain inserts into the vesicle membrane; the catalytic domain is delivered to the cytosol, resulting in inhibition of protein synthesis and death of the cell.

Host Defenses

Immunity to diphtheria involves an antibody response to diphtheria toxin following clinical disease or immunization with diphtheria toxoid.

Epidemiology

Before mass immunization of the U.S. population with diphtheria toxoid, diphtheria was typically a disease of children. A remarkable aspect of mass immunization with diphtheria toxoid is that as the percentage of the population with protective levels of antitoxin immunity (0.01 IU/ml) increases, the frequency of isolation of toxigenic strains from the population decreases. Today in the United States where there is an almost complete disappearance of clinical diphtheria, the isolation of toxigenic strains of *C diphtheriae* is rare. Since subclinical infection is no longer a source of diphtherial antigen exposure and, if not boosted, antitoxin immunity wanes, a large percentage of the adults (30 to 60%) have antitoxin levels that are below the protective level and are at risk. In the United States, Europe, and Eastern Europe recent outbreaks of diphtheria have occurred largely among alcohol and/ or drug abusers. Within this group, carriers of toxigenic *C diphtheriae* have moderately high levels of antitoxic immunity. The recent breakdown of public health measures in Russia has resulted in diphtheria becoming epidemic. By the end of 1994, Russia recorded more than 80,000 cases and greater than 2,000 deaths.

Focal outbreaks of diphtheria are almost always associated with an immune carrier who has returned from a region where diphtheria is endemic. Indeed, recent outbreaks of clinical diphtheria in the United States and Europe have been associated with travelers returning from Russia and Eastern Europe. Toxigenic strains of *C diphtheriae* spread directly from person to person by droplet infection. It is known that toxigenic strains may directly colonize the nasopharyngeal cavity. In addition, the *tox* gene may be spread indirectly by release of toxigenic corynebacteriophage and lysogenic conversion of nontoxigenic, autochthonous *C diphtheriae in situ.*

In addition to the determinationof biotype and lysotype of*C diphtheriae* isolates, it is now possible to use molecular biologic techniques in the study of diphtheria outbreaks. Restriction endonuclease digestion patterns of *C diphtheriae* chromosomal DNA, as well as the use of cloned corynebacterial insertion sequences as a genetic probe have been used in the study of clinical outbreaks of disease.

The Schick test has been used for many years to assess immunity to diphtheria toxin, although today it has been replaced in many regions by a serologic test for specific antibodies to diphtheria toxin. In the Schick test, a small amount of diphtheria toxin (ca. 0.8 ng in 0.2 ml) is injected intradermally into the forearm (test site) and 0.0124 μg of diphtheria toxoidin 0.2 ml is injected intradermally at a control site. After 48 and 96 hours, readings are made. Nonspecific skin reactions generally peak by 48 hours. At 96 hours, an erythematous reaction with some possible necrosis at the test site indicates that there is insufficient antitoxic immunity to neutralize the toxin (0.03 IU/ml). Inflammation at both the test and control sites at 48 hours indicates a hypersensitivity reaction to the antigen preparation. In many instances, diphtheria toxin is only partially purified prior to inactivation with formaldehyde, and as a result preparations of toxoid may contain other corynebacterial products, which may elicit a hypersensitivity reaction in some individuals.

Diagnosis

The clinical diagnosis of diphtheria requires bacteriologic laboratory confirmation of toxigenic *C diphtheriae* in throat or lesion cultures. For primary isolation, a variety of media may be used: Loeffler agar, Mueller-Miller tellurite agar, or Tinsdale tellurite agar. Sterile cotton-tipped applicators are used to swab the pharyngeal tonsils or their beds. Calcium alginate swabs may be inserted through both nares to collect nasopharyngeal samples for culture. Since diphtheritic lesions are often covered with a pseudomembrane, the surface of the lesion may have to be carefully exposed before swabbing with the applicator.

Following initial isolation, *C diphtheriae* may be identified as mitis, intermedius, or *gravis* biotype on the basis of carbohydrate fermentation patterns and hemolysis on sheep blood agar plates (Table 1). The toxigenicity of *C diphtheriae* strains is determined by a variety of *in vitro* and *in vivo* tests. The most common *in vitro* assay for toxigenicity is the Elek immunodiffusion test (Fig. 32-6). This test is based on the double diffusion of diphtheria toxin and antitoxin in an agar medium. A sterile, antitoxin-saturated filter paper strip is embedded in the culture medium, and *C diphtheriae* isolates are streak-inoculated at a 90° angle to the filter paper. The production of diphtheria toxin can be detected within 18 to 48 hours by the formation of a toxin-antitoxin precipitin band in the agar. Alternatively, many eukaryotic cell lines (e.g., African green monkey kidney, Chinese hamster ovary) are sensitive to diphtheria toxin, enabling *in vitro* tissue culture tests to be used for detection of toxin. Several sensitive *in vivo* tests for diphtheria toxin have also been described (e.g., guinea pig challenge test, rabbit skin test).

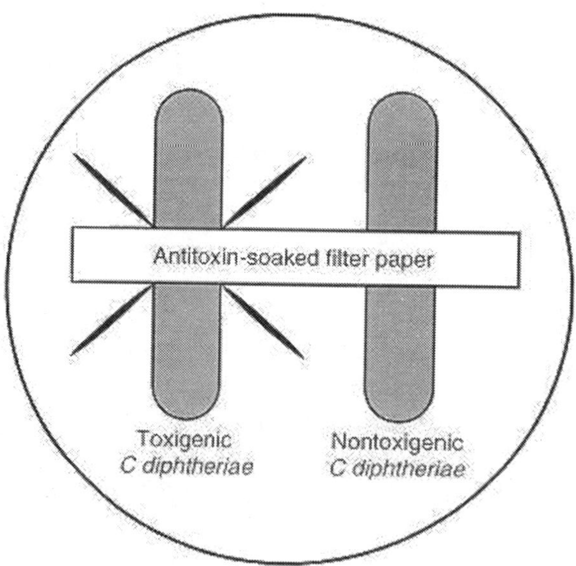

Figure 32-6 Elek immunodiffusion test. Sterile filter paper impregnated with diphtheria antitoxin is imbedded in agar culture medium. Isolates of *C diphtheriae* are then streaked across the plate at an angle of 90° to the antitoxin strip. Toxigenic *C diphtheriae* is detected because secreted toxin diffuses from the area of growth and reacts with antitoxin to form lines of precipitin.

Control

The control of diphtheria depends upon adequate immunization with diphtheria toxoid: formaldehyde-inactivated diphtheria toxin that remains antigenically intact. The toxoid is prepared by incubating diphtheria toxin with formaldehyde at 37° C under alkaline conditions. Immunization against diphtheria should begin in the second month of life with a series of three primary doses spaced 4 to 8 weeks apart, followed by a fourth dose approximately 1 year after the last primary inoculation. Diphtheria toxoid is widely used as a component in the DPT (diphtheria, pertussis, tetanus) vaccine. Epidemiologic surveys have shown that immunization against diphtheria is approximately 97% effective. Although, mass immunization against diphtheria is practiced in the United States and Europe and there is an adequate immunization rate in children, a large proportion of the adult population may have antibody titers that are below the protective level. The adult population should be reimmunized with diphtheria toxoid every 10 years. Indeed, booster immunization with diphtheria-tetanus toxoids should be administered to persons traveling to regions with high rates of endemic diphtheria (Central and South America, Africa, Asia, Russia and Eastern Europe). In recent years, the use of highly purified toxoid preparations for immunization has minimized the occasional severe hypersensitivity reaction.

Although, antibiotics (e.g., penicillin and erythromycin) are used as part of the treatment of patients who present with diphtheria, prompt passive immunization with diphtherial antitoxin is most effective in reducing the fatality rate. The long half-life of specific antitoxin in the circulation is an important factor in ensuring effective neutralization of diphtheria toxin; however, to be effective, the antitoxin must react with the toxin before it becomes internalized into the cell.

Other Corynebacterium Species

In addition to *C diphtheriae*, *C ulcerans* and *C pseudotuberculosis*, *C pseudodiphtheriticum* and *C xerosis* may occasionally cause infection of the nasopharynx and skin. The last two strains are recognized by their ability to produce pyrazinamidase. In veterinary medicine, *C renale* and *C kutscheri* are important pathogens and cause pyelonephritis in cattle and latent infections in mice, respectively.

Redesigning of Diphtheria Toxin for the Development of Eukarytoic Cell-Receptor Specific Cytotoxins

Protein engineering is a new and rapidly developing area within the field of molecular biology; it brings together recombinant DNA methodologies and solid phase DNA synthesis in the design and construction of chimeric genes whose products have unique properties. The study of diphtheria toxin structure/function relationships has clearly shown this toxin to be a three-domain protein: catalytic, transmembrane, and receptor binding (Fig. 32-4). It has been possible to genetically substitute the native diphtheria toxin receptor-binding domain with a variety of polypeptide hormones and cytokines (e.g., a-melanocyte-stimulating hormone [a-MSH], interleukin (IL) 2, IL-4, IL-6, IL-7, epidermal growth factor). The resulting chimeric proteins, or fusion toxins, combine the receptor-binding specificity of the cytokine with the transmembrane and catalytic domains of the toxin. In each instance, the fusion toxins have been shown to selectively intoxicate only those cells which bear the appropriate targeted receptor. The first of these genetically engineered fusion toxins, DAB389 IL-2, is currently being evaluated in human clinical trials for the treatment of refractory lymphomas and autoimmune diseases, in which cells with high affinity IL-2 receptors play a major role in pathogenesis. Administration of DAB389 IL-2 has been shown to be safe, well tolerated, and capable of inducing durable remission from disease in the absence of severe adverse effects. It is likely that the diphtheria toxin-based fusion toxins will be important new biological agents for the treatment of specific tumors or disorders in which specific cell surface receptors may be targeted.

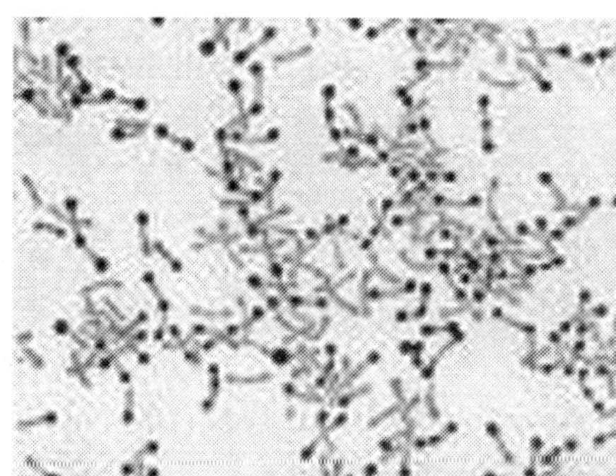

Cultural and Microscopic observation of *Cornybacterium*

REFERENCES

Bishai WR, Murphy JR: Bacteriophage gene products that cause human disease. In Calendar R (ed): The Bacteriophages. Plenum, New York, 1988.

Hesketh P, Caguioa P, Koh H, et al: Clinical activity of a cytotoxic fusion protein in the treatment of cutaneous T cell lymphoma. J Clin Oncol 11:1682, 1993

LeMaistre CF, Craig FE, Meneghetti C, et al: Phase I trial of a 90-minute infusion of the fusion toxin DAB486 IL-2 in hematological cancers. Cancer Res 53:3930, 1992

Pappenheimer AM, Jr: Diphtheria. In Germanier R (ed) Bacterial Vaccines. Academic Press, San Diego, 1984

Rappuoli R, Perugini M, Falsen E: Molecular epidemiology of the 1984-1986 outbreak of diphtheria in Sweden. N Engl J Med 318:12, 1988

Tao X, Schiering N, Zeng H-Y, et al: DtxR, iron and the regulation of diphtheria toxin expression. Mol Microbiol 14:191, 1994

VanderSpek J, Cosenza L, Woodworth T, et al: Diphtheria toxin-related cytokine fusion proteins: elongation factor 2 as a target for the treatment of neoplastic disease. Molec Cell Biochem 138:151, 1994.

Chapter **37**

Anaerobes: General Characteristics

General Concepts

Clinical Manifestations

Symptoms are related to the absence of oxygen from the affected area: hence, abscesses, devitalized tissue, and penetration of foreign matter lead to clinical infection.

Oxygen Toxicity

Low or undetectable levels of superoxide dismutase and catalase allow oxygen radicals to form in anaerobic bacteria and to inactivate other bacterial enzyme systems.

Pathogenic Anaerobes

Anaerobes are potentially pathogenic when displaced from normal environments (human colon, soil) and implanted in dead or dying tissue; abscesses, pneumonias, and oral and pelvic infections result.

Processing of Clinical Specimens

Anaerobic conditions are required for sample collection, culturing, and identification.

INTRODUCTION

The broad classification of bacteria as anaerobic, aerobic, or facultative is based on the types of reactions they employ to generate energy for growth and other activities. In their metabolism of energy-containing compounds, aerobes require molecular oxygen as a terminal electron acceptor and cannot grow in its absence (see Chapter 4). Anaerobes, on the other hand, cannot grow in the presence of oxygen. Oxygen is toxic for them, and they must therefore depend on other substances as electron acceptors. Their metabolism frequently is a fermentative type in which they reduce available organic compounds to various end products such as organic acids and alcohols. The facultative organisms are the most versatile. They preferentially utilize oxygen as a terminal electron acceptor, but also can metabolize in the absence of oxygen by reducing other compounds. Much more usable energy, in the form of high-energy phosphate, is obtained when a molecule of glucose is completely catabolized to carbon dioxide and water in the presence of oxygen (38 molecules of ATP) than when it is only partially catabolized by a fermentative process in the absence of oxygen (2 molecules of ATP). The ability to utilize oxygen as a terminal electron acceptor provides organisms with an extremely efficient mechanism for generating energy. Understanding the general characteristics of anaerobiosis provides insight into how anaerobic bacteria can proliferate in damaged tissue and why special care is needed in processing clinical specimens that may contain them.

Oxygen Toxicity

Several studies indicate that aerobes can survive in the presence of oxygen only by virtue of an elaborate system of defenses. Without these defenses, key enzyme systems in the organisms fail to function and the organisms die. Obligate anaerobes, which live only in the absence of oxygen, do not possess the defenses that make aerobic life possible and therefore cannot survive in air.

During growth and metabolism, oxygen reduction products are generated within microorganisms and secreted into the surrounding medium. The superoxide anion, one oxygen reduction product, is produced by univalent reduction of oxygen:

$$O_2 \; e \; ' \; O_2$$

It is generated during the interaction of molecular oxygen with various cellular constituents, including reduced flavins, flavoproteins, quinones, thiols, and iron-sulfur proteins. The exact process by which it causes intracellular damage is not known; however, it is capable of participating in a number of destructive reactions potentially lethal to the cell. Moreover, products of secondary reactions may amplify toxicity. For example, one hypothesis holds that the superoxide anion reacts with hydrogen peroxide in the cell:

$$O_2 + H_2O_2 \; ' \; OH + OH\cdot + O_2$$

This reaction, known as the Haber-Weiss reaction, generates a free hydroxyl radical (OH·), which is the most potent biologic oxidant known. It can attack virtually any organic substance in the cell. A subsequent reaction between the superoxide anion and the hydroxyl radical produces singlet oxygen (O_2*), which is also damaging to the cell:

$$O_2 + OH \; ' \; OH + O_2^*$$

The excited singlet oxygen molecule is very reactive. Therefore, superoxide must be removed for the cells to survive in the presence of oxygen.

Most facultative and aerobic organisms contain a high concentration of an enzyme called superoxide dismutase. This enzyme converts the superoxide anion into ground-state oxygen and hydrogen peroxide, thus ridding the cell of destructive superoxide anions:

$$2O_2 + 2H^+ \xrightarrow{\text{Superoxide Dismutase}} O_2 + H_2O_2$$

The hydrogen peroxide generated in this reaction is an oxidizing agent, but it does not damage the cell as much as the superoxide anion and tends to diffuse out of the cell. Many organisms possess catalase or peroxidase or both to eliminate the H_2O_2. Catalase uses H_2O_2 as an oxidant (electron acceptor) and a reductant (electron donor) to convert peroxide into water and ground-state oxygen:

$$H_2O_2 + H_2O_2 \xrightarrow{\text{Catalase}} 2H_2O + O_2$$

Peroxidase uses a reductant other than H_2O_2:

$$H_2O_2 + H_2R \xrightarrow{\text{Peroxidase}} 2H_2O + R$$

One study showed that facultative and aerobic organisms lacking superoxide dismutase possess high levels of catalase or peroxidase. High concentrations of these enzymes may alleviate the need for superoxide dismutase, because they effectively scavenge H_2O_2 before it can react with the superoxide anion to form the more active hydroxyl radical. However, most organisms show a positive correlation between the activity of superoxide dismutase and resistance to the toxic effects of oxygen.

In another study, facultative and aerobic organisms demonstrated high levels of superoxide dismutase. The enzyme was present, generally at lower levels, in some of the anaerobes studied, but was totally absent in others. The most oxygen-sensitive anaerobes as a rule contained little or no superoxide dismutase. In addition to the activity of superoxide dismutase, the rate at which an organism takes up and reduces

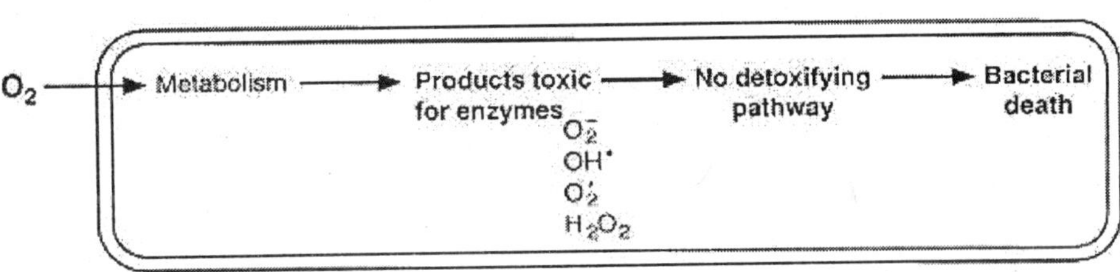

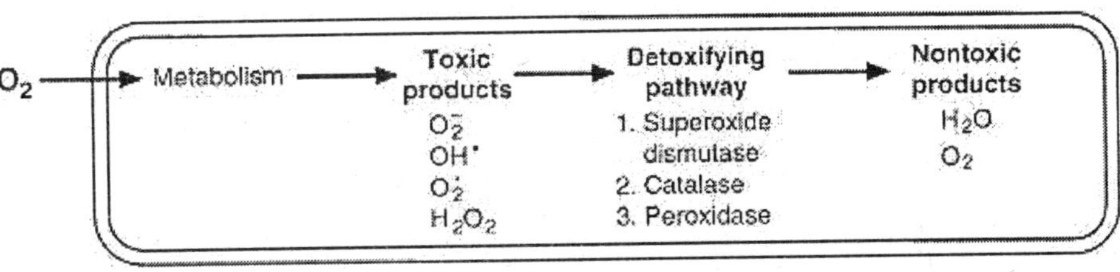

FIGURE 17-1 Effects of oxygen on aerobic, anaerobic, and facultative anaerobic bacteria.

oxygen was determined to be a factor in oxygen tolerance. Very sensitive anaerobes, which reduced relatively large quantities of oxygen and exhibited no superoxide dismutase activity, were killed after short exposure to oxygen. More tolerant organisms reduced very little oxygen or else demonstrated high levels of superoxide dismutase activity.

The continuous spectrum of oxygen tolerance among bacteria appears to be due partly to the activities of superoxide dismutase, catalase, and peroxidase in the cell and partly to the rate at which the cell takes up oxygen (Fig. 17-1). Clearly, other factors influence tolerance: the location of protective enzymes in the cell (surface versus cytoplasm), the rate at which cells form toxic oxygen products (e.g., the hydroxyl radical or singlet oxygen), and the sensitivities of key cellular components to the toxic oxygen products.

Pathogenic Anaerobes

Anaerobic bacteria are widely distributed in nature in oxygen-free habitats. Many members of the indigenous human flora are anaerobic bacteria, including spirochetes and Gram-positive and Gram-negative cocci and rods. For example, the human colon, where oxygen tension is low, contains large populations of anaerobic bacteria, exceeding 1011 organisms/g of colon content. Anaerobes in this region frequently outnumber facultative organisms by a factor of at least 100. Oxygen-sensitive organisms also are numerous in other areas of the body, such as the gingival crevices, tonsillar crypts, nasal folds, hair follicles, the urethra and vagina, and tooth surfaces.

Anaerobic indigenous flora components are potentially pathogenic if displaced from their normal habitat. Most anaerobic infections are endogenously acquired from members of the microflora, although Clostridium, found principally in the soil, also produces infections in humans. Proliferation of anaerobic bacteria in tissue depends on the absence of oxygen. Oxygen is excluded from the tissue when the local blood supply is impaired by trauma, obstruction, or surgical manipulation. Anaerobes multiply well in dead tissue. Multiplication of aerobic or facultative organisms in association with anaerobes in infected tissue also diminishes oxygen concentration and develops a habitat that supports growth of anaerobic bacteria.

Infections produced by anaerobic bacteria occur in all parts of the human body (Fig. 17-2). The infected tissues usually contain a mixture of several kinds of anaerobes and frequently also contain aerobic and facultative bacteria. The types of infections commonly produced by anaerobic bacteria are as follows:

1. **Intra-abdominal infections:** Abscesses, postoperative wound infections, and generalized peritonitis produced by anaerobes occur as a consequence of bowel perforation during surgery or injury.
2. **Pulmonary infections:** Anaerobic lung infections may originate in the bronchi or the blood. Aspirations from the upper respiratory tract, which contain

large numbers of anaerobic bacteria, are responsible for initiating infection in the bronchi.

3. **Pelvic infections:** Anaerobic infections of the vagina and uterus sometimes occur after gynecologic surgery or in association with malignancy of pelvic organs.
4. **Brain abscesses:** Anaerobes infrequently produce meningitis, but are a common cause of brain abscesses. The infecting organisms usually originate in the upper respiratory tract.
5. **Skin and soft tissue infections:** Combinations of anaerobes, aerobes, and facultative organisms often act synergistically to produce these infections.
6. **Oral and dental infections:** These local infections frequently extend to the face and neck and sometimes to other areas of the body such as the brain.
7. **Bacteremia and endocarditis:** Anaerobic bacteremia may follow disturbance in an area of the body where an established flora or an infection exists. Endocarditis, an inflammation of the endothelial lining of the heart cavities, is occasionally caused by anaerobic bacteria, especially anaerobic streptococci.

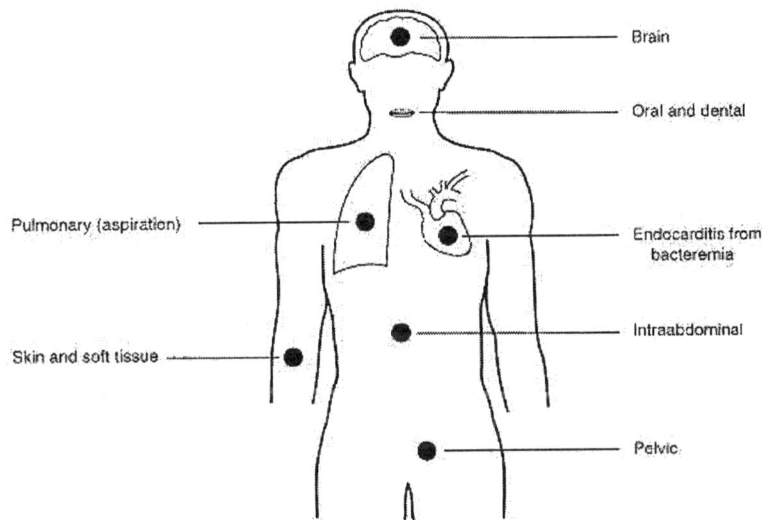

FIGURE 17-2 Types of infection commonly produced by anaerobic bacteria.

With the exception of the clostridia, which have been studied extensively, the mechanisms by which anaerobes cause infections in humans are not well understood. Clostridium species produce various toxins that destroy tissue cells, and two species, *C. botulinum* and *C. tetani*, release the neurotoxins responsible for botulism and tetanus, respectively. Enzymes excreted by other anaerobic bacteria, including proteases, lipases, hyaluronidase, chondroitin sulfatase, and neuraminidase, may play a role in infection by causing tissue cell destruction, and ß-lactamase may act as a virulence factor by inactivating antibiotics that possess a ß-lactam ring, such as the penicillins and cephalosporins. Inaddition, the capsules surrounding some anaerobic bacteria probably interfere with phagocytosis and act as a barrier against penetration by antimicrobial agents.

Processing of Clinical Specimens

When collecting specimens from patients for isolation and identification of anaerobic bacteria associated with infections, precautions must be taken to exclude air (Fig. 17-3). Materials for anaerobic culture are best obtained with a needle and syringe. Unless the specimen can be sent to the laboratory immediately, it is placed in an anaerobic transport tube containing oxygen-free carbon dioxide or nitrogen. The specimen is injected through the rubber stopper in the transport tube and remains in the anaerobic environment of the tube until processed in the bacteriology laboratory. If the specimen is collected with a swab, only a special commercially available anaerobic swab transport system is used.

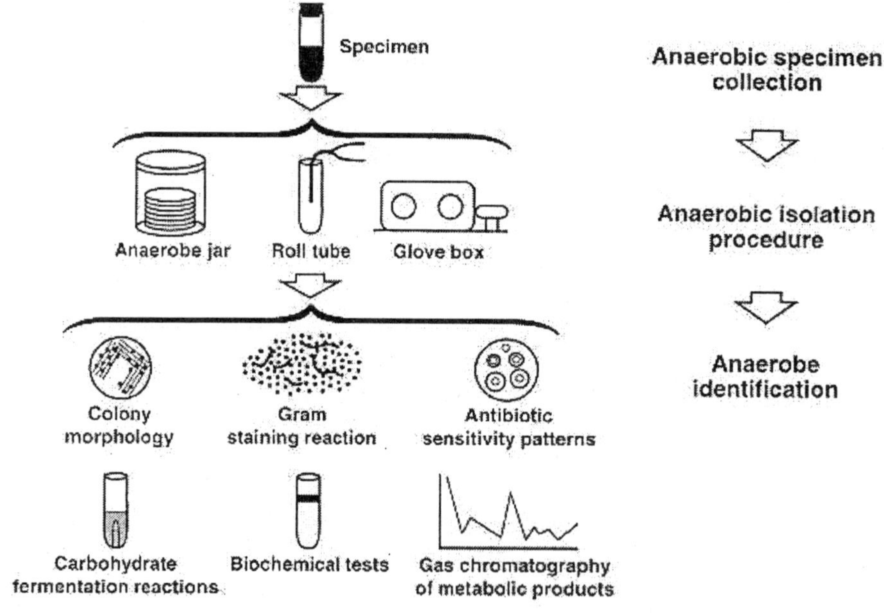

FIGURE 17-3 Isolation and identification of anaerobes.

Specimens should be free of contaminating bacteria. Material from sites that are normally sterile, such as blood, spinal fluid, or pleural fluid, poses no problem provided the usual precautions are taken to decontaminate the skin properly before puncturing it to obtain the specimen. Fecal specimens, sputum specimens, or vaginal secretions cannot be cultured routinely for pathogenic anaerobes because they normally contain other anaerobic organisms. Aspirates from abscesses or the specific sites of infections must be obtained in these cases to avoid undue contamination with indigenous flora components.

Although, several techniques are available for maintaining an oxygen-free environment during the processing of specimens for anaerobic culture, the anaerobic jar is the most common. It is a medium-sized glass or plastic jar with a tightly fitting lid containing palladium-coated alumina particles, which serve as a catalyst. It can be set up by two methods. The easiest uses a commercially available hydrogen and carbon dioxide generator envelope (GasPak) that is placed in the jar along with the culture plates. The generator is activated with water. Oxygen within the jar and the hydrogen that is generated are converted to water in the presence of the catalyst, thus producing anaerobic conditions. Carbon dioxide, which is also generated, is required for growth by some anaerobes and stimulates the growth of others. An alternative method for achieving anaerobiosis in the jar consists of evacuation and replacement. Air is evacuated from the sealed jar containing the culture plates and is replaced with an oxygen-free mixture of 80 percent nitrogen, 10 percent hydrogen, and 10 percent carbon dioxide.

More sophisticated procedures are used to isolate extremely oxygen-sensitive microorganisms that cannot be recovered by using the anaerobic jar. One, the roll tube method, consists of a stoppered test tube containing oxygen-free gas and a thin layer of prereduced agar medium on its inside surface. The medium in the tube is inoculated with a loop while the tube is rotated. This produces a spiral track on the agar surface. The tube is flushed with a stream of carbon dioxide to prevent entry of air while it is open during inoculation.

The anaerobic glove box isolator is another innovation developed for isolating anaerobic bacteria. It is essentially a large clear-vinyl chamber, with attached gloves, containing a mixture of 80 percent nitrogen, 10 percent hydrogen, and 10 percent carbon dioxide. A lock at one end of the chamber is fitted with two hatches, one leading to the outside and the other to the inside of the chamber. Specimens are placed in the lock, the outside hatch is closed, and the air in the lock is evacuated and replaced with the gas mixture. The inside hatch is then opened to introduce the specimen into the chamber. Conventional bacteriologic procedures are employed to process the specimen in the oxygen-free atmosphere.

Although, these complex systems are needed to isolate anaerobic flora components, studies have shown that the anaerobic jar is adequate to recover clinically significant anaerobes. The extremely oxygen-sensitive bacteria of the microflora apparently are not associated with infectious processes.

Procedures for cultivation and identification of anaerobic bacteria are well established (Fig. 17-3). A variety of selective and nonselective media is available for cultivation of anaerobes. A reliable, nonselective medium consists of Brucella agar supplemented with sheep blood, hemin, cysteine, sodium carbonate, and menadione. Usual bacteriologic procedures are used to identify anaerobes. These are based on Gram-staining reactions, cellular and colony morphology, antibiotic sensitivity patterns, carbohydrate fermentation reactions, and other biochemical tests. Analysis of metabolic end products, especially organic acids, provides additional information useful in classifying these organisms.

REFERENCES

Balows A, DeHaan RM, Dowell VR, Guze LB (eds): Anaerobic Bacteria. Charles C Thomas, Springfield, IL, 1974

Finegold SM: Anaerobic Bacteria in Human Disease. Academic Press, San Diego, 1977

Finegold SM, George WL (eds): Anaerobic Infections in Humans. Academic Press, San Diego, 1989

Holdeman LV, Cato EP, Moore WEC (eds): Anaerobe Laboratory Manual. 4th Ed. Virginia Polytechnic Institute and State University Anaerobe Laboratory, Blacksburg, VA, 1977

Lennette EH, Spaulding EH, Truant JP (eds): Manual of Clinical Microbiology. 2nd Ed. American Society for Microbiology, Washington, D.C., 1974

Morris JG: The physiology of obligate anaerobiosis. Adv Microb Physiol 12:169-246, 1975

Sutter VL, Citron DM, Edelstein MAC, Finegold SM: Wadsworth Anaerobic Bacteriology Manual, 4th Ed. Star Publishing, Belmont, CA, 1985

Chapter **38**

Anaerobic Gram-Negative Bacilli

General Concepts

Clinical Manifestations

Anaerobic Gram-negative bacilli are common elements of the mucous membrane flora throughout the body; they often act as secondary pathogens. They are the most common anaerobes involved in infection and include some of the most antibiotic-resistant species.

Structure, Classification, and Antigenic Types

Some are pleomorphic, whereas others have distinctive morphology. Most are obligate anaerobes. The key characters for classification are motility, arrangement of flagella, organic and volatile fatty acid metabolic end products and cellular fatty acid patterns. Antigens are not useful.

Pathogenesis

They usually invade as opportunistic pathogens through a break in the mucosa. A low redox potential favors infection. Some types produce virulence factors or interfere with host defenses (e.g., by inhibiting phagocytosis).

Host Defenses

Antibodies, complement (via both classic and alternative pathways), and cell-mediated immunity (involving both polymorphonuclear leukocytes and T lymphocytes) are important.

Epidemiology

Infections arise endogenously from the mucosal flora.

Diagnosis

Clues suggesting anaerobic infection include foul-smelling discharge, proximity of infection to mucosal surfaces, abscess formation, necrosis and gas in tissues, septic thrombophlebitis, various distinctive clinical pictures (e.g., actinomycosis, gas gangrene), and results of a Gram stain of clinical specimens. Collection of clinical specimens should avoid the mucosal flora, and transport must be anaerobic.

Control

Control involves (1) surgical drainage of abscesses and debridement of necrotic tissue; and (2) use of antimicrobials (particularly metronidazole, imipenem, chloramphenicol, or combinations of amoxicillin, ticarcillin, ampicillin or piperacillin with ß-lactamase inhibitors).

INTRODUCTION

At present there are over two dozen genera of Gram-negative anaerobic bacilli. In most clinical infections, only the genera Bacteroides, Prevotella, and Fusobacterium need be considered. These genera are prevalent in the body as members of the normal flora (Fig.20-1), constituting one-third of the total anaerobic isolates from clinical specimens, and may become involved in infections throughout the body (Fig.20-2). Within the Bacteroides group, *B fragilis* is the most common pathogen, followed by B thetaiotaomicron and other members of the *B fragilis* group. Among the bile-sensitive Prevotella species, the ones most commonly encountered clinically are *P melaninogenica, P oris,* and *P buccae.* Porphyromonas species seem to be much less pathogenic except in dental infections. *Fusobacterium nucleatum* is the Fusobacterium species most often found as a pathogen, but *F necrophorum* occasionally produces serious disease. These genera contain numerous other species that rarely or never infect humans.

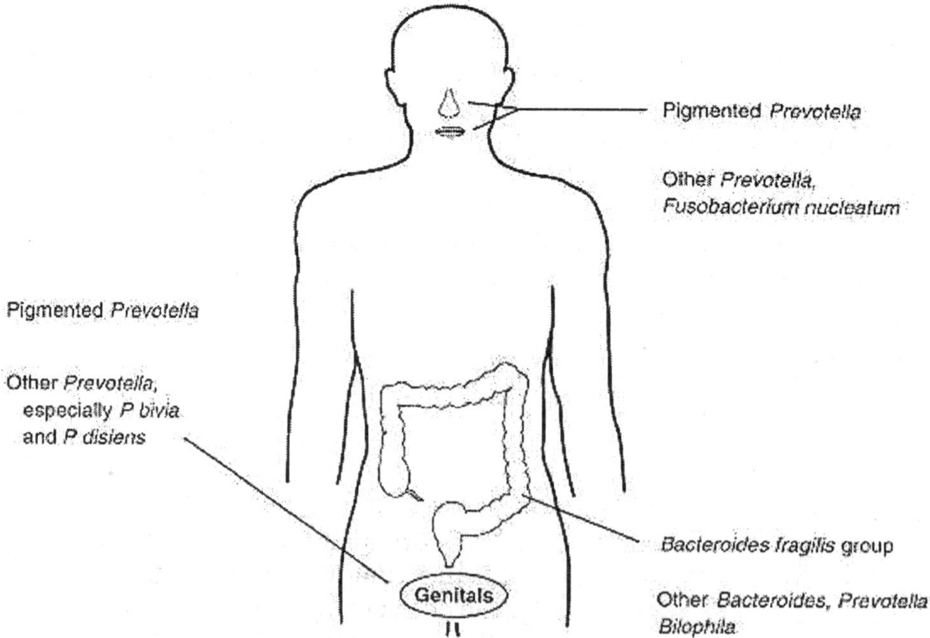

Figure 20-1 Predominant sites colonized by Bacteroides and other anaerobic bacilli.

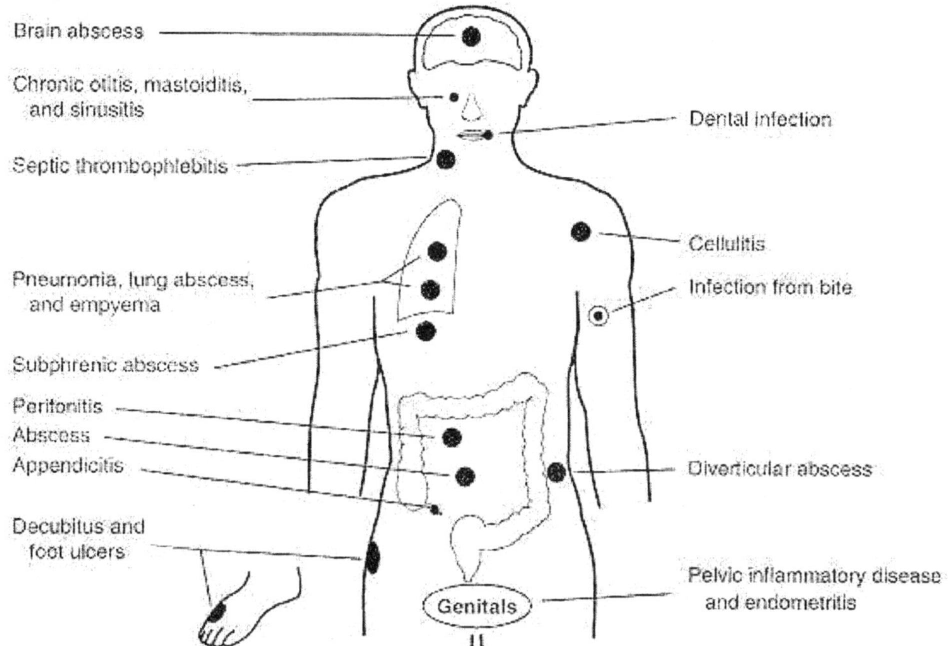

Figure 20-2 Sites of anaerobic infections.

Clinical Manifestations

Gram-negative anaerobic bacilli may cause infections anywhere in the body; the most common types are oral and dental, pleuropulmonary, intra-abdominal, female genital tract and skin, soft tissue and bone infections (Table 20-1). They may play a role in such diverse pathologic processes as periodontal disease and colon cancer. Bacteroides, Prevotella, Porphyromonas, and Fusobacterium produce enzymes (collagenase, neuraminidase, deoxyribonuclease, [DNase], heparinase, and proteinases) that may play a role in pathogenesis by helping the organisms to penetrate tissues and to set up infection after surgery or other trauma. The incidence of infection by these organisms can best be reduced or eliminated by avoiding conditions that decrease the redox potential of tissues and by preventing introduction of the anaerobes into compromised host tissues.

Table 20-1 Common Syndromes of Anaerobic Infection
Bite infections
Oral or dental infection
Aspriation pneumonia, lung abscess, empyema
Postabortion and puerperal infections
Infections following

Bowel and gall bladder surgery
Gynecologic surgery
Appendicitis, diverticulitis
Septicemia associated with

Cancer
Diabetes
Corticosteroids
"Negative" blood cultures
Septic thrombophlebitis
Gas-froming infection
Putirid infections

Structure, Classification, and Antigenic Types

Bacteroides fragilis (Fig.20-3), the most important of all anaerobes because of its frequency of occurrence in clinical infection and its resistance to antimicrobial agents, is a Gram-negative bacillus with rounded ends 0.5 to 0.8 µm in diameter and 1.5 to 4.5 µm long. Most strains are encapsulated. Vacuolization or irregular staining is common, particularly in broth media. Some pleomorphism also may be seen. By electron microscopy, the ultrastructure of *B fragilis* is similar to that of other Gram-negative bacteria. The guanine-plus-cytosine content is 42 percent. *Prevotella melaninogenica* and *Porphyromonas asaccharolytica* are short to coccoid Gram- negative rods; they produce a distinctive pigment (brown to black), which is a heme derivative that colors the colony (Figs.20-4 and 20-5). Many strains of *P melaninogenica* require vitamin K, or similar compounds, as well as heme.

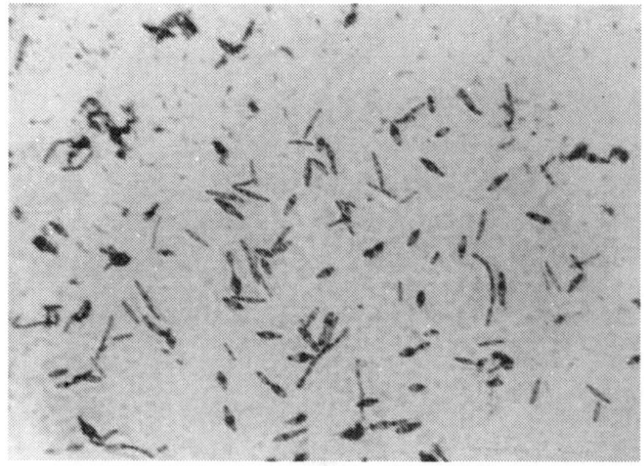

Figure 20-3 Microscopic morphology of *B fragilis* from broth culture. Note the irregular staining, rounded ends of bacilli, and some pleomorphism.

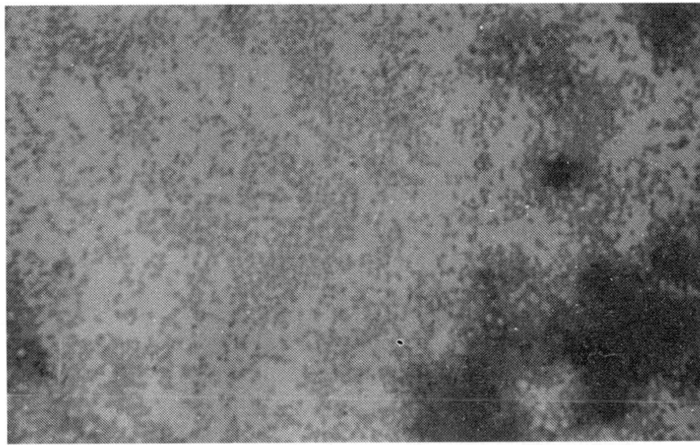

Figure 20-4 Microscopic morphology of *P melaninogenica*. Organisms are tiny coccobacilli that stain regularly.

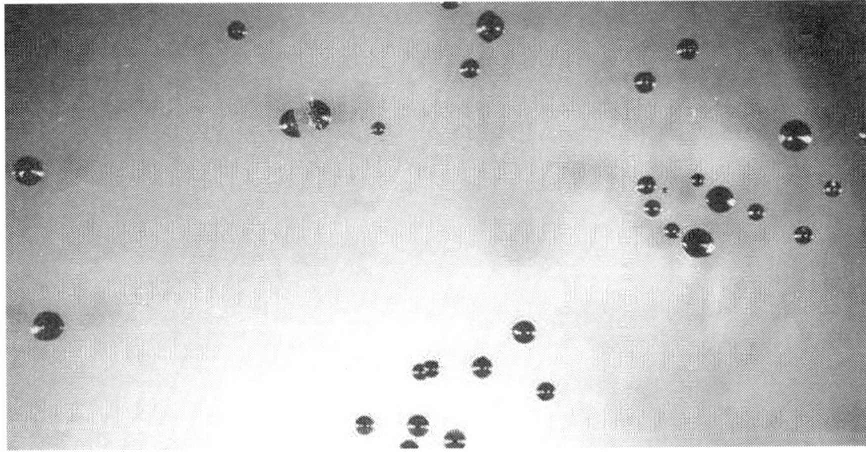

Figure 20-5 Colony morphology of *P melaninogenica*. Note the jet-black pigmented colonies.

Numerous studies of the endotoxin of Gram-negative anaerobic bacilli have determined that the *B fragilis* endotoxin contains little or no lipid A, 2-ketodeoxyoctanate, or heptose. It also lacks b-hydroxymyristic acid. This endotoxin exhibits little biologic activity in various test systems and little chemotactic activity; what activity there is, is complement-mediated by the alternative pathway. Poor biologic activity of endotoxin also has been demonstrated for the closely related species *B thetaiotaomicron, B ovatus, B vulgatus,* and *B distasonis. Prevotella melaninogenica* endotoxin contains no heptose or 2-ketodeoxyoctanate, and it and the endotoxin of *P oralis* both show weak biologic activity. Serologic methods have not been reliable for characterizing Gram-negative anaerobic rods.

Members of the genus Fusobacterium (Figs.20-6 and 20-7) may be spindle shaped or may have parallel sides and rounded ends. The guanine-plus-cytosine content ranges from 26 to 34 percent. Cells of *F necrophorum* often are elongated or filamentous, are curved, and possess spherical enlargements and large, free, round bodies. Fusobacterium nucleatum, although not producing infections as serious as those caused by *F necrophorum,* is a virulent organism and is much more common clinically. The cells of this species are usually spindle shaped, are 5 to 10 µm long, and are often seen in pairs, end to end.

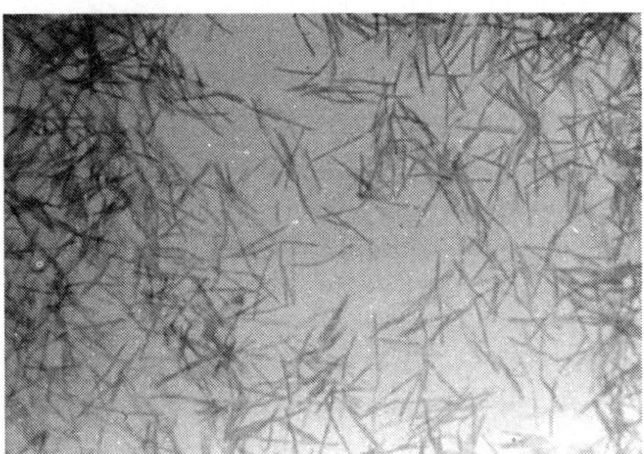

Figure 20-6 Microscopic morphology of *F nucleatum* from broth culture. Note the regular staining and thin, delicate bacilli with tapered ends. Organisms are sometimes found end to end.

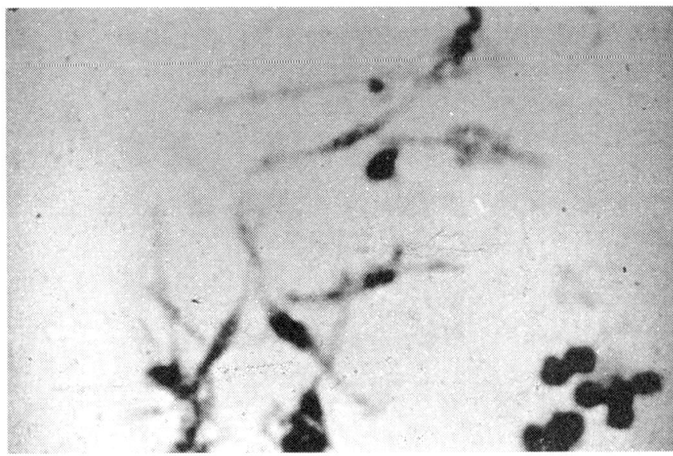

Figure 20-7 Microscopic morphology of *F mortiferum* from broth culture. Note the filaments with swollen central portions, large round bodies, and irregular staining.

The lipopolysaccharide of F *necrophorum* is located in a multilayered external coat. The endotoxin varies from strain to strain in its content of 2-ketodeoxyoctanate and sugars. Although, biologic activity varies also, many strains do show strong biologic activity, comparable to that of Salmonella enteritidis. The endotoxin of *F nucleatum* also is variable in its biologic activity, but often exhibits strong activity, comparable to that of S enteritidis.

Bacteriophages active against *B fragilis* are not uncommon. They are species specific and active against most strains. Bacteriocins also are produced by strains of *B fragilis* and *B thetaiotaomicron*. Plasmids have been found in about half the Bacteroides strains studied. For the most part, the biologic and clinical significance of these plasmids is not known; however, some code for resistance to such antimicrobial agents as clindamycin, erythromycin, tetracycline, chloramphenicol, ampicillin, and cephalothin. Plasmid-mediated antibiotic resistance has been transferred from strains of *B fragilis* to other strains of this species, to *B thetaiotaomicron*, and to *Escherichia coli*. Such resistance also has been transferred from *B distasonis* to *B fragilis*.

Most strains of the *B fragilis* group can deconjugate bile acids and are equally active whether the bile acid is conjugated with glycine or with taurine. Rarely, *P melaninogenica* may deconjugate bile acids, but in general this species, *P oralis*, and *F nucleatum* are inhibited by bile acids and do not deconjugate them. *Fusobacterium necrophorum* also is active in deconjugating bile acids but is active primarily on taurine conjugates. A few strains of Gram-negative anaerobic bacilli can convert primary bile acids to secondary bile acids. Bacteroides thetaiotaomicron can convert some lithocholic acid to its ethyl ester. Because lithocholic acid is toxic in humans and has been shown to exert tumor-promoting activity in animals, this reaction may be important. Bacteroides fragilis hydrolyzes the conjugated metabolites of benzpyrene. Glucuronidase produced by anaerobic Gram-negative bacilli may be of special significance in deconjugating compounds that had previously been detoxified in the liver by combination with glucuronide. There is speculation that this enzyme may be important in promoting bowel cancer. The activity of *B thetaiotaomicron, B distasonis*, and other members of the *B fragilis* group against plant polysaccharides, chondroitin, and mucin may be a factor in colon cancer and other disorders. Dietary fiber consists primarily of plant cell wall polysaccharides that are not digested in the stomach or small bowel.

Certain Bacteroides species possess distinguishing enzymes. Superoxide dismutase has been found in *B fragilis, B thetaiotaomicron, B vulgatus*, and *B ovatus*. In general, a good correlation exists between superoxide dismutase activity and oxygen tolerance. No consistent relationship has been found between catalase activity and oxygen tolerance, however, β-Lactamase activity has been demonstrated in several Bacteroides species, some Prevotella, and Bilophila; it accounts for most of the resistance to various ß-lactam antibiotics, such as penicillins and cephalosporins, although other mechanisms are responsible occasionally. Urease is produced by *Bilophila wadsworthia* and by *Bacteroides ureolyticus*. The latter organism also produces an agarase, which accounts for pitting of the agar by the colonies. A related pitting organism, *Sutterella wadsworthensis*, is much more pathogenic and is relatively resistant to antimicrobial agents.

Pathogenesis

Bacteroides, Prevotella, Porphyromonas, and Fusobacterium species are prevalent in the indigenous flora on all mucosal surfaces. They may have an opportunity to penetrate tissues and then to set up infection under certain circumstances such as surgical or other trauma or when tumors arise at the mucosal surface (Table 20-2). In certain cases, such as aspiration pneumonia, anaerobic bacteria from a site of normal carriage may move into another area that is normally free of organisms and infect that site. Tissue necrosis and poor blood supply lower the oxidation-reduction potential, thus favoring the growth of anaerobes. Accordingly, vascular disease, cold, shock, trauma, surgery, foreign bodies, cancer, edema, and gas production by bacteria may significantly predispose individuals to infection with anaerobes, as may prior infection with aerobic or facultative bacteria. Antimicrobial agents such as aminoglycosides, trimethoprim/sulfamethoxazole,and quinolones, to which anaerobes are notably resistant, may facilitate anaerobic infection. Conditions predisposing persons to anaerobic infection are summarized in Table 20-2. The more aerotolerant anaerobes are more likely to survive after the normally protective mucosal barrier is broken and until conditions are satisfactory for their multiplication and invasion. Once anaerobes begin to multiply, they can maintain their own reduced environment by excreting end products of fermentative metabolism. Infections involving Gram-negative anaerobic bacilli often are characterized by abscess formation and tissue destruction.

Table 20-2 Conditions Predisposing to Anaerobic Infection
General
Diabetes
Corticosteroids
Leukopenia
Hypogammaglobullnemia
Immunosuppression
Cytotoxis drugs
Splenectomy
Collagen diseases
Decreased redox potential
Tissue anoxia
Tissue destruction
Aerobic infection
Foreign body
Calcium salts
Burns
Peripheral vascular insufficiency
Specific clinical situations
Cancer

Colon, uterus, lung
Lukemia
Gastrointestinal and female pelvic surgery
Gastrointestinal trauma
Human and animal bites
Aminoglycoside therapy

Bacteroides, Prevotella, Porphyromonas, and Fusobacterium species produce enzymes that may play a role in pathogenesis. Prevotella melaninogenica is one of the few bacteria that produce collagenase, an enzyme of considerable importance. Cell extracts of P melaninogenica strains with collagenolytic activity, when given with a live Fusobacterium species, produce more severe lesions in rabbits than does the organism or the extract given alone. Porphyromonas gingivalis also produces collagenase and has trypsin-like activity. Neuraminidase may be important in the pathogenesis of Bacteroides infection. This enzyme alters neuraminic acid-containing glycoproteins of human plasma; Bacteroides strains isolated from clinical specimens have higher neuraminidase activity than do those isolated from stools, and strains of the *B fragilis* group have greater activity of this type than do strains of other Gram-negative anaerobic bacillary species. Hyaluronidase is produced by many strains of the *B fragilis* group and pigmented anaerobic Gram- negative rods. DNase is also produced by *B fragilis* and may be an important factor in infection. Many Gram-negative anaerobic bacilli produce phosphatase. A heparinase produced by *B fragilis* strains may contribute to intravascular clotting and hence increase the dosage of heparin needed to treat septic thrombophlebitis in infections caused by this organism. The lipopolysaccharides of *B fragilis, B vulgatus,* and *F mortiferum* activate the Hageman factor and thereby initiate the intrinsic pathway of coagulation. Fibrinolysin is produced by many *P melaninogenica* group strains and by a few *B fragilis* group strains. *Porphyromonas asaccharolytica* produces proteinases that render it capable of hydrolyzing gelatin, casein, coagulated protein, plasma protein, azacol, and collagen. Strains of Bacteroides and *P gingivalis* degrade complement factors and immunoglobulins G and M. A strain of *P melaninogenica* produces phospholipase A.

Fusobacteria necrophorum produces a leukocidin and hemolyses erythrocytes of humans, horses, rabbits, and, much less extensively, sheep and cattle. Certain *F necrophorum* cells hemagglutinate the erythrocytes of humans, chickens, and pigeons. A bovine isolate of F necrophorum demonstrates phospholipase A and lysophospholipase activity. *Fusobacterium gonidiaformans* produces an appreciable inflammatory reaction when inoculated into the skin of rabbits; when injected intraperitoneally into mice, it leads to liver abscesses and occasionally to death. A specific toxin has not yet been isolated.

Other factors may be involved in the continued growth and potential pathogenicity of certain anaerobes. For example, *P melaninogenica* can inhibit the growth of certain other organisms. Also, anaerobes such as *P melaninogenica* sometimes inhibit phagocytosis and killing of other organisms during mixed infection. Constituents of the cell envelope and cell surface may contribute to pathogenicity. The capsule of organisms such as *B fragilis* is an important virulence factor. Pili (fimbriae) and lectin like adhesins may also be important in the adherence of Bacteroides cells to epithelial surfaces. Butyrate and succinate produced by Bacteroides show a cytotoxic effect.

Host Defenses

Polymorphonuclear leukocytes have oxygen-dependent and oxygen-independent microbicidal systems. Components of both systems might be important in phagocytic killing of anaerobes under conditions of varying oxygen tension. Specifically, polymorphonuclear leukocytes normally kill *B fragilis* under anaerobic and aerobic conditions. Randommigration of polymorphonuclear leukocytes does not differ significantly under aerobic and anaerobic conditions. The same holds true for chemotaxis in response to factors generated by immune complexes in plasma; however, chemotaxis in response to factors generated by bacteria in plasma is markedly depressed under anaerobic conditions, and products of Gram-negative anaerobic bacilli may suppress neutrophil chemotaxis and phagocytic killing.

Studies of host defenses indicate that other interactions may occur between the bacteria and the host cells. Bacteroides fragilis, one organism used in the chemotaxis study described above, is more resistant to the normal bactericidal activity of serum than other members of the *B fragilis* group. *Fusobacterium mortiferum* is killed by serum alone or by serum plus leukocytes under aerobic and anaerobic conditions. Under anaerobic conditions, *B thetaiotaomicron* and *B fragilis* are phagocytosed and killed intracellularly by human polymorphonuclear leukocytes only in the presence of normal human serum. Similar results are obtained in an aerobic environment, except that *B fragilis* is phagocytosed and killed intracellularly to some extent in the absence of serum. There is evidence that the capsule of certain strains interferes with their phagocytosis.

Immunoglobulin and components of the classic and alternative complement pathways participate in chemotaxis, bacteriolysis, and opsonophagocytic killing of various Gram-negative anaerobic bacilli. Antibody to the capsular polysaccharide of *B fragilis* can be induced in animals by infection with encapsulated strains or by implantation of the capsular material itself along with outer membrane components that stimulate an antibody response. Such immunization of animals confers significant protection against subsequent abscess development from *B fragilis* strains. Furthermore, a study of women with acute pelvic inflammatory disease demonstrated antibody to the capsular antigen of *B fragilis* in women whose infecting flora contained *B fragilis*; the antibody was quantified by precipitin analysis. Immunodiffusion techniques have also been used on trichloracetic acid extracts from *B fragilis* in detecting precipitating antibodies against this organism in sera of immune rabbits. Data indicate that more than one serotype exists.

Fusobacterium necrophorum *persists for an extended period in the liver, where its proliferation* in Küpffer cells impairs macrophage function.

T cells are involved in immunity of humans to *B fragilis*, specifically linked to early stages of abscess formation.

Epidemiology

All infections involving anaerobic Gram-negative bacilli arise endogenously when mucosal damage related to surgery, trauma, or disease permits tissue penetration by members of the indigenous flora. Knowledge of the composition of the indigenous flora at various sites under different circumstances permits the clinician to anticipate the likely infecting species in acute infections at different locations. The pathogenicity of various species also must be taken into account. Ecologic determinants include the oxygen sensitivity of various organisms, the ability of organisms to adhere (discussed in Chapter 7), and microbial interrelationships. These interrelationships permit one organism to supply growth factors needed by the other, to provide assistance with adherence or motility to another organism, and to facilitate the production of inhibitory substances.

At birth, an infant's oral cavity usually is sterile; but by 12 months of age, *Fusobacterium* species can be cultured from 50 percent of infants and other Gram-negative anaerobic bacilli species from a smaller percentage. In the human gingival crevice area, Gram-negative anaerobic rods account for 16 to 20 percent of the total cultivable flora. *Prevotella melaninogenica* is seldom isolated before the age of 6 years, but by the early teens this organism can be isolated from the gingival crevice area of most individuals. Gram-negative anaerobic rods usually constitute 8 to 17 percent of the cultivable flora of human dental plaque. Selective localization is illustrated by the fact that *P melaninogenica* is found routinely in the gingival crevice but is not found, or is only rarely found, on the tongue, cheek, or coronal tooth surface.

The stomach normally has few organisms and, as a rule, no anaerobic bacteria; however, in the presence of pathologic conditions such as duodenal ulcer with bleeding or obstruction, abnormal colonization with *B fragilis* may occur in the stomach. In the terminal ileum, approximately equal numbers of facultative aerobes and anaerobes are present, with Bacteroides being one of the major anaerobes. Bacteroides species are almost invariably found in the feces of adult subjects; the mean count is 1011/g. Fusobacterium species are found in the feces of 18 percent of adults; the mean count is 108/g. *B thetaiotaomicron* and *B vulgatus* are the dominant species of Bacteroides encountered, followed by *B distasonis, B ovatus*, and *B fragilis*. In animal studies, Bacteroides protects against infection with Salmonella or Shigella.

Bacteroides, Prevotella, and Fusobacterium species are common in the vaginal flora.

In one quantitative study of the vaginal and cervical flora, Bacteroides and Prevotella species were recovered from half of the patients, with mean concentrations of 106/ g of material. Species recovered from the normal cervical flora of healthy women include *B fragilis, B capillosus, P oralis, P bivia, P disiens, P oris, P buccae*, and *B ureolyticus.*

Studies of the normal urethral flora are relatively limited, but Fusobacterium and other Gram-negative anaerobic bacilli have been isolated. *Fusiform bacilli* and *P melaninogenica* have been found regularly on the external genitalia.

Bacteroides cells placed on the forearms of human volunteers may persist for a few hours; strains placed on laboratory benches may survive even 10 hours after exposure to air, and Bacteroides has been recovered from the hospital environment on occasion. Clearly, however, the source of infection with these organisms is the indigenous flora of the body, particularly of mucosal surfaces.

Diagnosis

The clinical characteristics of infection with Bacteroides, Prevotella or Fusobacterium are primarily those seen with anaerobes in general. These characteristics include foul-smelling discharge, location of infection in proximity to mucosal surfaces, tissue necrosis, gas in tissues or discharges, association of infection with cancer, infection related to the use of aminoglycosides or other agents with poor activity against anaerobes, septic thrombophlebitis, infection following human or animal bites, and certain distinctive clinical features. The clinical presentation of sepsis may be distinctive in that onset is characterized by sore throat and fever often accompanied by chills. A membranous tonsillitis with foul odor to the breath may be noted, and in the absence of effective therapy, bacteremia and widespread metastatic infection occurs. Black discoloration of blood-containing exudates or red fluorescence of such exudates under ultraviolet light indicates infection with pigmented anaerobic Gram-negative bacilli.

A definitive diagnosis requires demonstration or isolation of the organisms responsible for the infection. Even direct Gram stain may be helpful because of the frequently unique morphology of Gram-negative anaerobic bacilli. In general, these organisms are pale staining and they may stain erratically. Fusobacterium cells may exhibit classic tapered ends and filamentous forms, with or without swollen areas and large round bodies. Direct gas-liquid chromatography of clinical specimens occasionally provides important clues to the presence of certain Gram-negative anaerobic bacilli. A large amount of butyric acid in the absence of isobutyric or isovaleric acid indicates the presence of Fusobacterium. The presence of succinic acid and only Gram-negative rods seen on Gram stain, or of both succinic and isobutyric acid in the specimen, indicates that Bacteroides is present. Both direct and indirect fluorescent antibody techniques may be useful for rapid detection of Bacteroides, Prevotella, and Fusobacterium in clinical material. False-positive reactions are sometimes a problem. Reagents are available commercially; they will undoubtedly be improved. Collection of clinical specimens should avoid the mucosal flora, and transport must be anaerobic. Use of selective and differential media may facilitate isolation and identification of different Gram-negative anaerobic bacilli. Tests for antibody development in response to the infection are not practical.

Control

There are two primary guidelines in preventing anaerobic infections: avoiding conditions that reduce the redox potential of the tissues and preventing the introduction of anaerobes of the normal flora to wounds, closed cavities, or other sites prone to infection. Prophylactic antimicrobial therapy is effective in selected situations. Patients with acute leukemia who are to be treated intensively with antitumor chemotherapy may be managed with a diet low in bacterial count and with administration of an antimicrobial regimen designed to reduce significantly the total body flora, including anaerobes. Some workers have advocated using antibacterial regimens that are relatively inactive against anaerobes; as a result, anaerobes persist in the bowel and provide colonization resistance against potential aerobic or facultative pathogens. There is some disagreement on this point.

Anaerobic bacteremia following dental manipulation may be managed effectively by administering an antibacterial agent 1 hour before the manipulation and continuing for a limited period (12 to 24 hours) afterward. The effectiveness of prophylactic antimicrobial therapy before bowel surgery is now well established. The physician may use oral neomycin plus erythromycin, giving the agents for a limited time before surgery to prevent overgrowth of other organisms. Prophylaxis also is effective before certain types of gynecologic surgery. When infection already is established but surgery is indicated (appendectomy, cholecystectomy), antimicrobial therapy just before surgery again may be helpful. Appropriate therapy of established infections such as chronic otitis media and sinusitis may prevent subsequent spread of infection that could lead to intracranial abscess. Precautions to minimize aspiration are helpful in preventing anaerobic pulmonary infection. Care must be observed in feeding feeble or confused patients and those who have difficulty swallowing. Good surgical technique minimizes the risk of postoperative infection. Minimizing injury and devitalization of tissue during surgery protects against infection. The use of closed methods of bowel resection, when feasible, decreases the likelihood of infection with members of the bowel flora.

Table 20-3 indicates the relative effectiveness of a number of drugs against Gram-negative anaerobic bacilli. Aminoglycosides such as gentamicin and amikacin are inactive against most anaerobes as are trimethoprim/sulfamethoxazole and fluoroquinolones. The activity of erythromycin varies significantly according to the testing procedure. Most penicillins and cephalosporins are less active than penicillin G; unfortunately, increasing numbers of anaerobes are showing resistance to penicillin, usually on the basis of ß-lactamase production. Ampicillin, carbenicillin, and penicillin V are roughly comparable to penicillin G on a weight basis, but the high blood levels safely achieved with carbenicillin and similar penicillins make them effective against 80 to 95 percent of *B fragilis* strains. Cefoxitin, which is resistant to penicillinase and cephalosporinase, used to be active against 95 percent of *B fragilis* strains, but now 25 percent of strains are resistant in some centers. Similarly, 20 percent of *B fragilis* strains are now resistant to clindamycin. The third-generation cephalosporins are usually less active than cefoxitin against the *B fragilis* group. Drugs active against essentially all Gram-negative (and other) anaerobes are metronidazole, imipenem, chloramphenicol, and combinations of ß-lactam drugs plus a ß-lactamase inhibitor.

Table 20-3 Antimicrobial Suceptibility of Gram-Negative Anaerobic Bacilli

Pathogen	Effectiveness of Antimicrobial Agent						
	Pencillin	Ch loramphenicol	Clindamycin	Metronidazole	Imipenem	Ticarcillin/ Clavvulanate	Cefoxcitin
B tragilis group	1	3	2	3	3	3	2
Pigmented *Prevotella and Porphyromonas spp*	2	3	3	3	3	3	3
F. varium	2-3	3	1-2	3	3	2-3	2-3
Other *Fusobacterium spp*	3	3	3	3	3	2-3	2-3

a Only drugs that might be used therapeutically are included.
b Numbers : 3, good activity; 2, moderate activity; 1, poor or incosistent activity
c Other combinations of a β-lactam drug and a b-lacternase inhibitor are comparable.
d A few strains are resistant

In addition to antimicrobial therapy, surgery is important in treating anaerobic infection. This includes drainage of abscesses, excision of necrotic tissue, relief of obstruction, and ligation or resection of infected veins. Percutaneous nonsurgical drainage may be effective in certain patients. Lung abscess, which responds well to medical therapy, is the primary exception to the rule that abscesses require surgical drainage.

Hyperbaric oxygen therapy is not useful in Gram-negative anaerobic rod infections. General supportive measures are, of course, important in managing any type of serious infection. Anticoagulation may be useful in patients with septic thrombophlebitis, along with appropriate antimicrobial therapy.

As noted above, the *B fragilis* group is among the most resistant of all anaerobes to antimicrobial agents. In part, this is related to ß-lactamase production by *B fragilis* and related strains, but other mechanisms of resistance exist as well. Sutterella wadsworthensis is often resistant to metronidazole and sometimes to penicillins (including piperacillin). Bilophila is also often resistant to penicillins by virtue of ß-lactamase production. Plasmid-mediated transferable resistance to several antimicrobial agents has been demonstrated with *B fragilis* and related Bacteroides species. Numerous other anaerobic Gram-negative bacillary species also produce ß-lactamases. Chloramphenicol acetyltransferase and nitroreductase have been demonstrated in Bacteroides but are not generally clinically significant. Among the fusobacteria, the primary organism manifesting resistance is *F varium*. Many strains of this species are resistant to clindamycin, and a number are resistant to penicillins and cephalosporins.

REFERENCES

Duerden BI, Drasar BS (eds): Anaerobes in Human Disease. Wiley-Liss, New York, 1991

Finegold SM: Anaerobic Bacteria in Human Disease. Academic Press, San Diego, 1977

Finegold SM, George WL (eds): Anaerobic Infections in Humans, Academic Press, San Diego, 1989

Finegold SM, Goldstein EJC, Mulligan ME: Proceedings of the 1994 Meeting of the Anaerobe Society of the Americas, Marina del Rey, CA. Clin Infect Dis 20 (Suppl.2):S111-383, 1995

Holdeman LV, Cato EP, Moore WEC: Anaerobic Laboratory Manual 4th ed. Virginia Polytechnic Institute and State University, Blacksburg, 1977

Kasper DL, Finegold SM (eds): Virulence factors of anaerobic bacteria (symposium). Rev Infect Dis 1:1-400, 1979

Rosebury T: Microorganisms Indigenous to Man. McGraw-Hill, New York, 1962

Smith LDS, Williams BL: The Pathogenic Anaerobic Bacteria. 3rd ed. Charles C Thomas, Springfield, IL, 1984

Styrt B, Gorbach SL: Recent developments in the understanding of the pathogenesis and treatment of anaerobic infections. N Engl J Med 321:240, 298, 1989

Summanen P, Baron EJ, Citron DM, Strong CA, Wexler HM, Finegold SM: Wadsworth Anaerobic Bacteriology Manual. 5th ed. Star Publishing Co., Belmont, CA 1993.

Chapter **39**

Anaerobic Cocci

General Concepts

Clinical Manifestations

A great variety of infections including abscesses, gangrene, cellulitis, bacteremia, pneumonia, peritonitis, bite wounds, and pelvic inflammatory disease.

Structure

Gram-positive and gram-negative cocci.

Classification and Antigenic Types

Diverse group of genera. Strictly anaerobic as well as aerotolerant species. Peptostreptococcus, Gemella, and Streptococcus are gram-positive genera of primary clinical importance. Veillonella is the gram-negative genus isolated most frequently from clinical specimens.

Pathogenesis

Infection usually results from invasion of damaged tissue by normal microbial flora. Most infections are polymicrobic. However, approximately 10%-15% of all clinical isolates come from pure culture infections.

Host Defenses

Undefined.

Epidemiology

Normal flora of skin, mouth, intestinal tract, and genitourinary tract.

Diagnosis

Laboratory isolation of organism from infected site with careful consideration of organism's role as a possible member of normal microbial flora of that particular site.

Control

Antibiotic therapy (e.g., penicillin, clindamycin), abscess drainage, debridement of necrotic tissue.

INTRODUCTION

Clinical Manifestations

Anaerobic cocci are not involved in any single specific disease process; rather, they may be present in a great variety of infections involving all areas of the human body (Fig. 19-1). These infections may range in severity from mild skin abscesses, which disappear spontaneously after incision and drainage, to more serious and life-threatening infections such as brain abscess, bacteremia, necrotizing pneumonia, and septic abortion. Infection by anaerobic cocci (and by anaerobes in general) usually involves invasion of devitalized tissue by organisms that are part of the normal flora of the affected tissue or of the surrounding areas.

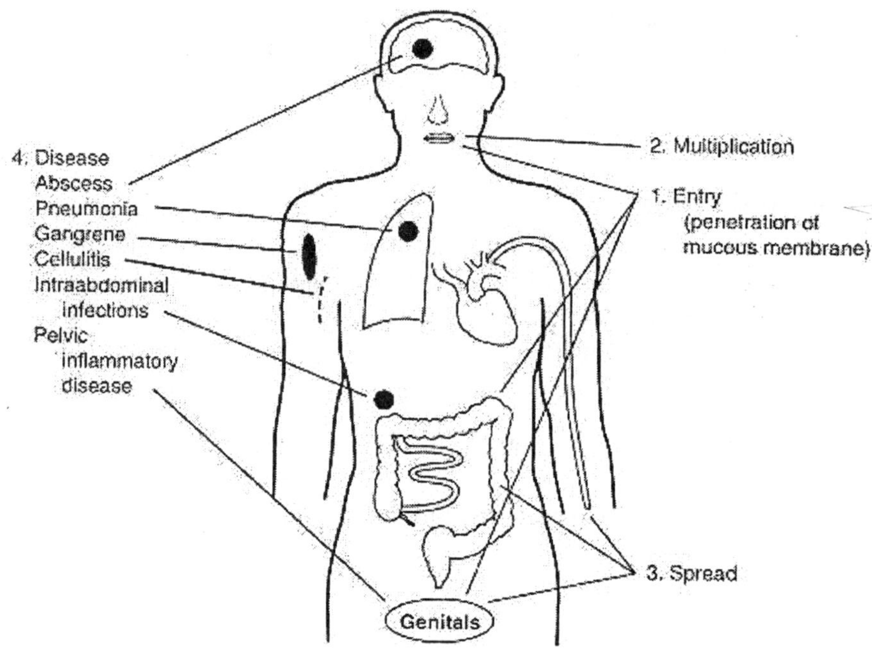

Figure 19-1 Pathogenesis of anaerobic cocci

Brain abscess, with a mortality rate of 40%, is one of the more serious infections involving anaerobic cocci. Anaerobes, rather than facultative or aerobic organisms, are a major cause; anaerobic cocci, Bacteroides and Fusobacterium, respectively, are the predominant groups isolated. Anaerobic cocci often have been isolated in pure culture from brain abscesses. Chronic otitis media or mastoiditis frequently is the primary source of the organisms and may result as a direct extension of the infection into the brain. Pleuropulmonary infection, sinusitis, congenital heart defects, and bacterial endocarditis are other conditions predisposing individuals to brain abscess by blood-borne metastases.

Pleuropulmonary infections in which anaerobic cocci may be etiologic agents are lung abscesses, necrotizing pneumonia, aspiration pneumonitis, and empyema. The incidence of anaerobes in these infections is 50-90%; anaerobic cocci account for about 40% of the anaerobic isolates. *F. nucleatum* and *P. melaninogenica* are often isolated concomitantly. These organisms are part of the normal microbial flora of the mouth and enter the lower respiratory tract as the result of aspiration, usually in association with altered consciousness.

Anaerobic pleuropulmonary infections frequently develop slowly and often are chronic. The mortality rate is about 15%.

Anaerobic cocci are involved in several skin and soft tissue infections that may be confused with clostridial myonecrosis (gas gangrene). These infections are anaerobic streptococcal myonecrosis, progressive bacterial synergistic gangrene, necrotizing fasciitis, crepitant cellulitis, chronic burrowing ulcer, and synergistic necrotizing cellulitis. These are severe infections, and the mortality rates may be as high as 75%. These conditions may be characterized by a purulent exudate, by varying degrees of tissue necrosis involving the skin, fascia, and/or underlying muscles, and sometimes by systemic toxicity. The infecting organisms often produce gas. Anaerobic cocci often are isolated with other organisms in these infections. They are characteristically found with *Staphylococcus aureus* and *Streptococcus pyogenes* in progressive bacterial synergistic gangrene, and also are found with gram-negative aerobic or facultative bacilli or Bacteroides or both in synergistic nonclostridial myonecrosis and synergistic necrotizing cellulitis. Diabetes mellitus and vascular insufficiency (often associated with trauma) are predisposing factors. Decubitus ulcers and postoperative wound infections are other soft-tissue infections from which anaerobic cocci have been isolated.

Anaerobic cocci have been recognized as significant pathogens in puerperal fever and septic abortion since the early 1900s. Other infections of the female genital tract in which anaerobic cocci have been implicated are pyometra, tuboovarian abscesses, postoperative wound infections following gynecologic surgery, and pelvic inflammatory disease, often in association with gonococci. Anaerobic cocci (*P prevotii, P anaerobius* and *S intermedius*) and *B fragilis* are the most frequently isolated anaerobes from these infections. Like the anaerobic cocci in other infections, these organisms are part of the normal flora of the affected area or of the surrounding tissues in this case, the vagina.

Periodontal disease, peritonitis, intraabdominal abscesses, and abscesses of the liver, spleen, and pancreas are types of intraabdominal infections from which anaerobic cocci have been isolated. Again, these are polymicrobic infections; concomitant isolates may be *Bacteroides sp., E. coli* and *Streptococcus* sp.

Structure, Classification and Antigenic Types

The anaerobic cocci are a physiologically diverse group that has recently undergone significant taxonomic changes. Anaerobic gram-positive cocci of clinical significance are found in three gram-positive genera (Peptostreptococcus, Gemella, and Streptococcus) and one gram-negative genus (Veillonella). There are other genera of anaerobic cocci, but they are rarely isolated from clinical specimens. Not all anaerobic cocci require stringent anaerobic conditions; for example, strains of Streptococcus intermedius are quite aerotolerant and may grow under reduced oxygen tension. Anaerobic cocci may be proteolytic or saccharolytic or both. They produce a variety of short-chain volatile fatty acids (i.e., acetic, propionic, butyric, caproic, and lactic acids) from the fermentation of simple sugars and amino acids. Both *P magnus* and *P anaerobius* possess species-specific cell wall antigens; in other anaerobic cocci, species-specific antigens have not yet been identified. Peptostreptococcus and Streptococcus are the most clinically important genera, with *P magnus* as the most frequent clinical isolate.

The anaerobic gram-positive cocci are difficult to speciate, but a few biochemical tests can be helpful. *P anaerobius* is the only species susceptible to sodium polyanethol sulfonate (SPS). *P asaccharolyticus* and *P hydrogenalis* are both indole positive, but alkaline phosphatase negative and positive, respectively. Of the indole-negative butyric acid producers, *P tetradius* is strongly saccharolytic and urease-positive, while *P prevotii* is weakly saccharolytic and usually urease-negative. *P magnus* and *P micros* are similar biochemically and are distinguished primarily on the basis of cell size and alkaline phosphatase reaction. The three prominent species of anaerobic cocci that are strongly saccharolytic and produce large amounts of lactic acid include *S intermedius, S constellatus, G morbillorum.* These latter species are either aerotolerant or become aerotolerant upon passage on laboratory media. Obligate anaerobic species in the genus Streptococcus are only rarely isolated from clinical specimens, but may be found in human feces, as can other genera of anaerobic gram-positive cocci.

Three genera of anaerobic gram-negative cocci can be found in human fecal flora: Veillonella, Acidominococcus, and Megosphora. Veillonella is considered the only clinically significant genus and V parvula is the species most frequently isolated from clinical specimens. Veillonella can be presumptively identified by the red fluorescence of colonies under ultraviolet light. This fluorescence is lost rapidly on exposure to oxygen.

Pathogenesis

Anaerobic cocci are opportunistic pathogens that cause a multitude of infections. They are part of the normal microbial flora of a healthy individual, but they can and do cause infections involving traumatized tissue or infections in the compromised host. They are isolated most often from a wide variety of polymicrobic infections (usually along with Bacteroides sp. or with facultative organisms or both), indicating a synergistic role in these infections. Approximately 10-15% of all isolates of anaerobic cocci come from pure culture infections, thus indicating that these organisms can be significant pathogens rather than innocuous commensals. In one series of 20 patients with anaerobic bacteremia, Peptostreptococcus species accounted for 21% of the 29 anaerobic isolates. The anaerobic cocci represent 25-30% of all anaerobic clinical isolates. Among anaerobes, they are second only to the gram-negative anaerobic bacilli in frequency of isolation from clinical specimens. The anaerobic cocci have received relatively little attention from microbiologists and clinicians. It is not known if anaerobic cocci produce toxins, capsules, or have other pathogenic attributes.

Host Defenses

P magnus infection (in pure culture) of hip prostheses has produced a serum antibody response; however, in most cases, specific immune responses to anaerobic cocci have not been investigated.

Epidemiology

Anaerobic cocci are part of the normal flora of the skin, the mouth, and the intestinal and genitourinary tracts of healthy individuals. Recently, with increasing study of the anaerobic cocci as pathogens, certain species are being associated with specific types of infection. As noted above, *P prevotii* and *P anaerobius* are associated with female genital tract and intraabdominal infections. *P magnus*, the most frequently isolated anaerobic coccus, is associated most often with chronic bone and joint infections and ankle ulcers. Pure cultures of this organism are not rare; they account for 15% of all *P magnus* isolates. The presence of foreign bodies, such as prosthetic joints, seems to be particularly significant in *P magnus* infections. In one study, anaerobic cocci were isolated in 15 (6%) of 246 cases of monomicrobial anaerobic bacteremia in cancer patients, indicating a relatively rare, but significant pathogenic potential for anaerobic cocci in this patient population. Veillonella and the anaerobic/aerotolerant *Streptococcus* are the anaerobic cocci isolated most frequently from infected human bites. These organisms are part of the normal oral flora. The microaerophilic gram-positive cocci are associated with abscesses and other purulent infections.

Diagnosis

Anaerobic infections generally occur in the compromised host; that is, in patients who have impaired host defense mechanisms. The primary host defense deficiency in these infections is the disruption of natural barriers (such as the skin and mucous membranes). Diabetes mellitus, connective tissue disorders, atherosclerotic disease, cancer (especially of the colon, uterus, and lung), irradiation damage, immunosuppressive treatment, and alcoholism are conditions that may disrupt these natural barriers.

To establish a definite role for anaerobic cocci in infections, the causative organism must be isolated from the affected tissue or the bloodstream. Because anaerobic cocci are a significant part of the normal flora, the proper choice of specimen is critical. For example, coughed sputum, feces, and vaginal swabs, all of which could be contaminated with normal microbial flora, are unacceptable.

Control

Treatment of infections caused by anaerobic cocci consists of antibiotic therapy and drainage, debridement, or both of necrotic tissue. In general, penicillin is the drug of choice, and clindamycin or metronidazole can be used for the patient allergic to penicillin. The clinician should be aware that in vitro antimicrobial susceptibility tests have shown that some strains of anaerobic cocci are resistant to penicillin or to clindamycin. Metronidazole is typically active against most strains of anaerobic cocci; however, aerotolerant species, such as Streptococcus spp. are uniformly resistant. Brain abscesses must be treated with an antimicrobial agent such as chloramphenicol or penicillin or metronidazole, sufficient doses of which can cross the blood barrier. Frequently *B fragilis*, an anaerobic gram-negative rod, is present in infections containing anaerobic cocci; this organism produces a ß-lactamase that can protect other organisms in the infection from the action of penicillin.

REFERENCES

Bourgault AM, Rosenblatt JE, Fitzgerald RH: *Peptococcus magnus*: A significant human pathogen. Ann Intern Med 93:244-248, 1980

Brook I: Peptostreptococcal infection in children. Scand J Infect Dis, 26(5):503, 1994

Brook I, Walker RI: Pathogenicity of anaerobic gram-positive cocci. Infect Immunol 320-324, 1984

Fainstein V, Elting LS, Bodey GF: Bacteremia caused by non-sporulating anaerobes in cancer patients - a 12-year experience. Medicine 3:151-162, 1989

Finegold SM, George WL (eds): Anaerobic infections in humans. Academic Press, New York, 1989

Kotiranta A, Haapasalo M, Lounatmaa K, Kari K: Crystalline surface protein of *Peptostreptococcus anaerobius*. Microbiology, 141(Pt. 5):1065, 1995

Peraino, VA, Cross SA, Goldstein EJ: Incidence and clinical significance of anaerobic bacteremia in a community hospital. Clin Infect Dis 16: S288, 1993

Summanen P, Baron EJ, Citron DM, Strong C, Wexler HM, Finegold, SM (eds): Wadsworth anaerobic bacteriology manual, 5th ed. Star Publishing Co., Belmont Calif, 1993

Summanen P: Recent taxonomic changes for anaerobic gram-positive and selected gram-negative organisms. Clin Infect Dis 16: S168, 1993

Taylor AG, Finham WJ, Golding MA et al.: Infection of total hip prostheses by *Peptococcus magnus*, an immuno-fluorescence and ELISA study of two cases. J Clin Pathol 32: 61-65, 1979

Chapter **40**

Miscellaneous Pathogenic Bacteria

General Concepts

Listeria Monocytogenes

This microorganism is a potential pathogen for both humans and animals. Most human cases occur in patients with debilitating disease or in prenatal or neonatal infants. Sepsis, meningitis, and disseminated abscesses occur in infected patients. Meat, vegetables, and various milk products are the most common sources of infection.

Erysipelothrix Rhusiopathiae

This species is transmitted occasionally from infected pigs to farmers or veterinarians, in whom it causes primarily inflammatory infections of the skin. Septicemia and endocarditis may develop secondarily.

Propionibacterium Acnes

A common inhabitant of the crypts of the skin, this species may contribute to acne.

Streptobacillus Moniliformis

This species may infect individuals through the bite of infected rodents. After local ulcerative inflammation, life-threatening septicemia may develop.

Calymmatobacterium Granulomatis

This microorganism causes granuloma inguinale, a venereal disease with local tissue destruction in the genital, inguinal, and perianal region.

INTRODUCTION

The microorganisms discussed in this chapter are taxonomically unrelated. The human infections they cause are rare except in the case of Listeria and Propionibacterium acnes. Some of these infections are fatal; others tend to be self-limited. Recognition depends largely on the proper use of bacteriological methods, which is important not only to ensure appropriate therapy but also to exclude other possible agents.

Listeria Monocytogenes

Clinical Manifestations

Listeriosis is a serious disease for humans, with a mortality greater than 25 percent. There are two main clinical manifestations, sepsis and meningitis (Fig. 16-1). Meningitis is often complicated by encephalitis, which is exceptional among bacterial infections. Occasionally, pyogenic infections of various organs have been found. Relapses may occur after apparent recovery.

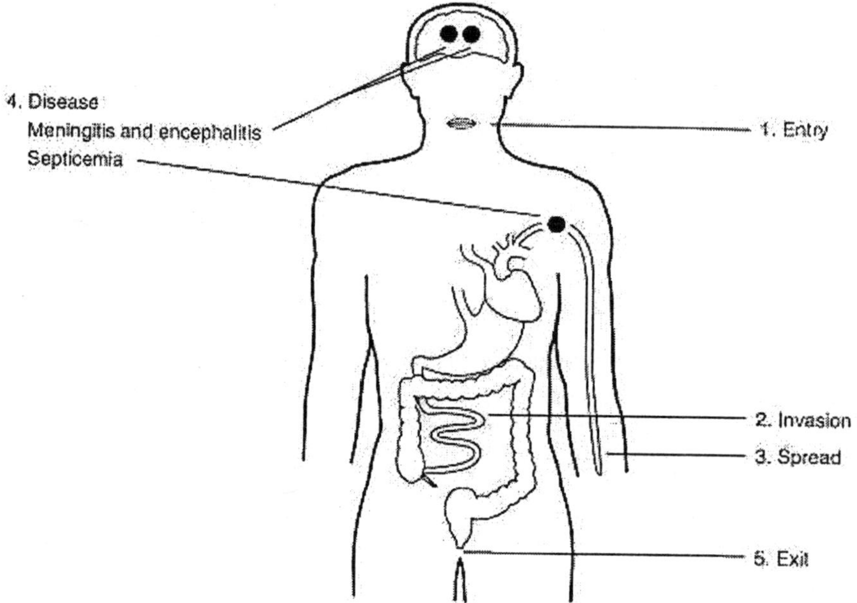

FIGURE 16-1 Pathogenesis of listeriosis.

Structure, Classification, and Antigenic Types

All Listeria species are small, Gram-positive rods, which are sometimes arranged in short chains. In direct smears they may be coccoid, so they can be mistaken for streptococci. Longer cells can be suggestive of corynebacteria. Flagella are produced at room temperature rather than at 37°C. Hemolysin production is an important marker for L monocytogenes, although it is not definitive, as L ivanovii and L seeligeri are likewise hemolytic on blood agar. Further biochemical characterization is necessary to distinguish between the different Listeria species.

It may be desirable for epidemiologic purposes to identify a particular strain by serotyping to characterize surface antigens, such as O antigens (teichoic acids) and H antigens (proteins). The serovars 1/2a and 4b are responsible for up to 90 percent of all cases of listeriosis.

A particular property of L monocytogenes is the ability to multiply at low temperatures (Fig. 16-2). Bacteria therefore can accumulate in contaminated food stored in the refrigerator.

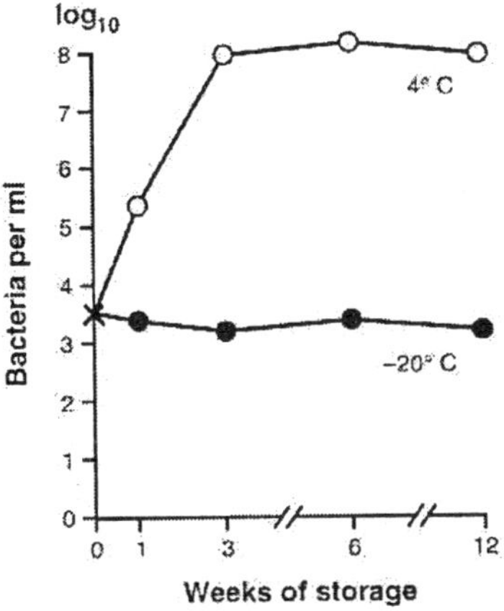

FIGURE 16-2 Multiplication of L monocytogenes in broth at low temperature.

Pathogenesis

Listeria monocytogenes is presumably ingested with raw, contaminated food (Fig. 16-2). An invasion factor secreted by the pathogenic bacteria enables them to penetrate host cells of the epithelial lining. Since this microorganism is widely distributed, this event may occur rather often. Normally, the immune system eliminates the infection before it spreads. Indeed, most adults who have no history of listeriosis have T lymphocytes primed specifically by Listeria antigens. If the immune system is compromised, however, systemic disease may develop. Listeria monocytogenes multiplies not only extracellularly but also intracellularly within macrophages after phagocytosis and even within parenchymal cells which are entered by induced phagocytosis. It therefore belongs to the large group of facultatively intracellular pathogens (Table 16-1).

TABLE 16-1 List of Some Facultative or Obligate Intercellular Microorganisms

Type of Microorganism	Agent	Disease
Bacteria	Mycobacterium tuberculosis	Tuberculosis
	Salmonella typhi	Typhoid fever
	Yorsinia postis	Plague
	Legionella pneumophlla	Legionellosis
	Listeria monocytogenes	Listeriosis
Fungi	Histoplasma capsulatum	Histoplasmosis
Protozoa	Toxoplasma gonchi	Toxoplasmosis
	Leishmania donovani	Kala azar
	Trypanosoma cruzi	Chagas disease

Survival within the phagosomes and eventual escape into the cytoplasm are mediated by a toxin, which also acts as a hemolysin. This toxin is one of the so-called SH-activated hemolysins, which are produced by a number of different bacteria such as serogroup A streptococci, pneumococci, and Clostridium perfringens. Obviously, nature has preserved the genetic code for this bacterial product in several species, and consequently the hemolysins from these different bacteria have common biochemical, biologic, and antigenic properties. Nonhemolytic variants of L monocytogenes are completely avirulent, as are the nonhemolytic species L innocua and L welshimeri. Hemolysin is not the only Listeria virulence factor, however, since the hemolytic Listeria species besides L monocytogenes (i.e., L seeligeri and L ivanovii) possess rather limited pathogenicity. The hemolysin gene is located on the chromosome within a cluster of other virulence genes which are all regulated by a common promotor. These additional genetic determinants are necessary for further steps in the intracellular life cycle of L monocytogenes. One particular gene product promotes the polymerization of actin, a component of the host cell cytoskeleton, on the bacterial surface. In this peculiar environment within host cells, surrounded by a sheet of actin filaments, the bacteria reside and even multiply. The growing actin sheet functions as a propulsive force which drives the bacterium across the intracellular pathways until it finally reaches the surface. Then, the host cell is urged to form slim, long protrusions containing living L. monocytogenes. Those cellular projections are engulfed by adjacent cells, even by non-professional phagocytes such as parenchymal cells. By such a mechanism a direct cell-to-cell spread in an infected organ may occur without an extracellular stage.

Host Defenses

Because it multiplies intracellularly, L monocytogenes is largely protected against humoral immune factors such as antibodies, and the effective host response is cell-mediated, involving both lymphokines (especially interferon) produced by CD4+ (T-helper) cells and direct lysis of infected cells by CD8+ (cytotoxic) T lymphocytes. Both of these fundamental defense mechanisms are expressed in the microenvironment of the infective foci. Histologically, these foci are organized as granulomas, characterized by a central accumulation of epitheliod cells (macrophages) with irregularly shaped nuclei and large, delicately structured cytoplasm and by peripheral lymphocytes recognizable by a round nucleus and a narrow border of intensely staining cytoplasm (Fig. 16-3).

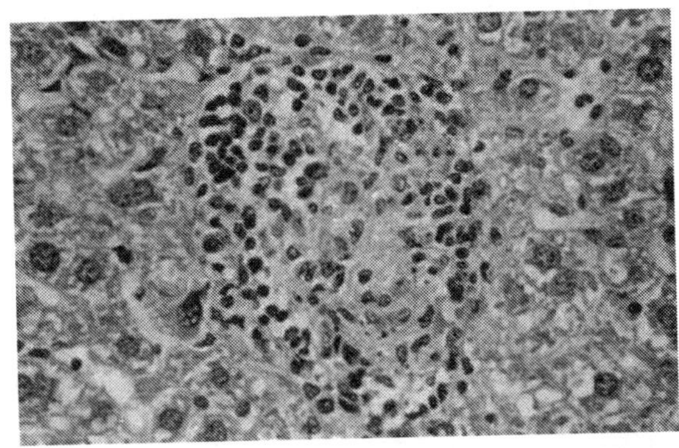

FIGURE 16-3 Infective focus in the liver of mice 7 days after infection with L monocytogenes. Note the granulomatous reaction characterized by central accumulation of epitheloid cells (macrophages) and the presence of some dark, round cells (lymphocytes) in the periphery.

Epidemiology

Listeria species are found in living and nonliving matter. Various foodstuffs of vegetable and animal origin are sources of infection. Animal and human carriers also have been described. Most human cases of listeriosis develop in immunocompromised hosts: newborns, old people, cancer patients, and transplant recipients. Reports of sporadic cases of listeriosis are becoming more frequent as the number of persons at risk, especially because of immunosuppression by medical therapy, increases. Outbreaks of listeriosis are due mainly to a common source of contaminated food.

Listeriosis also may be transmitted congenitally across the placenta. The immunocompetent mother suffers at worst a brief, flu-like febrile illness, but the fetus, whose defense system is still immature, becomes seriously ill. Depending on the stage of gestation, the fetus is either stillborn or born with signs of congenital infection. Typically, multiple pyogenic foci are found in several organs (granulomatosis infantiseptica). The onset of listeriosis is delayed (i.e., a few days after birth) when infection is acquired during labor by bacteria colonizing the genital tract of the mother.

Diagnosis

Listeria monocytogenes is implicated when monocytosis is observed in the peripheral blood as well as the cerebrospinal fluid. Early diagnosis may be obtained by finding pleocytosis with Gram-positive rods in a Gram stain of smears of the cerebrospinal fluid. Final proof is obtained by culture. Serologic tests are highly unreliable.

Control

Hygienic food processing and storage may reduce the risk of listeriosis. Individuals in high-risk groups (i.e., immunocompromised individuals and pregnant women) should avoid uncooked foodor should at least marinate salads for along time in a vinegar-based dressing to kill adherent bacteria.

Since a cell-mediated immune response (the most potent weapon against L monocytogenes) is induced only by injection of living antigen, vaccination is difficult. Even an attenuated living vaccine is dangerous for persons with impaired defenses, the proper target group. Completely avirulent live bacteria do not trigger an effective, cell-mediated immune response.

Antimicrobial agents are the mainstay of treatment. Most of the common antibiotics, except cephalosporins, are active against L monocytogenes in vitro. In practice, ampicillin combined with an aminoglycoside has given the best results. However, because infection occurs mainly in infirm patients and because intracellular bacteria are hardly accessible to most drugs, the cure rate is low. Furthermore, Listeria cells, although inhibited, are not killed by ampicillin. High doses for prolonged periods are indicated.

Erysipelothrix Rhusiopathiae

Clinical Manifestations

The most common human infection by E rhusiopathiae is erysipeloid, a well-defined, violet or wine-colored inflammatory lesion of the skin of the fingers or hand (Fig. 16-4). Itching is typical. Infrequently, septicemia develops, followed by various organ manifestations such as endocarditis or arthritis without fever.

Structure and Classification

Erysipelothrix rhusiopathiae is a slender, Gram-positive rod similar to L monocytogenes. In general, E rhusiopathiae rods are longer, especially in rough variants. They grow on routine culture media under aerobic conditions, but preferentially in a CO_2 atmosphere. In contrast to L monocytogenes, they are nonmotile, nonhemolytic, and catalase negative. The production of H_2S is highly indicative, since very few other Gram-positive bacteria have this property.

Pathogenesis

A minor skin injury may facilitate the penetration of E rhusiopathiae after contact with infected material. After an incubation of 1 to 4 days the local lesion develops; spontaneous recovery occurs in 2 to 3 weeks. Septicemia has been observed without previous local lesions so that an oral infection is assumed. Endocarditis may develop in a few cases.

Epidemiology

Erysipelothrix rhusiopathiae is found in mammals, poultry, and fish. Individuals who have occupational exposure to such animals (i.e., farmers, veterinarians, slaughterhouse workers, and fish handlers) are at risk.

Diagnosis

The typical, nonpyogenic lesions on occupationally exposed persons suggest erysipeloid. Since there is no wound, a swab is not useful. Bacteria can be cultured from a biopsy of the progressing, inflamed edge of the lesion. Blood culture is indicated in the setting of sepsis and endocarditis.

Control

Penicillin is the drug of choice to treat serious infections. Since local skin infection is self-limited, therapy is not essential.

Propionibacterium Acnes

Clinical Manifestations

The pathogenic role of Propionibacterium acnes is still disputed. Although, it is often detected in anaerobic blood cultures, it normally colonizes the skin crypts and is transported to cultures by pure chance (Fig. 16-4). Nevertheless, in compromised patients even this nonpathogenic species may induce pathologic reactions, such as endocarditis. In skin lesions P acnes is often found with other pathogenic bacteria, such as Staphylococcus aureus or actinomycetes, and is thought to support the damaging effect of those pathogens. It is doubtful whether P acnes alone is able to induce acne.

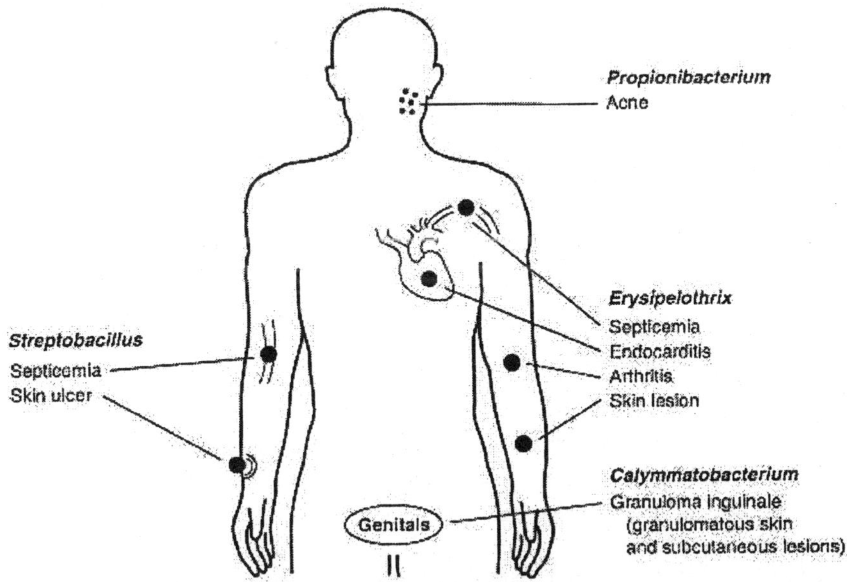

FIGURE 16-4 Disease manifestations of E rhusiopathiae, P acnes, C granulomatis, and S moniliformis.

Structure and Classification

The club-shaped, Gram-positive rods of P acnes resemble the diphtheroids but, unlike the latter, are slow-growing and anaerobic, so that their presence in blood cultures is detected after about a week.

Pathogenesis

Propionibacterium acnes produces several metabolic products, hemolysin, and various enzymes such as lipase and neuraminidase, which are excreted into the surroundings. These metabolites may clear the way for other bacteria. Furthermore, P acnes degrades sebaceous matter to produce fatty acids that stimulate an inflammatory reaction.

Epidemiology

Propionibacterium acnes normally colonizes the deep crypts of the skin, where the availability of oxygen is reduced. The same applies to mucous membranes of the oroanal areas. They may be transported to other sites by chance.

Control

Practically all common antibiotics, including penicillins, erythromycin, and tetracyclines, can be used to treat P acnes infections.

Streptobacillus Moniliformis

Clinical Manifestations

Streptobacillus moniliformis causes the clinical disease called rat bite fever. At the site of the rodent bite, an ulcer appears; this may heal spontaneously (Fig. 16-4). Occasionally, the infection spreads to the regional lymph nodes, and bacteremia has been observed. General malaise and fever may be present after a few days. This generalized disease may be fatal. Colonization of various parts of the body, such as joints or endocardium, may lead to chronic disease accompanied by local symptoms. Rat bite fever also is caused by Spirillum minus, a very different bacterium (see Ch. 35).

Structure and Classification

Streptobacillus moniliformis is a Gram-negative, nonmotile rod of variable length. The individual cells are not regularly shaped or stained, and thus pleomorphism is seen in smears. There is a tendency for spontaneous development of cell-wall-deficient L-forms (see Ch. 2). Consequently, growth on artificial media depends on certain additives, such as serum or ascitic fluid, which are not always present in common culture media. Growth is best under a CO_2 atmosphere.

Pathogenesis

Humans usually become infected with S moniliformis through the bite of an infected rat. Ingestion of contaminated food has rarely been incriminated as the source of infection.

Host Defenses

The nonspecific resistance mechanisms in the skin and draining lymph nodes prevent dissemination. The low pH in the stomach normally guarantees that S moniliformis cannot survive gastric passage.

Epidemiology

S moniliformis belongs to the common bacterial flora of the nasopharynx of rats, from which it reaches humans directly by a rat bite or indirectly via food.

Diagnosis

The coincidence of fever after a rat bite draws attention to this infection. Positive cultures can be obtained from blood or synovial fluid. Mice are highly susceptible to S moniliformis, exhibiting a rapid lethal infection after inoculation. Because of the existence of L-forms, special media must be used for culture, since on conventional agar plates L-form colonies hardly are visible and are likely to be overlooked.

Calymmatobacterium Granulomatis

Clinical Presentation

C granulomatis causes granuloma inguinale. This infection typically is localized in the genital region (Fig. 16-4). It spreads to adjacent areas, and the regional lymph nodes also may be inflamed. Persistent granulomatous lesions tend to ulcerate, destroying skin and subcutaneous tissue.

Structure and Classification

Calymmatobactcrium granulomatis is a Gram-negative, nonmotile rod. The capsule that surrounds the bacterial cell appears similar to that of Klebsiella. Addition of egg yolk and incubation in a CO_2 atmosphere are required for growth on artificial media.

Pathogenesis

Calymmatobacterium granulomatis is normally present in the gut flora and may be transmitted to the genital area by autoinoculation or sexual contact. After penetrating the skin the bacteria induce an inflammatory reaction, which may lead to destruction of the infected tissue. Within the inflammatory foci C granulomatis is found mainly intracellularly (Table 16-1) inside tissue macrophages (Donovan bodies). This is highly typical for granuloma inguinale. Superinfection of ulcers with other pathogenic organisms is possible.

Host Defenses

Antibodies against C granulomatis are produced during acute infection; their role in defense remains unclear. Cell-mediated defense mechanisms, expressed by a granulomatous reaction, are important in recovery.

Epidemiology

Granuloma inguinale occurs most frequently in people living under poor socioeconomic conditions (e.g., in the tropics). In the United States, infection of blacks is seven times more frequent than infection of whites. Transmission by sexual contacts is most common. Other sexually transmitted diseases, such as syphilis, may be associated.

Diagnosis

Microscopic evidence of intracellular Gram-negative encapsulated rods in ulcerative skin wounds of the genitoinguinal region is highly indicative for granuloma inguinale. Since experience with C granulomatis is lacking in most laboratories, cultural diagnosis probably will fail. Eventually, the yolk sac of 5-day-old chicken embryos can be inoculated directly with the infected material.

Control

Infection can be prevented by cleanliness or by avoiding sexual contacts with infected persons. Antibiotics active against intracellular bacteria, such as tetracycline or erythromycin, are effective in treatment.

REFERENCES

Davis MC: Granuloma inguinale. A clinical, histological and ultrastructural study.J Am Med Assoc 211:632, 1970

Edelson PJ: Intracellular parasites and phagocytic cells: cell biology and pathophysiology. Rev Infect Dis 4: 124, 1982

Gellin BG, Broome CV: Listeriosis. J Am Med Assoc 261:1313, 1989

Hahn H, Kaufmann SHE: The role of cell-mediated immunity in bacterial infection. Rev Infect Dis 3:1221, 1981

Lal S, Nicholas C: Epidemiological and clinical features in 165 cases of granuloma inguinale. Br J Vener Dis 46:461, 1970

Rogosa M: Streptobacillus moniliformis and Spirillum minus. p. 400. In Lenette EH, Balows A, Hausler WJ Jr, Shadomy HJ (eds): Manual of Clinical Microbiology. 4th Ed. American Society for Microbiology, Washington, DC, 1985

Sheehan B, Kocks C, Dramsi S, Gouin E, Klarsfeld AD, Mengaud J, Cossart P: Molecular and genetic determinants of the Listeria monocytogenes infectious process. Curr Top Microbiol Immunol 192:187, 1994

Weaver RE: Erysipelothrix. p. 209. In Lenette EH, Balows A, Hausler WJ Jr, Shadomy HJ (eds): Manual of Clinical Microbiology. 4th Ed. American Society for Microbiology, Washington, DC, 1985

Index

D

E

F

G

T

U

V

W

Y

Z

O

P

Q

R

S